Y0-AAX-910

# Methods in Enzymology

Volume 200

PROTEIN PHOSPHORYLATION

Part A

Protein Kinases: Assays, Purification, Antibodies, Functional Analysis, Cloning, and Expression

# METHODS IN ENZYMOLOGY

*Methods in Enzymology*

*Volume 200*

# *Protein Phosphorylation*

## *Part A*

## *Protein Kinases: Assays, Purification, Antibodies, Functional Analysis, Cloning, and Expression*

EDITED BY

*Tony Hunter*
*Bartholomew M. Sefton*

MOLECULAR BIOLOGY AND VIROLOGY LABORATORY
THE SALK INSTITUTE
SAN DIEGO, CALIFORNIA

ACADEMIC PRESS, INC.
**Harcourt Brace Jovanovich, Publishers**
San Diego New York Boston
London Sydney Tokyo Toronto

This book is printed on acid-free paper. ♾



Academic Press, Inc.
San Diego, California 92101

*United Kingdom Edition published by*
ACADEMIC PRESS LIMITED
24-28 Oval Road, London NW1 7DX

Library of Congress Catalog Card Number: 54-9110

ISBN 0-12-182101-3 (alk. paper)

PRINTED IN THE UNITED STATES OF AMERICA
91 92 93 94 9 8 7 6 5 4 3 2 1

# Table of Contents

## Section III. Purification of Protein Kinases

### A. General Methods



### B. Bacterial Protein Kinases



### C. Protein-Serine Kinases

## D. Protein-Tyrosine Kinases



## E. Protein-Histidine Kinases



## Section IV. Renaturation of Protein Kinases

## Section V. Antibodies against Protein Kinases



## Section VI. Affinity Labeling, Chemical Probing, and Crystallization of Protein Kinases



## Section VII. Cloning of Genes Encoding Protein Kinases

## Section VIII. Expression and Analysis of Protein Kinases Using cDNA Clones

# Contributors to Volume 200

Article numbers are in parentheses following the names of contributors.
Affiliations listed are current.

S. A. AARONSON (46), *Laboratory of Cellular and Molecular Biology, National Cancer Institute, National Institutes of Health, Bethesda, Maryland 20892*

C. DAVID ALLIS (4), *Department of Biology, Syracuse University, Syracuse, New York 13244*

NEIL G. ANDERSON (27), *Departments of Internal Medicine and Pharmacology, University of Virginia, Charlottesville, Virginia 22908*

MICHAEL ANOSTARIO, JR. (33), *Department of Medicinal Chemistry and Pharmacognosy, Purdue University, West Lafayette, Indiana 47907*

JEFFREY L. BENOVIC (28), *Department of Pharmacology, Jefferson Cancer Institute, Thomas Jefferson University, Philadelphia, Pennsylvania 19107*

J. MICHAEL BISHOP (53), *Departments of Microbiology and Immunology and Biochemistry and Biophysics, and G. W. Hooper Foundation, University of California, San Francisco, California 94143*

KATHERINE A. BORKOVICH (16), *Division of Biology, California Institute of Technology, Pasadena, California 91125*

ROBERT B. BOURRET (15), *Division of Biology, California Institute of Technology, Pasadena, California 91125*

NANCY A. BROWN (39), *Oregon Regional Primate Center, Beaverton, Oregon 97006*

JOSEPH A. BUECHLER (41), *Biosite, San Diego, California 92121*

DAVID CARLING (29), *Department of Pharmaceutical Chemistry, School of Pharmacy, University of London, London WC1 1AX, England*

MARIAN CARLSON (34), *Department of Genetics and Development and Institute of Cancer Research, Columbia University College of Physicians & Surgeons, New York, New York 10032*

DENNIS CARROLL (11), *Office of Health, Agency for International Development, Department of State, Washington, D.C. 20523*

JOHN E. CASNELLIE (9), *Department of Pharmacology, University of Rochester, Rochester, New York 14642*

JEAN-CLAUDE CAVADORE (23), *Institut National de la Santé et de la Recherche Médicale, 34033 Montpellier, France*

JOHN L. CELENZA (34), *Whitehead Institute for Biomedical Research, Cambridge, Massachusetts 02142*

TIMOTHY C. CHAMBERS (25), *Department of Pharmacology, Emory University School of Medicine, Atlanta, Georgia 30322*

LARS J. CISEK (24), *Department of Surgery, The Johns Hopkins University School of Medicine, Baltimore, Maryland 21205*

PAUL R. CLARKE (29), *EMBL, D-6900 Heidelberg, Germany*

GAIL M. CLINTON (39), *Department of Biochemistry and Molecular Biology, Oregon Health Sciences University, Portland, Oregon 97201*

JACKIE D. CORBIN (26), *Department of Molecular Physiology and Biophysics, Vanderbilt University School of Medicine, Nashville, Tennessee 37232*

JEFFRY L. CORDEN (24), *Howard Hughes Medical Institute and Department of Molecular Biology and Genetics, The Johns Hopkins University School of Medicine, Baltimore, Maryland 21205*

JEAN-CLAUDE CORTAY (17), *Institut de Biologie et Chimie des Protéines, C.N.R.S., Université de Lyon, 69622 Villeurbanne, France*

ALAIN-JEAN COZZONE (17), *Institut de Biologie et Chimie des Protéines, C.N.R.S., Université de Lyon, 69622 Villeurbanne, France*

MARCEL DORÉE (23), *Centre de Recherches de Biochimie Macromoleculaire, Centre National de la Recherche Scientifique, 34033 Montpellier, France*

LELAND ELLIS (54), *Howard Hughes Medical Institute and Department of Biochemistry, University of Texas Southwestern Medical Center, Dallas, Texas 75235*

ALAN K. ERICKSON (27), *Departments of Internal Medicine and Pharmacology, University of Virginia, Charlottesville, Virginia 22908*

ELEANOR ERIKSON (21), *Department of Pharmacology and Howard Hughes Medical Institute, University of Colorado School of Medicine, Denver, Colorado 80262*

R. L. ERIKSON (21), *Department of Cellular and Developmental Biology, Harvard University, Cambridge, Massachusetts 02138*

STEFANO FERRARI (12), *Friedrich Miescher-Institute, CH-4002 Basel, Switzerland*

JAMES E. FERRELL, JR. (35), *Department of Zoology, University of Wisconsin-Madison, Madison, Wisconsin 53706*

SHARRON H. FRANCIS (26), *Department of Molecular Physiology and Biophysics, Vanderbilt University School of Medicine, Nashville, Tennessee 37232*

YOKO FUJITA-YAMAGUCHI (5), *Department of Molecular Genetics, Beckman Research Institute of the City of Hope, Duarte, California 91010*

ROBERT L. GEAHLEN (33), *Department of Medicinal Chemistry and Pharmacognosy, Purdue University, West Lafayette, Indiana 47907*

CLAIBORNE V. C. GLOVER (4), *Department of Biochemistry, The University of Georgia, Athens, Georgia 30602*

DONALD J. GRAVES (36), *Department of Biochemistry and Biophysics, Iowa State University, Ames, Iowa 50011*

CHARLES GREENFIELD (52), *Department of Medicine, St. Mary's Hospital, London W2 1PG, England*

BOYD E. HALEY (40), *Department of Biochemistry, University of Kentucky Medical Center, Lexington, Kentucky 40536*

STEVEN K. HANKS (2, 44), *Department of Cell Biology, Vanderbilt University School of Medicine, Nashville, Tennessee 37232*

D. GRAHAME HARDIE (29), *Department of Biochemistry, The University, Dundee DD1 4HN, Scotland*

MARIETTA L. HARRISON (33), *Department of Medicinal Chemistry and Pharmacognosy, Purdue University, West Lafayette, Indiana 47907*

CARL-HENRIK HELDIN (30), *Ludwig Institute for Cancer Research, S-751 24 Uppsala, Sweden*

J. FRED HESS (15), *Department of Biochemistry, Merck Sharp & Dohme Research Laboratories, Rahway, New Jersey 07065*

JUSTIN HSUAN (31), *Structural Biology Laboratory, Ludwig Institute for Cancer Research, London W1P 8BT, England*

FREESIA L. HUANG (20, 38), *Section on Metabolic Regulation, Endocrinology and Reproduction Research Branch, National Institute of Child Health and Human Development, National Institutes of Health, Bethesda, Maryland 20892*

KUO-PING HUANG (20, 38), *Section on Metabolic Regulation, Endocrinology and Reproduction Research Branch, National Institute of Child Health and Human Development, National Institutes of Health, Bethesda, Maryland 20892*

TONY HUNTER (1), *Molecular Biology and Virology Laboratory, The Salk Institute, San Diego, California 92186*

JILL E. HUTCHCROFT (33), *Department of Biological Sciences, Purdue University, West Lafayette, Indiana 47907*

PAUL JENÖ (14), *Department of Biochemistry, Biocenter of the University of Basel, CH-4056 Basel, Switzerland*

KAREN E. JOHNSON (51), *DNAX, Palo Alto, California 94304*

JOSHUA M. KAPLAN (53), *Department of Biology, Massachusetts Institute of Technology, Cambridge, Massachusetts 02139*

SCOTT M. KEE (36), *Department of Chemistry, University of Texas, Austin, Texas 78731*

BRUCE E. KEMP (3, 10), *St. Vincent's Institute of Medical Research, Fitzroy, Victoria 3065, Australia*

USHIO KIKKAWA (18), *Department of Biochemistry, Kobe University School of Medicine, Kobe 650, Japan*

AKIRA KISHIMOTO (37), *Department of Biochemistry, Kobe University School of Medicine, Kobe 650, Japan*

DANIEL R. KNIGHTON (43), *Department of Chemistry, University of California, San Diego, La Jolla, California 92093*

M. H. KRAUS (46), *Laboratory of Cellular and Molecular Biology, National Cancer Institute, National Institutes of Health, Bethesda, Maryland 20892*

JEAN-CLAUDE LABBÉ (23), *Centre de Recherches de Biochimie Macromoleculaire, Centre National de la Recherche Scientifique, 34033 Montpellier, France*

HEIDI A. LANE (22), *Friedrich Miescher-Institute, CH-4002 Basel, Switzerland*

THOMAS A. LANGAN (25), *Department of Pharmacology, University of Colorado School of Medicine, Denver, Colorado 80262*

LONNY LEVIN (51), *Howard Hughes Medical Institute and Department of Molecular Biology, The Johns Hopkins University School of Medicine, Baltimore, Maryland 21205*

BARRY A. LEVINE (54), *Inorganic Chemistry Laboratory, University of Oxford, Oxford OX1 3QR, England*

SHU-LIAN LI (5), *Department of Molecular Genetics, Beckman Research Institute of the City of Hope, Duarte, California 91010*

RICHARD A. LINDBERG (44, 47), *Molecular Biology and Virology Laboratory, The Salk Institute, San Diego, California 92138*

JAMES L. MALLER (21), *Department of Pharmacology and Howard Hughes Medical Institute, University of Colorado School of Medicine, Denver, Colorado 80262*

R. M. MARAIS (19), *Eukaryotic Transcription Laboratory, Imperial Cancer Research Fund, London WC2A 3PX, England*

DANIEL R. MARSHAK (11), *Cold Spring Harbor Laboratory, Cold Spring Harbor, New York 11724*

G. STEVEN MARTIN (35), *Division of Biochemistry and Molecular Biology, Department of Molecular and Cell Biology, University of California, Berkeley, California 94720*

HARRY R. MATTHEWS (32), *Department of Biological Chemistry, University of California, Davis, California 95616*

MARIA L. MCGLONE (49), *Department of Chemistry, University of California, San Diego, La Jolla, California 92093*

W. TODD MILLER (42), *Department of Biology, Massachusetts Institute of Technology, Cambridge, Massachusetts 02139*

MICHAEL F. MORAN (56), *Samuel Lunenfeld Research Institute, Mount Sinai Hospital, Toronto, Ontario M5G 1X5, Canada*

DAVID O. MORGAN (53), *Department of Physiology, University of California, San Francisco, California 94143*

DIDIER NÈGRE (17), *Institut de Biologie et Chimie des Protéines, C.N.R.S., Université de Lyon, 69622 Villeurbanne, France*

YASUTOMI NISHIZUKA (18), *Department of Biochemistry, Kobe University School of Medicine, Kobe 650, Japan*

KOUJI OGITA (18, 37), *Department of Biochemistry, Kobe University School of Medicine, Kobe 650, Japan*

YOSHITAKA ONO (18), *Central Research Division, Takeda Chemical Industries, Osaka 532, Japan*

JOSEPH PARELLO (43), *Unité Associée 1111, C.N.R.S., Faculté de Pharmacie, 34060 Montpellier Cedex, France*

P. J. PARKER (19, 55), *Protein Phosphorylation Laboratory, Imperial Cancer Research Fund, London WC2A 3PX, England*

ELENA B. PASQUALE (47), *La Jolla Cancer Research Foundation, La Jolla, California 92037*

TONY PAWSON (56), *Samuel Lunenfeld Research Institute, Mount Sinai Hospital, Toronto, Ontario M5G 1X5, Canada*

RICHARD B. PEARSON (3, 10), *St. Vincent's Institute of Medical Research, Fitzroy, Victoria 3065, Australia*

ANNE MARIE QUINN (2), *Biocomputing Center, The Salk Institute, La Jolla, California 92037*

EFRAIM RACKER (7, 8), *Section of Biochemistry, Molecular and Cell Biology, Cornell University, Ithaca, New York 14853*

L. BRYAN RAY (27), *Departments of Internal Medicine and Pharmacology, University of Virginia, Charlottesville, Virginia 22908*

SYDONIA I. RAYTER (50), *Signal Transduction Laboratory, Imperial Cancer Research Fund, London WC2A 3PX, England*

GERT RIJKSEN (6), *Department of Hematology, Laboratory of Medical Enzymology, University Hospital Utrecht, 3508 GA Utrecht, The Netherlands*

LARS RÖNNSTRAND (30), *Ludwig Institute for Cancer Research, S-751 24 Uppsala, Sweden*

DINKAR SAHAL (5), *Department of Biophysics, University of Delhi South Campus, New Delhi 110021, India*

NAOAKI SAITO (37), *Department of Pharmacology, Kobe University School of Medicine, Kobe 650, Japan*

D. SCHAAP (55), *Dutch Cancer Institute, 1066 CX Amsterdam, The Netherlands*

PARIMAL C. SEN (8), *Section of Biochemistry, Molecular and Cell Biology, Cornell University, Ithaca, New York 14853*

MELVIN I. SIMON (15, 16), *Division of Biology, California Institute of Technology, Pasadena, California 91125*

LEE W. SLICE (49), *Virology Department, Hoffmann-La Roche Inc., Nutley, New Jersey 07110*

JANUSZ M. SOWADSKI (43), *Departments of Medicine and Biology, University of California, San Diego, La Jolla, California 92093*

GERARD E. J. STAAL (6), *Department of Hematology, Laboratory of Medical Enzymology, University Hospital Utrecht, 3508 GA Utrecht, The Netherlands*

S. STABEL (55), *Max-Delbruck-Labor in Der MPG, D-5000 Koln 30, Germany*

JAMES C. STONE (56), *Department of Biochemistry, University of Alberta, Edmonton, Alberta T6G 2H7, Canada*

THOMAS W. STURGILL (27), *Departments of Internal Medicine and Pharmacology, University of Virginia, Charlottesville, Virginia 22908*

SUSAN S. TAYLOR (41, 43, 49), *Department of Chemistry, University of California, San Diego, La Jolla, California 92093*

GEORGE THOMAS (12, 14, 22), *Friedrich Miescher-Institute, CH-4002 Basel, Switzerland*

JEAN A. TONER-WEBB (41), *Chiron Opthalmics, Irvine, California 92718*

BRIGIT A. VAN OIRSCHOT (6), *Department of Hematology, Laboratory of Medical Enzymology, University Hospital Utrecht, 3508 GA Utrecht, The Netherlands*

HAROLD E. VARMUS (53), *Departments of Microbiology and Immunology and Biochemistry and Biophysics, University of California, San Francisco, California 94143*

MICHAEL D. WATERFIELD (52), *Ludwig Institute for Cancer Research, London W1P 8BT, England*

YING-FEI WEI (32), *Laboratory of Toxicology, Harvard School of Public Health, Boston, Massachusetts 02115*

ANDREW F. WILKS (45) *Ludwig Institute for Cancer Research, PO Royal Hospital, Victoria 3050, Australia*

LYNN WOLFE (26), *Department of Biochemistry, University of Massachusetts Medical Center, Worcester, Massachusetts 01605*

JAMES ROBERT WOODGETT (13, 48), *Ludwig Institute for Cancer Research, London W1P 8BT, England*

PNINA YAISH (31), *Receptor Studies Laboratory, Ludwig Institute for Cancer Research, London W1P 8BT, England*

WES M. YONEMOTO (49, 51), *Department of Chemistry, University of California, San Diego, La Jolla, California 92093*

YASUYOSHI YOSHIDA (38), *Section on Metabolic Regulation, Endocrinology and Reproduction Research Branch, National Institute of Child Health and Human Development, National Institutes of Health, Bethesda, Maryland 20892*

CHIUN-JYE YUAN (36), *Department of Biochemistry and Biophysics, Iowa State University, Ames, Iowa 50011*

JIANHUA ZHENG (43), *Department of Chemistry, University of California, San Diego, La Jolla, California 92037*

MARK J. ZOLLER (51), *Department of Protein Engineering, Genentech, Inc., South San Francisco, California 94080*

# Preface

The field of protein phosphorylation has grown and changed considerably since it was covered in 1983 in Volume 99 on Protein Kinases in the *Methods in Enzymology* series. At that time fewer than five protein kinase amino acid sequences were known. The number of identified protein kinases and the number of processes known to be regulated by protein phosphorylation have both increased enormously since then, and the end is not yet in sight. Fundamental to this proliferation has been the ability to isolate novel genes encoding protein kinases using the techniques of molecular biology. Equally important is the fact that the similarity in amino acid sequence of the catalytic domains of the protein kinases allows the instantaneous realization that a molecular clone isolated on the basis of biological function or partial amino acid sequence encodes a protein kinase. It is now clear that many, and perhaps most, aspects of growth regulation are controlled by a complex network of protein kinases and phosphatases.

The techniques that have already defined the unexpectedly large size and degree of complexity of the protein kinase gene family, and will continue to do so, are described in Volumes 200 and 201. These two volumes were consciously entitled Protein Phosphorylation, rather than Protein Kinases. This decision had two origins. One was the emerging realization that the protein phosphatases may prove to be of as much regulatory significance as the protein kinases. The other was that the study of protein kinases is sterile in the absence of the identification and characterization of both upstream regulators and downstream polypeptide substrates, many of which will not be protein kinases.

Of necessity, the first protein kinases identified and studied were those whose activity was prominent in tissues that could be obtained in large quantities. Most of the protein kinases that are important in growth control, however, are present at extremely low levels in cells. The development of sensitive techniques to study nonabundant proteins was, therefore, imperative. Considerable attention is given in these volumes to the use of recombinant DNA techniques for the preparation of large quantities of protein kinases, to means by which to detect trace quantities of specific polypeptides in complex mixtures of proteins, and to techniques with which to perform protein chemistry on vanishingly small quantities of phosphoproteins.

What does the future of this field hold? A major "watershed" will be the determination of the three-dimensional structure of a protein kinase.

Techniques useful for the crystallization of cyclic AMP-dependent protein kinase are presented in Volume 200, but solution of the structure of the enzyme at atomic resolution has not yet been achieved. Knowledge of the structure of one or more protein kinases will almost certainly alter the study of these enzymes very significantly.

To date, with few exceptions, the study of protein phosphorylation has involved the study of the phosphorylation of proteins on serine, threonine, or tyrosine. The lack of attention paid to protein kinases generating acid-labile phosphoamino acids reflects not a lack of biological importance of these enzymes, since they clearly play a central role in bacterial chemotaxis, but rather the fact that methods for their study are few and poorly developed. Unanticipated and important roles for protein kinases may well become apparent if simple and reliable means with which to detect and study proteins containing labile phosphorylated amino acids are devised. No doubt the future will also hold other surprises, but we can only hope that four volumes are not needed the next time protein phosphorylation is covered in this series!

These volumes would never have seen "the light of day" without the diligence of Karen Lane. We thank her for her cheerful and tireless help.

BARTHOLOMEW M. SEFTON
TONY HUNTER

# METHODS IN ENZYMOLOGY

VOLUME XIII. Citric Acid Cycle
*Edited by* J. M. LOWENSTEIN

VOLUME XIV. Lipids
*Edited by* J. M. LOWENSTEIN

VOLUME XV. Steroids and Terpenoids
*Edited by* RAYMOND B. CLAYTON

VOLUME XVI. Fast Reactions
*Edited by* KENNETH KUSTIN

VOLUME XVII. Metabolism of Amino Acids and Amines (Parts A and B)
*Edited by* HERBERT TABOR AND CELIA WHITE TABOR

VOLUME XVIII. Vitamins and Coenzymes (Parts A, B, and C)
*Edited by* DONALD B. MCCORMICK AND LEMUEL D. WRIGHT

VOLUME XIX. Proteolytic Enzymes
*Edited by* GERTRUDE E. PERLMANN AND LASZLO LORAND

VOLUME XX. Nucleic Acids and Protein Synthesis (Part C)
*Edited by* KIVIE MOLDAVE AND LAWRENCE GROSSMAN

VOLUME XXI. Nucleic Acids (Part D)
*Edited by* LAWRENCE GROSSMAN AND KIVIE MOLDAVE

VOLUME XXII. Enzyme Purification and Related Techniques
*Edited by* WILLIAM B. JAKOBY

VOLUME XXIII. Photosynthesis (Part A)
*Edited by* ANTHONY SAN PIETRO

VOLUME XXIV. Photosynthesis and Nitrogen Fixation (Part B)
*Edited by* ANTHONY SAN PIETRO

VOLUME XXV. Enzyme Structure (Part B)
*Edited by* C. H. W. HIRS AND SERGE N. TIMASHEFF

VOLUME XXVI. Enzyme Structure (Part C)
*Edited by* C. H. W. HIRS AND SERGE N. TIMASHEFF

VOLUME XLI. Carbohydrate Metabolism (Part B)
*Edited by* W. A. WOOD

VOLUME XLII. Carbohydrate Metabolism (Part C)
*Edited by* W. A. WOOD

VOLUME XLIII. Antibiotics
*Edited by* JOHN H. HASH

VOLUME XLIV. Immobilized Enzymes
*Edited by* KLAUS MOSBACH

VOLUME XLV. Proteolytic Enzymes (Part B)
*Edited by* LASZLO LORAND

VOLUME XLVI. Affinity Labeling
*Edited by* WILLIAM B. JAKOBY AND MEIR WILCHEK

VOLUME XLVII. Enzyme Structure (Part E)
*Edited by* C. H. W. HIRS AND SERGE N. TIMASHEFF

VOLUME XLVIII. Enzyme Structure (Part F)
*Edited by* C. H. W. HIRS AND SERGE N. TIMASHEFF

VOLUME XLIX. Enzyme Structure (Part G)
*Edited by* C. H. W. HIRS AND SERGE N. TIMASHEFF

VOLUME L. Complex Carbohydrates (Part C)
*Edited by* VICTOR GINSBURG

VOLUME LI. Purine and Pyrimidine Nucleotide Metabolism
*Edited by* PATRICIA A. HOFFEE AND MARY ELLEN JONES

VOLUME LII. Biomembranes (Part C: Biological Oxidations)
*Edited by* SIDNEY FLEISCHER AND LESTER PACKER

VOLUME LIII. Biomembranes (Part D: Biological Oxidations)
*Edited by* SIDNEY FLEISCHER AND LESTER PACKER

VOLUME LIV. Biomembranes (Part E: Biological Oxidations)
*Edited by* SIDNEY FLEISCHER AND LESTER PACKER

VOLUME LV. Biomembranes (Part F: Bioenergetics)
*Edited by* SIDNEY FLEISCHER AND LESTER PACKER

VOLUME LVI. Biomembranes (Part G: Bioenergetics)
*Edited by* SIDNEY FLEISCHER AND LESTER PACKER

VOLUME LVII. Bioluminescence and Chemiluminescence
*Edited by* MARLENE A. DELUCA

VOLUME LVIII. Cell Culture
*Edited by* WILLIAM B. JAKOBY AND IRA PASTAN

VOLUME LIX. Nucleic Acids and Protein Synthesis (Part G)
*Edited by* KIVIE MOLDAVE AND LAWRENCE GROSSMAN

VOLUME LX. Nucleic Acids and Protein Synthesis (Part H)
*Edited by* KIVIE MOLDAVE AND LAWRENCE GROSSMAN

VOLUME 61. Enzyme Structure (Part H)
*Edited by* C. H. W. HIRS AND SERGE N. TIMASHEFF

VOLUME 62. Vitamins and Coenzymes (Part D)
*Edited by* DONALD B. MCCORMICK AND LEMUEL D. WRIGHT

VOLUME 63. Enzyme Kinetics and Mechanism (Part A: Initial Rate and Inhibitor Methods)
*Edited by* DANIEL L. PURICH

VOLUME 64. Enzyme Kinetics and Mechanism (Part B: Isotopic Probes and Complex Enzyme Systems)
*Edited by* DANIEL L. PURICH

VOLUME 65. Nucleic Acids (Part I)
*Edited by* LAWRENCE GROSSMAN AND KIVIE MOLDAVE

VOLUME 66. Vitamins and Coenzymes (Part E)
*Edited by* DONALD B. MCCORMICK AND LEMUEL D. WRIGHT

VOLUME 67. Vitamins and Coenzymes (Part F)
*Edited by* DONALD B. MCCORMICK AND LEMUEL D. WRIGHT

VOLUME 68. Recombinant DNA
*Edited by* RAY WU

VOLUME 69. Photosynthesis and Nitrogen Fixation (Part C)
*Edited by* ANTHONY SAN PIETRO

VOLUME 70. Immunochemical Techniques (Part A)
*Edited by* HELEN VAN VUNAKIS AND JOHN J. LANGONE

VOLUME 71. Lipids (Part C)
*Edited by* JOHN M. LOWENSTEIN

VOLUME 72. Lipids (Part D)
*Edited by* JOHN M. LOWENSTEIN

VOLUME 73. Immunochemical Techniques (Part B)
*Edited by* JOHN J. LANGONE AND HELEN VAN VUNAKIS

VOLUME 74. Immunochemical Techniques (Part C)
*Edited by* JOHN J. LANGONE AND HELEN VAN VUNAKIS

VOLUME 75. Cumulative Subject Index Volumes XXXI, XXXII, XXXIV–LX
*Edited by* EDWARD A. DENNIS AND MARTHA G. DENNIS

VOLUME 76. Hemoglobins
*Edited by* ERALDO ANTONINI, LUIGI ROSSI-BERNARDI, AND EMILIA CHIANCONE

VOLUME 77. Detoxication and Drug Metabolism
*Edited by* WILLIAM B. JAKOBY

VOLUME 78. Interferons (Part A)
*Edited by* SIDNEY PESTKA

VOLUME 79. Interferons (Part B)
*Edited by* SIDNEY PESTKA

VOLUME 80. Proteolytic Enzymes (Part C)
*Edited by* LASZLO LORAND

VOLUME 93. Immunochemical Techniques (Part F: Conventional Antibodies, Fc Receptors, and Cytotoxicity)
*Edited by* JOHN J. LANGONE AND HELEN VAN VUNAKIS

VOLUME 94. Polyamines
*Edited by* HERBERT TABOR AND CELIA WHITE TABOR

VOLUME 95. Cumulative Subject Index Volumes 61–74, 76–80
*Edited by* EDWARD A. DENNIS AND MARTHA G. DENNIS

VOLUME 96. Biomembranes [Part J: Membrane Biogenesis: Assembly and Targeting (General Methods; Eukaryotes)]
*Edited by* SIDNEY FLEISCHER AND BECCA FLEISCHER

VOLUME 97. Biomembranes [Part K: Membrane Biogenesis: Assembly and Targeting (Prokaryotes, Mitochondria, and Chloroplasts)]
*Edited by* SIDNEY FLEISCHER AND BECCA FLEISCHER

VOLUME 98. Biomembranes (Part L: Membrane Biogenesis: Processing and Recycling)
*Edited by* SIDNEY FLEISCHER AND BECCA FLEISCHER

VOLUME 99. Hormone Action (Part F: Protein Kinases)
*Edited by* JACKIE D. CORBIN AND JOEL G. HARDMAN

VOLUME 100. Recombinant DNA (Part B)
*Edited by* RAY WU, LAWRENCE GROSSMAN, AND KIVIE MOLDAVE

VOLUME 101. Recombinant DNA (Part C)
*Edited by* RAY WU, LAWRENCE GROSSMAN, AND KIVIE MOLDAVE

VOLUME 102. Hormone Action (Part G: Calmodulin and Calcium-Binding Proteins)
*Edited by* ANTHONY R. MEANS AND BERT W. O'MALLEY

VOLUME 103. Hormone Action (Part H: Neuroendocrine Peptides)
*Edited by* P. MICHAEL CONN

VOLUME 104. Enzyme Purification and Related Techniques (Part C)
*Edited by* WILLIAM B. JAKOBY

VOLUME 105. Oxygen Radicals in Biological Systems
*Edited by* LESTER PACKER

VOLUME 144. Structural and Contractile Proteins (Part D: Extracellular Matrix)
*Edited by* LEON W. CUNNINGHAM

VOLUME 145. Structural and Contractile Proteins (Part E: Extracellular Matrix)
*Edited by* LEON W. CUNNINGHAM

VOLUME 146. Peptide Growth Factors (Part A)
*Edited by* DAVID BARNES AND DAVID A. SIRBASKU

VOLUME 147. Peptide Growth Factors (Part B)
*Edited by* DAVID BARNES AND DAVID A. SIRBASKU

VOLUME 148. Plant Cell Membranes
*Edited by* LESTER PACKER AND ROLAND DOUCE

VOLUME 149. Drug and Enzyme Targeting (Part B)
*Edited by* RALPH GREEN AND KENNETH J. WIDDER

VOLUME 150. Immunochemical Techniques (Part K: *In Vitro* Models of B and T Cell Functions and Lymphoid Cell Receptors)
*Edited by* GIOVANNI DI SABATO

VOLUME 151. Molecular Genetics of Mammalian Cells
*Edited by* MICHAEL M. GOTTESMAN

VOLUME 152. Guide to Molecular Cloning Techniques
*Edited by* SHELBY L. BERGER AND ALAN R. KIMMEL

VOLUME 153. Recombinant DNA (Part D)
*Edited by* RAY WU AND LAWRENCE GROSSMAN

VOLUME 154. Recombinant DNA (Part E)
*Edited by* RAY WU AND LAWRENCE GROSSMAN

VOLUME 155. Recombinant DNA (Part F)
*Edited by* RAY WU

VOLUME 156. Biomembranes (Part P: ATP-Driven Pumps and Related Transport: The Na,K-Pump)
*Edited by* SIDNEY FLEISCHER AND BECCA FLEISCHER

VOLUME 170. Nucleosomes
*Edited by* PAUL M. WASSARMAN AND ROGER D. KORNBERG

VOLUME 171. Biomembranes (Part R: Transport Theory: Cells and Model Membranes)
*Edited by* SIDNEY FLEISCHER AND BECCA FLEISCHER

VOLUME 172. Biomembranes (Part S: Transport: Membrane Isolation and Characterization)
*Edited by* SIDNEY FLEISCHER AND BECCA FLEISCHER

VOLUME 173. Biomembranes [Part T: Cellular and Subcellular Transport: Eukaryotic (Nonepithelial) Cells]
*Edited by* SIDNEY FLEISCHER AND BECCA FLEISCHER

VOLUME 174. Biomembranes [Part U: Cellular and Subcellular Transport: Eukaryotic (Nonepithelial) Cells]
*Edited by* SIDNEY FLEISCHER AND BECCA FLEISCHER

VOLUME 175. Cumulative Subject Index Volumes 135–139, 141–167

VOLUME 176. Nuclear Magnetic Resonance (Part A: Spectral Techniques and Dynamics)
*Edited by* NORMAN J. OPPENHEIMER AND THOMAS L. JAMES

VOLUME 177. Nuclear Magnetic Resonance (Part B: Structure and Mechanism)
*Edited by* NORMAN J. OPPENHEIMER AND THOMAS L. JAMES

VOLUME 178. Antibodies, Antigens, and Molecular Mimicry
*Edited by* JOHN J. LANGONE

VOLUME 179. Complex Carbohydrates (Part F)
*Edited by* VICTOR GINSBURG

VOLUME 180. RNA Processing (Part A: General Methods)
*Edited by* JAMES E. DAHLBERG AND JOHN N. ABELSON

VOLUME 181. RNA Processing (Part B: Specific Methods)
*Edited by* JAMES E. DAHLBERG AND JOHN N. ABELSON

VOLUME 182. Guide to Protein Purification
*Edited by* MURRAY P. DEUTSCHER

VOLUME 183. Molecular Evolution: Computer Analysis of Protein and Nucleic Acid Sequences
*Edited by* RUSSELL F. DOOLITTLE

VOLUME 184. Avidin–Biotin Technology
*Edited by* MEIR WILCHEK AND EDWARD A. BAYER

VOLUME 185. Gene Expression Technology
*Edited by* DAVID V. GOEDDEL

VOLUME 186. Oxygen Radicals in Biological Systems (Part B: Oxygen Radicals and Antioxidants)
*Edited by* LESTER PACKER AND ALEXANDER N. GLAZER

VOLUME 187. Arachidonate Related Lipid Mediators
*Edited by* ROBERT C. MURPHY AND FRANK A. FITZPATRICK

VOLUME 188. Hydrocarbons and Methylotrophy
*Edited by* MARY E. LIDSTROM

VOLUME 189. Retinoids (Part A: Molecular and Metabolic Aspects)
*Edited by* LESTER PACKER

VOLUME 190. Retinoids (Part B: Cell Differentiation and Clinical Applications)
*Edited by* LESTER PACKER

VOLUME 191. Biomembranes (Part V: Cellular and Subcellular Transport: Epithelial Cells)
*Edited by* SIDNEY FLEISCHER AND BECCA FLEISCHER

VOLUME 192. Biomembranes (Part W: Cellular and Subcellular Transport: Epithelial Cells)
*Edited by* SIDNEY FLEISCHER AND BECCA FLEISCHER

VOLUME 193. Mass Spectrometry
*Edited by* JAMES A. MCCLOSKEY

VOLUME 194. Guide to Yeast Genetics and Molecular Biology
*Edited by* CHRISTINE GUTHRIE AND GERALD R. FINK

VOLUME 195. Adenylyl Cyclase, G Proteins, and Guanylyl Cyclase
*Edited by* ROGER A. JOHNSON AND JACKIE D. CORBIN

VOLUME 196. Molecular Motors and the Cytoskeleton
*Edited by* RICHARD B. VALLEE

VOLUME 197. Phospholipases
*Edited by* EDWARD A. DENNIS

VOLUME 198. Peptide Growth Factors (Part C)
*Edited by* DAVID BARNES, J. P. MATHER, AND GORDON H. SATO

VOLUME 199. Cumulative Subject Index Volumes 168–174, 176–194 (in preparation)

VOLUME 200. Protein Phosphorylation (Part A: Protein Kinases: Assays, Purification, Antibodies, Functional Analysis, Cloning, and Expression)
*Edited by* TONY HUNTER AND BARTHOLOMEW M. SEFTON

VOLUME 201. Protein Phosphorylation (Part B: Analysis of Protein Phosphorylation, Protein Kinase Inhibitors, and Protein Phosphatases)
*Edited by* TONY HUNTER AND BARTHOLOMEW M. SEFTON

VOLUME 202. Molecular Design and Modeling: Concepts and Applications (Part A: Proteins, Peptides, and Enzymes) (in preparation)
*Edited by* JOHN J. LANGONE

VOLUME 203. Molecular Design and Modeling: Concepts and Applications (Part B: Antibodies and Antigens, Nucleic Acids, Polysaccharides, and Drugs) (in preparation)
*Edited by* JOHN J. LANGONE

VOLUME 204. Bacterial Genetic Systems (in preparation)
*Edited by* JEFFREY H. MILLER

VOLUME 205. Metallobiochemistry (Part B: Metallothionein and Related Molecules) (in preparation)
*Edited by* JAMES F. RIORDAN AND BERT L. VALLEE

VOLUME 206. Cytochrome P450 (in preparation)
*Edited by* MICHAEL R. WATERMAN AND ERIC F. JOHNSON

VOLUME 207. Ion Channels (in preparation)
*Edited by* BERNARDO RUDY AND LINDA E. IVERSON

VOLUME 208. Protein–DNA Interactions (in preparation)
*Edited by* ROBERT T. SAUER

# Section I

## Classification of Protein Kinases and Phosphorylation Site Sequences

# [1] Protein Kinase Classification

*By* TONY HUNTER

Protein kinases are defined as enzymes that transfer a phosphate group from a phosphate donor onto an acceptor amino acid in a substrate protein. Generally the $\gamma$ phosphate of ATP, or another nucleoside triphosphate, is the donor, but individual enzymes may have other phosphate donors. Because most protein kinases have multiple substrates, it would seem reasonable to classify protein kinases based on the acceptor amino acid specificity rather than protein substrate specificity. In accordance with this idea, the Nomenclature Committee of the International Union of Biochemists has recommended that protein kinases be classified as follows:

1. Phosphotransferases with a protein alcohol group as acceptor called protein-serine/threonine kinases (i.e., serine- and threonine-specific protein kinases which generate phosphate esters)(E.C. 2.7.10)
2. Phosphotransferases with a protein phenolic group as acceptor called protein-tyrosine kinases (i.e., tyrosine-specific protein kinases which generate phosphate esters)(E.C. 2.7.11)
3. Phosphotransferases with a protein histidine, arginine, or lysine group as acceptor called protein-histidine kinases, etc. (i.e., enzymes that generate phosphoramidates at the 1- or 3-position of histidine, at the guanido group of arginine, or at the $\varepsilon$-$NH_2$ group of lysine)(E.C. 2.7.12)
4. Phosphotransferases with a protein cysteine group as acceptor called protein-cysteine kinases (i.e., enzymes that generate phosphate thioesters)(E.C. 2.7.13)
5. Phosphotransferases with a protein acyl group as acceptor called protein-aspartyl or glutamyl kinases (i.e., enzymes that generate mixed phosphate-carboxylate acid anhydrides)(E.C. 2.7.14)

Enzymes in the first two categories are well known,[1,2] whereas enzymes in the last three categories are less well characterized. Many enzymes that phosphorylate serine/threonine or tyrosine in proteins have been purified, and molecular clones for more than 100 of these protein kinases have been isolated from a variety of eukaryotic species.[3] In fact, the majority of the currently known protein kinase sequences were not first obtained from purified enzymes, but either as regulatory genes whose sequences proved to be related to the protein kinase family or else as a result of cloning strategies designed to isolate genes related to the protein kinase family. In many cases the acceptor amino acid specificity of these cloned genes has not been determined directly, but has been deduced indirectly from sequence comparison with protein kinases of known specificity. All of the known bona fide protein serine/threonine and protein-tyrosine kinases share a related catalytic domain of ~270 amino acids, which is discussed in detail by Hanks *et al.*,[4] and also described by Hanks and Quinn in the following chapter. Based on sequence comparisons within the catalytic domain, it is possible to place the protein-tyrosine kinases in a separate family, which can be further divided into subfamilies. There are several short sequence motifs within the catalytic domain of the protein-tyrosine kinase family that can be used to recognize this type of enzyme. Protein-tyrosine kinase activity has been reported in prokaryotes,[5–7] but so far none of these enzymes has been purified or molecularly cloned. Interestingly, a protein-tyrosine phosphatase closely related to the family of eukaryotic protein-tyrosine phosphatases has been cloned from the plague bacterium, *Yersinia,* but this gene is carried on a plasmid essential for virulence, and its evolutionary origins are obscure.[8]

The protein-serine/threonine family is more diverse than the protein-tyrosine kinase family.[4] One can recognize distinct subfamilies, and these families have sequence motifs that are distinct from the motifs found in the equivalent regions of the protein-tyrosine kinases. However, there are several protein kinases that are not closely related in sequence to any other protein kinase, while overall being closer to the protein-serine/threonine kinases than to the protein-tyrosine kinases. Some of these loners have been shown to be authentic protein-serine/threonine kinases, but a few of these protein kinases have been found to have the ability to phosphorylate both serine/threonine and tyrosine (e.g., wee1$^{+}$,[9,10] Clk and Nek,[11] YPK1,[12] and Spk1[13]) and it will be probably be necessary to create another group of protein kinases that have this dual amino acid acceptor specificity. Clearly this highlights the importance of determining acceptor amino acid specificity directly, rather than relying on sequence similarities. Protein-serine/threonine kinases have also been described in prokaryotic systems.[14] Apart from the isocitrate dehydrogenase kinase/phosphatase, rather little is known about most of these enzymes at the molecular level. The sequence of the isocitrate dehydrogenase kinase/phosphatase is largely unrelated to that of the eukaryotic protein kinase family.[15] It is not known whether there are similar bifunctional enzymes encoded by the nuclear genomes of eukaryotes. Given that the isocitrate dehydrogenase kinase/phosphatase is not related to other protein kinases, it should probably be put in a separate category. There may also be prokaryotic protein-serine/threonine kinases related to the eukaryotic enzymes, since a protein-serine/threonine kinase-like sequence has been cloned from myxobacteria.[16]

The existence of proteins phosphorylated on histidine, lysine, and arginine and enzymatic activities with these specificities has been reported in eukaryotic cells,[17–20] but so far none of the protein kinases responsible has been purified to homogeneity, and the molecular nature of these enzymes is currently unknown (see chapter by Matthews). It will be of obvious interest to determine whether the protein-histidine kinases are related in sequence to the protein-serine/threonine and protein-tyrosine kinases. Several two-member regulatory enzyme systems involved in stimulus–response coupling are known in prokaryotes (for review, see Ref. 21). In these systems, the stimulus causes one member to autophosphorylate on a histidine residue using ATP as a phosphate donor,[22,23] and this phosphate is then transferred onto an acceptor aspartyl residue in the second protein,[23] which is activated by phosphorylation to deliver the response. These enzymes should then strictly be called protein-aspartyl kinases, rather than protein-histidine kinases. Their phosphorylation mechanism is distinct from that of the protein-serine/threonine and protein-tyrosine kinases, which do not use a phosphoenzyme intermediate. These enzymes all have a related catalytic domain, which is totally distinct from that of the protein-serine/threonine and protein-tyrosine kinase families, and which contains three conserved regions including the acceptor histidine spaced over ~150 amino acids.[21] Whether such enzymes exist in eukaryotic cells is unknown. Protein-cysteine kinases have not been reported for eukaryotic cells. However, there is evidence that one of the proteins in the sugar transfer enzyme system in prokaryotes, which involves serial transfer of phosphate between four proteins starting with phosphoenolpyruvate as the initial phosphate donor,[24] can be phosphorylated on cysteine.[25]

It is more difficult to subdivide protein kinases within each major acceptor group. Because of the pleiotropic substrate specificity of most of the protein kinases, a system based on substrates does not seem useful. A classification based on the nature of the regulators of protein kinase activity will be valuable in some instances, but poses a problem in those cases where the nature of the regulator is unknown. For example, several of the protein-serine/threonine kinases can be classified according to their regulators. Thus, there is a family of cyclic nucleotide-regulated protein kinases (e.g., the cAMP-dependent protein kinases), a family of calmodulin-regulated protein kinases (e.g., the myosin light chain kinases), and a family of diacylglycerol-regulated protein kinases (i.e., the protein kinase Cs). In addition to their functional relationships, these groups of protein kinases prove to be closely related in primary sequence within their catalytic domains, and usually also within their regulatory domains. A number of the protein-tyrosine kinases are growth factor receptors. These enzymes are regulated by their growth factor ligands, and can be classified on this basis. Several of the second messenger-regulated protein-serine/threonine kinases have been given acronyms, which provides a convenient way to refer to them. However, there has not been totally consistent usage of these acronyms. The committee suggests that the use of PKA for the cAMP-dependent protein kinase, and PKC for protein kinase C, should be adopted wherever possible, since these are the most widely used.

Other protein kinases are not known to be regulated directly by second messengers, and are often known by trivial names derived from the substrate that was first used to identify them (e.g., the casein kinases). The use of such names is not ideal because most of these protein kinases have additional substrates. Ideally, these names should be changed, but this will be difficult because they are so widely accepted. In addition, there is no obvious alternative at present. The same holds true for the protein-tyrosine kinases that were first identified as the products of oncogenes, such as $pp60^{v\text{-}src}$, for which there are no known second messenger regulators. It seems reasonable to continue to describe these enzymes using the convention adopted for oncogene products, which uses the predicted or actual molecular weight of the protein ($\times 10^{-3}$) preceded by "p" for protein or "pp" for phosphoprotein, and followed by the three-letter italicized and superscripted acronym for the gene. Viral oncogene and cellular protooncogene products are distinguished by v- and c- prefixes, respectively. Cloned protein kinase genes, whose functions are not known, have generally been given three-letter acronyms. Human protein kinase genes are styled with three upper case italicized letters (e.g., *TRK*), and the proteins encoded by human genes are referred to as TRK, etc. For protein kinases from other species, the protein products are referred to as Trk, etc.

Most of the protein kinases that will be identified in the future are likely to be derived from gene cloning, either using directed search strategies or else as a result of sequencing individual genes or ultimately entire genomes. There are currently well over 100 bona fide or putative mammalian protein kinases, which have been cloned and which appear to be derived from distinct genes. The number is increased significantly when one includes protein kinase isoforms generated by alternate splicing, which for some genes can result in multiple distinct proteins. If one considers cloned protein kinases obtained from other species, only some of which are obviously homologs of the mammalian enzymes, then the number of protein kinases is over 200. From Fig. 1 it is apparent that the rate of discovery of new protein kinase sequences has not abated over the last 10 years. Indeed, with the advent of the polymerase chain reaction technique, which has enabled the cloning of protein kinases by using degenerate oligonucleotide primers corresponding to conserved motifs to amplify sequences lying

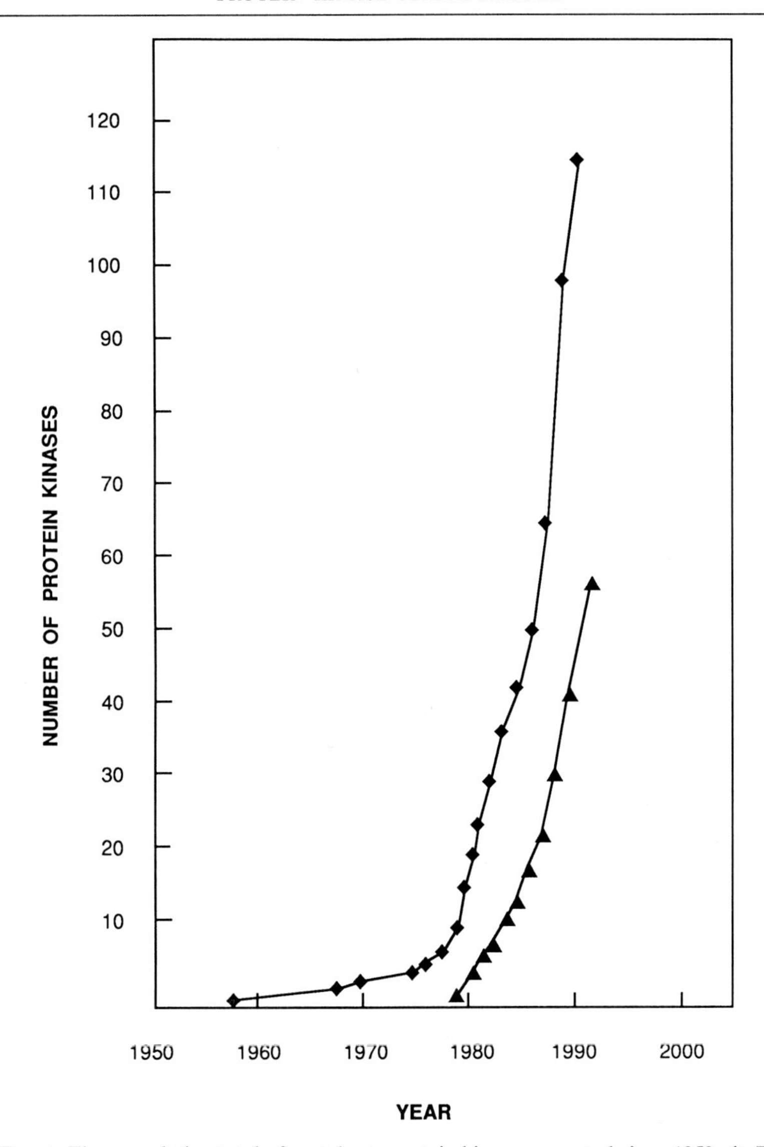


FIG. 1. The cumulative total of vertebrate protein kinases reported since 1959. ◆, Total protein kinases, plotted as the cumulative annual totals of protein-serine/threonine and protein-tyrosine kinases. ▲, Total protein-tyrosine kinases, plotted as the cumulative annual totals of protein-tyrosine kinases. Where homologs from different vertebrate species have been identified, only one example has been used. Partial catalytic domain sequences obtained by polymerase chain reaction (PCR) cloning have not been included. The final values for each curve were from March, 1990.

between such motifs, the rate of discovery of new protein kinases has accelerated. It is hard to predict where this curve will level off, but it could even reach the magic number of 1001[3]!

Clearly this explosion of protein kinase sequences poses a serious problem for classification. The best solution would appear to be the use of catalytic domain phylogenetic trees based on comparison of catalytic domain sequences, as outlined in the chapter by Hanks and Quinn ([2] in this volume). As discussed above this type of analysis groups successfully protein kinases that are known to have related functions. Thus the cNMP-dependent, calmodulin-dependent, and diacylglycerol-regulated protein-serine kinases all form clusters within the tree. Indeed, it is notable that these three groups of second messenger-regulated enzymes form a larger connected group in the tree. Other families of protein kinases also emerge, such as the cdc2 group. It is difficult to define precisely what level of similarity one should use to define true family members, but a level of identity above 60% would seem reasonable. This number is close to the degree of identity between the human and *S. cerevisiae* cdc2 proteins, which have been shown to be functionally interchangeable.

The chapter by Hanks and Quinn ([2], this volume) includes a list of 117 protein kinase catalytic domain sequences, an alignment of these sequences, and series of phylogenetic trees derived from these sequences. A database containing these sequences can be obtained electronically over Internet as described in that chapter. However, the database currently contains only one example of each homolog, when often there are many, and there are additional protein kinase sequences that have been reported since the chapter was written. Table I below is an updated, combined version of the tables of protein kinases to be found in Hanks *et al.*,[4] in which protein kinases were classified and grouped in subfamilies based on sequence relatedness. Some changes will be evident in classification due to the advent of new protein kinase sequences, which have enabled new families to be defined. Within the protein-serine/threonine kinase group, the families are listed in order of relatedness, starting with the cNMP-regulated protein kinases and moving clockwise as indicated in the phylogenetic tree in the chapter by Hanks and Quinn ([2], this volume). In addition, it should be noted that the dual-specificity protein kinases have for the moment been included under the protein-serine/threonine kinase heading.

Table I is not an exhaustive catalog of either protein kinases or citations. For instance, only protein-serine/threonine and protein-tyrosine kinases are listed. There are additional protein-serine kinase activities that have been reported (for review see Ref. 1), but only those for which there is some evidence that the catalytic entity is distinct from that of other protein kinases are included. The primary intention of Table I is to give pertinent references for the sequences of the protein-serine/threonine and protein-tyrosine kinase family. Where a complete or partial catalytic domain amino acid sequence has been reported, this is indicated as an inset entry below the main entry with the relevant references. If the sequence of an individual protein kinase from more than one species has been determined, these are listed as subheadings a, b, c, etc. Where there is more than one isolate of a viral or tumor oncogene-encoded protein kinase these are listed as subheadings i, ii, iii, etc. If no sequence data have been obtained, either a review for that protein kinase is cited or else an original reference is given. Where possible the protein kinases are grouped into families based on their mode of regulation and degree of sequence similarity in their catalytic domains. Where applicable the abbreviations for each protein kinase are those that were used in the sequence alignment in Hanks *et al.*[4] Table I is current as of November, 1990. Inevitably this list is incomplete, and apologies are due to those whose citations have inadvertently been omitted. Indeed, information concerning any omissions would be welcome, so that they can be corrected the next time this table is published.

TABLE I
PROTEIN KINASES

| Classification | References |
|---|---|
| **Protein-serine/threonine kinases** | |
| Vertebrate protein-serine/threonine kinases | |
| A. Cyclic nucleotide regulated | |
| 1. cAPK-$\alpha$: cAMP-dependent protein kinase catalytic subunit, $\alpha$ form | |
| a. Bovine cardiac muscle protein | 26 |
| b. Mouse S49 lymphoma cell cDNA | 27, 28 |
| c. Porcine LLC-$PK_1$ cell cDNA | 29 |
| d. Human HeLa cell cDNA | 30 |
| 2. cAPK-$\beta$: cAMP-dependent protein kinase catalytic subunit, $\beta$ form | |
| a. Bovine cardiac muscle cDNA | 31 |
| b. Mouse S49 lymphoma cell cDNA | 27, 28 |
| c. Porcine LLC-$PK_1$ cell cDNA | 29 |
| 3. cAPK-$\gamma$: cAMP-dependent protein kinase catalytic subunit, $\gamma$ form | |
| Human testis cDNA | 32 |
| 4. cGPK-$\alpha$: cGMP-dependent protein kinase, $\alpha$ form | |
| Bovine lung protein | 33 |

(*continued*)

TABLE I (*continued*)

| Classification | References |
|---|---|
| 5. cGPK-$\beta$: cGMP-dependent protein kinase, $\beta$ form (alternately spliced form of cGPK-$\alpha$ with a different N terminus) | |
| a. Bovine smooth muscle cDNA | 34 |
| b. Pig coronary artery protein | 35, 36 |
| B. Diacylglycerol regulated | |
| 1. PKC-$\alpha$: Protein kinase C, $\alpha$ form | |
| a. Bovine brain cDNA | 37 |
| b. Rabbit brain cDNA (designated $\gamma$) | 38 |
| c. Human brain cDNA | 39 |
| 2. PKC-$\beta$: Protein kinase C, $\beta$ form | |
| a. Bovine brain cDNA | 39 |
| b. Rat brain cDNA | 40, 41 |
| Two splice forms: II and III | |
| c. Rabbit brain cDNA | 38 |
| Two splice forms: $\beta$ and $\beta'$ | |
| d. Human brain cDNA | 39 |
| 3. PKC-$\gamma$: Protein kinase C, $\gamma$ form | |
| a. Bovine brain cDNA | 39 |
| b. Rat brain cDNA (designated form I) | 40 |
| c. Human brain cDNA | 39 |
| d. Rabbit brain cDNA (designated $\delta$) | 42 |
| 4. PKC-$\delta$: Protein kinase C, $\delta$ form | |
| Rat brain cDNA | 43, 44 |
| 5. PKC-$\varepsilon$: Protein kinase C, $\varepsilon$ form | |
| a. Rat brain cDNA | 41, 43, 44 |
| b. Rabbit brain cDNA (designated in nPKC) | 45 |
| 6. PKC-$\zeta$: Protein kinase C, $\zeta$ form | |
| Rat brain cDNA | 43, 44, 46 |
| 7. PKC-L: Protein kinase C-related gene (also called PKC-$\eta$) | |
| Human keratinocyte cDNA | 47, 48 |
| 8. PSK-K7: Protein kinase C-related gene | |
| Human cDNA | 49 |
| C. Calcium/calmodulin regulated | |
| 1. PhK-$\gamma$: Phosphorylase kinase, $\gamma$ subunit | |
| a. Rabbit skeletal muscle protein and cDNA | 50, 51 |
| b. Mouse muscle cDNA | 52 |
| c. Rat muscle cDNA | 53 |
| 2. PSK-C3: Putative protein-serine kinase related to phosphorylase kinase $\gamma$ | |
| Human HeLa cell cDNA | 54, 55 |
| 3. MLCK-K: Myosin light chain kinase, skeletal muscle | |
| Rabbit skeletal muscle protein | 56 |
| 4. MLCK-M: Myosin light chain kinase, smooth muscle | |
| Chicken gizzard cDNA | 57 |
| 5. nmMLCK: Myosin light chain kinase, nonmuscle type | |
| Chicken fibroblast cDNA | 58 |

TABLE I (*continued*)

| Classification | References |
|---|---|
| 6. CaMI: Calcium/calmodulin-dependent protein kinase Iβ | 59 |
| 7. CaMII-α: Calcium/calmodulin-dependent protein kinase II, α subunit | |
| Rat brain cDNA and protein | 60–62 |
| 8. CaMII-β: Calcium/calmodulin-dependent protein kinase II, β subunit | |
| Rat brain cDNA | 63 |
| 9. CaMII-γ: Calcium/calmodulin-dependent protein kinase II, γ subunit | |
| Rat brain cDNA | 64 |
| There is evidence for additional tissue-specific CaMII protein kinase isozymes. | |
| 10. CaMIII: Calcium/calmodulin-dependent protein kinase III | 65 |
| 11. CaMIV; λICM-1: Brain calmodulin-binding protein | |
| Mouse brain cDNA | 66 |
| 12. PSK-H1: Putative protein-serine kinase | |
| Human HeLa cell cDNA | 54, 67 |
| D. Ribosomal S6 protein kinases | |
| 1. S6KI: Ribosomal protein S6 kinase I. This type of protein kinase contains two complete protein kinase catalytic domains. It is not known whether one or both kinase domains are functional. The N-terminal domain is most closely related to the cAMP-dependent protein kinase family and the C-terminal domain to the calmodulin-regulated protein kinase family. | |
| a. *Xenopus* oocyte 90-kDa protein and ovarian cDNA. The S6KIIα cDNA corresponds to S6KI and not S6KII as originally reported. A second, related cDNA was found which may correspond to S6KII | 68, 69 |
| b. Mouse and chick fibroblast cDNAs (*rsk*-1) | 70 |
| c. Rat hepatoma cell cDNA | 71 |
| 2. S6KII: Ribosomal protein S6 kinase II | |
| a. *Xenopus* oocyte 92-kDa protein and ovarian cDNA (S6KIIβ) | 68, 69 |
| b. Mouse fibroblast cDNA (*rsk*-2) | 70 |
| 3. 70-kDa S6 kinase: This is distinct from S6KI and S6KII and has a single catalytic domain | |
| Rat liver and hepatoma cDNAs | 71, 72 |
| E. Serpentine receptor kinases | |
| 1. βARK: β-Adrenergic receptor protein kinase | |
| Bovine brain cDNA | 73 |
| 2. βARK-related protein kinase | |
| Bovine brain cDNA | 74 |
| 3. Rhodopsin kinase | 75 |
| F. Casein kinase II | |

(*continued*)

TABLE I (*continued*)

| Classification | References |
|---|---|
| 1. CKII-α: Casein kinase II-α | |
| a. Bovine lung α subunit protein | 76 |
| b. Human α subunit cDNA | 77 |
| c. Rat α subunit cDNA | 77 |
| d. Chicken α subunit cDNA | 78 |
| 2. CKIIα′: Casein kinase II α′ (closely related to CKII-α) | |
| a. Bovine α′ subunit cDNA | 79 |
| b. Chicken α′ subunit cDNA | 78 |
| 3. Nuclear protein kinase N2 (possibly the same as casein kinase II) | 80 |
| G. Glycogen synthase kinase 3 | |
| 1. GSK-3α: Glycogen synthase kinase 3α | |
| Rat brain cDNA | 81 |
| 2. GSK-3β: Glycogen synthase kinase 3β | |
| Rat brain cDNA | 81 |
| H. cdc2 family | |
| 1. cdc2Hs: Human functional homolog of yeast *cdc2⁺/CDC28* gene products. cdc2 is the catalytic subunit of histone H1 (growth associated) kinase and also the catalytic subunit of maturation-promoting factor (MPF). In both cases it is associated with a cyclin B subunit. | 82, 83, 83a |
| a. Human SV40-transformed fibroblast cDNA | 84 |
| b. Mouse fibroblast cDNA | 85, 85a, 85b |
| 2. cdc2′: cdc2-related protein kinase (does not complement the *cdc2* mutation) | |
| Human HeLa cell cDNA | 86 |
| 3. cdc2–eg1: cdc2-related cDNA (does not complement the *cdc2* mutation) | |
| a. *Xenopus* oocyte cDNA | 87 |
| b. Human HeLa cell cDNA | 86 |
| I. cdc2-related protein kinases | |
| 1. PSK-J3: Putative protein-serine kinase related to yeast cdc2⁺/CDC28 | |
| Human HeLa cell cDNA | 54 |
| 2. *mak* protein: Putative protein-serine kinase related to cdc2 | |
| Rat testis cDNA | 88 |
| 3. MO15: *Xenopus* gene product related to cdc2 that negatively regulates oocyte maturation | |
| *Xenopus* oocyte cDNA | 89 |
| 4. GTA kinase: cdc2-related protein kinase | |
| Hamster CHO cell cDNA | 90 |
| J. MAP kinases | |
| 1. MAPK-1: Microtubule-associated protein 2 (MAP-2) protein kinase (this is also known as the mitogen-activated protein (MAP) kinase (p42) and is the same as myelin basic protein kinase) | 91 |

TABLE I (*continued*)

| Classification | References |
|---|---|
| Rat brain cDNA (called ERK-1, this may be the 45-kDa form) (a similar cDNA has been cloned from starfish oocytes) | 92, 93 |
| 2. MAPK-2: Protein kinase related to MAP kinase (possibly MAPK p42) | |
| Rat brain cDNA (called ERK-2, this may be the 42-kDa form) | 92 |
| There are other ERK-1-related protein kinases, as well as MAP-2 protein kinases in other species which appear to be related to these enzymes. | 94 |
| 3. MAPK-3: 54-kDa MAP-2 protein kinase (it is not clear how closely related this enzyme is to MAPK-1 and MAPK-2) | |
| Rat liver protein | 95 |
| K. Mos/Raf protein kinases | |
| 1. c-*raf*/*mil* protein: Cellular homolog of v-*raf*/*mil* protein | |
| a. Human fetal liver cDNA | 96 |
| b. Chicken erythroblast cDNA | 97 |
| c. v-*raf*/*mil* retroviral oncogene products | |
| i. 3611 murine sarcoma virus | 98 |
| ii. Mill Hill 2 avian acute leukemia virus | 99, 100, 101 |
| 2. A-*raf* protein: Cellular oncogene closely related to c-*raf* protein | |
| a. Human T cell cDNA | 102 |
| b. Mouse spleen cDNA | 103 |
| c. Rat cDNA | 104 |
| 3. *pks* protein: Cellular oncogene product closely related to A-*raf* protein (may be identical to A-*raf*) | |
| Human fetal liver cDNA | 105 |
| 4. B-*raf* protein: Cellular oncogene product closely related to c-*raf* protein | |
| Human Ewing sarcoma-derived cDNA (activated during transfection by recombination) | 106 |
| 5. c-R-*mil* protein: Cellular oncogene product related to c-*raf*/*mil* protein | |
| v-R-*mil* retroviral oncogene product | |
| Avian oncogenic retrovirus | 107 |
| 6. c-*mos* protein: Cellular homolog of v-*mos* protein | |
| a. Human placenta genomic DNA | 108 |
| b. Mouse NIH/3T3 cell genomic DNA | 109 |
| c. Rat 3Y1 cell genomic DNA | 110 |
| d. Chicken genomic DNA | 111 |
| e. Monkey genomic DNA | 112 |
| f. *Xenopus* genomic DNA | 113, 114 |
| g. v-*mos* retroviral oncogene products | |
| i. Moloney murine sarcoma virus 124 | 109, 115, 116 |
| ii. Moloney murine sarcoma virus m1 | 117 |
| iii. Moloney murine sarcoma virus HT-1 | 118 |
| iv. Myeloproliferative sarcoma virus | 119 |

(*continued*)

TABLE I (*continued*)

| Classification | References |
|---|---|
| L. Casein kinase I | |
| 1. CKI: Casein kinase I | 120 |
| 2. Nuclear protein kinase N1 (possibly the same as casein kinase I) | 80 |
| M. Potential dual-specificity protein kinases | |
| 1. PYT: Putative protein kinase identified by screening an expression library with anti-phosphotyrosine antibodies | |
| Human fibroblast cDNA | 121, 122 |
| 2. *clk* protein: Putative protein kinase identified by screening an expression library with anti-phosphotyrosine antibodies | |
| Mouse erythroleukemia cell line cDNA | 11 |
| 3. *nek* protein: Putative protein kinase identified by screening an expression library with anti-phosphotyrosine antibodies | |
| Mouse erythroleukemia cell line cDNA | 11 |
| N. Other protein kinases with no close relatives | |
| 1. Double-stranded RNA-regulated protein kinase | |
| Human Daudi cell cDNA | 123 |
| 2. *cot* protein: Product of oncogene cloned from a human thyroid carcinoma | |
| Human thyroid carcinoma TC04-derived cDNA | 124 |
| 3. *pim-1* protein: Product of putative oncogene activated by murine leukemia virus integration | |
| a. Mouse BALB/c cell genomic DNA | 125 |
| b. Human K562 CML cell cDNA | 126, 127 |
| 4. PSK-G1: Putative protein-serine kinase related to PSK-H2 | |
| Human HeLa cell cDNA | 128 |
| 5. PSK-H2: Putative protein-serine kinase related to PSK-G1 | |
| Human HeLa cell cDNA | 128 |
| 6. Glycogen synthase kinase 4 | 129 |
| 7. Heme-regulated protein kinase | 129a |
| 8. Double-stranded DNA-regulated protein kinase | 130–132 |
| 9. Hydroxymethylglutaryl-CoA reductase kinase (identical to 5′ AMP-activated protein kinase) | 133, 134 |
| 10. Pyruvate dehydrogenase kinase | 135 |
| 11. Branched chain ketoacid dehydrogenase kinase | 136 |
| 12. Polypeptide-dependent protein kinase | 137 |
| 13. Polyamine-stimulated protein kinase | 138 |
| 14. Protease-activated kinase I (distinct from known protein kinase C isozymes) | 139, 140 |
| 15. Protease-activated kinase II may be the same as one of the protein kinase C isoforms | 141 |
| 16. Guanylate cyclase A: The negative regulatory domain of guanylate cyclase, which serves as the receptor for atrial natriuretic factor (ANF), has homology with protein-serine | |

(*continued*)

TABLE I (*continued*)

| Classification | References |
|---|---|
| kinases and may function by phosphorylating the catalytic cyclase domain. | |
| Rat brain cDNA (there are also sea urchin cDNAs) | 142–144 |
| 17. Guanylate cyclase B: ANF receptor-related guanylate cyclase | |
| Rat brain cDNA | 145 |
| 18. HSVK: Herpes simplex virus-*US3* gene product | |
| Herpes simplex virus genomic DNA | 146 |
| Three other putative herpesvirus-encoded protein kinases have been detected through sequence homology. | 147 |
| 19. hbx: Hepatitis B virus (HBV) transactivator protein (this protein is unrelated in sequence to other protein-serine/threonine kinases) | |
| HBV genomic sequence | 148 |
| *Drosophila* protein-serine/threonine kinases | |
| A. Cyclic nucleotide regulated | |
| 1. cAMP-dependent protein kinase *Drosophila* homolog (DC0) | |
| *D. melanogaster* cDNA | 149, 150 |
| 2. DC1: Novel protein kinase related to cAPK | |
| *D. melanogaster* cDNA | 150 |
| 3. DC2: Novel protein kinase related to cAPK | |
| *D. melanogaster* cDNA | 150 |
| 4. cGMP-dependent protein kinase *Drosophila* homolog (DG1) | |
| *D. melanogaster* cDNA | 149, 150 |
| 5. DG2: Novel protein kinase related to cGPK | |
| *D. melanogaster* cDNA | 150 |
| B. Diacylglycerol regulated | |
| 1. D-PKC: Protein kinase C homolog | |
| *D. melanogaster* cDNA | 151 |
| 2. D-PKC-53E: Protein kinase C homolog specifically expressed in *Drosophila* photoreceptor cells | |
| *D. melanogaster* cDNA | 152 |
| 3. D-PKC-98F: Protein kinase C homolog expressed in embryo and adult brain | |
| *D. melanogaster* cDNA | 152 |
| C. cdc2 family | |
| 1. CDC2: cdc2-related protein kinase that complements *Schizosaccharomyces pombe cdc2* mutants | |
| *D. melanogaster* cDNA | 153, 154 |
| 2. CDC2c: cdc2-related protein kinase that does not complement *S. pombe cdc2* mutants | |
| *D. melanogaster* cDNA | 153 |

(*continued*)

TABLE I (*continued*)

| Classification | References |
|---|---|
| D. Others | |
| 1. CKII$\alpha$-D: Casein kinase II, $\alpha$-subunit *Drosophila* homolog | |
| *D. melanogaster* cDNA | 155 |
| 2. D-*raf*-1 protein: Homolog of mammalian c-*raf* protein identical to l(1)polehole which is required for the function of the torso receptor protein-tyrosine kinase in development | 156 |
| *D. melanogaster* cDNA | 157, 158 |
| 3. D-*raf*-2 protein: Homolog of mammalian c-*raf* protein | 157 |
| 4. *ninaC* protein: Gene product essential for normal photoreceptor function, containing an N-terminal protein kinase catalytic domain joined to a C terminus related to myosin heavy chain | |
| *D. melanogaster* cDNA | 159 |
| 5. *fused* protein: Gene product essential for proper segmentation in *Drosophila* | |
| *D. melanogaster* genomic DNA | 160 |
| 6. *zeste-white 3* protein: Gene product essential for proper segmentation in *Drosophila*; identical to *shaggy* (homolog of GSK-3$\beta$) | |
| *D. melanogaster* cDNA and genomic DNA | 161, 162 |
| Yeast protein-serine/threonine kinases | |
| A. Cyclic AMP-dependent protein kinase related | |
| 1. TPK1 (also designated PK25 or SRA3): cAMP-dependent protein kinase yeast homolog, type 1 | |
| *Saccharomyces cerevisiae* genomic DNA | 163–165 |
| 2. TPK2: cAMP-dependent protein kinase yeast homolog, type 2 | |
| *S. cerevisiae* genomic DNA | 163 |
| 3. TPK3: cAMP-dependent protein kinase yeast homolog, type 3 | |
| *S. cerevisiae* genomic DNA | 163 |
| 4. SCH9: cAMP-dependent protein kinase-related gene product | |
| *S. cerevisiae* genomic DNA | 166 |
| B. Calmodulin regulated | |
| 1. CMK1: Calmodulin-dependent protein kinase II yeast homolog | |
| *S. cerevisiae* genomic DNA | 167 |
| 2. CMK2: Calmodulin-dependent protein kinase II yeast homolog | |
| *S. cerevisiae* genomic DNA | 167 |
| C. Diacylglycerol regulated | |
| 1. PKC1: Protein kinase C yeast homolog | |
| *S. cerevisiae* genomic DNA | 168 |
| 2. PKC2: Protein kinase C-like, $Ca^{2+}$/phosphatidylserine/diacylglycerol-dependent *S. cerevisiae* protein kinase | 167 |
| D. CDC28/cdc2$^{+}$ family | |
| 1. CDC28: "Cell-division-cycle" gene product—protein-serine kinase | |
| *S. cerevisiae* genomic DNA | 169 |

TABLE I (*continued*)

| Classification | References |
|---|---|
| 2. cdc2$^{+}$: "Cell-division-cycle" gene product—protein-serine kinase | |
| *S. pombe* genomic DNA | 170 |
| 3. KIN28: Putative protein-serine kinase related to CDC28/cdc2$^{+}$ | |
| *S. cerevisiae* genomic DNA | 171 |
| 4. KSS1: Putative protein-serine kinase—suppressor of SST2 | |
| *S. cerevisiae* genomic DNA | 172 |
| 5. PHO85: Putative protein-serine kinase—negative regulator of the PHO system (phosphate uptake) | |
| *S. cerevisiae* genomic DNA | 173 |
| 6. FUS3: Putative protein-serine kinase—negative regulator of mitosis and positive regulator of meiosis | |
| *S. cerevisiae* genomic DNA | 174 |
| 7. spk1$^{+}$: Putative protein-serine kinase related to KSS1 | |
| *S. pombe* genomic DNA | 175 |
| E. STE7 family | |
| 1. STE7: "Sterile" mutant wild-type allele gene product | |
| *S. cerevisiae* genomic DNA | 176 |
| 2. PBS2: Polymyxin B antibiotic resistance gene product | |
| *S. cerevisiae* genomic DNA | 177 |
| F. SNF1 family | |
| 1. SNF1: "Sucrose nonfermenting" gene product | |
| *S. cerevisiae* genomic DNA | 178 |
| 2. KIN1: Unusual yeast protein-serine kinase, type 1 | |
| a. *S. cerevisiae* genomic DNA | 179 |
| b. *S. pombe* genomic DNA | 180 |
| 3. KIN2: Unusual yeast protein-serine kinase, type 2 | |
| *S. cerevisiae* genomic DNA | 179 |
| 4. nim1$^{+}$: "Novel inducer of mitosis": suppressor of *cdc25* mutants | |
| *S. pombe* genomic DNA | 181 |
| G. Casein kinase family | |
| 1. Casein kinase II $\alpha 1$ yeast homolog (CKA1) | |
| *S. cerevisiae* genomic DNA | 182 |
| 2. Casein kinase II $\alpha 2$ (CKA2) | |
| *S. cerevisiae* genomic DNA | 183 |
| H. Putative dual-specificity protein kinases | |
| 1. wee1$^{+}$: "Reduced size at division" mutant wild-type allele gene product | |
| *S. pombe* genomic DNA | 184 |
| 2. mik1$^{+}$: Novel yeast protein-serine kinase related to wee1$^{+}$ | |
| *S. pombe* genomic DNA | 185 |
| 3. SPK1: Novel yeast protein kinase gene essential for sporulation isolated by screening with anti-phosphotyrosine antibodies | |
| *S. cerevisiae* genomic DNA | 13 |

(*continued*)

TABLE I (*continued*)

| Classification | References |
|---|---|
| 4. HRR25: Novel yeast protein-serine kinase gene needed for meiotic recombination and progression through the cell cycle | |
| *S. cerevisiae* genomic DNA | 186 |
| I. Others | |
| 1. ran1$^{+}$: "Meiotic bypass" mutant wild-type allele gene product | |
| *S. pombe* genomic DNA | 187 |
| 2. CDC7: "Cell-division-cycle" mutant wild-type allele gene product | |
| *S. cerevisiae* genomic DNA | 188 |
| 3. STE11: "sterile" mutant wild-type allele gene product | |
| *S. cerevisiae* genomic DNA | 189 |
| 4. GCN2: Product of gene essential for translational derepression of GCN4. The protein has a predicted $tRNA^{His}$-binding site and may be regulated by uncharged tRNA | 190 |
| *S. cerevisiae* genomic DNA | 190, 191 |
| 5. byr1$^{+}$: Suppressor of sporulation defects | |
| *S. pombe* genomic DNA | 192 |
| 6. YPK1: Novel yeast protein-serine kinase | |
| *S. cerevisiae* genomic DNA | 193 |
| 7. KIN3: Novel yeast protein-serine kinase | |
| *S. cerevisiae* genomic DNA | 194 |
| 8. YKR2: Novel yeast protein-serine kinase | |
| *S. cerevisiae* genomic DNA | 195 |
| 9. YAK1: Novel yeast protein-serine kinase | |
| *S. cerevisiae* genomic DNA | 196 |
| 10. DBF2: Novel yeast protein-serine kinase needed for progression through the cell cycle (there is a second, closely related gene product) | |
| *S. cerevisiae* genomic DNA | 197 |
| 11. SME1: Novel yeast protein-serine kinase needed for initiation of meiosis and sporulation | |
| *S. cerevisiae* genomic DNA | 198 |
| 12. VPS15: Novel yeast protein-serine kinase needed for vacuolar protein sorting | |
| *S. cerevisiae* genomic DNA | 199 |
| 13. RNA polymerase kinase 58-kDa subunit | |
| *S. cerevisiae* genomic DNA | 200 |
| Nematode protein-serine/threonine kinases | |
| 1. cAMP-dependent protein kinase: *Caenorhabditis elegans* homolog of cAMP-dependent protein kinase | |
| *C. elegans* genomic DNA and cDNA | 201 |
| 2. Casein kinase II$\alpha$: *C. elegans* homolog of CKII-$\alpha$ | |
| *C. elegans* genomic DNA and cDNA | 202 |
| 3. *tpa-1* protein: *C. elegans* protein kinase C homolog | |
| *C. elegans* genomic DNA and cDNA | 203 |

TABLE I (*continued*)

| Classification | References |
|---|---|
| 4. *daf-1* gene product: *C. elegans* protein kinase required for larval development | |
| *C. elegans* genomic DNA and cDNA | 204 |
| 5. *daf-4* gene product: *C. elegans* protein kinase required for larval development | |
| *C. elegans* genomic DNA and cDNA | 205 |
| 6. *unc-22* gene product: *C. elegans* protein kinase required for proper control of muscle contraction. This protein, known as twitchin, has a protein kinase domain near the C terminus and many copies of two different repeats. | |
| *C. elegans* genomic DNA and cDNA | 206 |
| *Aplysia* protein-serine/threonine kinases | |
| 1. $C_{APL-A}$: *Aplysia* cAMP-dependent protein kinase homolog. There are two alternate spliced forms, A-1 and A-2, differing in a 42-amino acid exon. | |
| *Aplysia* cDNA | 207 |
| 2. $C_{APL-B}$: *Aplysia* cAMP-dependent protein kinase-related gene | |
| *Aplysia* cDNA | 208 |
| Plant protein-serine kinases | |
| 1. PVPK-1: Bean (*Phaseolus vulgaris*) cAMP-dependent protein kinase related protein kinase | |
| Bean cDNA | 209 |
| 2. G11A: Rice (*Oryzae sativa*) cAMP-dependent protein kinase related protein kinase | |
| Rice cDNA | 209 |
| 3. ZmPK1: Maize receptor-like protein-serine kinase related to *raf* protein | |
| Maize cDNA | 210 |
| 4. cdc2: Pea homolog of yeast cdc2 | |
| Pea cDNA | 211 |
| Other protein-serine/threonine kinases | |
| 1. nimA: Cell cycle control protein kinase from *Aspergillus nidulans* | |
| *A. nidulans* cDNA | 212 |
| 2. cAMP-dependent protein kinase: *Dictyostelium discoideum* cAMP-dependent protein kinase homolog | |
| *D. discoideum* cDNA | 213 |
| 3. aceK: Isocitrate dehydrogenase kinase/phosphatase from *Escherichia coli* | |
| *E. coli* genomic DNA | 15 |
| **Protein-tyrosine kinases** | |
| Vertebrate protein-tyrosine kinases | |
| A. *src* gene family | |
| 1. c-*src* protein: Cellular homolog of v-*src* protein. There are 3 alternate splice forms, with 2 neuronal forms differing from the | |

(*continued*)

TABLE I (*continued*)

| Classification | References |
|---|---|
| fibroblast form by inclusion of a 6-amino acid exon (NI) or the NI exon plus an 11-amino acid exon (NII) at position-114 in chicken fibroblast form. | |
| a. Human fetal liver genomic DNA | 214 |
| b. Mouse brain cDNA; neuronal forms | 215–217 |
| c. Chicken genomic DNA | 218 |
| d. *Xenopus* ovary cDNA. There are two distinct *src* genes. | 219, 220 |
| e. v-*src* avian retroviral oncogene product | |
| i. Rous sarcoma virus Prague C | 221 |
| ii. Rous sarcoma virus Schmidt–Ruppin A | 222–224 |
| iii. Rous sarcoma virus B77 | 225 |
| iv. Rous sarcoma virus Bryan | 226 |
| v. Avian sarcoma viruses S1 and S2 | 227 |
| 2. c-*yes* protein: Cellular homolog of v-*yes* | |
| a. Human embryo fibroblast cDNA | 228 |
| b. v-*yes* retroviral oncogene product | |
| Yamaguchi 73 avian sarcoma virus | 229 |
| 3. c-*fgr* protein: Cellular homolog of v-*fgr* protein | |
| a. Human genomic DNA (designated c-*src2*), embryo fibroblast cDNA, and mononuclear cell cDNA | 230–236 |
| b. Murine monocyte cDNA | |
| c. v-*fgr* retroviral oncogene product | |
| Gardner–Rasheed feline sarcoma virus | 237 |
| 4. *fyn* protein: Protein-tyrosine kinase related to *fgr* and *yes* proteins. There are two alternately spliced forms differing by the use of an alternate exon 7 in the ATP-binding domain—one found exclusively in thymocytes and the other in brain. | 238 |
| a. Human fibroblast cDNA | 239, 240 |
| b. Mouse myelomonocytic cDNA | 238 |
| c. *Xenopus* oocyte cDNA | 241 |
| 5. *lyn* protein: Protein-tyrosine kinase related to *lck* and *yes* proteins | |
| Human placenta cDNA | 242 |
| 6. *lck* protein: Lymphoid T cell protein-tyrosine kinase | |
| a. Human T cell leukemia line cDNA | 243, 244 |
| b. Mouse T cell lymphoma line cDNA | 245, 246 |
| 7. *hck* protein: Hematopoietic cell protein-tyrosine kinase | |
| a. Human placenta and peripheral leukocyte cDNAs | 247, 248 |
| b. Mouse macrophage cDNA | 249, 250 |
| 8. *tkl* protein: Putative protein-tyrosine kinase (possibly chicken homolog of *lck* protein) | |
| Chicken spleen cDNA | 251 |
| 9. *blk* protein: Lymphoid B cell protein-tyrosine kinase | |
| Murine B cell cDNA | 252 |

TABLE I (*continued*)

| Classification | References |
|---|---|
| B. *abl* gene family | |
| 1. c-*abl* protein: Cellular homolog of v-*abl* protein (there are alternately spliced forms differing solely by a short N-terminal coding exon) | |
| a. Human fetal liver and K562 CML cDNAs | 253, 254 |
| b. Mouse testis cDNA | 255 |
| c. v-*abl* retroviral oncogene product | |
| Abelson murine leukemia virus | 256 |
| 2. *arg* protein: Protein-tyrosine kinase related to *abl* protein | |
| Human genomic and K562 cell cDNAs | 257, 258 |
| C. *fps*/*fes* gene family | |
| 1. c-*fes*/*fps* protein: Cellular homolog of v-*fes*/*fps* protein | |
| a. Human genomic DNA and cDNA | 259, 260 |
| b. Feline genomic DNA | 261 |
| c. Chicken genomic DNA | 262 |
| d. Murine cDNA | 263 |
| e. v-*fps*/*fes* retroviral gene product | |
| i. Gardner–Arnstein and Snyder–Theilen feline sarcoma virus strains | 264 |
| ii. Fujinami avian sarcoma virus | 265 |
| iii. PRCII avian sarcoma virus | 266 |
| 2. p94 c-*fes*-related protein expressed in myeloid cells | 267, 268 |
| a. Rat brain cDNA (designated *flk*) | 269 |
| b. Human fibroblast (designated FER) and human B cell cDNAs (designated TYK3) | 270, 271 |
| D. Growth factor receptors | |
| Epidermal growth factor receptor subfamily | |
| 1. EGFR: Epidermal growth factor receptor; cellular homolog of v-*erbB* protein | |
| a. Human placenta and A431 cell line cDNAs | 272–274 |
| b. Chicken fibroblast cDNA | 275 |
| c. *Xiphophorus* genomic DNA | 276 |
| d. v-*erbB* retroviral oncogene product | |
| i. AEV-H avian erythroblastosis virus | 277 |
| ii. AEV-ES4 avian erythroblastosis virus | 278 |
| 2. *neu* protein: Product of oncogene activated in ENU-induced rat neuroblastoma (also designated *erbB2* and *HER2*) | |
| a. Human placenta and gastric cancer cell line cDNAs | 279, 280 |
| b. Rat cDNA (from mouse B104-1-1 cell line transformed with DNA from a rat neuroblastoma cell line) | 281 |
| 3. *erbB3* protein: Protein related to the EGF receptor (also designated *HER3*) | |
| Human epithelial cell cDNAs | 282, 283 |
| 4. *Xmrk* protein: Normal homolog of the dominant *Tu* gene product in *Xiphophorus* which causes malignant melanoma | |
| *Xiphophorus* genomic and cDNA clones | 284 |

(*continued*)

TABLE I (*continued*)

| Classification | References |
|---|---|
| Insulin receptor subfamily | |
| 1. INS.R: Insulin receptor | |
| Human placenta cDNA | 285, 286 |
| 2. IGF1R: Insulin-like growth factor I receptor | |
| Human placenta cDNA | 287 |
| 3. IRRK: Insulin receptor-related kinase | |
| Human and guinea pig genomic DNA | 288 |
| 4. c-*ros* protein: Cellular homolog of v-*ros* protein. There is at least one alternately spliced form of c-*ros*-1 protein lacking the protein kinase domain, but containing the extracellular and transmembrane domains. | |
| a. Human placenta genomic DNA | 289, 290 |
| b. Chicken genomic DNA | 291 |
| c. Rat c-*ros*-1 genomic DNA and cDNA | 291a |
| d. v-*ros* retroviral oncogene product | |
| UR2 avian sarcoma virus | 292 |
| 5. *trk* protein: Product of cellular gene that gave rise to the *trk* oncogene (NGF receptor) | |
| a. Human K562 CML cell line cDNA | 293 |
| b. *trk* oncogene product | |
| i. Human colon carcinoma cDNA (activated by mutation) | 294 |
| ii. Human breast carcinoma-derived cDNA (activated during transfection by recombination) | 295 |
| 6. *trk*-B protein: Cellular protein related to *trk* protein; there are several alternately spliced forms of *trk*-B protein, some lacking the protein kinase domain | |
| a. Mouse brain cDNA | 296, 297 |
| b. Rat brain cDNA | 298 |
| 7. *met* protein: Product of cellular protein gene that gave rise to the *met* oncogene | |
| a. Human MNNG-HOS cell line cDNA (activated by mutation) | 299 |
| b. Mouse fibroblast cDNA | 300 |
| 8. c-*sea* protein: cellular homolog of v-*sea* protein | |
| v-*sea* retroviral oncogene product | |
| S13 avian erythroleukemia virus | 301 |
| Platelet-derived growth factor receptor subfamily | |
| 1. PDGFR-B: Platelet-derived growth factor receptor B type (recognizes only PDGF B chain) | |
| a. Mouse fibroblast cell line cDNA | 302 |
| b. Human fibroblast cDNA | 303, 304 |
| 2. PDGFR-A: Platelet-derived growth factor receptor A type (recognizes both PDGF A and B chains) (the product of the *patch* locus in mice) | |
| a. Human embryo fibroblast and glioblastoma cDNAs | 305, 306 |
| b. Rat olfactory bulb, and partial mouse placental cDNAs | 307 |
| c. Mouse fibroblast cDNA | 307a |

TABLE I (*continued*)

| Classification | References |
|---|---|
| 3. CSF1R: Colony stimulating factor-1 receptor; cellular homolog of v-*fms* protein | |
| a. Human placenta cDNA and genomic DNA | 308, 309 |
| b. Mouse pre-B cell line cDNA | 310 |
| c. v-*fms* retroviral oncogene product | |
| McDonough feline sarcoma virus | 311 |
| 4. c-*kit* protein: Cellular homolog of v-*kit* protein and the product of the developmental locus W in the mouse | |
| a. Human placenta cDNA | 312 |
| b. Mouse brain cDNA | 313 |
| c. v-*kit* retroviral oncogene product | |
| Hardy–Zuckerman 4 feline sarcoma virus | 314 |
| 5. *flt* protein: Putative protein-tyrosine kinase | |
| Human genomic DNA and cDNA | 315, 316 |
| *eph* gene subfamily | |
| 1. *eph* protein: Receptor-like protein-tyrosine kinase detected in an erythropoietin-producing hepatoma | |
| Human hepatoma cDNA | 317 |
| 2. *eck* protein: Receptor-like protein-tyrosine kinase detected in epithelial cells | |
| HeLa cell cDNA | 317a |
| 3. *elk* protein: Receptor-like protein-tyrosine kinase detected in brain | |
| Rat brain cDNA | 269 |
| 4. *eek* protein: Receptor-like protein-tyrosine kinase detected in brain | |
| Rat brain cDNA | 318 |
| Related genes have been identified in the chicken and mouse and their analysis suggests that there are several additional members of this family, including Cek5 and tyro-1, 4, 5, 6, and 11. | 319 |
| Fibroblast growth factor receptor subfamily | |
| 1. FGFR: Basic fibroblast growth factor receptor (*flg* endothelial protein-tyrosine kinase gene product) | |
| a. Human endothelial cell cDNA | 320, 321 |
| b. Chicken embryo fibroblast cDNA | 322, 323 |
| c. Mouse neuroepithelial cell and mouse brain cDNAs (there are two alternately spliced forms with different external domains) | 324, 325 |
| 2. FGFR2: Basic fibroblast growth factor receptor-related protein (*bek* gene product) | |
| Mouse liver cDNA | 326 |
| The K-*sam* human cellular oncogene product may be the human homolog of *bek* (human gastric carcinoma KATOIII cell line cDNA). | 327 |
| There are additional FGF receptor-related proteins—Cek2 and Cek3; JTK2(FGFR4), JTK4, JTK12, and JTK17; and FGFR3. | 328–330 |

(*continued*)

TABLE I (*continued*)

| Classification | References |
|---|---|
| E. Others | |
| 1. p75 liver-specific protein-tyrosine kinase | 331 |
| 2. *ltk* protein: Lymphocyte receptor-like protein-tyrosine kinase | |
| a. Mouse pre-B cell line cDNA | 332 |
| b. Human B cell line cDNA and genomic DNA | 333 |
| 3. *ret* protein: Putative receptor-like protein-tyrosine kinase and cellular oncogene product. There are two alternately spliced products differing at their C termini. | |
| a. Human leukemic T cell cDNA (activated by recombination) | 334, 335 |
| b. Human colonic cancer cDNA (activated by recombination) | 336 |
| 4. JAK1: Mouse growth factor-dependent hematopoietic cell gene product (formerly known as FD17) related to JAK2 | |
| Mouse cDNA | 337 |
| 5. JAK2: Mouse growth factor-dependent hematopoietic cell gene product (formerly known as FD22) related to JAK1 | |
| Mouse cDNA | 337 |
| 6. TYK2: Putative protein-tyrosine kinase related to JAK1 and JAK2 | |
| Human B cell cDNA | 271, 338 |
| 7. D-src28 homolog | |
| Mouse cDNA | 337 |
| 8. CSK/OB19: Potential $pp60^{c\text{-}src}$ (Tyr-527) protein-tyrosine kinase | |
| a. Human Hela cell cDNA | 121 |
| b. Rat brain cDNA | 339, 340 |
| 9. TKR11: Putative protein-tyrosine kinase (possibly chicken homolog of human *elk* protein) | |
| Chicken genomic DNA | 341 |
| 10. TKR16: Putative protein-tyrosine kinase | |
| Chicken genomic DNA | 341 |
| 11. *tec* gene product: Putative protein-tyrosine kinase (this may be the same as the liver p75 enzyme) | |
| Mouse liver cDNA | 342 |
| 12. *flk-1* protein: Putative receptor-like protein-tyrosine kinase | |
| Mouse fetal liver cDNA | 343 |
| 13. *flk-2* protein: Putative receptor-like protein-tyrosine kinase | |
| Mouse fetal liver cDNA | 343 |

Multiple novel protein-tyrosine kinase partial cDNAs have been isolated by PCR amplification of RNA and DNA from various vertebrate cell sources.

TABLE I (*continued*)

| Classification | References |
| --- | --- |
| *Drosophila* protein-tyrosine kinases | |
| A. *src* gene family | |
| 1. D*src*64 protein: *Drosophila* gene product; polytene locus 64B | |
| *D. melanogaster* genomic DNA | 344, 345 |
| 2. D*src*28 protein: *Drosophila* gene product; polytene locus 28C | |
| *D. melanogaster* adult female cDNA | 346 |
| B. *abl* gene family | |
| D*ash* protein: *Drosophila* gene product related to *abl* protein | |
| *D. melanogaster* genomic DNA | 345, 347 |
| C. *fps/fes* gene family | |
| D*fps* protein: *Drosophila* gene product related to *fps* protein | |
| *D. melanogaster* genomic DNA | 348 |
| D. Growth factor receptors | |
| 1. DER: *Drosophila* gene product related to EGFR; DER is allelic with *faint little ball, torpedo,* and *ellipse* | 349–351 |
| *D. melanogaster* genomic DNA | 352 |
| 2. DILR: *Drosophila* gene product related to INS.R | |
| *D. melanogaster* embryo cDNA | 353 |
| 3. 7less: *sevenless* gene product essential for R7 photoreceptor cell development in *Drosophila* (related to vertebrate c-*ros* protein) | |
| a. *D. melanogaster* eye imaginal disk cDNA | 354 |
| b. *D. virilis* eye imaginal disk cDNA | 355 |
| 4. *torso* protein: *torso* gene product essential for determination of anterior and posterior structures in the embryo | |
| *D. melanogaster* genomic DNA and cDNA | 356 |
| 5. D-TRK: *Drosophila* homolog of the mammalian *trk* protein | |
| *D. melanogaster* genomic DNA and cDNA | 357 |
| 6. D-RET: *Drosophila* homolog of the mammalian *ret* protein | |
| *D. melanogaster* genomic DNA | 358 |
| 7. FGF receptor: *Drosophila* homolog of one of the mammalian FGF receptor gene products | |
| *D. melanogaster* genomic DNA and cDNA | 359 |
| E. Others | |
| 1. PCRTK1: Putative *D. melanogaster* protein-tyrosine kinase which may correspond to the *basket* gene product | |
| *D. melanogaster* genomic DNA | 358 |
| 2. PCRTK2: Putative *D. melanogaster* protein-tyrosine kinase | |
| *D. melanogaster* genomic DNA | 358 |
| A number of novel protein-tyrosine kinase partial cDNAs have been isolated by PCR amplification of RNA from eye disks. | 360 |

(*continued*)

TABLE I (*continued*)

| Classification | References |
|---|---|
| Yeast protein-tyrosine kinases | |
| YPK1: The reported protein-tyrosine kinase activity in yeast has been purified and cloned. The cloned gene is more closely related to the protein-serine kinases than to the protein-tyrosine kinases and YPK1 could be a dual-specificity protein kinase. The cloned protein has not yet been shown to have protein-tyrosine kinase activity. | 361 |
| *S. cerevisiae* genomic DNA | 12 |
| Other protein-tyrosine kinases | |
| N-*abl* protein: *C. elegans* gene product related to *abl* protein | 362 |
| *let-23* protein: *C. elegans* putative receptor-like protein-tyrosine kinase needed for vulval induction | 363 |
| kin-5–kin-9: Five novel *C. elegans* putative protein-tyrosine kinases identified by low-stringency screening with v-*ros* sequences | 364 |
| tyr-kin I–V: Five novel *C. elegans* putative protein-tyrosine kinases identified by PCR | 365 |
| DPYK1 and DPYK2: Novel protein-tyrosine kinases from *D. discoideum* | 366 |
| *Dictyostelium* cDNA | |
| STK: *src*-related gene product in *Hydra* | 367 |

## References


[1] A. M. Edelman, D. K. Blumenthal, and E. G. Krebs, *Annu. Rev. Biochem.* **56,** 567–613 (1987).

[2] T. Hunter and J. A. Cooper, *Annu. Rev. Biochem.* **54,** 897–930 (1985).

[3] T. Hunter, *Cell* **50,** 823–829 (1987).

[4] S. K. Hanks, A. M. Quinn, and T. Hunter, *Science* **241,** 42–52 (1988).

[5] R. H. Vallejos, L. Holuigue, H. A. Lucero, and M. Torruella, *Biochem. Biophys. Res. Commun.* **126,** 685–691 (1985).

[6] J. C. Cortay, B. Duclos, and A. J. Cozzone, *J. Mol. Biol.* **187,** 305–315 (1986).

[7] T. M. Chiang, J. Reizer, and E. H. Beachey, *J. Biol. Chem.* **264,** 2957–2962 (1989).

[8] K. Guan and J. E. Dixon, *Science* **249,** 553–556 (1990).

[9] C. Featherstone and P. Russell, *Nature (London)* **349,** 808–811 (1991).

[10] H. Piwnica-Worms, unpublished observations (1990).

[11] Y. Ben-David, K. Letwin, L. Tannock, A. Bernstein, and T. Pawson, *EMBO J.*, in press. (1991).

[12] D. Dailey, G. L. Schieven, M. Y. Lim, H. Marquardt, T. Gilmore, J. Thorner, and G. S. Martin, *Mol. Cell. Biol.* **10,** 6244–6256 (1990).

[13] D. F. Stern, P. Zheng, D. R. Beidler, and C. Zerillo, *Mol. Cell. Biol.* **11,** 987–1001 (1991).

[14] A. J. Cozzone, *Annu. Rev. Microbiol.* **42,** 97–125 (1988).

[15] D. J. Klumpp, D. W. Plank, L. J. Bowdin, C. S. Stueland, T. Chung, and D. C. LaPorte, *J. Bacteriol.* **170,** 2763–2769 (1988).

[16] M. Inouye, unpublished observations (1991).

[17] C.-C. Chen, D. L. Smith, B. B. Bruegger, R. M. Halpern, and R. A. Smith, *Biochemistry* **13,** 3785 (1974).

[18] V. D. Huebner and H. R. Matthews, *J. Biol. Chem.* **260,** 16106–16113 (1985).
[19] K. H. Pesis, Y. Wei, M. Lewis, and H. R. Matthews, *FEBS Lett.* **239,** 151–154 (1988).
[20] F. Levy-Favatier, M. Delpech, and J. Kruh, *Eur. J. Biochem.* **166,** 617–621 (1987).
[21] J. B. Stock, A. J. Ninfa, and A. M. Stock, *Microbiological Rev.* **53,** 450–490 (1989).
[22] J. F. Hess, R. B. Bourret, and M. I. Simon, *Nature (London)* **336,** 139 (1988).
[23] V. Weiss and B. Magasanik, *Proc. Natl. Acad. Sci. U.S.A.* **85,** 8919–8923 (1988).
[24] M. H. Saier, Jr., L.-F. Wu, and J. Reizer, *TIBS* **15,** 391–395 (1990).
[25] H. H. Pas and G. T. Robillard, *Biochemistry* **27,** 5835–5839 (1988).
[26] S. Shoji, L. H. Ericsson, K. A. Walsh, E. H. Fischer, and K. Titani, *Biochemistry* **22,** 3702–3709 (1983).
[27] M. D. Uhler, J. C. Chrivia, and G. S. McKnight, *J. Biol. Chem.* **261,** 15360–15363 (1986).
[28] J. C. Chrivia, M. D. Uhler, and G. S. McKnight, *J. Biol. Chem.* **263,** 5739–5744 (1988).
[29] S. R. Adavani, M. Schwarz, M. O. Showers, R. Maurer, and B. A. Hemmings, *Eur. J. Biochem.* **167,** 221–226 (1987).
[30] F. Maldonado and S. K. Hanks, *Nucleic Acids Res.* **16,** 8189–8190 (1988).
[31] M. O. Showers and R. A. Maurer, *J. Biol. Chem.* **261,** 16288–16291 (1986).
[32] S. J. Beebe, O. Oyen, M. Sandberg, A. Froysa, V. Hansson, and T. Jahnsen, *Mol. Endocrinol.* **4,** 465–475 (1990).
[33] K. Takio, R. D. Wade, S. B. Smith, E. G. Krebs, K. A. Walsh, and K. Titani, *Biochemistry* **23,** 4207–4218 (1984).
[34] W. Wernet, V. Flockerzi, and F. Hofmann, *FEBS Lett.* **251,** 191–196 (1989).
[35] L. Wolfe, J. D. Corbin, and S. H. Francis, *J. Biol. Chem.* **264,** 7734–7741 (1989).
[36] S. H. Francis, T. A. Woodford, L. Wolfe, and J. D. Corbin, *Sec. Mess. Phosphoproteins* **12,** 301–310 (1989).
[37] P. J. Parker, L. Coussens, N. Totty, L. Rhee, S. Young, E. Chen, S. Stabel, M. D. Waterfield, and A. Ullrich, *Science* **233,** 853–859 (1986).
[38] S. Ohno, H. Kawasaki, S. Imajoh, K. Suzuki, M. Inagaki, H. Yokokura, T. Sakoh, and H. Hidaka, *Nature (London)* **325,** 161–166 (1987).
[39] L. Coussens, P. J. Parker, L. Rhee, T. L. Yang-Feng, E. Chen, M. D. Waterfield, U. Francke and A. Ullrich, *Science* **233,** 859–866 (1986).
[40] J. L. Knopf, M.-H. Lee, L. A. Sultzman, R. W. Kriz, C. R. Loomis, R. M. Hewick, and R. M. Bell, *Cell* **46,** 491–502 (1986).
[41] G. M. Housey, C. A. O'Brian, M. D. Johnson, P. Kirschmeier, and I. B. Weinstein, *Proc. Natl. Acad. Sci. U.S.A.* **84,** 1065–1069 (1987).
[42] S. Ohno, H. Kawasaki, Y. Konno, M. Inagaki, H. Hidaka, and K. Suzuki, *Biochemistry* **27,** 2083–2087 (1988).
[43] Y. Ono, T. Fujii, K. Ogita, U. Kikkawa, K. Igarishi, and Y. Nishizuka, *FEBS Lett.* **226,** 125–128 (1987).
[44] Y. Ono, T. Fujii, K. Ogita, U. Kikkawa, K. Igarishi, and Y. Nishizuka, *J. Biol. Chem.* **263,** 6927–6932 (1988).
[45] S. Ohno, Y. Aleita, Y. Konno, S. Imajoh, and K. Suzuki, *Cell* **53,** 731–741 (1988).
[46] Y. Ono, T. Fujii, K. Ogita, U. Kikkawa, K. Igarishi, and Y. Nishizuka, *Proc. Natl. Acad. Sci. U.S.A.* **86,** 3099–3103 (1989).
[47] N. Bacher, Y. Zisman, E. Berent, and E. Livneh, *Mol. Cell. Biol.,* **11,** 126–133 (1991).
[48] S. I. Osada, K. Mizuno, T. C. Saido, Y. Akita, K. Suzuki, T. Kuroki, and S. Ohno, *J. Biol. Chem.* **265,** 22434–22440 (1990).
[49] P. Coffer and J. R. Woodgett, unpublished observations (1990).
[50] E. M. Reimann, K. Titani, L. H. Ericsson, R. D. Wade, E. H. Fischer, and K. A. Walsh, *Biochemistry* **23,** 4185–4192 (1984).
[51] E. da Cruz e Silva and P. T. Cohen, *FEBS Lett.* **220,** 36–42 (1987).

[52] J. S. Chamberlain, P. Van Tuinen, A. A. Reeves, B. A. Philip, and C. T. Caskey, *Proc. Natl. Acad. Sci. U.S.A.* **84,** 2886–2890 (1987).
[53] K. C. Cawley, C. Ramachandran, F. A. Gorin, and D. A. Walsh, *Nucleic Acids Res.* **16,** 2355–2356 (1988).
[54] S. K. Hanks, *Proc. Natl. Acad. Sci. U.S.A.* **84,** 388–392 (1987).
[55] S. K. Hanks, *Mol. Endocrinol.* **3,** 110–116 (1989).
[56] K. Takio, D. K. Blumenthal, A. M. Edelman, K. A. Walsh, E. G. Krebs, and K. Titani, *Biochemistry* **24,** 6028–6037 (1985).
[57] V. Guerriero, M. A. Russo, N. J. Olson, J. A. Putkey, and A. R. Means, *Biochemistry* **25,** 8372–8381 (1986).
[58] M. O. Shoemaker, W. Lau, R. L. Shattuck, A. Kwiatowski, P. Matrisian, L. Guerra-Santos, E. Wilson, T. J. Lukas, L. Van Eldik, and D. M. Watterson, *J. Cell Biol.* **111,** 1107–1125 (1990).
[59] A. C. Nairn and P. Greengard, *J. Biol. Chem.* **262,** 7273–7281 (1987).
[60] R. M. Hanley, A. R. Means, T. Ono, B. E. Kemp, K. E. Burgin, N. Waxham, and P. T. Kelly, *Science* **237,** 293–297 (1987).
[61] C. R. Lin, M. S. Kapiloff, S. Durgerian, K. Tatemoto, A. F. Russo, P. Hanson, H. Schulman, and M. G. Rosenfeld, *Proc. Natl. Acad. Sci. U.S.A.* **84,** 5962–5966 (1987).
[62] H. LeVine III, D. F. Hunt, N.-Z. Zhu, and J. Shabanowitz, *Biochem. Biophys. Res. Commun.* **148,** 1104–1109 (1987).
[63] M. K. Bennett and M. B. Kennedy, *Proc. Natl. Acad. Sci. U.S.A.* **84,** 1794–1798 (1987).
[64] T. Tobimatsu, I. Kameshita, and H. Fujisawa, *J. Biol. Chem.* **263,** 16082–16086 (1988).
[65] A. C. Nairn, B. Bhagat, and H. C. Palfrey, *Proc. Natl. Acad. Sci. U.S.A.* **82,** 7939–7943 (1985).
[66] J. M. Sikela and W. E. Hahn, *Proc. Natl. Acad. Sci. U.S.A.* **84,** 3038–3042 (1987).
[67] J. R. Woodgett, unpublished observations (1990).
[68] E. Erikson and J. L. Maller, *J. Biol. Chem.* **261,** 350–355 (1986).
[69] S. W. Jones, E. Erikson, J. Blenis, J. L. Maller, and R. L. Erikson, *Proc. Natl. Acad. Sci. U.S.A.* **85,** 3377–3381 (1988).
[70] D. A. Alcorta, C. M. Crews, L. J. Sweet, L. Bankston, S. W. Jones, and R. L. Erikson, *Mol. Cell. Biol.* **9,** 3850–3859 (1989).
[71] P. Banerjee, M. R. Ahmad, J. R. Grove, C. Kozlosky, D. J. Price, and J. Avruch, *Proc. Natl. Acad. Sci. U.S.A.* **87,** 8550–8554 (1990).
[72] S. Kozma, S. Ferrari, P. Bassand, M. Siegmann, N. Totty, and G. Thomas, *Proc. Natl. Acad. Sci. U.S.A.* **87,** 7365–7369 (1990).
[73] J. L. Benovic, A. DeBlasi, W. C. Stone, M. G. Caron, and R. J. Lefkowitz, *Science* **246,** 235–240 (1989).
[74] J. Benovic, unpublished observations (1990).
[75] H. Kuhn, *Biochemistry* **17,** 4389–4395 (1978).
[76] K. Takio, E. A. Kuenzel, K. A. Walsh, and E. G. Krebs, *Proc. Natl. Acad. Sci. U.S.A.* **84,** 4851–4855 (1987).
[77] H. Meisner, R. Heller-Harrison, J. Buxton, and M. P. Czech, *Biochemistry* **28,** 4072–4082 (1989).
[78] G. Maridor, W. Park, W. Krek, and E. A. Nigg, *J. Biol. Chem.,* **266,** 2362–2368 (1991).
[79] F. J. Lozeman, D. W. Litchfield, C. Peining, K. Takio, K. A. Walsh, and E. G. Krebs, *Biochemistry,* **29,** 8436–8447 (1990).
[80] H. R. Matthews and V. D. Huebner, *Mol. Cell. Biochem.* **59,** 81–99 (1984).
[81] J. R. Woodgett, *EMBO J.* **9,** 2431–2438 (1990).
[82] T. C. Chambers and T. A. Langan, *J. Biol. Chem.* **265,** 16940–16947 (1990).
[83] W. G. Dunphy, L. Brizuela, D. Beach, and J. Newport, *Cell* **54,** 423–431 (1988).
[83a] J. Gautier, M. Newbury, M. Lohka, P. Nurse, and J. Maller, *Cell* **54,** 433–439 (1988).

[84] M. G. Lee and P. Nurse, *Nature (London)* **327,** 31–35 (1987).
[85] L. J. Cisek and J. L. Corden, *Nature (London)* **339,** 679 (1989).
[85a] J. P. H. Th'ng, P. S. Wright, J. Hamaguchi, M. G. Lee, C. J. Norbury, P. Nurse, and E. M. Bradbury, *Cell* **63,** 313–324 (1990).
[85b] N. Spurr, A. Gough, and M. Lee, *DNA Sequence 1* 49–54 (1990).
[86] J. Pines and T. Hunter, unpublished observations (1990).
[87] J. Paris, R. Le Guellec, A. Couturier, K. Le Guellec, F. Omilli, J. Camonis, S. MacNeill, and M. Philippe, *Proc. Natl. Acad. Sci. U.S.A.* **88,** 1039–1043 (1991).
[88] H. Matsushime, A. Jinno, N. Tagaki, and M. Shibuya, *Mol. Cell. Biol.* **10,** 2261–2268 (1990).
[89] J. Shuttleworth, R. Godfrey, and A. Colman, *EMBO J.* **9,** 3233–3240 (1990).
[90] B. A. Bunnell, L. S. Heath, D. E. Adams, J. M. Lahti, and V. J. Kidd, *Proc. Natl. Acad. Sci. U.S.A.* **87,** 7467–7471 (1990).
[91] L. B. Ray and T. W. Sturgill, *J. Biol. Chem.* **263,** 12721–12727 (1988).
[92] T. G. Boulton, G. D. Yancopoulos, J. S. Gregory, C. Slaughter, C. Moomaw, J. Hsu, and M. H. Cobb, *Science* **249,** 64–67 (1990).
[93] S. Pelech, unpublished observations (1990).
[94] M. Cobb, unpublished observations (1990).
[95] J. M. Kyriakis and J. Avruch, *J. Biol. Chem.* **265,** 17355–17363 (1990).
[96] T. I. Bonner, H. Oppermann, P. Seeburg, S. B. Kerby, M. A. Gunnell, A. C. Young, and U. R. Rapp, *Nucleic Acids Res.* **14,** 1009–1015 (1986).
[97] M. Koenen, A. E. Sippel, C. Trachmann, and K. Bister, *Oncogene* **2,** 179–185 (1988).
[98] G. E. Mark and U. R. Rapp, *Science* **224,** 285–289 (1984).
[99] F. Galibert, S. Dupont de Dinechin, M. Rishi, and D. Stehelin, *EMBO J.* **3,** 1333–1338 (1984).
[100] N. C. Kan, C. S. Flordellis, G. E. Mark, P. H. Duesberg, and T. S. Papas, *Science* **223,** 813–816 (1984).
[101] P. Sutrave, T. I. Bonner, U. R. Rapp, H. W. Jansen, T. Patschinsky, and K. Bister, *Nature (London)* **309,** 85–88 (1984).
[102] T. W. Beck, M. Huleihel, M. Gunnell, T. I. Bonner, and U. R. Rapp, *Nucleic Acids Res.* **15,** 595–609 (1987).
[103] M. Huleihel, M. Goldsborough, J. Cleveland, M. Gunnell, T. Bonner, and U. R. Rapp, *Mol. Cell. Biol.* **6,** 2655–2662 (1986).
[104] F. Ishikawa, F. Takaku, M. Nagao, and T. Sugimura, *Oncogene Res.* **1,** 243–253 (1987).
[105] G. E. Mark, T. W. Seeley, T. B. Shows, and J. D. Mountz, *Proc. Natl. Acad. Sci. U.S.A.* **83,** 6312–6316 (1986).
[106] S. Ikawa, M. Fukui, Y. Ueyama, N. Tamaoki, T. Yamamoto, and K. Toyoshima, *Mol. Cell. Biol.* **8,** 2651–2654 (1988).
[107] M. Marx, A. Eychene, D. Laugier, C. Becahde, P. Crisanti, P. Dezelee, B. Pessac, and G. Calothy, *EMBO J.* **7,** 3369–3373 (1988).
[108] R. Watson, M. Oskarsson, and G. F. Vande Woude, *Proc. Natl. Acad. Sci. U.S.A.* **79,** 4078–4082 (1982).
[109] C. Van Beveren, F. van Straaten, J. A. Galleshaw, and I. M. Verma, *Cell* **27,** 97–108 (1981).
[110] F. A. Van der Hoorn and J. Firzlaff, *Nucleic Acids Res.* **12,** 2147–2156 (1984).
[111] M. Schmidt, M. K. Oskarsson, J. K. Dunn, D. G. Blair, S. Hughes, F. Propst, and G. F. Vande Woude, *Mol. Cell. Biol.* **8,** 923–929 (1988).
[112] R. S. Paules, F. Propst, K. J. Dunn, D. G. Blair, K. Kaul, A. E. Palmer, and G. F. Vande Woude, *Oncogene* **3,** 59–68 (1988).
[113] N. Sagata, M. Oskarsson, T. Copeland, J. Brumbaugh, and G. F. Vande Woude, *Nature (London)* **335,** 519–525 (1988).

[114] R. S. Freeman, K. M. Pickham, J. P. Kanki, B. A. Lee, S. V. Pena, and D. J. Donoghue, *Proc. Natl. Acad. Sci. U.S.A.* **86,** 5805–5809 (1989).
[115] E. P. Reddy, M. J. Smith, and S. A. Aaronson, *Science* **214,** 445–450 (1981).
[116] D. J. Donoghue, *J. Virol.* **42,** 538–546 (1982).
[117] M. A. Brow, A. Sen, and J. G. Sutcliffe, *J. Virol.* **49,** 579–582 (1984).
[118] A. Seth and G. Vande Woude, *J. Virol.* **56,** 114–152 (1985).
[119] A. Stacey, C. Arbuthnot, R. Kollek, L. Coggins, and W. Ostertag, *J. Virol.* **50,** 725–732 (1984).
[120] G. M. Hathaway and J. A. Traugh, *Curr. Top. Cell. Regul.* **21,** 101–127 (1982).
[121] R. A. Lindberg and T. Hunter, unpublished observations (1990).
[122] G. Mills, unpublished observations (1990).
[123] E. Meurs, K. Chong, J. Galabru, N. S. B. Thomas, I. M. Kerr, B. R. G. Williams, and A. G. Hovanessian, *Cell* **62,** 379–390 (1990).
[124] J. Miyoshi, T. Higashi, H. Mukai, T. Ohuchi, and T. Kakunaga, *Mol. Cell. Biol.,* in press. (1991).
[125] G. Selten, H. T. Cuypers, W. Boelens, E. Robanus-Maandag, J. Verbeek, J. Domen, C. Van Beveren, and A. Burns, *Cell* **46,** 603–611 (1986).
[126] T. C. Meeker, L. Nagarajan, A. Ar-Rushdi, G. Rovera, K. Huebner, and C. M. Croce, *Oncogene Res.* **1,** 87–101 (1987).
[127] J. Domen, M. Von Lindern, A. Hermans, M. Breuer, G. Grosveld, and A. Berns, *Oncogene Res.* **1,** 103–112 (1987).
[128] S. K. Hanks, unpublished observations (1990).
[129] P. Cohen, D. Yellowlees, A. Aitken, A. Donella-Deana, B. A. Hemmings, and P. J. Parker, *Eur. J. Biochem.* **145,** 21–35 (1982).
[129a] C. G. Proud, *TIBS* **11,** 73–77 (1986).
[130] A. I. Walker, T. Hunt, R. J. Jackson, and C. W. Anderson, *EMBO J.* **4,** 139–145 (1985).
[131] T. Carter, I. Vancurova, I. Sun, W. Lou, and S. DeLeon, *Mol. Cell. Biol.* **10,** 6460–6471 (1990).
[132] S. P. Lees-Miller, Y.-R. Chen, and C. W. Anderson, *Mol. Cell. Biol.* **10,** 6472–6481 (1990).
[133] D. M. Gibson and R. A. Parker, *The Enzymes* **18,** 180–215 (1987).
[134] D. G. Hardie, D. Carling, and A. T. R. Sim, *TIBS* **14,** 20–23 (1989).
[135] L. J. Reed and S. J. Yeaman, *The Enzymes* **18,** 77–95 (1987).
[136] P. J. Randle, P. A. Patston, and J. Espinal, *The Enzymes* **18,** 97–121 (1987).
[137] M. Abdel-Ghany, C. Riegler, and E. Racker, *Proc. Natl. Acad. Sci. U.S.A.* **81,** 7388–7391 (1984).
[138] V. J. Atmar and G. D. Kuehn, *Proc. Natl. Acad. Sci. U.S.A.* **78,** 5518 (1981).
[139] S. U. Tahara and J. A. Traugh, *J. Biol. Chem.* **256,** 11558–11564 (1981).
[140] R. Wettenhall, unpublished observations (1990).
[141] T. H. Lubben and J. A. Traugh, *J. Biol. Chem.* **258,** 13992–13997 (1983).
[142] M. Chinkers, D. L. Garbers, M.-S. Chang, D. G. Lowe, H. Fhin, D. V. Goeddel, and S. Schulz, *Nature (London)* **338,** 78–83 (1989).
[143] S. Singh, D. G. Lowe, D. S. Thorpe, H. Rodriguez, W. J. Kuang, L. J. Dangott, M. Chinkers, D. V. Goeddel, and D. L. Garbers, *Nature (London)* **334,** 708–712 (1988).
[144] D. S. Thorpe and D. L. Garbers, *J. Biol. Chem.* **264,** 6545–6549 (1989).
[145] S. Schulz, S. Singh, R. A. Bellet, G. Singh, D. J. Tubb, H. Chin, and D. L. Garbers, *Cell* **58,** 1155–1162 (1989).
[146] D. J. McGeoch and A. J. Davison, *Nucleic Acids Res.* **14,** 1765–1777 (1986).
[147] R. F. Smith and T. Smith, *J. Virol.* **63,** 450–455 (1989).
[148] J. W. Wu, Z.-Y. Zhou, A. Judd, C. A. Cartwright, and W. S. Robinson, *Cell* **63,** 687–695 (1990).

[149] J. L. Foster, G. C. Higgins, and F. R. Jackson, *J. Biol. Chem.* **263,** 1676–1685 (1988).
[150] D. Kalderon and G. M. Rubin, *Genes Devel.* **2,** 1539–1556 (1988).
[151] A. Rosenthal, L. Rhee, R. Yadegari, R. Paro, A. Ullrich, and D. V. Goeddel, *EMBO J.* **6,** 433–441 (1987).
[152] E. Schaefer, D. Smith, G. Mardon, W. Quinn, and C. Zuker, *Cell* **57,** 403–412 (1989).
[153] C. F. Lehner and P. H. O'Farrell, *EMBO J.* **9,** 3573–3581 (1990).
[154] J. Jiminez, L. Alphey, P. Nurse, and D. M. Glover, *EMBO J.* **9,** 3565–3571 (1990).
[155] A. Saxena, R. Padmanabha, and C. V. C. Glover, *Mol. Cell. Biol.* **7,** 3409–3418 (1987).
[156] L. Ambrosio, A. P. Mahowald, and N. Perrimon, *Nature (London)* **342,** 288–290 (1989).
[157] G. E. Mark, R. J. MacIntyre, M. E. Digan, L. Ambrosio, and N. Perrimon, *Mol. Cell. Biol.* **7,** 2134–2140 (1987).
[158] Y. Nishida, M. Hata, T. Ayaki, H. Ryo, M. Yamagata, K. Shimizu, and Y. Nishizuka, *EMBO J.* **7,** 775–781 (1988).
[159] C. Montell and G. M. Rubin, *Cell* **52,** 757–772 (1988).
[160] T. Preat, P. Therond, C. Lamour-Isnard, B. Limbourg-Bouchon, H. Tricoire, I. Erk, M.-C. Mariol, and D. Busson, *Nature (London)* **347,** 87–89 (1990).
[161] E. Siegfried, L. A. Perkins, T. M. Capaci, and N. Perrimon, *Nature (London)* **345,** 825–829 (1990).
[162] M. Bourouis, P. Moore, L. Ruel, Y. Grau, P. Heitzler, and P. Simpson, *EMBO J.* **9,** 2877–2884 (1990).
[163] T. Toda, S. Cameron, P. Sass, M. Zoller, and M. Wigler, *Cell* **50,** 277–287 (1987).
[164] J. Lisziewicz, A. Godany, H.-H. Forster, and H. Kuntzel, *J. Biol. Chem.* **262,** 2549–2553 (1987).
[165] J. F. Cannon and K. Tatchell, *Mol. Cell. Biol.* **7,** 2653–2663 (1987).
[166] T. Toda, S. Cameron, and M. Wigler, *Genes Develop.* **2,** 517–527 (1988).
[167] J. Thorner, unpublished observations (1990).
[168] D. E. Levin, F. O. Fields, R. Kunisawa, J. M. Bishop, and J. Thorner, *Cell* **62,** 213–224 (1990).
[169] A. T. Lorincz and S. I. Reed, *Nature (London)* **307,** 183–185 (1984).
[170] J. Hindley and G. A. Phear, *Gene* **31,** 129–134 (1984).
[171] M. Simon, B. Seraphin, and G. Faye, *EMBO J.* **5,** 2697–2701 (1986).
[172] W. E. Courchesne, R. Kunisawa, and J. Thorner, *Cell* **58,** 1107–1119 (1989).
[173] A. Toh-e, K. Tanaka, Y. Uesomo, and R. B. Wickner, *Mol. Gen. Genet.* **214,** 162–164 (1988).
[174] E. A. Elion, P. L. Grisell, and G. R. Fink, *Cell* **60,** 649–664 (1990).
[175] T. Toda, M. Shimanuki, and M. Yanagida, *Genes Devel.* **5,** 60–73 (1991).
[176] M. A. Teague, D. T. Chaleff, and B. Errede, *Proc. Natl. Acad. Sci. U.S.A.* **83,** 7371–7375 (1986).
[177] G. Buguslawski and J. O. Polazzi, *Proc. Natl. Acad. Sci. U.S.A.* **84,** 5848–5852 (1987).
[178] J. L. Celenza and M. Carlson, *Science* **233,** 1175–1180 (1986).
[179] D. E. Levin, C. I. Hammond, R. O. Ralston, and J. M. Bishop, *Proc. Natl. Acad. Sci. U.S.A.* **84,** 6035–6039 (1987).
[180] D. Levin, unpublished observations (1990).
[181] P. Russell and P. Nurse, *Cell* **49,** 569–576 (1987).
[182] J. L.-P. Chen-Wu, R. Padmanabha, and C. V. C. Glover, *Mol. Cell. Biol.* **8,** 4981–4990 (1988).
[183] R. Padmanabha, J. L.-P. Chen-Wu, D. E. Hanna, and C. V. C. Glover, *Mol. Cell. Biol.* **10,** 4089–4099 (1990).
[184] P. Russell and P. Nurse, *Cell* **49,** 559–567 (1987).
[185] D. Beach, unpublished observations (1990).
[186] M. Hoekstra, unpublished observations (1990).

[187] M. McLeod and D. Beach, *EMBO J.* **5,** 3665–3671 (1986).
[188] M. Patterson, R. A. Sclafani, W. L. Fangman, and J. Rosamond, *Mol. Cell. Biol.* **6,** 1590–1598 (1986).
[189] N. Rhodes, L. Connell, and B. Errede, *Genes Develop.* **4,** 1862–1874 (1990).
[190] R. C. Wek, B. M. Jackson, and A. G. Hinnebusch, *Proc. Natl. Acad. Sci. U.S.A.* **86,** 4579–4583 (1989).
[191] I. Roussou, G. Thireos, and B. M. Hauge, *Mol. Cell. Biol.* **8,** 2132–2139 (1988).
[192] S. A. Nadin-Davis and A. Nasim, *EMBO J.* **7,** 985–992 (1988).
[193] R. A. Maurer, *DNA* **7,** 469–474 (1988).
[194] D. G. Jones and J. Rosamund, *Gene* **90,** 87–92 (1990).
[195] K. Kubo, S. Ohno, S. Matsumoto, I. Yahara, and K. Suzuki, *Gene* **76,** 177–180 (1989).
[196] S. Garrett and J. Broach, *Genes Devel.* **3,** 1336–1348 (1989).
[197] L. H. Johnston, S. Eberly, J. W. Chapman, H. Araki, and A. Sugino, *Mol. Cell. Biol.* **10,** 1358–1366 (1990).
[198] M. Yoshida, H. Kawaguchi, Y. Sakata, K.-I. Kominami, M. Hirano, H. Shima, R. Akada, and I. Yamashita, *Mol. Gen. Genet.* **221,** 176–186 (1990).
[199] P. K. Herman, J. H. Stack, J. A. DeModena, and S. D. Emr, *Cell,* **64,** 425–437 (1991).
[200] A. Greenleaf, unpublished observations (1990).
[201] R. E. Gross, S. Bagchi, X. Lu, and C. S. Rubin, *J. Biol. Chem.* **265,** 6896–6907 (1990).
[202] E. Hu and C. S. Rubin, *J. Biol. Chem.* **265,** 5072–5080 (1990).
[203] Y. Tabuse, K. Nishiwaki, and J. Miwa, *Science* **243,** 1713–1716 (1989).
[204] L. L. Georgi, P. S. Albert, and D. L. Riddle, *Cell* **61,** 635–645 (1990).
[205] M. Estevez, D. S. Albert, and D. L. Riddle, unpublished observations (1990).
[206] G. M. Benian, J. E. Kiff, N. Neckelmann, D. G. Moerman, and R. H. Waterston, *Nature* (*London*) **342,** 45–50 (1989).
[207] S. Beushausen, P. Bergold, S. Sturner, A. Elste, V. Roytenberg, J. H. Schwartz, and H. Bayley, *Neuron* **1,** 853–864 (1988).
[208] S. Beushausen and H. Bayley, *Mol. Cell. Biol.,* **10,** 6775–6780. (1990).
[209] M. A. Lawton, R. T. Yamamoto, S. K. Hanks, and C. J. Lamb, *Proc. Natl. Acad. Sci. U.S.A.* **86,** 3140–3144 (1989).
[210] J. C. Walker and R. Zhang, *Nature* (*London*) **345,** 743–746 (1990).
[211] H. S. Feiler and T. W. Jacobs, *Proc. Nat. Acad. Sci. U.S.A.* **87,** 5397–5401 (1990).
[212] S. A. Osmani, R. T. Pu, and R. Morris, *Cell* **53,** 237–244 (1988).
[213] M. Veron, R. Mutzel, M. L. Lacombe, M. N. Simon, and V. Wallet, *Dev. Genet.* **9,** 247–258 (1988).
[214] S. K. Anderson, C. P. Gibbs, A. Tanaka, H.-J. Kung, and D. J. Fujita, *Mol. Cell. Biol.* **5,** 1122–1129 (1985).
[215] R. Martinez, B. Mathey-Prevot, A. Bernards, and D. Baltimore, *Science* **237,** 411–414 (1987).
[216] J. B. Levy, T. Dorai, L. H. Wang, and J. S. Brugge, *Mol. Cell. Biol.* **7,** 4142–4145 (1987).
[217] J. M. Pyper and J. B. Bolen, *Mol. Cell. Biol.* **10,** 2035–2040 (1990).
[218] T. Takeya and H. Hanafusa, *Cell* **32,** 881–890 (1983).
[219] R. E. Steele, *Nucleic Acids Res.* **13,** 1747–1761 (1985).
[220] R. Steele, T. F. Unger, M. J. Mardis, and J. B. Fero, *J. Biol. Chem.* **264,** 10649–10653 (1989).
[221] D. Schwartz, R. Tizard, and W. Gilbert, *Cell* **32,** 853–869 (1983).
[222] T. Takeya and H. Hanafusa, *J. Virol.* **44,** 12–18 (1982).
[223] A. P. Czernilofsky, A. D. Levinson, H. E. Varmus, J. M. Bishop, E. Tischler, and H. M. Goodman, *Nature* (*London*) **287,** 198–203 (1980).

[224] A. P. Czernilofsky, A. D. Levinson, H. E. Varmus, J. M. Bishop, E. Tischler, and H. M. Goodman, *Nature (London)* **301,** 736–738 (1983).
[225] G. Mardon and H. E. Varmus, *Cell* **32,** 871–879 (1983).
[226] B. J. Mayer, R. Jove, J. F. Krane, F. Poirier, G. Calothy, and H. Hanafusa, *J. Virol.* **60,** 858–867 (1986).
[227] S. Ikawa, K. Hagino-Yamagishi, S. Kawai, T. Yamamoto, and K. Toyoshima, *Mol. Cell. Biol.* **6,** 2420–2428 (1986).
[228] J. Sukegawa, K. Semba, Y. Yamanashi, M. Nishizawa, N. Miyajima, T. Yamamoto, and K. Toyoshima, *Mol. Cell. Biol.* **7,** 41–47 (1987).
[229] N. Kitamura, A. Kitamura, K. Toyoshima, Y. Hirayama, and M. Yoshida, *Nature (London)* **297,** 205–208 (1982).
[230] R. C. Parker, G. Mardon, R. V. Lebo, H. E. Varmus, and J. M. Bishop, *Mol. Cell. Biol.* **5,** 831–838 (1985).
[231] M. Nishizawa, K. Semba, M. C. Yoshida, T. Yamamoto, M. Sasaki, and K. Toyoshima, *Mol. Cell. Biol.* **6,** 511–517 (1986).
[232] M. Patel, S. J. Leevers, and P. M. Brickell, *Oncogene* **5,** 201–206 (1990).
[233] K. Inoue, S. Ikawa, K. Semba, J. Sukegawa, T. Yamamoto, and K. Toyoshima, *Oncogene* **1,** 301–304 (1987).
[234] S. Katamine, V. Notario, C. D. Rao, T. Miki, M. S. C. Cheah, S. R. Tronick, and K. C. Robbins, *Mol. Cell. Biol.* **8,** 259–266 (1988).
[235] T.-L. Yi and C. L. Willman, *Oncogene* **4,** 1081–1087 (1989).
[236] F. J. King and M. D. Cole, *Oncogene* **5,** 337–344 (1990).
[237] G. M. Naharra, K. C. Robbins, and E. P. Reddy, *Science* **223,** 63–66 (1984).
[238] M. P. Cooke and R. M. Perlmutter, *New Biol.* **1,** 66–74 (1989).
[239] K. Semba, M. Nishizawa, N. Miyajima, M. C. Yoshida, J. Sukegawa, Y. Yamanishi, M. Sasaki, T. Yamamoto, and K. Toyoshima, *Proc. Natl. Acad. U.S.A.* **83,** 5459–5463 (1986).
[240] T. Kawakami, C. Y. Pennington, and K. C. Robbins, *Mol. Cell. Biol.* **6,** 4195–4201 (1986).
[241] R. Steele, J. C. Deng, C. R. Ghosn, and J. B. Fero, *Oncogene* **5,** 369–376 (1990).
[242] Y. Yamanashi, S. I. Fukushige, K. Semba, J. Sukegawa, N. Miyajima, K. I. Matsubara, T. Yamamoto, and K. Toyoshima, *Mol. Cell. Biol.* **7,** 237–243 (1987).
[243] J. M. Trevillyan, Y. Lin, S. J. Chen, C. A. Phillips, C. Canna, and T. J. Linna, *Biochim. Biophys. Acta* **888,** 286–295 (1986).
[244] Y. Koga, N. Caccia, B. Toyonaga, R. Spolski, Y. Yanagi, Y. Yoshikai, and T. W. Mak, *Eur. J. Immunol.* **16,** 1643–1646 (1986).
[245] J. D. Marth, R. Peet, E. G. Krebs, and R. M. Perlmutter, *Cell* **43,** 393–404 (1985).
[246] A. F. Voronova and B. M. Sefton, *Nature (London)* **319,** 682–685 (1986).
[247] N. Quintrell, R. Lebo, H. Varmus, J. M. Bishop, M. J. Pettenati, M. M. LeBeau, M. O. Diaz, and J. D. Rowley, *Mol. Cell. Biol.* **7,** 2267–2275 (1987).
[248] S. F. Ziegler, J. D. Marth, D. B. Lewis, and R. M. Perlmutter, *Mol. Cell. Biol.* **7,** 2276–2285 (1987).
[249] D. A. Holtzman, W. D. Cook, and A. R. Dunn, *Proc. Natl. Acad. Sci. U.S.A.* **84,** 8325–8329 (1987).
[250] M. J. Klemz, S. R. McKercher, and R. A. Maki, *Nucleic Acids Res.* **15,** 9600 (1987).
[251] K. Strebhart, J. I. Mullins, C. Bruck, and H. Rübsamen-Waigmann, *Proc. Natl. Acad. Sci. U.S.A.* **84,** 8778–8782 (1988).
[252] S. Dymecki, J. Niederhuber, and S. V. Desiderio, *Science* **247,** 332–336 (1990).
[253] E. Shtivelman, B. Lifshitz, R. P. Gale, B. A. Roe, and E. Canaani, *Cell* **47,** 277–284 (1986).
[254] A. M. Mes-Masson, J. McLaughlin, G. Q. Daley, M. Paskind, and O. N. Witte, *Proc. Natl. Acad. Sci. U.S.A.* **83,** 9768–9772 (1986).

[255] C. Oppi, S. K. Shore, and E. P. Reddy, *Proc. Natl. Acad. Sci. U.S.A.* **84,** 8200–8204 (1987).
[256] E. P. Reddy, M. J. Smith, and A. Srinivasan, *Proc. Natl. Acad. Sci. U.S.A.* **80,** 3623–3627 (1983).
[257] G. D. Kruh, C. R. King, M. H. Kraus, N. C. Popescu, S. C. Amsbaugh, W. O. McBride, and S. A. Aaronson, *Science* **234,** 1545–1548 (1986).
[258] G. D. Kruh, R. Perego, T. Miki, and S. A. Aaronson, *Proc. Natl. Acad. Sci. U.S.A.* **87,** 5802–5806 (1990).
[259] A. J. M. Roebroek, J. A. Schalken, J. S. Verbeek, A. M. W. Van den Ouweland, C. Onnekink, H. P. J. Bloemers, and W. J. M. Van de Ven, *EMBO J.* **4,** 2897–2903 (1985).
[260] M. Alcalay, F. Antolini, W. J. Van de Ven, L. Lanfrancone, F. Grignani, and P. G. Pelicci, *Oncogene,* **5,** 267–275 (1990).
[261] A. J. M. Roebroek, J. A. Schalken, C. Onnekink, H. P. J. Bloemers, and W. J. M. Van de Ven, *J. Virol.* **61,** 2009–2016 (1987).
[262] C.-C. Huang, C. Hammond, and J. M. Bishop, *J. Mol. Biol.* **181,** 175–186 (1985).
[263] A. F. Wilks and R. R. Kurban, *Oncogene* **3,** 289–294 (1988).
[264] A. Hampe, I. Laprevotte, F. Galibert, L. A. Fedele, and C. J. Sherr, *Cell* **30,** 775–785 (1982).
[265] M. Shibuya and H. Hanafusa, *Cell* **30,** 787–795 (1982).
[266] C.-C. Huang, C. Hammond, and J. M. Bishop, *J. Virol.* **50,** 125–131 (1984).
[267] R. A. Feldman, J. L. Gabrelove, J. P. Tam, M. A. Moore, and H. Hanafusa, *Proc. Natl. Acad. Sci. U.S.A.* **82,** 2379–2383 (1985).
[268] I. MacDonald, T. Levy, and T. Pawson, *Mol. Cell Biol.* **5,** 2543–2553 (1985).
[269] K. Letwin, S. P. Yee, and T. Pawson, *Oncogene* **3,** 621–627 (1988).
[270] Q.-L. Hao, N. Heisterkamp, and J. Groffen, *Mol. Cell. Biol.* **9,** 1587–1593 (1989).
[271] J. Krolewski, R. Lee, R. Eddy, T. B. Shows, and R. Dalla-Favera, *Oncogene* **5,** 277–282 (1990).
[272] A. Ullrich, L. Coussens, J. S. Hayflick, T. J. Dull, A. Gray, A. W. Tam, J. Lee, Y. Yarden, T. A. Liberman, J. Schlessinger, J. Downward, E. L. V. Mayes, M. D. Waterfield, M. Whittle, and P. H. Seeburg, *Nature* (*London*) **309,** 418–425 (1984).
[273] C. R. Lin, W. S. Chen, W. Kruijer, L. S. Stolarsky, W. Weber, R. M. Evans, I. M. Verma, G. N. Gill, and M. G. Rosenfeld, *Science* **224,** 843–848 (1984).
[274] G. T. Merlino, Y.-H. Xu, S. Ishii, A. J. L. Clark, K. Semba, K. Toyoshima, T. Yamamoto, and I. Pastan, *Science* **224,** 417–419 (1984).
[275] I. Lax, A. Johnson, R. Howk, J. Sap, F. Bellot, M. Winkler, A. Ullrich, B. Vennstrom, J. Schlessinger, and D. Givol, *Mol. Cell. Biol.* **8,** 1970–1978 (1988).
[276] C. Zechel, U. Schleenbecker, A. Anders, and F. Anders, *Oncogene* **3,** 605–617 (1988).
[277] T. Yamamoto, T. Nishida, N. Miyajima, S. Kawai, T. Ooi, and K. Toyoshima, *Cell* **35,** 71–78 (1983).
[278] M. Privalsky, R. Ralston, and J. M. Bishop, *Proc. Natl. Acad. Sci. U.S.A.* **81,** 704–707 (1984).
[279] L. Coussens, T. L. Yang-Feng, Y.-C. Liao, E. Chen, A. Gray, J. McGrath, P. H. Seeburg, T. A. Libermann, J. Schlessinger, U. Francke, A. Levinson, and A. Ullrich, *Science* **230,** 1133–1139 (1985).
[280] T. Yamamoto, S. Ikawa, T. Akiyama, K. Semba, N. Nomura, N. Miyajima, T. Saito, and K. Toyoshima, *Nature* (*London*) **319,** 230–234 (1986).
[281] C. I. Bargmann, M.-C. Hung, and R. A. Weinberg, *Nature* (*London*) **319,** 226–230 (1986).
[282] M. H. Kraus, W. Issing, T. Miki, N. C. Popescu, and S. A. Aaronson, *Proc. Natl. Acad. Sci. U.S.A.* **86,** 9193–9197 (1989).

[283] G. D. Plowman, G. S. Whitney, M. G. Neubauer, J. M. Green, V. L. McDonald, G. J. Todaro, and M. Shoyab, *Proc. Natl. Acad. Sci. U.S.A.* **87,** 4905–4909 (1990).
[284] J. Wittbrodt, D. Adam, B. Malitschek, W. Mäueler, F. Raulf, A. Telling, S. M. Robertson, and M. Schartl, *Nature (London)* **341,** 415–421 (1989).
[285] A. Ullrich, J. R. Bell, E. Y. Chen, R. Herrara, L. M. Petruzzelli, T. J. Dull, A. Gray, L. Coussens, Y.-C. Liao, M. Tsubokawa, A. Mason, P. H. Seeburg, C. Grunfeld, O. M. Rosen, and J. Ramachandran, *Nature (London)* **313,** 756–761 (1985).
[286] Y. Ebina, L. Ellis, K. Jarnagin, M. Edery, L. Graf, E. Clauser, J.-H. Ou, F. Masiarz, Y. W. Kan, I. D. Goldfine, R. A. Roth, and W. J. Rutter, *Cell* **40,** 747–758 (1985).
[287] A. Ullrich, A. Gray, A. W. Tam, T. Yang-Feng, M. Tsubokawa, C. Collins, W. Henzel, T. Le Bon, S. Kathuria, E. Chen, S. Jacobs, U. Francke, J. Ramachandran, and Y. Fujita-Yamaguchi, *EMBO J.* **5,** 2503–2512 (1986).
[288] P. Shier and V. M. Watt, *J. Biol. Chem.* **264,** 14605–14608 (1989).
[289] H. Matsushime, L.-H. Wang, and M. Shibuya, *Mol. Cell. Biol.* **6,** 3000–3004 (1986).
[290] C. Birchmeier, D. Birnbaum, G. Waitches, O. Fasano, and M. Wigler, *Mol. Cell. Biol.* **6,** 3109–3116 (1986).
[291] W. S. Neckameyer, M. Shibuya, M. Hsu, and L.-H. Wang, *Mol. Cell. Biol.* **6,** 1478–1486 (1986).
[291a] H. Matsushime and M. Shibuya, *J. Virol.* **64,** 2117–2125 (1990).
[292] W. S. Neckameyer and L.-H. Wang, *J. Virol.* **53,** 879–884 (1985).
[293] D. Martin-Zanca, R. Osham, G. Mitra, T. Copeland, and M. Barbacid, *Mol. Cell. Biol.* **9,** 24–33 (1989).
[294] D. Martin-Zanca, S. H. Hughes, and M. Barbacid, *Nature (London)* **319,** 743–748 (1986).
[295] S. C. Kozma, S. M. S. Redmond, F. Xiao-Chang, S. M. Saurer, B. Groner, and N. E. Hynes, *EMBO J.* **7,** 147–154 (1988).
[296] R. Klein, L. F. Parada, F. Coulier, and M. Barbacid, *EMBO J.* **8,** 3701–3709 (1989).
[297] R. Klein, D. Conway, L. F. Parada, and M. Barbacid, *Cell* **61,** 647–656 (1990).
[298] D. S. Middlemas, R. A. Lindberg, and T. Hunter, *Mol. Cell. Biol.* **11,** 143–153 (1991).
[299] M. Park, M. Dean, K. Kaul, M. J. Braun, M. A. Gonda, and G. Vande Woude, *Proc. Natl. Acad. Sci. U.S.A.* **84,** 6379–6383 (1987).
[300] A. M.-L. Chan, H. W. S. King, E. A. Deakin, P. R. Tempest, J. Hilkens, V. Kroezen, D. R. Edwards, A. J. Wills, P. Brookes, and C. S. Cooper, *Oncogene* **1,** 229–335 (1987).
[301] D. R. Smith, P. K. Vogt, and M. J. Hayman, *Proc. Natl. Acad. Sci. U.S.A.* **86,** 5291–5295 (1989).
[302] Y. Yarden, J. A. Escobedo, W.-J. Kuang, T. L. Yang-Feng, T. O. Daniel, P. M. Tremble, E. Y. Chen, M. E. Ando, R. N. Harkins, U. Francke, V. A. Fried, A. Ullrich, and L. T. Williams, *Nature (London)* **323,** 226–232 (1986).
[303] L. Claesson-Welsh, A. Eriksson, A. Moren, L. Severinsson, B. Ek, A. Ostman, C. Betsholtz, and C. H. Heldin, *Mol. Cell. Biol.* **8,** 3476–3486 (1988).
[304] R. G. K. Gronwald, F. J. Grant, B. A. Haldeman, C. E. Hart, P. J. O'Hara, F. S. Hagen, R. Ross, D. F. Bowen-Pope, and M. J. Murray, *Proc. Natl. Acad. Sci. U.S.A.* **85,** 3435–3439 (1988).
[305] T. Matsui, M. Heidaran, T. Miki, N. Popescu, W. La Rochelle, M. Kraus, J. Pierce, and S. Aaronson, *Science* **243,** 800–804 (1989).
[306] L. Claesson-Welsh, A. Eriksson, B. Westermark, and C. H. Heldin, *Proc. Natl. Acad. Sci. U.S.A.* **86,** 4917–4921 (1989).
[307] K.-H. Lee, D. F. Bowen-Pope, and R. R. Reed, *Mol. Cell. Biol.* **10,** 2237–2246 (1990).
[307a] C. Wang, J. Kelly, D. F. Bowen-Pope, and C. D. Stiles, *Mol. Cell. Biol.* **10,** 6781–6784 (1990).
[308] L. Coussens, C. Van Beveren, D. Smith, E. Chen, R. L. Mitchell, C. M. Isacke, I. M. Verma, and A. Ullrich, *Nature (London)* **320,** 277–280 (1986).

[309] A. Hampe, B.-M. Shamoon, M. Gobet, C. J. Sherr, and F. Galibert, *Oncogene Res.* **4,** 9–17 (1989).

[310] V. M. Rothwell and L. R. Rohrschneider, *Oncogene Res.* **1,** 311–324 (1987).

[311] A. Hampe, M. Gobet, C. J. Scherr, and F. Galibert, *Proc. Natl. Acad. Sci. U.S.A.* **81,** 85–89 (1984).

[312] Y. Yarden, W.-J. Kuang, T. Yang-Feng, L. Coussens, S. Munemitsu, T. J. Dull, E. Chen, J. Schlessinger, U. Francke, and A. Ullrich, *EMBO J.* **6,** 3341–3350 (1987).

[313] F. Qiu, P. Ray, K. Brown, P. E. Barker, S. Jhanwar, F. H. Ruddle, and P. Besmer, *EMBO J.* **7,** 1003–1011 (1988).

[314] P. Besmer, J. E. Murphy, P. C. George, F. Qiu, P. J. Bergold, L. Lederman, H. W. J. Snyder, D. Brodeur, E. E. Zuckerman, and W. D. Hardy, *Nature* (*London*) **320,** 415–421 (1986).

[315] H. Matsushime, M. D. Yoshida, M. Sasaki, and M. Shibuya, *Japanese J. Cancer Res.* **78,** 655–661 (1987).

[316] M. Shibuya, S. Yamaguchi, A. Yamane, T. Ikeda, A. Tojo, H. Matsushime, and M. Sato, *Oncogene* **5,** 519–524 (1990).

[317] H. Hirai, Y. Maru, K. Hagiwara, J. Nishida, and F. Takaku, *Science* **238,** 1717–1720 (1987).

[317a] R. A. Lindberg and T. Hunter, *Mol. Cell. Biol.* **10,** 6316–6324 (1990).

[318] L. Chan and V. M. Watt, *Oncogene,* in press. (1991).

[319] E. Pasquale, S. K. Hanks, and C. Lai, unpublished observations (1990).

[320] M. Ruta, R. Howk, G. Ricca, W. Drohan, M. Zabelshansky, G. Laureys, D. E. Barton, U. Francke, J. Schlessinger, and D. Givol, *Oncogene* **3,** 9–15 (1988).

[321] M. Ruta, W. Burgess, D. Givol, J. Epstein, J. Kaplow, G. Crumley, C. Dionne, M. Jaye, and J. Schlessinger, *Proc. Natl. Acad. Sci. U.S.A.* **86,** 8722–8726 (1989).

[322] P. L. Lee, D. E. Johnson, L. S. Cousens, V. A. Fried, and L. T. Williams, *Science* **245,** 57–60 (1989).

[323] E. Pasquale and S. J. Singer, *Proc. Natl. Acad. Sci. U.S.A.* **86,** 5449–5453 (1989).

[324] H. H. Reid, A. F. Wilks, and O. Bernard, *Proc. Natl. Acad. Sci. U.S.A.* **87,** 1596–1600 (1990).

[325] A. Safran, A. Avivi, N. Orr-Urtereger, G. P. Lonai, D. Givol, and Y. Yarden, *Oncogene* **5,** 635–643 (1990).

[326] S. Kornbluth, K. E. Paulson, and H. Hanafusa, *Mol. Cell. Biol.* **8,** 5541–5544 (1988).

[327] Y. Hattori, H. Odagiri, H. Nakatani, K. Miyagawa, K. Naito, H. Sakamoto, O. Katoh, T. Yoshida, T. Sugimura, and M. Terada, *Proc. Natl. Acad. Sci. U.S.A.* **87,** 5983–5987 (1990).

[328] E. Pasquale, *Proc. Natl. Acad. Sci. U.S.A.* **87,** 5812–5816 (1990).

[329] J. Partanen, T. P. Mäkelä, R. Alitalo, H. Lehväslaiho, and K. Alitalo, *Proc. Natl. Acad. Sci. U.S.A.* **87,** 8913–8917 (1990).

[330] K. Keegan, D. E. Johnson, L. T. Williams, and M. J. Hayman, *Proc. Natl. Acad. Sci. U.S.A.* **88,** 1095–1099 (1991).

[331] T. W. Wong and A. R. Goldberg, *Proc. Natl. Acad. Sci. U.S.A.* **80,** 2529–2533 (1983).

[332] Y. Ben-Neriah and A. R. Bauskin, *Nature* (*London*) **333,** 672–676 (1988).

[333] Y. Maru, H. Hirai, and F. Takaku, *Oncogene* **5,** 199–204 (1990).

[334] M. Takahashi and G. M. Cooper, *Mol. Cell. Biol.* **7,** 1378–1385 (1987).

[335] M. Takahashi, Y. Buma, T. Iwamoto, Y. Inaguma, H. Ikeda, and H. Hiai, *Oncogene* **3,** 571–578 (1988).

[336] Y. Ishizaka, T. Tahira, M. Ochiai, I. Ikeda, T. Sugimura, and M. Nagao, *Oncogene Res.* **3,** 193–197 (1988).

[337] A. F. Wilks, *Proc. Natl. Acad. Sci. U.S.A.* **86,** 1603–1607 (1989).

[338] I. Firmbach-Kraft, unpublished observations (1990).

[339] S. Nada, M. Okada, A. MacAuley, J. A. Cooper, and H. Nakagawa, *Nature (London)*, in press. (1990).
[340] D. Middlemas and T. Hunter, unpublished observations (1990).
[341] D. A. Foster, J. B. Levy, G. Q. Daley, M. C. Simon, and H. Hanafusa, *Mol. Cell. Biol.* **6,** 325–331 (1986).
[342] H. Mano, F. Ishikawa, J. Nishida, H. Hirai, and F. Takaku, *Oncogene,* **5,** 1781–1786 (1990).
[343] I. Lemischka, unpublished observations (1990).
[344] M. Simon, B. Drees, T. Kornberg, and J. M. Bishop, *Cell* **42,** 831–840 (1985).
[345] F. M. Hoffman, L. D. Fresco, H. Hoffman-Falk, and B.-Z. Shilo, *Cell* **35,** 393–340 (1983).
[346] R. J. Gregory, K. L. Kammermeyer, W. S. I. Vincent, and S. G. Wadsworth, *Mol. Cell. Biol.* **7,** 2119–2127 (1987).
[347] M. J. Henkemeyer, R. L. Bennett, F. B. Gertler, and F. M. Hoffmann, *Mol. Cell. Biol.* **8,** 843–853 (1988).
[348] A. Katzen, D. Montarras, J. Jackson, R. F. Paulson, T. Kornberg, and J. M. Bishop, *Mol. Cell. Biol.* **11,** 226–239 (1991).
[349] J. V. Price, R. J. Clifford, and T. Schüpbach, *Cell* **56,** 1085–1092 (1989).
[350] E. D. Schejter and B.-Z. Shilo, *Cell* **56,** 1093–1104 (1989).
[351] N. E. Baker and G. M. Rubin, *Nature (London)* **340,** 150–153 (1989).
[352] E. Livneh, L. Glazer, D. Segal, J. Schlessinger, and B.-Z. Shilo, *Cell* **40,** 599–607 (1985).
[353] Y. Nishida, M. Hata, Y. Nishizuka, W. J. Rutter, and Y. Ebina, *Biochem. Biophys. Res. Commun.* **141,** 474–481 (1986).
[354] E. Hafen, K. Basler, J.-E. Edstroem, and G. M. Rubin, *Science* **236,** 55–63 (1987).
[355] W. M. Michael, D. D. Bowtell, and G. M. Rubin, *Proc. Natl. Acad. Sci. U.S.A.* **87,** 5351–5353 (1990).
[356] F. Sprenger, L. M. Stevens, and C. Nüsslein-Volhard, *Nature (London)* **338,** 478–483 (1989).
[357] M. Barbacid, unpublished observations (1990).
[358] E. Hafen, unpublished observations (1990).
[359] B. Shilo, unpublished observations (1990).
[360] G. M. Rubin, unpublished observations (1990).
[361] G. Schieven, J. Thorner, and G. S. Martin, *Science* **231,** 390–393 (1986).
[362] J. M. Goddard, J. J. Weiland, and M. R. Capecchi, *Proc. Natl. Acad. Sci. U.S.A.* **83,** 2172–2176 (1986).
[363] R. V. Aroian, M. Koga, J. E. Mendel, Y. Ohshima, and P. W. Sternberg, *Nature (London)* **348,** 693–699 (1990).
[364] M. Koga and Y. Ohshima, *Worm Breeder's Gazette* **11,** 37–38 (1989).
[365] A. Kamb, M. Weir, B. Rudy, H. Varmus, and C. Kenyon, *Proc. Natl. Acad. Sci. U.S.A.* **86,** 4372–4376 (1989).
[366] J. L. Tan and J. A. Spudich, *Mol. Cell. Biol.* **10,** 3578–3583 (1990).
[367] T. C. G. Bosch, T. F. Unger, D. A. Fischer, and R. E. Steele, *Mol. Cell. Biol.* **9,** 4141–4151 (1989).

## Acknowledgments

I thank Steve Hanks for beginning this compilation and for helping keep the new protein kinase sequence entries current. I am also grateful to those who provided unpublished papers and sequence information, which will ensure that this list does not go out of date quickly. I am indebted to Karen Lane for the many hours of painstaking work she put in preparing this chapter.

# [2] Protein Kinase Catalytic Domain Sequence Database: Identification of Conserved Features of Primary Structure and Classification of Family Members

*By* STEVEN K. HANKS and ANNE MARIE QUINN

## Introduction

In recent years it has become apparent that the eukaryotic protein kinases make up an unusually large protein family related on the basis of a homologous catalytic domain.[1,2] In mammals, over 60 distinct protein kinase family members have been described at the sequence level, while many others have been identified from a wide variety of different eukaryotic organisms. Indications are that this family will continue to expand during the next several years. It is now rare for a month to pass without the appearance of research reports describing genes or cDNAs that encode novel protein kinases.

To help cope with the rapidly expanding protein kinase family, we have established a database of the catalytic domain amino acid sequences and are making it available on request (see below). This database should be a useful resource for the initial classification of novel protein kinases and for other studies that require extensive sequence comparisons. The catalytic domain database is updated frequently and often includes new sequences before they can be found in the Genbank/EMBL/PIR resources. More importantly, the availability of this large group of sequences in a single file saves investigators from the tedious task of collecting them themselves.

In this chapter we describe the current makeup of the catalytic domain database and present two examples of its use: (1) analysis and graphic display of conserved catalytic domain residues using conservation plots and (2) classification of protein kinases by phylogenetic mapping. A prerequisite for both of these tasks is a multiple sequence alignment. The one we used is shown in Fig. 1.

## Protein Kinase Catalytic Domain Database

The protein kinase catalytic domain database file can be obtained electronically over Internet using the standard network file transfer pro-

[1] T. Hunter, *Cell (Cambridge, Mass.)* **50,** 823 (1987).
[2] S. K. Hanks, A. M. Quinn, and T. Hunter, *Science* **241,** 42 (1988).

gram (FTP). Computers linked to the national network run FTP locally to connect with node SALK-SC2.SDSC.EDU. Follow the steps listed below.

1. Log into the local networked computer system
2. Type the command FTP
3. Type SALK-SC2.SDSC.EDU at the FTP>
4. Type LOGIN at the SALK-SC2.SDSC.EDU>
5. Type ANONYMOUS at the Foreign username:
6. TYPE GUEST at the Password:
7. Type GET PKINASES.IG at the SALK-SC2.SDSC.EDU>
8. Type a filename for your copy at the To local file:
9. Type EXIT when the file transfer is complete

The database file PKINASES.IG is organized according to Intelligenetics format.[3] This format can be converted to Wisconsin format using the UWGCG program FROMIG. Brief descriptions of the kinases and references for the sequences are included in the database file.

One hundred and seventeen distinct sequences had been entered by mid-February, 1990. Seventy-five of these are taken from protein-serine/threonine kinases and 42 from protein-tyrosine kinases. Sixty-eight of the sequences are from vertebrate species, 24 from yeasts (both budding and fission), 18 from *Drosophila,* 2 from nematode, and 1 each from *Aplysia, Aspergillus, Hydra,* bean plant (*Phaseolus*), and avian erythroblastosis virus S13. The individual entries are listed in Table I.

Some catalytic domain sequences (e.g., from the *src* protooncogene product) have been determined from several different vertebrate species but, since these are very similar, we have included only one listing to represent them all. If available, the human sequence is used. In fact, 43 of the 68 current vertebrate entries are taken from human sources. Twenty-two of the remaining vertebrate sequences derive from 4 other mammals: bovine, rabbit, rat, and mouse. In contrast to the vertebrates, presumed functional homologs from lower eukaryotes are given separate entries as a reflection of their greater evolutionary distance. We have not included listings for protein kinases encoded by viral genomes (an exception is the retroviral oncoprotein v-Sea, whose cellular homolog has not been reported).

Boundaries for the catalytic domains are defined on the basis of homology, i.e., by determining the extreme amino- and carboxyl-terminal residues conserved throughout the family. At the amino terminus, the bound-

[3] Intelligenetics Suite Reference Manual, Intelligenetics Inc., Mountain View, California, January 1989.

**A**

```
cAPKα     FERIKTLGTGSFGRVMLVKHKE--------------TGNHYAMKILDKQKVVKLK----------QIEHTLNEKRILQAV---------NFPFLVKLEFSFKDN------SNLYMVMEYVPGG-EMFSHLRRIGR-------
cAPKβ     FERKKTLGTGSFGRVMLVKHKA--------------TEQYYAMKILDKQKVVKLK----------QIEHTLNEKRILQAV---------NFPFLVRLEYAFKDN------SNLYMVMEYVPGG-EMFSHLRRIGR-------
cAPKγ     FERLRTLGMGSFGRVMLVRHQE--------------TGGHYAMKILNKQKVVKMK----------QVEHILNEKRILQAI---------DFPFLVKLQFSFKDN------SYLYLVMEYVPGG-EMFSRLQRVGR-------
DC0       FERIKTLGTGSFGRVMIVQHK---------------PTKDYAMKILDKQKVVKLK----------QVEHTLNEKRILQAI---------QFPFLVSLRYHFKDN------SNLYMVLEYVPGQ-EMFSHLRKVGR-------
DC1       YITRAVLGNGSFGTVMLVREKS--------------GKNYYAAKMMSKEDLVRLK----------QVAHVHNEKHVLNAA---------RFPFLIYLVDSTKCF------DYLYLILPLVNGG-ELESYHRRVRK-------
DC2       YQIIKTVGTGTFGRVCLCRVRI--------------SEKYCAMKILAMTEVIRLK----------QIEHVKNERNILREI---------RHPFVISLEWSTKDD------SNLYMIFDYVCGG-ELFTYLRNAGK-------
AplC      FNRIKTLGTGSFGRVMLVQHKGE-------------SRNFYAMKILDKQKVVKLK----------QVEHTLNEKKILQSI---------DFPFLVKLEYSFKDN------SNLYMVLEFVTGG-EMFSHLRRIGR-------
TPK1      FQILRTLGTGSFGRVHLIRSRH--------------NGRYYAMKVLKKEIVVRLK----------QVEHTNDERLMLSIV---------THPFIIRMWGTFQDA------QQIFMIMDYIEGG-ELFSLLRKSQR-------
TPK2      FQIMRTLGTGSFGRVHLVRSVH--------------NGRYYAIKVLKKQQVVKMK----------QVEHTNDERRMLKLV---------EHPFLIRMWGTFQDA------RNIFMVMDYIEGG-ELFSLLRKSQR-------
TPK3      FQILRTLGTGSFGRVHLIRSNH--------------NGRFYALKTLKKHTIVKLK----------QVEHTNDERRMLSIV---------SHPFIIRMWGTFQDS------QQVFMVMDYIEGG-ELFSLLRKSQR-------
cGPK      FNIIDTLGVGGFGRVELVQLKSE-------------ESKTFAMKILKKRHIVDTR----------QQEHIRSEKQIMQGA---------HSDFIVRLYRTFKDS------KYLYMLMEACLGG-ELWTILRDRGS-------
DG1       LEVVSTLGIGGFGRVELVKAHHQD------------RVDIFALKCLKKRHIVDTK----------QEEHIFSERHIMLSS---------RSPFICRLYRTFRDE------KYVYMLLEACMGG-EIWTMLRDRGS-------
DG2       LRVIATLGVGGFGRVELVQTNGD-------------SSRSFALKQMKKSQIVETR----------QQQHIMSEKEIMGEA---------NCQFIVKLFKTFKDK------KYLYMLMESCLGG-ELWTILRDKGN-------

PKCα      FNFLMVLGKGSFGKVMLADRKG--------------TEELYAIKILKKDVVIQDD----------DVECTMVEKRVLALLD--------KPPFLTQLHSCFQTV------DRLYFVMEYVNGG-DLMYHIQQVGK-------
PKCβ      FNFLMVLGKGSFGKVMLSERKG--------------TDELYAVKILKKDVVIQDD----------DVECTMVEKRVLALPG--------KPPFLTQLHSCFQTM------DRLYFMMEYVNGG-DLMYHIQQVGR-------
PKCγ      FSFLMVLGKGSFGKVMLAERRG--------------SDELYAIKILKKDVIVQDD----------DVDCTLVEKRVLALGGRGPGG---RPHFLTQLHSTFQTP------DRLYFVMEYVTGG-DLMYHIQQLGK-------
PKCδ      FTFQKVLGKGSFGKVLLAELKG--------------KERYFAIKYLKKDVVLIDD----------DVECTMVEKRVLALAW--------ENPFLTHLICTFQTK------DHLFFVMEFLNGG-DLMFHIQDKGR-------
PKCε      FNFIKVLGKGSFGKVMLAELKG--------------KDEVYAVKVLKKDVILQDD----------DVDCTMTEKRILALAR--------KHPYLTQLYCCFQTK------DRLFFVMEYVNGG-DLMFQIQRSRK-------
PKCζ      FDLIRVIGRGSYAKVLLVRLKKN-------------DQIIYAMKVVKKELVHDDE----------DIDWVQTEKHVFEQAS--------SNPFLVGLHSCFQTT------SRLFLVIEYVNGG-DLMFHMQRQRK-------
DPKC53b   FNFIKVLGKGSFGKVLLAERKG--------------SEELYAIKILKKDVIIQDD----------DVECTMIEKRVLALGE--------KPPFLVQLHSCFQTM------DRLFFVMEYVNGG-DLIFQIQQFGK-------
DPKC98    FNFIKVLGKGSFGKVMLAEKKG--------------TDEIYAIKVLKKDAIIQDD----------DVDCTMTEKRILALAA--------NHPFLTALHSCFQTP------DRLFFVMEYVNGG-DLMFQIQKARR-------
DPKC53e   FNFVKVIGKGSFGKVLLWERRG--------------TDELYAVKVLRKDVIIQTD----------DMELPMNEKKILALSG--------RPPFLVSMHSCFQTM------DRLFFVMEYCKGG-DLMYHMQQYGK-------
TPA1      FNLLKVLGKGSFGKVMLVELKG--------------KNEFYAMKCLKKDVILEDD----------DTECTYIERRVLILAS--------QCPFLCQLFCSFQTN------EYLFFVMEYLNGG-DLMHHIQQIKK-------
YPK1      FDLLKVIGKGSFGKVMQVRKKD--------------TQKVYALKAIRKSYIVSKS----------EVTHTLAERTVLARV---------DCPFIVPLKFSFQSP------EKLYFVLAFINGG-ELFYHLQKEGR-------
YKR2      FDLLKVIGKGSFGKVMQVRKKD--------------TQKIYALKALRKAYIVSKC----------EVTHTLAERTVLARV---------DCPFIVPLKFSFQSP------EKLYLVLAFINGG-ELFYHLQHEGR-------
SCH9      FEVLRLLGKGTFGQVYQVKKKD--------------TQRIYAMKVLSKKVIVKKN----------EIAHTIGERNILVTTASK------SSPFIVGLKFSFQTP------TDLYLVTDYMSGG-ELFWHLQKEGR-------

RSK1N     FELLKVLGQGSFGKVFLVRKVTRPD-----------SGHLYAMKVLKKATLKVR-----------DRVRTKMERDILADV---------NHPFVVKLHYAFQTE------GKLYLILDFLRGG-DLFTRLSKEVM-------
PVPK1     FRLLKKLGCGDIGSVYLAELSG--------------TRTSFAMKVMNKTELANRK----------KLLRAQTEREILQSL---------DHPFLPTLYTHFETE------IFSCLVMEFCPGG-DLHALRQRQPGKY-----
βARK      FSVHRIIGRGGFGEVYGCRKAD--------------TGKMYAMKCLDKKRIKMKQ----------GETLALNERIMLSLVSTG------DCPFIVCMSYAFHTP------DKLSFILDLMNGG-DLHYHLSQHGV-------

CAMIIα    YQLFEELGKGAFSVVRRCVKVL--------------AGQEYAAKIINTKKLSAR-----------DHQKLEREARICRLL---------KHPNIVRLHDSISEE------GHHYLIFDLVTGG-ELFEDIVAREY-------
CAMIIβ    YQLYEDIGKGAFSVVRRCVKLC--------------TGHEYAAKIINTKKLSAR-----------DHQKLEREARICRLL---------KHSNIVRLHDSISEE------GFHYLVFDLVTGG-ELFEDIVAREY-------
CAMIIγ    YQLFEELGKGAFSVVRRCVKKT--------------STQEYAAKIINTKKLSAR-----------DHQKLEREARICRLL---------KHPNIVRLHDSISEE------GFHYLVFDLVTGG-ELFEDIVAREY-------
PHKγ      YEPKEILGRGVSSVVRRCIHKP--------------TCKEYAVKIIDVTGGGSFSAEEVQE----LREATLKEVDILRKVS--------GHPNIIQLKDTYETN------TFFFLVFDLMKKG-ELFDYLTEKVT-------
PSK-C3    YDPKDVIGRGVSSVVRRCVHRA--------------TGHEFAVKIMEVTAERLSPEQLEE-----VREATRRETHILRQVA--------GHPHIITLIDSYESS------SFMFLVFDLMRKG-ELFDYLTEKVA-------
RSK1C     YVVKETIGVGSYSVCKRCVHKA--------------TNMEYAVKVIDK-----------------SKRDPSEEIEILLRYG--------QHPNIITLKDVYDDG------KHVYLVTELMRGG-ELLDKILRQF--------
RSK2C     YEVKEDIGVGSYSVCKRCIHKA--------------TNMEFAVKIIDK-----------------SKRDPTEEIEILLRYG--------QHPNIITLKDVYDDG------KYVYVVTELMKGG-ELLDKILRQF--------
PSK-H1    YDIKALIGRGSFSRVVRVEHRA--------------TRQPYAIKMIETKYRE-------------GREVCESELRVLRRV---------RHANIIQLVEVFETQ------ERVYMVMELATGG-ELFDRIIAKGS-------
MLCK-K    MNSKEALGGGKFGAVCTCTEKS--------------TGLKLAAKVIKKQTPK-------------DKEMVMLEIEVMNQL---------NHRNLIQLYAAIETP------HEIVLFMEYIEGG-ELFERIVDEDYH------
MLCK-M    YNIEERLGSGKFGQVFRLVEKK--------------TGKVWAGKFFKAYSAK-------------EKENIRDEISIMNCL---------HHPKLVQCVDAFEEK------ANIVMVLEMVSGG-ELFERIIDEDFE------
KIN1      WEFVETVGAGSMGKVKLAKHRY--------------TNEVCAVKIVNRATKAFLHKEQMLP-{}-RDKRTIREASLGQIL---------YHPHICRLFEMCTLS------NHFYMLFEYVSGG-QLLDYIIQHGS-------
KIN2      WEFLETVGAGSMGKVKLVKHRQ--------------TKEICVIKIVNRASKAYLHKQHSLP-{}-RDKRTVREASLGQIL---------YHPHICRLFEMCTMS------NHFYMLFEYVSGG-QLLDYIIQHGS-------
SNF1      YQIVKTLGEGSFGKVKLAYHTT--------------TGQKVALKIINKKVLAKSD----------MQGRIEREISYLRLL---------RHPHIIKLYDVIKSK------DEIIMVIEYIAGN-ELFDYIVQRDK-------
nim1      WRLGKTLGIGSTSCVRLAKHAK--------------TGDLAAIKIIPIRYAS-------------------IGMEILMMRLL---------RHPNILRLYDVWTDH------QHMYLALEYVPDG-ELFHYIRKHGP-------

                 I                                  II                       III                   IV                      V
```

```
CDC2Hs  YTKIEKIGEGTYGVVYKGRHKT-------------TGQVVAMKKIRLESEEEG-----------VPSTAIREISLLKEL---------RHPNIVSLQDVLMQD------SRLYLIFEFLSM--DLKKYLDSIPPGQY----
cdc2    YQKVEKIGEGTYGVVYKARHKL-------------SGRIVAMKKIRLEDESEG-----------VPSTAIREISLLKEVNDEN-----NRSNCVRLLDILHAE------SKLYLVFEFLDM--DLKKYMDRISETGATS--
CDC28   YKRLEKVGEGTYGVVYKALDLRPGQ----------GQRVVALKKIRLESEDEG-----------VPSTAIREISLLKEL---------KDDNIVRLYDIVHSDA-----HKLYLVFEFLDL--DLKRYMEGIPKDQP----
PHO85   FKQLEKLGNGTYATVYKGLNKT-------------TGVYVALKEVKLDSEEG------------TPSTAIREISLMKEL---------KHENIVRLYDVIHTE------NKLTLVFEFMDN--DLKKYMDSRTVANTPRG-
PSK-J3  YEPVAEIGVGAYGTVYKARDPH-------------SGHFVALKSVRVPNGGGGGGG--------LPISTVREVALLRRLEAF------EHPNVVRLMDVCATSRTDRE-IKVTLVFEHVDQ--DLRTYLDKAPPPG-----
KIN28   YTKEKKVGEGTYAVVYLGCQHS-------------TGRKIAIKEIKTSEFKDG-----------LDMSAIREVKYLQEM---------QHPNVIELIDIFMAY------DNLNLVLEFLPT--DLEVVIKDKSIL------
KSS1    YKLVDLIGEGAYGTVCSAIHKP-------------SGIKVAIKKIQPFSKKL------------FVTRTIREIKLLRYFH--------EHENIISILDKVRPVSIDKL-NAVYLVEELMET--DLQKVINNQNSGFST---
MAK     YTTMRQLGDGTYGSVLMGKSNE-------------SGELVAIKRMKRKFYSW------------DECMNLREVKSLKKL---------NHANVIKLKEVIREN------DHLYFIFEYMKE--NLYQLMKDRNKL------
YAK1    YLVLDILGQGTFGQVVKCQNLL-------------TKEILAVKVVKSRTE--------------YLTQSITEAKILELLNQKIDPT--NKHHFLRMYDSFVHK------NHLCLVFELLSN--NLYELLKQNKFHG-----
GSK3    YTDIKVIGNGSFGVVYQARLAE-------------TRELVAIKKVLQDKRF-----------------KNRELQIMRKL---------DHCNIVRLRYFFYSSGEKKDELYLNLVLEYVPE--TVYRVARHFTKAKLI---
CKII    YQLVRKLGRGKYSEVFEAINIT-------------NNEKVVVKILKPV----------------KKKKIKREIKILENLR--------GGPNADIVKDPVS--------RTPALVFEHVNNT-DFKQLYQT----------
DCKII   YQLVRKLGRGKYSEVFEAINIT-------------TTEKCVVKILKPV----------------KKKKIKREIKILENLR--------GGTNLAVVKDPVS--------RTPALIFEHVNNT-DFKQLYQT----------
CKAI    YEIENKVGRGKYSEVFQGVKLD-------------SKVKIVIKMLKPV----------------KKKKIKREIKILTDLS--------GHANIIHLFDIIKDPIS----KTPALVFEYVDNV-DFRILYPK----------
PSK-H2  YEVLKVIGKGSFGQVVKAYDHK-------------VHQHVAIKMVRNEKRF--------------HRQAAEEIRILEHLRKQDKD---NTMNVIHMLENFTFR------NHICMTFELLSM--NLYELIKKNKFQG-----
PSK-G1  YEIVSTLGEGTFGRVVQSVDHRR------------GGARVAIKIIKNVEKY--------------KEAARIEINVLEKINEKDPD---NKNLCVQMFDWFDYH------GHMCISFELLGL--STFDFLKDNNYLP-----

PBS2    LEFLDELGHGNYGNVSKVLHKP-------------TNVIMATKEVRLELDEA------------KFRQILMELEVLHKC---------NSPYIVDFYGAFFIE------GAVYMCMEYMDGG-SLDKIYDESSEIGG----
byr1    LEVVRHLGEGNGGAVSLVKHRN---------------IFMARKTVYVGSDSK------------LQKQILRELGVLHHC---------RSPYIVGFYGAFQYK------NNISLCMEYMDCG-SLDAILR-----------
STE7    LVQLGKIGAGNSGTVVKALHVP-------------DSKIVAKKTIPVEQNNST-----------IINQLVRELSIVKNVK--------PHENIITFYGAYYNQHIN---NEIIILMEYSDCG-SLDKILSVYKRFVQRGTV
ninaC   FEIYEEIAQGVNAKVFRAKELD-------------NDRIVALKIQHYD----------------EEHQVSIEEEYRTLRDYC------DHPNLPEFYGVYKLSKPNGP-DEIWFVMEYCAGG-TAVDMVNKLLKLDRR---
nimA    YEVLEKIGCGSFGIIRKVKRKS-------------DGFILCRKEINYIKMSTK-----------EREQLTAEFNILSSL---------RHPNIVAYYHREHLKAS----QDLYLYMEYCGGG-DLSMVIKNLKRTNKY---
PIM1    YQVGPLLGSGGFGSVYSGIRVS-------------DNLPVAIKHVEKDRISDWGELP-------NGTRVPMEVVLLKKVSS-------GFSGVIRLLDWFERP------DSFVLILERPEPVQDLFDFITERGA-------
ran1    LRFVSIIGAGAYGVVYKAEDIY-------------DGTLYAVKALCKDGLNEK-----------QKKLQARELALHARVS--------SHPYIITLHRVLETE------DAIYVVLQYCPNG-DLFTYITEKKVYQ-----
CDC7    YKLIDKIGEGTFSSVYKAKDITGKITKKFASHFWNYGSNYVALKKIYVTSSP---------------QRIYNELNLLYIMT--------GSSRVAPLCDAKRVR------DQVIAVLPYYPHE-EFRTFYRD----------
wee1    FRNVTLLGSGEFSEVFQVEDPVE------------KTLKYAVKKLKVKFSGPK-----------ERNRLLQEVSIQRALK--------GHDHIVELMDSWEHG------GFLYMQVELCENG-SLDRFLEEQGQLSR----
PYT     YSILKQIGSGGSSKVFQVLNEK--------------KQIYAIKYVNLEEADNQ-----------TLDSYRNEIAYLNKLQQ-------HSDKIIRLYDYEITD------QYIYMVMECGNI--DLNSWLKKKKS-------
COT     NIGSDFIPRGAFGKVYLAQDIK-------------TKKRMACKLIPVDQF----------------KPSDVEIQACF-----------RHENIAELYGAVLWG------ETVHLFMEAGEGG-SVLEKLESCGP-------
GCN2    LKRLNFSGQGAFGQVVKARNAL-------------DSRYYAIKKIRNTEE--------------KLSTMISEVMLLASLNHQYV-{}-RRRNFVKPMTAVKKK------STLFIQMEYCENR-TLYDLIHSENLN------

c-raf   VMLSTRIGSGSFGTVYKGKWHG----------------DVAVKILKVVDPTPE-----------QFQAFRNEVAVLRKT---------RHVNILLFMGYMTK-------DNLAIVTQWCEGS-SLYKHLHVQETK------
Araf    VQLLKRIGTGSFGTVFRGRWHG----------------DVAVKVLKVSQPTAE-----------QAQAFKNEMQVLRKT---------RHVNILLFMGFMTR-------PGFAIITQWCEGS-SLYHHLHVADTR------
Braf    ITVGQRIGSGSFGTVYKGKWHG----------------DVAVKMLNVTAPTPQ-----------QLQAFKNEVGVLRKT---------RHVNILLFMGYSTK-------PQLAIVTQWCEGS-SLYHHLHIIETK------
Draf    ILIGPRIGSGSFGTVYRAHWHG----------------PVAVKTLNVKTPSPA-----------QLQAFKNEVAMLKKT---------RHCNILLFMGCVSK-------PSLAIVTQWCEGS-SLYKHVHVSETK------
c-mos   VCLLQRLGAGGFGSVYKATYRG---------------VPVAIKQVNKCTKNRLA----------SRRSFWAELNVARL----------RHDNIVRVVAASTRTPAGSN-SLGTIIMEFGGNV-TLHQVIYGAAGHPEGDAG

                 I                          II                           III                 IV                          V
```

FIG. 1. Multiple amino acid sequence alignment of the catalytic domains of (A) 75 protein-serine/threonine kinases and (B) 42 protein-tyrosine kinases. The dashes represent gaps that have been inserted in the sequence to align conserved regions. Highly conserved amino acids with three or less variant residues at any given position are shown by black boxes with white lettering. Long insertions which are not conserved are indicated by replacement braces. Roman numerals along the bottom refer to subdomains conserved across the protein kinase family.

```
cAPKα     ---------FSEPHARFYAAQIVLTFEYLHSL---------DLIYRDLKPENLLIDQQG----------YIQVTDFGFAKRVKGR-----------------TWTLCGTPEYLAPEII-------------LSKGYNK
cAPKβ     ---------FSEPHARFYAAQIVLTFEYLHSL---------DLIYRDLKPENLLIDHQG----------YIQVTDFGFAKRVKGR-----------------TWTLCGTPEYLAPEII-------------LSKGYNK
cAPKγ     ---------FSEPHACFYAAQVVLAVQYLHSL---------DLIHRDLKPENLLIDQQG----------YLQVTDFGFAKRVKGR-----------------TWTLCGTPEYLAPEII-------------LSKGYNK
DC0       ---------FSEPHSRFYAAQIVLAFEYLHYL---------DLIYRDLKPENLLIDSQG----------YLKVTDFGFAKRVKGR-----------------TWTLCGTPEYLAPEII-------------LSKGYNK
DC1       ---------FNEKHARFYAAQVALALEYMHKM---------HLMYRDLKPENILLDQRG----------YIKITDFGETKRVDGR-----------------TSTLCGTPEYLAPEIV-------------QLRPYNK
DC2       ---------FTSQTSNFYAAEIVSALEYLHSL---------QIVYRDLKPENLLINRDG----------HLKITDFGFAKKLRDR-----------------TWTLCGTPEYIAPEII-------------QSKGHNK
AplC      ---------FSEPHSRFYAAQIVLVLEYLHHL---------DIMYRDLKPENLLIDSYG----------YLKVTDFGFAKRVKGR-----------------TWTLCGTPEYLAPEII-------------LSKGYDK
TPK1      ---------FPNPVAKFYAAEVCLALEYLHSK---------DIIYRDLKPENILLDKNG----------HIKITDFGFAKYVPDV-----------------TYTLCGTPDYIAPEVV-------------STKPYNK
TPK2      ---------FPNPVAKFYAAEVILALEYLHAH---------NIIYRDLKPENILLDRNG----------HIKITDFGFAKEVQTV-----------------TWTLCGTPDYIAPEVI-------------TTKPYNK
TPK3      ---------FPNPVAKFYAAEVCLALEYLHSK---------DITYRDLKPENILLDKNG----------HIKITDFGFAKYVPDV-----------------TYTLCGTPDYIAPEVV-------------STKPYNK
cGPK      ---------FEDSTTRFYTACVVEAFAYLHSK---------GIIYRDLKPENLILDHRG----------YAKLVDFGFAKKIGFG---------------KKTWTFCGTPEYVAPEII-------------LNKGHDI
DG1       ---------FEDNAAQFIIGCVLQAFEYLHAR---------GIIYRDLKPENLMLDERG----------YVKIVDFGFAKQIGTS---------------SKTWTFCGTPEYVAPEII-------------LNKGHDR
DG2       ---------FDDSTTRFYTACVVEAFDYLHSR---------NIIYRDLKPENLLLNERG----------YGKLVDFGFAKKLQTG---------------RKTWTFCGTPEYVAPEVI-------------LNRGHDI

PKCα      ---------FKEPQAVFYAAEISIGLFFLHKR---------GIIYRDLKLDNVMLDSEG----------HIKIADFGMCKEHMMD---------------GVTTRTFCGTPDYIAPEII-------------AYQPYGK
PKCβ      ---------FKEPHAVFYAAEIAIGLFFLQSK---------GIIYRDLKLDNVMLDSEG----------HIKIADFGMCKENIWD---------------GVTTKTFCGTPDYIAPEII-------------AYQPYGK
PKCγ      ---------FKEPHAAFYAAEIAIGLFFLHNQ---------GIIYRDLKLDNVMLDAEG----------HIKITDFGMCKENVFP---------------GSTTRTFCGTPDYIAPEII-------------AYQPYGK
PKCδ      ---------FELYRATFYAAEIICGLQFLHGK---------GIIYRDLKLDNVMLDKDG----------HIKIADFGMCKENIFG---------------ENRASTFCGTPDYIAPEIL-------------QGLKYSF
PKCε      ---------FDEPRSGFYAAEVTSALMFLHQH---------GVIYRDLKLDNILLDAEG----------HSKLADFGMCKEGILN---------------GVTTTTFCGTPDYIAPEIL-------------QELEYGP
PKCζ      ---------LPEEHARFYAAEICIALNFLHER---------GIIYRDLKLDNVLLDADG----------HIKLTDYGMCKEGLGP---------------GDTTSTFCGTPNYIAPEIL-------------RGEEYGF
DPKC53b   ---------FKEPVAVFYAAEIAAGLFFLHTK---------GILYRDLKLDNVLLDADG----------HVKIADFGMCKENIVG---------------DKTTKTFCGTPDYIAPEII-------------LYQPYGK
DPKC98    ---------FEASRAAFYAAEVTLALQFLHTH---------GVIYRDLKLDNILLDQEG----------HCKLADFGMCKEGIMN---------------GMLTTTFCGTPDYIAPEIL-------------KEQEYGA
DPKC53e   ---------FKESVAIFYAVEVAIALFFLHER---------DIIYRDLKLDNILLDGEG----------HVKLVDFGLSKEGVTE---------------RQTTRTFCGTPNYMAPEIV-------------SYDPYSI
TPA1      ---------FDEARTRFYACEIVVALQFLHTN---------NIIYRDLKLDNVLLDCDG----------HIKLADFGMAKTEMNR---------------ENGMASTFCGTPDYISPEII-------------KGQLYNE
YPK1      ---------FDLSRARFYTAELLLALDNLHKL---------DVVYRDLKPENILLDYQG----------HIALCDFGLCKLNMKD---------------DDKTDTFCGTPEYLAPELL-------------LGLGYTK
YKR2      ---------FSLARSRFYIAELLCALDSLHKL---------DVIYRDLKPENILLDYQG----------HIALCDFGLCKLNMKD---------------NDKTDTFCGTPEYLAPEIL-------------LGQGYTK
SCH9      ---------FSEDRAKFYIAELVLALEHLHDN---------DIVYRDLKPENILLDANG----------NIALCDFGLSKADLKD---------------KDRTNTFCGTTEYLAPELLL------------DETGYTK

RSK1N     ---------FTEEDVKFYLAELALGLDHLHSL---------GIIYRDLKPENILLDEEG----------HIKLTDFGLSKEAIDH---------------EKKAYSFCGTVEYMAPEVV-------------NRQGHTH
PVPK1     ---------FSEHAVRFYVAEVLLSLEYLHML---------GIIYRDLKPENVLVREDG----------HIMLSDFDLSLRCSVSPTLVKSSNNLQT-{}-NARSMSFVGTHEYLAPEII-------------KGEGHGS
βARK      ---------FSEADMRFYAAEIILGLEHMHNR---------FVVYRDLKPANILLDEHG----------HVRISDIGLACDFSKK-----------------KPHASVGTHGYMAPEVLQ-------------KGVAYDS

CAMIIα    ---------YSEADASHCIQQILEAVLHCHQM---------GVVHRDLKPENLLLASKLKGA-------AVKLADFGLAIEVEGE---------------QQAWFGFAGTPGYLSPEVL-------------RKDPYGK
CAMIIβ    ---------YSEADASHCIQQILEAVLHCHQM---------GVVHRDLKPENLLLASKCKGA-------AVKLADFGLAIEVQGD---------------QQAWFGFAGTPGYLSPEVL-------------RKEAYGK
CAMIIγ    ---------YSEADASHCIHQILESVNHIHQH---------DIVHRDLKPENLLLASKCKGA-------AVKLADFGLAIEVQGE---------------QQAWFGFAGTPGYLSPEVL-------------RKDPYGK
PHKγ      ---------LSEKETRKIMRALLEVICALHKL---------NIVHRDLKPENILLDDDM----------NIKLTDFGFSCQLDPG---------------EKLREVCGTPSYLAPEIIECSMND--------NHPGYGK
PSK-C3    ---------LSEKETRSIMRSLLEAVSFLHAN---------NIVHRDLKPENILLDDNM----------QIRLSDFGFSCHLEPG---------------EKLRELCGTPGYLAPEILKCSMDE--------THPGYGK
RSK1C     ---------FSEREASFVLHTISKTVEYLHSQ---------GVVHRDLKPSNILYVDESGNPE------CLRICDFGFAKQLRAE---------------NGLLMTPCYTANFVAPEVL-------------KRQGYDE
RSK2C     ---------FSEREASAVLFTITKTVEYLHAQ---------GVVHRDLKPSNILYVDESGNPE------SIRICDFGFAKQLRAE---------------NGLLMTPCYTANFVAPEVL-------------KRQGYDA
PSK-H1    ---------FTERDATRVLQMVLDGVRYLHAL---------GITHRDLKPENLLYYHPGTDS-------KIIITDFGLASARKKG---------------DDCLMKTTCGTPEYIAPEVL-------------VRKPYTN
MLCK-K    ---------LTEVDTMVFVRQICDGILFMHKM---------RVLHLDLKPENILCVNTTGH--------LVKIIDFGLARRYNPN---------------EKLKVNFGTPEFLSPEVV-------------NYDQISD
MLCK-M    ---------LTERECIKYMRQISEGVEYIHKQ---------GIVHLDLKPENIMCVNKTGT--------SIKLIDFGLARRLESA---------------GSLKVLFGTPEFVAPEVI-------------NYEPIGY
KIN1      ---------IREHQARKFARGIASALIYLHAN---------NIVHRDLKIENIMISDSS----------EIKIIDFGLSNIYDSR---------------KQLHTFCGSLYFAAPELL-------------KANPYTG
KIN2      ---------LKEHHARKFARGIASALQYLHAN---------NIVHRDLKIENIMISSSG----------EIKIIDFGLSNIFDYR---------------KQLHTFCGSLYFAAPELL-------------KAQPYTG
SNF1      ---------MSEQEARRFFQQIISAVEYCHRH---------KIVHRDLKPENLLLDEHL----------NVKIADFGLSNIMTDG---------------NFLKTSCGSPNYAAPEVI-------------SGKLYAG
nim1      ---------LSEREAAHYLSQILDAVAHCHRF---------RFRHRDLKLENILIKVNEQ---------QIKIADFGMATVEPND---------------SCLENYCGSLHYLAPEIV-------------SHKPYRG

                 VIa                      VIb                       VII                              VIII
```

```
CDC2Hs  ---------MDSSLVKSYLYQILQGIVFCHSR---------RVLHRDLKPQNLLIDDKG-----------TIKLADFGLARAFGIP-----------------IRVYTHEVVTLWYRSPEVLL-------------GSARYST
cdc2    ---------LDPRLVQKFTYQLVNGVNFCHSR---------RIIHRDLKPQNLLIDKEG-----------NLKLADFGLARSFGVP-----------------LRNYTHEIVTLWYRAPEVLL-------------GSRHYST
CDC28   ---------LGADIVKKFMMQLCKGIAYCHSH---------RILHRDLKPQNLLINKDG-----------NLKLGDFGLARAFGVP-----------------LRAYTHEIVTLWYRAPEVLL-------------GGKQYST
PHO85   ---------LELNLVKYFQWQLLQGLAFCHEN---------KILHRDLKPQNLLINKRG-----------QLKLGDFGLARAFGIP-----------------VNTFSSEVVTLWYRAPEVLM-------------GSRTYST
PSK-J3  ---------LPAETIKDLMRQFLRGLDFLHAN---------CIVHRDLKPENILVTSGG-----------TVKLADFGLARIYSYQ------------------MALTPVVVTLWYRAPEVLL--------------QSTYAT
KIN28   ---------FTPADIKAWMLMTLRGVYHCHRN---------FILHRDLKPNNLLFSPDG-----------QIKVADFGLARAIPAP-----------------HEILTSNVVTRWYRAPELLF-------------GAKHYTS
KSS1    ---------LSDDHVQYFTYQILRALKSIHSA---------QVIHRDIKPSNLLLNSNC-----------DLKVCDFGLARCLASSSDSRETL----------VGFMTEYVATRWYRAPEIML-------------TFQEYTT
MAK     ---------FPESVIRNIMYQILQGLAFIHKH---------GFFHRDMKPENLLCMGPE-----------LVKIADFGLARELRSQ------------------PPYTDYVSTRWYRAPEVLL-------------RSSVYSS
YAK1    ---------LSIQLIRTFTTQILDSLCVLKES---------KLIHCDLKPENILLCAPDKP---------ELKIIDFGSSCEEA--------------------RTVYTYIQSRFYRAPEIIL--------------GIPYST
GSK3    ---------IPIIYVKVYMYQLFRSLAYIHSQ---------GVCHRDIKPQNLLVDPDTA----------VLKLCDFGSAKQLVRG------------------EPNVSYICSRYYRAPELIF-------------GATDYTS
CKII    ---------LTDYDIRFYMYEILKALDYCHSM---------GIMHRDVKPHNVMIDHEHR----------KLRLIDWGLAEFYHPG------------------QEYNVRVASRYFKGPELLV-------------DYQMYDY
DCKII   ---------LTDYEIRYYLFELLKALDYCHSM---------GIMHRDVKPHNVMIDHENR----------KLRLIDWGLAEFYHPG------------------QEYNVRVASRYFKGPELLV-------------DYQMYDY
CKAI    ---------LTDLEIRFYMFELLKALDYCHSM---------GIMHRDVKPHNVMIDHKNK----------KLRLIDWGLAEFYHVN------------------MEYNVRVASRFFKGPELLV-------------DYRMYDY
PSK-H2  ---------FSLPLVRKFAHSILQCLDALHKN---------RIIHCDLKPENILLKQQGRS---------GIKVIDFGSSCYEHQR--------------------VYTYIQSRFYRAPEVIL--------------GPWYGM
PSK-G1  ---------YPIHQVRHMAFQLCQAVKFLHDN---------KLTHTDLKPENILFVNSDYELTYNL-{}-AARVVDFGSATFDHEH--------------------HSTIVSTRHYRAPEVIL--------------ELGWSQ

PBS2    ---------IDEPQLAFIANAVIHGLKELKEQH--------NIIHRDVKPTNILCSANQG----------TVKLCDFGVSGNLVAS-------------------LAKTNIGCQSYMAPERIKSLNP---------DRATYTV
byr1    ---------EGGPIPLDILGKIINSMVKGLIYLYNVL----HIIHRDLKPSNVLINSKG-----------EIKLCDFGVSGELVNS-------------------VAQTFVGTSTYMSPERI--------------RGGKYTV
STE7    SSKKTW---FNELTISKIAYGVLNGLDHLYRQY--------KIIHRDIKPSNVLINSKG-----------QIKLCDFGVSKKLINS-------------------IADTFVGTSTYMSPERI--------------QGNVYSI
ninaC   ---------MREEHIAYIIRETCRAAIELNRN---------HVLHRDIRGDNILLTKNG-----------RVKLCDFGLSRQVDST-----------------LGKRGTCIGSPCWMAPEVVSAME----------SEPDITV
nimA    ---------AEEDFVWRILSQLVTALYRCHYGTDPAE-{}-TILHRDLKPENIFLGSDN-----------TVKLGDFGLSKLMHSH------------------DFASTYVGTPFYMSPEIC--------------AAEKYTL
PIM1    ---------LQEELARSFFWQVLEAVRHCHNC---------GVLHRDIKDENILIDLNRG----------ELKLIDFGSGALLKDT-------------------VYTDFDGTRVYSPPEWIR-------------YHRYHGR
ran1    ---------GNSHLIKTVFLQLISAVEHCHSV---------GIYHRDLKPENIMVGNDGN----------TVYLADFGLATTEPYS--------------------SDFGCGSLFYMSPECQREVKKLSSLS-{}-SSSFATA
CDC7    ---------LPIKGIKKYIWELLRALKFVHSK---------GIIHRDIKPTNFLFNLELG----------RGVLVDFGLAEAQMDYKSMISSQNDY-{}-RIKNANANRAGTRGFRAPEVLM-------------KCGAQST
wee1    ---------LDEFRVWKILVEVALGLQFIHHK---------NYVHLDLKPANVMITFEG-----------TLKIGDFGMASVWPVP-------------------RGMEREGDCEYIAPEVL--------------ANHLYDK
PYT     ---------IDPWERKSYWKNMLEAVHTIHQH---------GIVHSDLKPANFLIVDG------------MLKLIDFGIANQMQPD---------------TTSVVKDSQVGTVNYMPPEAIKDMSSSRENGK---SKSKISP
COT     ---------MREFEIIWVTKHVLKGLDFLHSK---------KVIHHDIKPSNIVFMST------------KAVLVDFGLSVQMTED-----------------VYFPKDLRGTEIYMSPEVI--------------LCRGHST
GCN2    ---------QQRDEYWRLFRQILEALSYIHSQ---------GIIHRDLKPKNIFIDESR-----------NVKIGDFGLAKNVHRSLDILKDSQNLPGS--SDNLLTSAIGTAMYVATEVLD-------------GTGHYNE

c-raf   ---------FQMFQLIDIARQTAQGMDYLHAK---------NIIHRDMKSNNIFLHEGL-----------TVKIGDFGLATVKSRW---------------SGSQQVEQPTGSVLWMAPEVIRMQ-----------DNNPFSF
Araf    ---------FDMVQLIDVARQTAQGMDYLHAK---------NIIHRDLKSNNIFLHEGL-----------TVKIGDFGLATVKTRW---------------SGAQPLEQPSGSVLWMAAEVIRMQ-----------DPNPYSF
Braf    ---------FEMIKLIDIARQTAQGMDYLHAK---------SIIHRDLKSNNIFLHEDL-----------TVKIGDFGLATVKSRW---------------SGSHQFEQLSGSILWMAPEVIRMQ-----------DKNPYSF
Draf    ---------FKLNTLIDIGRQVAQGMDYLHAK---------NIIHRDLKSNNIFLHEDL-----------SVKIGDFGLATAKTRW---------------SGEKQANQPTGSILWMAPEVIRMQ-----------ELNPYSF
c-mos   EPHCRTGGQLSLGKCLKYSLDVVNGLLFLHSQ---------SIVHLDLKPANILISEQD-----------VCKISDFGCSEKLEDL--------------LCFQTPSYPLGGTYTHRAPELL--------------KGEGVTP

                   VIa                      VIb                         VII                                   VIII
```

FIG. 1. (*continued*)

```
cAPKα    -AVDWWALGVLIYEMAA-GYPPFFA-------DQPIQIYEKIVSGKVRFPSH----------------------------------------------------FSSDLKD-LLRNLLQVDLTKRFGNLKNGVNDIKN-HKWF
cAPKβ    -AVDWWALGVLIYEMAA-GYPPFFA-------DQPIQIYEKIVSGKVRFPSH----------------------------------------------------FSSDLKD-LLRNLLQVDLTKRFGNLKNGVSDIKT-HKWF
cAPKγ    -AVDWWALGVLIYEMAV-GFPPFYA-------DQPIQIYEKIVSGRVRFPSK----------------------------------------------------LSSDLKD-LLRSLLQVDLTKRFGNLRNGVGDIKN-HKWF
DC0      -AVDWWALGVLVYEMAA-GYPPFFA-------DQPIQIYEKIVSGKVRFPSH----------------------------------------------------FGSDLKD-LLRNLLQVDLTKRYGNLKAGVNDIKN-QKWF
DC1      -SVDWWAFGILVYEFVA-GRSPFAIHN-----RDVILMYSKICICDYKMPSY----------------------------------------------------FTSQLRS-LVESLMQVDTSKRLGNSNDGSSDVKS-HPWF
DC2      -AVDWWALGVLIYEMLV-GYPPFYD-------EQPFGIYEKILSGKIEWERH----------------------------------------------------MDPIAKD-LIKKLLVNDRTKRLGNMKNGADDVKR-HRWF
AplC     -AVDWWALGVLIYEMAA-GYPPFFA-------DEPIQIYEKIVSGKVRFPSH----------------------------------------------------FSSDLKD-EERNLLQVDLTKRFGDLKDGVNDIKN-HKWF
TPK1     -SIDWWSFGILIYEMLA-GYTPFYD-------SNTMKTYEKILNAELRFPPF----------------------------------------------------FNEDVKD-LLSRLITRDLSQRLGNLQNGTEDVKN-HPWF
TPK2     -SVDWWSLGVLIYEMLA-GYTPFYD-------TTPMKTYEKILQGKVVYPPY----------------------------------------------------FQPDVVD-LLSKLITADLTRRIGNLQSGSRDIKA-HPWF
TPK3     -SVDWWSFGVLIYEMLA-GYTPFYN-------SNTMKTYENILNAELKFPPF----------------------------------------------------FHPDAQD-LLKKLITRDLSERLGNLQNGSEDVKN-HPWF
cGPK     -SADYWSLGILMYELLT-GSPPFSG-------PDPMKTYNIILRGIDMIEFPKK--------------------------------------------------IAKNAAN-LIKKLCRDNPSERLGNLKNGVKDIQK-HKWF
DG1      -AVDYWALGILIHELLN-GTPPFSA-------PDPMQTYNLILKGIDMIAFPKH--------------------------------------------------ISRWAVQ-LIKRLCRDVPSERLGYQTGGIQDIKK-HKWF
DG2      -SADYWSLGVLMFELLT-GTPPFTG-------SDPMRTYNIILKGIDAIEFPRN--------------------------------------------------ITRNASN-LIKKLCRDNPAERLGYQRGGISEIQK-HKWF

PKCα     -SVDWWAYGVLLYEMLA-GQPPFDG-------EDEDELFQSIMEHNVSYPKS----------------------------------------------------LSKEAVS-ICKGLMTKHPAKRLGCGPEGERDVRE-HAFF
PKCβ     -SVDWWAFGVLLYEMLA-GQAPFEG-------EDEDELFQSIMEHNVAYPKS----------------------------------------------------MSKEAVA-ICKGLMTKHPGKRLGCGPEGERDIKE-HAFF
PKCγ     -SVDWWSFGVLLYEMLA-GQPPFDG-------EDEEELFQAIMEQTVTYPKS----------------------------------------------------LSREAVA-ICKGFLTKHPGKRLGSGPDGEPTIRA-HGFF
PKCδ     -SVDWWSFGVLLYEMLI-GQSPFHG-------DDEDELFESIRVDTPHYPRW----------------------------------------------------ITKESKD-IMEKLFERDPAKRLGVTGNIRL-----HPFF
PKCε     -SVDWWALGVLMYEMMA-GQPPFEA-------DNEDDLFESILHDDVLYPVW----------------------------------------------------LSKEAVS-ILKAFMTKNPHKRLGCAAQNGEDAIKQHPFF
PKCζ     -SVDWWALGVLMFEMMA-GRSPFDIITDNPDMNTEDYLFQVILEKPIRIPRF----------------------------------------------------LSVKASH-VLKGFLNKDPKERLGCPQTGFSDIKS-HAFF
DPKC53b  -SVDWWAYGVLLYEMLV-GQPPFDG-------EDEEELFAAITDHNVSYPKS----------------------------------------------------LSKEAKE-ACKGFLTKQPNKRLGCGSSGEEDVRL-HPFS
DPKC98   -SVDWWALGVLMYEMMA-GQPPFEA-------DNEDELFDSIMHDDVLYPVW----------------------------------------------------LSREAVS-ILKGFLTKNPEQRLGCGDENEIRK---HPFF
DPKC53e  -AADWWSFGVLLFEFMA-GQAPFEG-------DDETTVFRNIKDKKAVFPKH----------------------------------------------------FSVEAMD-IITSFLTKKPNNRLGAGRYARQEITT-HPFF
TPA1     -AVDFWSFGVLMYEMLV-GQSPFHG-------EGEDELFDSILNERPYFPKT----------------------------------------------------ISKEAAK-CLSALFDRNPNTRLGMPECPDGPIRQ-HCFF
YPK1     -AVDWWTLGVLLYEMLT-GLPPYYR-------EDVPKIYKKILQEPLVFPDG----------------------------------------------------FDRDAKD-LLIGLLSRDPTRRLGYNGADEIRN---HPFF
YKR2     -TVDWWTLGILLYEMMT-GLPPYSD-------ENVPVMYKKILQQPLLFPDG----------------------------------------------------FDPAAKD-LLIGLLSRDPSRRLGVNGTDEIRN---HPFF
SCH9     -MVDFWSLGVLIFEMCC-GWSPFFA-------ENNQKMYQKIAFGKVKFPRDV---------------------------------------------------LSQEGRS-FVKGLLNRNPKHRLGAIDDGRELRA--HPFF

RSK1N    -SADWWSYGVLMG-------------------KDRKETMTLILKAKLGMPQF----------------------------------------------------LSTEAQS-LLRALFKRNPANRLGSGPDGAEEIKR-HIFY
PVPK1    -AVDWWTFGIFLYELLF-GRTPFKG-------ANRATLFNVIGQPLRFPESPT---------------------------------------------------VSFAARD-LIRGLLVKEPQHRLAYRRGATEIKQ--HPFF
βARK     -SADWFSLGCMLFKLLR-GHSPFRQ-------HKTKDKHEIDRMTLTMAVELPDS-------------------------------------------------FSPELRS-LLEGLLQRDVNRRLGCLGRGAEVKE--SPFF

CAMIIα   -PVDLWACGVILYILLV-GYPPFWD-------EDQHRLYQQIKAGAYDFPSPEWDT------------------------------------------------VTPEAKD-LINKMLTINPSKRITAAEALK------HPWI
CAMIIβ   -PVDIWACGVILYILLV-GYPPFWD-------EDQHKLYQQIKAGAYDFPSPEWDT------------------------------------------------VTPEAKN-LINQMLTINPAKRITAHEALK------HPWV
CAMIIγ   -PVDIWACGVILYILLV-GYPPFWD-------EDQHKLYQQIKAGAYDFPSPEWDT------------------------------------------------VTPEAKN-LINQMLTINPAKRITADQALK------HPWV
PHKγ     -EVDMWSTGVIMYTLLA-GSPPFWH-------RKQMLMLRMIMSGNYQFGSPEWDD------------------------------------------------YSDTVKD-LVSRFLVVQPQKRYTAEEALA------HPFF
PSK-C3   -EVDLWACGVILFTLLA-GSPPFWH-------RRQILMLRMIMEGQYQFSSPEWDD------------------------------------------------RSSTVKD-LISRLLQVDPEARLTAEQALQ------HPFF
RSK1C    -GCDIWSLGILLYTMLA-GYTPFANGPS----DTPEEILTRIGSGKFTLSGGNWNT------------------------------------------------VSETAKD-LVSKMLHVDPHQRLTAKQVLQ------HPWI
RSK2C    -ACDIWSLGVLLYTMLT-GYTPFANGPD----DTPEEILARIGSGKFSLSGGYWNS------------------------------------------------VSDTAKD-LVSKMLHVDPHQRLTAALVLR------HPWI
PSK-H1   -SVDMWALGVIAYILLS-GTMPFED-------DNRTRLYRQILRGKYSYSGEPWPS------------------------------------------------VSNLAKD-FIDRLLTVDPGARMTALQALR------HPWV
MLCK-K   -KTDMWSLGVITYMLLS-GLSPFLG-------DDDTETLNNVLSGNWYFDEETFEA------------------------------------------------VSDEAKD-FVSNLIVKEQGARMSAAQCLA------HPWL
MLCK-M   -ETDMWSIGVICYILVS-GLSPFMG-------DNDNETLANVTSATWTFDDEAFDE------------------------------------------------ISDDAKD-FISNLLKKDMKSRLNCTQCLQ------HPWL
KIN1     PEVDVWSFGVVLFVLVC-GKVPFDD-------ENSSVLHEKIKQGKVEYPQH----------------------------------------------------LSIEVIS-LLSKMLVVDPKRRATLKQVVE------HHWM
KIN2     PEVDIWSFGIVLYVLVC-GKVPFDD-------ENSSILHEKIKKGKVDYPSH----------------------------------------------------LSIEVIS-LLTRMIVVDPLRRATLKNVVE------HPWM
SNF1     PEVDVWSCGVILYVMLC-RRLPFDD-------ESIPVLFKNISNGVYTLPKF----------------------------------------------------LSPGAAG-LIKRMLIVNPLNRISIHEIMQ------DDWF
nim1     APADVWSCGVILYSLLS-NKLPFGG-------QNTDVIYNKIRHGAYDLPSS----------------------------------------------------ISSAAQD-LLHRMLDVNPSTRITIPEFFS------HPFL

            IX                            X                                                         XI
```

```
CDC2Hs  -PVDIWSIGTIFAELAT-KKPLFHGDSEI---DQLFRIFRALGTPNNEVWPEVESLQD-----------------YKNTFPKWKPGSLASHVKN--LDENGLD-LLSKMLIYDPAKRISGKMALN------HPYF
cdc2    -GVDIWSVGCIFAEMIR-RSPLFPGDSEI---DEIFKIFQVLGTPNEEVWPGVTLLQD-----------------YKSTFPRWKRMDLHKVVPN--GEEDAIE-LLSAMLVYDPAHRISAKRALQ------QNYL
CDC28   -GVDTWSIGCIFAEMCN-RKPIFSGDSEI---DQIFKIFRVLGTPNEAIWPDIVYLPD-----------------FKPSFPQWRRKDLSQVVPS--LDPRGID-LLDKLLAYDPINRISARRAAI------HPYF
PHO85   -SIDIWSCGCILAEMIT-GKPLFPGTNDE---EQLKLIFDIMGTPNESLWPSVTKLPKYNPN-------------IQQRPPRDLRQVLQPHTHEP-LDGNLMD-FLHGLLQLNPDMRLSAKQALH------HPWF
PSK-J3  -PVDMWSVGCIFAEMFR-RKPLFCGNSEA---DQLGKIFDLIGLPPEDDWPRDVSLP-------------------RGAFPPRGPRPVQSVVPE--MEESGAQ-LLLEMLTFNPHKRISAFRALQ------HSYL
KIN28   -AIDIWSVGVIFAELML-RIPYLPGQNDV---DQMEVTFRALGTPTDRDWPEVSSFMTYNK--------------LQIYPPPSRDELRKRFIA---ASEYALD-FMCGMLTMNPQKRWTAVQCLE------SDYF
KSS1    -AMDIWSCGCILAEMVS-GKPLFPGRDYH---HQLWLILEVLGTPSFEDFNQ-IKSKRAKEYIAN----------LPMRPPLPWETVWSKTD----LNPDMID-LLDKMLQFNPDKRISAAEALR------HPYL
MAK     -PIDVWAVGSIMAELYT-FRPLFPGTSEV---DEIFKICQVLGTPKKSDWPEGYQLASS----------------MNFRFPQCIPINLKTLIPN--ASSEAIQ-LMTEMLNWDPKKRPTASQALK------HPYF
YAK1    -SIDMWSIGCIVAELFL-GIPIFPGASEY---NQLTRIIDTLGYPPSWMIDMGKNSGKFMKKLAPEESSSS-{}-RNYRYPKSIQNSQELIDQEMQNRECLIH-FLGGVLNLNPLERWTPQQAML------HPFI
GSK3    -SIDVWSAGCVLAELLL-GQPIFPGDSGV---DQLVEIIKVLGTPTREQIREMNPNYTEFKF-------------PQIKAHPWTKVFKSR------TPPEAIA-LCSSLLEYTPSSRLSPLEACA------HSFF
CKII    -SIDMWSLGCMLASMIFRKEPFFHGHDNY---DQLVRIAKVLGTEDLYDYIDKYNIELDPRFNDIL---------QRHSRKRWERFVHSENQHL--VSPEALD-FLDKLLRYDHQSRLTAREAME------HPYF
DCKII   -SIDMWSLGCMLASMIFRKEPFFHGHDNY---DQLVRIAKVLGTEELYAYLDKYNIDLDPRFHDIL---------QRHSRKRWERFVHSDNQHL--VSPEALD-FLDKLLRYDHVDRLTAREAMA------HPYF
CKAI    -SIDLWSFGTMLASMIFKREPFFHGTSNT---DQLVKIVKVLGTSDFEKYLLKYEITLPREFYDM----------DQYSRKPWHRFINDGNKHLS-GNDEIID-LIDKLLRYDHQERLTAKEAMG------HPWF
PSK-H2  -PIDMWSLGCILAELLT-GYPLLPGEDEGDQLACMIELLGMPSQKLLDASKRAKNFVSSKGYPRYCTVTTL-{}-GKLRGPPESREWGNALKGC--DDPLFLG-FLKQCLGWDPAVRMTPGQALR------HPWL
PSK-G1  -PODVWSIGCIIFEYYV-GFTLFQTHDNREHLAMMERILGPIPSRMIRKTRKQKYFYRGRLDWDENTS-------AGRYVRENCKPLRRYLTSEAEEHHQLFD-LIESMLEYEPAKRLTLGEALQ------HPFF

PBS2    -QSDIWSLGLSILEMAL-GRYPYPP-------ETYDNIFSQLSAIVQEPPPRLPSDK---------------------------------------FSSDAQD-FVSLCLQKIPERRPTYAALTE------HPWL
byr1    -KSDIWSLGISIIELAT-QELPWSFSNI----DDSIGILDLLHCIVNEEPPRLPSS----------------------------------------FPEDLRL-FVDACLHKDPTLRASPQQLCA------MPYF
STE7    -KGDVWSLGLMIIELVT-GEFPLGGHN-----DTPDGILDLLQRIPSPRLPKDRI-----------------------------------------YSKEMTD-FVNRCCIKNERERSSIHELLH------HDLI
ninaC   -RADVWALGITTIELAD-GKPPFADMHPT---RAMFQIIRNPPPTLMRPTN---------------------------------------------WSQQIND-FISESLEKNAENRPMMVEMVE------HPFL
nimA    -RSDIWAVGCIMYELCQ-REPPFNA-------RTHIQLVQKIREGKFAPLPDF-------------------------------------------YSSELKN-VIASCLRVNPDHRPDTATLIN------TPVI
PIM1    -SADVWSLGILLYDMVC-GDIPF---------EHDEEIIRGQVFFVDGRQR---------------------------------------------VSSECQH-LIRWCLALRPSDRPTFEEIQN------HPWM
ran1    -PNDVWALGIILINLCCKRNPWK---------RACSQTDGTYRSYVHNPSTLLSILP---------------------------------------ISRELNS-LLNRIFDRNPKTRITLPELST------LVSN
CDC7    -KIDIWSVGVILLSLLG-RRFPMFQSL-----DDADSLLELCTIFGWKELRKCAALHGLGFEASGLIWDKP-{}-KTNMDAVDAYELKKYQEEIWSDHYWCFQ-VLEQCFEMDPQKRSSAEDLLK------TPFF
wee1    -PADIFSLGITVFEAAA-NIVLPDN-------GQSWQKLRSGDLSPRLSSTD-----------------------NGSSLTSSSRETPANSII---GQGGLDR-VVEWMLSPEPRNRPTIDQILATD----EVCW
PYT     -KSDVWSLGCILYYMTY-GKTPFQQII-----NQISKLHAIIDDAPNHEIEFPDI-----------------------------------------PEKDLQD-VVKCCLKRDPKQRISIPELLA------HPYV
COT     -KADIYSLGATLIHMQT-GTPPWVK-------RYPRSAYPSYLYIIHKQAPPLEDIADD-------------------------------------CSPGMRE-LIEASLERNPNHRPRAADLLK------HEAL
GCN2    -KIDMYSLGIIFFEMIY---PFSTG-------MERVNILKKLRSVSIEFPPDFDDN----------------------------------------KNKVEKK-IIRLLIDHDPNKRPGARTLLN------SGWL

c-raf   -QSDVYSYGIVLYELMT-GELPYSHI------NNRDQIIFMVGRGYASPDLSKLYKN---------------------------------------CPKAMKR-LVADCVKKVKEERPLFPQI--------LSSI
Araf    -QSDVYAYGVVLYELMT-GSLPYSHI------GCRDQIIFMVGRGYLSPDLSKISSN---------------------------------------CPKAMRR-LLSDCLKFQREERPLFPQI--------LATI
Braf    -QSDVYAFGIVLYELMT-GQLPYSNI------NNRDQIIFMVGRGYLSPDLSKVRSN---------------------------------------CPKAMKR-LMAECLKKKRDERPLFPQI--------LASI
Draf    -QSDVYAFGIVMYELLA-ECLPYGHI------SNKDQILFMVGRGLLRPDMSQVRSD---------------------------------------ARRHSKR-LAEDCIKYTPKDRPLFRPL--------LNML
c-mos   -KADIYSFAITLWQMTT-KQAPYSG-------ERQHILYAVVAYDLRPSLSAAVFEDSL-------------------------------------PGQRLGD-VIQRCWRPSAAQRPSARLL--------LVDL

              IX                        X                                                                                XI
```

FIG. 1 (*continued*)

**B**

```
c-src    LRLEVKLGQGCFGEVWMGTWNG--------TTRVAIKTLKPGTM--SPEAFLQEAQVMKKL-RHEKLVQLYAVVSE--EPIYIVTEYMSKGSLLDFLKGETGKY-------------------------LRLPQLVDMAAQIASGMAYVERM
FYN      LQLIKRLGNGQFGEVWMGTWNG--------NTKVAIKTLKPGTM--SPESFLEEAQIMKKL-KHDKLVQLYAVVSE--EPIYIVTEYMNKGSLLDFLKDGEGRA-------------------------LKLPNLVDMAAQVAAGMAYIERM
c-yes    LRLEVKLGQGCFGEVWMGTWNG--------TTKVAIKTLKPGTM--MPEAFLQEAQIMKKL-RHDKLVPLYAVVSE--EPIYIVTEFMSKGSLLDFLKEGDGKY-------------------------LKLPQLVDMAAQIADGMAYIERM
c-fgr    ITLERRLGTGCFGDVWLGTWNG--------STKVAVKTLKPGTM--SPKAFLEEAQVMKLL-RHDKLVQLYAVVSE--EPIYIVTEFMCHGSLLDFLKNPEGQD-------------------------LRLPQLVDMAAQVAEGMAYMERM
LYN      IKLVKRLGAGQFGEVWMGYYNN--------STKVAVKTLKPGTM--SVQAFLEEANLMKTL-QHDKLVRLYAVVTRE-EPIYIITEYMAKGSLLDFLKSDEGGK-------------------------VLLPKLIDFSAQIAEGMAYIERK
HCK      LKLEKKLGAGQFGEVWMATYNK--------HTKVAVKTMKPGSM--SVEAFLAEANVMKTL-QHDKLVKLHAVVTK--EPIYIITEFMAKGSLLDFLKSDEGSK-------------------------QPLPKLIDFSAQIAEGMAFIEQR
LCK      LKLVERLGAGQFGRCGWGTTTG--------TTKVAVKSLKQGSM--SAGRLPAEANLMKQL-QHQRLVRLYAVVTQ--EPIYIITEYMENGSLVDFLKTPSGIK-------------------------LTINKLLDMAAQIAEGMAFIEER
TKL      LKLVEKLGAGQFGEVWMGFYNG--------HTKVAIKNLKQGSM--SPSAFLAEANLMKNL-QHPRLVRLYAVVTK--EPIYIITEYMEKGSLVDFLKTSEGIK-------------------------LSINKLLDMAAQIAEGMAFIEAK
BLK      LKLVRKLGSGQFGEVWMGYYKN--------NMKVAIKTLKEGTM--SPEAFLGEANVMKTL-QHERLVRLYAVVTR--EPIYIVTEYMARGCLLDFLKTDEGSR-------------------------LSLPRLIDMSAQVAEGMAYIERM
STK      LRLNRKLGAGQFGEVWAGVWNN--------TTQVAVKTLKPGTM--SPASFLDEAGVMKKL-RHKHLVQLTAICSDR-EPIYIVTEYMSGGSLLDYLSKGEGVN-------------------------LQLPTLIDMAAQVASGMAFLEAQ
Dsrc64   IQLLRKLGRGNFGEVFYGKWRN--------SIDVAVKTLREGTM--STAAFLQEAAIMKKF-RHNRLVALYAVCSQE-EPIYIVQEYMSKGSLLDFLREGDGRY-------------------------LHFEDLIYIATQVASGMEYLESK
Dsrc28   LMLMEELGSGQFGVVRRGKWRG--------SIDTAVKMMKEGTM--SEDDFIEEAKVMTKL-QHPNLVQLYGVCTKH-RPIYIVTEYMKHGSLLNYLRRHEKTLI------------------------GNMGLLLDMCIQVSKGMTYLERH
c-abl    ITMKHKLGGGQYGEVYEGVWKK-------YSLTVAVKTLKEDTM--EVEEFLKEAAVMKEI-KHPNLVQLLGVCTRE-PPFYIITEFMTYGNLLDYLRECNRQE-------------------------VNAVVLLYMATQISSAMEYLEKK
Dabl     IMMKHKLGGGQYGEVYEAVWKR-------YGNTVAVKTLKEDTM--ALKDFLEEAAIMKEM-KHPNLVQLIGVCTRE-PPFYIITEFMSHGNLLDFLRSAGRET-------------------------LDAVALLYMATQIASGMSYLESR
Nabl     IIMHNKLGGGQYGDVYEGYWKR-------HDCTIAVKALKEDAM--PLHEFLAEAAIMKDL-HHKNLVRLLGVCTHE-APFYIITEFMCNGNLLEYLRRTDKSL-------------------------LPPIILVQMASQIASGMSYLEAR

ECK      VTRQKVIGAGEFGEVYKGMLKTSSGK---KEVPVAIKTLKAGYTEKQRVDFLGEAGIMGQF-SHHNIIRLEGVISKY-KPMMIITEYMENGALDKFLREKDGE--------------------------FSVLQLVGMLRGIAAGMKYLANM
ELK      VKIEEVIGAGEFGEVYKGRLKLPGK----REIYVAIKTLKAGYSEKQRRDFLSEASIMGQF-DHPNIIRLEGVVTKS-RPVMIITEFMENGALDSFLRQNDGQ--------------------------FTVIQLVGMLRGIAAGMKYLSEM
EPH      LMVDTVIGEGEFGEVYRGTLRLPSQ----DCKTVAIKTLKDTSPGGQWWNFLREATIMGQF-SHPHILHLEGVVTKR-KPIMIITEFMENGALDAFLREREDQ--------------------------LVPGQLVAMLQGIASGMNYLSNH
FER      VILGELLGKGNFGEVYKGTLKD--------KTSVAVKTCKEDLPQELKIKFLQEAKILKQY-DHPNIVKLIGVCTQR-QPVYIIMELVSGGDFLTFLRRKKDE--------------------------LKLKQLVKFSLDAAAGMLYLESK
c-fes    LVLGEQIGRGNFGEVFSGRLRAD-------NTLVAVKSCRETLPPDIKAKFLQEAKILKQY-SHPNIVRLIGVCTQK-QPIYIVMELVQGGDFLTFLRTEGAR--------------------------LRVKTLLQMVGDAAAGMEYLESK

INS.R    ITLLRELGQGSFGMVYEGNARDIIKGE--AETRVAVKTVNESASLRERIEFLNEASVMKGF-TCHHVVRLLGVVSKG-QPTLVVMELMAHGDLKSYLRSLRPEAENNPGRPP-----------------PTLQEMIQMAAEIADGMAYLNAK
IRR      ISIIRELGQGSFGMVYEGLARGLEAGE--ESTPVALKTVNELASPRECIEFLKEASVMKAF-KCHHVVRLLGVVSQG-QPTLVIMELMTRGDLKSHLRSLRPEAENNPGLPQ-----------------PALGEMIQMAGEIADGMAYLAAN
IGF1R    ITMSRELGQGSFGMVYEGVAKGVVKDE--PETRVAIKTVNEAASMRERIEFLNEASVMKEF-NCHHVVRLLGVVSQG-QPTLVIMELMTRGDLKSYLRSLRPEMENNPVLAP-----------------PSLSKMIQMAGEIADGMAYLNAN
DILR     IIQLAPIGQGSFGMVYEGILKSFPPNG--VDRECAIKTVNENATDRERTNFLSEASVMKEF-DTYHVVRLLGVCSRG-QPALVVMELMKKVFLKSYLRAHRPRSGMRPDDVSLIAWMTGNVQP------PTYGRIYQVAIEIADGMAYLAAK
LTK      VTLLRALGHGAFGEVYEGLVTGLPGDS--SPLPVAIKTLPELCSHQDELDFLMEALIISKF-SHQNIVRCVGLSFRS-APILILLELMSGGDMKSFLRHSRPHPGQLAP--------------------LTMQDLLQLAQDIAQGCHYLEEN
c-ros    LTLRLLLGSGAFGEVYEGTAVDILGVGS-GEIKVAVKTLKKGSTDQEKIEFLKEAHLMSKF-NHPNILKQLGVCLLN-EPQYIILELMEGGDLLTYLRKARMATFYGPL--------------------LTLVDLVDLCVDISKGCVYLERM
7LESS    LKLLRFLGSGAFGEVYEGQLKTEDSE---EPQRVAIKSLRKGAS--EFAELLQEAQLMSNF-KHENIVRLVGICFDT-ESISLIMEHMEAGDLLSYLRAARATSTQEPQPTAG----------------LSLSELLAMCIDVANGCSYLEDM
TRK-A    IVLKWELGEGAFGKVFLAECHNLLPEQ--DKMLVAVKALKEASES-ARQDFQREVELLTML-QHQHIVRFFGVCTEG-RPLLMVFEYMRHGDLNRFLRSHGPDAKLLAGGEDVAPGP------------LGLGQLLAVASQVAAGMVYLAGL
TRK-B    IVLKRELGEGAFGKVFLAECYNLCPEQ--DKILVAVKTLKDASDN-ARKDFHREAELLTNL-QHEHIVKFYGVCVEG-DPLIMVFEYMKHGDLNKFLRAHGPDAVLMAEGNPPTE--------------LTQSQMLHIAQQIAAGMVYLASQ
MET      VHFNEVIGRGHFGCVYHGTLLDNDG----KKIHCAVKSLNRITDIGEVSQFLTEGIIMKDF-SHPNVLSLLGICLRSEGSPLVVLPYMKHGDLRNFIRNETHN--------------------------PTVKDLIGFGLQVAKAMKYLASK
SEA      THRSRVIGRGHFGSVYHGTYMDPLL----GNLHCAVKSLHRITDLEEVEEFLREGILMKGF-HHPQVLSLLGVCLPRHGLPLVVLPYMRHGDLRHFVRAQERS--------------------------PTVKELIGFGLQVAIGMEYLAQK

PDGFR-B  LVLGRTLGSGAFGQVVEATAHGLSHSQ--ATMKVAVKMLKSTARSSEKQALMSELKIMSHLGPHLNVVNLLGACTKG-GPIYIITEYCRYGDLVDYLHRNKHTFLQHHSDKRRPPSAELYSNALPV{}-LSYMDLVGFSYQVANGMEFLASK
PDGFR-A  LVLGRVLGSGAFGKVVEGTAYGLSRSQ--PVMKVAVKMLKPTARSSEKQALMSELKIMTHLGPHLNIVNLLGACTKS-GPIYIITEYCFYGDLVNYLHKNRDSFLSHHPEKPKKELDIFGLNPADE{}-LTLLDLLSFTYQVARGMEFLASK
CSF1R    LQFGKTLGAGAFGKVVEATAFGLGKED--AVLKVAVKMLKSTAHADEKEALMSELKIMSHLGQHENIVNLLGACTHG-GPVLVITEYCCYGDLLNFLRRKAEAMLGPSLSPGQDPEPGVDYKNIHL{}-LELRDLLHFSSQVAQGMAFLASK
c-kit    LSFGKTLGAGAFGKVVEATAYGLIKSD--AAMTVAVKMLKPSAHLTEREALMSELKVLSYLGNHMNIVNLLGACTIG-GPTLVITEYCCYGDLLNFLRRKRDSFICSKQEDHAEAALYKNLLHSKE{}-LDLEDLLSFSYQVAKGMAFLASK
FGFR     LVLGKPLGEGCFGQVVLAEAIGLDKDKPNRVTKVAVKMLKSDATEKDLSDLISEMEMMKMIGKHKNIINLLGACTQE-GPLYVIVEYASKGNLREYLQARRPRGLEYCYNPSHNPEEQ-----------LSSKDLVSCAYQVARGMEYLASK
RET      LVLGKTLGEGEFGKVVKATAFHLKGRA--GYTTVAVKMLKENASPSELRDLLSEFNVLKQV-NHPHVIKLYGACSQD-GPLLLIVEYAKYGSLRGFLRESRKVGPGYLGSGGSRNSSSLDHPDERA---LTMGDLISFAWQISQGMQYLAEM
TORSO    VLLQDVLGEGAFGIVRRGVYK---------KRQVAVKLLKDEPNDDEVYAFKCEIQMLKAVGKHPNIVGIVGYSTRFSNQMMLLIEYCSLGSLQNFLREEWKFRQEQNAIGLKKNLEQNVDNRREN{}-LKYADLLDIAQQVAVGMEFLAQN

EGFR     FKKIKVLGSGAFGTVYKGLWIPEGEK---VKIPVAIKELREATSPKANKEILDEAYVMASV-DNPHVCRLLGICLT--STVQLITQLMPFGCLLDYVREHKDN--------------------------IGSQYLLNWCVQIAKGMNYLEDR
NEU      LRKVKVLGSGAFGTVYKGIWIPDGEN---VKIPVAIKVLRENTSPKANKEILDEAYVMAGV-GSPYVSRLLGICLT--STVQLVTQLMPYGCLLDHVRENRGR--------------------------LGSQDLLNWCMQIAKGMSYLEDV
Xmrk     FKKDRVLGSGAFGTVYKGLWNPDGEN---IRIPVAIKVLREATSPKVNQEVLDEAYVMASV-DHPHVCRLLGICLT--SAVQLVTQLMPYGCLLDYVRQHQER--------------------------ICGQWLLNWCVQIAKGMNYLEER
DER      LRKGGVLGMGAFGRVYKGVWVPEGEN---VKIPVAIKELLKSTGAESSEEFLREAYIMASE-EHVNLLKLLAVCMS--SQMMLITQLMPLGCLLDYVRNNRDK--------------------------IGSKALLNWSTQIAKGMSYLEEK

                 I                          II          III         IV              V                                                       VIa
```

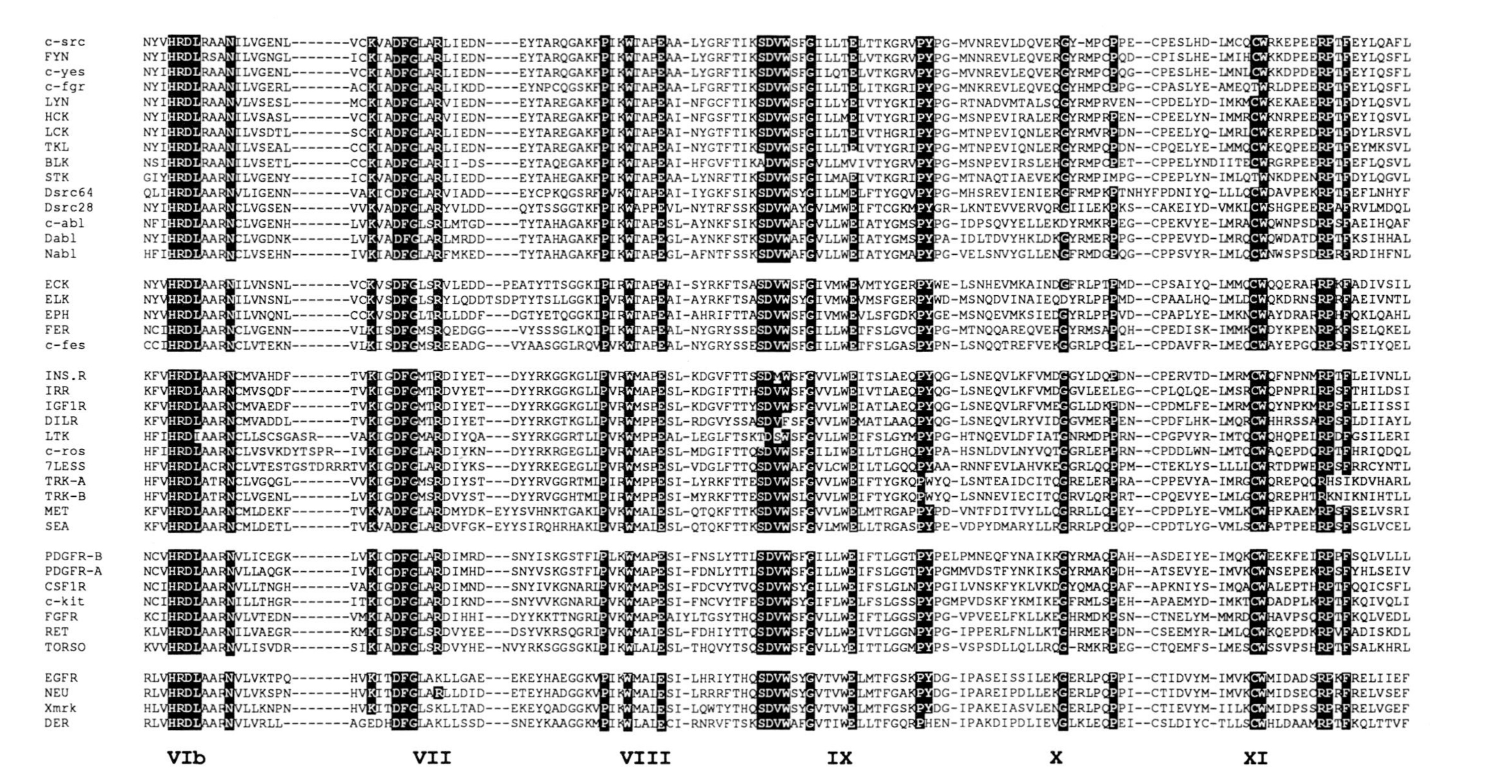
c-src
FYN
c-yes
c-fgr
LYN
HCK
LCK
TKL
BLK
STK
Dsrc64
Dsrc28
c-abl
Dabl
Nabl
ECK
ELK
EPH
FER
c-fes
INS.R
IRR
IGF1R
DILR
LTK
c-ros
7LESS
TRK-A
TRK-B
MET
SEA
PDGFR-B
PDGFR-A
CSF1R
c-kit
FGFR
RET
TORSO
EGFR
NEU
Xmrk
DER
VIb
VII
VIII
IX
X
XI

FIG. 1 (continued)

TABLE I
ENTRIES IN CATALYTIC DOMAIN DATABASE[a]

| Abbreviated form | Source and reference |
|---|---|
| cAPK-$\alpha$ | Human cAMP-dependent protein kinase catalytic subunit, $\alpha$ form [F. Maldonado and S. K. Hanks (1988). *Nucleic Acids Res.* **16,** 8189–8190] |
| cAPK-$\beta$ | Human cAMP-dependent protein kinase catalytic subunit, $\beta$ form [S. J. Beebe *et al.* (1990). *Mol. Endocrinol.* **4,** 465–475] |
| cAPK-$\gamma$ | Human cAMP-dependent protein kinase catalytic subunit, $\gamma$ form [S. J. Beebe *et al.* (1990). *Mol. Endocrinol.* **4,** 465–475] |
| DCO | *Drosophila* cAPK homolog [J. L. Foster *et al.* (1988). *J. Biol. Chem.* **263,** 1676–1685; D. Kalderon and G. M. Rubin (1988). *Genes Dev.* **2,** 1539–1556] |
| DC1 | *Drosophila* protein related to cAPK [D. Kalderon and G. M. Rubin (1988). *Genes Dev.* **2,** 1539–1556] |
| .DC2 | *Drosophila* protein related to cAPK [D. Kalderon and G. M. Rubin (1988). *Genes Dev.* **2,** 1539–1556] |
| Apl-C | Marine mollusk (*Aplysia californica*) cAPK homolog [S. Beushausen *et al.* (1988). *Neuron* **1,** 853–864] |
| TPK1 | Yeast (*Saccharomyces cerevisiae*) cAPK homolog, type 1 [T. Toda *et al.* (1987). *Cell (Cambridge, Mass.)* **50,** 277–287; J. Lisziewicz *et al.* (1987). *J. Biol.Chem.* **262,** 2549–2553; J. F. Cannon and K. Tatchell (1987). *Mol. Cell. Biol.* **7,** 2653–2663] |
| TPK2 | Yeast (*S. cerevisiae*) cAPK homolog, type 2 [T. Toda *et al.* (1987). *Cell (Cambridge, Mass.)* **50,** 277–287] |
| TPK3 | Yeast (*S. cerevisiae*) cAPK homolog, type 3 [T. Toda *et al.* (1987). *Cell (Cambridge, Mass.)* **50,** 277–287] |
| cGPK | Bovine cGMP-dependent protein kinase [K. Takio *et al.* (1984). *Biochemistry* **23,** 4207–4218] |
| DG2 | *Drosophila* protein related to cGMP-dependent protein kinase [D. Kalderon and G. M. Rubin (1989). *J. Biol. Chem.* **264,** 10738–10748] |
| DG1 | *Drosophila* protein related to cGMP-dependent protein kinase [D. Kalderon and G. M. Rubin (1989). *J. Biol. Chem.* **264,** 10738–10748] |
| PKC-$\alpha$ | Rat protein kinase C, $\alpha$ form [U. Kikkawa *et al.* (1987). *FEBS Lett.* **223,** 212–216] |
| PKC-$\beta$ | Rat protein kinase C, $\beta$ form [J. L. Knopf *et al.* (1986). *Cell (Cambridge, Mass.)* **46,** 491–502; G. M. Housey *et al.* (1987). *Proc. Natl. Acad.* |

TABLE I (*continued*)

| Abbreviated form | Source and reference |
|---|---|
| | *Sci. U.S.A.* **84,** 1065–1069; U. Kikkawa *et al.* (1987). *FEBS Lett.* **223,** 212–216] |
| PKC-$\gamma$ | Rat protein kinase C, $\gamma$ form [J. L. Knopf *et al.* (1986). *Cell (Cambridge, Mass.)* **46,** 491–502; U. Kikkawa *et al.* (1987). *FEBS Lett.* **223,** 212–216] |
| PKC-$\delta$ | Rat protein kinase C, $\delta$ form [Y. Ono *et al.* (1988). *J. Biol. Chem.* **263,** 6927–6932] |
| PKC-$\varepsilon$ | Rat protein kinase C, $\varepsilon$ form [G. M. Housey *et al.* (1987). *Proc. Natl. Acad. Sci. U.S.A.* **84,** 1065–1069; Y. Ono *et al.* (1988). *J. Biol. Chem.* **263,** 6927–6932] |
| PKC-$\zeta$ | Rat protein kinase C, $\zeta$ form [Y. Ono *et al.* (1988). *J. Biol. Chem.* **263,** 6927–6932] |
| DPKC-53E(br) | *Drosophila* protein kinase C homolog, polytene locus 53E [A. Rosenthal *et al.* (1987). *EMBO J.* **6,** 433–441; E. Schaeffer *et al.* (1989). *Cell (Cambridge, Mass.)* **57,** 403–412] |
| DPKC-98F | *Drosophila* protein kinase C homolog, polytene locus 98F [E. Schaeffer *et al.* (1989). *Cell (Cambridge, Mass.)* **57,** 403–412] |
| DPKC-53E(ey) | *Drosophila* protein kinase C homolog specifically expressed in photoreceptor cells, polytene locus 53E [E. Schaeffer *et al.* (1989). *Cell (Cambridge, Mass.)* **57,** 403–412] |
| TPA1 | Nematode (*Caenorhabditis elegans*) protein kinase C homolog [Y. Tabuse *et al.* (1989). *Science* **243,** 1713–1716] |
| YPK1 | Yeast (*S. cerevisiae*) cAPK-related gene product [R. A. Maurer (1988). *DNA* **7,** 469–474] |
| YKR2 | Yeast (*S. cerevisiae*) cAPK-related gene product [K. Kubo *et al.* (1989). *Gene* **76,** 177–180] |
| SCH9 | Yeast (*S. cerevisiae*) cAPK-related gene product, complements cdc25 [T. Toda *et al.* (1988). *Genes Dev.* **2,** 517–527] |
| PVPK1 | Bean (*Phaseolus vulgaris*) putative protein kinase [M. A. Lawton *et al.* (1989). *Proc. Natl. Acad. Sci. U.S.A.* **86,** 3140–3144] |
| $\beta$ARK | Bovine $\beta$-adrenergic receptor kinase [J. L. Benovic *et al.* (1989). *Science* **246,** 235–240] |
| RSK1-N | Mouse ribosomal protein S6 kinase, type 1, amino domain [D. A. Alcorta *et al.* (1989). *Mol. Cell. Biol.* **9,** 3850–3859] |
| CaMII-$\alpha$ | Rat calcium/calmodulin-dependent protein kinase II, $\alpha$ subunit [C. R. Lin *et al.* (1987). *Proc. Natl. Acad. Sci. U.S.A.* **84,** 5962–5966; H. LeVine *et al.* (1987). *Biochem. Biophys. Res. Commun.* **148,** 1104–1109] |

(*continued*)

TABLE I (*continued*)

| Abbreviated form | Source and reference |
|---|---|
| CaMII-β | Rat calcium/calmodulin-dependent protein kinase II, β subunit [M. K. Bennett and M. B. Kennedy (1987). *Proc. Natl. Acad. Sci. U.S.A.* **84,** 1794–1798] |
| CaMII-γ | Rat calcium/calmodulin-dependent protein kinase II, γ subunit [T. Tobimatsu *et al.* (1988). *J. Biol. Chem.* **263,** 16082–16086] |
| PhK-γ | Rabbit skeletal muscle phosphorylase kinase, γ subunit [E. M. Reimann *et al.* (1984). *Biochemistry* **23,** 4185–4192; E. F. da Cruz e Silva and P. Cohen (1987). *FEBS Lett.* **220,** 36–42] |
| PSK-C3 | Human protein related to PhK-γ, transcript abundant in testis [S. K. Hanks (1989). *Mol. Endocrinol.* **3,** 110–116] |
| MLCK-K | Rabbit myosin light chain kinase, skeletal muscle [K. Takio *et al.* (1985). *Biochemistry* **24,** 6028–6037] |
| MLCK-M | Chicken myosin light chain kinase, smooth muscle [V. Guerriero *et al.* (1986). *Biochemistry* **25,** 8372–8381] |
| PSK-H1 | Human putative protein-serine kinase [J. R. Woodgett, unpublished data] |
| RSK1-C | Mouse ribosomal protein S6 kinase, type 1, carboxyl domain [D. A. Alcorta *et al.* (1989). *Mol. Cell. Biol.* **9,** 3850–3859] |
| RSK2-C | Mouse ribosomal protein S6 kinase, type 2, carboxyl domain [D. A. Alcorta *et al.* (1989). *Mol. Cell. Biol.* **9,** 3850–3859] |
| SNF1 | Yeast (*S. cerevisiae*) sucrose-nonfermenting mutant wild-type gene [J. L. Celenza and M. Carlson (1986). *Science* **233,** 1175–1180] |
| $nim1^+$ | Yeast (*Schizosaccharomyces pombe*) "new inducer of mitosis" gene product [P. Russell and P. Nurse (1987). *Cell (Cambridge, Mass.)* **49,** 569–576] |
| KIN1 | Yeast (*S. cerevisiae*) unusual protein kinase, type 1 [D. E. Levin *et al.* (1987). *Proc. Natl. Acad. Sci. U.S.A.* **84,** 6035–6039] |
| KIN2 | Yeast (*S. cerevisiae*) unusual protein kinase, type 2 [D. E. Levin *et al.* (1987). *Proc. Natl. Acad. Sci. U.S.A.* **84,** 6035–6039] |
| CDC2Hs | Human functional homolog of yeast $cdc2^+$/CDC28 [M. G. Lee and P. Nurse (1987). *Nature (London)* **327,** 31–35] |
| CDC28 | Yeast (*S. cerevisiae*) "cell division control" gene product [A. T. Lorincz and S. I. Reed (1984). *Nature (London)* **307,** 183–185] |
| $cdc2^+$ | Yeast (*S. pombe*) "cell division control" gene product [J. Hindley and G. A. Phear (1984). *Gene* **31,** 129–134] |

TABLE I (*continued*)

| Abbreviated form | Source and reference |
|---|---|
| PHO85 | Yeast (*S. cerevisiae*) negative regulator of PHO system [A. Toh-e *et al.* (1988). *Mol. Gen. Genet.* **214,** 162–164] |
| PSK-J3 | Human putative protein-serine kinase related to CDC2 [S. K. Hanks, unpublished data] |
| KIN28 | Yeast (*S. cerevisiae*) putative protein kinase [M. Simon *et al.* (1986). *EMBO J.* **5,** 2697–2701] |
| MAK | Rat putative protein-serine kinase related to CDC2 [H. Matsushime *et al.* (1990). *Mol. Cell. Biol.* **10,** 2261–2268] |
| KSS1 | Yeast (*S. cerevisiae*) suppressor of sst2 mutant [W. E. Courchesne *et al.* (1989). *Cell (Cambridge, Mass.)* **58,** 1107–1119] |
| YAK1 | Yeast (*S. cerevisiae*) ras suppressor [S. Garrett and J. Broach (1989). *Genes Dev.* **3,** 1336–1348] |
| GSK-3 | Rat glycogen synthase kinase, type 3 [J. R. Woodgett, unpublished data] |
| CKII-$\alpha$ | Human casein kinase II $\alpha$ (catalytic) subunit [H. Meisner *et al.* (1989). *Biochemistry* **28,** 4072–4082] |
| DCKII | *Drosophila* casein kinase II $\alpha$ subunit [A. Saxena *et al.* (1987). *Mol. Cell. Biol.* **7,** 3409–3418] |
| CKA1 | Yeast (*S. cerevisiae*) casein kinase II, $\alpha_1$ subunit [J. L. Chen-Wu *et al.* (1988). *Mol. Cell. Biol.* **8,** 4981–4990] |
| PSK-G1 | Human putative protein-serine kinase [S. K. Hanks, unpublished data] |
| PSK-H2 | Human putative protein-serine kinase [S. K. Hanks, unpublished data] |
| PIM-1 | Human homolog of oncogene activated by murine leukemia virus [T. C. Meeker *et al.* (1987). *Oncogene Res.* **1,** 87–101; J. Domen *et al.* (1987). *Oncogene Res.* **1,** 103–112] |
| PYT | Human protein kinase isolated using anti-phosphotyrosine antibodies [R. A. Lindberg and T. Hunter, unpublished data] |
| STE7 | Yeast (*S. cerevisiae*) "sterile" mutant wild-type gene product [M. A. Teague *et al.* (1986). *Proc. Natl. Acad. Sci. U.S.A.* **83,** 7371–7375] |
| PBS2 | Yeast (*S. cerevisiae*) polymyxin B antibiotic resistance gene product [G. Buguslawski and J. O. Polazzi (1987). *Proc. Natl. Acad. Sci. U.S.A.* **84,** 5848–5852] |
| byr1$^+$ | Yeast (*S. pombe*) suppressor of sporulation defects [S. A. Nadin-Davis and A. Nasim (1988). *EMBO J.* **7,** 985–992] |
| nimA | *Aspergillus nidulans* cell cycle control protein kinase [S. A. Osmani *et al.* (1988). *Cell (Cambridge, Mass.)* **53,** 237–244] |

(*continued*)

TABLE I (*continued*)

| Abbreviated form | Source and reference |
|---|---|
| ninaC | *Drosophila* gene product essential for normal photoreceptor function [C. Montell and G. M. Rubin (1988). *Cell (Cambridge, Mass.)* **52,** 757–772] |
| ran1$^+$ | Yeast (*S. pombe*) "meiotic bypass" mutant wild-type gene product [M. McLeod and D. Beach (1986). *EMBO J.* **5,** 3665–3671] |
| wee1$^+$ | Yeast (*S. pombe*) "reduced size at division" mutant wild-type gene product [P. Russell and P. Nurse (1987). *Cell (Cambridge, Mass.)* **49,** 559–567] |
| GCN2 | Yeast (*S. cerevisiae*) product of gene essential for translational derepression of *GCN4* [I. Roussou *et al.* (1988). *Mol. Cell. Biol.* **8,** 2132–2139] |
| cot | Human oncogene cloned from a human thyroid carcinoma [J. Miyoshi *et al.*, unpublished data] |
| CDC7 | Yeast (*S. cerevisiae*) "cell division control" gene product [M. Patterson *et al.* (1986). *Mol. Cell. Biol.* **6,** 1590–1598] |
| Mos | Human cellular homolog of viral oncogene v-*mos* product [R. Watson *et al.* (1982). *Proc. Natl. Acad. Sci. U.S.A.* **79,** 4078–4082] |
| 2Raf | Human cellular homolog of viral oncogene v-*raf/mil* product [T. I. Bonner *et al.* (1986). *Nucleic Acids Res.* **14,** 1009–1015] |
| Araf | Human cellular oncogene closely related to Raf [T. W. Beck *et al.* (1987). *Nucleic Acids Res.* **15,** 595–609] |
| Braf | Human cellular oncogene closely related to Raf [S. Ikawa *et al.* (1988). *Mol. Cell. Biol.* **8,** 2651–2654] |
| Draf | *Drosophila Raf* homolog [G. E. Mark *et al.* (1987). *Mol. Cell. Biol.* **7,** 2134–2140; Y. Nishida *et al.* (1988). *EMBO J.* **7,** 775–781] |
| Src | Human cellular homolog of viral oncogene v-*src* product [S. K. Anderson *et al.* (1985). *Mol. Cell. Biol.* **5,** 1122–1129] |
| Yes | Human cellular homolog of viral oncogene v-*yes* product [J. Sukegawa *et al.* (1987). *Mol. Cell. Biol.* **7,** 41–47] |
| Fgr | Human cellular homolog of viral oncogene v-*fgr* product [R. C. Parker *et al.* (1985). *Mol. Cell. Biol.* **5,** 831–840; M. Nishizawa *et al.* (1986). *Mol. Cell. Biol.* **6,** 511–517; K. Inoue *et al.* (1987). *Oncogene* **1,** 301–304; S. Katamine *et al.* (1988). *Mol. Cell. Biol.* **8,** 259–266] |
| FYN | Human protein-tyrosine kinase related to Fgr and Yes [K. Semba *et al.* (1986). *Proc. Natl. Acad. Sci. U.S.A.* **83,** 5459–5463; T. Kawakami *et al.* (1986). *Mol. Cell. Biol.* **6,** 4195–4201] |

TABLE I (*continued*)

| Abbreviated form | Source and reference |
| --- | --- |
| LYN | Human protein-tyrosine kinase related to LCK and Yes [Y. Yamanashi *et al.* (1987). *Mol. Cell. Biol.* **7,** 237–243] |
| LCK | Human lymphoid cell protein-tyrosine kinase [J. M. Trevillyan *et al.* (1986). *Biochim. Biophys. Acta* **888,** 286–295; Y. Koga *et al.* (1986). *Eur. J. Immunol.* **16,** 1643–1646] |
| HCK | Human hematopoietic cell protein-tyrosine kinase [N. Quintrell *et al.* (1987). *Mol. Cell. Biol.* **7,** 2267–2275; S. F. Ziegler *et al.* (1987). *Mol. Cell. Biol.* **7,** 2276–2285] |
| TKL | Chicken gene product related to Src [K. Strebhart *et al.* (1988). *Proc. Natl. Acad. Sci. U.S.A.* **84,** 8778–8782] |
| BLK | Mouse B lymphoid cell protein-tyrosine kinase [S. M. Dymecki *et al.* (1990). *Science* **247,** 332–336] |
| Dsrc64 | *Drosophila* gene product related to Src, polytene locus 64B [M. A. Simon *et al.* (1985). *Cell (Cambridge, Mass.)* **42,** 831–840; F. M. Hoffman *et al.* (1983). *Cell (Cambridge, Mass.)* **35,** 393–401] |
| Dsrc28 | *Drosophila* gene product related to Src, polytene locus 28C [R. J. Gregory *et al.* (1987). *Mol. Cell. Biol.* **7,** 2119–2127] |
| STK | *Hydra* gene product related to Src [T. C. G. Bosch *et al.* (1989). *Mol. Cell. Biol.* **9,** 4141–4151] |
| Abl | Human cellular homolog of viral oncogene v-*abl* product [E. Shtivelman *et al.* (1986). *Cell (Cambridge, Mass.)* **47,** 277–284; A. M. Mes-Masson *et al.* (1986). *Proc. Natl. Acad. Sci. U.S.A.* **83,** 9768–9772] |
| Dabl | *Drosophila* gene product related to Abl [F. M. Hoffman *et al.* (1983). *Cell (Cambridge, Mass.)* **35,** 393–401; M. J. Henkemeyer *et al.* (1988). *Mol. Cell. Biol.* **8,** 843–853] |
| Nabl | Nematode (*C. elegans*) gene product related to Abl [J. M. Goddard *et al.* (1986). *Proc. Natl. Acad. Sci. U.S.A.* **83,** 2172–2176] |
| Fes | Human cellular homolog of viral oncogene v-*fes/fps* product [A. J. M. Roebroek *et al.* (1985). *EMBO J.* **4,** 2897–2903] |
| FER | Human Fes-related protein expressed in myeloid cells [Q. L. Hao *et al.* (1989). *Mol. Cell. Biol.* **9,** 1587–1593] |
| EPH | Human kinase detected in erythropoietin-producing hepatoma [H. Hirai *et al.* (1982). *Science* **238,** 1717–1720] |
| ELK | Rat protein-tyrosine kinase detected in brain [K. Letwin *et al.* (1988). *Oncogene* **3,** 621–627] |
| ECK | Human protein-tyrosine kinase detected in epithelial cells [R. A. Lindberg and T. Hunter, unpublished data] |

(*continued*)

TABLE I (*continued*)

| Abbreviated form | Source and reference |
|---|---|
| INS.R | Human insulin receptor [A. Ullrich *et al.* (1985). *Nature* (*London*) **313,** 756–761; Y. Ebina *et al.* (1985). *Cell* (*Cambridge, Mass.*) **40,** 747–758] |
| IRR | Human protein related to insulin receptor [P. Shier and V. M. Watt (1989). *J. Biol. Chem.* **264,** 14605–14608] |
| IGR1R | Human insulin-like growth factor I receptor [A. Ullrich *et al.* (1986). *EMBO J.* **5,** 2503–2512] |
| Ros | Human cellular homolog of viral oncogene v-*ros* product [H. Matsushime *et al.* (1986). *Mol. Cell. Biol.* **6,** 3000–3004; C. Birchmeier *et al.* (1986). *Mol. Cell. Biol.* **6,** 3109–3116] |
| LTK | Mouse lymphocyte protein-tyrosine kinase [Y. Ben-Neriah and A. R. Bauskin (1988). *Nature* (*London*) **333,** 672–676] |
| TRK-A | Human cellular oncogene product [D. Martin-Zanca *et al.* (1989). *Mol. Cell. Biol.* **9,** 24–33] |
| TRK-B | Mouse cellular oncogene product related to TRK-A [R. Klein *et al.* (1989). *EMBO J.* **8,** 3701–3709] |
| MET | Human cellular oncogene product [M. Park *et al.* (1987). *Proc. Natl. Acad. Sci. U.S.A.* **84,** 6379–6383] |
| Sea | Oncogene product of avian erythroblastosis virus S13 [D. R. Smith *et al.* (1989). *Proc. Natl. Acad. Sci. U.S.A.* **86,** 5291–5295] |
| DILR | *Drosophila* gene product related to INS.R [Y. Nishida *et al.* (1986). *Biochem. Biophys. Res. Commun.* **141,** 474–481] |
| 7LESS | *Drosophila* "sevenless" gene product essential for R7 photoreceptor cell development [E. Hafen *et al.* (1987). *Science* **236,** 55–63] |
| PDGFR-A | Human platelet-derived growth factor receptor, A type [T. Matsui *et al.* (1989). *Science* **243,** 800–804] |
| PDGFR-B | Human platelet-derived growth factor receptor, B type [L. Claesson-Welsh *et al.* (1988). *Mol. Cell. Biol.* **8,** 3476–3486; R. G. Gronwald *et al.* (1988). *Proc. Natl. Acad. Sci. U.S.A.* **85,** 3435–3439] |
| CSF1R | Human colony-stimulating factor-1 receptor; cellular homolog of viral oncogene v-*fms* product [L. Coussens *et al.* (1986). *Nature* (*London*) **320,** 277–280] |
| Kit | Human cellular homolog of viral oncogene v-*kit* product and product of developmental locus W in mouse [Y. Yarden *et al.* (1987). *EMBO J.* **6,** 3341–3350] |

TABLE I (*continued*)

| Abbreviated form | Source and reference |
|---|---|
| RET | Human cellular oncogene product [M. Takahashi and G. M. Cooper (1987). *Mol. Cell. Biol.* **7,** 1378–1385; M. Takahashi *et al.* (1988). *Oncogene* **3,** 571–578] |
| FGFR | Human basic fibroblast growth factor receptor [M. Ruta *et al.* (1988). *Oncogene* **3,** 9–15] |
| TORSO | *Drosophila* gene product essential for determination of anterior and posterior structures in the embryo [F. Sprenger *et al.* (1989). *Nature (London)* **338,** 478–483] |
| EGFR | Human epidermal growth factor receptor; cellular homolog of viral oncogene v-*erbB* product [A. Ullrich *et al.* (1984). *Nature (London)* **309,** 418–425; C. R. Lin *et al.* (1984). *Science* **224,** 843–848; G. T. Merlino *et al.* (1984). *Science* **224,** 417–419] |
| NEU | Human oncogene product (also designated ERBB2 and HER2) [L. Coussens *et al.* (1985). *Science* **230,** 1132–1139; T. Yamamoto *et al.* (1986). *Nature (London)* **319,** 230–234] |
| Xmrk | Fish (*Xiphophorus*) dominant oncogene product [J. Wittbrodt *et al.* (1989). *Nature (London)* **341,** 415–421] |
| DER | *Drosophila* homolog of EGFR [E. Livneh *et al.* (1985). *Cell (Cambridge, Mass.)* **40,** 599–607] |

[a] Through February, 1990.

ary is set at the position lying seven residues upstream from the first Gly in the Gly-Xxx-Gly-Xxx-Xxx-Gly consensus, where hydrophobic residues are present in all but 2 of the 117 sequences. At the carboxyl terminus, the boundary is set at a hydrophobic residue that lies a short but variable distance downstream from the invariant Arg. For all protein-tyrosine kinases, this residue lies 10 residues downstream of the invariant Arg. For the protein-serine/threonine kinases, it lies 12–18 residues downstream from the invariant Arg. The four-residue stretch His-Pro-(aromatic)-(hydrophobic) is often found at the protein-serine/threonine kinase catalytic domain carboxyl terminus.

The sequences currently entered in the database range from 247 ($nim1^+$) to 437 (CDC7) amino acid residues. About 90% fall within the range of 247–290 residues. The few catalytic domains that exceed this length can be characterized as having large insert regions consisting of stretches of 20 or more residues conserved in few, if any, of the other sequences. The size distribution of the catalytic domains is shown in Fig. 2.

## Aligning Sequences and Identifying Conserved Residues

Catalytic domain amino acid residues conserved throughout the entire protein kinase family are strongly implicated as playing important roles in the mechanism of phospho transfer. The identification of such residues arises from a multiple amino acid sequence alignment. In producing an alignment of the catalytic domains, we have relied primarily on the old-fashioned "eyeballing" technique. Computer programs developed for generating multiple alignments have difficulty projecting past the large inserts found in many of the catalytic domains. What is needed is an algorithm

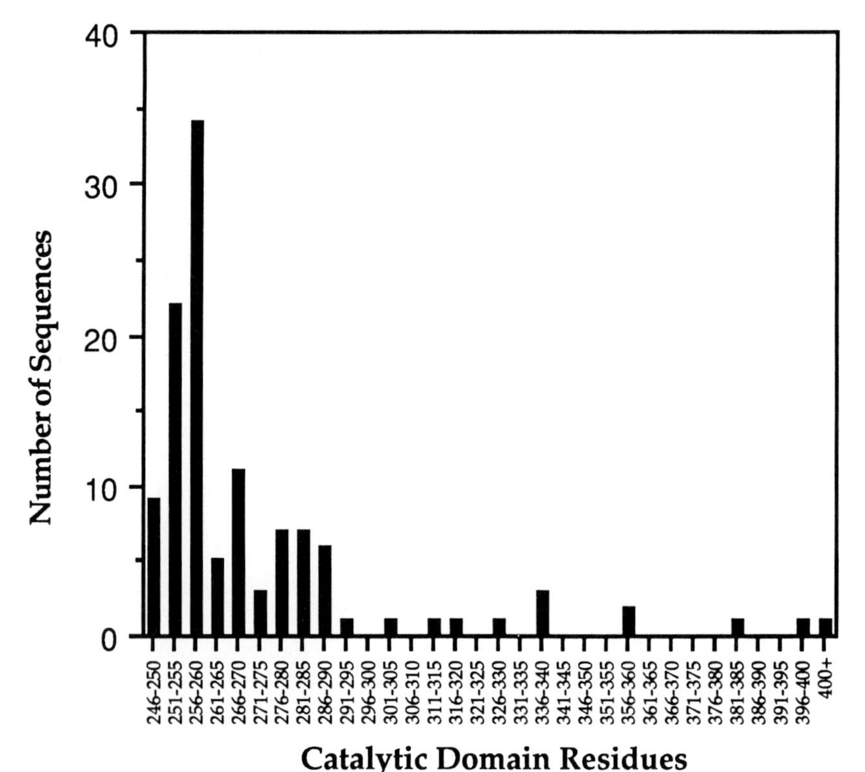


FIG. 2. Size distribution of protein kinase catalytic domains.

that first identifies and aligns the regions conserved throughout the entire family and leaves the more divergent regions, including the gap/insert segments, for last.

The conserved features of the protein kinase catalytic domains can be displayed through a "conservation profile" that represents a quantitative analysis of the extent of conservation at each position in the alignment. To quantitate structural similarities in amino acids, we made use of the structure–genetic scoring matrix described by Feng *et al.*[4] The positional scores plotted in the conservation profile were based on comparisons against the most prevalent amino acid at each position. The scores are weighted against a "perfect" score, arbitrarily set at 100, obtained for invariant residues. If a position contained a gap for 10 or more of the sequences, it was not included in the profile.

The conservation profile obtained from the 117 catalytic domain alignment (Fig. 3A) indicates 8 residues that are invariant throughout the family. These residues, along with some others that are almost as well conserved, are shown at the top of Fig. 3A. Viewed as a whole, this pattern of conserved amino acids represents a protein kinase consensus. Many of these most highly conserved residues have already been implicated as playing essential roles in catalysis (reviewed by Taylor[5]).

The catalytic domain can be further divided into a number of smaller subdomains that represent localized regions of high conservation. The larger gap/insert regions serve to define boundaries for these subdomains. Twelve different subdomains are evident from the 117 catalytic domain alignment and these are indicated by the roman numerals in the conservation profile shown in Fig. 3A. Note that this represents an increase from the 11 subdomains we previously defined in the 65 catalytic domain alignment[2] as a new gap/insert region was identified. This region falls within old subdomain VI, dividing it into new subdomains VI-A and VI-B. Subdomains I, II, III, VI-B, VII, VIII, IX, and XI can each be characterized by the prominent invariant or nearly invariant residues they contain. All 12 subdomains, however, contain positions where structurally similar residues (e.g., hydrophobic, hydrophilic, or small neutral R groups) are conserved throughout the family.

When the conservation profile analysis is repeated using certain subsets of the sequences in the catalytic domain database, conserved features characteristic of smaller protein kinase subgroups can be displayed. For example, separately profiling protein-serine/threonine kinases (Fig. 3B)

[4] D. F. Feng, M. S. Johnson, and R. F. Doolittle, *J. Mol. Evol.* **21,** 112 (1985).

[5] S. S. Taylor, *J. Biol. Chem.* **264,** 8443 (1989).

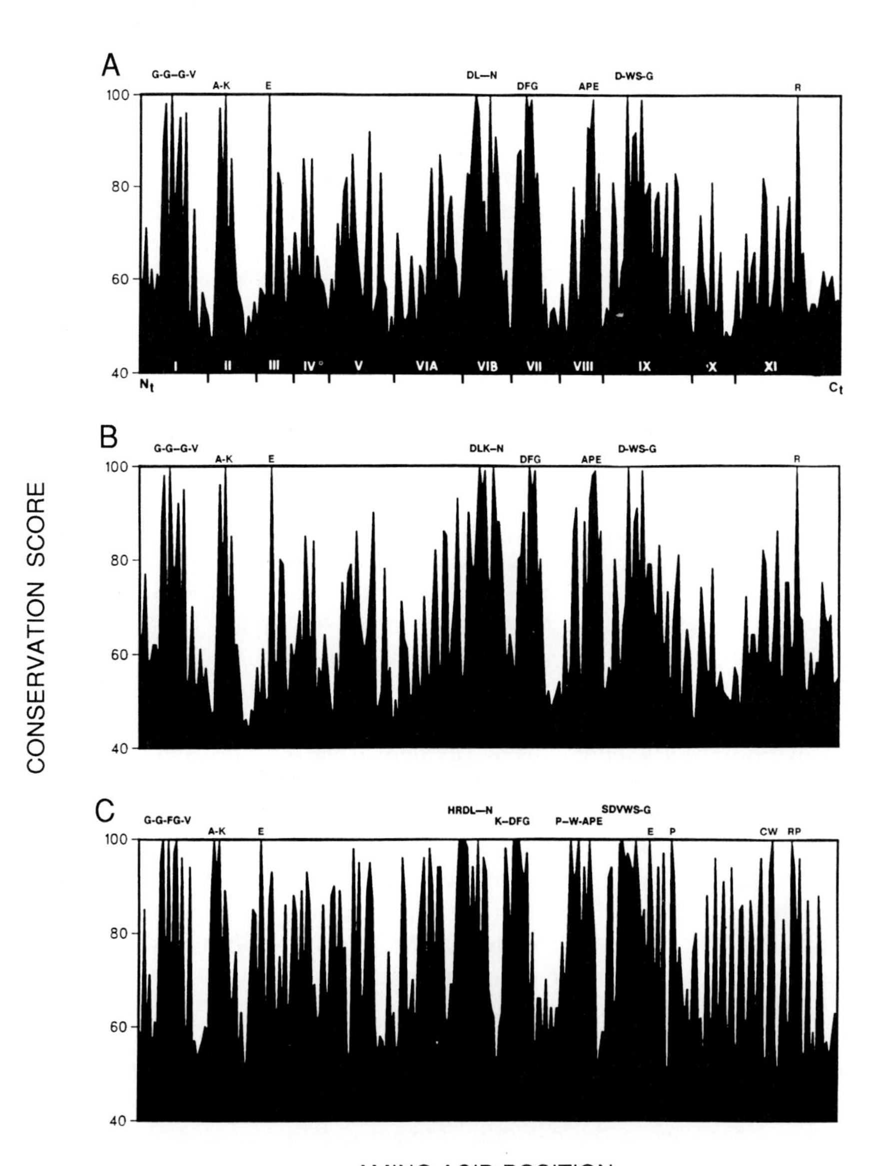


FIG. 3. Conservation profiles for protein kinase catalytic domains: (A) all protein kinase catalytic domains; (B) protein-serine/threonine kinase catalytic domains only; (C) protein-tyrosine kinase catalytic domains only.

and protein-tyrosine kinases (Fig. 3C) reveals residues specifically conserved in one or the other of these broad subgroups. One can speculate, therefore, that some of these residues may play a role in recognition of the correct hydroxyamino acid in the protein substrate.

Another benefit from analyzing the catalytic domains for conserved features is in the design of degenerate oligonucleotides for use in homology-based approaches for identifying novel family members. Highly conserved short stretches of amino acids indicate choice targets for hybridization probes or PCR primers. These aspects are covered in other chapters in this volume.[6,7]

## Protein Kinase Classification

A common use for the catalytic domain database will be in the initial classification of a new family member whose amino acid sequence has just been deduced. Once it has been determined that the new sequence includes the structural features characteristic of a protein kinase catalytic domain (by using the program PROFILESCAN,[8] for example), the investigator will want to know whether the new sequence represents a novel family member or one that has been previously identified. If the sequence is unique, the investigator will be curious to learn if there are any close relatives—hoping, perhaps, that an interesting structure–function relationship will emerge. Using the entries in the catalytic domain database, one can rapidly compare the new sequence against all other known or putative protein kinases and obtain a ranking of the database entries based on a quantitative analysis of sequence similarities. One database-searching program that can be used for this purpose is FASTA.[9] It is available in most of the popular sequence analysis software packages, including Intelligenetics[3] and UWGCG.[10]

Phylogenetic trees based on multiple sequence alignments reveal the overall structural relationships among the members of a protein family and can serve as a useful tool in classification. With this in mind, we have constructed a phylogenetic tree for the protein kinases based on our 117 catalytic domain alignment (Fig. 4). To accomplish this, we used a software package developed by Feng and Doolittle[11] that makes use of the tree-

[6] A. F. Wilks, this volume [45].

[7] S. K. Hanks and R. A. Lindberg, this volume [44].

[8] M. Gribskov, R. Luthy, and D. Eisenberg, this series, Vol. 183, p. 146.

[9] W. R. Pearson and D. J. Lipman, *Proc. Natl. Acad. Sci. U.S.A.* **85,** 2444 (1988).

[10] University of Wisconsin Genetics Computer Group, Sequence Analysis Software Package, Program Manual, Madison, Wisconsin, February 1989.

[11] D. F. Feng and R. F. Doolittle, *J. Mol. Evol.* **25,** 351 (1987).

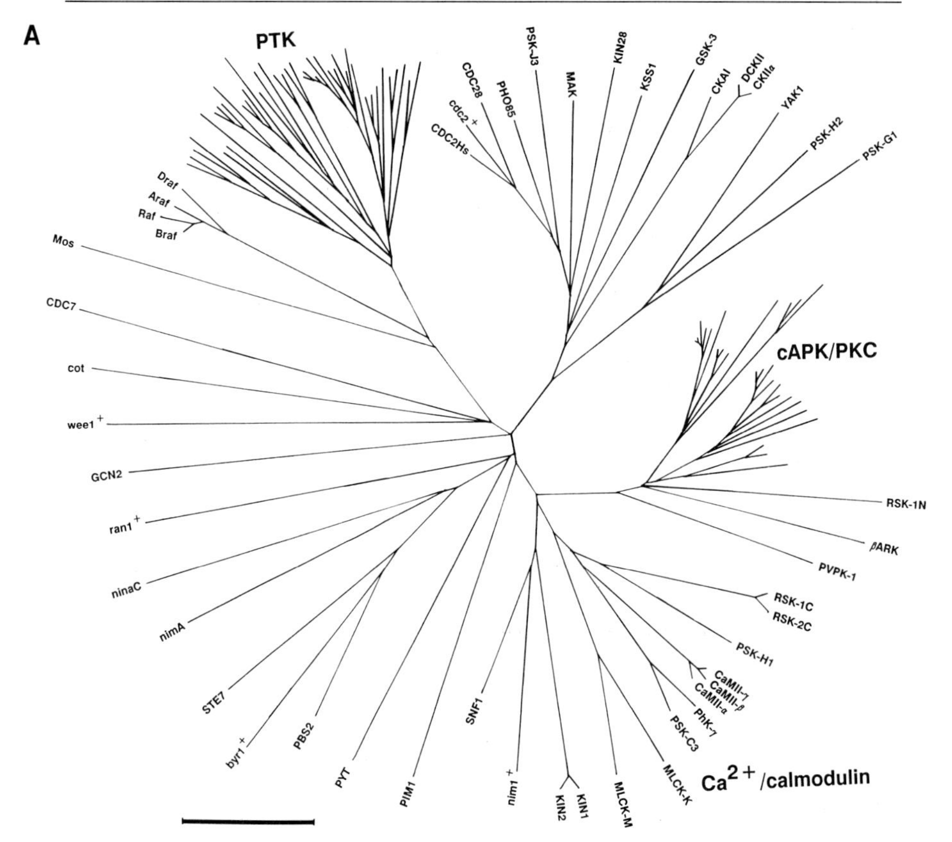


FIG. 4. Protein kinase catalytic domain phylogenetic tree. (A) All 117 catalytic domains are represented. The individual protein kinases are indicated by the abbreviated names used in Table I. For two large branch clusters, the protein-tyrosine kinase subfamily (PTK) and the cyclic nucleotide-dependent and protein kinase C subfamilies (cAPK/PKC), individual branches are not labeled but these clusters are redrawn with the branches labeled in panels (B) and (C) for PTK and cAPK/PKC subfamilies, respectively. Scale bars in all panels represent a branch length corresponding to a relative difference score of 25.

building concept of Fitch and Margoliash.[12] According to this strategy, a matrix of difference scores is calculated from the multiple amino acid sequence alignment and a best fit approach is then taken to calculate branch lengths and tree topology. The tree-building software we used is available on request from Dr. R. F. Doolittle, Department of Chemistry, University of California at San Diego. The upper limit of sequences that can be analyzed using this software is 52. We therefore constructed three

[12] W. M. Fitch and E. Margoliash, *Science* **155,** 279 (1967).

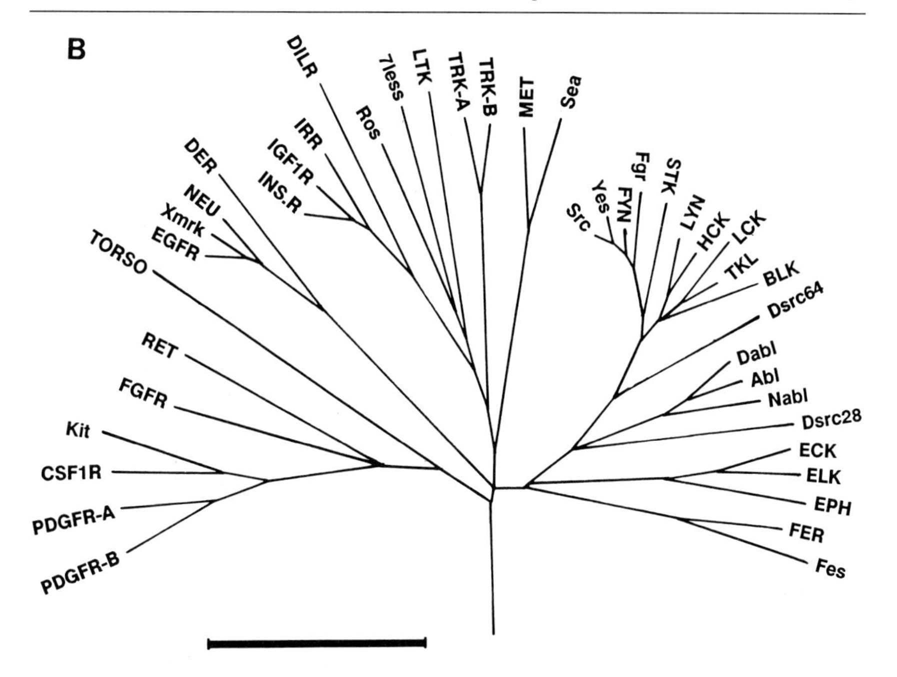


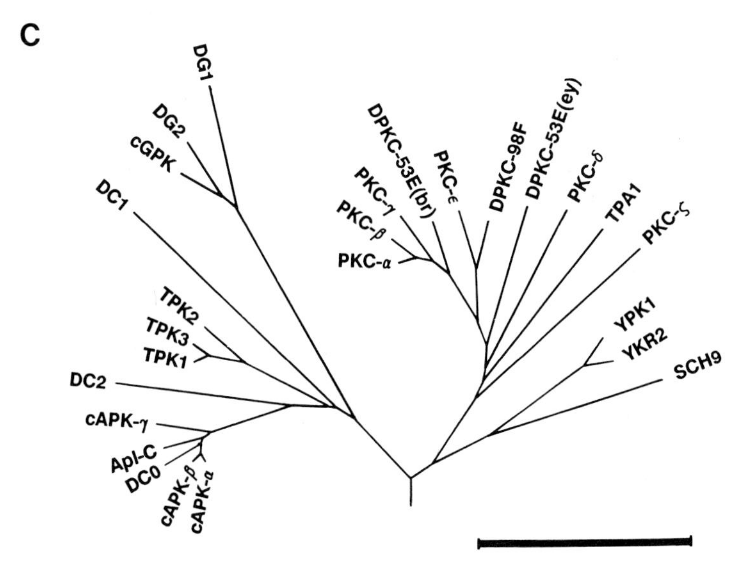


FIG. 4. (*continued*)

smaller trees from which a composite tree that includes all 117 sequences was drawn (Fig. 4). These smaller trees, each containing about 40–45 sequences, were calculated in approximately three central processing unit hours on a VAX-6220 equipped with 64 megabytes of memory running the virtual memory operating system (VMS).

The potential usefulness of the catalytic domain phylogenetic tree as a tool for classification is illustrated by the clustering of protein kinases that have similar properties. This clustering can be used to define protein kinase subfamilies. The densest cluster is composed of the 42 known or putative protein-tyrosine kinases (Fig. 4A and B). This large subfamily includes both cytoplasmic enzymes (Src, Abl, Fes, etc.) as well as the receptors for growth and differentiation factors. Two distinct protein-serine/threonine kinase clusters are recognized as containing members whose activities are under similar modes of regulation. These are the cyclic nucleotide-dependent protein kinase subfamily (13 members) and the protein kinase C subfamily (10 members). They map quite near one another within the tree (Fig. 4A and C). In addition, the protein-serine/threonine kinases regulated by $Ca^{2+}$/calmodulin map near one another and fall within a broad cluster. It is uncertain, however, if all members of this cluster (e.g., RSK-1C, RSK-2C, PSK-H1) are regulated by $Ca^{2+}$/calmodulin.

Clearly, some structure–function relationships have been maintained during the course of catalytic domain evolution. We predict other such relationships will emerge with the continued identification and functional characterization of novel protein kinases.

## Acknowledgments

S.K.H. was supported by United States Public Health Service Grant GM-38793. We thank Tony Hunter for continued encouragement and Lisa Caballero for many useful comments.

# [3] Protein Kinase Phosphorylation Site Sequences and Consensus Specificity Motifs: Tabulations

*By* RICHARD B. PEARSON and BRUCE E. KEMP

## Introduction

The phosphorylation site sequences for protein-serine/threonine and protein-tyrosine kinases are presented in Table I of this chapter. Over 240 phosphorylation site sequences are included with the phosphorylated residue(s) indicated by an asterisk. Plant and prokaryote phosphorylation site sequences have not been included. References are listed at the end of the chapter. Anyone wanting an annual update of this tabulation and/or wanting to contribute a new sequence should contact the authors.

Note that residue numbers followed by a # in the tabulation have been obtained from SWISSPROT protein database. (P) denotes a phosphorylated residue which acts as a substrate-specificity determinant. The question marks indicate potential sites within a phosphorylated peptide or sites inferred from mutagenesis studies or the stoichiometry of phosphorylation (see Table I, footnotes *c* and *h*, respectively).

Table II contains consensus phosphorylation site motifs for each enzyme, where applicable. The frequencies listed are derived from Table I unless indicated otherwise. The $S:T$ ratio is for the total number of phosphorylation sites. Asterisks indicate the phosphorylated residue and specificity determinants are shown in boldface type.

The one-letter code for amino acids in Tables I and II is as follows: A, alanine; C, cysteine; D, aspartic acid; E, glutamic acid; F, phenylalanine; G, glycine; H, histidine; I, isoleucine; K, lysine; L, leucine; M, methionine; N, asparagine; P, proline; Q, glutamine; R, arginine; S, serine; T, threonine; V, valine; W, tryptophan; Y, tyrosine.

TABLE I
PROTEIN KINASE PHOSPHORYLATION SITE SEQUENCES[a]

| Protein kinase | Phosphorylation site sequence | Protein | Refs.[b] |
|---|---|---|---|
| **Protein-serine/ threonine kinases** | | | |
| AMP-activated protein kinase (previously called acetyl-CoA carboxylase kinase-3, HMG-CoA reductase kinase, hormone-sensitive lipase kinase) | $M_{74}$ R S S M S* G L H L | Acetyl-CoA carboxylase (rat) | 42 |
| | $L_{1196}$ N R M S* F A S N | Acetyl-CoA carboxylase (rat) | 42 |
| | $G_{1209}$ M TH V A S* V S D V L L D | Acetyl-CoA carboxylase (rat) | 49 |
| | $M_{560}$ R R S V S* E A A L | Hormone-sensitive lipase (rat) | 42, 49, 91 |
| | $H_{865}$ M V H N R S* K I N L Q D L | HMG-CoA reductase (rat) | 49 |
| β-Adrenergic receptor kinase[c] | $G_{353}$ Y $S^{?*}$ $S^{?*}$ N G N $T^{?*}$ G E Q $S^{?*}$ | β-Adrenergic receptor# | 50 |
| | $E_{379}$ D L P G $T^{?*}$ E D | β-Adrenergic receptor# | 50 |
| | $G_{392}$ G $T^{?*}$ V P $S^{?*}$ D N I D $S^{?*}$ Q G R N C $S^{?*}$ $T^{?*}$ N D $S^{?*}$ L L-COOH | β-Adrenergic receptor# | 50 |

(*continued*)

TABLE I (*continued*)

| Protein kinase | Phosphorylation site sequence | Protein | Refs.[b] |
|---|---|---|---|
| Branched chain α-ketoacid dehydrogenase kinase | $T_{326}$ Y R I G H H S* T S D D S S | Branched chain α-ketoacid dehydrogenase site 1# | 45 |
| | $A_{340}$ Y R S* V D E V N Y W D K | Branched chain α-ketoacid dehydrogenase site 2# | 45 |
| Calmodulin-dependent protein kinase I | $N_2$ Y L R R R L S* D S N F M | Synapsin I site 1 | 14, 100 |
| Calmodulin-dependent protein kinase II | $T_{564}$ R Q T S* V S G Q A P P K | Synapsin I site 2 (bovine) | 14, 100 |
| | $T_{601}$ R Q A S* Q A G P M P R | Synapsin I site 3 (bovine) | 14, 100 |
| | $R_{13}$ R A S T* I E M P Q Q A R | Phospholamban | 15, 78 |
| | $P_1$ L S R T L S* V S S | Glycogen synthase site 2 | 16 |
| | $K_{191}$ M A R V F S* V L R E | Calcineurin# | 17, 139 |
| | $R_{15}$ R A V S* E Q D A K | Tyrosine hydroxylase (monooxygenase) site C | 18 |
| | $R_{37}$ R Q S* L I E D A R K | Tyrosine hydroxylase (monooxygenase) site A | 18 |
| | $R_{13}$ K L S* D/N F G E/Q | Phenylalanine hydroxylase (monooxygenase) | 19 |
| | $Y_7$ L R R A S* V A Q L T* Q E | Pyruvate kinase# | 20 |
| | $R_{10}$ S K Y L A S* A S T M | Myelin basic protein | 21 |
| | $R_{64}$ T T H Y G S* L P Q K | Myelin basic protein | 21 |
| | $K_{91}$ N I V T* P R T P P | Myelin basic protein | 21 |
| | $R_{113}$ F S* W G A E G Q K | Myelin basic protein | 21 |
| | R/K T A S* F S E S R | ATP-citrate lyase | 43 |
| | $F_{19}$ I I G S V S* E D N | Acetyl-CoA carboxylase# | 28 |
| | $M_{560}$ R R S V S* E A A L | Hormone-sensitive lipase | 70 |
| | $R_{808}$ A I G R L S S* M A M | Smooth muscle myosin light chain kinase | 130, 131 |
| | $R_{283}$ Q E T* V D C L K K F N A R R K L K | Autophosphorylation (α subunit) | 96–98 |
| | $R_{284}$ Q E T* V E C L K K F N A R R K L K | Autophosphorylation (β subunit) | 98, 99 |

TABLE I (*continued*)

| Protein kinase | Phosphorylation site sequence | Protein | Refs.[b] |
|---|---|---|---|
| Calmodulin-dependent protein kinase III | $A_{51}$G E T$^{?*}$ R F T$^{?*}$ D T$^{?*}$ R | Elongation factor 2 | 22 |
| Casein kinase I | $V_{22}$ S(P) S(P) S(P) E E S* I I S | $\alpha_{s2}$-Casein# | 29 |
| | $L_{143}$ S(P) T S(P) E E N S* K K | $\alpha_{s2}$-Casein# | 29 |
| | $L_{31}$ S(P) S(P) S(P) E E S* I T R | $\beta$-Casein# | 29 |
| | $V_{37}$ N E L S* K D I | $\alpha_{s1}$-Casein# | 29 |
| | $G_{45}$ S(P) E S(P) T* E D Q | $\alpha_{s1}$-Casein# | 29 |
| | $P_1$ L S* R T L S*(P) V S S* L P G L | Glycogen synthase | 30, 118 |
| | $K_{23}$ R S G S* V/I Y E P L K | Phosphorylase kinase ($\beta$ subunit)# | 102 |
| Casein kinase II[d] | $R_{40}$ L S* E H S* S P E E E A | DARPP-32 (bovine) | 25 |
| | $E_{98}$ N Q A S* E E E D E L G E | DARPP-32 (bovine) | 25 |
| | $E_{225}$ R D K E V S* D D E A E E | $\alpha$-Heart shock protein-90 ($\alpha$-Hsp-90) | 26 |
| | $E_{225}$ R E K E I S* D D E A E E | $\beta$-Hsp-90 (human) | 26 |
| | $E_{257}$ I E D V G S* D E E E E | $\alpha$-Hsp 90 (human) | 26 |
| | $K_{249}$ I E D V G S* D E E D D | $\beta$-Hsp 90 (human) | 26 |
| | $G_{130}$ D R F T* D E E V D E | Myosin regulatory light chain, chicken smooth muscle | 27 |
| | $S_{23}$ V S E D N S* E D E I S N L | Acetyl-CoA carboxylase# | 28 |
| | $S_1^*$ D E E V E | Troponin T | 29 |
| | $P_{654}$ H Q S* E D E E E | Glycogen synthase (residue numbers based on human muscle sequence) | 29, 41, 150 |
| | $A_{72}$ D S* E S* E D E E D | cAMP-dependent protein kinase regulatory subunit $R_{II}$# | 108 |
| | $V_5$ E E D A E S* E D E D E E D | Nucleolar protein B23# | 29 |
| | $P_{85}$ A S* E D E D E E E D | Nucleolar protein C23# | 29 |

(*continued*)

TABLE I (*continued*)

| Protein kinase | Phosphorylation site sequence | Protein | Refs.[b] |
|---|---|---|---|
| | L E L S* D D D D E S K | Myosin heavy chain (brain) | 56 |
| | $D_{80}$ D D D A Y S* D T E T T E | Phosphatase inhibitor-2 | 71 |
| | $E_{117}$ Q E S* S* G E E D S D L | Phosphatase inhibitor-2 | 71 |
| | $R_3$ R P R H S I Y S* S* D D D E E D | c-Myb | 87 |
| | $P_{245}$ P T?* T?* S?* S?* D S?* E E E Q E D E E E | Myc | 88 |
| | $S_{344}$ P R S?* S?* D T E E N | Myc | 88 |
| | $S_9$ S S* E S* G A P E A A E E D | Clathrin light chain $LC_b$ | 109 |
| | $S^*_{89}$ D E E D E E | Elongation factor 1$\beta$ | 110 |
| | $D_{78}$ T* D S E E E I R E | Calmodulin | 116 |
| | $D_{95}$ G D G Y I S* A A E L R H | Calmodulin | 116 |
| | $E_{26}$ Q L N D S* S* E E E D E I D | Human papillomavirus E7 oncoprotein (HPV E7) | 117 |
| | $T_{301}$ G S* D D E D E S N E Q | Ornithine decarboxylase[#] | 29 |
| | $E_{84}$ E S P A S* D E A E E K | High mobility group 14 protein | 140 |
| Crystallin kinase(s) | $R_{116}$ R Y R L P S* N V D | $\alpha_A$-Crystallin | 58 |
| | $R_{12}$ P F F P F H S* P S R | $\alpha_B$-Crystallin | 59 |
| | $P_{39}$ A S T S L S* P F Y L R P P | $\alpha_B$-Crystallin | 59 |
| cAMP-dependent protein kinase (mammalian)[e] | $Y_7$ L R R A S* L/V A Q L T | Pyruvate kinase[#] | 1 |
| | $F_1$ R R L S* I S T | Phosphorylase kinase $\alpha$ chain | 1 |
| | $Q_{692}$ W P R R A S* C T S | Glycogen synthase site 1a (residue numbers based on human muscle sequence) | 1, 150 |
| | $G_{706}$ S K R S N S* V D T | Glycogen synthase site 1b (residue numbers based on human muscle sequence) | 1, 150 |
| | $R_{21}$ T K R S G S* V/I Y E | Phosphorylase kinase $\beta$ chain[#] | 1 |

TABLE I (*continued*)

| Protein kinase | Phosphorylation site sequence | Protein | Refs.[b] |
|---|---|---|---|
| | $A_{29}$ G A R R K A S* G P P | Histone H1[#] (residue numbers from bovine sequence) | 1 |
| | $K_{13}$ A K T R S S* R A | Histone H2A[#] (residue numbers from bovine sequence) | 1 |
| | $G_{26}$ K K R K R S* R K E S* Y S | Histone H2B[#] (residue numbers from bovine sequence) | 1 |
| | E R R K S* K S G A G | cAMP regulated phosphoprotein $M_r$ = 21,000 (ARPP-21) | 55 |
| | $Y_3$ L R R R L S* D S N F | Synapsin I site 1 | 100 |
| | $N_{19}$ Y R G Y S* L G N Y V | Reduced carboxymethylated maleylated (RCMM)-lysozyme | 76 |
| | $R_{27}$ A S* F G S R G S* G S | Desmin | 86 |
| | $S_{47}$ R T S* A V P T | Desmin | 86 |
| | $P_1$ L S R T L S* V S S | Glycogen synthase site 2 | 1 |
| | $A_{16}$ V R R S* D R A | Troponin I (cardiac) | 69 |
| | $M_{560}$ R R S* V S E A A L | Hormone-sensitive lipase | 42 |
| | $M_{74}$ R S S* M S G L H L | Acetyl-CoA carboxylase | 42 |
| | $S_{12}$ Q R R R S* L E P P D | $pp60^{c\text{-}src}$ | 79 |
| | $K_{23}$ R K R K S?* S?* Q C L V K | c-erbA | 134 |
| | $K_{11}$ H K R K S?* S?* Q C L V K | v-erbA | 134 |
| | $R_{45}$ N T D G S* T D Y G I | RCMM-lysozyme | 76 |
| | $R_{212}$ R K G T* D V | Lipocortin I (p35, calpactin II) | 81 |
| | $I_{29}$ R R R R P T* P A T | Phosphatase inhibitor-1[#] | 1 |
| | K P R R K D T* P A L | G substrate | 1 |
| | $K_{188}$ R V K G R T W T* L C G T | Autophosphorylation of catalytic subunit | 1, 95 |
| | $R_{89}$ F D R R V S* V C A | Autophosphorylation of regulatory subunit $R_{II}$ | 1 |
| Yeast | $K_{218}$ R K Y L K K L T R R A S* F S A | ADR1 | 2 |

(*continued*)

TABLE I (*continued*)

| Protein kinase | Phosphorylation site sequence | Protein | Refs.[b] |
|---|---|---|---|
| | $P_7$ R R D S* T E G F | Fructose-1,6-bisphosphatase | 128 |
| | $Q_{141}$ R R T S* V S G E | Autophosphorylation of regulatory subunit | 129 |
| cGMP-dependent protein kinase | $F_1$ R R L S* I S T E | Phosphorylase kinase $\alpha$ chain | 3 |
| | $D_{25}$ G K K R K R S* R K E S* | Histone H2B | 3 |
| | $R_{92}$ R R R G A I S* A E V Y | cAMP-dependent protein kinase regulatory subunit $R_I$ | 3 |
| | K K P R R K D T* P A L H | G substrate site 1 | 3 |
| | Q K P R R K D T* P A L H | G substrate site 2 | 3 |
| | $R_{29}$ R R R P T* P A M L | DARPP-32[#] | 3 |
| | $Q_{28}$ I R R R R P T* P A T L | Phosphatase inhibitor-1 | 3 |
| | $P_1$ K R K V S* S A E G | High mobility group 14 protein | 3 |
| | $K_{17}$ R R S A R L S* A K P A | High mobility group 14 protein | 3 |
| | $M_{560}$ R R S* V S E A A L | Hormone-sensitive lipase | 91 |
| | $A_{11}$ I R R A S* T I E M | Phospholamban | *f* |
| | $V_{45}$ L P V P S* T H I G P | Autophosphorylation | 3 |
| | $I_{53}$ G P R T T* R A Q G I | Autophosphorylation | 3 |
| | $P_{67}$ Q T Y R S* F H D L R | Autophosphorylation | 3 |
| | $A_{79}$ F R K F T* K S E R S | Autophosphorylation | 3 |
| Double-stranded DNA-activated protein kinase | $P_1$ E E T* Q T* Q D Q P M E | Heat-shock protein 90$\alpha$ (human) | 37 |
| Double-stranded RNA-activated protein kinase (p68 kinase) | $I_{45}$ L L S E L S* R R | eIF-2$\alpha$ | 38 |
| Endogenous eIF-4E kinase | $K_{49}$ N D K S* K T W Q A N L R | eIF-4E | 40 |
| Glycogen synthase kinase-3 | $P_{637}$ R P A S* V P P S* P S L S* R H S S* P Q H S(P) | Glycogen synthase sites 3a, b, c and 4 (residue numbers based on human muscle sequence) | 35, 103, 150 |

(*continued*)

TABLE I (*continued*)

| Protein kinase | Phosphorylation site sequence | Protein | Refs.[b] |
|---|---|---|---|
| | $D_{68}$ E P S T* P Y H S M I G D D D D A Y S(P) D | Phosphatase inhibitor-2 | 71 |
| | $L_{39}$ R E A R S* R A S* T P P | cAMP-dependent protein kinase regulatory subunit $R_{II}$ | 72 |
| | $K_{36}$ P G F S* P Q P S* R R G S(P) | Protein phosphatase-1 G subunit[j] | 36 |
| | $A_{343}$ P V S?* C L G E H H H C T?* P S?* P P V D H G C L | c-Myb[#] (residue numbers based on chicken sequence) | 119 |
| | $E_{238}$ T* P P L S* P I D M E S* Q E R | c-Jun | 120 |
| Glycogen synthase kinase-4 | $P_1$ L S R T L S* V S S | Glycogen synthase site 2 | 89 |
| | $M_{560}$ R R S V S* E A A L | Hormone-sensitive lipase | 42, 90 |
| Growth-associated H1 histone kinase (MPF, cdc2+/CDC28 protein kinases)[g] | $T_{10}$ S E/Q P A K T* P V K | Histone H1 (calf thymus) | 53, 54 |
| | $K_{130}$ A T G A A T* P K | Histone H1 (calf thymus) | 53, 54 |
| | $K_{152}$ T* P K | Histone H1 (calf thymus) | 53, 54 |
| | $V_{177}$ A K S* P K | Histone H1 (calf thymus) | 53, 54 |
| | $F_{29}$ P A S Q T* P N K T A | pp60[c-src] | 104 |
| | $P_{41}$ D T H R T* P S R S F | pp60[c-src] | 104 |
| | $S_{67}$ D T V T S* P Q R A G | pp60[c-src] | 104 |
| | $D_{562}$ A P D T* P E L L H T K | c-Abl type IV | 137, 138 |
| | $S_{583}$ E P A V S* P L L P R | c-Abl type IV | 137, 138 |
| | $H_{122}$ S T* P P K K K R K | Large T antigen | 137 |
| | $S_{313}$ S S* P Q P | p53 | 137 |
| | $S^*_{16}$ P T R | Lamin B | 137 |
| | $S^*_1$ S K R A K A K T* T K K | Myosin regulatory light chain, nonmuscle | 137 |
| | K R S* P K K | SW15 | 137 |
| | Y S?* P T S?* P S | RNA polymerase II | 105 |
| Growth factor-regulated kinase | $P_{667}$ L T* P S G E A | EGF receptor | 34 |
| Heme-regulated eIF-2α kinase | $I_{45}$ L L S* E L S* R R | eIF-2α | 38, 39 |
| Histone H4 kinase I | $V_{43}$ K R I S* G L | Histone H4 | 48 |

(*continued*)

TABLE I (*continued*)

| Protein kinase | Phosphorylation site sequence | Protein | Refs.[b] |
|---|---|---|---|
| Histone H4 kinase II | Ac-$S_1^*$ G R G K G G | Histone H4 | 48 |
| Insulin receptor-associated serine kinase (IRSK) | K/$R_{1292}$ $S^{?*}$ $S^{?*}$ H C Q R | Insulin receptor | 115 |
| Isocitrate dehydrogenase kinase | $T_{104}$ T P V G G G I R S* L N V A | Isocitrate dehydrogenase | 47 |
| Mitogen-activated protein kinase (MAP kinase) | $T_{94}$ P R T* P P P | Myelin basic protein | 107 |
| Mitogen-activated S6 kinase ($M_r$ 70,000) | $R_{232}$ R L S* S* L R A S* T S K S* E S S* Q K | Ribosomal protein S6 | 147, 148 |
| Myosin I heavy chain kinase | R/K A G T* T Y A L N L N K | Myosin 1A heavy chain | 32 |
| | $G_{307}$ G A G A K K M S* T Y N V | Myosin 1B heavy chain | 32 |
| | $G_{303}$ E Q G R G R S S* V Y S C | Myosin 1C heavy chain | 32 |
| Myosin light chain kinase<br>Skeletal muscle | $K_5$ R R A A E G S S* N V F | Myosin regulatory light chain, chicken skeletal muscle | 6, 80 |
| Smooth muscle | $K_{11}$ K R P Q R A T* S* N V F | Myosin regulatory light chain, chicken smooth muscle | 5, 60 |
| Phosphorylase kinase | $D_6$ Q E K R K Q I S* V R G | Phosphorylase | 4 |
| | $P_1$ L S R T L S* V S S | Glycogen synthase site 2 | 4 |
| | $R_{21}$ T K R S G S* V/I Y E | Autophosphorylation ($\beta$ chain)# | 1, 101 |
| Proline-directed protein kinase | $P_2$ T P S A P S* P Q P K | Tyrosine hydroxylase (monooxygenase) | 33 |
| | $T_{546}$ R P P A S* P S P Q R | Synapsin I (bovine) | 100 |
| Protease-activated kinases I and II | $A_{229}$ K R R R L S S* L R A | Ribosomal protein S6 | 52 |
| Protein kinase C[e] | $Q_4$ K R P S* Q R S K Y L | Myelin basic protein | 7 |
| | $K_{1329}$ K N G R V L T* L P R S | Insulin receptor (rat) | 8 |
| | $K_{1329}$ K N G R I L T* L P R S | Insulin receptor (human) | 121, 122 |
| | $Y_{20}$ T R F S* L A R | Transferrin receptor | 9 |
| | $P_1$ L S R T L S* V S S | Glycogen synthase site 2 | 135 |
| | $P_{691}$ Q W P R R A S* C T S | Glycogen synthase site 1a (residue numbers based on human muscle sequence) | 135, 150 |

TABLE I (*continued*)

| Protein kinase | Phosphorylation site sequence | Protein | Refs.[b] |
|---|---|---|---|
| | $K_{37}$ I Q A S* F R G H I T R K K | Neuromodulin | 136 |
| | $S_{I}^{*}$ S* K R A K A K T* T K K R | Myosin regulatory light chain, chicken smooth muscle | 10 |
| | $K_{339}$ K K K K R F S* F K K S* F K L S* G F S* F K K N K K | MARCKS protein | 11 |
| | $G_{97}$ T G A S G S* F K | Histone H1 | 12 |
| | $F_{43}$ F G S* D R G | Myelin basic protein | 7 |
| | $G_{148}$ T L S* K I F | Myelin basic protein | 7 |
| | $R_{645}$ R R H I V R K R T* L R R L | EGF receptor | 23 |
| | $R_{89}$ R R H I V R K R T* L R R L | erb B | 126, 127 |
| | $A_{229}$ K R R R L S S* L R A | Ribosomal protein S6 | 24 |
| | $Y_{21}$ V Q T* V K S* S* K G G P G | Lipocortin I (p35, calpactin II) | 81, 82 |
| | $S_{21}$ A Y G S* V K P Y T N F D | Lipocortin II (p36, calpactin I heavy chain) | 83 |
| | $G_{555}$ K S* S* S* Y S K | Fibrinogen | 84 |
| | $H_{594}$ E G T H S* T K R | Fibrinogen | 84 |
| | $D_{354}$ L K L R R $S^{?*}$ $S^{?*}$ $S^{?*}$ V G Y | Acetylcholine receptor | 85 |
| | $R_{9}$ V S S* Y R R T F G | Desmin | 86 |
| | $R_{27}$ A S* F G S R G S G S S* V T S R | Desmin | 86 |
| | $P_{53}$ T L S* T F R T T R | Desmin | 86 |
| | Ac-$S_{I}^{*}$ L K D H L I H N V H K | Lactate dehydrogenase | 123 |
| | $W_{240}$ Q R R Q R K S* R R T I | Interleukin-2 receptor | 124 |
| | $G_{2}$ S S K S K P K D P S* Q R R R S | pp60$^{src}$ | 125 |
| | $R_{75}$ S S* M S G L H | Acetyl-CoA carboxylase | 42, 141 |
| | $R_{88}$ D R K K I D S* F A S N | Acetyl-CoA carboxylase | 42, 141 |
| Pyruvate dehydrogenase kinase | $R_{226}$ Y G M G T S* V E R | E1$\alpha$ pyruvate dehydrogenase[#] (residue numbers based on human sequence) | 51 |
| | $R_{287}$ Y H G H S* M S D P G V S* Y R | E1$\alpha$ pyruvate dehydrogenase[#] (residue numbers based on human sequence) | 51 |

(*continued*)

TABLE I (*continued*)

| Protein kinase | Phosphorylation site sequence | Protein | Refs.[b] |
|---|---|---|---|
| Rhodopsin kinase[h] | $D_{330}$ D E A S* T* T* V S* K T?* E T?* S* Q V A P | Rhodopsin | 44 |
| S6 kinase II ($M_r$ 92,000) | $A_{229}$ K R R R L S?* S?* L R A | Ribosomal protein S6 | 31 |
| | $I_{35}$ G R R Q S* L I E D A | Tyrosine hydroxylase (monooxygenase) | 31 |
| | $P_1$ L S R T L S* V S S L P G | Glycogen synthase site 2 | 31 |
| | $R_{339}$ G R A S S* H S S | Lamin C# (residue numbers based on human sequence) | 31 |
| Sperm-specific histone kinase | $P_1$ G S* P Q K R A A S* P R K S* P K K S* P R K A S A S* P R | Histone H1 (sea urchin sperm) | 57 |
| | $K_8$ R S* P T K R S* P Q K G | Histone H2B$_1$ (sea urchin sperm) | 57 |
| | $R_{10}$ K G S* P R K G S* P K R G | Histone H2B$_2$ (sea urchin sperm) | 57 |
| Tropomyosin kinase | $D_{275}$ H A L N D M T S* I-COOH | $\alpha$-Tropomyosin | 46 |
| | $D_{275}$ N A L N D I T S* L-COOH | $\beta$-Tropomyosin | 46 |
| Unknown kinase(s) | $K_{85}$ K R R L S* F S E T F | Interleukin-1$\alpha$ | 144 |
| | $R_{662}$ E L V E P L T* P S* G E A P | EGF receptor | 145 |
| | $F_{1041}$ L Q R Y S* S* D P T G A L | EGF receptor | 142, 145 |
| | $T_{82}$ F P P A P G S* P E P P | Adenovirus type 5 289R E1A protein | 146 |
| | $R_{215}$ R P T S* P V S R E C N S S T D S* C D S G | Adenovirus type 5 289R E1A protein | 146 |
| **Protein-tyrosine kinase**[i] | | | |
| Colony-stimulating factor 1 receptor kinase | $Q_{689}$ D S E G D S S Y* K N I H | Autophosphorylation | 132 |
| | $K_{698}$ N I H L E K K Y* V R R D | Autophosphorylation | 132, 133 |
| | $R_{799}$ D I M N D S N Y* V V K G | Autophosphorylation | 132, 133 |
| EGF-receptor kinase | $S_{1166}$ T R E N A E Y* L R V A P Q S | Autophosphorylation | 61 |
| | $I_{1141}$ S L D N P D Y* Q Q D F F P K | Autophosphorylation | 61 |

(*continued*)

TABLE I (*continued*)

| Protein kinase | Phosphorylation site sequence | Protein | Refs.[b] |
|---|---|---|---|
| | $T_{1061}$ F L P V P E Y* I N Q S V P K | Autophosphorylation | 61 |
| | $S_{1080}$ V Q N P V Y* H N Q P L N | Autophosphorylation | 94 |
| | $D_{979}$ E E D M D D V V D A D E Y* L I P Q Q | Autophosphorylation | 111, 142 |
| | $D_{135}$ E E V D E M Y* R E A P I D K | Myosin regulatory light chain, smooth muscle | 61 |
| | $D_{148}$ V K G N F N Y* V E F T R I L | Myosin regulatory light chain, smooth muscle | 61 |
| | $I_{1}$ D N E E Q E Y* I K T V K G S | Lipocortin I (p35, calpactin II), porcine | 61, 63 |
| | $I_{14}$ E N E E Q E Y* V Q T V K S S | Lipocortin I (p35, calpactin II), human | 81 |
| | $L_{466}$ A E G S A Y* E E | Phospholipase C-γ | 92 |
| | $A_{767}$ E P D Y* G A L Y E | Phospholipase C-γ | 92 |
| | $N_{779}$ P G F Y* V E A N | Phospholipase C-γ | 92 |
| | $E_{1251}$ A R Y* Q Q P F E D F R | Phospholipase C-γ | 92 |
| | $T_{1}$ D V E T T Y* A D F I A S G | Protein kinase inhibitor protein | 61, 73 |
| Insulin receptor | $R_{1145}$ D I Y* E T D Y* Y* R K G G K G | Autophosphorylation (mouse) | 61, 64 |
| | $R_{1143}$ D I Y* E T D Y* Y* R K G G K G | Autophosphorylation (human) | 64, 113 |
| | $K_{1313}$ R S Y* E E H I P Y* T H M N G G K | Autophosphorylation (human) | 64, 113 |
| | E N F D D Y* M K E | pp15 [422(aP2) protein] | 114 |
| | $D_{93}$ K D G N G Y* I S A A E | Calmodulin | 67, 68 |
| PDGF receptor | $D_{746}$ E S V D Y* V P M L D M K | Autophosphorylation | 65 |
| | $D_{854}$ S N Y* I S K | Autophosphorylation | 65 |
| pp50$^{v\text{-}abl}$ | $I_{14}$ E N E E Q E Y* V Q T V K S S | Lipocortin I (p35, calpactin II), Human | 81 |
| pp60$^{c\text{-}src}$ | $L_{308}$ E E E E E E Y* M P M E D L Y | Polyomavirus middle T antigen | 61, 93 |
| | $S_{243}$ L L S N P T Y* S V M R S H S | Polyomavirus middle T antigen[#] | 61 |
| | $I_{14}$ E N E E Q E Y* V Q T V K S S | Lipocortin I (p35, calpactin II), human | 81 |
| | $R_{409}$ L I E D N E Y* T A R Q G A K | Autophosphorylation | 143 |

(*continued*)

TABLE I (*continued*)

| Protein kinase | Phosphorylation site sequence | Protein | Refs.[b] |
|---|---|---|---|
| pp60$^{c\text{-}src}$ and/or unknown kinase(s) | $F_{520}$ T S T E P Q Y* Q P G E N L | pp60$^{c\text{-}src}$ | 61, 66, 143 |
| pp60$^{v\text{-}src}$ | $R_{409}$ L I E D N E Y* T A R Q G A K | Autophosphorylation | 61 |
| | $S_{336}$ G G K G G S Y* S Q A A C S D | HLA-B7 (α chain)# | 61 |
| | $S_{290}$ D R K G G S Y* S Q A A S S D | HLA-A2 (α chain)# | 61 |
| | $H_{16}$ S T P P S A Y* G S V K A Y T | Lipocortin II (p36, calpactin I heavy chain) | 61, 62 |
| p90$^{gag\text{-}yes}$ | $R_{417}$ L I E D N E Y* T A R Q G A K | Autophosphorylation# (residue numbers based on human sequence) | 61 |
| p56$^{lck}$ | $R_{387}$ L I E D N E Y* T A R E G A K | Autophosphorylation | 61 |
| p140$^{gag\text{-}fps}$ | $R_{417}$ Q E E D G V Y* A S T G G M K | Autophosphorylation | 61 |
| | $K_{231}$ Q V V E S A Y* E V I R L K G | Lactate dehydrogenase# (residue numbers based on chicken sequence) | 61 |
| | $S_{36}$ G A S T G I Y* E A L E L R | Enolase# (residue numbers based on human sequence) | 61 |
| p110$^{gag\text{-}fes}$ | $R_{493}$ E A A D G I Y* A A S G G L R | Autophosphorylation | 61, 149 |
| p85$^{gag\text{-}fes}$ | $R_{361}$ E E A D G V Y* A A S G G L R | Autophosphorylation | 61, 149 |
| p120$^{gag\text{-}abl}$ | $R_{386}$ L M T G D T Y* T A H A G A G | Autophosphorylation# (residue numbers based on human sequence) | 61 |
| Endogenous kinase (?p40) | $M_{1}$ E E L Q D D Y* E D D M E E N | Band 3 | 61, 74, 75 |
| p40 | $E_{669}$ E D G E R Y* D E D E E | Glycogen synthase (residue numbers based on human muscle sequence) | 74, 150 |
| Unknown kinase(s) | $F_{498}$ T A T E G Q Y* Q P Q P | p56$^{lck}$ | 143 |
| Unknown kinase | $K_{36}$ K R K S* G N S R E R | Avian retrovirus nucleocapsid protein (pp12) | 77 |
| Unknown kinase | $K_{9}$ I G E G T Y* G V V Y K A R H K | cdc2$^{+}$ (pp34) | 106 |

[a] #, Residue numbers obtained from SWISSPROT protein data base; (P) denotes a phosphorylated residue which acts as a substrate specificity determinant.

[b] References for Tables I and II are combined at the end of the chapter.

[c] The phosphorylation sites for the $\beta$-adrenergic receptor kinase are inferred from mutagenesis studies.[50]

[d] Artificial protein substrates have not been included for casein kinase II.

[e] Only selected phosphorylation sites for cAMP-dependent protein kinase and protein kinase C have been included to illustrate the various recognition motifs.

[f] P. J. Robinson (unpublished result, 1990).

[g] Although the growth-associated H1 histone kinases appear to require proline residues for substrate recognition, so do a number of other kinases, including glycogen synthase kinase-3, growth factor-regulated kinase, mitogen-activated protein kinase, proline-directed protein kinase, and sperm-specific histone kinase.

[h] Threonine residues 340 and 342 are assumed sites based on stoichiometry of rhodopsin phosphorylation.

[i] Some doubt exists as to whether tyrosine phosphorylation of the various tyrosine kinases is due to autophosphorylation *in vivo*.

[j] Residue numbers for protein phosphatase-1 G subunit obtained from P. Tang, J. Bondor, and A. A. DePaoli-Roach (personal communication, 1990).

TABLE II
CONSENSUS PHOSPHORYLATION SITES: SPECIFICITY MOTIFS FOR PROTEIN KINASES

| Protein kinase | S:T ratio[a] | Motif[b] | Frequency |
|---|---|---|---|
| Calmodulin-dependent protein kinase II | 15:5 | X**R**XXS*/T* | 13/20[c] |
| | | X**R**XXS*/T***V** | 6/20 |
| Casein kinase I | 8:1 | S**(P)**XXS*/T* | 5/9[d] |
| Casein kinase II | 28:2 | S*/T*XX**E**X | 23/30 |
| | | S*/T*XX**D**X | 3/20[e] |
| cAMP-dependent protein kinase[f] | 40:6 | **R**XS* | 21/46 |
| | | **RR**XS* | 12/46 |
| | | **R**XXS* | 11/46 |
| | | **KR**XXS* | 2/46 |
| cGMP-dependent protein kinase | 7:3[g] | **R/K**XS*/T* | 9/10 |
| | | **R/K**XXS*/T* | 8/10 |
| | | **R/KR/K**XS*/T* | 7/10 |
| | | **R/K**XXXS*/T* | 5/10 |
| | | S*/T*X**R/K** | 2/10 |
| Glycogen synthase kinase-3 | 10:2 | S*XXX**S(P)** | 6/12[h] |
| Growth-associated histone H1 kinase (MPF, cdc2+/CDC28 protein kinases) | 7:8 | S*/T***P**X**K/R** | 6/15 |
| | | **K/R**S*/T***P** | 5/15 |
| | | S*/T***PK/R** | 4/15 |
| Phosphorylase kinase | 3:0 | **K/R**XXS***V/I** | 3/3 |
| Protein kinase C | 31:6 | S*/T*X**K/R** | 20/37[i] |
| | | **K/R**XXS*/T* | 13/37 |
| | | **K/R**XXS*/T*X**K/R** | 7/37 |
| | | **K/R**XS*/T* | 10/37 |
| | | **K/R**XS*/T*X**K/R** | 6/37 |

*(continued)*

TABLE II (*continued*)

| Protein kinase | S:T ratio[a] | Motif[b] | Frequency |
|---|---|---|---|
| Tyrosine kinase[j] | | | |
| EGF-receptor kinase | Tyrosine | X**E**/**D**Y*X | 7/14 |
| | | X**E**/**D**Y***I**/**L**/**V** | 5/14 |

[a] S:T ratio is for the total number of phosphorylation sites.

[b] Asterisks indicate the phosphorylated residue. Specificity determinants are shown in bold type.

[c] Three of 20 phosphorylation sites for calmodulin-dependent protein kinase II are on threonine residues, including both autophosphorylation sites.

[d] Assuming phosphorylation of Ser-3 in glycogen synthase directs Ser-7 phosphorylation.

[e] Eighteen of 30 Casein kinase II phosphorylation site sequences contain 3 consecutive acidic residues following the phosphorylated residue.

[f] cAMP-dependent protein kinase motifs and frequency data are derived from O. Zetterqvist, U. Ragnarsson, and L. Engström, *in* "Peptides and Protein Phosphorylation" (B. E. Kemp, ed.), p. 171. Uniscience CRC Press, Boca Raton, Florida, 1990, and from Table I in this chapter. The most striking feature of the cAMP-dependent protein kinase phosphorylation site sequences is the variability, with less than one-third corresponding to the **RR**XS* motif.

[g] The only autophosphorylation site included was Thr-58. Autophosphorylation at Ser-50, Ser-72,and Thr-84 occurs only following activation with cAMP.

[h] Assuming sequential phosphorylation of glycogen synthase and protein phosphatase-1 G subunit by glycogen synthase kinase-3.

[i] Twenty-three of 37 protein kinase C phosphorylation site sequences contain an adjacent hydrophobic residue on the COOH-terminal side of the phosphorylated residue.

[j] Apart from the EGF-receptor kinase, the tyrosine kinase substrate sequences do not reveal consensus recognition motifs. Relatively few phosphorylation site sequences are known for exogenous proteins and autophosphorylation sites may not reflect the specificity determinants required in substrates. A more informative indication of likely phosphorylation site arrangements may be drawn from this volume [10].

## References


[1] O. Zetterqvist, U. Ragnarsson, and L. Engström, *in* "Peptides and Protein Phosphorylation" (B. E. Kemp, ed.), p. 171. Uniscience CRC Press, Boca Raton, Florida, 1990.

[2] J. R. Cherry, T. R. Johnson, C. Dollard, J. R. Shuster, and C. L. Denis, *Cell* (*Cambridge, Mass.*) **56,** 409 (1989).

[3] D. B. Glass, *in* "Peptides and Protein Phosphorylation" (B. E. Kemp, ed.), p. 209. Uniscience CRC Press, Boca Raton, Florida, 1990.

[4] K.-F. J. Chan, M. O. Hurst, and D. J. Graves, *J. Biol. Chem.* **257,** 3655 (1982).

[5] R. B. Pearson, R. Jakes, M. John, and J. Kendrick-Jones, *FEBS Lett.* **168,** 108 (1984).

[6] G. Matsuda, Y. Suzuyama, T. Maita, and T. Umegane, *FEBS Lett.* **84,** 53 (1977).

[7] A. Kishimoto, K. Nishiyama, H. Nakanishi, Y. Uratsuji, H. Nomura, Y. Takeyama, and Y. Nishizuka, *J. Biol. Chem.* **260,** 12492 (1985).

[8] O. Koshio, Y. Akanuma, and M. Kasuga, *FEBS Lett.* **254,** 22 (1989).

[9] R. J. Davis, G. L. Johnson, D. J. Kelleher, J. K. Anderson, J. E. Mole, and M. P.Czech, *J. Biol. Chem.* **261,** 9034 (1986).

[10] M. Ikebe, D. J. Hartshorne, and M. Elzinga, *J. Biol. Chem.* **262,** 9569 (1987).

[11] J. M. Graff, D. J. Stumpo, and P. J. Blackshear, *J. Biol. Chem.* **264,** 11912 (1989).

[12] S. Jakes, T. G. Hastings, E. M. Reimann, and K. K. Schlender, *FEBS Lett.* **234,** 31 (1988).

[13] Reference deleted in proof.

[14] A. J. Czernik, D. T. Pang, and P. Greengard, *Proc. Natl. Acad. Sci. U.S.A.* **84,** 7518 (1987).

[15] H. K. B. Simmerman, J. H. Collins, J. L. Theibert, A. D. Wegener, and L. R. Jones, *J. Biol. Chem.* **261,** 13333 (1986).

[16] J. R. Woodgett, M. T. Davison, and P. Cohen, *Eur. J. Biochem.* **136,** 481 (1983).

[17] Y. Hashimoto and T. R. Soderling, *J. Biol. Chem.* **264,** 16524 (1989).

[18] D. J. Campbell, D. G. Hardie, and P. R. Vulliet, *J. Biol. Chem.* **261,** 10489 (1986).

[19] A. P. Doskeland, C. M. Schworer, S. O. Doskeland, T. D. Chrisman, T. R. Soderling, J. R. Corbin, and T. Flatmark, *Eur. J. Biochem.* **145,** 31 (1984).

[20] T. R. Soderling, C. M. Schworer, M. R. El-Maghrabi, and S. J. Pilkis, *Biochem. Biophys. Res. Commun.* **139,** 1017 (1986).

[21] S. Shoji, J. Ohnishi, T. Funakoshi, K. Fukunaga, E. Miyamoto, H. Ueki, and Y. Kubota, *J. Biochem.* (*Tokyo*) **102,** 1113 (1987).

[22] A. C. Nairn and H. C. Palfrey, *J. Biol. Chem.* **262,** 17299 (1987).

[23] T. Hunter, N. Ling, and J. A. Cooper, *Nature* (*London*) **311,** 480 (1984).

[24] C. House, R. E. H. Wettenhall, and B. E. Kemp, *J. Biol. Chem.* **262,** 772 (1987).

[25] J.-A. Girault, H. C. Hemmings, K. R. Williams, A. C. Nairn, and P. Greengard, *J. Biol. Chem.* **264,** 21748 (1989).

[26] S. P. Lees-Miller and C. W. Anderson, *J. Biol. Chem.* **264,** 2431 (1989).

[27] Y. Tashiro, S. Matsumura, N. Murakami, and A. Kumon, *Arch. Biochem. Biophys.* **233,** 540 (1984).

[28] T. A. J. Haystead, D. G. Campbell, and D. G. Hardie, *Eur. J. Biochem.* **175,** 347 (1988).

[29] L. A. Pinna, F. Meggio, and F. Marchiori, *in* "Peptides and Protein Phosphorylation" (B. E. Kemp, ed.), p. 145. Uniscience CRC Press, Boca Raton, Florida, 1990.

[30] H. Flotow and P. J. Roach, *J. Biol. Chem.* **264,** 9126 (1989).

[31] E. Erikson and J. L. Maller, *Second Messengers Phosphoproteins* **12,** 135 (1988).

[32] H. Brzeska, T. J. Lynch, B. Martin, and E. D. Korn, *J. Biol. Chem.* **264,** 19340 (1989).

[33] P. R. Vulliet, F. L. Hall, J. P. Mitchell, and D. G. Hardie, *J. Biol. Chem.* **264,** 16292 (1989).

[34] J. L. Countaway, I. C. Northwood, and R. J. Davis, *J. Biol. Chem.* **264,** 10828 (1989).

[35] D. B. Rylatt, A. Aitken, T. Bilham, G. D. Condon, N. Embi, and P. Cohen, *Eur. J. Biochem.* **107,** 529 (1980).

[36] P. Dent, D. G. Campbell, M. J. Hubbard, and P. Cohen, *FEBS Lett.* **248,** 67 (1989).

[37] S. P. Lees-Miller and C. W. Anderson, *J. Biol. Chem.* **264,** 17275 (1989).

[38] D. R. Colthurst, D. G. Campbell, and C. G. Proud, *Eur. J. Biochem.* **166,** 357 (1987).

[39] W. Kudlicki, R. E. H. Wettenhall, B. E. Kemp, R. Szyszka, G. Kramer, and B. Hardesty, *FEBS Lett.* **215,** 16 (1987).

[40] W. Rychlik, M. A. Russ, and R. E. Rhoads, *J. Biol. Chem.* **262,** 10434 (1987).

[41] P. Cohen, D. Yellowlees, A. Aitken, A. Donella-Deana, B. A. Hemmings, and P. J. Parker, *Eur. J. Biochem.* **124,** 21 (1982).

[42] D. G. Hardie, D. Carling, and A. T. R. Sim, *Trends Biochem. Sci.* **14,** 20 (1989).

[43] D. G. Hardie, D. Carling, S. Ferrari, P. S. Guy, and A. Aitken, *Eur. J. Biochem.* **157,** 553 (1986).

[44] P. Thompson and J. B. C. Findlay, *Biochem. J.* **220,** 773 (1984).
[45] R. Paxton, M. Kuntz, and R. A. Harris, *Arch. Biochem. Biophys.* **244,** 187 (1986).
[46] K. Montgomery and A. S. Mak, *J. Biol. Chem.* **259,** 5555 (1984).
[47] A. C. Borthwick, W. H. Holmes, and H. G. Nimmo, *FEBS Lett.* **174,** 112 (1984).
[48] R. A. Masaracchia, B. E. Kemp, and D. A. Walsh, *J. Biol. Chem.* **252,** 7109 (1977).
[49] P. R. Clarke and D. G. Hardie, *EMBO J.* **9,** 2439 (1990).
[50] W. P. Hausdorff, M. Bouvier, B. F. O'Dowd, G. P. Irons, M. G. Caron, and R. J. Lefkowitz, *J. Biol. Chem.* **264,** 12657 (1989).
[51] A. M. Edelman, D. K. Blumenthal, and E. G. Krebs, *Annu. Rev. Biochem.* **56,** 567 (1987).
[52] N. A. Morrice and R. E. H. Wettenhall, personal communication (1990).
[53] T. A. Langan, *Methods Cell Biol.* **19,** 127 (1978).
[54] T. A. Langan, J. Gautier, M. Lohka, R. Hollingsworth, S. Moreno, P. Nurse, J. Maller, and R. A. Sclafani, *Mol. Cell. Biol.* **9,** 3860 (1989).
[55] H. C. Hemmings, J.-A. Girault, K. R. Williams, M. B. LoPresti, and P. Greengard, *J. Biol. Chem.* **264,** 7726 (1989).
[56] N. Murakami, G. Healy-Louie, and M. Elzinga, *J. Biol. Chem.* **265,** 1041 (1990).
[57] C. S. Hill, L. C. Packman, and J. O. Thomas, *EMBO J.* **9,** 805 (1990).
[58] C. E. M. Voorter, J. W. M. Mulders, H. Bloemendal, and W. W. de Jong, *Eur. J. Biochem.* **160,** 203 (1986).
[59] C.E. M. Voorter, W. A. de Haard-Hoekman, E. S. Roersma, H. E. Meyer, H. Bloemendal, and W. W. de Jong, *FEBS Lett.* **259,** 50 (1989).
[60] M. Ikebe, D. J. Hartshorne, and M. Elzinga, *J. Biol. Chem.* **261,** 36 (1986).
[61] R. L. Geahlen and M. L. Harrison, *in* "Peptides and Protein Phosphorylation" (B. E. Kemp, ed.), p. 239. Uniscience CRC Press, Boca Raton, Florida, 1990.
[62] J. R. Glenney and B. F. Tack, *Proc. Natl. Acad. Sci. U.S.A.* **82,** 7884 (1985).
[63] B. K. De, K. S. Misono, T. J. Lukas, B. Mroczkowski, and S. Cohen, *J. Biol. Chem.* **261,** 13784 (1986).
[64] J. R. Flores-Riveros, E. Sibley, T. Kastelic, and M. D. Lane, *J. Biol. Chem.* **264,** 21557 (1989).
[65] A. Kazlauskas and J. A. Cooper, *Cell (Cambridge, Mass.)* **58,** 1121 (1989).
[66] M. Okada and H. Nakagawa, *J. Biol. Chem.* **264,** 20886 (1989).
[67] J. P. Laurino, J. C. Colca, J. D. Pearson, D. B. DeWald, and J. M. McDonald, *Arch. Biochem. Biophys.* **265,** 8 (1988).
[68] D. M. Watterson, F. Sharief, and T. C. Vanaman, *J. Biol. Chem.* **255,** 962 (1980).
[69] B. E. Kemp, *J. Biol. Chem.* **254,** 2638 (1979).
[70] S. J. Yeaman, *Biochim. Biophys. Acta* **1052,** 128 (1990).
[71] C. F. B. Holmes, J. Kuret, A. A. K. Chisholm, and P. Cohen, *Biochim. Biophys. Acta* **870,** 408 (1986).
[72] B. A. Hemmings, A. Aitken, P. Cohen, M. Rymond, and F. Hofmann, *Eur. J. Biochem.* **127,** 473 (1982).
[73] S. M. Van Patten, G. J. Heisermann, H.-C. Cheng, and D. A. Walsh, *J. Biol. Chem.* **262,** 3398 (1987).
[74] A. M. Mahrenholz, P. Votaw, P. J. Roach, A. A. Depaoli-Roach, T. F. Zioncheck, M. L. Harrison, and R. L. Geahlen, *Biochem. Biophys. Res. Commun.* **155,** 52 (1988).
[75] S. A. Dekowski, A. Rybicki, and K. Drickamer, *J. Biol. Chem.* **258,** 2750 (1983).
[76] D. B. Bylund and E. G. Krebs, *J. Biol. Chem.* **250,** 6355 (1975).
[77] X. Fu, P. T. Tuazon, J. A. Traugh, and J. Leis, *J. Biol. Chem.* **263,** 2134 (1988).
[78] J. Fujii, A. Ueno, K. Kitano, S. Tanaka, M. Kadoma, and M. Tada, *J. Clin. Invest.* **79,** 301 (1987).
[79] T. Patschinsky, T. Hunter, and B. M. Sefton, *J. Virol.* **591,** 73 (1986).

[80] C. H. Michnoff, B. E. Kemp, and J. T. Stull, *J. Biol. Chem.* **261,** 8320 (1986).
[81] L. Varticovski, S. B. Chahwala, M. Whitman, L. Cantley, D. Schindler, E. P. Chow, L. K. Sinclair, and R. B. Pepinsky, *Biochemistry* **27,** 3682 (1988).
[82] D. D. Schlaepfer and H. T. Haigler, *Biochemistry* **27,** 4253 (1988).
[83] K. L. Gould, J. R. Woodgett, C. M. Isacke, and T. Hunter, *Mol. Cell. Biol.* **6,** 2738 (1986).
[84] P. Heldin and E. Humble, *Arch. Biochem. Biophys.* **252,** 49 (1987).
[85] A. Safran, R. Sagi-Eisenberg, D. Neumann, and S. Fuchs, *J. Biol. Chem.* **262,** 10506 (1987).
[86] S. Kitamura, S. Ando, M. Shibata, K. Tanabe, C. Sato, and M. Inagaki, *J. Biol. Chem.* **264,** 5674 (1989).
[87] B. Luscher, E. Christensen, D. W. Litchfield, E. G. Krebs, and R. N. Eisenman, *Nature (London)* **344,** 517 (1990).
[88] B. Luscher, E. A. Kuenzel, E. G. Krebs, and R. N. Eisenman, *EMBO J.* **8,** 1111 (1989).
[89] P. Cohen, *Nature (London)* **296,** 613 (1982).
[90] H. Olsson, P. Stralfors, and P. Belfrage, *FEBS Lett.* **209,** 175 (1986).
[91] A. J. Garton, D. G. Campbell, D. Carling, D. G. Hardie, R. J. Colbran, and S. J. Yeaman, *Eur. J. Biochem.* **179,** 249 (1989).
[92] J. W. Kim, S. S. Sim, U.-H. Kim, S. Nishibe, M. I. Wahl, G. Carpenter, and S. G. Rhee, *J. Biol. Chem.* **265,** 3940 (1990).
[93] D. A. Talmage, R. Freund, A. T. Young, J. Dahl, C. J. Dawe, and T. L. Benjamin, *Cell (Cambridge, Mass.)* **59,** 55 (1989).
[94] B. L. Margolis, I. Lax, R. Kris, M. Dombalagian, A. M. Honegger, R. Howk, D. Givol, A. Ullrich, and J. Schlessinger, *J. Biol. Chem.* **264,** 10667 (1989).
[95] S. Shoji, D. C. Parmelee, R. D. Wade, S. Kumar, L. H. Ericsson, K. A. Walsh, H. Neurath, G. L. Long, J. G. Demaille, E. H. Fischer, and K. Titani, *Proc. Natl. Acad. Sci. U.S.A.* **78,** 848 (1981).
[96] C. R. Lin, M. S. Kapiloff, S. Durgerian, K. Tatemoto, A. F. Russo, P. Hanson, H. Schulman, and M. G. Rosenfeld, *Proc. Natl. Acad. Sci. U.S.A.* **84,** 5962 (1987).
[97] G. Thiel, A. J. Czernik, F. Gorelick, A. C. Nairn, and P. Greengard, *Proc. Natl. Acad. Sci. U.S.A.* **85,** 6337 (1987).
[98] C. M. Schworer, R. J. Colbran, J. R. Keefer, and T. R. Soderling, *J. Biol. Chem.* **263,** 13486 (1988).
[99] M. K. Bennett and M. B. Kennedy, *Proc. Natl. Acad. Sci. U.S.A.* **84,** 1794 (1987).
[100] F. L. Hall, J. P. Mitchell, and P. R. Vulliet, *J. Biol. Chem.* **265,** 6944 (1990).
[101] M. M. King, T. J. Fitzgerald, and G. M. Carlson, *J. Biol. Chem.* **258,** 9925 (1983).
[102] T. J. Singh, A. Akutsuka, and K.-P. Huang, *J. Biol. Chem.* **257,** 13379 (1982).
[103] C. J. Fiol, A. Wang, R. W. Roeske, and P. J. Roach, *J. Biol. Chem.* **265,** 6061 (1990).
[104] S. Shenoy, J.-K. Choi, S. Bagrodia, T. D. Copeland, J. L. Maller, and D. Shalloway, *Cell (Cambridge, Mass.)* **57,** 763 (1989).
[105] L. J. Cisek and J. L. Corden, *Nature (London)* **339,** 679 (1989).
[106] K. L. Gould and P. Nurse, *Nature (London)* **342,** 39 (1989).
[107] A. K. Erickson, D. M. Payne, P. A. Martino, A. J. Rossomando, J. Shabanowitz, M. J. Weber, D. F. Hunt, and T. W. Sturgill, *J. Biol. Chem.* **265,** 19728 (1990).
[108] D. F. Carmichael, R. L. Geahlen, S. M. Allen, and E. G. Krebs, *J. Biol. Chem.* **257,** 10440 (1982).
[109] B. L. Hill, K. Drickamer, F. M. Brodsky, and P. Parham, *J. Biol. Chem.* **263,** 5499 (1988).
[110] G. M. C. Janssen, G. D. F. Maessen, R. Amons, and W. Moller, *J. Biol. Chem.* **263,** 11063 (1988).
[111] G. M. Walton, W. S. Chen, M. G. Rosenfeld, and G. N. Gill, *J. Biol. Chem.* **265,** 1750 (1990).

[112] Reference deleted in proof.

[113] H. E. Tornqvist, J. R. Gunsalus, R. A. Nemenoff, A. R. Frackelton, M. W. Pierce, and J. Avruch, *J. Biol. Chem.* **263,** 350 (1988).

[114] R. C. Hresko, M. Bernier, R. D. Hoffman, J. R. Flores-Riveros, K. Liao, D. M. Laird, and M. D. Lane, *Proc. Natl. Acad. Sci. U.S.A.* **85,** 8835 (1988).

[115] R. E. Lewis, G. P. Wu, R. G. MacDonald, and M. P. Czech, *J. Biol. Chem.* **265,** 947 (1990).

[116] S. Nakajo, Y. Masuda, K. Nakaya, and Y. Nakamura, *J. Biochem.* (*Tokyo*) **104,** 946 (1988).

[117] M. S. Barbosa, C. Edmonds, C. Fisher, J. T. Schiller, D. R. Lowy, and K. H. Vousden, *EMBO J.* **9,** 153 (1990).

[118] J. Kuret, J. R. Woodgett, and P. Cohen, *Eur. J. Biochem.* **151,** 39 (1985).

[119] T. Hunter, P. Angel, W. J. Boyle, R. Chiu, E. Freed, K. L. Gould, C. M. Isacke, M. Karin, R. A. Lindberg, and P. van der Geer, *Cold Spring Harbor Symp. Quant. Biol.* **53,** 131 (1988).

[120] W. J. Boyle, T. Smeal, L. H. K. Defize, P. Angel, J. R. Woodgett, M. Karin, and T. Hunter, *Cell* (*Cambridge,Mass.*) **64,** 573 (1991).

[121] R. E. Lewis, L. Cao, D. Perregaux, and M. P. Czech, *Biochemistry* **29,** 1807 (1990).

[122] A. Ullrich, J. R. Bell, E. Y. Chen, R. Herrera, R. M. Petruzzelli, T. J. Dull, A. Gray, L. Coussens, Y.-C. Liao, M. Tsubokawa, A. Mason, P. H. Seeburg, C. Grunfeld, O. M. Rosen, and J. Ramachandran, *Nature* (*London*) **313,** 756 (1985).

[123] J. R. Woodgett, K. L. Gould, and T. Hunter, *Eur. J. Biochem.* **161,** 177 (1986).

[124] D. A. Shackelford and I. S. Trowbridge, *J. Biol. Chem.* **261,** 8334 (1986).

[125] K. L. Gould, J. R. Woodgett, J. A. Cooper, J. E. Buss, D. Shalloway, and T. Hunter, *Cell* (*Cambridge, Mass.*) **42,** 849 (1985).

[126] S. J. Decker, B. Dorai, and S. Russell, *J. Virol.* **62,** 3649 (1988).

[127] J. Downward, Y. Yarden, E. Mayes, G. Scrace, N. Totty, P. Stockwell, A. Ullrich, J. Schlessinger, and M. D. Waterfield, *Nature* (*London*) **307,** 521 (1984).

[128] J. Rittenhouse, L. Moberly, and F. Marcus, *J. Biol. Chem.* **262,** 10114 (1987).

[129] J. Kuret, K. E. Johnson, C. Nicolette, and M. J. Zoller, *J. Biol. Chem.* **263,** 9149 (1988).

[130] Y. Hashimoto and T. R. Soderling, *Arch. Biochem. Biophys.* **278,** 41 (1990).

[131] N. J. Olson, R. B. Pearson, D. S. Needleman, M. Y. Hurwitz, B. E. Kemp, and A. R. Means, *Proc. Natl. Acad. Sci. U.S.A.* **87,** 2284 (1990).

[132] P. Tapley, A. Kazlauskas, J. A. Cooper, and L. R. Rohrschneider, *Mol. Cell. Biol.* **10,** 2528 (1990).

[133] P. van der Geer and T. Hunter, *Mol. Cell. Biol.* **10,** 2991 (1990).

[134] Y. Goldberg, C. Glineur, J.-C. Gasquiere, A. Ricouart, J. Sap, B. Vennstrom, and J. Ghysdael, *EMBO J.* **7,** 2425 (1988).

[135] Z. Ahmad, F.-T. Lee, A. DePaoli-Roach, and P. J. Roach, *J. Biol. Chem.* **259,** 8743 (1984).

[136] E. D. Apel, M. F. Byford, D. Au, K. A. Walsh, and D. R. Storm, *Biochemistry* **29,** 2330 (1990).

[137] J. Pines and T. Hunter, *New Biol.* **2,** 389 (1990).

[138] E. T. Kipreos and J. Y. J. Wang, *Science* **248,** 217 (1990).

[139] T. M. Martensen, B. M. Martin, and R. L. Kincaid, *Biochemistry* **28,** 9243 (1989).

[140] G. M. Walton, J. Spiess, and G. N. Gill, *J. Biol. Chem.* **260,** 4745 (1985).

[141] T. A. J. Haystead and D. G. Hardie, *Eur. J. Biochem.* **175,** 339 (1988).

[142] A. Ullrich, L. Coussens, J. S. Hayflick, T. J. Dull, A. Gray, A. W. Tam, J. Lee, Y. Yarden, T. A. Libermann, J. Schlessinger, J. Downward, E. L. V. Mayes, N. Wittle, M. D. Waterfield, and P. H. Seeburg, *Nature* (*London*) **309,** 418 (1984).

[143] J. A. Cooper, *in* "Peptides and Protein Phosphorylation" (B. E. Kemp, ed.), p. 85. Uniscience CRC Press, Boca Raton, Florida, 1990.

[144] H. U. Beuscher, M. W. Nickells, and H. R. Colten, *J. Biol. Chem.* **263,** 4023 (1988).

[145] G. J. Heisermann and G. N. Gill, *J. Biol. Chem.* **263,** 13152 (1988).

[146] M. L. Tremblay, C. J. McGlade, G. E. Gerber, and P. E. Branton, *J. Biol. Chem.* **263,** 6375 (1988).

[147] P. Jeno, L. M. Ballou, I. Novak-Hofer, and G. Thomas, *Proc. Natl. Acad. Sci. U.S.A.* **85,** 406 (1988).

[148] J. Krieg, J. Hofsteenge, and G. Thomas, *J. Biol. Chem.* **263,** 11473 (1988).

[149] A. Hampe, I. Laprevotte, F. Galibert, L. A. Fedele, and C. J. Sherr, *Cell (Cambridge, Mass.)* **30,** 775 (1982).

[150] M. F. Browner, K. Nakano, A. G. Bang, and R. J. Fletterick, *Proc. Natl. Acad. Sci. U.S.A.* **86,** 1443 (1989).

# Section II

## Assays of Protein Kinases

# [4] Enzyme Activity Dot Blots for Assaying Protein Kinases

*By* CLAIBORNE V. C. GLOVER and C. DAVID ALLIS

Assay of protein kinase activity generally requires two distinct steps: (1) transfer of the (labeled) terminal phosphoryl group of the nucleoside triphosphate donor to the protein or peptide substrate and (2) separation of the phosphorylated product from unutilized nucleotide. Step 1 is generally carried out in solution, with both the enzyme and the substrate in the liquid phase. Step 2 is usually accomplished by trichloroacetic acid (TCA) precipitation, by sodium dodecyl sulfate (SDS) gel electrophoresis, or by binding the labeled product to a solid support such as phosphocellulose paper[1] or nitrocellulose membrane.[2]

Perhaps not surprisingly, step 1 can also be carried out with either the enzyme or the substrate immobilized on a solid support. For example, complex protein mixtures can be fractionated by SDS gel electrophoresis, blotted onto nitrocellulose, and then tested as potential substrates by incubating the sheet with a nonspecific blocking agent followed by the desired protein kinase plus labeled ATP.[3] Carrying this idea one step further, we have designed a protein kinase assay (enzyme activity dot blot) in which both the kinase *and* its substrate are immobilized on nitrocellulose.[4] This assay, originally worked out for casein kinase II, is comparable to the standard liquid assay with respect to sensitivity, linearity, etc., and has a number of advantageous features. Related solid-phase assays are described elsewhere in this volume [5], [34], and [35].

## Principle of Assay

Samples containing the protein kinase activity to be assayed are immobilized on a nitrocellulose filter, either by manual spotting or by vacuum filtration using a commercial slot or dot blotter. The filter is then incubated in an appropriate protein substrate, which binds to remaining binding sites on the filter, and then with radioactive ATP. The reaction is stopped and

[1] D. B. Glass, R. A. Masaracchia, J. R. Feramisco, and B. E. Kemp, *Anal. Biochem.* **87,** 566 (1978).

[2] J. D. Buxbaum and Y. Dudai, *Anal. Biochem.* **169,** 209 (1988).

[3] F. Valtorta, W. Schiebler, R. Jahn, B. Ceccarelli, and P. Greengard, *Anal. Biochem.* **158,** 130 (1986).

[4] C. D. Allis, L. G. Chicoine, C. V. C. Glover, E. M. White, and M. A. Gorovsky, *Anal. Biochem.* **159,** 58 (1986).

unutilized ATP simultaneously removed simply by washing the filter. Incorporation is determined by autoradiography and/or liquid scintillation counting. While perhaps counterintuitive, it appears that the bound kinase and/or its bound substrate retain sufficient mobility on nitrocellulose to allow phosphorylation of the substrate to take place. Indeed, simple calculations indicate that each bound enzyme molecule catalyzes multiple phosphorylation events.

*Solutions*

Buffer A: 25 m*M* Tris, 1 m*M* EDTA, 100 m*M* NaCl, pH 8.0
Buffer B: 25 m*M* Tris, 1 m*M* EDTA, 200 m*M* NaCl, 10% (v/v) glycerol, 1 m*M* dithiothreitol, pH 8.0
Substrate for casein kinase II: Hydrolyzed and partially dephosphorylated casein (Sigma, St. Louis, MO; C4765) at 10 mg/ml in buffer A
Reaction mix for casein kinase II: 25 m*M* Tris, pH 8.5, 100 m*M* NaCl, 10 m*M* $MgCl_2$, 1 m*M* dithiothreitol, 0.1 $\mu M$ [$\gamma$-$^{32}$P]ATP (100 Ci/mmol). Essentially identical results are achieved using 10 $\mu M$ [$\gamma$-$^{32}$P]ATP at 1 Ci/mmol
TE: 10 m*M* Tris, 1 m*M* EDTA, pH 8.0

## Procedures

At no point during the following manipulations should the nitrocellulose membrane be allowed to air dry once it has been wet. The filter should always be either submerged in buffer or maintained on a sheet of buffer-saturated filter paper, and all transfers should be made with reasonable haste. This caveat does not apply to the one-sided "drying" which occurs when nitrocellulose is held on damp filter paper. The latter presents no problems, and bound kinase activity appears to be quite stable in this situation.

### *Manual Method*

1. Cut a sheet of nitrocellulose (Schleicher and Schuell, Keene, NH; BA-85) to the desired size and rule into 1-cm squares using a straight edge and a No. 2 pencil.

2. Wet the nitrocellulose with buffer A and place it on a sheet of Whatman 3MM filter paper saturated with the same buffer. Allow any puddles on the nitrocellulose to drain through into the filter paper.

3. Dilute samples to be assayed to an appropriate concentration in buffer B containing 1 mg/ml bovine serum albumin (BSA; Sigma A7030).

Each dot applied to the nitrocellulose should contain from 0.5 to 50 p*M* units of protein kinase activity (1 p*M* unit represents that amount of enzyme transferring 1 pmol of $^{32}P$ to substrate/min). Assuming a typical protein kinase specific activity of 1 $\mu$mol/min/mg, this represents approximately 0.5 to 50 ng of pure kinase per dot.

4. Using a pipetter, spot 1 to 5 $\mu$l of each sample onto the nitrocellulose. Place each dot at the center of a ruled square. Wait until all dots have drained in before proceeding to step 5.

5. Briefly rinse the nitrocellulose in buffer A and place it face up on a sheet of Parafilm. The latter should be taped by its edges onto a glass plate so as to obtain a completely flat surface. Cover the nitrocellulose with a solution (25 $\mu l/cm^2$) of a suitable protein substrate (1 to 10 mg/ml in buffer A) and incubate in a moist chamber for 30 min at 23°. For large sheets of nitrocellulose, it may be preferable to conduct this and the following incubation in a heat-sealed bag.

6. Wash the strip twice in 100 ml of buffer A, place on a fresh sheet of Parafilm, and cover with the reaction mix (25 $\mu l/cm^2$). Incubate in a moist chamber for 5 to 30 min at 23°.

7. Wash the strip four times in 100 ml of buffer A, once in 100 ml of TE, and air dry. Wrap the dried strip in Saran wrap and autoradiograph on Cronex 4 (Du Pont, Wilmington, DE) or XAR-5 (Kodak, Rochester, NY) X-ray film for 15 min to 2 hr at room temperature using a Lightening Plus intensifying screen. The autoradiograph provides a qualitative record of the results and is useful for diagnosing background problems, bleeding of high-activity dots, etc.

8. To quantitate the assay, cut individual squares (dots) from the nitrocellulose, place them in scintillation vials with 0.2 ml of Scintiverse E (Fisher, Pittsburgh, PA), and count in a liquid scintillation counter. Cerenkov counts can be measured instead (by omitting the fluor) but at a cost of a 40% reduction in counting efficiency. If available, an area detector (e.g., Betascope 603 blot analyzer; Betagen Corp., Waltham, MA) can be used to provide both qualitative and quantitative analysis of the results.

9. If desired, the identity of the protein substrate(s) phosphorylated in a dot assay can be ascertained by sodium dodecyl sulfate (SDS) gel electrophoresis, essentially as described by Allis *et al.* for histone acetyltransferase activity.[4] Briefly, cut out the labeled dots, boil them (including the paper) in 30 $\mu$l of SDS sample buffer, and analyze the boiled samples by SDS gel electrophoresis, Coomassie Blue staining, and autoradiography. For autoradiography, wrap the wet gel in Saran wrap and autoradiograph at room temperature with an intensifying screen. Alternatively, dry the gel on a commercial gel drier, wrap the dried gel in Saran wrap, and autoradiograph at $-80°$ with screen.

*Filtration Method*

A commercial dot- or slot-blotting apparatus (e.g., VacuDot-VS; American Bionetics, Inc., Hayward, CA) can be used instead of a manual pipetter to apply samples to the nitrocellulose. In this case, steps 1 through 4 are modified as follows.

1. Cut a sheet of nitrocellulose to the size required to fit the apparatus. It is not necessary to rule the nitrocellulose because the sample wells leave an imprint which is a sufficient guide for cutting out the samples at step 8.
2. Wet the nitrocellulose with buffer A and assemble the apparatus as described by the manufacturer. Apply gentle suction to remove any liquid present in the wells. Do not overdry. Disconnect the vacuum.
3. Dilute samples to be assayed to an appropriate concentration in buffer B plus 10 $\mu$g/ml BSA. As above, each dot applied to the nitrocellulose should contain from 0.5 to 50 p$M$ units of protein kinase activity.
4. Pipette 100 to 500 $\mu$l of each sample into a well of the blotting apparatus. When all wells have been filled, apply gentle suction to pull samples through the nitrocellulose. As soon as all wells are empty, disassemble the apparatus and proceed immediately to step 5 above.

## Potential Problems and Limitations

Most problems with the technique can be traced to overloading the nitrocellulose paper, either with total protein or with enzymatic activity. Nitrocellulose binds BSA linearly up to 50 $\mu g/cm^2$ and has an absolute capacity of 500 $\mu g/cm^2$.[5] The concentration of BSA in the diluents described above yields a maximum of 5 $\mu$g/slot or dot, which is equivalent to about 50 $\mu g/cm^2$. This load appears to leave more than adequate binding capacity for the protein substrate. If the samples themselves contribute significantly to the protein load, it may be necessary to reduce the concentration of BSA used in the diluent. Care must also be taken to avoid excessive input of protein kinase activity because the dot assay begins to deviate from linearity beyond about 50 p$M$ units/dot (as does the standard liquid assay). In addition, bleeding of high-activity dots may contaminate adjacent dots on the nitrocellulose.

An appropriate substrate for the enzyme to be assayed obviously must be available in adequate quantity. Substrates which are in short supply may be bound to nitrocellulose at concentrations well below 1 mg/ml but at the cost of a reduced signal intensity. For most protein substrates, binding to nitrocellulose should not be a problem. However, binding of

[5] D. C. Hinkle and M. J. Chamberlin, *J. Mol. Biol.* **70,** 157 (1972).

peptide substrates may present difficulties, and the feasibility of using such substrates has not yet been investigated.

## Applicability to Other Protein Kinases

In addition to casein kinase II, the catalytic (C) subunit of cAMP-dependent protein kinase from bovine heart (Sigma P2645) can also be assayed by enzyme activity dot blotting (C. Glover, unpublished observations). With hydrolyzed, partially dephosphorylated casein as substrate, the only variation required in the method is adjustment of the various buffers to accommodate the stability and activity requirements of the C subunit (the pH of all solutions is set at 7.5 and NaCl is omitted from the reaction mix). With these modifications the background, sensitivity, and efficiency of the solid-phase assay are remarkably similar to those of the standard liquid assay, as reported earlier for casein kinase II.[4] Given that these two kinases are distantly related[6] and, furthermore, that casein kinase II is heterotetrameric while the C subunit is monomeric, it appears probable that many protein kinases can be efficiently assayed by enzyme activity dot blotting. We note that a variety of other enzymatic activities have been found to assay efficiently following immobilization on nitrocellulose, including histone acetyltransferase[4] and $\beta$-1,3- and $\beta$-1,4-glucan synthases.[7]

## Applications

Since enzyme activity dot blots are comparable to standard liquid assays with respect to sensitivity, range, linearity, etc., they can be substituted for the latter in most routine assays. In certain circumstances, dot blots will be superior to the liquid assay because of the ease with which enzyme activity can be concentrated from dilute solution and the ease with which buffer exchange can be carried out once the enzyme is immobilized on the filter. These features make dot blots particularly useful for monitoring enzyme purification, where fractions are often of low activity and/or contaminated with buffer salts, inhibitors or activators used to elute activity from affinity columns, etc. The nitrocellulose-based assay can also be used to analyze the distribution of protein kinase activity in a native gel following gel electrophoresis.[4] The gel is blotted by standard techniques onto a nitrocellulose membrane, and the membrane is then incubated with a substrate followed by labeled ATP, exactly as described

[6] S. K. Hanks, A. M. Quinn, and T. Hunter, *Science* **241,** 42 (1988).

[7] P. Nodet, J. Grange, and M. Ferre, *Anal. Biochem.* **174,** 662 (1988).

above. This technique has been successfully used to detect protein kinase activity (casein kinase II) in crude extracts of whole cells[4] and could potentially be used to monitor tissue- and developmental stage-specific expression, to identify and characterize electrophoretic variants, etc. The method can also potentially be used to identify cDNA clones encoding protein kinase catalytic subunits in cDNA expression libraries or to screen for variants following mutagenesis of previously cloned protein kinase cDNAs.

### Acknowledgments

This work was supported by NIH Grants HD 16259 and GM 40922 to C.D.A. and NIH Grant GM 33237 to C.V.C.G.

# [5] Solid-Phase Protein-Tyrosine Kinase Assay

*By* DINKAR SAHAL, SHU-LIAN LI, and YOKO FUJITA-YAMAGUCHI

## Introduction

Protein kinases play an important role in signal transduction mechanisms by regulating a variety of cellular enzyme activities. Since tyrosine-specific protein kinase (TPK) activity is much less abundant than serine/threonine (Ser/Thr)-specific protein kinase activity in cells, determination of TPK activity in crude cell extracts or column chromatography eluates has been difficult. Protein kinase assays generally require either trichloroacetic acid washing to remove excess [$\gamma$-$^{32}$P]ATP,[1] while keeping all phosphorylated macromolecules precipitated on a filter paper, or washing with mild acid[2] to allow all basic protein or peptide substrates to bind to phosphocellulose paper, while the anionic [$\gamma$-$^{32}$P]ATP is washed away. Synthetic tyrosine-containing polymers are used by many investigators since they are specific substrates for TPKs and are much less expensive than a peptide substrate such as angiotensin II. However, polymers have disadvantages over small peptides since trichloroacetic acid precipitates endogenously phosphorylated proteins along with the phosphorylated substrate polymers. Thus, it would be helpful to develop a rapid, economical method

[1] J. D. Corbin and E. M. Reimann, *Methods in Enzymology* **38,** 287–290 (1974).
[2] R. Roskoski, Jr., *Methods in Enzymology* **99,** 3–6 (1983).

for the separation of high endogenous protein phosphorylation background from phosphorylation of added tyrosine-containing substrates.

A new solid-phase TPK assay has been developed to measure TPK activity of crude enzyme preparations.[3] The phosphorylation reaction is performed in a well of a polyacrylamide gel into which a TPK specific polymeric substrate such as $(Glu,Tyr)_n$, 4:1 has been immobilized by copolymerization. Incorporation of $^{32}P$ into the immobilized TPK-specific substrate is detected by either autoradiography or liquid scintillation counting of the gel pieces excised from the bottom of the well.

This solid-phase TPK assay is different from the assays previously described by others[4-6] in that previous methods used polyacrylamide gel electrophoresis to separate enzymes and detected the enzyme activity in the gel by overlaying $[\gamma\text{-}^{32}P]ATP$ and protein substrates on the gels.[7,8] In design of external appearance, the gels used here look like ouchterlony wells or the wells used in radial enzyme diffusion methods. However, by way of providing an immobilized substrate on the gel wall to an enzyme in solution, the wells in this assay represent the actual sites of enzymatic reaction.

## Materials and Methods

### *Materials*

The following are required to perform the solid-phase assay.

Bottle caps [5.4 cm (inside diameter) × 1 cm; see Fig. 1A]
Flexible microtiter 96-well assay plates (Falcon 3912, Becton Dickinson, Lincoln Park, NJ) trimmed to the size shown in Fig. 1
Random copolymer [$(Glu,Tyr)_n$, 4:1 (P0275; Sigma, St. Louis, MO); molecular weight range, 20,000–50,000]
Acrylamide [Bethesda Research Laboratories (BRL), Gaithersburg, MD]
Bis (*N,N*-methylenebisacrylamide; Mallinckrodt, St. Louis, MO)
$[\gamma\text{-}^{32}P]ATP$ (New England Nuclear, Boston, MA)
Horizontal gel electrophoresis apparatus (e.g., BRL model H5)

All other chemicals are reagent grade.

[3] D. Sahal and Y. Fujita-Yamaguchi, *Anal. Biochem.* **182,** 37 (1989).
[4] O. Gabriel, this series, Vol. 22, p. 565.
[5] R. L. Geahlen, M. Anostario, Jr., P. S. Low, and M. L. Harrison, *Anal. Biochem.* **153,** 151 (1986).
[6] R. I. Glazer, G. Yu, and M. C. Knode, *Anal. Biochem.* **164,** 214 (1987).
[7] G. F. B. Schumancher and W. B. Schill, *Anal. Biochem.* **48,** 9 (1972).
[8] H. Lowenstein and A. Ingild, *Anal. Biochem.* **71,** 204 (1976).

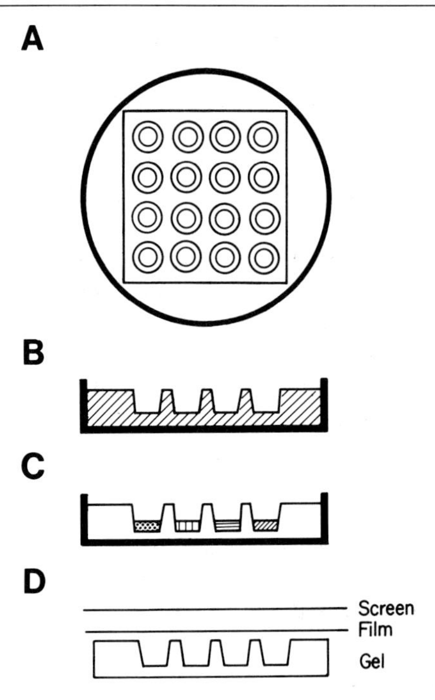


FIG. 1. A schematic representation of the solid-phase TPK assay. (A) A bottle cap with a microtiter plate mold. (B) A cross-section of a uniform-substrate gel. Hatched lines indicate an immobilized TPK substrate. (C) A cross-section of a substrate-layered gel. Each well indicates either different concentrations of a TPK substrate or different TPK substrates. (D) How to produce an autoradiogram of the gel after the kinase reaction.

### *Preparation of Substrate Gels*

Two types of substrate gels are described. Uniform-substrate gels are easy to prepare and suitable, for example, to screen TPK inhibitors and activators, whereas substrate-layered gels allow more versatile experiments, including determination of $K_m$ values and substrate specificity since different concentrations of substrates or different substrates can be prepared in the wells of the same gel.

*Uniform-Substrate Gel (Fig. 1B).* To prepare a 7.5% (w/v) acrylamide gel containing 2 m*M*[9] substrate, 3.4 ml of 22.2% acrylamide–0.6% Bis, 6.4 ml of 50 m*M* Tris-HCl buffer, pH 7.4, containing 15.4 mg of $(Glu,Tyr)_n$, 4 : 1 and 0.1 ml of 10% ammonium persulfate, are mixed. Ten microliters of TEMED is added just before this solution is poured into the bottle-cap

[9] The molecular weight of a unit $(Glu,Tyr)_n$, 4 : 1 has been used for calculating the concentration of tyrosine-containing polymers.

cast. The titer plate mold is placed on the gel solution as seen in Fig. 1A. Butanol is layered on top of the gel solution to prevent air contact and ~1 g of weight is placed on the titer plate to keep the plastic well embedded in the gel solution. The gel is kept for 2 hr at 25° to allow for polymerization and overnight at 4° to ensure all reactive groups are inactivated.

*Substrate-Layered Gel (Fig. 1C).* First, a 7.5% acrylamide gel is prepared by mixing 6.7 ml of 22.2% acrylamide–0.6% Bis, 12.9 ml of 50 m*M* Tris-HCl buffer, pH 7.4, and 0.2 ml of 10% ammonium persulfate. Four milliliters of the acrylamide solution is mixed with 4 μl of TEMED and poured into the bottle cap. After ~15 min at 25°, ~1 ml of unpolymerized solution is removed. The titer plate mold is placed on the gel bed with ~4 g of weight, and then 10 ml of the acrylamide solution mixed with 10 μl of TEMED is layered. Butanol is added on top of the gel solution to prevent air contact. The polymerization reaction is allowed to proceed for ~20 min at 25°. The titer plate mold is taken out and the wells are washed with the Tris-HCl buffer.

In each well, 100 μl of an acrylamide solution containing TPK substrates of varying concentrations are polymerized. For example, the two-layered gel containing different substrate concentrations shown in Fig. 3 is prepared as follows:

| Final substrate concentration (m*M*) | Substrate + gel solution | |
|---|---|---|
| 2 | 0.5 ml of 6 m*M* $(Glu,Tyr)_n$, 4:1 | + 1 ml of 7.5% gel solution |
| 1 | 0.75 ml of 2 m*M* solution | + 0.75 ml of 7.5% gel solution |
| 0.5 | 0.75 ml of 1 m*M* solution | + 0.75 ml of 7.5% gel solution |
| 0.25 | 0.75 ml of 0.5 m*M* solution | + 0.75 ml of 7.5% gel solution |
| 0.125 | 0.75 ml of 0.25 m*M* solution | + 0.75 ml of 7.5% gel solution |

It is important to put a reference mark on the gel before proceeding further. A mechanical cut, a puncture by a thin needle, or entrapment of a dye by microinjection and sealing the hole can be used as a reference mark.

*Storage of Acrylamide Gels.* The gels can be preapred as above and stored in a sealed bag at −20° for ~50 days or at 4° for ~10 days.

### *Kinase Assay*

*Enzyme Reaction Mixtures.* Enzyme solution is prepared in 30 μl of 50 m*M* Tris-HCl buffer, pH 7.4, containing 40 μ*M* [γ-$^{32}$P]ATP (~12,000 cpm/pmol), 15 m*M* $MgCl_2$, 2 m*M* $MnCl_2$, and 0.001% (w/v) Bromphenol Blue.[10]

[10] Bromphenol blue helps deliver the solution into a well accurately.

*Reaction.* The phosphorylation reaction is initiated by placing the enzyme solution in a well made of the acrylamide gel in a bottle cap. After 40 min at 25°, the reaction is terminated by the addition of 10 $\mu$l of 20 m$M$ unlabeled ATP.

*Washing Gels.* The enzyme reaction mixtures are removed from the wells. The gel, while still in the bottle cap, is washed for 10 min four times in 0.5 liter of water. The gel is placed on the flat platform of a horizontal gel electrophoresis apparatus (e.g., BRL model H5). The wells are filled with 50 $\mu$l of Laemmli's sample buffer containing 5 m$M$ ATP.[11] The gel is electrophoresed in 25 m$M$ Tris–glycine buffer, pH 8.3, containing 0.1% (w/v) sodium dodecyl sulfate (SDS) for ~2 hr at 100 V until the dye marker completely moves away. The gel is then fixed in 50% methanol–10% acetic acid (v/v) for 30 min, and washed in 25% methanol–10% acetic acid (v/v) or water for 4 hr to overnight.

*Autoradiography.* The gel is exposed to an X-ray film [Kodak (Rochester, NY) XAR5] with an intensifying screen as shown in Fig. 1D. The solution in the wells should be removed to allow efficient autoradiography. If a film is placed at the bottom of the gel, autoradiography cannot occur due to shielding by the acrylamide gel.

*Quantitation.* After autoradiography, the gel is frozen on dry ice and the bottoms of the wells are cut out using a cork borer that has a diameter slightly smaller than that of the well. The gel pieces are placed in vials containing 4 ml of scintillation solution (Ecoscint; National Diagnostics, Manville, NJ). Incorporation of $^{32}$P into the immobilized substrate is quantitated by liquid scintillation counting. A quantitative assessment of the gel can also be obtained by scanning the gel on a radioisotopic scanner.

## Solid-Phase Tyrosine-Specific Protein Kinase Assays Using Human Placenta Cytosolic Preparation

### *Uniform-Substrate Gel*

Figure 2 represents a typical autoradiogram showing the results of the solid-phase TPK assay. The gel contained 2 m$M$ (Glu,Tyr)$_n$, 4:1. TPK activity of the crude placenta enzyme preparation at different concentrations was measured in triplicate. Groups 1, 2, 3, and 4 show the results of 1, 2, 5, and 0 $\mu$l of the enzyme fraction. Quantitative determination of $^{32}$P incorporation into the immobilized substrate revealed 1,669 ± 127 cpm, 5480 ± 356 cpm, and 15,243 ± 1306 cpm for 1, 2, and 5 $\mu$l of the enzyme fraction, respectively.

[11] U. K. Laemmli, *Nature (London)* **227,** 680 (1970).

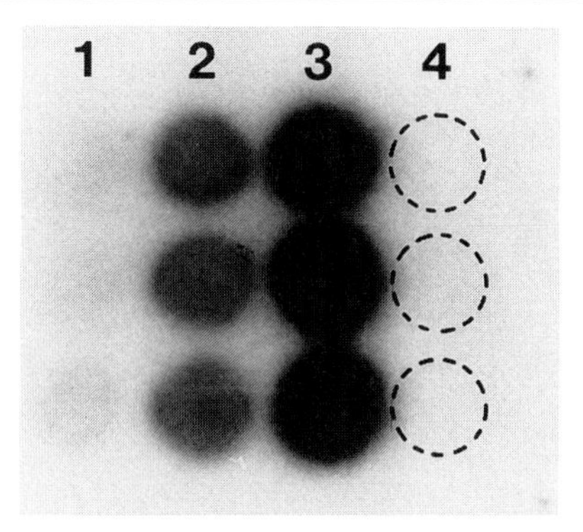


FIG. 2. Phosphorylation of immobilized TPK-specific substrate by crude placenta TPKs using a uniform-substrate gel containing 2 m$M$ (Glu,Tyr)$_n$, 4 : 1. Correlation between enzyme concentration and $^{32}$P incorporation into the substrate was measured. Enzyme preparation used was 30% (w/v) $(NH_4)_2SO_4$ supernatant of human placental 100,000 *g* cytosolic fraction [Y. Fujita-Yamaguchi *et al., J. Biol. Chem.* **258,** 5045 (1983)]. Three wells each of 1, 2, 3, and 4 represent the results of phosphorylation reaction by 1, 2, 5, and 0 $\mu$l of the crude enzyme, respectively. The phosphorylation reaction was performed as described in the text. The gel was exposed to a film for 2 hr at −70° with an intensifying screen.

### *Substrate-Layered Gel*

Uniform-substrate gels have a drawback in that all the gel wells have a uniform concentration of substrate. In addition, a considerable gel mass which lies outside the well area consumes more substrate than the amount of substrate actually needed for the experiments. In order to overcome these problems, a 2-tier substrate-layered gel has been developed. Substrate-free wells are first made in the gel. Then into each well the acrylamide solutions containing the desired substrate or the desired concentration is overlayered and polymerization allowed to continue.

Figure 3 shows the result of the determination of a $K_m$ value using placenta crude cytosolic enzyme. In the acrylamide gel shown in Fig. 3A, each well contained a different concentration of (Glu,Tyr)$_n$, 4 : 1. Therefore, the same amount of enzyme reaction mixture produced a different level of $^{32}$P incorporation into the immobilized substrate in each well; well 2 (0 m$M$), 456 cpm; wells 3 (0.1 m$M$), 2017 cpm; wells 4 (0.125 m$M$), 2209 cpm; wells 5 (0.25 m$M$), 2666 cpm; wells 6 (0.5 m$M$), 3862 cpm; wells 7 (1 m$M$), 4675 cpm; wells 8 (2 m$M$), 4722 cpm. The results are expressed as a double-reciprocal plot (Fig. 3B), from which a $K_m$ value of 0.21 m$M$ is obtained.[12]

[12] Similarly, a $K_m$ value of 0.22 m$M$ is obtained for another immobilized substrate, (Glu,Ala,Tyr)$_n$, 6 : 3 : 1.

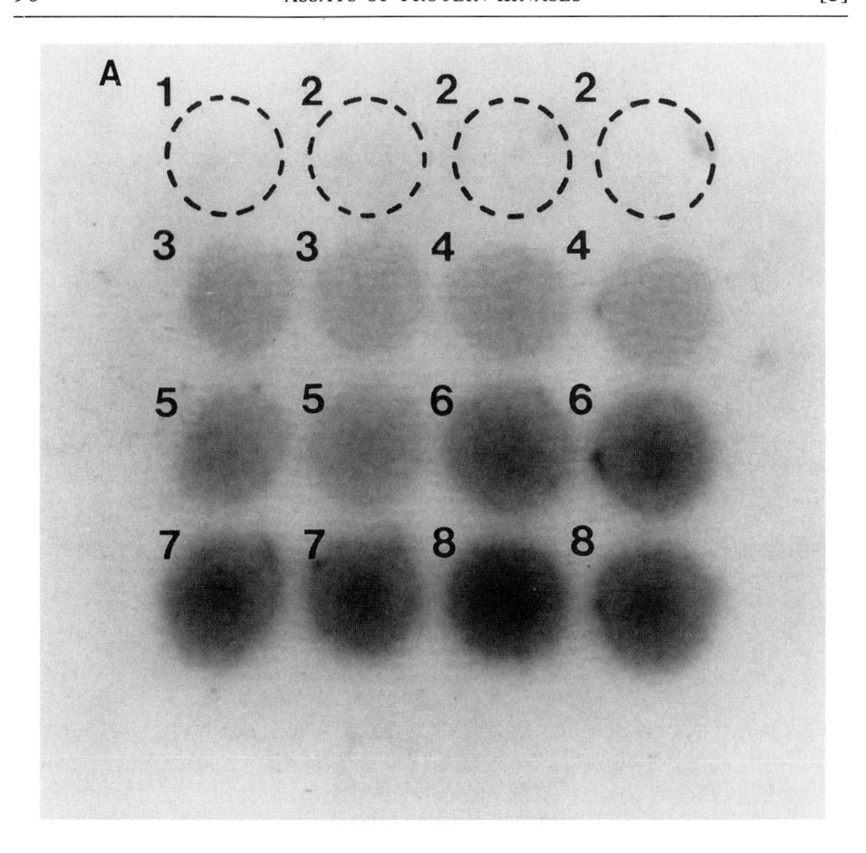


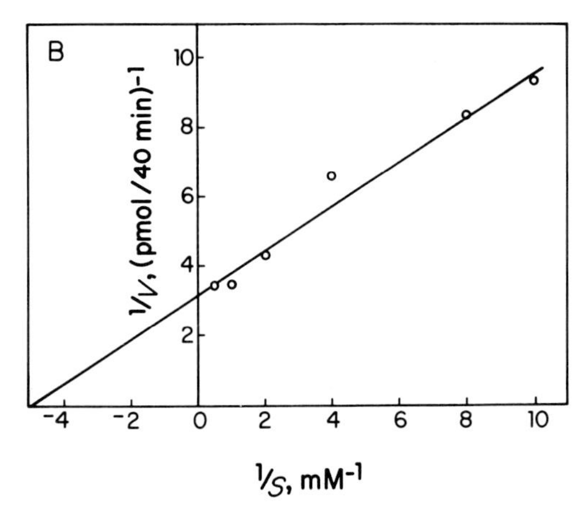


FIG. 3. Phosphorylation of immobilized TPK-specific substrate by crude placenta TPKs using a substrate-layered gel containing different concentrations of (Glu,Tyr)$_n$, 4 : 1 in wells.

## Application and Limitation

The major advantages of the solid-phase TPK assay are that multiple assays can be conveniently performed and that TPK activity can be detected in crude enzyme preparations without interference by Ser/Thr kinases whose catalytic activity accounts for over 99% of cellular protein phosphorylation. The presence of Ser/Thr kinase activity in the crude enzyme extracts used in our experiments has been confirmed.[3] A larger gel than that described here can be used for up to 96 sample wells as long as an appropriate size cast is available, e.g., an agarose gel tray for BRL model H5. However, the fragility of the gel must be overcome.

The solid-phase TPK assay can be useful for screening inhibitors and activators of TPKs as well as examining substrate specificity. The same method can be applied for assaying Ser/Thr kinases after immobilization of judiciously chosen substrates. The solid-phase assay may also be useful for assaying protein phosphatases since the product of the protein kinase reaction in the gels can perhaps be used as substrates for the protein phosphatase reaction.

The disadvantage of the solid-phase assay is low sensitivity due to the limited accessibility of the substrate to the enzymes; for instance, the level of $^{32}P$ incorporation into the immobilized substrate was ~1/2000 of the value when compared with an equivalent solution assay in which the polymer substrate was separated by SDS–PAGE as shown.[13,14] This can lead to some difficulties in detection of TPK peaks from column eluates when TPK activity is not very high.

[13] D. Sahal, J. Ramachandran, and Y. Fujita-Yamaguchi, *Arch. Biochem. Biophys.* **260,** 416 (1988).

[14] However, it seems that a $K_m$ value measured by the solution assay[11] is ~10 times higher than that measured by the solid-phase assay.

(A) Well 2 contains no substrate (0 m*M*). Wells 3, 4, 5, 6, 7, and 8 contained 0.1, 0.125, 0.25, 0.5, 1, and 2 m*M* substrate, respectively. Enzyme reaction mixture was prepared and 30 μl each of the mixtures which contained 2 μl of the human placental enzyme as used in Fig. 2 was delivered to each well except well 1. The phosphorylation reaction was performed as described in the text. The gel was exposed to a film for 1.5 hr at −70° with an intensifying screen. (B) Double-reciprocal plot of velocity ($V$) versus substrate concentration ($S$) of data shown in (A). The bottoms of the wells were cut out with a cork borer and radioactivity associated with the area was quantitated by liquid scintillation counting.

## Acknowledgments

We thank Drs. T. R. LeBon and C. M. Rotella for valuable discussions. D. S. thanks Drs. R. B. Wallace and Dan Y. Wu for valuable discussions and for help in the use of the radioisotopic scanner. This work was supported by NIH Grant DK34427. D. S. was a recipient of the American Diabetes Association, Southern California Affiliate's Research Fellowship.

# [6] Nonradioactive Assays of Protein-Tyrosine Kinase Activity Using Anti-phosphotyrosine Antibodies

*By* GERT RIJKSEN, BRIGIT A. VAN OIRSCHOT, and GERARD E. J. STAAL

## Principle of Methods

In general, protein-tyrosine kinase activity (PTK) is assayed using [$^{32}$P]ATP and macromolecular or peptide substrates. The incorporation of [$^{32}$P]phosphate is measured by either precipitating the polypeptide substrate on filter paper with trichloroacetic acid, extensive washing and counting for radioactivity,[1,2] or by using phosphocellulose paper in the case of smaller peptides. We described a new method, which is based on the detection of phosphorylated tyrosyl residues by using monoclonal antibodies to phosphotyrosine.[3] This principle can be applied to a liquid assay system for quantitative determinations[3] as well as to a solid-phase variant, which is particularly useful for rapid semiquantitative screening of, e.g., elution fractions obtained in purification procedures. In the liquid-phase assay the PTK sample is incubated in a reaction vessel with the substrates poly(GluNa,Tyr) (4 : 1) (PGT) and unlabeled ATP. After termination of the reaction an aliquot is transferred to a poly(vinylidene difluoride) (PVDF) membrane. The extent of tyrosine phosphorylation is measured by probing the membrane with anti-phosphotyrosine antibodies, followed by detection by the immunogold silver-staining procedure. The signal is quantified by densitometry.

The solid-phase assay procedure consists of a combination of the above method and the enzyme activity dot-blot method of Allis *et al.*[4] (see [4] in

[1] D. Sahal and Y. Fujita-Yamaguchi, *Anal. Biochem.* **167,** 23 (1987).

[2] S. Braun, W. E. Raymond, and E. Racker, *J. Biol. Chem.* **259,** 2051 (1984).

[3] G. Rijksen, B. A. van Oirschot, and G. E. J. Staal, *Anal. Biochem.* **182,** 98 (1989).

[4] C. D. Allis, L. G. Chicoine, C. V. C. Glover, E. M. White, and M. A. Gorovsky, *Anal. Biochem.* **159,** 58 (1986).

this volume). PTK samples are spotted to PVDF membrane and incubated consecutively with PGT and ATP. Thus the phosphorylation reaction proceeds on the membrane. Detection of phosphorylated tyrosine residues is again achieved with anti-phosphotyrosine antibodies.

## Method A: Liquid Assay System

### *Preparation of PTK Samples*

1. Tissues or cells are dispersed mechanically in 4 vol of extraction buffer.

*Extraction Buffer*

Tris-HCl (pH 7.4), 10 m*M*
$MgCl_2$, 1 m*M*
EDTA-$Na_2$, 1 m*M*
Dithiothreitol, 1 m*M*
Sucrose, 0.25 *M*
Phenylmethylsulfonyl fluoride (PMSF), 1 m*M*: From a 500 m*M* stock solution kept in ethanol
Aprotinin, 50 kallikrein inhibitor units (KIU)

*Note:* Aprotinin and PMSF are added immediately before use. After these additions the pH is again adjusted to 7.4 (4°).

2. Centrifuge the suspension at 800 *g* for 10 min at 4°. The pellet containing intact cells, nuclei, and debris is discarded.
3. Centrifuge the supernatant at 150,000 *g* for 1 hr at 4°. The resulting supernatant is collected as the cytosolic fraction.
4. Resuspend the pellet in a volume of solubilizing buffer equivalent to ~0.2 vol of the original lysate. Solubilize the membrane fraction for 1 hr on ice. Sonicate the suspension twice during this period for 10 sec with 1 min of cooling in ice between each sonication.

*Solubilizing Buffer*

Tris-HCl (pH 7.5), 50 m*M*
Magnesium acetate, 20 m*M*
NaF, 5 m*M*
EDTA-$Na_2$, 0.2 m*M*
EGTA, 0.8 m*M*
Dithiothreitol (DTT), 1 m*M*
$Na_3VO_4$, 30 $\mu M$
Nonidet P-40 (NP-40), 0.5%

5. Centrifuge at 150,000 *g* for 1 hr at 4°. The supernatant is collected and contains the solubilized membrane fraction. The pellet containing detergent-insoluble material is discarded.

*Preparation of Phosphorylated PGT Standard*

1. Select a tissue or cell type with a high PTK activity in both membrane and cytosol. Breast cancer specimens and lymphoid cells appeared to be very useful. Prepare a total cell lysate in extraction buffer; immediately before use add 1 m*M* PMSF and 50 KIU/ml aprotonin. Centrifuge at 800 *g* for 10 min at 4° and dilute the supernatant with solubilizing buffer to ~1 mg protein/ml. After 1 hr of solubilization, centrifuge the suspension at 150,000 *g* for 1 hr at 4°.

*Incubation Buffer*

Tris-HCl (pH 7.5), 50 m*M*
Magnesium acetate, 20 m*M*
NaF, 5 m*M*
EDTA-$Na_2$, 0.2 m*M*
EGTA, 0.8 m*M*
Dithiothreitol, 1 m*M*
$Na_3VO_4$, 30 $\mu M$

2. Incubate the supernatant overnight at 37° in the presence of 1 mg/ml PGT and 0.5 m*M* ATP. The incubation is terminated by adding 25 m*M* EDTA and 25 m*M* EGTA (final concentrations).

3. Centrifuge the suspension at 150,000 *g* for 1 hr. The supernatant is collected and stored in small quantities at −20° for later use as a standard.

4. Quantification of phosphate incorporated in the PGT is performed by running a microscale experiment in parallel with [$\gamma$-$^{32}$P]ATP under identical conditions. Incorporation of [$^{32}$P]phosphate is quantified by using the conventional trichloroacetic acid (TCA) filter paper assay.[3] The stoichiometry of the incorporation varies from approximately 2 to 25 mmol/mol tyrosyl residues, depending on the source of the PTK activity.

*Protein-Tyrosine Kinase Assay*

1. Immediately before starting the preparation of the assay mix the PTK sample is diluted with incubation buffer (cytosolic fraction) or solubilizing buffer (solubilized membrane fraction) to 0.6 mg protein/ml. Higher dilutions may be appropriate if high PTK activities are to be expected.

2. For each sample to be assayed, mix in duplicate:

| | Total phosphorylation mix | Endogenous phosphorylation mix |
|---|---|---|
| Incubation buffer (see above) | — | 40 μl |
| PGT | 40 μl | — |
| PTK sample | 10 μl | 10 μl |

*PGT Solution*

Poly(GluNa,Tyr) (4 : 1) (Sigma, St. Louis, MO), 2 mg/ml in incubation buffer: Due to batch-to-batch variations it is advisable to buy a relatively large amount of PGT from one batch. Store it in small aliquots at −20°. Before opening, allow the bottle to warm to room temperature

3. Start the phosphorylation reaction by adding 10 μl ATP (3 m*M*) and incubate for 1 hr at 37°. ATP is dissolved in incubation buffer, and the pH must be readjusted.

4. Terminate the reaction by adding 10 μl stop reagent. If desired, the assay mix can now be stored at −20° for later analysis.

*Stop Reagent*

EDTA-$Na_2$ (175 m*M*) + 175 m*M* EGTA in incubation buffer: Adjust pH to 7.5; any precipitate dissolves during adjusting

### *Detection and Quantification of Phosphotyrosyl Residues with Immunogold Silver Staining Procedure*

1. Pretreat a poly(vinylidene difluoride) membrane (PVDF, trade name Immobilon; Millipore, Bedford, MA) by rinsing it in methanol for 3 sec and in doubly distilled water for 1 min. Put the membrane on a wet filter paper, remove air bubbles, and wipe off excess fluid with a tissue, or mount it in a dot–blot apparatus.

2. Spot 5-μl aliquots of the assay mix in duplicate onto the membrane manually or using the dot–blot apparatus (in the latter case it is advisable to dilute the assay mix five times and to spot 25 μl). Allow the spots to dry (do not let the membrane become too dry), or, if using the dot–blot apparatus, slowly pull through the assay mix by applying vacuum. It is advisable to rinse the well with incubation buffer and also pull this through the membrane to remove protein from the template.

3. Construct a standard curve by spotting a series of 5-μl aliquots containing 0–30 pmol of phosphotyrosyl residues by proper dilutions of the stock phosphorylated PGT in incubation or solubilizing buffer (see above). In this series of standards the total amount of PGT (both phosphor-

ylated and unphosphorylated) is adjusted to 5.7 μg/5 μl in order to create identical conditions for standards and assay mixtures.

4. Block the membrane by incubating it for 10–30 min in buffer I; then rinse it three times for 5 min in buffer II.

| *Buffer I* | *Buffer II* |
|---|---|
| NaCl, 0.137 *M* | NaCl, 0.137 *M* |
| KCl, 2.7 m*M* | KCl, 2.7 m*M* |
| $Na_2HPO_4 \cdot 2H_2O$, 8.0 m*M* | $Na_2HPO_4 \cdot 2H_2O$, 8.0 m*M* |
| $KH_2PO_4$, 1.5 m*M* | $KH_2PO_4$, 1.5 m*M* |
| Sodium azide, 10 m*M* | Sodium azide, 10 m*M* |
| Bovine serum albumin, 50 mg/ml | Bovine serum albumin, 1 mg/ml |

*Notes:* Use fatty acid-free bovine serum albumin (Sigma, St. Louis, MO). Make all solutions in doubly distilled water.

5. Incubate the membrane 1–2 hr with anti-phosphotyrosine at 20° under gentle shaking.

*Anti-Phosphotyrosine Solution*

Anti-phosphotyrosine* (2 μg/ml) in buffer II
Normal goat serum, 1%

6. Rinse three times by incubating 5 min with buffer II.

7. Incubate the membrane 1–2 hr with the secondary gold-labeled antibody at 20° under gentle shaking.

*Secondary Antibody Solution*

Auroprobe BL plus goat anti-mouse IgG + IgM (Amersham) diluted 100 times in buffer II plus 0.05 vol gelatin

8. Rinse the membrane consecutively two times in buffer II (5 min) and two times in phosphate-buffered saline (1 min).

9. Fix by incubating for 10 min in 2% glutaraldehyde (v/v) in phosphate-buffered saline.

10. Rinse four times with doubly distilled water for 5 min.

### *Silver Enhancement of Gold-Labeled Secondary Antibody*

1. Dissolve 0.055 g silver lactate in 7.5 ml in doubly distilled water in a dark tube, and stir for at least 3 min on a magnetic stirrer. Dissolve 0.425 g hydroquinone in 7.5 ml; keep the tube dark. Mix 30 ml doubly distilled

* Monoclonal antibodies to phosphotyrosine, clone 1G2, described by Huhn *et al.*,[5] were purchased from Amersham (London, England).

water plus 5 ml citrate buffer with the silver lactate and hydroquinone solutions. (*Note:* Silver lactate and hydroquinone must be added simultaneously and mixed very quickly.) The mixture must remain clear. Use immediately. Incubate the membrane in a rolling tube with the mixture for exactly 3 min. Protect from direct light. Use doubly distilled water for all solutions.

*Citrate Buffer*

Adust the pH of a 0.1 *M* citric acid solution with 0.1 *M* sodium citrate to pH 4.5

2. Rinse with doubly distilled water for 5 sec. Remove the membrane from the tube and incubate with 50 ml of a common photographic fixative, e.g., Hypam (Ilford, England) diluted 1 : 10 in doubly distilled water for exactly 3 min. Keep as dark as possible; shake regularly.

3. Rinse with doubly distilled water for 30 min. Refresh after 1, 5, 10, 15, 20, and 25 min.

### *Quantification*

The optical density of the entire spot on the wet membrane can be quantified with a densitometer. The availability of a densitometry with beam dimensions covering at least the entire dot is a prerequisite for accurate quantitation of the signal. PTK activity in the samples is calculated from the standard curve. If a signal is obtained in the absence of PGT (endogenous phosphorylation mix), it must be subtracted from the total phosphorylation signal. If present it probably originates from the tyrosine phosphorylation of endogenous proteins or from aspecific interactions. Using PTK samples from various human tumor biopsies, background staining was usually negligible. The PTK activity is expressed in picomoles of phosphate incorporated per minute.

*Note:* When other anti-phosphotyrosine antibodies are used, the extent of a background signal may be more significant. The antibody used here (1G2 clone, originally described by Huhn *et al.*[5]) is not very well suited for Western blotting of endogenous tyrosylphosphorylated proteins. For unknown reasons, however, it is very well suited for the immunoblotting of phosphorylated PGT, as shown by the present method.

[5] R. D. Huhn, M. R. Posner, S. I. Rayter, J. G. Foulkes, and A. R. Frackelton, *Proc. Natl. Acad. Sci. U.S.A.* **84,** 4408 (1987).

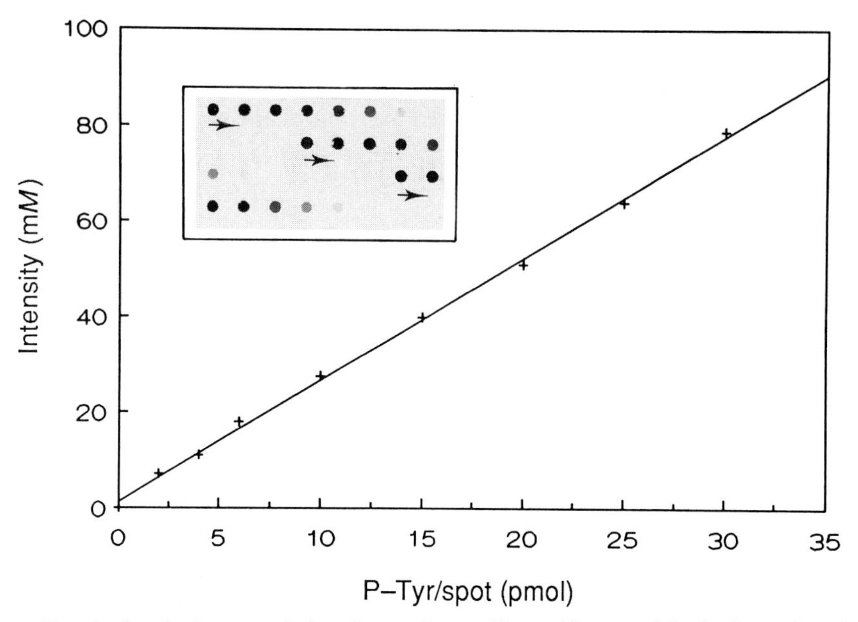


FIG. 1. Standard curve of phosphotyrosine residues. Aliquots of 5 μl of a series of phosphotyrosine standards containing 0–30 pmol of phosphotyrosyl residues in a total amount of 5.7 μg PGT were spotted to a PVDF membrane and stained by the immunogold silver-staining procedure. The inset shows the silver-stained membrane from which the data were derived by densitometry. The standard curve is spotted in triplicate. The arrows indicate the starting point of each series.

## *Assay Performance and Comparison with Other Methods*

The assay described here produces a linear response up to 30 pmol of phosphotyrosine residues/spot, as shown by a standard curve depicted in Fig. 1. No background signal of unphosphorylated PGT is observed. The sensitivity of the method allows the quantitative detection of less than 2.5 pmol of phosphotyrosine/spot. Binding of the anti-phosphotyrosine antibody is diminished by the addition of 2 m*M* phosphotyrosine or 10 m*M* phenylphosphate during the incubation of the PVDF membrane with antibody solution, proving the specificity of the signal. The sensitivity and accuracy evidenced by the very low standard deviation of the curve shown in Fig. 1 allow kinetic studies to establish optimal substrate concentrations for PTKs from various sources.[3] The assay is linear in a broad range of enzyme activities and is linear during an incubation time of 2–3 hr, depending on the activity of the sample.

A comparison of the present method with the conventional filter-paper assay, in which the incorporation of [$^{32}$P]phosphate is determined by TCA

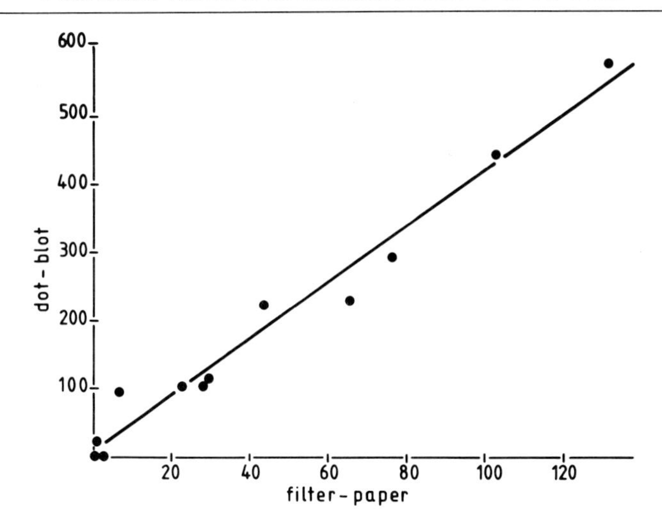


FIG. 2. Comparison of the conventional filter-paper assay and the present nonradioactive dot–blot assay. Cytosolic PTK activities of normal breast tissue, benign breast disease, and breast carcinoma were assayed by both assays. The results were analyzed by linear regression: slope = 4.1 ± 0.2 (SE), correlation coefficient = 0.98. The activities are expressed in picomoles of phosphate incorporated per minute per milligram of protein.

precipitation and liquid scintillation counting,[1,2] is shown in Fig. 2. PTK activities of a series of cytosolic extracts from normal human breast and breast tumors were determined by both methods. A linear correlation was obtained with a correlation coefficient of 0.98. Activities measured by the new assay were approximately four times higher than in the filter-paper assay. This difference appeared to be caused mainly by differences in the incubation temperature and only to a smaller extent by differences in the substrate concentrations. The lower detection limit of both assays was comparable. However, the sensitivity of the filter-paper assay was greatly dependent on the extent of background phosphorylation of endogenous protein (probably mainly on serine residues). As mentioned before, no background signal was present in the present assay.

The liquid assay method is at least equally sensitive and accurate and is even easier to perform. The fundamental differences in favor of the present assay are the absence of noise and the extremely low level of background derived from endogenous phosphoproteins. In the TCA precipitation assay the main problem is to get rid of the noise, which constitutes over 99% of the total radioactivity.[1] Furthermore, the background of endogenous protein phosphorylation is considerably higher in the latter assay as it originates mainly from proteins phosphorylated on serine. In fact, the lower detection limit in this assay is determined by the level of background phosphorylation and is therefore variable. These differences are probably the main reason for the relatively high sensitivity and accuracy of the nonradioactive assay presented here. Another advantage of the new method is the absence of restrictions in the choice of the ATP concentration in the assay mixture. In the conventional filter-paper assay the choice of the ATP concentration is a compromise between the need for a high specific activity of the [$^{32}$P]ATP and the amount of radioactivity which can be handled at the same time. As a consequence the ATP concentration is usually rather low, thus underestimating the activity of higher $K_m$ PTKs and limiting the time in which the reaction is linear. Finally, the main advantage of the new method is probably the avoidance of the use of radioactivity, which enables the simultaneous assay of a large number of samples without the need for special facilities and safety precautions. Therefore, this method seems to be useful to facilitate further studies on PTKs from various sources and, moreover, makes the PTK activity determination more readily accessible for routine diagnostic purposes.

## Method B: Solid-Phase Assay System

1. Pretreat a PVDF membrane according to the procedure described in Method A.
2. Spot 5-$\mu$l aliquots of PTK samples directly on the membrane and allow the spots to dry (do not let the membrane become too dry).
3. Incubate the membrane with PGT (2 mg/ml in incubation buffer, see Method A) for 30 min at room temperature.
4. Rinse the membrane for 10 sec in doubly distilled water or incubation buffer.
5. Incubate the membrane with ATP (500 $\mu M$ in incubation buffer) for 60 min at 37°.
6. Proceed with the detection and quantification of phosphotyrosyl residues by the immunogold silver-staining procedure as described in Method A.

### *Evaluation of Method*

Although the solid-phase variant of the PTK assay described here provides only semiquantitative reşults, it is very suitable when a rapid screening of many PTK samples is desired, for instance, in purification procedures. It should be noted, however, that the assay conditions are quite different from the liquid-phase assay: The PTK protein is adsorbed to the membrane before the phosphorylation reaction is started and excess PGT, not bound to the enzyme or the membrane, is washed away before adding ATP as a second substrate. As a consequence PTK activities from the two assay systems cannot be compared directly. Some PTKs may be relatively inactive in the solid-phase assay, although they are very active in the liquid-phase assay or vice versa.

# [7] Use of Synthetic Amino Acid Polymers for Assay of Protein-Tyrosine and Protein-Serine Kinases

*By* EFRAIM RACKER

## Introduction

Synthetic random polymers of amino acids with an average molecular weight of about 30,000 to 60,000 serve as excellent substrates for many protein kinases (PK). A polymer of glutamate (E) and tyrosine (Y) with a ratio of 4 : 1 is a substrate for all protein-tyrosine kinases (PTK) thus far tested.[1-3] A large number of protein-serine kinases (PSK) phosphorylate either poly(arginine–serine), RS (3 : 1), or poly(glutamate–threonine), ET (4.4 : 1), at rates comparable to those observed with natural substrates.[4] Insertion of other amino acids into glutamate–tyrosine polymers such as alanine (A) or lysine (K) markedly alters the rates of phosphorylation catalyzed by different PTKs,[1] as will be described in the section, Substrate Specificity. Nevertheless, the fact that EY (4 : 1), RS (3 : 1), and ET (4.4 : 1) serve as excellent substrates for many protein kinases[1-4] raises the question of the significance of so-called "consensus sequences." Although convincing evidence for the importance of the amino acid sequences has been demonstrated in numerous laboratories with synthetic low-molecular-weight peptides as substrates,[5] the question of how much it contributes to the susceptibility of large proteins to phosphorylation remains to be established. For example, RAS 2 of yeast, lacking the classical consensus

[1] S. Braun, W. E. Raymond, and E. Racker, *J. Biol. Chem.* **259,** 2051 (1984).

[2] S. Braun, M. Abdel-Ghany, J. A. Lettieri, and E. Racker, *Arch. Biochem. Biophys.* **247,** 424 (1986).

[3] S. Nakamura, S. Braun, and E. Racker, *Arch. Biochem. Biophys.* **252,** 538 (1987).

[4] M. Abdel-Ghany, D. Raden, E. Racker, and E. Katchalski-Katzir, *Proc. Natl. Acad. Sci. U.S.A.* **85,** 1408 (1988).

[5] P. D. Boyer and E. G. Krebs, eds., "The Enzymes," Vols. 17 and 18. Academic Press, Orlando, Florida, 1987.

sequence for protein kinase A (PK-A), is a better substrate for this enzyme than histone 1.[6] It was proposed[7,8] that charge distribution plays an important role in substrate susceptibility and that the distance between a tyrosine and a charged amino acid (glutamate or aspartate), or between serine and arginine or lysine, is a key signal for attack by protein kinases. An appropriate charge constellation may be provided by the primary sequence or induced by the three-dimensional conformation of proteins or even by the addition of charged polymers that are not substrates.[9,10] A remarkable feature of these polymer substrates, particularly in the case of PTK, is that their $K_m$ values may be orders of magnitude lower than those of small consensus sequence peptides. A second advantage is that random polymers are easily synthesized and many are commercially available at a reasonable price. Moreover, a tyrosine polymer can be used as substrate in crude cell extracts without interference by the large number of protein-serine kinases. A disadvantage of the currently available polymers is that they are very heterogeneous with respect to molecular weight and therefore not very useful for SDS–PAGE analysis. They can be used, however, in gel electrophoresis in the absence of SDS.[11] Some basic polymers do not even enter the SDS–polyacrylamide gel. A most attractive feature is that the polymers can be used for studies of signal transduction between either tyrosine and serine kinases or between serine and tyrosine kinases, as will be described below.

## Substrates

Synthetic random amino acid polymers are synthesized as described.[12] Several tyrosine-containing polymers, poly(arginine–serine), and poly(lysine–serine), can be purchased from Sigma (St. Louis, MO). They are widely used for the assay of protein kinases.[2–4,11,13–15]

[6] R. Resnick and E. Racker, *Proc. Natl. Acad. Sci. U.S.A.* **85,** 2474 (1988).

[7] E. Racker, *JNCI, J. Natl. Cancer Inst.* **81,** 247 (1989).

[8] M. Abdel-Ghany, H. K. Kole, M. Abou El Saad, and E. Racker, *Proc. Natl. Acad. Sci. U.S.A.* **86,** 6072 (1989).

[9] E. Racker, *in* "Current Topics in Cellular Regulation." (G. R. Welch, ed.) Academic Press, in press.

[10] M. Abdel-Ghany, K. El-Gendy, S. Zhang, and E. Racker, *Proc. Natl. Acad. Sci. U.S.A.* **87,** 7061 (1990).

[11] G. Schieven, J. Thorner, and G. S. Martin, *Science* **231,** 390 (1986).

[12] M. Sela and L. A. Steiner, *Biochemistry* **2,** 416 (1963).

[13] N. Sasaki, R. W. Rees-Jones, Y. Zick, S. P. Nissley, and M. M. Rechler, *J. Biol. Chem.* **260,** 9793 (1985).

[14] K. Yonezawa and R. A. Roth, *FASEB J.* **4,** 194 (1990).

[15] Y. Yanagita, M. Abdel-Ghany, D. Raden, N. Nelson, and E. Racker, *Proc. Natl. Acad. Sci. U.S.A.* **84,** 925 (1987).

## Assays

Assays are performed in Eppendorf tubes in the presence of appropriate activators. For each protein kinase optimal conditions of buffer, pH, cations, reducing agents, activators, ATP, substrate concentration, and linear time course need to be established. With preparations that contain phosphatase activity, inhibitors are added (e.g., 100 $\mu M$ vanadate) provided they do not interfere with the PK activity. Sometimes lowering the temperature to 4° is used to reduce the relative effectiveness of phosphatases and proteases. Optimal concentrations of the polymers are critical. The $K_m$ values between polymers for, e.g., the v-p140 PTK of Fujinami virus vary between 6 $\mu$g and 1.5 mg/ml.[1] Amounts of polymer in excess of optimum sometimes inhibit PTK activities. However, optimal substrate concentrations may vary in the presence of charged activators. Protein kinase P (PK-P) is a membranous casein kinase similar to but not identical to casein kinase II.[15] PK-P phosphorylation of some protein substrates has an absolute requirement for poly(lysine). With the ET polymer, poly(lysine) shows only marginal stimulation of PK-P activity when optimal concentrations of the acidic polymers are used, but stimulation is pronounced in the presence of suboptimal concentrations of the polymer substrate.[4]

A typical experiment is performed in a final volume of 50 $\mu$l containing 20 m$M$ HEPES buffer (pH 7.4), enzyme, polymer, 10 m$M$ $MgCl_2$, and 10 $\mu M$ [$\gamma$-$^{32}$P]ATP (specific radioactivity of 3000 to 8000 cpm/pmol). After incubation at room temperature (5 to 60 min), 30-$\mu$l samples are removed and placed on 2-cm filter paper squares (Whatman 3 MM), held by a forceps. Each square is immediately immersed into a plastic round container (~6 in. in diameter) containing about 50 ml of 10% trichloroacetic acid and 10 m$M$ sodium pyrophosphate, resulting in the precipitation of the polymers. The container is closed and gently shaken for 1 to 2 hr (depending on background), changing the fluid every 15 min. After removal of the squares by forceps, they are dried in an oven or on a hot plate and counted in a scintillation counter.

## Substrate Specificity

### *Tyrosine Polymers*

There are distinct differences with respect to substrate specificity with various PTKs. For the insulin receptor PTK the best substrate is EY (4 : 1); less active is EAY (6 : 3 : 1). Most other polymers, including EKAY

(36 : 24 : 35 : 5), are less than 10% as active. The EGF receptor and v-*abl* PTKs prefer EAY (6 : 3 : 1) over EY (4 : 1) as substrate. For the v-*fps* PTK EKAY is by far the best substrate, whereas EY (1 : 1) is about 30% as active and is slightly superior to EY (4 : 1). The latter is a suitable substrate for all the other PTKs tested, including two enzymes from placenta, one from brain,[2] and one from Ehrlich ascites tumor cells.[3]

Of particular interest are the polymers EY (1 : 1) and EAY (1 : 1 : 1). EY (1 : 1) is a substrate for the v-*fps* PTK (1), but it is very poorly phosphorylated by the insulin and EGF receptor kinases. It was shown that EY (1 : 1) is an excellent substrate for insulin receptor PTK if large amounts of bovine serum albumin (BSA) (up to 10 mg/ml) are present.[14] We have shown that the same is true for c- and v-*src* kinases. Stimulation by BSA is most pronounced with EAY (1 : 1 : 1) as substrate.[10] From plasma membrane of bovine brain an activator was extracted and partially purified that stimulates phosphorylation of EY (4 : 1) by c-*src* kinase. Of special interest is that stimulation of v-*src* activity by this activator was small by comparison with c-*src*. The activation was separated from an inhibitor that preferentially inhibits c-*src*.[10]

These examples document again the phenomenon of chaperones or substrate modulators, proteins that are not kinase substrates themselves but greatly influence phosphorylation of other proteins.[7,8,9]

### *Use of Synthetic Amino Acid Polymers for Study of Signal Transduction between Protein Kinases*

With the availability of synthetic polymers that contain either tyrosine or serine or threonine as the only amino acid susceptible to phosphorylation, the effect of a tyrosine kinase on serine kinase or of serine kinases on tyrosine kinases can be explored *in vitro*. However, it is essential to assay under conditions of appropriate kinetics to show cross-talk effects. As a rule the converter enzyme must be present in large excess, whereas the enzyme to be converted must be present at small concentrations within the linear range of activity and preferably with its substrate at saturating concentrations. Attention must be drawn to the fact that very acidic polymers, e.g., EY (4 : 1), may inhibit serine kinases, e.g., casein kinase II or PK-P, and basic polymers in excess may inhibit PTKs.[8] Activators of PK-C ($Ca^{2+}$ and lipids) stimulate *src* PTK activity.[10] It is therefore essential that signal transduction experiments of this kind are performed in the same assay medium so that controls with single enzymes are meaningful.

With cell extracts or partially purified protein kinase preparations that contain multiple kinases, it would be of great advantage to have specific PK inhibitors available. Unfortunately, very few specific inhibitors are generally available. In our hands, using synthetic polymers as substrate, only the peptide inhibitor of PK-A has proved to be a specific inhibitor. Staurosporine inhibits protein kinase C at 5 n$M$ concentration without affecting several other PSKs. However, it is also a potent inhibitor of PTKs, e.g., src. Other inhibitors such as sphingosine[16] and H7, were found to lack specificity. Our knowledge of specific consensus sequences required for the phosphorylation of small polypeptides should, however, become very valuable for the design of specific kinase inhibitors and are slowly becoming available. However, in contrast to cell-permeable inhibitors, such as staurosporine, or activators, such as epinephrine or phorbol esters, these polypeptides are not suitable for experiments on signal transduction in intact cells and specific permeable inhibitors would be of great value.

### *Use of Synthetic Polymers for Purification of Protein Kinases*

Affinity columns with synthetic polymers are easily prepared and useful for the purification of protein kinases. Several unidentified PTKs were prepared from human placenta, bovine brain,[2] and Ehrlich ascites tumors.[3] In some cases the extent of purification was modest with one polymer affinity column and much better with another. For example, in the case of c-src and v-src purification, two polymer columns used sequentially yielded a virtually homogeneous kinase protein.[17] On these affinity columns c-src and v-src kinases eluted at different salt concentrations. Exploration of the marked differences both in substrate specificity and $K_m$ values described above should greatly increase the potential usefulness of such affinity columns.

### Acknowledgments

This investigation was supported by Public Health Service Grant CA-08964, awarded by the National Cancer Institute, DHHS, and in part by a grant from the Cornell Biotechnology Program which is sponsored by the New York State Science and Technology Foundation, a consortium of industries, the U.S. Army Research Office, and the National Science Foundation.

[16] Y. Igarashi, S. Hakomori, T. Toyokuni, B. Dean, S. Fujita, M. Sugimoto, T. Ogawa, K. El-Ghendy, and E. Racker, *Biochemistry* **28,** 6796 (1989).

[17] S. Zhang, M. El-Gendy, M. Abdel-Ghany, and R. Clark, F. McCormick, and E. Racker *Cell. Physiol. Biochem.*, in press (1991).

# [8] Assay of Phosphorylation of Small Substrates and of Synthetic Random Polymers That Interact Chemically with Adenosine 5′-Triphosphate

*By* EFRAIM RACKER and PARIMAL C. SEN

## Introduction

Several procedures have been published to measure the transfer of [$\gamma$-$^{32}$P]ATP to proteins by kinases. The most popular is the small filter paper squares procedure,[1] which is rapid and allows for simultaneous assay of large numbers of samples. For peptides or proteins with a suitable charge, ion-exchange filter papers are equally effective.[2] High-pressure liquid chromatography (HPLC)[3] and electrophoresis[4,5] are less suitable when large numbers of samples are analyzed, e.g., during purification.

A general and rapid method for the assay of phosphorylated substrates of low molecular weight that are not precipitated by trichloroacetic acid has been described.[6] It is based on the lability of [$\gamma$-$^{32}$P]ATP to 1 *N* HCl at 100°, whereas phosphothreonine/phosphoserine residues in proteins are quite stable. A variant of this method that includes measurements of phosphotyrosine residues as well[7] is described below.

Basic proteins and particularly some very basic synthetic polypeptides react nonenzymatically with ATP to form a covalent intermediate.[8] With concentrations of ATP of 10 to 100 $\mu M$ and high specific radioactivity the filter paper assays are unsuitable because of high background counts. This is particularly the case with random polypeptides that contain proline that are used in our laboratory to measure p34$^{cdc28}$ activity in the absence of protein kinase C (or in the presence of 10 *nM* staurosporine). We will

[1] J. D. Corbin and E. M. Reimann, this series, Vol. 38, p. 287.

[2] D. B. Glass, R. B. Masaracchia, J. R. Feramisco, and B. E. Kemp, *Anal. Biochem.* **87,** 566 (1978).

[3] J. E. Casnelli, M. L. Harrison, L. J. Pike, K. E. Hellström, and E. G. Krebs, *Proc. Natl. Acad. Sci. U.S.A.* **79,** 282 (1982).

[4] T. Hunter, *J. Biol. Chem.* **257,** 4843 (1982).

[5] T. W. Wong and A. R. Goldberg, *J. Biol. Chem.* **258,** 1022 (1983).

[6] F. Meggio, A. Donella, and L. A. Pinna, *Anal. Biochem.* **71,** 583 (1976).

[7] S. Braun, M. Abdel-Ghany, and E. Racker, *Anal. Biochem.* **135,** 369 (1983).

[8] M. Abdel-Ghany, D. Raden, E. Racker, and E. Katchalski-Katzir, *Proc. Natl. Acad. Sci. U.S.A.* **85,** 1408 (1988).

describe a simple method based on differential acid stability to eliminate the high background counts which allows the use of the standard filter paper assay with these polymers.

## Assays of Protein Kinase Activity with Low-Molecular-Weight Substrates[7]

### *Principle*

After phosphorylation of substrate with [$\gamma$-$^{32}$P]ATP, the samples are hydrolyzed at 100° in the presence of 1 *N* HCl, which converts all residual ATP to [$^{32}$P]$P_i$ and AMP without hydrolyzing the phosphoester bond in the substrate. [$^{32}$P]$P_i$ is then extracted as an ammonium molybdate complex with organic solvents (upper phase) while the phosphorylated substrate remains in the lower water phase.

The reaction mixture (25 $\mu$l) contains 20 m*M* HEPES (pH 7.4), 10 m*M* thioglycerol, divalent cations ($Mg^{2+}$, $Mn^{2+}$), [$\gamma$-$^{32}$P]ATP (3000 cpm/pmol), protein kinase, and substrate. After a suitable time for incubation the mixture is deproteinized. When concentrations of proteins are used that do not give rise to turbidity, deproteinization is not required and the reaction is stopped directly by addition of 175 $\mu$l of 1.14 *N* HCl and the mixture is immediately subjected to hydrolysis. With low amounts of protein that cause turbidity on addition of trichloroacetic acid, bovine serum albumin (BSA, 1 mg/ml) is added prior to trichloroacetic acid to permit complete precipitation.

With high concentrations of proteins the reaction is stopped by addition of 100 $\mu$l of 25% (w/v) trichloroacetic acid. The mixture is kept on ice for 30 min to complete protein precipitation and then centrifuged at 4° for 10 min at top speed in an Eppendorf microfuge. Supernatants (100 $\mu$l) are collected, transferred to Pyrex tubes (18 × 150 mm), and 2 *N* HCl (100 $\mu$l) is added. The tubes are covered with glass marbles and incubated in a boiling water bath for 20 min, which results in the complete release of a [$\gamma$-$^{32}$P]phosphate of ATP.

After addition to the sample of 4 ml of 1.15 *N* perchloric acid, 4 ml of isobutanol–benzene 1 : 1 (v/v), and 1 ml of 5% ammonium molybdate, the mixtures are shaken vigorously in a multitube Vortexer (Kraft Appar. Inc., Mineola, NY) twice for 1 min (with few samples the tubes are shaken in a Vortex). The upper phase is removed by aspiration. The water layer is reextracted twice with 4 ml of water–saturated isobutanol and finally with 2 ml of ether.[7,9]

The amount of phosphorylated substrate is determined by counting an

[9] M. E. Pullman, this series, Vol. 10, p. 59.

aliquot of the water layer in a liquid scintillation counter. The background value is determined by subjecting a sample of [$\gamma$-$^{32}$P]ATP to hydrolysis and the extraction procedure as described above.

This method is used successfully not only for peptides but also for low-molecular-weight substrates such as glycerol, tyramine, *N*-acetyltyrosine, and others.[7] We have used it for assay of a small synthetic polypeptide (donated by Dr. P. Brautigan at Brown University, Providence, RI) that is phosphorylated by p34$^{cdc28}$.

## Phosphorylation by Protein Kinases of Large Synthetic Random Polypeptides that React Chemically with ATP

### *Principle*

Polypeptides that are phosphorylated by [$\gamma$-$^{32}$P]ATP, enzymatically as well as chemically, are hydrolyzed for 15 min in 1 *N* HCl to remove $^{32}$P incorporated nonenzymatically.[8]

For studies of p34$^{cdc28}$ kinase we searched for synthetic random polypeptides that substitute for histones as substrate. The latter contain tyrosine residues that are phosphorylated by src kinases and other protein-tyrosine kinases and are therefore not suitable for studies of signal transduction in reconstituted systems. We tested several random proline-containing polypeptides that were synthesized for us by Dr. S. Deshmane of Sigma Chemical Company (St. Louis, MO). We selected as most suitable a random polymer poly(arginine–threonine–proline) (RTP) (1 : 1 : 1) which was phosphorylated by p34$^{cdc28}$ purified from *Saccharomyces cerevisiae* (P. C. Sen and E. Racker, unpublished observations, 1990).

To eliminate high background counts the following procedure is used. The reaction mixture (50 $\mu$l) contained, for measurement of p34$^{cdc28}$, 20 m*M* HEPES, Tris (pH 7.4), 15 m*M* $MgCl_2$, 10 m*M* thioglycerol, 0.01% Brij 35, 1 mg/ml of RTP (1 : 1 : 1) (or other basic polypeptides), various concentrations of protein kinase, and 50 $\mu M$ [$\gamma$-$^{32}$P]ATP (2500–3000 cpm/pmol). The reaction is started by addition of radioactive ATP and after 20 min at room temperature is terminated by the addition of 5 $\mu$l of 12 *N* HCl. The reaction mixture is heated for 15 min at 90° and an aliquot of 25 $\mu$l is assayed by the filter paper method.[1]

## Acknowledgment

This investigation was supported by Public Health Service Grant CA08964, awarded by the National Cancer Institute, DHHS; by a grant from the Alzheimer's Disease Foundation; and by a grant from the Cornell Biotechnology Program, which is sponsored by the New York State Science and Technology Foundation, a consortium of industries, the United States Army Research Office, and the National Science Foundation.

# [9] Assay of Protein Kinases Using Peptides with Basic Residues for Phosphocellulose Binding

*By* JOHN E. CASNELLIE

Studies of the function and regulation of protein kinases demand sensitive and specific assays for their activity. The early observation that denatured proteins such as histones and casein are phosphorylated by these enzymes has resulted in the use of these proteins as substrates in assays for protein kinase activity. However, the very fact that such proteins are phosphorylated by many different protein kinases can make it difficult to use them to assay for the activity of a specific protein kinase. The inability to separate readily the $^{32}P$-labeled protein product from other proteins phosphorylated during the reaction can add to the problems of using proteins as the phosphate acceptor in assays for protein kinase activity, especially when assaying for this activity in crude extracts.

Small synthetic peptide substrates suffer less from these disadvantages. The sequence of peptide substrates can be designed to optimize the assay of specific enzymes. Their relatively low molecular weight plus the ability to modify the sequence at will provide a variety of approaches for separating the phosphorylated peptide product from the other radioactive components in the reaction mixture. The highly defined nature of synthetic peptides and the presence of only a single hydroxyl-containing residue make it possible to determine kinetic parameters that permit comparisons of results from different investigations. Small peptide substrates are especially suited for certain unique situations such as assays of protein kinase activation in permeabilized cells and for studies of enzyme kinetics. A synthetic peptide with the sequence of an *in vivo* site of protein phosphorylation can be very useful in identifying the protein kinase responsible for that phosphorylation.

Although a number of different techniques are available to separate the phosphorylated peptide product from other $^{32}P$-labeled material present in assay mixtures, the most efficient method is to make use of the ability of peptides containing several basic charges to bind via ionic interactions to phosphocellulose paper.[1,2] This technique originated with assays using histones as substrates. Even after phosphorylation histones are still highly

[1] J. J. Witt and R. Roskoski, Jr., *Anal. Biochem.* **66,** 253 (1975).

[2] D. B. Glass, R. A. Masaracchia, J. R. Feramisco, and B. E. Kemp, *Anal. Biochem.* **87,** 566 (1978).

positively charged and bind to phosphocellulose. The pieces of phosphocellulose can be washed in water and the phosphorylated histones are retained while the [$\gamma$-$^{32}$P]ATP and any [$^{32}$P]P$_i$ are washed away. Because some peptide substrates contain only two basic residues it is necessary to modify the assay by using acid washes that suppress the negative charges of the phosphate group.[2] The phosphocellulose technique has been extended to enzymes that recognize acidic residues in their peptide substrates instead of basic residues.[3,4] In this situation basic residues were added to the peptides at positions that did not interfere with recognition by the enzymes but still permitted the use of the phosphocellulose assay, again using acid washes to suppress the negative charges of the acidic groups on the peptides. Although the phosphocellulose assay can be used with peptide substrates that have only a single basic residue, quantitative retention on the phosphocellulose paper after phosphorylation requires the presence of at least two basic residues and a free amino terminus.[2]

## Synthetic Peptides That Can Be Used with Phosphocellulose Technique to Assay for Protein-Serine/Threonine Kinases

The sequences of peptides* listed in Table I have been used to assay for the activity of protein-serine/threonine kinases using the phosphocellulose assay. Table I includes only those peptides that have actually been used to monitor enzyme activity rather than a comprehensive list of all peptides that serve as substrates for these kinases. Several of these peptides, especially those that contain multiple basic residues, have been found to undergo phosphorylation by more than one enzyme. For example, the peptide used to assay for the S6 kinase and the peptide used to assay for the cGMP-dependent protein kinase are also phosphorylated by the ubiquitous cAMP-dependent protein kinase.[5,6] Thus when using these peptides with impure preparations it is necessary to include the peptide inhibitor of the cAMP-dependent protein kinase in order to suppress the activity of this enzyme. The S6 peptide and the peptide used to assay for $Ca^{2+}$/calmodulin-dependent protein kinase II are probably also substrates for protein kinase C. In general the identification of an enzyme in a crude

[3] J. E. Casnellie, M. L. Harrison, L. J. Pike, K. E. Hellström, and E. G. Krebs, *Proc. Natl. Acad. Sci. U.S.A.* **79,** 282 (1982).

[4] E. A. Keunzel and E. G. Krebs, *Proc. Natl. Acad. Sci. U.S.A.* **82,** 737 (1985).

* A, Alanine; D, aspartic acid; E, glutamic acid; F, phenylalanine; G, glycine; H, histidine; I, isoleucine; K, lysine; L, leucine; N, asparagine; P, proline; Q, glutamine; R, arginine; S, serine; T, threonine; V, valine; and Y, tyrosine.

[5] B. Gabrielli, R. E. H. Wettenhall, B. E. Kemp, M. Quinn, and L. Bizonova, *FEBS Lett.* **175,** 219 (1984).

[6] D. B. Glass and E. G. Krebs, *J. Biol. Chem.* **257,** 1196 (1982).

TABLE I
PEPTIDE SUBSTRATES USED TO ASSAY FOR SERINE/THREONINE PROTEIN KINASE ACTIVITY

| Peptide | Enzyme | $K_m$ ($\mu M$) | $V_m$ ($\mu$mol/min/mg) | Ref. |
|---|---|---|---|---|
| LRRASLG | cAMP-dependent protein kinase | 16 | 20 | *a* |
| RKRSRAE | cGMP-dependent protein kinase | 22 | 4 | *b* |
| KRTLRR | Protein kinase C | nd[c] | nd | *d* |
| RRLSSLRA | Ribosomal S6 kinase | 180 | nd | *e* |
| KKRPQRATSNVFS | Myosin light-chain kinase | 20 | 2.5 | *f* |
| PLARTLSVAGLPGKK | $Ca^{2+}$/calmodulin-dependent | 12 | 2.8 | *g, h* |
| PLRRTLSVAA | protein kinase II | 4 | 11 | |
| RRREEETEEE | Casein kinase II | 500 | 0.5 | *i* |
| RRRDDDSDDD | | 60 | 2.2 | *i, j* |
| PAPAAPSPQPKG | Proline-directed protein kinase | 50 | nd | *k* |

[a] B. E. Kemp, D. J. Graves, E. Benjamini, and E. G. Krebs, *J. Biol. Chem.* **252,** 4888 (1977).
[b] D. B. Glass and E. G. Krebs, *J. Biol. Chem.* **257,** 1196 (1982).
[c] nd, Not determined.
[d] R. J. Davis and M. P. Czech, *J. Biol. Chem.* **262,** 6832 (1987).
[e] S. L. Pelech, B. B. Olwin, and E. G. Krebs, *Proc. Natl. Acad. Sci. U.S.A.* **83,** 5968 (1986).
[f] B. E. Kemp, R. B. Pearson, and C. House, *Proc. Natl. Acad. Sci. U.S.A.* **80,** 7471 (1983).
[g] Y. Hashimoto and T. R. Soderling, *Arch. Biochem. Biophys.* **252,** 418 (1987).
[h] R. B. Pearson, J. R. Woodgett, P. Cohen, and B. E. Kemp, *J. Biol. Chem.* **260,** 14471 (1985).
[i] E. A. Kuenzel, J. A. Mulligan, J. Sommercorn, and E. G. Krebs, *J. Biol. Chem.* **262,** 9136 (1987).
[j] D. W. Litchfield, F. J. Lozeman, C. Piening, J. Sommercorn, K. Takio, K. A. Walsh, and E. G. Krebs, *J. Biol. Chem.* **265,** 7638 (1990).
[k] P. R. Vulliet, F. L. Hall, J. P. Mitchell, and D. G. Hardie, *J. Biol. Chem.* **264,** 16292 (1989).

extract must rely on either immunological techniques or the use of specific activators or inhibitors rather than simply the ability to phosphorylate a particular peptide.

## Synthetic Peptide Substrates for Protein-Tyrosine Kinases That Can Be Used with Phosphocellulose Method

Synthetic peptides potentially provide ideal substrates for assaying the activity of protein-tyrosine kinases. Because peptides can be made with only the tyrosine as the phosphorylatable residue they provide a means to

TABLE II
PEPTIDE SUBSTRATES FOR TYROSINE PROTEIN KINASES

| Peptide | Source of sequence | Ref. |
|---|---|---|
| DRVYIHPF | Angiotensin II | *a* |
| DRVYVHPF | [Val[5]]Angiotensin II | *a* |
| RRLIEDAEYAARG | pp60$^{src}$ | *b, c* |
| EDAEYAARRRG | | |
| TRDIYETDYYRK | Insulin receptor | *d* |
| KGSTAENAEYLRV | Epidermal growth factor receptor | *e* |

[a] T. W. Wong and A. R. Goldberg, *J. Biol. Chem.* **258,** 1022 (1983).
[b] J. E. Casnellie, M. L. Harrison, K. E. Hellström, and E. G. Krebs, *Proc. Natl. Acad. Sci. U.S.A.* **79,** 282 (1982).
[c] G. Swarup, J. D. Dasgupta, and D. L. Garbers, *J. Biol. Chem.* **258,** 10341 (1983).
[d] L. Stadtmauer and O. Rosen, *J. Biol. Chem.* **261,** 10000 (1986).
[e] A. Honegger, T. J. Dull, D. Szapary, A. Komoriya, R. Kris, A. Ullrich, and J. Schlessinger, *EMBO J.* **7,** 3053 (1988).

assay specifically for the activity of these enzymes even in the presence of a large amount of protein-serine/threonine kinase activity. The specificity of protein-tyrosine protein kinases toward synthetic peptides has not been well defined although there seems to be a general requirement for acidic residues on the amino-terminal side of the tyrosine. One disadvantage in using synthetic peptides to assay for protein-tyrosine kinase activity is that the $K_m$ values of the enzymes for the peptides are relatively high (0.1 to 2 m*M*), resulting in the inefficient use of these expensive reagents.

The sequences of the peptides listed in Table II have been used to assay for protein-tyrosine kinase activity using the phosphocellulose method. Angiotensin II and [Val[5]]angiotensin II have the advantage of being comparatively inexpensive although this consideration must be counterbalanced by the fact that the $K_m$ values for these peptides are usually considerably higher than for peptides whose sequences are based on physiologically signficiant sites of tyrosine phosphorylation. The other peptides have sequences based on sites of tyrosine phosphorylation in various protein-tyrosine kinases. The angiotensin peptides and the peptides based on the site of tyrosine phosphorylation in pp60$^{src}$ are commercially available. Note that all the peptides listed in Table II have at least one acidic residue on the amino side of the tyrosine and at least two basic residues to permit the peptides to be used with the phosphocellulose assay.

### Assay Procedures Using Phosphocellulose Paper

Assays are set up in 0.5-ml microfuge tubes. Assay volumes are kept at 50 μl or less in order to conserve reagents. The specific conditions of the assay will vary depending on the enzyme preparation. In the case of impure preparations of protein-tyrosine kinases it is important to include 0.1 m*M* sodium orthovanadate and/or 5 m*M* *p*-nitrophenol phosphate in order to inhibit phosphatases that can dephosphorylate the phosphorylated peptide. The use of $MnCl_2$ in the assay may also serve to suppress phosphotyrosine phosphatase activity.[3]

If there is no appreciable background protein phosphorylation then the reaction can be stopped by spotting an aliquot of the reaction mixture onto the piece of phosphocellulose and dropping it into 75 m*M* phosphoric acid. Alternatively, the reactions can first be stopped with 30% (v/v) acetic acid and then an aliquot spotted onto the piece of phosphocellulose. In situations where it is necessary to remove endogenously phosphorylated proteins that would otherwise contribute to the background, the reactions are stopped by adding an equal volume of 5–10% (v/v) trichloroacetic acid. A 10-$\mu$l aliquot of 10 mg/ml bovine serum albumin (BSA) is added as carrier and the tubes are incubated for 10 min on ice to allow the proteins to precipitate. Alternatively the BSA can simply be incorporated into the assay mixture. The precipitated protein is removed by centrifugation and one-half of the supernatants are spotted onto 2 $\times$ 2 cm pieces of p81 phosphocellulose (Whatman, Clifton, NJ) numbered with pen or pencil.

The phosphocellulose pieces are washed in a beaker on a shaker. The pieces are washed five times with 500 ml of 75 m*M* phosphoric acid [5 ml 85% (v/v) phosphoric acid/liter]. There is no advantage to wash times longer than 5 min. The progress and completeness of the washing can be readily monitored by removing one of the blanks and checking it with a Gieger counter.

## Problems Encountered with Phosphocellulose Assay

The phosphocellulose assay is straightforward and highly sensitive. The only problem that is encountered with the assay is high blanks. Even when assaying for activity in crude extracts the blanks should be 0.1% or less of the total input counts. The most common cause of high blanks is the presence of $^{32}P$-labeled contaminants in commercial [$\gamma$-$^{32}P$]ATP that have a net positive charge in acid and thus bind to phosphocellulose. Such contaminants can result in blanks of 1% or more of the input counts. High blanks due to the [$\gamma$-$^{32}P$]ATP can be diagnosed by running a blank without enzyme. The suggested solution to this problem is to obtain a replacement shipment of [$\gamma$-$^{32}P$]ATP. Although less convenient but probably more successful, the problem can also be alleviated by repurifying the ATP by ion-exchange chromatography. If the peptide substrate has three or more basic residues and no acidic residues then it will be retained on the phosphocellulose even if a neutral solution is used in the washes and this substitution may improve the background.[7]

## Alternative Methods for Assaying for Peptide Phosphorylation

### *Anion-Exchange Column Method*

This method[8] has the advantage that it can be used even when the peptide does not have any basic residues. Since the principle of this method is essentially the same as that for the phosphocellulose paper method, the anion-exchange method generally gives the same level of peptide phosphorylation and background as the phosphocellulose assay.

### *Thin-Layer Electrophoresis*

Several investigators[9,10] have used this method to monitor the phosphorylation of peptides. It may be a useful alternative when high blanks with the phosphocellulose assay obscure enzyme activity.

### *Sodium Dodecyl Sulfate Electrophoresis*

Honegger *et al.*[11] reported the use of 5 to 20% (w/v) polyacrylamide gradient gels to monitor the autophosphorylation of the epidermal growth factor (EGF) receptor and the phosphorylation of an exogenous peptide substrate simultaneously. A layer of 20% acrylamide at the bottom of the gel provided for an effective means to separate the peptide from ATP and $P_i$. Because separation is presumably at least in part due to sieving by the gel, this approach may be of general utility regardless of the charge of the peptide.

### *Acid Hydrolysis Followed by Extraction of $^{32}P$*[12]

Braun *et al.* developed a general method for monitoring protein kinase activity by taking advantage of the relative acid stability of phosphoesters. In this procedure the [$\gamma$-$^{32}$P]ATP is hydrolyzed by heating the samples with 1 *N* HCl. The $^{32}$P is effectively separated from the phosphoesters by extraction as the molybdate complex (see [8] in this volume).

[7] L. E. Heasley and G. L. Johnson, *J. Biol. Chem.* **264,** 8646 (1986).

[8] B. E. Kemp, E. Benjamini, and E. G. Krebs, *Proc. Natl. Acad. Sci. U.S.A.* **73,** 1038 (1976).

[9] T. Hunter, *J. Biol. Chem.* **257,** 4843 (1982).

[10] R. J. Davis and M. P. Czech, *J. Biol. Chem.* **262,** 6832 (1987).

[11] A. Honegger, T. J. Dull, D. Szapary, A. Komoriya, R. Kris, A. Ullrich, and J. Schlessinger, *EMBO J.* **7,** 3050 (1988).

[12] S. Braun, M. A. Ghany, and E. Racker, *Anal. Biochem.* **135,** 369 (1983).

# [10] Design and Use of Peptide Substrates for Protein Kinases

*By* BRUCE E. KEMP and RICHARD B. PEARSON

Synthetic peptide substrates have played an important role in the study of protein kinase (PK) substrate specificity as well as in the measurement of protein kinase activities in cell extracts. Their great utility as experimental reagents became apparent from studies of the substrate specificity of the cAMP-dependent protein kinases and phosphorylase kinase.[1] It was found that the principal substrate specificity determinants for these enzymes were located in short segments of the primary sequence around phosphorylation sites. Arginine residues were identified as important specificity determinants for both of these enzymes by studies using genetic variants of protein substrates[1] and synthetic peptides.[2] Some short synthetic peptides of 7–10 residues were phosphorylated with kinetic properties comparable to the parent protein. The sequencing of a number of phosphorylation sites from a variety of protein substrates for the cAMP-dependent protein kinase provided evidence for the concept of a recognition motif, RRXS or KRXXS, indicating that all of the features necessary for phosphorylation could be combined in a short peptide.[2a] While these findings certainly reinforced the concept of a recognition motif it should be noted that only approximately one-third of the known phosphorylation sites for the cAMP-dependent protein kinase actually conform to the RRXS or KRXXS motif.[3] The liver pyruvate kinase peptide LRRASLG[3] (kemptide) proved to be an excellent substrate for the cAMP-dependent protein kinase with a $K_m$ of approximately 5 $\mu M$ and a $V_{max}$ of 16 $\mu$mol min$^{-1}$ mg$^{-1}$. Subsequently it was found that synthetic peptides corresponding to other phosphorylation site sequences for this enzyme varied widely in their capacity to act as substrates with $K_m$ values in the range 0.1 to 2 m$M$. At present this phenomenon is understood in terms of the local phosphorylation site

[1] B. E. Kemp, D. B. Bylund, T. S. Huang, and E. G. Krebs, *Proc. Natl. Acad. Sci. U.S.A.* **72,** 3448 (1975).

[2] B. E. Kemp, D. J. Graves, E. Benjamini, and E. G. Krebs, *J. Biol. Chem.* **252,** 4888 (1977).

[2a] Single-letter abbreviations are used for amino acids: A, alanine; C, cysteine; D, aspartic acid; E, glutamic acid; F, phenylalanine; G, glycine; H, histidine; I, isoleucine; K, lysine; L, leucine; M, methionine; N, asparagine; P, proline; Q, glutamine; R, arginine; S, serine; T, threonine; V, valine; W, tryptophan; X, unknown; Y, tyrosine.

[3] O. Zetterqvist, U. Ragnarsson, and L. Engström, *in* "Peptides and Protein Phosphorylation" (B. E. Kemp, ed.), p. 171. Uniscience CRC Press, Boca Raton, Florida, 1990.

sequence being the principal determinant of substrate specificity, but its context in the parent protein structure is also important. It seems likely that nearby residues required to accommodate the phosphorylation site sequence into the parent protein structure may influence the capacity of the corresponding synthetic peptide to act as a substrate either positively or negatively.

The major goals of designing peptide substrates for protein kinases are to construct peptides that have excellent kinetic properties and a high degree of specificity. Although some synthetic peptides are phosphorylated with $K_m$ and $V_{max}$ values comparable to their parent proteins, many are not and the reason for this is poorly understood. Peptide substrates that are phosphorylated with $K_m$ values of less than 50 $\mu M$ and preferably in the range 1 to 10 $\mu M$ are typically the most useful. A low $K_m$ value improves the likelihood that the peptide substrate will be relatively specific for that enzyme and has a cost benefit if large numbers of assays are required using an expensive synthetic substrate. On the other hand, from a practical viewpoint the $V_{max}$ value is more important than the $K_m$ value because it determines the detection sensitivity of the peptide phosphorylation reaction. The greater the $V_{max}$ value, the greater the latitude available for varying the peptide substrate concentration. Overlapping specificities of some protein kinases occurs, such as for phosphorylase kinase, protein kinase C, and the multifunctional calmodulin-dependent protein kinase, all of which phosphorylate Ser-7 in the glycogen synthase peptide (see below). Examples will be given below of ways in which the sensitivity and specificity of protein kinase peptide substrates have been enhanced. Sensitivity is a function of $V_{max}/K_m$ and the relative specificity for a peptide substrate toward two enzymes is a function of the ratio, $[V_{max}/K_m$ (enzyme 1)]/[$V_{max}/K_m$ (enzyme 2)], or coefficient of specificity.

## Peptide Substrate Synthesis

Previously Glass[4] has reported procedures for the synthesis of oligopeptides for the study of cyclic nucleotide-dependent protein kinases in an earlier volume of this series. For this reason a detailed account of the chemistry of oligopeptide synthesis will not be covered here. The Merrifield solid-phase synthesis procedure with either tBoc or Fmoc chemistries is used widely.[5] For the chemical synthesis of phosphorylated peptides, see [18] and [19] (Volume 201 in this series). The advent of modern automated peptide synthesizers such as the model 430 (Applied Biosystems, Foster

[4] D. B. Glass, this series, Vol. 99, p. 119.

[5] S. B. H. Kent, *Annu. Rev. Biochem.* **57,** 957 (1988).

City, CA) has meant that the assembly of peptides is no longer rate limiting. However, peptide purification and characterization remain areas of great importance. The assembled peptides are typically cleaved from the resin using anhydrous HF[6] or trifluoromethanesulfonic acid. The resultant crude peptide is then extracted from the resin using one of several suitable volatile buffers, depending on the amino acid composition of the peptide. Most basic and neutral peptides are soluble in 60% acetonitrile (v/v) containing 0.1% trifluoroacetic acid (v/v) and acidic peptides are soluble in 60% acetonitrile (v/v) containing 0.1 *M* $NH_4HCO_3$. Either acetic acid (30%, v/v) or trifluoroacetic acid (50%, v/v) can also be used for more hydrophobic peptides. The completeness of peptide extraction from the resin is readily monitored using the resin ninhydrin test.[7]

In general, protein kinase substrates are relatively polar and solubility is not normally limiting. The most reliable way to purify synthetic peptide substrates routinely is to use both ion-exchange and reversed-phase chromatography. Alternatively, a two-step reversed-phase chromatography using ion pairing at two pH values has been widely used for other synthetic peptides.[8] Reversed-phase chromatography carried out at a single pH can be insufficient, particularly in cases where hydrophobic and hydrophilic residues are separated to different ends of the peptide. Since many protein kinases utilize either acidic or basic residues as specificity determinants an ion-exchange step in the purification of peptide substrates is particularly effective. The procedures outlined below are designed for a single person simultaneously synthesizing and purifying peptides in a small laboratory with limited instrument resources.

Ion-exchange chromatography is conveniently carried out using a commercial purification system such as fast protein liquid chromatography (Pharmacia Piscataway, NJ FPLC) or Waters (Milford, MA) 650 protein purification system (K. I. Mitchelhill and B. E. Kemp, unpublished). Basic peptides are purified on a S-Sepharose HP (Pharmacia) cation-exchange column (1 × 25 cm) using a 0 to 1.0 *M* NaCl gradient in trifluoroacetic acid 0.1% (v/v) with 20% (v/v) acetonitrile. The inclusion of acetonitrile in the ion-exchange chromatography buffer is to enhance peptide solubility and recovery. The flow rate is 1 ml/min for 1140 min (overnight). The peptide peak is located using ninhydrin staining of dot blots of the fractions followed by analytical reversed-phase high-performance liquid chromatography (HPLC) of the fractions containing peptide. The analytical separa-

[6] J. M. Stewart and J. D. Young, "Solid Phase Peptide Synthesis," pp. 44 and 66. Freeman, San Francisco, California, 1966.

[7] V. K. Savin, S. B. H. Kent, J. P. Tam, and R. B. Merrifield, *Anal. Biochem.* **117,** 147 (1981).

[8] J. Rivier, *J. Liq. Chromatogr.* **1,** 343 (1978).

tions are routinely carried out on an RP-300 guard column (4.6 × 30 mm, 7-$\mu$m particle size) using a linear gradient from 0 to 60% (v/v) acetonitrile in 0.1% (v/v) trifluoroacetic acid. This gradient may be modified to optimize the elution time and resolution, facilitating a larger throughput of samples. The peptide is detected by monitoring absorbance at 210 nm. The fractions are pooled and the acetonitrile removed by rotary evaporation. The ion-exchange purified peptide is chromatographed on a hand-packed preparative reversed-phase low-pressure column (2.5 × 100 cm) containing Vydac $C_{18}$ resin (218TPB 2030). The combination of a large column and overnight chromatography provides resolution equivalent to an analytical HPLC column. An acetonitrile gradient in trifluoroacetic acid (0.1%, v/v) is run at 1.5 ml/min for 14 hr. The acetonitrile gradient is tailored to the properties of the particular peptide based on its retention on analytical reversed-phase HPLC. Again the fractions containing peptide are located by dot blot developed with ninhydrin followed by analytical reversed-phase HPLC of the fractions containing peptide. On-line UV monitoring of preparative column eluants is an advantage, but in many cases the concentration of eluted peptide may be off scale on the detector unless customized preparative flow cells are used. The fractions containing pure peptide are pooled and lyophilized as their trifluoroacetic acid salts.

Acidic peptides are purified on a Q-Sepharose HP (Pharmacia) anion-exchange column (1.6 × 10 cm) using a 0 to 0.4 *M* NaCl gradient in 20 m*M* $NH_4HCO_3$ with 20% acetonitrile (v/v) and a flow rate of 1 ml/min for 1140 min. The acetonitrile is removed by rotary evaporation prior to reversed-phase chromatography of the acidic peptides in the presence of 20 m*M* $NH_4HCO_3$. The anion-exchange column fractions can be screened for peptide using the ninhydrin dot blot procedure provided the blot is dipped in diisopropylethylamine (30%, v/v, in dichloromethane) and dried to removed $NH_4^+$ prior to staining with ninhydrin. By using a combination of ion-exchange and reversed-phase chromatography as described, it is possible to process approximately 500 mg crude peptide.

The amino acid composition of the synthetic peptide typically is determined following acid hydrolysis at 110° for 24 hr in 200 $\mu$l 6 *M* HCl containing 0.1% phenol (v/v) and 1% thioglycol (v/v) to maximize recovery of tyrosine and methionine, respectively. The samples are dried under vacuum, dissolved in 0.2 *M* sodium citrate buffer, and run on a Beckman 6300 automated amino acid analyzer. For a purified synthetic peptide, molar ratios are typically within 2 to 5% of unity. If the method of hydrolysis and amino acid analysis is less precise than this, it has doubtful value as a quality control criteria for peptide purity. Amino acid analysis does not reveal the presence of residual protective groups since these are removed by the acid hydrolysis procedure. However, the presence of most protective groups is readily detected from the UV spectrum and generally result in distinct reversed-phase chromatography. Capillary electrophoresis[9] is likely to become a more widely used criterion of purity for synthetic peptides since this effectively complements analytical reversed-phase HPLC.

## Measurement of Peptide Phosphorylation

There have been a variety of protein kinase assays developed that utilize synthetic peptide substrates. These include spectrophotometric,[10] fluorescent,[11] and radioisotopic methods.[12] Quantitation of the transfer of $^{32}P_i$ to the peptide requires a simple separation system to remove [$\gamma$-$^{32}$P]ATP. The most widely used method has been the P81 phosphocellulose cation-exchange paper assay.[12,13] This procedure depends on the presence of multiple basic residues in the synthetic peptide for binding to the P81 paper. For other peptides it is possible to use the anion-exchange column procedure,[14] gel electrophoresis,[15] thin-layer and paper electrophoresis,[16,17] and isoelectric focusing.[18] A tandem column procedure combining both cation- and anion-exchange chromatography has been developed by Egan *et al.*[19] that gives very low backgrounds, but it is restricted to use with basic synthetic peptides. A number of workers have added arginine residues to nonbasic synthetic peptide substrates to adapt them to the P81 paper assay (see [9] in this volume). Both the column separation procedures and the P81 paper assays can employ Cerenkov counting so there is no need for the continued use of scintillation fluids in many cases.

It is important when characterizing a new peptide substrate to measure the stoichiometry of phosphorylation. In addition to providing information about the specificity of the enzyme this also acts as a quality control for the synthetic peptide. The site of phosphorylation can be determined either

[9] W. G. Kuhr, *Anal. Chem.* **62,** 403 (1990).
[10] H. N. Bramson, E. T. Kaiser, and A. S. Mildvan, *CRC Crit. Rev. Biochem.* **15,** 93 (1984).
[11] D. E. Wright, E. S. Noiman, P. B. Chock, and V. Chau. *Proc. Natl. Acad. Sci. U.S.A.* **78,** 6028 (1981).
[12] R. Roskoski, Jr, this series, Vol. 99, p. 3.
[13] D. B. Glass, R. A. Masaracchia, J. R. Feramisco, and B. E. Kemp, *Anal. Biochem.* **87,** 566 (1978).
[14] B. E. Kemp, E. Benjamini, and E. G. Krebs, *Proc. Natl. Acad. Sci. U.S.A.* **73,** 1038 (1976).
[15] T. Hunter, N. Ling, and J. A. Cooper, *Nature (London)* **311,** 480 (1984).
[16] T. W. Wong and A. R. Goldberg, *J. Biol. Chem.* **258,** 1022 (1983).
[17] T. Hunter, *J. Biol. Chem.* **257,** 4843 (1982).
[18] C. J. Fiol, A. Wang, R. W. Roeske, and P. J. Roach, *J. Biol. Chem.* **265,** 6061 (1990).
[19] J. J. Egan, M. K. Chang, and C. Londos, *Anal. Biochem.* **175,** 552 (1988).

by the direct sequencing procedures,[20] mass spectrometry,[21] or by more classical approaches.[22] The introduction of capillary electrophoresis has also provided an important analytical tool for characterizing phosphopeptides[9] due to its great sensitivity and resolving power with polar peptides (Applied Biosystems, bioseparations application notes).

## Design of Peptide Substrates

There are three approaches to the design of protein kinase peptide substrates. The most widely used approach is to synthesize analogs of known phosphorylation site sequences. These may be from natural substrates, autophosphorylation sites, or phosphorylation sites in exogenous substrates such as histone, myelin basic protein, or caseins. The second approach is to use random polymers of amino acids such as Tyr and Glu for tyrosine kinases (see [7] in this volume). While random polymers may be excellent substrates the disadvantage of this approach is that the phosphorylation site is less well defined. In principle, degenerate random peptide sequences could be used in a cocktail to study the specificity of a protein kinase. The most rapidly phosphorylated peptides could then be isolated and their sequence determined. It would be possible to optimize the kinetic properties of a given peptide substrate by "shotgun" substitution of all other amino acids at every residue. Analogous approaches have already been used to study antibody specificity and to epitope map proteins. Geysen *et al.*[23] and Houghton[24] have developed procedures for making small amounts of large numbers of peptides that could be employed for studying protein kinase specificity in this way. The third approach is to prepare substrate analogs of the pseudosubstrate autoregulatory regions (see below) that have been found in a number of protein kinases. A report of engineering phosphorylation sites into recombinant proteins[25] has been made; however, this is beyond the scope of this chapter.

A number of attempts have also been made to understand phosphoryla-

[20] R. E. H. Wettenhall, R. H. Aebersold, L. E. Hood, and S. B. H. Kent, this series, Vol. 201 [15].

[21] H. E. Meyer, E. Hoffmann-Posorske, H. Korte, and L. M. Heilmeyer, *FEBS Lett.* **204,** 61 (1986).

[22] R. B. Pearson, R. Jakes, M. John, J. Kendrick-Jones, and B. E. Kemp, *FEBS Lett.* **168,** 108 (1984).

[23] H. M. Geysen, R. H. Meloen, and S. J. Barteling, *Proc. Natl. Acad. Sci. U.S.A.* **81,** 3998 (1984).

[24] R. A. Houghton, *Proc. Natl. Acad. Sci. U.S.A.* **82,** 5131 (1985).

[25] K. Nakai and M. Kanehisa, *J. Biochem.* (*Tokyo*) **104,** 693 (1988).

tion site sequences in terms of their possible secondary structures.[25–29] Although it is not yet possible to design potent synthetic peptides purely from theoretical structural considerations alone, the rapid expansion in the phosphorylation site sequence database, improved structure predictions, and increased knowledge of protein kinase specificity may make this possible in the future. An alternative to using synthetic peptide substrates directly is to conjugate them to carrier proteins or even immobilize them on various resins. Little has been done in this area except that tandem repeat peptide substrates have been used with the cell cycle kinase,[30] cdc2, and peptide substrates have been conjugated to antibodies as a means of labeling the antibody. It is possible that the kinetics of peptide phosphorylation may be enhanced by conjugation to a carrier protein. A difficulty with chemically conjugated peptide substrates is that, depending on the strategy used, the product may be heterogenous. Peptide substrates can be conjugated to proteins or solid supports through cysteine residues at the carboxyl- or amino-terminal end without interfering with side chains required for recognition by the protein kinase. It is desirable to use a heterobifunctional maleimide cross-linking reagent to give a thioether linkage that is stable in the presence of sulfhydryl reagents. A wide variety of chemical cross-linking procedures have been used to conjugate peptides to proteins to enhance their immunogenicity and some of these could be applied to conjugation and immobilization of protein kinase synthetic peptide substrates (see this series).

A list of synthetic peptide substrates for a number of protein kinases prepared from the local phosphorylation site sequences of known substrates is given in Table I.[31–63] The degree of specificity varies widely with

[25a] B. L. Li, J. A. Langer, B. Schwartz, and S. Pestka, *Proc. Natl. Acad. Sci. U.S.A.* **86,** 558 (1989).

[26] R. E. Williams, *Science* **192,** 473 (1976).

[27] J. F. Leszczynski and G. D. Rose, *Science* **234,** 849 (1986).

[28] C. Radziejewski, W. T. Miller, S. Mobashery, A. R. Goldberg, and E. T. Kaiser, *Biochemistry* **28,** 9047 (1989).

[29] D. A. Tinker, E. G. Krebs, I. C. Felthan, S. K. Attah-Poku, and V. S. Ananthanarayanan, *J. Biol. Chem.* **263,** 5024 (1988).

[30] L. J. Cisek and J. L. Corden, *Nature (London)* **339,** 679 (1989).

[31] D. B. Glass and E. G. Krebs, *J. Biol. Chem.* **254,** 9728 (1979).

[32] R. B. Pearson, J. R. Woodgett, P. Cohen, and B. E. Kemp, *J. Biol. Chem.* **260,** 14471 (1985).

[33] Y. Hashimoto and T. R. Soderling, *Arch. Biochem. Biophys.* **252,** 418 (1987).

[34] B. E. Kemp and R. B. Pearson, *J. Biol. Chem.* **260,** 3355 (1985).

[35] C. H. Michnoff, B. E. Kemp, and J. T. Stull, *J. Biol. Chem.* **261,** 8320 (1986).

[36] B. E. Kemp and M. John, *Cold Spring Harbor Conf. Cell Proliferation* **8,** 331 (1980).

[37] C. House, R. E. H. Wettenhall, and B. E. Kemp, *J. Biol. Chem.* **262,** 772 (1987).

[38] I. Yasuda, A. Kishimoto, S. Tanaka, M. Tominaga, A. Sakurai, and Y. Nishizuka, *Biochem. Biophys. Res. Commun.* **166,** 1220 (1990).

a number of the peptides acting as substrates for multiple protein kinases. In the case of the glycogen synthase kinase-3 substrate peptide listed, prior phosphorylation with casein kinase II is required[51] before it can act as a substrate because of the specificity requirement for an $n + 4$ acidic residue which can be provided by phosphorylated serine [Ser(P)] in this case. In some cases multiple synthetic peptide substrates have been prepared by different laboratories for particular protein kinases. Generally only the one with the most favorable kinetics of phosphorylation has been listed. The protein kinase C peptide substrate PLSRTLS*VAAKK derived from glycogen synthase has the most favorable kinetics but the peptide QKRPS*QRSKYL derived from myelin basic protein is probably more

[39] J. R. Woodgett, K. L. Gould, and T. Hunter, *Eur. J. Biochem.* **161,** 177 (1986).

[40] A. D. Blake, R. A. Mumford, H. V. Strout, E. E. Slater, and C. D. Strader, *Biochem. Biophys. Res. Commun.* **147,** 168 (1987).

[41] B. Gabrielli, R. E. H. Wettenhall, B. E. Kemp, M. Quinn, and L. Bizonova, *FEBS Lett.* **175,** 219 (1984).

[42] S. L. Pelech, B. B. Olwin, and E. G. Krebs, *Proc. Natl. Acad. Sci. U.S.A.* **83,** 5968 (1986).

[43] S. P. Davies, D. Carling, and D. G. Hardie, *Eur. J. Biochem.* **186,** 123 (1989).

[44] P. Hohmann and R. S. Greene, *Biochem. Biophys. Res. Commun.* **168,** 763 (1990).

[45] M. Peter, J. Nakagawa, M. Dorée, J. C. Labbé, and E. A. Nigg, *Cell (Cambridge, Mass.)* **60,** 791 (1990).

[46] P. R. Vulliet, F. L. Hall, J. P. Mitchell, and D. G. Hardie, *J. Biol. Chem.* **264,** 16292 (1989).

[47] J. L. Countaway, I. C. Northwood, and R. J. Davis, *J. Biol. Chem.* **264,** 10828 (1989).

[48] F. Meggio, J. W. Perich, H. E. Meyer, E. Hoffmann-Posorske, D. P. W. Lennon, R. B. Johns, and L. A. Pinna, *Eur. J. Biochem.* **186,** 459 (1989).

[49] P. Agostinis, L. A. Pinna, F. Meggio, O. Marin, J. Goris, J. R. Vandenheede, and W. Merlevede, *FEBS Lett.* **259,** 75 (1989).

[50] E. A. Kuenzel, J. A. Mulligan, J. Sommercorn, and E. G. Krebs, *J. Biol. Chem.* **262,** 9136 (1987).

[51] C. J. Fiol, A. M. Mehrenholz, Y. Wang, R. W. Roeske, and P. J. Roach, *J. Biol. Chem.* **262,** 14042 (1987).

[52] K. J. Chan, *Biochem. Biophys. Res. Commun.* **165,** 93 (1989).

[53] R. A. Masaracchia, B. E. Kemp, and D. A. Walsh, *J. Biol. Chem.* **252,** 7109 (1977).

[54] K. Palczewski, A. Arendt, J. H. McDowell, and P. A. Hargrave, *Biochemistry* **28,** 8764 (1989).

[55] M. H. Watson, A. K. Taneja, R. S. Hodges, and A. S. Mak, *Biochemistry* **27,** 4506 (1988).

[56] G. S. Baldwin, A. W. Burgess, and B. E. Kemp, *Biochem. Biophys. Res. Commun.* **109,** 656 (1982).

[57] C. House, G. S. Baldwin, and B. E. Kemp, *Eur. J. Biochem.* **140,** 363 (1984).

[58] W. Weber, P. J. Bertics, and G. Gill, *J. Biol. Chem.* **259,** 14631 (1984).

[59] J. E. Casnellie and E. G. Krebs, *Adv. Enzyme Regul.* **22,** 501 (1984).

[60] M. L. Harrison, P. S. Low, and R. L. Geahlen, *J. Biol. Chem.* **259,** 9248 (1984).

[61] L. A. Stadtmauer and O. M. Rosen, *J. Biol. Chem.* **258,** 6682 (1983).

[62] J. G. Foulkes, B. Mathey-Prévot, B. C. Guild, R. Prynes, and D. Baltimore, *in* "Cancer Cells/3, Growth Factors and Transformation" (J. Feramisco, B. Ozanne, and C. Stiles, eds.), p. 319. Cold Spring Harbor Lab., Cold Spring Harbor, New York, 1985.

[63] J. A. Cooper, F. S. Esch, S. S. Taylor, and T. Hunter, *J. Biol. Chem.* **259,** 7835 (1984).

TABLE I

PEPTIDE SUBSTRATES BASED ON PROTEIN PHOSPHORYLATION SITE ANALOGS IN NATURAL AND EXOGENOUS SUBSTRATES

| Enzyme | Peptide sequence[a] | $K_m$ ($\mu M$) | $V_{max}$ ($\mu$mol/min/mg) | Ref.[b] |
|---|---|---|---|---|
| Serine/threonine kinases | | | | |
| cAMP-PK (mammalian) | LRRAS*LG | 4.5 | 16 | 1 |
| cAMP-PK (yeast) | VKRKYLKKLTRRAS*FSAQ | 4.3 | 28 | *c* |
| cGMP-PK | MDKVQYLTRSAIRRAS*TIE-MPQQARQNLQNL | 7.0 | 5 | *d* |
| cGMP-PK | RKRS*RAE | 29 | 20 | 31 |
| Cam II PK[e] | PLRRTLS*VAA | 3.5 | 11.3 | 32 |
| | PLARTLS*VAGLPGKK | 12 | 2.75 | 33 |
| Sm MLCK | KKRAARATS*NVFA | 7.5 | 1.4 | 34 |
| Sk. MLCK | AKRAARATS*NVFS | 10 | 31 | 35 |
| Phosphorylase kinase | KAKQIS*VRGSL | 900 | 2.9 | 36 |
| Protein kinase C (mixed) | PLSRTLS*VAAKK | 4 | 12 | 37 |
| | QKRPS*QRSKYL | 7 | d | 38 |
| | AKRRRLSS*LRA | 0.51 | 1.1 | 37 |
| | VRKRT*LRRL | 48 | 0.99 | 39 |
| | YQRRQRKS*RRTI | 24 | 0.92 | 39 |
| | myr-GSSKSKPKDPS*QRRRSLE | 48 | 0.56 | 39 |
| | CNle-RRSSSKAYG | 4.1 | 5.0 | 40 |
| S6 kinase | RRLSS*LRA | 180 | *f* | 41, 42 |
| AMP-PK | HMRSAMS*GLHLVKRR | 30 | 1 | 43 |
| $p34^{cdc2}$ | KS*PAKT*PVK | *f* | *f* | 44 |
| | $(S^*PTS^*PSY)_6$ | 200 | *f* | 30 |
| | AVT*PAKKAAT*PAKKA | 20[g] | *f* | 45 |
| Proline-dependent PK | PTPSAPS*PQPKG | ~50 | *f* | 46 |
| Growth factor PK (p45) | ELVEPLT*PSGEAPNQALLR | *f* | *f* | 47 |
| Casein kinase I | DDDEES*ITRR | 1000 | *h* | 48 |
| Casein kinase II | ESLS*SSEE-NHMe | 11 | *f* | 49 |
| | RRREEES*EEE | 180 | 2.13 | 50 |
| | RRRDDDS*DDD | 60 | 2.19 | 50 |
| Mammary gland PK | ESLSS*SEE-NHMe | 38 | *f* | 49 |
| Glycogen synthase kinase-3 | PRPAS*VPPS*PSLS*RH SS*PHQSEDEEEP | 2 | *f* | 51 |
| Ganglioside PK | RFS*WGAEGQK | *f* | *f* | 52 |
| Histone $H_4$ kinase I | VKRIS*GLG | 43 | 0.016 | 53 |
| Rhodopsin kinase | -DEASTTVSKTETSQVAP- | 1400 | 0.008 | 54 |
| Tropomyosin kinase | KLKYKAISEELDHALNDMTS*I | 500 | 0.034 | 55 |
| Protein-tyrosine kinases | | | | |
| EGF-receptor kinase | RRLEEEEEAY*G | 150 | 0.002 | 56 |
| | LIEDAEYTA | 440 | 0.006 | 57 |
| | DRVY*IHPF (angiotensin II) | 800 | 0.011 | 58 |
| | RREELQDDY*EDD | 90 | 0.001 | *i* |

(*continued*)

TABLE I (*continued*)

| Enzyme | Peptide sequence[a] | $K_m$ ($\mu M$) | $V_{max}$ ($\mu$mol/min/mg) | Ref.[b] |
|---|---|---|---|---|
| p56$^{lck}$ | RRLIADAEY*AARG | 1300 | 0.006 | 59 |
| | DRVY*IHPFHL (angiotensin I) | 2300 | 0.005 | 60 |
| pp60$^{v\text{-}src}$ | DRVY*IHPFHL (angiotensin I) | 1540 | 0.003 | 16 |
| | DRVY*IHPF (angiotensin II) | 2000 | 0.010 | 16 |
| | DRVY*VHPF [V$^5$]angiotensin II) | 870 | 0.007 | 16 |
| | IENEEQEY*VQTVK | 440 | 0.010 | 28 |
| | Raytide | 100 | *f* | *j* |
| Insulin receptor | RRLIEDAEY*ARG | 1200 | *f* | 61 |
| | DRVY*IHPF (angiotensin II) | 2600 | *f* | 61 |
| | RVY*VHPF (angiotensin III inhibitor) | 8000 | *f* | 61 |
| pt$^{abl}$50 | DRVY*IHPFHL (angiotensin I) | 3700 | 1.25 | 62 |
| p140$^{gag\text{-}fps}$ | AAVPSGASTGIY*EALELR (enolase peptide) | 200 | *f* | 63 |

[a] *, Phosphate acceptor site.
[b] Numbers refer to text footnotes.
[c] C. Denis *et al.* (unpublished, 1990); J. R. Cherry, T. R. Johnson, C. Dollard, J. R. Shuster, and C. L. Denis, *Cell (Cambridge, Mass.)* **56,** 409 (1989).
[d] P. J. Robinson, B. Michell, K. I. Mitchelhill, and B. E. Kemp (unpublished, 1990).
[e] Cam, calmodulin; Sm, smooth; Sk, skeletal; MLCK, myosin light chain kinase; myr, myristate; Nle, norleucine; Me, methylester.
[f] Not reported or not given per unit protein.
[g] Inferred from $K_i$ value.
[h] $V_{max}$ equivalent to $\beta$-casein $A^2$.
[i] C. House and B. E. Kemp (erythrocyte band 3 site, unpublished).
[j] Raytide is a model peptide substrate supplied by Oncogene Sciences, Inc. Although the structure of this peptide is not provided by the manufacturer, it has a $K_m$ value similar to the gastrin and band 3 peptides (see above).

specific (Table I). The data concerned with protein kinase C refers to the preparation from brain consisting of a mixture of $\alpha$, $\beta$, and $\gamma$ isoenzymes. There are not yet any isoenzyme-specific synthetic peptide substrates available for protein kinase C, but work in this area is being undertaken.[64] The S6 kinase refers to the growth factor-sensitive enzyme activity; how-

[64] R. M. Marais and P. J. Parker, *Eur. J. Biochem.* **182,** 129 (1989).

TABLE II
PEPTIDE SUBSTRATES BASED ON AUTOPHOSPHORYLATION SITES

| Enzyme | Peptide | $K_m$ ($\mu M$) | $V_{max}$ ($\mu$mol/min/mg) | Ref.[a] |
|---|---|---|---|---|
| Cam II-PK | MHRQET*VDC | 10 | 3.15 | 66 |
| cGMP-PK | IGPRTT*RAQGI | 578 | 0.069 | 67 |
| EGF-receptor PK | RRKGSTAENAEY*LRV | 160 | 0.009 | 68 |
| | RRISKDNPDY*QQD | 340 | 0.010 | 68 |
| | RRDDTFLPVPEY*INQS | 410 | 0.011 | 68 |
| Insulin-receptor PK | TRDIY*ETDY*Y*RK | 240 | ND[b] | 69 |
| pp60$^{src}$ | EDNEY*TARQG | 6250 | 0.001 | 16 |
| pp60$^{src}$ | EDNEY*VARQG | 5890 | 0.001 | 16 |
| p56$^{lck}$ | PRLIEDAEY*AARG | 1160 | 0.01 | 29 |

[a] Numbers refer to text footnotes and these contain the names of the parent substrate proteins.
[b] ND, Not reported.

ever, it is now recognized that there are multiple S6 kinases. There are a number of protein kinases capable of phosphorylating Thr/Ser-Pro sites in addition to p34$^{cdc2}$ and further work is required to understand their comparative specificities. A detailed account of tyrosine kinase synthetic peptide substrates has recently been prepared by Geahlen and Harrison.[65] For several protein kinases consensus phosphorylation site sequences are known, such as RRXS*X for the cyclic AMP-dependent protein kinase and RXXS*XR for protein kinase C. The corresponding synthetic peptides tend to be good substrates, but it should be noted that the consensus sequence is usually recognized by comparing multiple phosphorylation site sequences as well as taking into account peptide substrate structure/function data in which the roles of key specificity determinants have been studied.

Many protein kinases undergo autophosphorylation and there are a number of examples where autophosphorylation site sequences have been used to construct synthetic peptide substrates (Table II).[66–69] The kinetics of phosphorylation of these peptides is generally no better, and frequently worse, than for peptides based on the local phosphorylation site sequences

[65] R. J. Geahlen and M. L. Harrison, *in* "Peptides and Protein Phosphorylation" (B. E. Kemp, ed.), p. 239. Uniscience CRC Press, Boca Raton, Florida, 1990.
[66] R. J. Colbran, Y. L. Fong, C. M. Schworer, and T. R. Soderling, *J. Biol. Chem.* **263,** 18145 (1988).
[67] D. B. Glass and S. B. Smith, *J. Biol. Chem.* **258,** 14797 (1983).
[68] J. Downward, M. D. Waterfield, and P. J. Parker, *J. Biol. Chem.* **260,** 14538 (1985).
[69] L. Stadtmauer and O. M. Rosen, *J. Biol. Chem.* **261,** 10000 (1986).

TABLE III
PEPTIDE SUBSTRATES BASED ON PSEUDOSUBSTRATE PROTOTOPES

| Enzyme | Peptide | $K_m$ ($\mu M$) | $V_{max}$ ($\mu$mol/min/mg) | Ref.[a] |
|---|---|---|---|---|
| $\alpha$PK-C(19–31), $S_{25}$ | RFARKGS*LRQKNV | 0.2 | 8 | 70 |
| $\alpha$PK-C(15–31), $S_{25}$ | DVANRFARKGS*LRQKNV | 18 | 4.5 | 64 |
| $\beta_1$PK-C(15–31), $S_{25}$ | ESTVRFARKGS*LRQKNV | 7.2 | 1.9 | 64 |
| $\gamma$PK-C(15–31), $S_{25}$ | GPRPLFCRKGS*LRQKVV | 9.6 | 2.3 | 64 |
| $\varepsilon$PK-C(149–164), $S_{159}$ | ERMRPRKRQGS*VRRRV | 68 | 1.9 | 71 |
| PKI(14–22), $S_{22}$ | GRTGRRNS*I | 0.11 | 9.2 | 72 |

[a] Numbers refer to text footnotes.

of substrates. Accordingly none of these is used routinely as model substrates.

Potent synthetic peptide substrates have been constructed using the pseudosubstrate sequences found in protein kinase C and the cAMP-dependent protein kinase inhibitor protein. Some of these peptides have $K_m$ values in the submicromolar range (Table III).[70–72] On the other hand, the pseudosubstrate-based peptide substrates for the myosin light chain kinases from both smooth muscle and skeletal muscle are exceedingly poor substrates (not listed). Substrate analogs of the pseudosubstrate regions have been prepared for protein kinase C isoenzyme forms $\alpha$, $\beta_1$, and $\gamma$ and all act as effective substrates (Table III), but are not isoenzyme specific.

Schaap and Parker[71] exploited the pseudosubstrate idea to design a peptide substrate for recombinant $\varepsilon$ protein kinase C that does not phosphorylate histone. As more protein kinases are identified by cDNA cloning techniques, particularly through the use of the polymerase chain reaction, the demand for peptide substrates designed using the pseudosubstrate regulatory concept will increase.

The specificity of synthetic peptide substrates toward different protein kinases may be modulated by making amino acid substitutions. An early example of this was shown for the phosphorylase peptide,[2,36] KRKQIS$_{14}$VRGL, where replacement of Arg-16 with Ala made the peptide a potent substrate for the cAMP-dependent protein kinase and a poor

[70] C. House and B. E. Kemp, *Science* **238,** 1726 (1987).
[71] D. Schaap and P. J. Parker, *J. Biol. Chem.* **265,** 7301 (1990).
[72] D. B. Glass, H. C. Cheng, L. Mende-Mueller, J. Reed, and D. A. Walsh, *J. Biol. Chem.* **264,** 8802 (1989).

TABLE IV
MODULATION OF PEPTIDE SUBSTRATE SELECTIVITY BY SUBSTITUTION

| Peptide sequence [Phosphorylase(9–18)[a]] | Phosphorylase kinase $K_m$ ($\mu M$) | Phosphorylase kinase $V_{max}$ ($\mu$mol/min/mg) | cAMP-PK $K_m$ ($\mu M$) | cAMP-PK $V_{max}$ ($\mu$mol/min/mg) |
|---|---|---|---|---|
| KRKQIS*VRGL | 900 | 2.9 | 3900 | 4.1 |
| KRKQIS*VAGL | 2500 | 0.18 | 36 | 21.4 |
| KAKQIS*VRGL | 900 | 2.7 | 2200 | 0.04 |

[a] From Ref. 2.

substrate for phosphorylase kinase (Table IV). Substitution of Arg-10 with Ala had the opposite effect, enhancing specificity toward phosphorylase kinase and suppressing phosphorylation by the cAMP-dependent protein kinase. Similar switches in specificity have been engineered for the glycogen synthase peptide,[32] PLSRTLS$_7$VAA, where substitution with Arg instead of Ser-3 favors phosphorylation with the calmodulin-dependent kinase II. Substitution of Thr-5 with Arg, however, favors phosphorylation by the cAMP-dependent protein kinase.

## Applications

Synthetic peptide substrates have proved extremely useful reagents in the study of protein kinases across a wide spectrum of studies, extending from those carried out with crude extracts to structural studies with NMR and X-ray crystallography. By using peptides containing a single phosphorylatable residue it has been possible to detect protein kinases in cell extracts containing multiple protein kinase activities. The sensitivity and specificity of synthetic peptide substrates has made them the substrate of choice in the study of hormonal regulation of protein kinases.[73] Stability and chemical purity are also major benefits in using synthetic peptide substrates. Since quantities are not usually limiting they can often be used at saturating concentrations, making it possible to obtain maximum rates of phosphorylation. Synthetic peptides may be used effectively in the confirmation of natural phosphorylation sites. Often the available protein substrate is limiting; synthesis of peptides corresponding to a number of potential phosphorylation sites in the protein can be used to determine the

[73] S. A. Livesey, B. E. Kemp, C. A. Re, N. C. Partridge, and T. J. Martin, *J. Biol. Chem.* **257,** 14983 (1982).

most likely site.[74] The power of this approach can be further enhanced by making complementary point mutations in the protein substrate. With improvements in the potency of peptide substrates it is likely that applications of affinity purification will also increase ([13], this volume).

In the past the development of synthetic peptide substrates has lagged behind the discovery of protein kinases. This situation is changing due to the widespread use of many synthetic peptides, such as the kemptide and the ribosomal S6 peptide, which have been used to detect protein kinases with unforeseen overlapping specificities to the cAMP-dependent protein kinase.[75,76] The availability of relatively specific inhibitors such as the cyclic AMP-dependent protein kinase inhibitor peptide PKI(5–22)[77] and other pseudosubstrate inhibitors (this series, Volume, 201 [24]) as well as calcium chelators for calcium-dependent protein kinases greatly facilitates attempts to detect new protein kinase activities using peptide substrates capable of being phosphorylated by multiple protein kinases.

It can be expected that in the forthcoming years there will be even greater synergy between the use of recombinant protein expression and synthetic peptides to create a variety of protein kinase substrates in order to explore the mechanisms of regulation by protein phosphorylation.

[74] B. Luscher, E. Christenson, D. W. Litchfield, E. G. Krebs, and R. N. Eisenman, *Nature (London)* **344,** 517 (1990).
[75] E. Erikson and J. L. Maller, *Second Messengers Phosphoproteins* **12,** 135 (1988).
[76] J. K. Klarlund, A. P. Bradford, M. G. Milla, and M. P. Czech, *J. Biol. Chem.* **265,** 227 (1990).
[77] B. E. Kemp, H. C. Cheng, and D. A. Walsh, this series, Vol. 159, p. 173.

# [11] Synthetic Peptide Substrates for Casein Kinase II

*By* DANIEL R. MARSHAK and DENNIS CARROLL

## Casein Kinase II

The enzyme casein kinase II (CKII) is a protein-serine/threonine kinase found in all eukaryotic cells.[1] Its ubiquitous distribution among species and tissues implies a function central to all nucleated cells. CKII was first identified from rabbit reticulocyte lysates,[2] and subsequently isolated from hypotonic extracts of mammalian liver and lung tissue.[3] Following

[1] G. M. Hathaway and J. A. Traugh, *Curr. Top. Cell. Regul.* **21,** 101 (1982).
[2] G. M. Hathaway and J. A. Traugh, *J. Biol. Chem.* **254,** 762 (1979).

anion-exchange chromatography of these extracts on DEAE-cellulose and assay of the resulting fractions using partially hydrolyzed casein as substrate, two major peaks of casein kinase activity were identified. The species eluting at a higher salt concentration was designated CKII. Purification of CKII has been accomplished from various sources, including mammals, worms, flies, and yeast. The enzyme consists of two subunits, $\alpha$ and $\beta$, with molecular weight ranges of 37K–44K and 24K–28K, respectively, with an apparent subunit composition of $\alpha_2\beta_2$. The $\alpha$ subunit appears to be catalytic based on homology to other protein kinases, particularly in the consensus ATP-binding region, by affinity labeling the ATP-binding site, and by expression of the cDNA for the $\alpha$ subunit.[4] There appears to be a second form of the $\alpha$ subunit, designated $\alpha'$, but the significance of this alternative structure is not yet clear. The role of the $\beta$ subunit appears to be that of a regulatory element, as it is phosphorylated by exogenous kinases and contains sites of autophosphorylation.[5] The complete amino acid sequences of these subunits is available, either from protein structural determinations or inferred from cDNA structures. The enzyme is capable of using either ATP or GTP as substrate with similar $K_m$ values (10 and 30 $\mu M$, respectively). The enzyme activity can be activated by polycationic materials, such as polyamines, and inhibited with low concentrations of polyanionic compounds, such as heparin. In recent years, CKII has been found to be stimulated by insulin[6] or epidermal growth factor[7] in cultured cells, by serum stimulation of quiescent fibroblasts,[8] by virus infection of epithelial cells,[9] and by phorbol ester stimulation of primary cultures of kidney cells.[10] These studies indicate that CKII is regulated in various cells, both by hormonal stimulation and by pharmacological agents.

Many substrates have been identified for CKII, although the physiological relevance of the phosphorylation is not clear in all cases. Casein, the

[3] A. M. Edelman, D. K. Blumenthal, and E. G. Krebs, *Annu. Rev. Biochem.* **56,** 567 (1987). The characteristics of CKII referred to are summarized in this review.

[4] R. Padmanabha, J. L. Chen-Wa, D. E. Hanna, and C. V. Glover, *Mol. Cell. Biol.* **10,** 4089 (1990).

[5] P. Ackerman, C. V. C. Glover, and N. Osheroff, *Proc. Natl. Acad. Sci. U.S.A.* **87,** 821 (1990).

[6] J. Sommercorn and E. G. Krebs, *J. Biol. Chem.* **262,** 3839 (1987).

[7] P. Ackerman and N. Osheroff, *J. Biol. Chem.* **264,** 11958 (1989).

[8] D. Carroll and D. R. Marshak, *J. Biol. Chem.* **264,** 7345 (1989).

[9] D. Carroll, E. Moran, and D. R. Marshak, unpublished observations (1988).

[10] D. Carroll, N. Santoro, and D. R. Marshak, *Cold Spring Harbor Symp. Quant. Biol.* **53,** 91 (1988). Two points in this reference require clarification. The $V_{max}$ values quoted in this reference are in nmol/10 min/mg, and are therefore consistent with the results presented here. Also, the sequence of the Fos peptide was the 19-mer described here (Table I).

substrate first used for the enzyme, appears to be a nonphysiological substrate. The CKII enzyme activity is found both in cytosol and in nuclei, and there are substrates identified in both locations. Cytosolic substrates include proteins involved in translational control[1] (e.g., eukaryotic initiation factor (eIF)-2, -3, -4B, and -5), metabolic regulation (e.g., glycogen synthase[11] and calmodulin[12]), and the cytoskeleton (e.g. nonmuscle myosin,[13] $\beta$-tubulin[14]). Substrates found in the nucleus include DNA topoisomerase II,[15] RNA polymerases I and II,[16] oncoproteins[10,17,18] (e.g., adenovirus E1a, SV40 large T antigen, Myc, Fos, Myb), and transcription factors[19] (e.g., serum response factor, cAMP regulatory element-binding protein). The extraordinary range of substrates for this enzyme supports the contention that CKII plays a significant role in cell physiology.

The specificity of CKII, like many other protein kinases, requires only a short recognition sequence that can be identified from the primary structure, including a serine or threonine, followed by a cluster of at least three acidic amino acids.[20,21] Structures containing predicted $\beta$ turns generally have a lower $K_m$ than those without such structures.[20–22] For example, one or two proline residues preceding the phosphorylated serine or threonine are common in CKII substrates. Basic residues immediately adjacent to the site of phosphorylation reduce the performance of a substrate, but basic residues distal to the site may have some beneficial effect. This may relate to the activation properties of polycationic substances.

In this chapter we review the use of new peptide substrates for CKII based on several of the nuclear oncoproteins. This class of CKII substrates is of special interest in transcriptional control, cell proliferation, tissue differentiation, and the pathophysiology of cancer. The methods presented permit the quantitative analysis of the phosphorylation of these model

[11] C. Picton, J. Woodgett, B. Hemmings, and P. Cohen, *FEBS Lett.* **150,** 191 (1982).
[12] F. Meggio, A. M. Brunati, and L. A. Pinna, *FEBS Lett.* **215,** 241 (1987).
[13] N. Murakami, G. Healy-Louie, and M. Elzinga, *J. Biol. Chem.* **265,** 1041 (1990).
[14] D. S. Kohtz and S. Puszkin, *J. Neurochem.* **52,** 285 (1989).
[15] P. Ackerman, C. V. C. Glover, and N. Osheroff, *J. Biol. Chem.* **263,** 12653 (1988).
[16] D. A. Stetler and K. M. Rose, *Biochemistry* **21,** 3721 (1982).
[17] B. Lüscher, E. A. Kuenzel, E. G. Krebs, and R. N. Eisenman, *EMBO J.* **8,** 1111 (1989).
[18] B. Lüscher, E. Christenson, D. W. Litchfield, E. G. Krebs, and R. N. Eisenman, *Nature (London)* **344,** 517 (1990).
[19] J. R. Manak, N. de Bisschop, R. M. Kris, and R. Prywes, *Genes Dev.* **4,** 955 (1990).
[20] F. Meggio, F. Marchiori, G. Borin, G. Chessa, and L. A. Pinna, *J. Biol. Chem.* **259,** 14576 (1984).
[21] E. A. Kuenzel, J. A. Mulligan, J. Sommercorn, and E. G. Krebs, *J. Biol. Chem.* **262,** 9136 (1987).
[22] D. Small, P. Y. Chou, and G. D. Fasman, *Biochem. Biophys. Res. Commun.* **79,** 341 (1977).

substrates. Throughout such studies, we advocate a multidisciplinary approach[23,24] utilizing a variety of methodologies, such as peptide synthesis, high-performance liquid chromatography (HPLC), thin-layer chromatography, chemical modification, protein sequencing, amino acid analysis, and mass spectrometry. Through such an approach, complex questions can be addressed, such as the level of phosphorylation at adjacent sites, the order of addition of phosphates at multiple sites, and the identification of peptide contaminants in synthetic products.

## Synthetic Peptide Substrates Based on Oncoproteins

Oncoproteins refer to the proteins products of oncogenes; that is, genes defined by mutations that affect the ability of those genes to transform cells in culture or to produce tumors in animals when introduced through viruses, artifically transfected cells, or external mutagenesis.[25] Many of the known cytoplasmic oncoproteins affect some aspect of signal transduction mechanisms at the level of growth factors, receptors, G proteins, or protein kinases. The nuclear oncoproteins of viral and cellular origins are often DNA-binding proteins, transcription factors, and/or regulators of DNA replication.

Through genetic analysis, nuclear oncoproteins have been found to have discrete domains that are responsible for cell transformation, cell proliferation, DNA binding, transcriptional control, or protein–protein interactions.[26,27] In several cases, these sites include sequences that meet the requirements for candidate CKII phosphorylation sites. Labeling studies using [$^{32}$P]phosphate indicate that most, if not all of these proteins are phosphorylated *in vivo,* and in some cases, peptide maps are consistent with the notion that CKII phosphorylates the expected site.[17,18] Several transcription factors that are either oncoproteins themselves or are involved in oncogene expression contain consensus sequences for CKII phosphorylation sites. These include the serum response factor (srf) and the cAMP response element-binding protein (creb) that regulate the *fos* gene,[28] the fos/jun protein families themselves,[29] and the yeast transcrip-

[23] D. R. Marshak and B. A. Fraser, *in* "Brain Peptides Update" (J. B. Martin, M. J. Brownstein, and D. T. Krieger, eds.), Vol. 1, p. 13. Wiley, New York, 1987.

[24] D. R. Marshak and B. A. Fraser, *in* "High Performance Liquid Chromatography in Biotechnology" (W. Hancock, ed.), pp. 531. Wiley, New York.

[25] J. M. Bishop, *Annu. Rev. Biochem.* **52,** 301 (1983).

[26] J. Ma and M. Ptashne, *Cell (Cambridge, Mass.)* **48,** 847 (1987).

[27] E. Moran and M. B. Matthews, *Cell (Cambridge, Mass.)* **48,** 177 (1987).

[28] M. Gilman, R. N. Wilson, and R. A. Weinberg, *Mol. Cell. Biol.* **6,** 4305 (1986).

[29] B. R. Franza, Jr., F. J. Rauscher, III, S. F. Josephs, and T. Curran, *Science* **239,** 1150 (1988).

TABLE I
SYNTHETIC PEPTIDE SUBSTRATES OF CASEIN KINASE II

| Protein | $K_m$ ($\mu M$) | $V_{max}$ (nmol/min/mg) | Sequence[a] |
|---|---|---|---|
| E1a | 25 | 45 | H-HEAGFPPSDDEDEEG-$NH_2$ |
| LTag | 35 | 54 | H-SEEMPSSDDEATAD-$NH_2$ |
| Myc | 27 | 31 | H-EEETPPTTSSDSEEEQEDEEE-OH |
| Fos | 2 | 6 | H-RRGKVEQLSPEEEEKRRIRR-$NH_2$ |
| Nef | 0.65 | 30 | H-MDDVDSDDDD-$NH_2$ |
| ETE | 720 | 38 | H-RRREEETEEE-OH |
| p53 | *b* | *b* | H-RRTEEE-OH |

[a] A, Alanine; D, aspartic acid; E, glutamic acid; F, phenylalanine; G, glycine; H, histidine; I, isoleucine; L, leucine; M,methionine; P, protine; Q, glutamine; R, arginine; S, serine; T, threonine; V, valine.

[b] The p53 peptide did not show significant incorporation of phosphate using bovine liver casein kinase II at peptide concentrations up to 10 m*M*.

tion factors GAL4 and GCN4.[26] In the case of srf, CKII phosphorylation appears to cause a more avid interaction with DNA,[19] while Myb,[18] Myc,[30] and LTag[31] display decreased binding to DNA after CKII phosphorylation. Whether these effects mimic the *in vivo* function of CKII is not yet established, but it is likely that each protein will have to be analyzed independently for the effects of CKII phosphorylation.

We have used synthetic peptides as models of the phosphorylation sites for CKII, based on the primary sequences of adenovirus E1a, simian virus 40 large T antigen (LTag), the human immunodeficiency virus *nef* protein, the human antioncoprotein p53, Myc, Fos, and Myb. The sequence of these peptides are shown in Table I, along with the $K_m$ and $V_{max}$ values for the peptides using purified bovine liver CKII. For comparison, the peptide substrate developed by Kuenzel and Krebs[32] (RRREEETEEE) is shown, designated the ETE peptide. The kinetic parameters for the peptides vary over three orders of magnitude, but all appear to be specific substrates of CKII and give values similar to those previously reported.[10] Of note are the Fos and *nef* peptides that have quite low $K_m$ values. The *nef* peptide is an outstanding substrate, displaying submicromolar $K_m$ values, and is similar to the DSD peptide used by Lüscher *et al.*[18] (RRRDDDSDDD). The Fos peptide shows a low $V_{max}$ value, as well as a low $K_m$ value, and displays inhibition of CKII phosphorylation at high

[30] D. Carroll, W.-K. Chan, M. T. Vandenberg, D. Spector, and D. R. Marshak, in preparation.
[31] D.Carroll, D. McVey, and D. R. Marshak, (1989) unpublished observations.
[32] E. A. Kuenzel and E. G. Krebs, *Proc. Natl. Acad. Sci. U.S.A.* **82,** 737 (1985).

concentrations. This observation led us to begin to examine peptides closely for the possibility of contamination with adducts or other modified synthetic materials. In addition, the reported $K_m$ value for the ETE model peptide varies slightly from that originally described,[32] suggesting inconsistencies either in assay procedures or in the synthetic products from different laboratories. It is of interest to compare the ETE peptide with the p53 peptide, which is not a good substrate for CKII. The major differences in the two peptides is the overall size and the proximity of the arginine residues to the threonine residue. Therefore, we felt it necessary to eliminate the possibility that termination, deletion or other artifacts of the synthesis were not altering the apparent kinetic parameters for the peptide. Below we address problems in synthetic procedures for CKII substrate peptides, followed by a summary of assay protocols that have given reproducible results in our laboratory.

## Synthesis of Peptides

Chemical synthesis of peptides using automated instrumentation is convenient, rapid, and cost effective. However, the synthesis of peptide substrates for an enzyme, in this case CKII, requires attention to the details of the synthetic chemistry and significant purification and characterization of the product. It is essential to remove minor contaminants such as deletion products, termination products, incompletely deprotected analogs, scavenger adducts, and chemically modified forms. Such contaminants can lead to erroneous results and/or lot-to-lot variability when measuring the kinetics of peptide phosphorylation. For example, a modified peptide might act as a competitive inhibitor, and even a contaminant of 1% would seriously affect the kinetics if the $K_i$ value is below the $K_m$ value of the major product. Therefore, crude peptides produced by automated synthesis without purification are not suitable for detailed kinetic analysis.

### *Automated Synthesis*

The synthetic chemistry employed for peptides in our laboratory utilizes an Applied Biosystems (Foster City, CA) 430A instrument to couple preformed, symmetric anhydrides or 1-hydroxybenzotriazole (HOBt)-activated esters of $N^{\alpha}$-*t*-Boc amino acids on phenylacetamidomethyl-derivatized or benzhydrylamine-derivatized polystyrene resins (1% divinylbenzene cross-linked) at a substitution level >0.5 mmol/g. These procedures are based on the methods of Merrifield for solid-phase peptide synthesis.[33]

[33] G. Barany and R. B. Merrifield, *in* "The Peptides: Structure, Function, Biology" (E. Gross and J. Meienhofer, eds.), Vol. 2, p. 1. Academic Press, New York, 1980.

Other chemistries, particularly that using 9-fluorenylmethyloxycarbonyl (Fmoc)-protected amino acids, are quite suitable for synthesis. Choice of solvent system is important to optimize yields of the desired peptide. Dichloromethane, dimethylformamide, *N*-methylpyrrolidone, and dimethyl sulfoxide are the most frequently employed. When using dichloromethane, it is useful to recouple each amino acid in a more polar solvent, such as dimethylformamide. Our work has been with *N*-methylpyrrolidone and dichloromethane, including an additional coupling period in dimethyl sulfoxide. Although reaction times are increased to about 90–120 min, these solvents promote swelling of the polystyrene, allowing a higher coupling yield. In addition, we routinely treat the resin with acetic anhydride following the completion of the coupling reaction to block any unreacted $N^{\alpha}$-amino groups and prevent further reaction. This technique results in the formation of a series of termination products with $N^{\alpha}$-acetylations rather than a large number of deletion products. Acetylated termination products are more easily separated by chromatography than are deletion products because termination peptides generally have shorter retention times compared to the desired product.

*Deprotection*

Another area of concern is the modifications and side reactions that can occur during acidolytic cleavage of the peptide from the resin and deprotection of the side chains. Synthetic peptide substrates for CKII always include several aspartic and/or glutamic acid residues, and there are particular side reactions that plague these syntheses. Failure to compensate for these reactions will lead to peptide products that yield erroneous results on phosphorylation. Cleavage and deprotection of the peptide from the resin involve acidolytic treatment, usually in either anhydrous hydrogen fluoride or trifluoromethanesulfonic acid.[34] The latter treatment is extremely dehydrating and can lead to the cyclization of acidic or amidated amino acids. In particular, the use of the *O*-benzyl-protecting group on aspartic acid can lead to the cyclization of the side chain to form the aspartimide. Using *O*-cycloalkanes as protecting groups for aspartic acid residues this side reaction because these esters are poor leaving groups compared to *O*-benzyl, and hydrolysis is favored over cyclization by the peptide backbone $N^{\alpha}$ moiety. Therefore, we routinely use $N^{\alpha}$-*t*-Boc-($\beta$-*O*-cyclohexyl)aspartic acid. The second most frequent problem with synthesis of acidic peptides is adduct formation between the peptide

[34] H. Yajima and N. Fujii, *in* "The Peptides: Analysis, Synthesis, Biology" (E. Gross and J. Meienhofer, eds.), Vol. 5, p. 66. Academic Press, New York, 1983.

glutamate and glutamine side chains and scavenger molecules, such as anisole or cresol. These chemicals are necessary in acidolytic deprotection to guard against reattachment of protecting groups. However, anisole adducts with glutamyl residues can significantly alter the charge of the product and introduce an unwanted modification. Such adducts are not readily detectable on amino acid analysis because aqueous acid hydrolysis at elevated temperatures destroys the adduct, producing free glutamic acid. Thus, amino acid analysis alone is not adequate documentation of the purity of a synthetic peptide.

*Solubilization and Desalting*

Following deprotection, scavengers and other organic reaction products are removed by ether extraction and precipitation of the peptide. Solubilization of the precipitated peptide can be difficult, particularly for acidic CKII substrates. If the product is to be lyophilized, neutralizing the precipitate with a volatile base, such as ammonium hydroxide or triethylamine, is necessary. Variable proportions of an organic solvent, such as methanol or acetonitrile, are often required for solubilization of the neutralized peptide in aqueous media. Alternatively, the product can be neutralized with sodium hydroxide or Tris (trishydroxymethylaminomethane), followed by solubilization in aqueous 6 *M* guanidine-hydrochloride or 8 *M* urea. The choice of solubilization conditions depends on the desalting step to follow. Three options are available, all of which have advantages and disadvantages for CKII substrates. First, large-scale gel filtration can be used to remove small-molecule contaminants. We have used columns of 5.0 × 100 cm or greater in size containing polyacrylamide resin, BioGel P-2 (Bio-Rad, Richmond, CA). Solvents for these columns are either 0.05 *M* ammonium acetate or 0.1% acetic acid containing 20% (v/v) acetonitrile. Choice of solvents depends on the solubility of the particular peptide. When using gel filtration, it is usually necessary to concentrate the solubilized, ether-precipitated peptide through lyophilization. Therefore, a volatile solvent system must be used in solubilization.

Second, large-scale, reversed-phase HPLC is very useful for desalting the peptide. We have used the Waters (Milford, MA) Delta Prep 3000 preparative HPLC equipped with the Prep-Pak module that permits cartridge columns of 4.8 × 30 cm. Flow rates are generally 25–75 ml/min with solvent consisting of 0.1% aqueous trifluoroacetic acid and increasing proportions of acetonitrile. The choice of $C_4$- or $C_{18}$-modified silica depends on the size of the peptide; peptides <5000 atomic mass units (amu) are run on $C_{18}$ material. One advantage of large-scale, reversed-phase desalting is its rapidity and tolerance of peptides solubilized in guanidine hydrochloride. Large amounts of urea are not recommended for preparative HPLC because urea is uncharged and can interfere with the chromatographic separation. Guanidine hydrochloride is a charged species and falls through the column with the unbound material. We routinely solubilized peptides by stirring overnight at 4° in 6 *M* guanidine hydrochloride at neutral pH.

The third method for desalting that is quite useful for CKII substrates is the use of Dowex anion-exchange resins. The peptide is solubilized in 0.2% triethylamine-hydrochloride, pH 9–11, and the solution is stirred for 60 min at room temperature with Dowex AG1-X2 (200–400 mesh). The unbound material is washed with the equilibration buffer, and the peptide is eluted with 10 m*M* HCl or 0.1% (w/v) trifluoroacetic acid. Urea can be included in the buffer if necessary to solubilize the peptide, and does not interfere with ion exchange. In this case guanidine hydrochloride is not desirable because it will interfere with ion exchange, unless the pH is elevated beyond the $pK_a$ of the guanidinium group. Extremely high pH is not recommended for peptides, because deamidation of asparagine and glutamine can occur, as well as base hydrolysis of the peptide chain. Elevated pH may be used for short periods (<2 hr) at low temperatures, particularly when no amidated side chains are present in the peptide. Ion exchange on Dowex resins is a rapid and simple method often overlooked as a procedure for desalting crude peptides.

*Purification*

Final purification by reversed-phase HPLC is done preparatively on a column (2.2 × 25 cm) of octyldecylsilanyl silica using 300-Å pore size, synthetic, spherical silica, 10 μm in diameter (Vydac, The Separations Group, Hesperia, CA). This column gives very high resolution with reasonable back-pressure characteristics. A guard column is recommended to protect the resin bed from particulates and irreversibly bound materials. In addition, we recommend filtration of the sample through 0.22-μm filters of nylon 66 (Schleicher and Schuell, Keene, NH). We typically load 400 mg of peptide in 0.1% trifluoroacetic acid on this column and elute with a linear gradient of increasing proportions of acetonitrile. It is useful to monitor the effluent at several wavelengths, typically 210, 220, 235, 260, and 280 nm to obtain a useful chromatographic profile. Multiwavelength or diode-array detectors are available for this purpose. Clearly, peptides without aromatics or sulfhydryls will not be expected to show absorption at 275–282 nm. In these cases, peaks displaying significant optical density at 280 nm are likely to be adducts of anisole or other aromatic or sulfhydryl scavengers. Another useful method for optimizing the purification is complex gradient programs. Adducts generally elute at a retention time later than the desired product, while salts and termination products elute at earlier retention times. Introduction of an isocratic step at low organic solvent and another isocratic step at high (60–80%, v/v) organic solvent serves to batch elute these unwanted products. In between, one runs a shallow gradient or isocratic "plateau" at a proportion of organic solvent that yields the best separation. This level must be determined empirically using several analytical injections of the partially purified peptide (0.5–1.0 mg). Contaminants that are vey similar to the desired peptide chemically, such as deletion and dehydration products, are often separated only by isocratic or shallow gradient elutions. This is the primary reason for the acetylation of unreacted species during synthesis; the elimination of deletion products is a much more difficult separation problem than that for a set of nested termination products. Similarly, elimination of dehydration products by the use of $\beta$-*O*-cyclohexyl-protected aspartic acid simplifies the separation.

*Analysis*

Critical to the analysis of the synthetic product is a multidisciplinary approach including analytical HPLC, amino acid analysis, sequencing, and mass spectrometry. We use plasma desorption, time-of-flight, mass spectrometry[35] to screen fractions from the HPLC separation. If the mass of the molecular ion $(M + H)^+$ is incorrect, it is not worthwhile to pursue further characterization of the fraction. Usually, it is possible to tentatively assign the contaminants based solely on their measured mass and knowledge of coupling yields in the synthesis. For example, a product with a molecular mass that is 90 mass units above the expected mass is likely to be an anisole adduct, arising from the addition of anisole (108 mass units) and the elimination of the elements of water (18 mass units). Products of mass lower than expected are often assigned by calculating the termination product mass and adding 42 for the acetylation (43 mass units for acetyl minus 1 mass unit for the hydrogen atom replaced). A helpful tool in this pursuit is a computer program for calculating predicted molecular weights of a peptide.[36] It is often possible to assign the residue location of the adduct or the cyclization from the fragment ion pattern. Fast atom bom-

[35] B. Sundqvist, P. Roepstorff, J. Fohlman, A. Hedin, P. Häkansson, I. Kamensky, M. Lindberg, M. Salehpour, and G. Säwe, *Science* **226,** 696 (1984).

[36] We use PEPTOP v 3.0 available from Dr. Blair A. Fraser, Food and Drug Administration, Bethesda, MD 20892.

bardment mass spectrometry yields excellent fragmentation patterns,[37] and $^{252}$Cf-fission fragment bombardment also generates useful fragments for peptides of 1000–4000 mass units.[38] However, analysis of the fragmentation pattern for every side fraction is time consuming, and it is usually sufficient to characterize the $(M + H)^+$ to identify the correct fraction. Occasionally, modified peptides are useful as inhibitors of the enzyme.

After the desired product has been identified by mass spectrometry, further analysis can be performed by amino acid analysis and protein sequencing. Amino acid analysis is particularly useful for determining the exact peptide concentration of a solution. For kinetic analysis exact concentration measurements are necessary. Dissolve the peptide in water or a volatile buffer such as 0.1 *M* ammonium bicarbonate or 0.1% acetic acid, and store as a stock solution of 10 m*M* peptide. Dilutions of peptide are then made volumetrically based on the peptide content as measured by amino acid analysis. This is the only acceptable method for determining the peptide concentration accurately. Sequence analysis can be helpful in confirming a peptide structure. However, the yields of phenylthiohydantoin derivatives of serine and threonine tend to be low due to $\beta$ elimination during the Edman degradation, so it is difficult to determine if such dehydration has occurred in the peptide product or during the sequencing process. Dehydration due to aspartimide formation can be analyzed on the sequencer. Aspartic acid yields are low on cycles where an aspartimide has occurred, but losses due to repetitive yield, washout, and lag complicate the analysis. Most dehydration products will have been identified by mass spectrometry and discarded. It is usual that only the final, purified product undergoes verification of structure by sequencing. Thus, sequencing, amino acid analysis, and mass spectrometry complement each other as methods to characterize synthetic substrates of CKII.

## Mass Spectrometric Analysis of ETE Peptide

To emphasize the importance of proper care in synthesis, purification, and characterization of peptide substrates, we present the analysis of the purification of a batch of the ETE peptide of Kuenzel and Krebs.[32] The sequence of this peptide, shown in Table I, contains three arginine residues, six glutamic acid residues, and one threonine residue. The peptide is synthesized as described above using HOBt-activated esters of $N^{\alpha}$-*t*-

[37] W. H. Burgess, T. Mehlman, D. R. Marshak, B. A. Fraser, and T. Maciag, *Proc. Natl. Acad. Sci. U.S.A.* **83,** 7216 (1986).

[38] D. R. Marshak and G. E. Binns, *in* "Current Research in Protein Chemistry" (J. J. Villafranca, ed.), p. 127. Academic Press, San Diego, California, 1990.

Boc amino acids in *N*-methylpyrrolidone and dimethyl sulfoxide. The side chains of glutamic acid and threonine residues are protected as the *O*-benzyl derivatives and the arginines are protected as the tosyl derivative. The peptide is removed from the resin with liquid HF : anisole : dimethyl sulfide (26.6 : 1 : 1) for 2 hr at $-10°$. Following precipitation in diethyl ether, the crude peptide is redissolved in 0.1% (w/v) trifluoroacetic acid and subjected to chromatography on a column (4.8 × 30 cm) of $C_{18}$, 300-Å pore size, 15-$\mu$m diameter silica (Waters Prep-Pak). The column is eluted with a linear gradient of acetonitrile (0–60%, v/v) over 30 min at a flow rate of 25 ml/min. Fractions are collected based on absorbance at 214 nm, and aliquots (20 $\mu$l) of each fraction are analyzed by plasma desorption mass spectrometry on a Bio-Ion spectrometer (Bio-Ion Division, Applied Biosystems, Uppsala, Sweden).

The results of the analysis are shown in Fig. 1. The expected isotopically averaged mass of the peptide protonated molecular ion $(M + H)^+$ is 1363.24 mass units. Figure 1A shows the fraction containing the desired product, with measured *m/z* for the $(M + H)^+$ of 1363.9. The smaller peaks at lower *m/z* correspond to sequence fragment ions. Figure 1B and C shows apparently dehydrated products: loss of one molecule of water, 18 mass units, to give the *m/z* 1845.7 species (Fig. 1B) and loss of two molecules of water, 36 mass units, to give the *m/z* 1327.5 species (Fig. 1C). The spectra in Fig. 1D, E, and F show apaprent anisole adducts, with additions of 90 mass units. One anisole addition gives *m/z* 1454.0, two gives *m/z* 1543.8, and three gives *m/z* 1633.4 mass units. The fraction of interest, that giving the result in Fig. 1A, was further purified by reversed-phase HPLC on a column (2.2 × 25 cm) of $C_{18}$, 300-Å, 10-$\mu$m silica (Vydac, The Separations Group, Hesperia, CA). The yield of peptide from 0.5 mmol substituted resin was 430 mg, a final overall yield of 63.1%. These results demonstrate that it is possible to get excellent yield of active, CKII substrate peptides that are free of dehydration and adduct contaminants using high-resolution, preparative HPLC methods and mass spectrometry to screen fractions. Three peptide fractions were tested as substrates for CKII: the purified ETE peptide after the second HPLC, the crude peptide shown in Fig. 1A and the mixture of dehydrated and desired product shown in Fig. 1B. All peptide samples were tested at 0.64 m*M*, as measured by amino acid analysis, for 30 min at 37° with purified bovine brain CKII. The purified material showed incorporation of 9.52 pmol of $[^{32}P]$phosphate. The material in Fig. 1A, incorporated 7.61 pmol, and the material in Fig. 1B incorporated 4.69 pmol. Thus, the pure peptide, after high-resolution HPLC, incorporated twice the phosphate of the material contaminated with the dehydration product, and significantly more than the material after only one round of HPLC purification. All three pools of

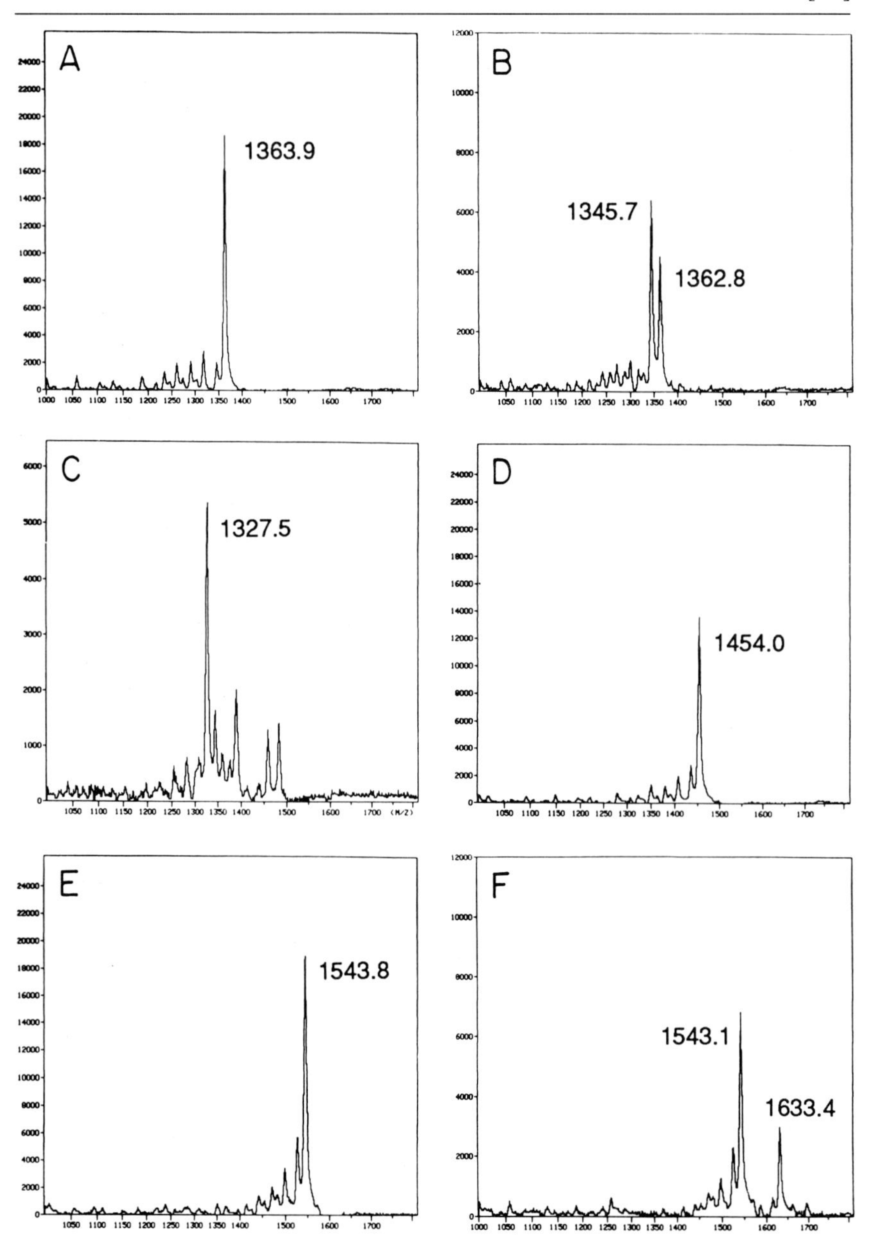


FIG. 1. Mass spectrometric analysis of ETE peptide. The ETE peptide (sequence shown in Table I) was synthesized as described in the text, and desalted by preparative reversed-phase HPLC. Fractions were collected and analyzed by plasma desorption, time-of-flight mass spectrometry (A–F). The ordinate shows the number of positive ions detected, and the abscissa shows the mass-to-charge ratio, $m/z$. The apparent molecular weights of the major peaks are marked.

peptide were indistinguishable by amino acid analysis. It is clearly imperative to purify synthetic peptide substrates for CKII to true chemical homogeneity to obtain accurate measurements of protein kinase activity.

## Quantitative Assay Methods

The methods of assay of peptides for CKII phosphorylation can be either similar to assays for other protein kinases or quite different, dependent upon the structure of the peptides to be analyzed. CKII substrate peptides have the distinction of containing a large negative charge, due to the acidic nature of the phosphorylation site, and, of course, to the subsequent addition of phosphate to the peptide. Because the other reactants and products, ATP and ADP, of the reaction are also acidic, separation of the peptide from the nucleotides can be difficult. In the cases of the cAMP-dependent protein kinase, protein kinase C, and calmodulin kinases, the substrate peptides have a basic nature, containing at least two arginine or lysine residues, which facilitates the separation of the peptide from the ATP by ion-exchange methods.[39] One of the simplest methods involves the binding of the phosphorylated peptide to phosphocellulose paper strips, followed by washing in dilute phosphoric acid to remove ATP. This method can be applied to CKII substrate peptides only when the peptide has a basic portion consisting of two or more basic residues, preferably arginine. The ETE peptide is a good example of such a peptide with three arginine residues added to acidic clusters surrounding the phosphorylated threonine. If the arginine residues are adjacent to the serine or threonine to be phosphorylated, the $K_m$ value may be seriously affected. For example, the p53 peptide shown in Table I has two arginine residues adjacent to the substrate threonine, and the peptide is not phosphorylated by CKII. In some cases, the native sequence of the protein substrate does not contain any basic residues, as in the case of the Myc peptide (Table I). It is possible to artifically add basic residues to the peptide during the synthesis, but this might alter the substrate characteristics of the peptide. To maintain the native sequence, it is necessary to find an alternative method to separate the phosphorylated peptide from ATP. We have employed a thin-layer procedure using reversed-phase silica plates. Although somewhat more time consuming and less convenient than the phosphocellulose paper method, thin layer is nonetheless useful for assaying multiple samples. Below are listed the reagents and

[39] R. Roskoski, this series, Vol. 99, p. 3.

procedures for both assays: the phosphocellulose paper-binding method and the thin-layer method for acidic peptides.

### Method I: Binding to Phosphocellulose Paper[39]

*Reagents*

Buffer: 20 m*M* HEPES, pH 7.5; 150 m*M* KCl; 10 m*M* $MgCl_2$; 1 m*M* dithiothreitol; 1 m*M* EGTA; 30 m*M* *p*-nitrophenylphosphate; 10% (v/v) glycerol
Casein kinase II (bovine liver, lung, or brain)
[$\gamma$-$^{32}$P]ATP, 0.5 m*M*, 100–500 cpm/pmol
Peptide substrate, stock solution, 10 m*M* in buffer
Bovine serum albumin, 0.1 g/ml
40% (w/v) Trichloroacetic acid
Phosphocellulose paper (Whatman P81)
Phosphoric acid, 75 m*M* (5 ml of 85%/liter)

*Procedure*. The reaction mixture has a total volume of 50 $\mu$l containing 30 $\mu$l buffer, 10 $\mu$l ATP, 5 $\mu$l substrate peptide, 5 $\mu$l enzyme. The enzyme is added last to start the reaction. The final concentrations of the reactants are 0.1 m*M* ATP and 1 m*M* peptide substrate. The mixture is incubated for 30 min at 37°. The reaction is terminated by the addition of 10 $\mu$l of 0.1 g/ml bovine serum albumin immediately followed by 20 $\mu$l of ice cold 40% (w/v) trichloroacetic acid. The resulting mixture contains a final concentration of 10% (w/v) trichloroacetic acid, and is incubated on ice for 30 min, followed by centrifugation at 12,000 *g* for 5 min at 25° in a table-top microcentrifuge. An aliquot (20 $\mu$l) of the supernatant is applied to a strip of phosphocellulose paper, and the strip is washed four times for 5 min each in 2 liters of 75 m*M* phosphoric acid with occasional stirring. The paper can be dried using a hair dryer or an oven, or by using additional washes in acetone followed by diethyl ether. The individual assays are measured for radioactivity using liquid scintillant, water, or without scintillant (Čerenkov).

*Commentary*. The trichloroacetic acid step is designed to precipitate large proteins in the sample that may be phosphorylated by CKII and add to the background.[40] The albumin is included as carrier protein for the precipitation. When using purified CKII in catalytic amounts, the precipitation step can be eliminated. For column fractions or for crude mixtures, this step is essential for low background. The peptide substrate should

[40] G. M. Hathaway and J. A. Traugh, this series, Vol. 99, p. 317.

remain soluble under conditions of precipitation of proteins, but this should be tested empirically for every peptide substrate to be used.

### Method II: Reversed-Phase Thin-Layer Chromatography

*Reagents*

Buffer: 20 m*M* HEPES, pH 7.5; 150 m*M* KCl; 10 m*M* $MgCl_2$; 1 m*M* dithiothreitol; 1 m*M* EGTA; 30 m*M* *p*-nitrophenylphosphate; 10% (v/v) glycerol
Casein kinase II (bovine liver, lung, or brain)
[$\gamma$-$^{32}$P]ATP, 0.5 m*M*, 100–500 cpm/pmol
Peptide substrate stock solution, 10 m*M* in buffer
ATP, 10 m*M* (unlabeled)
$C_{18}$ silica thin-layer chromatography plates, 20 cm (Whatman KC18F)
Chromatography solvent: 3% (v/v) acetonitrile; 0.1% (v/v) phosphoric acid; 0.28% (v/v) triethylamine

*Procedure.* The assays are performed exactly as described above for the phosphocellulose method. Following trichloroacetic acid precipitation, an aliquot of the entire reaction mixture is applied to the origin of the thin-layer plate that has been prespotted with 10 $\mu$l of unlabeled 10 m*M* ATP. The spot should not be allowed to dry before applying the sample. The plate is then placed in a chromatography tank, and the solvent front is allowed to reach the top of the plate. The plate is allowed to dry and is exposed to X-ray film (Kodak XAR-5). After development, the film is used to identify the boundaries of the radioactive spot at the origin, and the silica in the radioactive circle surrounding the origin is scraped off the plate into a scintillation vial. Each sample is measured for radioactivity by scintillation counting.

*Commentary.* In this procedure, the radioactive ATP is eluted rapidly from the origin and appears near the solvent front at the top of the plate. Figure 2 shows an example of such an assay using the Myc peptide (Table I). It is essential to use the unlabeled ATP to block nonspecific adsorption of the radioactive ATP to the origin. Without this procedure, or if the ATP solution is allowed to dry at the origin before applying the sample, there is high background. In some cases, it is possible to perform a second chromatography step using the same solvent containing 10–20% acetonitrile to elute the peptide from the origin and move it toward the center of the plate. This tends to give a less uniform spot shape and decreases the reproducibility of the method. By using the unlabeled competitor ATP, it is unnecessary to do a second chromatography step.

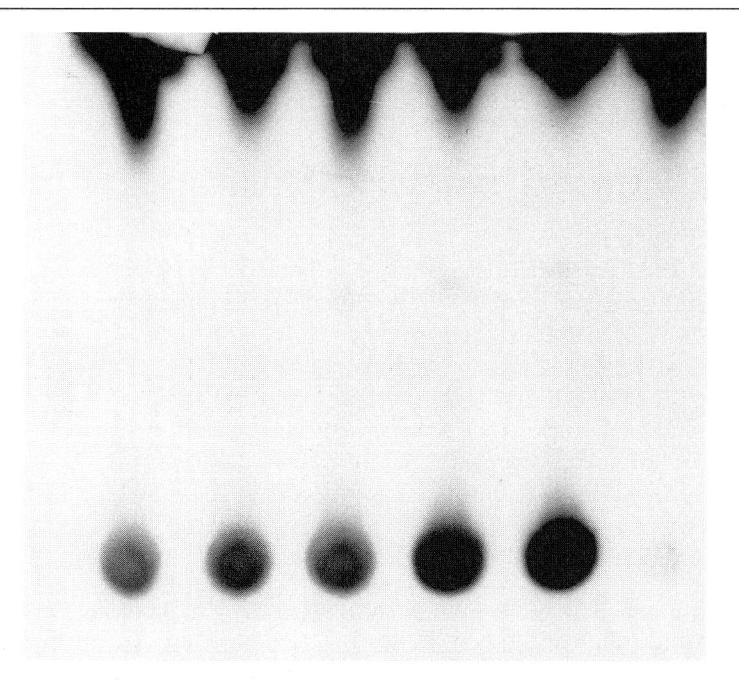

FIG. 2. Thin-layer chromatography of CKII phosphorylation of Myc peptide (sequence shown in Table I). Assays were performed with bovine liver CKII using increasing concentrations of peptide. From left: 0.063, 0.125, 0.250, 0.500, and 1.00 m*M*. The control without peptide is shown on the final lane on the right. Chromatography on $C_{18}$ silica plates was performed as described in the text.

## Qualitative Assay Methods

New methods have become available to identify phosphorylated peptides qualitatively, either as synthetic peptides, or as proteolytic fragments of proteins phosphorylated *in vivo* or *in vitro*. The two methods used frequently in our laboratory employ the gel electrophoresis method of Schägger and von Jagow[41] and plasma desorption mass spectrometry.[35] The electrophoresis method utilizes a Tris-Tricine buffer system in polyacrylamide gels containing urea. The resolution of these gels extends to 1000 mass units, and is suitable for relatively small peptides. Following autoradiography of the gels, bands containing $^{32}P$ label are excised, and the peptide is eluted in buffer containing 0.2 *M* ammonium bicarbonate, pH 7.4, 0.1% (w/v) sodium dodecyl sulfate, and 1 m*M* dithiothreitol. We find it important to concentrate, dialyze, and acetone precipitate the

[41] H. Schägger and G. von Jagow, *Anal. Biochem.* **166,** 368 (1987).

peptide after elution. The peptide can be further purified by reversed-phase HPLC and subjected to Edman degradation on automated sequencing machines. We have used this method successfully to identify phosphorylation sites on LTag[42] and Myc.[30] This technique can be combined with mass spectrometry to analyze phosphopeptides isolated from electrophoretic separations of proteolytic digestion products of phosphorylated proteins.

The mass spectrometer can be used independently to monitor the progress of a phosphorylation reaction and to assay the number of phosphate groups added to a peptide. A major problem with CKII substrate peptides on mass spectrometry is their acidic nature. Most mass spectrometric analysis is done in the positive ionization mode in which the peptide must acquire a proton during the ionization process to be identified as the protonated molecular ion, $(M + H)^+$. In many cases, an extremely acidic peptide often shows low yield under positive ionization. Plasma desorption mass spectrometry on a time-of-flight instrument is quite flexible and, under negative ionization, one can identify the negatively charged, deprotonated molecular ion, $(M - H)^-$. Figure 3 shows the results of a negative ionization, time-of-flight mass spectrum of a synthetic CKII substrate peptide, based on the LTag sequence. This peptide was phosphorylated using bovine brain CKII for various amounts of time, and an aliquot of each time point was used for the analysis. The peptide solution was applied to an aluminized Mylar foil, previously electrosprayed with 50 $\mu$g nitrocellulose dissolved in acetone. The peptide is adsorbed to the nitrocellulose matrix, and the buffer and salts are washed away with 50 $\mu$l of water. As shown in panel A, the $(M - H)^-$ ion can be clearly identified. During the time course of the reaction, the unphosphorylated substrate peak decreases, and the product peaks containing an integral number of phosphates (in this case, 1, 2, or 3) begin to appear. The signal-to-noise ratio for the phosphorylated products is low, but detectable. Further refinements of the technique including new matrices and washing procedures should make this a routine method in the near future.

## Kinetics of Multiple Phosphorylation Sites

One of the merits of using synthetic peptides as models of *in vivo* phosphorylation sites is the ability to elucidate complex kinetic phenomena, such as multisite or sequential phosphorylation. Several known CKII sites are intimately involved with neighboring phosphorylation sites. For example, glycogen synthase kinase 3 and CKII collaborate in the phos-

[42] D. McVey, L. Brizuela, I. Mohr, D. R. Marshak, Y. Gluzman, and D. Beach, *Nature (London)* **341,** 503 (1989).

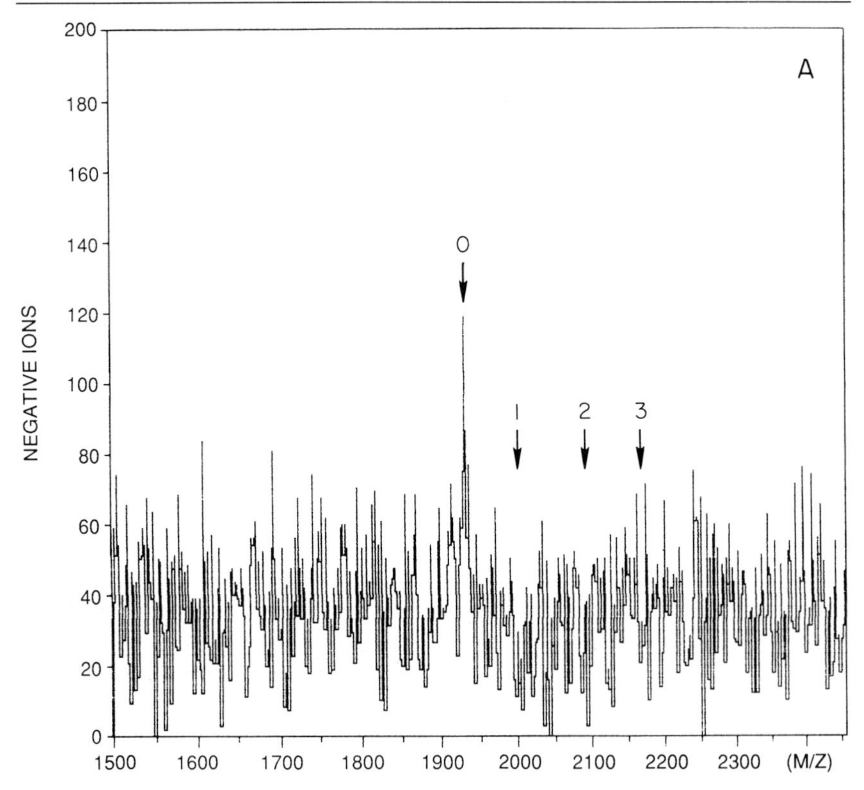


FIG. 3. Negative ionization mass spectrometry of phosphorylation of a synthetic peptide by CKII. A synthetic peptide based on the LTag sequence was synthesized as described in the text. The sequence was H-EENLFCSEEMPSSDDEA-$NH_{12}$. The peptide was phosphorylated using bovine brain CKII at 37° for 0 hr (A), 1 hr (B), and 4 hr (C). The expected position of the deprotonated molecular ion $(M - H)^-$ is shown by the arrow labeled 0. The positions of phosphorylated derivatives, with shifts of +80, +160, and +240 mass units by addition of 1, 2, and 3 phosphates, are shown.

phorylation of a particular domain on glycogen synthase.[11] In LTag, there are three potential CKII sites in a short region at serine residues 106, 111, and 112. Serine-106 has been identified as a site of CKII phosphorylation *in vivo,* and there is a suggestion of phosphorylation at 111 and/or 112 from both *in vivo* and *in vitro* experiments.[43] The results of the mass spectrometric analysis of a peptide including all three sites (Fig. 3) suggests that it may be possible to incorporate phosphate into all three sites. We

[43] F. A. Grässer, K. H. Scheidtmann, P. T. Tuazon, J. A. Traugh, and G. Walter, *Virology* **165,** 13 (1988).

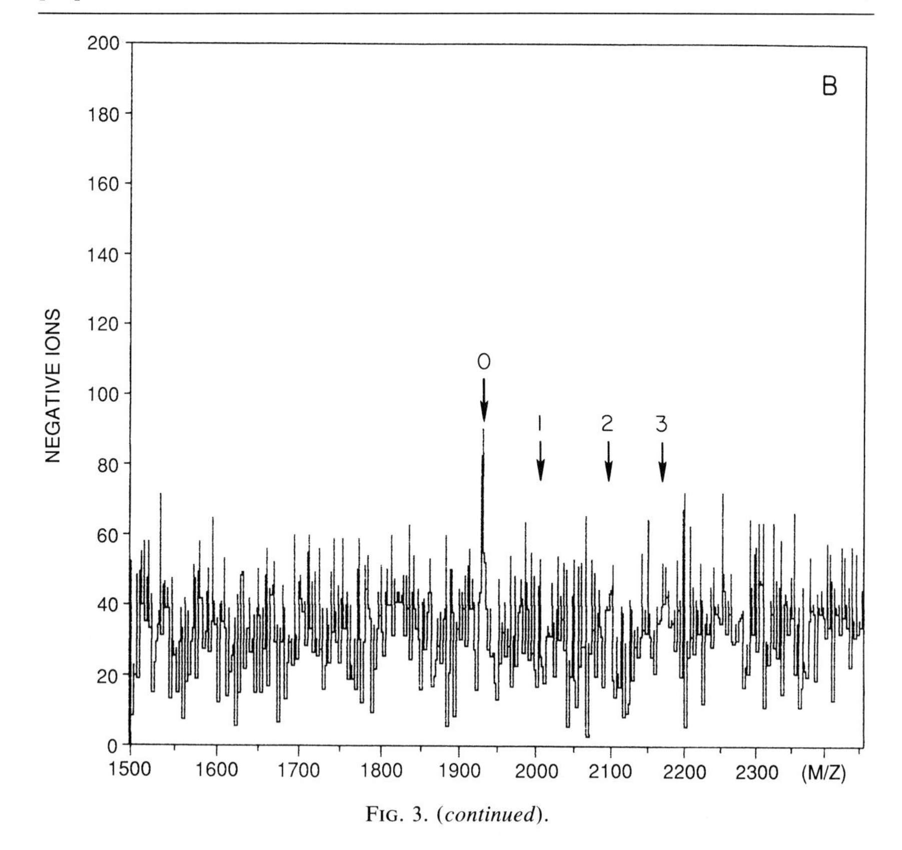


FIG. 3. (*continued*).

utilized a shorter LTag analog peptide as substrate for CKII that contains serine-106 as the N-terminal amino acid residue (sequence shown in Table I). In this case, the amino terminal serine is not phosphorylated, but the two downstream serines, at the analogous positions to residues 111 and 112 do incorporate phosphate. This analysis is based on results obtained by using the method of Heilmeyer,[44] employing $\beta$ elimination of phosphate with base $[Ba(OH)_2]$[45] and derivatizing the ensuing dehydroalanine derivative with ethanethiol to form the *S*-ethylcysteine derivative. Sequence analysis of the final derivative permitted quantitation of the phosphate incorporated in each of the serine residues. Shown in Fig. 4 are the results of the sequencer analysis for several time points during the incubation.

[44] H. E. Meyer, E. Hoffman-Posorske, and L. M. G. Heilmeyer, Jr., this series, Vol. 201 [14].

[45] The use of $Ba(OH)_2$ was suggested by M. Byford and K. Walsh, University of Washington, Seattle.

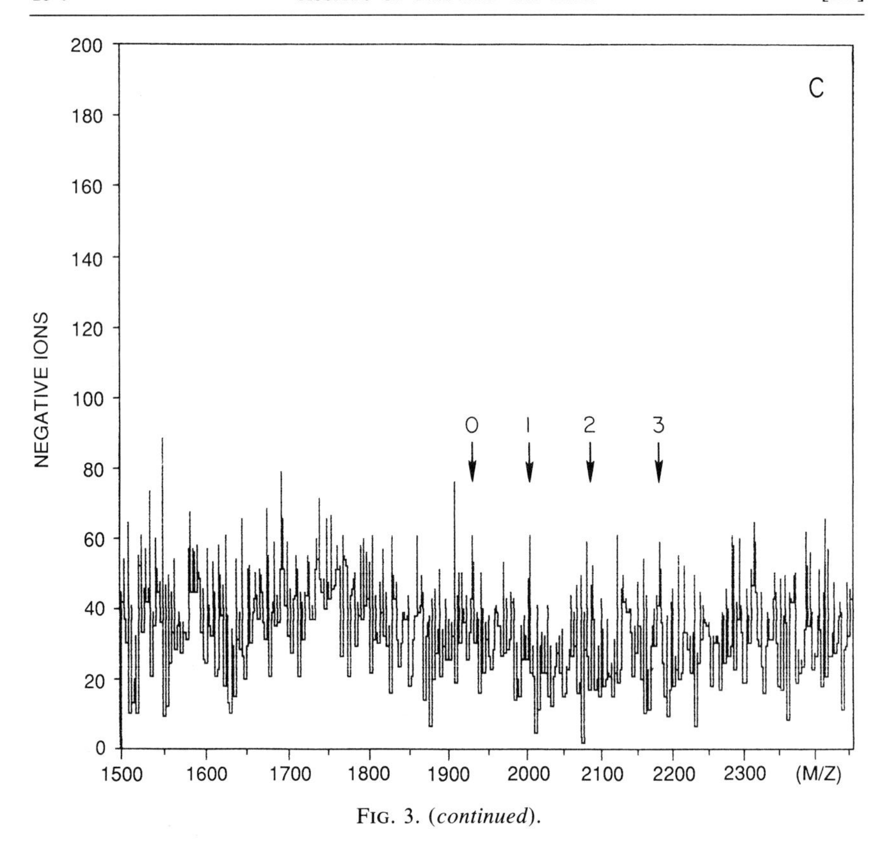


FIG. 3. (*continued*).

The serine at position 7, corresponding to residue 112, acquires phosphate with linear kinetics, while the serine at position 6, corresponding to residue 111, displays a lag, followed by a similar rate of phosphorylation. These results suggest the sequential addition of phosphate to these adjacent position. Once serine-112 is phosphorylated, the preceding serine becomes a substrate for CKII. There may be important implications of this ordered phosphorylation, particularly with respect to DNA replication and neighboring phosphorylation sites. A separate analysis of this domain on LTag revealed a downstream phosphorylation site at threonine-124 for the cell division cycle-regulated protein kinase, p34$^{cdc2}$. Phosphorylation at that site increases DNA replication by altering binding of LTag to the viral origin of replication.[42] The role of the neighboring CKII sites is not full understood. However, the use of synthetic peptide substrates, in conjunction with biochemical and genetic analysis of the intact protein, provides

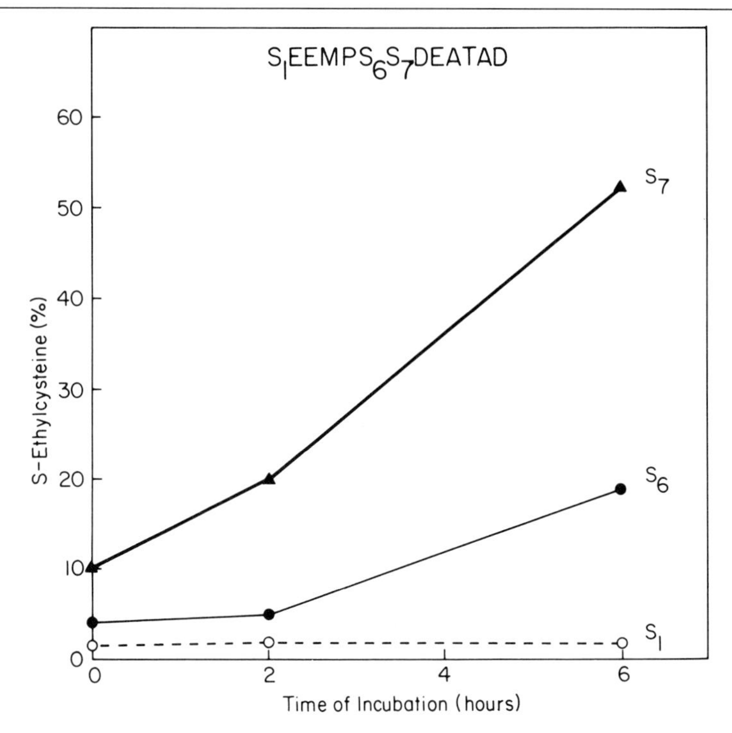


FIG. 4. Ordered addition of phosphate to an LTag peptide (sequence shown in Table I). The peptide was phosphorylated with bovine liver CKII for various amounts of time as shown on the abscissa. Following the phosphorylation, the peptide was treated with 0.1 *M* $Ba(OH)_2$ and 1 *M* ethanethiol at 30° for 1 hr. Another aliquot of ethanethiol was added to bring the final concentration to 2 *M*, and the reaction was incubated for an additional 2 hr. The reaction was terminated by lowering the pH to 4.0 with acetic acid, and the peptide was desalted by reversed-phase HPLC. The modified peptide was then subjected to automated sequencing on an Applied Biosystems 470A instrument equipped with an on-line HPLC and data system. The Phenylthiohydantoin (Pth)-*S*-ethylcysteine at each cycle was quantitated by comparison to a standard synthesized in our laboratory.

a powerful approach to understanding the phosphorylation domains of nuclear oncoproteins.

## Summary

Synthetic peptide substrates for CKII are useful reagents in the analysis of phosphorylation sites when used in conjunction with biochemical and genetic analysis of the protein substrates for the enzyme. A multidiscipli-nary approach should be applied to the characterization of the synthetic peptide products, including amino acid analysis, sequencing, and mass spectrometry. Synthetic procedures for CKII substrate peptides often result in anisole adducts and dehydrated forms. Mass spectrometry is invaluable in identifying these contaminants, and preparative HPLC can be used to separate them from the desired product. Quantitative analysis of the CKII phosphorylation of peptides can utilize phosphocellulose paper if the peptide has a basic sequence, or thin-layer chromatography, if the peptide has no basic portion. Qualitative analysis using electrophoresis and mass spectrometry help to establish the stoichiometry of phosphorylation. Sequence analysis of phosphoserine after $\beta$ elimination and derivitization is useful in quantifying adjacent phosphorylation sites. Overall, application of a variety of techniques permits detailed analysis of CKII phosphorylation sites on synthetic peptides that are model substrates.

## Acknowledgments

We are indebted to M. Vandenberg, G. Binns, M. Meneilly, and N. Santoro for expert technical assistance. This work was supported by NIH Grants CA13106, CA45508, and CA09311 and NSF Grant BNS8707558.

# Section III

## Purification of Protein Kinases

A. General Methods
*Articles 12 through 14*

B. Bacterial Protein Kinases
*Articles 15 through 17*

C. Protein-Serine Kinases
*Articles 18 through 29*

D. Protein-Tyrosine Kinases
*Articles 30 and 31*

E. Protein-Histidine Kinases
*Article 32*

# [12] Micro- and Macropurification Methods for Protein Kinases

*By* STEFANO FERRARI and GEORGE THOMAS

## Introduction

Phosphorylation of key proteins or enzymes has been implicated as the major intracellular regulatory mechanism in many biological processes.[1,2] Protein kinases are often activated by hormones either directly, as in the case of receptor-tyrosine kinases, or indirectly, through the generation of second messengers or through kinase cascades. Protein kinases were initially discovered during classical investigations of metabolic processes such as glycogen breakdown.[3-5] More recently, many new members have been added through studies of growth factor action,[6,7] oncogenesis,[8] development, and the control of the cell cycle.[9-12] In addition, molecular biology now offers a new way to identify and to isolate novel kinases by screening cDNA libraries using oligonucleotides based on conserved protein sequences of either serine/threonine kinases or tyrosine kinases.[13] This latter approach will undoubtedly lead to the identification of many new kinases. However, when a novel protein kinase activity is detected, it may not be possible to identify the enzyme by molecular cloning but instead it may be necessary to purify the enzyme using classical biochemical methods. Furthermore, if the protein kinase was initially identified in cell cultures, it may be present in too small an amount to obtain either protein sequence data or antibodies. It would then be necessary to search for a second source capable of providing enough material for these studies. While

[1] S. K. Hanks, A. M. Quinn, and T. Hunter, *Science* **241,** 42 (1988).
[2] A. M. Edelman, D. K. Blumenthal, and E. G. Krebs, *Annu. Rev. Biochem.* **56,** 567 (1987).
[3] E. H. Fisher and E. G. Krebs, *J. Biol. Chem.* **216,** 121 (1955).
[4] E. W. Sutherland and W. D. Wosilait, *Nature (London)* **175,** 169 (1955).
[5] E. G. Krebs, D. J. Graves, and E. H. Fisher, *J. Biol. Chem.* **234,** 2867 (1959).
[6] S. Cohen, G. Carpenter and L. King, Jr., *J. Biol. Chem.* **255,** 4834 (1980).
[7] M. Kasuga, F. A. Karlsson, and C. R. Kahn, *Science* **215,** 185 (1982).
[8] R. L. Erikson, A. F. Purchio, E. Erikson, M. S. Collett, and J. S. Brugge, *J. Cell Biol.* **87,** 319 (1980).
[9] P. H. O'Farrel, B. A. Edgar, D. Lakich, and C. F. Lehner, *Science* **246,** 635 (1989).
[10] D. Arion, L. Meijer, L. Briyuela, and D. Beach, *Cell (Cambridge, Mass.)* **55,** 371 (1988).
[11] J. Gautier, C. Norbury, M. Lohka, P. Nurse, and J. Maller, *Cell (Cambridge, Mass.)* **54,** 433 (1988).
[12] J. C. Labbé, A. Picard, E. Karsenti, and M. Dorée, *Dev. Biol.* **127,** 157 (1988).
[13] S. K. Hanks, *Proc. Natl. Acad. Sci. U.S.A.* **84,** 388 (1987).

purifying the mitogen-activated S6 kinase we faced a number of these problems.[14] From our experience and that of others we have attempted to describe methods that can be used to identify and purify a rare protein kinase, assuming its mode of activation is known, and the problems that may be encountered in scaling up from micro- to macropurification. In this chapter we emphasize the use of Pharmacia-LKB (Piscataway, NJ) hardware because we have the most experience with it. Each investigator should consider a number of products or systems which may be more suitable for his or her own particular needs.

## Assay

The first step in working with a protein kinase is to establish a specific and rapid assay for the enzyme. The most important component in the assay is the protein substrate, which can be (1) an endogenous cellular target protein, (2) a synthetic peptide based on the sequence phosphorylated in this protein, or (3) a model substrate protein. The best choice is an endogenous substrate, but this may be difficult or very costly to obtain in the amounts required for routine assays. If a synthetic peptide is used, it should contain the sites phosphorylated in the endogenous substrate as well as the kinase recognition determinants. Large quantities of peptides can be easily synthesized and their purification normally requires only one step. Such peptides are extremely useful if they are specifically recognized by the protein kinase of interest. However, in general, protein kinases have a relatively low affinity for peptide substrates and peptides in many cases are phosphorylated by a number of other protein kinases. Therefore, if using a peptide, it is essential to find out whether it is recognized by the protein kinase in the same way as the native substrate. If the peptide is phosphorylated by a number of enzymes it may still prove useful at later stages of purification when other protein kinases have been removed. A model substrate is an abundant protein, usually commercially available, that can substitute for the native protein substrate in the same way a synthetic peptide can. A model substrate contains both the phosphorylatable amino acids and kinase recognition determinants which are either acidic, as are casein and phosvitin, or basic, as are histones and protamines. There are several advantages in using model substrates, but they can be subjected to the same criticisms as applied to synthetic peptides above. Having chosen a substrate one must work out the other parameters of the phosphorylation reaction. These include (1) the nucleotide triphosphate donor, usually ATP but in rare cases GTP, (2) the ions required to

[14] H. A. Lane and G. Thomas, this volume [22].

produce maximal activity, usually $Mg^{2+}$, although $Mn^{2+}$ for most tyrosine kinases, as well as other salts, and (3) the optimum pH and temperature at which the reaction should be carried out. Once the protein kinase is purified to homogeneity, these parameters can be more rigorously defined.

Following the phosphorylation reaction, one of three widely used methods can be employed to separate the substrate from nucleotide triphosphates: (1) trichloroacetic acid (TCA) precipitation, (2) spotting of positively charged substrates on phosphocellulose paper, and (3) detection of phosphorylated substrates by gel electrophoresis.

### *TCA Precipitation*

This method involves the denaturation and then the precipitation of the substrate in 10–20% (w/v) TCA, leaving the free nucleotide phosphates in solution. This technique can be applied to both basic and acidic protein substrates and, contrary to common belief, can also be applied to peptides, depending on their length and physical makeup. In addition, "carriers" such as bovine serum albumin or deoxycholate can be added to the reaction prior to the TCA in order to render the precipitation more efficient. In general, substrates are allowed to precipitate on ice for 5 to 10 min. Then they are filtered on 2.5-cm Whatman (Clifton, NJ) glass fiber (GF/C) filters,[15] washed three times with 5 ml of ice-cold 5% TCA (w/v), dried, and counted in scintillation fluid. Several devices are available from Millipore (Bedford, MA) to filter one or more samples at a time. An alternative to filtration on glass fiber is to centrifuge the tubes in a bench microfuge. The pellets are washed three times with 1 ml of ice-cold 5% TCA, dissolved in scintillation fluid, and counted.[16]

### *Phosphocellulose Paper*

In this method the kinase reaction mixture is spotted onto a piece of phosphocellulose paper (Whatman P81). The ionic interaction between positively charged residues in the substrate and negative phosphate groups on the paper keeps the peptide bound to the paper, but [$\gamma$-$^{32}$P]ATP can be readily washed away.[17] The reactions are stopped by adding phosphoric acid to a 5% final concentration and the samples are then spotted onto numbered filter paper squares and allowed to absorb for a few seconds. The paper squares are then immersed in a beaker containing 0.5% phos-

[15] J. D. Corbin and E. M. Reimann, this series, Vol. 38, p. 287.

[16] N. Embi, D. B. Rylatt, and P. Cohen, *Eur. J. Biochem.* **100,** 339 (1979).

[17] D. B. Glass, R. A. Masaracchia, J. R. Feramisco, and B. E. Kemp, *Anal. Biochem.* **87,** 566 (1978).

phoric acid and stirred gently for 5 min. The washing process is repeated twice more, and the filters are rinsed in acetone, dried, and counted as above. The method is very useful for processing large numbers of samples but limited to peptide substrates that contain at least two or three basic residues (see [9] in this volume).

### *Polyacrylamide Gel Electrophoresis*

When the protein substrate is impure or is a component of a multienzyme complex, the phosphorylated protein of interest can be separated from the others by sodium dodecyl sulfate (SDS)–polyacrylamide gel electrophoresis. This procedure can be performed relatively quickly by using short slab gels ($20 \times 8 \times 0.5$ cm) that require 25–30 min to run and 10 min for staining and electrophoretic destaining. The gel is then dried and autoradiographed. Single bands can be cut out and counted in the presence of liquid scintillation fluid. The volume of radioactive waste is small if the dye front, where [$\gamma$-$^{32}$P]ATP migrates, is not allowed to run off the gel and is removed at the end of the run. For analytical purposes, Phast-System (Pharmacia) is an alternative since the gel is electrophoresed, stained, and destained in a single chamber within 1 hr with no handling of the gel.

## Micropurification

To elucidate the mechanisms that regulate basic biological processes, many investigators use cell lines grown in defined culture conditions. These model systems have several advantages over animals or specific tissues: (1) clonal cell lines represent a homogeneous cell type; (2) they can be synchronized in the cell cycle; (3) intracellular components can be labeled to a high specific activity with radioactive markers; and (4) specific factors can be used to induce the intracellular pathway of interest. However, it may be extremely difficult to purify and identify a protein kinase from cells grown in culture because of the small amount of starting material. It is possible to increase the number of cells by growing them on beads, roller bottles, or large plastic trays, or by using a cell line that grows in suspension. Although one must consider the cost of maintaining and growing large numbers of cells, this approach may be essential for the initial identification and characterization of the protein kinase. Based on this knowledge, it may be possible to obtain the same enzyme from a less costly source that provides the larger amounts of protein required for sequencing and antibody production.

*Strategy*

Information about the physical properties of a new protein kinase is often not available. Thus, there is not a straightforward protocol to follow for its purification. To purify a protein kinase from a limited source, the strategy we found most useful was to (1) remove abundant proteins by classical chromatographic techniques before applying affinity steps; (2) remove inhibitors or inactivators at an early stage in the purification; (3) keep sample volumes to a minimum while avoiding dialysis; and (4) move through the purification rapidly without freezing the enzyme. The classical chromatographic techniques used to remove bulk protein include cation- and anion-exchange, hydrophobic, and gel-filtration chromatography. These techniques are described in detail in most manuals on protein purification. In an ideal protocol of enzyme purification one should take advantage of the chromatography steps such that an eluate can be directly loaded onto the next column with no need of dialysis. For example, ion-exchange steps should be followed by hydrophobic interaction chromatography or vice versa. In the case of gel-permeation chromatography the volumes loaded must be small to achieve optimal resolution. This is done by running an ion-exchange step prior to the gel filtration step, which in turn makes possible the separation of the enzyme from the salt. However, it is usually recommended that a gel-filtration chromatography be run in the presence of 0.2–0.5 *M* salt to avoid unspecific binding of the enzyme to the gel beads or aggregation of proteins. Thus, if this method is employed at low salt concentrations the behavior of the protein kinase at high salt concentrations should also be examined. The easiest way to test both binding and elution on either ion-exchange or hydrophobic columns is to mix a small portion of the protein kinase with the resin in a specific buffer at a suitable pH, wash the resin with buffers with increasing ionic strength, and assay the supernatants. Again, compatible buffers along with their physicochemical properties are listed in many manuals on purification. After having exploited classical approaches one can then turn to affinity steps which are based on either a general biochemical feature of protein kinases or a specific property of the protein kinase under investigation.[18] Without a good affinity step in the purification, rare protein kinases which require $1 \times 10^4$- to $1 \times 10^6$-fold purification, will not be purified to homogeneity. Useful steps employed in the purification of protein kinases are shown in Table I.

For purifying protein kinases from tissue culture we have largely relied

[18] P. Jenö and G. Thomas, this volume [14].

TABLE I
PURIFICATION OF KINASES BY CHROMATOGRAPHY

| Chromatography step | Matrix | Ligand | Comment |
|---|---|---|---|
| Ion exchange | | | |
| Cation exchange | BioGel A, cellulose, Sephadex, Sepharose | Carboxymethyl (CM) | Remove bulk proteins<br>Remove phosphatases |
| | Sephadex | Phospho | |
| | Sepharose | Sulfonate (S) | |
| Anion exchange | BioGel A, cellulose, Sephacel, Sephadex, Sepharose | Diethylaminoethyl (DEAE) | Remove bulk proteins |
| | Sepharose | Quaternary amine (Q) | |
| Hydrophobic interaction | Sepharose | Octyl | Remove salt<br>Remove hydrophilic proteins |
| | Sepharose | Phenyl | |
| Gel filtration | BioGel P, Sephacryl, Sephadex, Sepharose, Superose | | Select molecular weight interval<br>Remove salt<br>Reequilibrate sample |
| Affinity | Activated → agarose | ATP | Homogeneously purify protein kinase |
| | Activated → Sepharose | Peptide substrate effector | |

on the fast protein liquid chromatography (FPLC) system of Pharmacia. The advantages are that the system is largely inert and designed for preserving the activity of biological molecules, it contains all of the required hardware in a single unit, and it can be kept and operated in the cold room as well as at room temperature. In addition, gradient runs are rapid (40 to 45 min) and a wide range of high-resolution resins in prepacked columns is available as well as compatible columns in which one may pack any resin of choice. We found that cation-exchange chromatography at neutral pH on either prepacked Mono S columns or self-packed Fast Flow S columns is an ideal first step in purification for the mitogen-activated S6 kinase. This is because most proteins do not bind to the resin under these conditions, including phosphatases that inactivate the S6 kinase. Surprisingly, a number of protein kinases bind to Mono S at neutral pH even though they have apparent p*I* values $\leq 5.5$. Possibly the sulfonate groups mimic some molecular aspect of ATP or the substrate. In micropurification methods one may be faced with large volumes in making extracts from large numbers of cells. Moreover, it may be sometimes necessary to have two ion-exchange steps following one other, and when this is the case, a successful loading of high salt pools can be achieved by dilution to large volumes. To circumvent the problem of loading these large volumes it is possible to apply the sample directly through the pumps. Although this is not recommended by Pharmacia, this method saves a great deal of time. For example, in loading volumes of between 500 and 1000 ml on a Pharmacia HR 10/10 column (1 × 10 cm) packed with Fast Flow S it is possible to load continuously through the pumps at 12 ml/min, rather than stopping to reload the Superloop, which has a maximum capacity of only 50 ml. Furthermore, contrary to common belief, the FPLC system is also compatible with many HPLC columns, such as the Phenyl TSK column distributed by Beckman. All that is required are suitable adaptors (compression screw unions; Pharmacia). We have obtained superior resolution of proteins employing these Beckman columns (Palo Alto, CA).

Prepacked FPLC columns for gel filtration can be developed in about 30 min at flow rates much higher than those using classical gels. However, due to the small size of these columns, the sample volume should not exceed 200 μl and it should contain no more than 5–10 mg of protein. These parameters are sometimes difficult to achieve and it may be necessary to carry out multiple runs. Pharmacia introduced a new FPLC-compatible matrix for gel filtration (Sephacryl HR) that can be run at flow rates five times higher than those of Sephacryl Superfine. These resins could greatly decrease the time required for gel-filtration chromatography.

In affinity chromatography the possible ligands for a protein kinase are (1) nucleotide triphosphates, (2) substrates, and (3) effector molecules.[18] We have successfully used three affinity ligands to purify the mitogen-activated S6 kinase: ATP-type IV agarose, threonine-agarose, and a 32-amino acid peptide substrate coupled to agarose. All three resins have been packed into Pharmacia HR 5/5 columns and run on the FPLC system at flow rates similar to those used for Mono S and Mono Q HR 5/5 columns, i.e., 0.5 to 1.5 ml/min. In all three cases we have found little or no measurable back pressure. Finally, it should be noted that protein kinases do not elute as sharply from the affinity columns as from ion-exchange columns. Since the affinity steps are employed at the end of the purification, the protein kinase at this point is in a dilute solution. To concentrate the enzyme we have removed ~100 μl of packed Mono Q anion-exchange resin without allowing the material to break up, and repacked the resin into a Pharmacia HR 5/2 column (5 mm × 0.5 cm). The column is fitted at both ends with a top assembly, allowing compression of the resin between both plungers. This column can be loaded on the FPLC at 1 ml/min and then developed at 200 μl/min, collecting 100-μl fractions into siliconized Eppendorf tubes. This allows one to concentrate the kinase into 200–300 μl, which may be important for subsequent storage or for visualizing the enzyme on silver-stained gels. There is virtually no loss of kinase activity on such a column, although a carrier protein is included to prevent nonspecific binding.[14] The carrier protein should be incubated at 56° to destroy any protease activity and its molecular weight should be sufficiently different from that of the protein kinase so that it can be easily removed by gel filtration or SDS gel electrophoresis. Moreover, it should not interfere with the binding of the protein kinase to resins when added before the last step of purification. All tubes, Eppendorf tips, or other plasticware which come into contact with the protein kinase at the later stages of purification should be siliconized. Siliconization is achieved by leaving the plastics for 30 min under vacuum in vapors of dimethyldichlorosilane. They are then rinsed thoroughly in water and autoclaved at 110° for 15 min. In addition, chemicals and solvents used throughout the purification and especially during storage should be of the highest quality to ensure that the protein kinase is not inactivated by heavy metal ions or by oxidizing agents.

At the end of a micropurification the purity of the protein kinase should be checked on a silver-stained SDS–polyacrylamide gel. If few contaminants are present, the band corresponding to the kinase can be indirectly identified by plotting the intensity of the stained band in each fraction versus the peak of activity. The property of autophosphorylation, shared by most protein kinases, can also be used to localize the band of interest. More direct evidence can be obtained by running the sample on an SDS–polyacrylamide gel and reactivating the protein kinase after extraction from the gel[19] or by labeling the protein kinase with radioactive ATP affinity analogs followed by gel electrophoresis.[20] At the final stage of purification it is also necessary to determine the protein concentration of the sample in order to calculate the specific activity and the level of purification. Conventional methods such as the modified Lowry[21] or the Bradford[22] are not applicable due to the low amount of protein. In this case the two possible alternatives are quantitation by either fluorometric methods[23] or amino acid analysis.

[19] S. A. Lacks and S. S. Springhorn, *J. Biol. Chem.* **255,** 7467 (1980).

[20] B. E. Haley, this volume [40].

[21] G. L. Peterson, *Anal. Biochem.* **83,** 346 (1977).

[22] M. Bradford, *Anal. Biochem.* **72,** 248 (1976).

[23] V. A. Fried, M. E. Ando, and A. J. Bell, *Anal. Biochem.* **146,** 271 (1985).

Macropurification

As pointed out above, the amount of protein kinase required to make the initial identification and characterization is not usually limiting. For amino acid sequencing, antibody production and further characterization studies, however, a specific tissue or organ will probably have to be used as starting material. Even then, the yield of enzyme may be limited. For instance, 200 $\mu$g of protein kinase C can be easily purified from 30 g of bovine brain with a recovery of 15%, while only 20 to 30 $\mu$g of active S6 kinase can be obtained from 1500 g of rat liver with a recovery of 7%.[14] Thus, depending on the abundance of the kinase, a number of practical considerations must be taken into account: (1) availability and cost of the source, (2) if required, the best protocol for activating the protein kinase *in vivo*, and (3) demonstrating that the purified protein kinase is the same enzyme first identified in tissue culture or the model system.

*Strategy*

In scaling up the purification of a protein kinase the first problems encountered are the heterogeneity of proteins present in tissue extracts, the higher amount of starting material, and the large volumes of crude extract. One can use ammonium sulfate precipitation to reduce volumes and gain some purification. However, we found that this protocol can lead to large losses in activity without much gain in purification. In the end, the strategy we adopted was similar to that outlined for the micropurification above. It may seem that using larger amounts of material directly contradicts the requirement of keeping volumes to a minimum and the need to move through the purification rapidly. As outlined below, these limitations can largely be overcome by using preparative columns at high flow rates and introducing additional affinity steps.

By using a combination of low-speed centrifugation followed by cation- and anion-exchange chromatography on Fast Flow S and Fast Flow Q resins, it is possible in a few hours to reduce 20 to 30 liters of extract containing 1.5 kg of tissue to 200 mg of protein in 100 ml. This is accomplished by using BioProcess columns (Pharmacia) in combination with a Watson–Marlow 501 U peristaltic pump. The columns are made of oxirane glass, allowing easy visual control during packing and running. The smaller diameter columns (11.3 cm) are produced in lengths of 15, 30, and 60 cm. However, they can be bolted to one another to produce any desirable length. We have packed the 15- and 30-cm columns with 300 to 1000 ml of Fast Flow Q or Fast Flow S resin, respectively.[14] The columns are coupled with $\frac{1}{4}$-in. polypropylene tubing to the Watson–Marlow pump using Jaco connectors (Pharmacia). To avoid leakages we line the threads of all screw connections with Teflon tape. The pumps can run at speeds as low as 10 ml/min and as high as 300 to 350 ml/min. We have been able to load these columns at maximum speed, which means that we can load 20 or 30 liters of extract in 1 to 1.5 hr. In general, the columns are developed at 50 to 100 ml/min with linear salt gradients. On-line conductivity meters and UV flow cells are available which are very useful when one carries out repetitive purifications. The UV flow cells fit into the standard UV recorder of the FPLC system so the conductivity and the absorbance can be monitored simultaneously on the FPLC chart recorder. Along with the pump the only other piece of equipment in the FPLC which is not compatible with large volumes is the fraction collector. In this case, one can substitute another fraction collector such as the Super Rac fraction collector (LKB) which comes with racks which can accommodate 50-ml collecting tubes. One run allows the collection of 84 fractions or ~4200 ml, but the fraction collector can be programmed to carry out multiple runs, requiring only the addition of new racks. We found the purification to be highly reproducible and we now simply pool fractions based on conductivity and the $OD_{280}$ profile. It should also be noted that even though the column is developed with a salt gradient, it is still much faster than loading and stepping off the kinase using standard batch procedures. Running the two ion-exchange steps alone we typically gain a 100-fold purification of the mitogen-activated S6 kinase. At this stage there may be 200 or 300 mg of protein in a volume of 100 or 200 ml of high salt buffer. As with the micropurification, at this point we have turned to hydrophobic interaction columns. Beckman markets a preparative TSK phenyl column which can be directly mounted, with the proper connectors described above, to the FPLC system. In general, a 10-fold purification is obtained at this step which can be run on the same day as the first two steps. In our hands a 1000-fold purification is gained on the first day and the total amount of protein is reduced to a few milligrams, which is compatible with affinity purification. Moreover, by the use of a combination of affinity ligands an additional 100- to 1000-fold purification can be obtained, which gives an overall $1 \times 10^5$- to $1 \times 10^6$-fold purification. At this level most kinases should be pure.

Obviously, in working with large amounts of material the help of colleagues may be required in the initial step. With this one consideration, and even the possibility that an additional step of purification may be required in the macro- versus the micropurification, both tasks can be completed in the same amount of time, 2 to 3 days.

## Storage

Kinases can be stored under a variety of conditions. However, no method can be *a priori* recommended as the best and kinase stability over time should be tested in each individual case. If the primary purpose is to obtain sequence data or raise polyclonal antibodies, carrier proteins or peptide protease inhibitors should not be added to the homogeneous kinase preparation and the enzyme should be stored in liquid nitrogen or at −70°. If the aim is to characterize the biochemical properties of the kinase, storage at −20° in the presence of 50% glycerol should be compared with the methods above. In the case of freezing, it should be noted that the pH of some buffer systems is unstable on freezing and thawing. Protein kinase C has been stored at 4° in the presence of protease inhibitors, but in this case all subsequent experiments were carried out within 1 week.[24] Sucrose at a final concentration of 34% (w/v) has also been used to stabilize protein kinase C at 4°.[25] Lyophilization, as well as freezing and thawing, has been reported to inactivate or denature some kinases,[26] while cAMP-dependent protein kinase is commercially available in the lyophilized form.

## Acknowledgments

We thank Drs. Lisa Ballou, Greg Goodall, and Dennis Keefe for their critical reading of the manuscript, and Carol Wiedmer for typing it.

[24] C. J. Le Peuch, R. Ballester, and O. M. Rosen, *Proc. Natl. Acad. Sci. U.S.A.* **80,** 6858 (1983).

[25] B. C. Wise, R. L. Raynor, and J. F. Kuo, *J. Biol. Chem.* **257,** 8481 (1982).

[26] U. Kikkawa, Y. Takai, R. Minakuchi, S. Inohara, and Y. Nishizuka, *J. Biol. Chem.* **257,** 13341 (1982).

# [13] Use of Synthetic Peptides Mimicking Phosphorylation Sites for Affinity Purification of Protein-Serine Kinases

*By* JAMES ROBERT WOODGETT

The parameters used by protein kinases to recognize, bind, and phosphorylate particular residues in proteins depend largely on the amino acid sequences around the target amino acids. Thus synthetic peptides corresponding to sites phosphorylated by cyclic AMP-dependent protein kinase, protein kinase C, casein kinase II, calmodulin-dependent protein kinase II, and ribosomal S6 kinase, among others, are relatively effective substrates compared to their protein substrates.[1-4] The importance of protein kinases in signal transduction and cellular regulation has become apparent in the past decade. However, study of these enzymes is complicated by their relatively low abundance in tissue, necessitating extensive purification. We have utilized the affinity of two protein-serine kinases for a synthetic peptide substrate to effect their purification. One of the kinases, glycogen synthase kinase 3 (GSK-3), exhibits an unusual specificity requirement. Certain substrates must be prephosphorylated by a distinct protein kinase before they are recognized by GSK-3.[5-8] A purification strategy has been developed to exploit this unusual prerequisite using a peptide derived from glycogen synthase, which exhibits high-affinity binding to GSK-3 only when prephosphorylated by casein kinase II. The utility of synthetic peptides for the affinity purification of other protein kinases is discussed.

## Peptide Assay of Casein Kinase II and GSK-3

### *Peptide Synthesis*

A 25-residue peptide termed GS-1, derived from a region of glycogen synthase containing the residues phosphorylated by casein kinase II and GSK-3 (Fig. 1),[9,10] was synthesized on an Applied Biosystems (Foster City, CA) model 430A using 9-fluorenylmethyloxycarbonyl (FMOC) chemistry and a free carboxyl terminus. The peptide contains two arginine residues at the amino terminus to facilitate binding to phosphocellulose paper and hence simplify the assay of phosphorylation. The peptide was purified by reversed-phase high-performance liquid chromatography (HPLC) and authenticity confirmed by amino acid analysis and Edman peptide se-

[1] B. E. Kemp, E. Benjamini, and E. G. Krebs, *Proc. Natl. Acad. Sci. U.S.A.* **73,** 1038 (1976).
[2] O. Zetterqvist, U. Ragnarsson, E. Humble, L. Berglund, and L. Engström, *Biochem. Biophys. Res. Commun.* **70,** 696 (1976).
[3] L. Engström, P. Ekman, E. Humble, U. Ragnarsson, and O. Zetterqvist, this series, Vol. 107, p. 130.
[4] B. E. Kemp and R. B. Pearson, this volume [10].
[5] C. Picton, J. R. Woodgett, B. A. Hemmings, and P. Cohen, *FEBS Lett.* **150,** 191 (1982).
[6] A. A. DePaoli-Roach, *J. Biol. Chem.* **259,** 12144 (1984).
[7] C. J. Fiol, A. M. Mahrenholz, Y. Wang, R. W. Roeske, and P. J. Roach, *J. Biol. Chem.* **262,** 14042 (1987).
[8] J. R. Woodgett and P. Cohen, *Biochim. Biophys. Acta* **788,** 339 (1984).
[9] C. Picton, A. Aitken, T. Bilham, and P. Cohen, *Eur. J. Biochem.* **124,** 37 (1982).
[10] L. Poulter, S. G-. Ang, B. W. Gibson, D. H. Williams, C. F. B. Holmes, F. B. Caudwell, J. Pitcher, and P. Cohen, *Eur. J. Biochem.* **175,** 497 (1988).

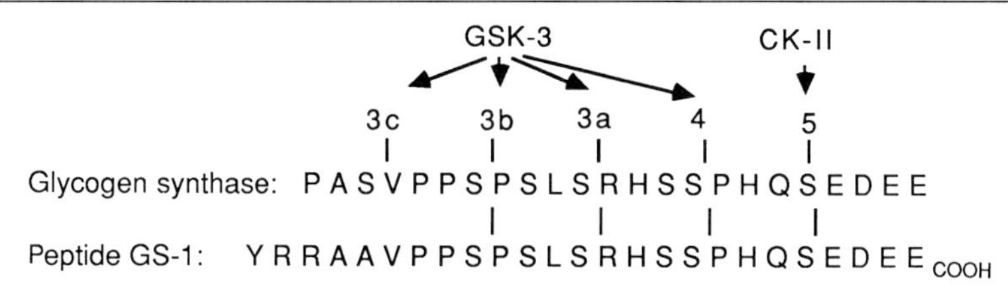


FIG. 1. Amino acid sequence of synthetic peptide GS-1 and the region to which it corresponds in glycogen synthase.[9,10] The sites targeted by GSK-3 and casein kinase II are indicated.

quencing. Aliquots of the peptide (5 m*M* in water) are stored in the dark at 4 or −20°. Concentration is determined by amino acid analysis.

*Procedure for Casein Kinase II Assay*

Peptide GS-1 (50 $\mu M$) is incubated with 20 m*M* HEPES–OH (pH 7.5), 0.2 m*M* EGTA, 100 m*M* NaCl, and casein kinase II for 2 min at 30°. The reaction (final volume 0.05 ml) is initiated by the addition of $MgCl_2$ to 4 m*M* and [$\gamma$-$^{32}$P]ATP (100 cpm/pmol) to 10 $\mu M$. After 2–10 min, reactions are terminated by spotting an aliquot (0.04 ml) onto 1-cm$^2$ squares of phosphocellulose paper (Whatman Clifton, NJ; P-81) which are then immersed into 7.5% phosphoric acid (200 ml) and stirred. Following three changes of the acid washing solution, the papers are allowed to dry and associated radioactivity determined in the presence of scintillant.

*Procedure for GSK-3 Assay*

Since phosphorylation of GS-1 by GSK-3 requires prephosphorylation of the peptide by casein kinase II, this protein kinase is more easily assayed using glycogen synthase as a substrate, since the casein kinase II site on this protein is not dephosphorylated during isolation.[11] However, synthetic peptide prephosphorylated by casein kinase II in the presence of unlabeled $Mg^{2+} \cdot$ ATP can be prepared using the scaled up conditions described below for preparation of the affinity columns, but using 10 $\mu M$ unlabeled ATP rather than 1 m*M* ATP. Casein kinase II activity is inactivated after the prephosphorylation by the addition of 0.1 vol of 20 $\mu$g/ml heparin and heating to 50° for 10 min. Assays can then be performed by the addition of radiolabeled $Mg^{2+} \cdot$ ATP and initiated with GSK-3 as described for the assay of casein kinase II (see above). Control reactions are performed in the absence of GSK-3 to confirm inhibition of casein kinase II. For

[11] B. A. Hemmings and P. Cohen, this series, Vol. 99, p. 337.

quantitation of activity and stoichiometry, the dilution in the specific activity of the radiolabeled ATP by the unlabeled nucleotide added for the casein kinase II preincubation must be considered.

### *Definition of Unit*

One unit of GSK-3 or casein kinase II is that amount that catalyzes the incorporation of 1.0 $\mu$mol of phosphate into glycogen synthase or GS-1, respectively, per minute from [$\gamma$-$^{32}$P]ATP at 30°.

## Preparation of Peptide Affinity Columns

### *Preparation of Phosphorylated GS-1*

Peptide GS-1 (10 mg) is incubated in a final volume of 0.2 ml in the presence of 4 m*M* $MgCl_2$, 0.2 m*M* EGTA, 20 m*M* HEPES–OH (pH 7.5), 100 m*M* NaCl, 2 mU casein kinase II, and 10 m*M* [$\gamma$-$^{32}$P]ATP (100 cpm/nmol) for 12 hr at 30°. Additional 2-mU aliquots of casein kinase II and ATP are added after each 60-min period. Final stoichiometry of phosphate incorporation should be 0.3–0.4 mol phosphate/mol GS-1 peptide.

### *Preparation of Peptide Agarose*

Affi-Gel 15 (8 ml; Bio-Rad, Richmond, CA) is washed with water and resuspended in 5 ml 100 m*M* sodium bicarbonate, pH 8.0. Peptide GS-1 or phospho-GS-1 (10 mg in water) is added and rotated end over end for 16 hr at room temperature. Unreacted *N*-hydroxysuccinimide ester groups on the agarose are blocked by the addition of 1.0 *M* ethanolamine hydrochloride, pH 8.0, for 2 hr at room temperature. The agarose is washed with 25 m*M* Tris-HCl, 1 m*M* EDTA, 15 m*M* 2-mercaptoethanol, 5% (v/v) glycerol, pH 7.5 (buffer A) containing 0.5 *M* NaCl to remove noncovalently linked peptide and is stored at 4° in buffer A containing 3 m*M* sodium azide. The coupling efficiency is monitored by the inclusion of trace amounts of [$^{32}$P]phosphate-labeled peptide and should be >50%.

## Use of Peptide Affinity Columns to Purify Casein Kinase II and GSK-3

All procedures are carried out at 4°. The initial steps in the purification of casein kianse II and GSK-3 are similar, permitting their simultaneous isolation. Skeletal muscle from the hind limbs and back of female New Zealand White rabbits, killed by intravenous injection of a lethal dose of sodium pentabarbitone, is rapidly excised and placed into ice. The muscle is coarsely ground and homogenized in 2.5 vol of 4 m*M* EDTA, 15 m*M* 2-mercaptoethanol, 0.1 m*M* phenylmethylsulfonyl fluoride, 1 m*M* benzamidine, pH 7.0, for two 30-sec low-speed pulses in a Waring commercial blender. Following centrifugation at 6000 *g* for 40 min, the supernatant is decanted through glass wool to remove lipid deposits. Optionally at this stage, the extract can be acidified to pH 6.1 to precipitate glycogen particles as previously described.[12] This allows subsequent isolation of glycogen synthase from the pellet.[13] However, this step only marginally affects the purification of GSK-3 and results in a 50% reduction in the yield of casein kinase II. Therefore, this acidification step is usually omitted. The pH of the extract is adjusted to 7.5 with 15 *M* ammonium hydroxide before addition of solid ammonium sulfate to 33% saturation (195 g/liter). The precipitate is left to form for 20 min and is removed by centrifugation at 6000 *g* for 40 min. Solid ammonium sulfate is added to the supernatant to achieve 55% saturation (a further 142 g/liter). After standing for 20 min, the precipitate is collected by centrifugation at 6000 *g* for 40 min. The pellet is resuspended in 400 ml 25 m*M* Tris, 1 m*M* EDTA, 15 m*M* 2-mercaptoethanol, pH 7.0, and dialyzed against two 8-liter changes of this buffer overnight. Following centrifugation at 10,000 *g* for 30 min to remove insoluble material, the supernatant is applied to a column of phosphocellulose (Whatman P-11, 5 × 6 cm) equilibrated in buffer A containing 50 m*M* NaCl. After extensive washing with buffer A + 100 m*M* NaCl (4 liter), GSK-3 is eluted with buffer A + 200 m*M* NaCl (2 liter). The column is then washed with buffer A + 400 m*M* NaCl (2 liter) and casein kinase II eluted with buffer A + 1 *M* NaCl (500 ml).

### *Further Purification of GSK-3*

Fractions containing GSK-3 activity from the 200 m*M* NaCl phosphocellulose elution are dialyzed for 2 hr against buffer A and applied to a 2.6 × 50 cm column of Blue-Sepharose CL-6B (Pharmacia/LKB, Piscataway, NJ) equilibrated in buffer A + 50 m*M* NaCl. GSK-3 is eluted with a 400-ml linear 50–1000 m*M* NaCl gradient in buffer A, the active fractions eluting around 400 m*M* NaCl are pooled, dialyzed overnight against buffer A, and concentrated on a column of CM-Sepharose CL-4B (1 × 10 cm; Pharmacia/LKB) in the same buffer. The column is eluted with buffer A + 200 m*M* NaCl and the peak of protein dialyzed against two changes of buffer A (1 liter) for 3 hr. The preparation is applied to a column of GS-1 agarose (1 × 8 cm), which is then washed with buffer A (20 ml). GSK-3 is not retained by the column (Fig. 2A). Flow-through and wash

[12] P. Cohen, this series, Vol. 99, p. 243.

[13] H. G. Nimmo, C. G. Proud, and P. Cohen, *Eur. J. Biochem.* **68,** 21 (1976).

fractions containing GSK-3 activity are pooled and directly applied to a column of phospho-GS-1 agarose (1 × 4 cm) at a flow rate of 0.5 ml/min. After a brief (20 ml) wash with buffer A, GSK-3 is eluted with buffer A + 50 m*M* NaCl (Fig. 2B). The active fractions are immediately applied to a Mono S fast protein liquid chromatography (FPLC) column (Pharmacia/LKB) equilibrated in buffer A and eluted with a linear 0–0.5 *M* NaCl gradient. Active fractions are dialyzed against buffer A containing 200 m*M* NaCl and 50% rather than 5% glycerol and stored at −20°. Activity is stable for several years. A summary of the purification is given in Table I.[14,15]

*Further Purification of Casein Kinase II*

The peak of protein eluted by 1 *M* NaCl from phosphocellulose column is dialyzed for 2 hr against buffer A containing 200 m*M* NaCl and applied to heparin-Sepharose (1.5 × 10 cm; Pharmacia/LKB) equilibrated in the dialysis buffer. After washing with buffer A containing 0.4 *M* NaCl, the casein kinase II is step eluted by increasing the NaCl concentration to 1.0 *M*. The peak of activity is dialyzed overnight against buffer A containing 100 m*M* NaCl. Following centrifugation at 10,000 *g* for 20 min to remove precipitated material, the preparation (approximately 30 ml) is applied to GS-1 agarose (1 × 6 cm) at a flow rate of 0.5 ml/min. The column is washed with buffer A containing 100 m*M* NaCl and the casein kinase II is completely eluted by increasing the NaCl concentration to 200 m*M* (Fig. 2C). The enzyme purified thus has a specific activity of ~600 mU/mg using peptide GS-1 and is >40% pure. Purification to homogeneity can be readily accomplished by FPLC chromatography on Mono Q beads (Pharmacia/LKB), elution being achieved with a 0–600 m*M* NaCl gradient. Fractions containing activity, eluting at 350 m*M* NaCl, are pooled and dialyzed

[14] M. M. Bradford, *Anal. Biochem.* **72,** 248 (1976).
[15] J. R. Woodgett, *Anal. Biochem.* **180,** 237 (1989).

FIG. 2. Chromatography of GSK-3 and casein kinase II on peptide-agarose. (A) GSK-3 from the CM-Sepharose step was applied to a column (1 × 8 cm) of GS-1 agarose at a flow rate of 30 ml/hr. Protein kinase activity (◆) and protein (⊡) were monitored. Fractions of 2 ml were collected. (B) The activity from the GS-1 agarose column [essentially the breakthrough fractions in (A)] was applied to a column of phospho-GS-1 (1 × 4 cm). After washing with load buffer (20 ml), the column was eluted by increasing the NaCl concentration to 50 m*M* (at X) and 200 m*M* (at Y). Symbols as in (A). (C) Casein kinase II from the heparin-Sepharose step was applied to a column (1 × 8 cm) of GS-1 agarose at a flow rate of 30 ml/hr. The column was eluted with 200 m*M* (X), 500 m*M* (Y), or 1 *M* NaCl (Z). Symbols as in (A). Results are adapted from Woodgett.[15]

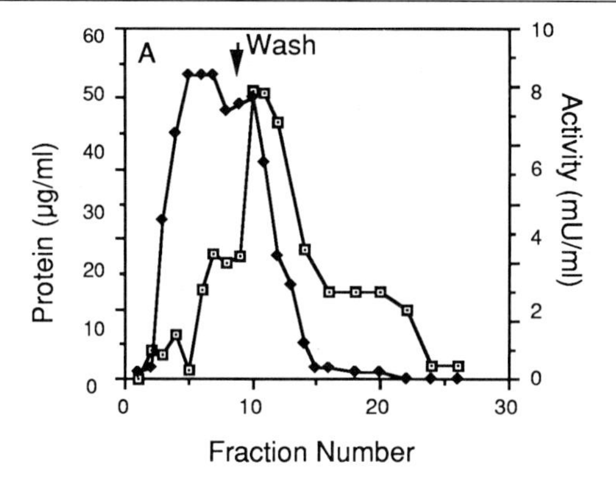
A
Wash
Protein (μg/ml)
Activity (mU/ml)
Fraction Number
60
50
40
30
20
10
0
10
8
6
4
2
0
0
10
20
30

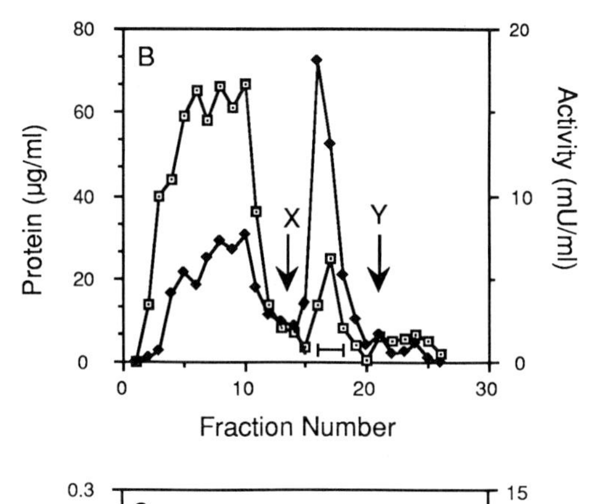
B
X
Y
Protein (μg/ml)
Activity (mU/ml)
Fraction Number
80
60
40
20
0
20
10
0
0
10
20
30

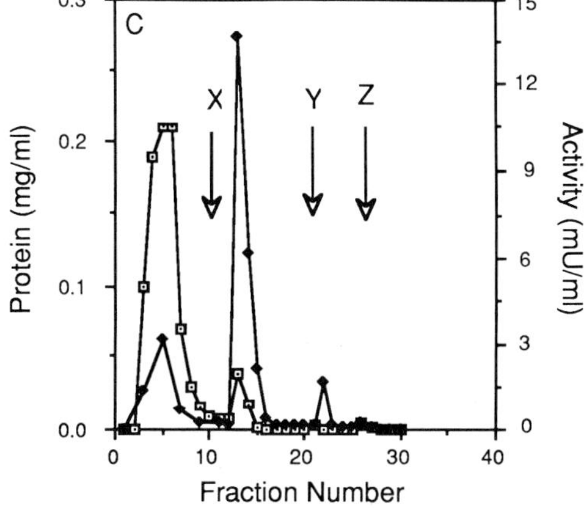
C
X
Y
Z
Protein (mg/ml)
Activity (mU/ml)
Fraction Number
0.3
0.2
0.1
0.0
15
12
9
6
3
0
0
10
20
30
40

TABLE I
PURIFICATION OF GSK-3 FROM RABBIT SKELETAL MUSCLE[a]

| Step | Protein (mg) | Activity (mU) | Specific activity (mU/mg) | Purification (-fold) | Yield (%) |
|---|---|---|---|---|---|
| Phosphocellulose | 110 | 950 | 8.6 | 290 | 30 |
| Blue-Sepharose | 36 | 696 | 19.3 | 643 | 23 |
| CM-Sepharose | 10.6 | 430 | 41 | 1,360 | 15 |
| GS-1 agarose (breakthrough) | 4.2 | 350 | 83 | 2,800 | 12 |
| Phospho-GS-1 agarose | 0.09 | 120 | 1,333 | 44,450 | 4 |
| Mono S | 0.04 | 84 | 2,100 | 70,000 | 2.8 |

[a] Starting material was 3000 g muscle. Protein was determined according to Bradford.[14] The specific activity in crude muscle extracts was taken from Woodgett and Cohen[8] as 0.03 mU/mg in 100,000 mg of starting protein. The results are taken from Woodgett.[15]

against buffer A containing 200 m*M* NaCl and 50% (v/v) glycerol. Storage at $-20°$ maintains activity for several years. A summary of the purification is given in Table II.

## Comments on Purification

The passage of GSK-3 through GS-1 agarose results in only a twofold purification. However, omission of this step greatly reduces the efficiency of the subsequent purification on the phosphorylated GS-1 column. Indeed, the subtractive nature of the two columns results in a 30-fold enrichment, whereas chromatography on phospho-GS-1 agarose alone effects

TABLE II
PURIFICATION OF CASEIN KINASE II FROM RABBIT SKELETAL MUSCLE[a]

| Step | Protein (mg) | Activity (mU) | Specific activity (mU/mg) | Purification (-fold) | Yield (%) |
|---|---|---|---|---|---|
| Phosphocellulose | 32 | 154 | 4.8 | — | 100 |
| Heparin-Sepharose | 6.5 | 106 | 16 | 3.4 | 68 |
| GS-1-agarose | 0.25 | 70 | 280 | 59 | 45 |
| Mono Q | 0.1 | 59 | 590 | 110 | 38 |

[a] Yields and purification are given from the phosphocellulose step due to interference in assays of crude extracts. Starting material was 3000 g of rabbit skeletal muscle. Protein was determined according to Bradford.[14] Results are adapted from Woodgett.[15]

only a 5-fold purification. Only 50% of the GSK-3 is retained by the phosphorylated column. This is apparently not due to overloading, since reapplication of the breakthrough to a fresh column fails to cause significant retention. The reason for the lower affinity for phospho-GS-1 of a proportion of the GSK-3 is presently unclear. However, preliminary Western blotting experiments suggest that the breakthrough activity has a slightly lower molecular weight (48K) to that of the retained activity (51K), perhaps representing posttranslational modification or distinct gene products.[16] Chromatography on Mono S is recommended following the peptide affinity steps since this step acts to concentrate the dilute enzyme and removes minor contaminants, resulting in the isolation of homogeneous GSK-3 (51K). Ability to chemically synthesize the phosphorylated form of peptide GS-1 may improve the performance of the chromatography and would greatly simplify preparation of the phospho-GS-1 agarose.[17]

## General Use of Peptide Columns as Affinity Matrices

While the peptide used in these studies exploited the unusual specificity of GSK-3 for prephosphorylated substrates, the ability of the peptide to effect a reasonable purification of casein kinase II suggests that substrate peptides may be generally useful for affinity chromatography. The most critical requirement is for a peptide that exhibits high affinity for a protein kinase; that is, with a $K_m$ value lower than 10 $\mu M$, a condition now satisfied by a number of peptide substrates for a variety of protein kinases such as protein kinase C and myosin light-chain kinase, for example.[18–20] The $K_m$ values of peptides GS-1 and phospho-GS-1 for casein kinase II and GSK-3 are 10 and 5 $\mu M$, respectively. A 32-residue peptide derived from the C terminus of ribosomal peptide S6 has been effective in the purification of a hepatic cycloheximide-induced S6 kinase, facilitating a 94-fold enrichment.[21] The $K_m$ value of the S6 kinase for this peptide is 12 $\mu M$. There are several advantages in the use of peptides as affinity columns. The peptides can be synthesized in large quantities (100 mg–1 g), whereas protein substrates are often difficult to obtain in sufficient amounts for affinity applications. The large size of proteins endows them with regions of charge and/or hydrophobicity that cause nonspecific interaction with other

[16] J. R. Woodgett, this volume [48].

[17] J. Perich, this series, Vol. 201 [18], [19].

[18] B. E. Kemp, R. B. Pearson, and C. House, *Proc. Natl. Acad. Sci. U.S.A.* **80,** 7471 (1983).

[19] J. R. Woodgett, K. L. Gould, and T. Hunter, *Eur. J. Biochem.* **161,** 177 (1986).

[20] B. E. Kemp, R. B. Pearson, V. Guerriero, Jr., I. C. Bagchi, and A. R. Means, *J. Biol. Chem.* **262,** 2542 (1987).

[21] D. J. Price, R. A. Nemenoff, and J. Avruch, *J. Biol. Chem.* **264,** 13825 (1989).

proteins, reducing their efficacy in purification. Often proteins are substrates for several protein kinases, further reducing their specificity. Peptides, in contrast, can be designed to selectively associate with a particular kinase by varying the composition of amino acids around the target phosphorylation site. Idealized synthetic peptides could be designed exhibiting greater affinity than the parent protein. In many ways the advantages of peptides for purification of protein kinases are analogous to their benefits in the generation of antibodies.

Acknowledgments

The technical expertise of Rob Philp in synthesizing the synthetic peptide GS-1 is gratefully appreciated.

# [14] Affinity Purification of Protein Kinases Using Adenosine 5′-Triphosphate, Amino Acid, and Peptide Analogs

*By* PAUL JENÖ and GEORGE THOMAS

The first well-characterized protein kinases were investigated in conjunction with classical hormonal signaling pathways mediated by cyclic AMP or $Ca^{2+}$.[1] Initially it was clear that posttranslational modification of proteins by phosphorylation served as one mechanism for transducing intracellular signals. More recent studies, however, indicate that the spectrum of protein kinases controlling specific intracellular processes is immense,[2] arguing that phosphorylation–dephosphorylation may be the major intracellular regulatory mechanism operating in eukaryotic cells. Many of the more recently described protein kinases were discovered using molecular biological strategies, and the number of new protein kinases discovered in this way will continue to increase at an exponential pace. The success of these cloning strategies lies in the fact that all known protein kinases possess conserved sequences which allow the construction of oligonucleotide probes for screening purposes.[3,4] Although such a procedure can lead rapidly to the identification of a new protein kinase, without

[1] A. M. Edelman, D. K. Blumenthal, and E. G. Krebs, *Annu. Rev. Biochem.* **56,** 567 (1987).

[2] T. Hunter, *Cell (Cambridge, Mass.)* **50,** 823 (1987).

[3] S. K. Hanks, *Proc. Natl. Acad. Sci. U.S.A.* **84,** 388 (1987).

[4] S. K. Hanks, A. M. Quinn, and T. Hunter, *Science* **241,** 42 (1988).

a complementary genetic system,[5,6] it is not a simple task to then establish the physiological importance or the substrate specificity of the protein kinase. The other approach is to identify a kinase by biochemical procedures, exploiting a substrate which is known to be phosphorylated under specific physiological conditions. This may also prove to be a difficult task since many protein kinases appear to be rare enzymes requiring both a sufficient source for their purification and a knowledge of their mode of activation. The two approaches, homology probing and biochemical purification, are not mutually exclusive but complementary. Indeed, protein purification and sequencing provided the basis for homology probing.[4] Thus, when identifying a novel protein kinase one is initially faced with the necessity of carrying out a biochemical purification. When sufficient amounts of the material are available to produce antibodies and/or protein sequence, gene cloning can be used to determine the complete primary structure of the protein kinase.

In establishing a strategy for the identification of a protein kinase the aim is to purify the enzyme as efficiently and rapidly as possible. Since many protein kinases are relatively rare enzymes, classical procedures such as anion-exchange, cation-exchange, hydrophobic, and gel-filtration chromatography, although useful in removing bulk proteins, have proved inadequate in purifying these enzymes to homogeneity. To achieve this end and to recover the protein kinase in a relatively high yield requires one or more affinity steps. These steps are based on the interaction of the protein kinase with either its two substrates, ATP and the protein to be phosphorylated, or with molecules which modulate its activity, such as cofactors, effectors, or inhibitors. This chapter is intended to compile some of the affinity methods which have been successfully applied to the purification of protein kinases. It is not meant as a complete overview of the field but rather as a survey of possible approaches and illustrates the potential of the technique by presenting several examples.

## ATP and ATP Analogs

In most cases ATP is utilized by protein kinases as the phosphate donor during the phosphorylation of specific substrates. However, it is also the major intracellular energy source for driving enzymatic reactions. Thus, it has a long history of being employed as an immobilized affinity ligand for the purification of a number of enzymes. The interaction of enzymes with immobilized ATP is very much dependent on the moiety which is

[5] J. Hayles, D. Beach, B. Durkacz, and P. Nurse, *Mol. Gen. Genet.* **202,** 291 (1986).
[6] K. Basler and E. Hafen, *Cell (Cambridge, Mass.)* **54,** 299 (1988).

coupled to the resin, as this can lead to steric constraints. Therefore, when exploring this technique it is important to test the binding characteristic of the protein kinase with the different types of immobilized ATP.

ATP and many of its analogs can be coupled via the N-6 or C-8 positions of adenine, through the terminal phosphate group, or through the hydroxyl groups of the ribose ring by oxidation to the resin, usually agarose. These forms of ATP are referred to as types 1 through type 4, respectively, and are commercially available from Pharmacia (Piscataway, NJ). They can also be easily prepared in the laboratory at significantly less cost by following established protocols.[7] In general, these protocols involve coupling ATP to the support matrix via hexyl side chains. These side chains act as spacer arms, allowing the enzyme to interact more readily with the affinity ligand. The $N^6$-substituted analog is synthesized by quaternizing the N-1 position of the adenine moiety, followed by rearrangement of the substituent onto the 6-amino group. Alternatively, it has been prepared as an $N^6$-ε-aminocaproyl-ATP. The most common approach for linking ATP at the C-8 position is by derivatizing this site with an aminohexyl group. A number of ATP analogs have been prepared that are immobilized through the terminal phosphate group, such as the 6-amino-1-hexyl nucleotide phosphodiester of ADP prepared by condensation of the phosphorylimidazolide of AMP with 6-amino-1-hexyl pyrophosphate. Finally, the ribose-linked ATP analog is obtained by oxidizing the ring with sodium periodate and then reacting this with a dihydrazide, followed by coupling to the activated matrix. In general, approximately 5–10 μmol of analog can be coupled per milliliter of swollen matrix. An extensive discussion of immobilized nucleotide affinity chromatography analogs and methods for their preparation has been reviewed by Mosbach.[7]

An alternative to nucleotide affinity chromatography is the use of immobilized dyes. It was simultaneously discovered during the purification of pyruvate kinase and phosphofructokinase on gel-filtration columns that both kinases eluted in the void volume with high-molecular-weight dextran, the marker used to calibrate the exclusion volume of gel-filtration columns.[8,9] To make the dextran visible it is coupled to Cibacron blue F3GA (Pharmacia). When the blue marker was omitted from the gel-filtration step, the enzymes eluted after the void volume, at their correct molecular weights. This finding rapidly led to the use of this dye for the purification of many other enzymes, especially dehydrogenases, and to the generally accepted idea that most enzymes which bind purine nucleotides

[7] K. Mosbach, *Adv. Enzymol.* **46,** 205 (1978).

[8] R. Haeckel, B. Hess, W. Lauterborn, and K. Wurster, *Hoppe-Seyler's Z. Physiol. Chem.* **349,** 699 (1968).

[9] G. Kopperschläger, R. Freyer, W. Diezel, and E. Hofmann, *FEBS Lett.* **1,** 137 (1968).

show affinity toward Cibacron blue F3GA. Because this and other dyes have few structural features in common with the true substrate, they have been referred to as pseudoaffinity substrates. In the case of Cibacron blue F3GA, the groups which appear to contribute to the binding are the planar ring structure and the negatively charged moieties. By X-ray crystallography it has been shown that Cibacron blue F3GA binds to the nicotinamide adenine dinucleotide binding site of alcohol dehydrogenase, at positions corresponding to the adenine and ribose rings.[10] The dye, therefore, serves as an analog of ADP-ribose and for this reason binds strongly to the ATP binding site of kinases. It should be noted that, in contrast to ATP, attachment of the dye directly to the matrix instead of through a spacer arm has no effect on binding.

Although Cibacron blue F3GA has been utilized in many purification protocols, other reactive triazinyl dyes have been only recently applied to protein purification. Amicon (Danvers, MA) has introduced a Dyematrex kit, which includes a number of these dyes. The kit has the advantage in that it can be employed as a rapid screening method to assess the suitability of this approach in the purification of a specific kinase. In many cases these dyes are employed as another purification step. However, proper application of the principles of affinity elution and more extensive screening of different dyes have led to the development of powerful purification steps.

Proteins bound to either ATP or dye affinity columns can be eluted by increasing the salt concentration or by applying micro- or millimolar concentrations of substrates or specific eluants. Obviously, elution with specific substrates yields the highest purification factors. Nevertheless, in the case of strong interactions one may have to turn to salt elution. In the case of dyes exhibiting hydrophobic interactions, proteins are best eluted with chaotropic agents such as thiocyanate, guanidinium hydrochloride, or ethylene glycol. (A more thorough discussion of these approaches can be found in Scopes.[11]) Two examples of kinases purified by (1) ATP affinity chromatography or (2) dye affinity chromatography are given below.

### *Affinity Chromatography of Mitogen-Activated S6 Kinase on ATP Type 4 Agarose*

Purification of the mitogen-activated S6 kinase is carried out in sequential steps on Fast Flow S, Mono Q, Sephacryl 200, and ATP-agarose type

[10] J. F. Biellmann, J. P. Samama, C. I. Bränden, and H. Eklund, *Eur. J. Biochem.* **102,** 107 (1979).

[11] R. K. Scopes, "Protein Purification Principles and Practice." Springer-Verlag, New York, 1987.

4.[12,13] The last step results in a 43% recovery and a 20-fold purification of the enzyme. More recently, the efficiency of this step has been dramatically improved by developing the column with ATP rather than salt.[14] It should be noted that binding of the kinase to the resin requires low ionic strength and that ATP coupled through either the N-6 (type 2) or the C-8 (type 3) position of adenine fails to bind the enzyme under identical conditions.

*Affinity Chromatography of Casein Kinase I on Affi-Gel Blue*

The purification of casein kinase I is carried out by sequential chromatography on DEAE-cellulose, phosphocellulose, CM-cellulose, hydroxyapatite, and Affi-Gel Blue.[15] The last step results in a 68% recovery and an approximately three- to fourfold purification of the enzyme. Elution of the enzyme requires not only a strong salt, ammonium sulfate, but also the presence of 10 m*M* ATP.

## Affinity Chromatography of Kinases on Immobilized Amino Acids, Proteins, and Peptide Substrates

Immobilization of protein substrates has gained widespread application as a means to purify protein kinases present in low amounts. One of three approaches is usually used. First, when there is no information regarding the sequence surrounding the sites of phosphorylation in the substrate the amino acids which are known to be phosphorylated *in vivo* can be used as affinity ligands (see Table I[16–19]). Obviously, if this method is chosen the protein kinase of interest must be first purified away from other protein kinases. Although affinity chromatography on an amino acid support may appear too general, it can be a powerful technique. Second, if the substrate has been identified and can be obtained in large amounts, it can be attached

[12] P. J. Jenö, L. M. Ballou, I. Novak-Hofer, and G. Thomas, *Proc. Natl. Acad. Sci. U.S.A.* **85,** 406 (1988).

[13] P. J. Jenö, N. Jäggi, H. Luther, M. Siegmann, and G. Thomas, *J. Biol. Chem.* **264,** 1293 (1989).

[14] L. M. Ballou, M. Siegmann, and G. Thomas, *Proc. Natl. Acad. Sci. U.S.A.* **85,** 7145 (1988).

[15] M. E. Dahmus, *J. Biol. Chem.* **256,** 3319 (1981).

[16] U. Kikkawa, J. Koumoto, and Y. Nishizuka, *Biochem. Biophys. Res. Commun.* **135,** 636 (1986).

[17] H. A. Lane and G. Thomas, this volume [22].

[18] B. A. Hemmings, D. Yellowlees, J. C. Kernohan, and P. Cohen, *Eur. J. Biochem.* **119,** 443 (1981).

[19] J. Woodgett, this volume [13].

TABLE I
AFFINITY LIGANDS

| Affinity ligand | Kinase | Type of elution | Ref. |
|---|---|---|---|
| ATP-agarose type IV | S6 Kinase | Salt or ATP | 13 |
| Affi-Gel Blue | Casein kinase I | Salt or ATP | 15 |
| L-Threonine | Protein kinase C | Salt | 16 |
| S6 Peptide | S6 kinase | Salt | 17 |
| Glycogen synthase-agarose | Glycogen synthase kinase-3 | Salt | 18 |
| Peptide GS-1-agarose (± phosphorylated) | Glycogen synthase kinase-3 | Salt | 19 |
| Heat-stable protein kinase inhibitor (inhibitory domain) | cAMP-dependent protein kinase ($C_\alpha$, $C_\beta$ isoforms) | Salt or L-arginine | 22 (20) |
| p13 | Maturation-promoting factor | SDS p13 | 27 |

to the support matrix. However, in many cases it may not be possible to isolate the protein substrate in amounts large enough to achieve the high ligand densities required for affinity chromatography. In addition, due to the high number of charged amino acid side chains, such columns may act as ion-exchange resins leading to nonspecific protein–protein interactions. In order to suppress nonspecific binding, high salt conditions must be employed during the loading of the kinase to the column. Even with these drawbacks, this method has been applied with great success. The final approach has been the use of synthetic peptides or amino acids corresponding to the phosphorylation sites in the substrate. This approach has been applied to a number of protein kinases and is rapidly becoming the method of choice. The use of peptides requires that the sites of phosphorylation be known, that the kinase recognition determinants be identified, and that the peptides have affinities similar to those of the parent substrate. Taking these considerations into account, the main advantage is that one can easily produce sufficient amounts of peptide which can be readily coupled to preactivated resins at high densities. Application of peptide substrates might even be further extended in light of the identification of pseudosubstrate sequences within a number of protein kinases, including protein kinase C[20] and the myosin light chain kinase.[21]

Amino acids, proteins, or peptides can be coupled to a number of resins

[20] C. House and B. E. Kemp, *Science* **238,** 1726 (1987).

[21] R. B. Pearson, R. E. H. Wettenhall, A. R. Means, P. J. Hartshorne, and B. E. Kemp, *Science* **241,** 970 (1988).

available from either Pharmacia or Bio-Rad (Richmond, CA). In general, these resins have spacer arms and allow coupling through either the free carboxyl group (AH-Sepharose 4B; Pharmacia) or through primary amino groups (CH-Sepharose 4B and activated CH-Sepharose 4B from Pharmacia or Affi-Gel 10 and 15 from Bio-Rad). Coupling is straightforward, especially with the preactivated resins. However, in the case of coupling through free amino groups, it is important that the pH be between 5 and 10 with the optimum at around pH 8. At lower pH values the hydrolysis of active $\alpha$-amino groups is minimized. It is also important to note that the preactivated resins react preferentially with $\alpha$-amino groups of proteins or peptides, but will also react with the $\varepsilon$-amino groups of lysine and guanidinium groups of arginine. Since many peptides contain basic residues that define the protein kinase recognition determinants, efficient binding of the enzyme to the resin will be achieved only when the ligand concentration is in severalfold excess of the active ester groups of the resin. If the ligand concentration is too low, many of the peptides will be incorrectly linked to the resin, leading to masking of recognition determinants and the production of affinity resins with reduced binding capacities. The ligand concentration required can be quite high if one considers that the ester groups are present at 10 $\mu$mol/ml swollen resin. This is also one reason why peptides may prove more efficient in the purification of kinases, since, as mentioned above, it is difficult to reach ideal concentrations of coupling with proteins. For instance, to couple a 30-amino acid peptide, at an excess of 2-fold, to 1 ml of resin containing 10 $\mu$mol of active ester would require ~90 mg of peptide. In contrast, to couple an equivalent amount of a protein substrate with a molecular weight of 50,000 would require almost 1.5 g of protein. It should be noted that even though high ligand concentrations are necessary, the total amount of affinity resin required for the purification of the protein kinase may be significantly less than stated in the example above. In our experience, for the purification of the 70,000 S6 kinase, a small amount of resin, 50 to 200 $\mu$l, can be coupled just as efficiently as 1 ml. If this volume is insufficient to pack into a column, the modified resin can be diluted with Sepharose 4B. Below are three examples in which either (1) amino acids, (2) proteins, or (3) peptides attached to an affinity support have been used for the purification of a protein kinase.

*Affinity Chromatography of Protein Kinase C on Threonine-Sepharose*

L-Threonine (150 mg) is coupled, with 600 mg 1-ethyl-3-(3-dimethylaminopropyl)carbodiimide, to 50 ml of AH-Sepharose 4B in a volume of 120 ml.[16] The resin is then washed and packed into a 1 × 15 cm column. A protein kinase C fraction derived from 30 g of bovine brain, previously purified on a DEAE-52 column, is passed over the affinity column. Elution of the kinase is achieved with a gradient from 0 to 1.0 *M* NaCl. The recovery from the threonine-Sepharose 4B column is 53% with a 36-fold increase in specific activity.

### *Affinity Purification of Glycogen Synthase Kinase-3 on Glycogen Synthase-Agarose*

A 12.5-ml solution of Affi-Gel 15 is washed with water and resuspended in 5 ml of 100 m*M* sodium bicarbonate, pH 8.0, containing 15 mg of glycogen synthase.[18] After mixing for 10 hr at 4°, 10 ml of 10 m*M* glycine is added to block unreacted esters. Subsequent analysis indicates that 95% of the glycogen synthase is covalently attached to the Affi-Gel 15. The protein kinase is first purified by ammonium sulfate fractionation followed by chromatography on DEAE-cellulose, phosphocellulose, and Affi-Gel Blue. The pooled active fractions are concentrated, dialyzed, and applied to the Affi-Gel 15-glycogen synthase column. At this step 95% of the protein is removed with a recovery of 35% of the activity and a 10-fold purification.

### *Affinity Purification of $C_\alpha$ and $C_\beta$ Isoforms of Catalytic Subunit of cAMP-Dependent Protein Kinase on Synthetic Peptide*

This is a spectacular example of a single-step purification from cells overexpressing the catalytic subunits of the cyclic AMP-dependent protein kinase based on the use of a synthetic peptide corresponding to the inhibitory domain of the heat-stable protein kinase inhibitor (PKI).[22] PKI contains a recognition site for the cyclic AMP-dependent kinase in which alanine is present rather than serine. The peptide employed lacks the two C-terminal amino acids from the synthetic peptide inhibitor described by Cheng *et al.*[23] Although this reduces the inhibitory properties of the peptide 100-fold, (50% inhibition of enzymatic activity is obtained at 200 n*M*), it makes this peptide a good candidate for an affinity ligand. The peptide (15 mg) is coupled to 22 ml of preswollen Affi-Gel 10 in 0.1 *M* 4-morpholinopropane sulfonic acid (MOPS) buffer, pH 8.5, for 4 hr at 4°. After blocking unreacted esters the resin is equilibrated in column buffer, and a total cell extract containing 100 mg of protein is mixed with the resin at 4° for 16 hr with agitation. The mixture is then packed into a column, washed first

[22] S. R. Olsen and M. D. Uhler, *J. Biol. Chem.* **264,** 18662 (1989).

[23] H. C. Cheng, B. E. Kemp, R. B. Pearson, A. J. Smith, L. Miconi, S. M. Van Patten, and D. A. Walsh, *J. Biol. Chem.* **261,** 989 (1986).

with equilibration buffer containing 250 $\mu M$ ATP, and then with 200 m$M$ NaCl. The catalytic subunit is released by adding 200 m$M$ L-arginine to the column. Starting with 100 mg of cell extract protein, 100–200 $\mu$g of purified subunit is obtained with a yield of 30 to 50% in a single step. It should be noted that the degree of purity achieved is dependent on the extent of catalytic subunit expression induced by $Zn^{+}$. In either case this represents an exquisite example of affinity chromatography.

## Affinity Chromatography Based on Effector Molecules

In this approach one exploits effector molecules which are known to interact directly with the kinase. Many examples of effector molecules exist, such as calmodulin binding and activation of a number of kinases, including myosin light chain kinase, calmodulin-dependent kinases I–IV, and the more recently described $Ca^{2+}$/calmodulin-dependent translation elongation factor (EF-2) kinase.[24] Other positive and negative effector molecules have been exploited for affinity purification, including peptides based on PKI[20] (see above) and analogs of diacylglycerol or phospholipids in the purification of protein kinase C.[25,26] More recently, a similar approach was employed for the purification of the maturation-promoting factor (MPF) kinase in *Xenopus* oocytes.[27] Advantage was taken of the fact that this enzyme is equivalent to the cdc2 kinase of yeast. A number of gene products are known to interact with cdc2 and one of them, the *suc1* gene product or p13, is known to exist in a complex with cdc2 and cyclin. Based on this knowledge, it was found that yeast p13 is a potent inhibitor of MPF activation in cell-free extracts, leading to the use of this molecule as an affinity analog for the purification of MPF from *Xenopus* eggs.[27] Below we describe two cases in which either (1) a positive or (2) a negative effector molecule is utilized in the affinity purification of a kinase.

### *Affinity Purification of Myosin Light Chain Kinase on Calmodulin Sepharose 4B*

For this purification an affinity column is developed based on the fact that myosin light chain kinase is activated by $Ca^{2+}$ and calmodulin. Calmodulin is isolated from rabbit skeletal muscle according to Yazawa

[24] A. G. Ryazanou, E. A. Shestakova, and P. G. Natapov, *Nature* (*London*) **334,** 170 (1988).
[25] R. C. Schatzman, R. L. Raynor, R. B. Fritz, and J. F. Kuo, *Biochem. J.* **209,** 435 (1983).
[26] T. Vehida and C. R. Filburn, *J. Biol. Chem.* **259,** 12311 (1984).
[27] W. G. Dunphy, L. Brizuela, D. Beach, and J. Newport, *Cell* (*Cambridge, Mass.*) **54,** 433 (1988).

*et al.*[28] The purified calmodulin (51 mg) dissolved in 38 ml of 0.1 *M* $NaHCO_3$ is mixed with 21 g of preswollen activated Sepharose 4B for 16 hr at 4°.[29] The yield of coupling is reported to be 0.73 mg of calmodulin/ml of resin. Partially purified myosin light chain kinase containing 0.55 m*M* $CaCl_2$ is applied to the calmodulin-Sepharose 4B column (2.6 × 5 cm) equilibrated with 10 m*M* Tris-HCl, pH 8, and 0.2 m*M* $CaCl_2$. After applying the enzyme, the column is washed with equilibration buffer until the UV absorbance returns to basal levels. This procedure removes ≥97% of the proteins present. The protein kinase is eluted by washing the column with a buffer containing 1 m*M* EGTA.

### *Affinity Purification of Maturation Promoting Factor from Xenopus Eggs Employing suc1 Gene Product from Yeast*

For the purification of the MPF kinase from *Xenopus* egg extracts p13 is purified from an overproducing strain of *Escherichia coli* by gel filtration on Sepharose CL-6B, followed by either chromatography on a Mono Q anion-exchange column or a BioGel P-100 column. p13 is coupled to either CNBr-activated Sepharose 4B or Affi-Gel 10 at a final concentration of 5 mg/ml of resin. All of the MPF activity present in the cell extract is bound to the column and can be released only by SDS. However, more recently, a 3 m*M* solution of p13 was used to release the protein kinase from the column. The protein kinase is then separated from p13 on a Mono S column.[30]

## Conclusion

We have tried to review the most commonly used affinity steps in the purification of protein kinases. Obviously, the more one knows about the protein kinase the more affinity approaches can be exploited in its purification. Although it may take some effort to establish an affinity step for purification, the value of this investment cannot be overestimated.

## Acknowledgments

We thank Dr. L. M. Ballou and H. A. Lane for critical reading of the manuscript and C. Wiedmer for patience during the typing of this chapter.

[28] M. Yazawa, H. Kuwayama, and K. Yagi, *J. Biochem.* (*Tokyo*) **84,** 1253 (1978).

[29] M. Yazawa and K. Yagi, *J. Biochem.* (*Tokyo*) **84,** 1259 (1978).

[30] J.-C. Labbé, J.-P. Capony, D. Caput, J.-C. Covadore, J. Derancourt, M. Kaghad, J.-M. Lelias. A. Piccard, and M. Dorée, *EMBO J.* **8,** 3053 (1989).

# [15] Phosphorylation Assays for Proteins of the Two-Component Regulatory System Controlling Chemotaxis in *Escherichia coli*

*By* J. FRED HESS, ROBERT B. BOURRET, and MELVIN I. SIMON

## Introduction

### *Chemotaxis and Two-Component Regulatory Systems*

*In vitro* phosphorylation assays using purified proteins have been critical to success in unraveling the signal transduction mechanism which controls bacterial chemotaxis. The known reactions of the excitation pathway are summarized in Fig. 1. The CheA protein autophosphorylates[1-3] and transfers the phosphoryl group to the CheY protein.[2,3] CheY–phosphate is believed to interact with the flagellar switch to regulate the swimming behavior of the cell. Activation of CheA–phosphate and CheY–phosphate formation is coupled to the ligand occupancy state of the transmembrane receptor Tar by the CheW protein.[4,4a] CheY–phosphate spontaneously autodephosphorylates[5] to liberate $P_i$; this reaction is greatly accelerated by the CheZ protein.[2,5] The CheB protein, which is involved in adaptation, is also phosphorylated by CheA; however, dephosphorylation of CheB is not affected by the presence of CheZ.[2] From the perspective of this chapter CheB is little different than CheY and will not be discussed further. This chapter focuses on the central reactions involving CheA, CheY, and CheZ, whereas [16] in this volume describes the coupling of these reactions to Tar and CheW.

Chemotaxis is only one example of a widespread paradigm in bacteria for sensing and responding appropriately to various environmental sig-

[1] J. F. Hess, K. Oosawa, P. Matsumura, and M. I. Simon, *Proc. Natl. Acad. Sci. U.S.A.* **84,** 7600 (1987).
[2] J. F. Hess, K. Oosawa, N. Kaplan, and M. I. Simon, *Cell (Cambridge, Mass.)* **53,** 79 (1988).
[3] D. Wylie, A. M. Stock, C.-Y. Wong, and J. B. Stock, *Biochem. Biophys. Res. Commun.* **151,** 891 (1988).
[4] K. A. Borkovich, N. Kaplan, J. F. Hess, and M. I. Simon, *Proc. Natl. Acad. Sci. U.S.A.* **86,** 1208 (1989).
[4a] K. A. Borkovich and M. I. Simon, *Cell* **63,** 1339 (1990).
[5] J. F. Hess, R. B. Bourret, K. Oosawa, P. Matsumura, and M. I. Simon, *Cold Spring Harbor Symp. Quant. Biol.* **53,** 41 (1988).

nals.[6,6a] Two-component regulatory systems consist of a CheA-like protein (termed the sensor) paired with a CheY-like protein (termed the regulator). There is now substantial experimental evidence Table I[1–3,5,7–24a]) supporting a common mechanism of action, although each two-component regulatory system has unique properties. The proposed scheme involves autophosphorylation of a histidine residue in the sensor protein with the $\gamma$-phosphoryl group of ATP, and subsequent transfer of the phosphoryl group to an aspartate residue in the regulator protein, which modifies regulator activity. Thus, formation of regulator–phosphate is believed to be controlled environmentally. The system is reset by a specific phosphatase activity which hydrolyzes regulator–phosphate.

### *General Considerations*

Chemotaxis is one of the best understood of the two-component regulatory systems. The phosphorylation assays we have employed to study the chemotaxis proteins are described below as a guide for investigating re-

[6] J. B. Stock, A. J. Ninfa, and A. M. Stock, *Microbiol. Rev.* **53,** 450 (1989).
[6a] R. B. Bourret, K. A. Borkovich, and M. I. Simon, *Ann. Rev. Biochem.* **60,** 401 (1991).
[7] J. F. Hess, R. B. Bourret, and M. I. Simon, *Nature (London)* **336,** 139 (1988).
[8] D. A. Sanders, B. L. Gillece-Castro, A. M. Stock, A. L. Burlingame, and D. E. Koshland, Jr., *J. Biol. Chem.* **264,** 21770 (1989).
[9] A. J. Ninfa and B. Magasanik, *Proc. Natl. Acad. Sci. U.S.A.* **83,** 5909 (1986).
[10] J. Keener and S. Kustu, *Proc. Natl. Acad. Sci. U.S.A.* **85,** 4976 (1988).
[11] V. Weiss and B. Magasanik, *Proc. Natl. Acad. Sci. U.S.A.* **85,** 8919 (1988).
[12] M. M. Igo and T. J. Silhavy, *J. Bacteriol.* **170,** 5971 (1988).
[13] M. M. Igo, A. J. Ninfa, and T. J. Silhavy, *Genes Dev.* **3,** 598 (1989).
[14] M. M. Igo, A. J. Ninfa, J. B. Stock, and T. J. Silhavy, *Genes Dev.* **3,** 1725 (1989).
[15] H. Aiba, T. Mizuno, and S. Mizushima, *J. Biol. Chem.* **264,** 8563 (1989).
[16] H. Aiba, F. Nakasai, S. Mizushima, and T. Mizuno, *J. Biol.Chem.* **264,** 14090 (1989).
[17] S. Forst, J. Delgado, and M. Inouye, *Proc. Natl. Acad. Sci. U.S.A.* **86,** 6052 (1989).
[18] K. Makino, H. Shinagawa, M. Amemura, T. Kawamoto, M. Yamada, and A. Nakata, *J. Mol. Biol.* **210,** 551 (1989).
[19] M. Amemura, K. Makino, H. Shinagawa, and A. Nakata, *J. Bacteriol.* **172,** 6300 (1990).
[20] S. Jin, T. Roitsch, R. G. Ankenbauer, M. P. Gordon, and E. W. Nester, *J. Bacteriol.* **172,** 525 (1990).
[21] Y. Huang, P. Morel, B. Powell, and C. I. Kado, *J. Bacteriol.* **172,** 1142 (1990).
[22] S. Jin, R. K. Prusti, T. Roitsch, R. G. Ankenbauer, and E. W. Nester, *J. Bacteriol.* **172,** 4945 (1990).
[23] M. Perego, S. P. Cole, D. Burbulys, K. Trach, and J. A. Hoch, *J. Bacteriol.* **171,** 6187 (1989).
[23a] K. Mukai, M. Kawata, and T. Tanaka, *J. Biol. Chem.* **265,** 20000 (1990).
[23b] W. R. McCleary and D. R. Zusman, *J. Bacteriol.* **172,** 6661 (1990).
[24] A. J. Ninfa, E. G. Ninfa, A. N. Lupas, A. Stock, B. Magasanik, and J. B. Stock, *Proc. Natl. Acad. Sci. U.S.A.* **85,** 5492 (1988).
[24a] G. Olmedo, E. G. Ninfa, J. Stock, and P. Youngman, *J. Mol. Biol.* **215,** 359 (1990).

TABLE I
*In Vitro* PHOSPHORYLATION DATA FOR TWO-COMPONENT REGULATORY SYSTEM PROTEINS

| | | Sensor[a] | | | | | Regulator | | | | |
|---|---|---|---|---|---|---|---|---|---|---|---|
| | | | Labeled with | | | | | | | | |
| Function regulated | Organism | Protein | [γ-$^{32}$P]ATP | [γ-$^{32}$P]GTP | [α-$^{32}$P]ATP | Phosphorylated amino acid | Protein | Phosphorylated by sensor[b] | Phosphorylated amino acid | Regulator dephosphorylation stimulated by | Refs. |
| Chemotaxis | *E. coli, Salmonella typhimurium* | CheA | Yes | No | No | His-48 | CheY | Yes | Asp-57 | CheZ | 1–3, 5, 7, 8 |
| | | | | | | | CheB | Yes | ND | [c] | |
| Nitrogen assimilation | *E. coli* | NtrB | Yes | ND | No | Histidine | NtrC | Yes | Aspartate | NtrB/$P_{II}$/ATP | 9–11 |
| Outer membrane protein expression | *E. coli* | EnvZ[d] | Yes | No | No | Histidine?[e] | OmpR | Yes | Aspartate? | EnvZ/ATP | 12–17 |
| Phosphate assimilation | *E. coli* | PhoR[d], | Yes | ND | No | Histidine? | PhoB | Yes | Aspartate? | ND | 18, 19 |
| | | PhoM[d] | Yes | ND | No | Histidine | PhoMORF2 | Yes | Aspartate? | ND | |
| Virulence | *Agrobacterium tumefaciens* | VirA[d] | Yes | No | No | Histidine | VirG | Yes | Asp-52 | [f] | 20–22 |
| Sporulation | *Bacillus subtilis* | SpoIIJ | Yes | ND | ND | ND | SpoOA | Yes | ND | ND | 23 |
| | | | | | | | SpoOF | Yes | ND | ND | |
| Degradative enzymes | *B. subtilis* | DegS | Yes | ND | No | Histidine? | DegU | Yes | Aspartate? | ND | 23a |
| Motility/ development | *Myxococcus xanthus* | FrzE[g] | Yes | ND | ND | [g] | FrzE | [g] | Aspartate? | ND | 23b |

[a] ND, No data available.

[b] Sensor proteins have also been shown to correctly phosphorylate noncognate regulator proteins *in vitro*, although to a much lesser extent than cognate regulators. CheA can phosphorylate NtrC,[24] OmpR,[14] and SpoOA[24a]; NtrB can phosphorylate SpoOA[10] and CheY[24]; and EnvZ can phosphorylate NtrC.[14] PhoM can phosphorylate PhoB, but phosphorylation of PhoMORF2 by PhoR has not been observed.[19]

[c] No effect of CheZ observed.[2]

[d] Whereas CheA, NtrB, and SpoIIJ are soluble cytoplasmic proteins, most known sensors are transmembrane proteins[6] which have proved difficult or impossible to purify in large amounts. This obstacle has been circumvented for EnvZ,[12,15,17] PhoM,[19] PhoR,[18] and VirA[20,21] by construction of truncated or fusion proteins which lack transmembrane segments but retain the biochemically active carboxy-terminal portion of the protein. Antibodies raised against fusion proteins were used to confirm activity of the full-length wild-type EnvZ[13] and PhoM[19] proteins.

[e] A "?" signifies chemical properties which are consistent with the indicated phosphoamino acid, but conclusive identification has not been made.

[f] No effect observed with three different carboxy-terminal fragments of VirA, in either the presence or absence of ATP.[22]

[g] FrzE contains both sensor and regulator elements fused together in a single protein.[23b]

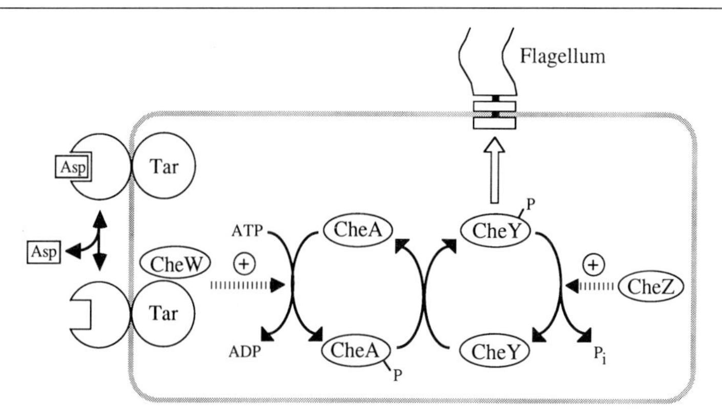


FIG. 1. The excitation signaling pathway of chemotaxis in *Escherichia coli*. See text for details.

lated systems. Where available, relevant data from similar systems are included for comparison. The assays are especially useful for characterizing qualitative features of the various proteins, such as the nature and regulation of their biochemical activities with regard to phosphorylation, the functional relationships between proteins, and the altered properties of mutant proteins. There is great flexibility to customize the assay to emphasize the particular aspect of the reaction under consideration. Each assay incorporates different segments of the signal transduction pathway. Furthermore, the prominent radioactive species observed (ATP, sensor–phosphate, regulator–phosphate, or $P_i$) depends on the balance between the phosphorylation and dephosphorylation rates of the sensor and regulator proteins, and can thus be altered by adjusting the reaction time and the proportions of the various reaction components.

One practical consequence of the apparently common reaction mechanism is the necessity of devising assay conditions under which the phosphoprotein linkages in question are stable. Boiling greatly accelerates hydrolysis of CheA–phosphate and especially CheY–phosphate; therefore samples are not heated at any point during analysis. In addition, phosphohistidine is acid labile, whereas phosphoaspartate is both acid and base labile.[25] Routine laboratory procedures which utilize acid [e.g., staining protein gels in acetic acid, or trichloroacetic acid (TCA) precipitation of proteins] result in significant loss of signal, but when limited in duration have nevertheless been usefully employed.

The CheA, CheY, and CheZ proteins are purified and stored at −20°

[25] J. M. Fujitaki and R. A. Smith, this series, Vol. 107 [2].

in TEDG buffer (see below); thus all phosphorylation reactions are performed in TEDG supplemented with the appropriate reagents. All reactions are performed at room temperature. Reaction components are thoroughly mixed and allowed to equilibrate at room temperature prior to addition of the component which initiates the reaction. Reaction volumes are governed by the concentration of purified proteins available and the method of analysis employed, but typically are ~10–20 μl.

*Solutions*

LB: 1% tryptone, 0.5% yeast extract, 1% NaCl
3β-Indoleacrylic acid: 20 mg/ml stock solution in 100% ethanol
SDS sample buffer: 0.125 *M* Tris/pH 6.8, 4% SDS, 20% sucrose, 10% (v/v) 2-mercaptoethanol, 0.02% Bromphenol Blue
Stain: 45% (v/v) methanol, 10% (v/v) acetic acid, 0.15% Coomassie Blue
Destain: 15% (v/v) methanol, 7.5% (v/v) acetic acid
TEDG: 50 m*M* Tris/pH 7.5, 0.5 m*M* EDTA, 2 m*M* Dithiothreitol, 10% (v/v) glycerol
TEDG20: TEDG made with 20% (v/v) glycerol
[$\gamma$-$^{32}$P]ATP: A 5- to 10-fold concentrated stock solution is made by mixing the appropriate small amount of high specific activity [$\gamma$-$^{32}$P]ATP (7000 Ci/mmol, ICN) with unlabeled ATP. The actual specific activity of the stock [$\gamma$-$^{32}$P]ATP thus prepared can be determined by diluting and counting an aliquot, which then permits expression of results in terms of picomoles of labeled protein

## Purification of *Escherichia coli* CheA, CheY, and CheZ Proteins

A wide variety of protein phosphorylation reactions occurs in bacteria.[6,26,27,27a] In most cases, the experimental advantages of working with purified proteins in a defined system, rather than with crude cell extracts, justify the additional effort necessary to purify the proteins in question.

High-level expression of the chemotaxis proteins from plasmids made by Matsumura and co-workers was critical to development of simple purification procedures. CheA (and CheW) are expressed under the control of the *Serratia marcescens trp* promoter in pDV4,[28] whereas CheY

[26] A. J. Cozzone, *Annu. Rev. Microbiol.* **42,** 97 (1988).

[27] A. J. Cozzone, ed., "Protein Phosphorylation in Prokaryotes," Special issue of *Biochimie* **71,** 987 (1989).

[27a] J.-C. Cortay, D. Negre, and A. J. Cozzone, this volume [17].

[28] J. Stader, P. Matsumura, D. Vacante, G. E. Dean, and R. M. Macnab, *J. Bacteriol.* **166,** 244 (1986).

and CheZ are similarly expressed in pRL22.[29] The chemotaxis proteins are produced at sufficiently high levels from these plasmids that purification progress can be monitored simply by following a prominent protein species that is (1) of the correct molecular weight and (2) observed only in the presence of the plasmid. Thus, at appropriate points in each purification, aliquots are subjected to sodium dodecyl sulfate-polyacrylamide gel electrophoresis (SDS–PAGE) on a 12.5% polyacrylamide gel and Coomassie Blue staining to identify fractions containing the protein of interest. CheA is 71K,[30] CheZ is 24K,[31] and CheY is 14K.[29,31]

To prepare cell lysates, plasmid-containing bacteria are grown in LB with ampicillin to $OD_{600}$ = 1. The *trp* promoter is induced by the addition of 3β-indoleacrylic acid to 100 μg/ml, and the culture is incubated until saturation. All subsequent steps are performed at 0–4° in TEDG, except as noted. The culture is harvested by centrifugation for 10 min at 6000 *g*. The pellet is resuspended in a smaller volume of TEDG (TEDG20 for CheA) to wash the cells, and the centrifugation repeated. The pellet is drained, and may be stored indefinitely at −20° if desired. The pellet is resuspended in 5 to 15 ml of TEDG (TEDG20 for CheA) and lysed by sonication. Debris and membranes are removed by a 10-min centrifugation at 8000 *g*, followed by a 1-hr centrifugation at 165,000 *g*.

Purification of *E. coli* CheA, CheY, and CheZ from cell lysates is described below; alternate purification schemes for the closely related *Salmonella typhimurium* proteins have been reported.[32–34] After the final gel-filtration step in each procedure, fractions are concentrated as desired by ultrafiltration and stored at −20°. Purified CheA, CheY, and CheZ retain activity during storage for a period of years.

### *Purification of CheA*

CheA is purified using ammonium sulfate precipitation, dye–ligand chromatography, and gel filtration. TEDG20 is used instead of TEDG in the early steps of the procedure to facilitate CheA solubility. CheA is precipitated from the cell lysate by the addition of $(NH_4)_2SO_4$ to 35% saturation. CheA is one of the first proteins to precipitate. The precipitate

[29] P. Matsumura, J. J. Rydel, R. Linzmeier, and D. Vacante, *J. Bacteriol.* **160,** 36 (1984).

[30] E. C. Kofoid and J. S. Parkinson, *J. Bacteriol.* **173,** in press (1991).

[31] N. Mutoh and M. I. Simon, *J. Bacteriol.* **165,** 161 (1986).

[32] A. M. Stock, T. Chen, D. Welsh, and J. B. Stock, *Proc. Natl. Acad. Sci. U.S.A.* **85,** 1403 (1988).

[33] A. M. Stock, D. E. Koshland, Jr., and J. B. Stock, *Proc. Natl. Acad. Sci. U.S.A.* **82,** 7989 (1985).

[34] A. M. Stock and J. B. Stock, *J. Bacteriol.* **169,** 3301 (1987).

is collected by a 15-min centrifugation at 19,000 *g*, resuspended in 5 ml TEDG20, and dialyzed against TEDG20 to remove salt.

The material is loaded onto a 10-ml Affi-Gel Blue (100–200 mesh, Bio-Rad, Richmond, CA) column equilibrated with TEDG and the column washed with two to three column volumes of TEDG. CheA is eluted using a 0.25 to 1 *M* NaCl/TEDG gradient developed over 10 column volumes. Column fractions containing CheA are pooled, concentrated by precipitation with $(NH_4)_2SO_4$ added to 45% saturation, and the precipitate resuspended in 200 $\mu$l TEDG.

The blue column pool is loaded onto a Sepharose 12 fast protein liquid chromatography (FPLC) gel-filtration column (Pharmacia, Piscataway, NJ) equilibrated with TEDG and chromatographed at room temperature. CheA elutes as a single peak of approximately tetrameric size (~300 kDa). This procedure yields ~1 to 2 mg CheA from a 1-liter culture.

*Purification of CheY*

CheY is purified by a modification of the method of Matsumura *et al.*[29] The procedure consists of dye–ligand chromatography, anion-exchange chromatography, and gel filtration. Cell lysate is loaded onto a 10-ml Affi-Gel Blue column equilibrated with TEDG and the column washed with two to three column volumes of TEDG. CheY is eluted using a gradient of 0 to 0.6 *M* NaCl in TEDG, developed over eight column volumes. Fractions containing CheY are pooled and dialyzed against TEDG to remove salt.

The blue column pool is loaded onto a 10-ml DEAE-Sepharose CL-6B (Pharmacia) column equilibrated with TEDG, the column washed with one column volume of TEDG, and CheY eluted with 0.1 *M* NaCl in TEDG. CheY binds weakly to this column, so it is important that salt is removed from the sample prior to loading the column. In fact, many mutant CheY proteins containing a 2+ charge change with respect to wild-type CheY do not bind to the DEAE column. Fractions containing CheY are pooled and concentrated by ultrafiltration.

The DEAE pool is loaded onto a Superose 12 FPLC column (Pharmacia) equilibrated with TEDG and chromatographed at room temperature. CheY elutes as a single peak of monomeric size (~14 kDa). This procedure yields ~5 mg CheY from a 1-liter culture.

*Purification of CheZ*

CheZ may be purified from the same cells as CheY, using dye–ligand chromatography, ammonium sulfate precipitation, anion-exchange chromatography, and gel filtration. CheZ does not bind to the blue column (see above) and is recovered in the wash effluent. Fractions containing CheZ are pooled and CheZ precipitated by the addition of $(NH_4)_2SO_4$ to 43% saturation. This step results in a substantial purification. The precipitate is collected by a 15-min centrifugation at 19,000 *g*, resuspended in TEDG, and dialyzed against TEDG to remove salt.

The material is loaded onto a Mono Q FPLC (Pharmacia) column equilibrated with TEDG and the column washed with two to three column volumes of 0.1 *M* NaCl/TEDG. CheZ is eluted with a linear gradient of 0.1 to 0.35 *M* NaCl in TEDG, developed over ~10 column volumes. Fractions containing CheZ are pooled, concentrated by precipitation with $(NH_4)_2SO_4$ added to 45% saturation, and the precipitate resuspended in TEDG.

The Mono Q pool is loaded onto a Superose 12 FPLC column (Pharmacia) equilibrated with TEDG and chromatographed at room temperature. CheZ elutes as a peak of approximately tetrameric size (~110 kDa). Very large aggregates of CheZ are sometimes observed in the void volume. This procedure yields ~1 to 2 mg CheZ from a 1-liter culture.

### *Small-Scale Purification of Mutant Proteins*

Small-scale purification of mutant CheA or CheY proteins can be accomplished from 100-ml cultures using the basic strategy outlined above with minor modifications. The bed volumes are reduced to 3 ml for the Blue column and 1.5 ml for the DEAE column. Protein is eluted from the columns with steps of increasing salt concentration rather than a linear gradient. Buffer is applied in aliquots of the indicated size and the column drained by gravity flow, with manual collection of fractions. This allows one person to operate four columns simultaneously. The step gradient schemes are as follow:

| | |
|---|---|
| CheA, Blue column: | 5 × 1 ml TEDG wash |
| | 5 × 1 ml 0.25 *M* NaCl/TEDG wash |
| | 10 × 1 ml 1 *M* NaCl/TEDG elution |
| CheY, Blue column: | 5 × 1 ml TEDG wash |
| | 9 × 1 ml 0.6 *M* NaCl/TEDG elution |
| CheY, DEAE column: | 4 × 1 ml TEDG wash |
| | 7 × 0.75 ml 0.1 *M* NaCl/TEDG elution |

The yield of pure protein per unit volume of culture is reduced approximately two- to three-fold with respect to the large-scale procedures; however, it is more than sufficient to perform the assays described in the remainder of this chapter.

Autophosphorylation of CheA

Sensor autophosphorylation is the first step in the signal transduction pathway, and an obligate intermediate in regulator phosphorylation. To measure CheA autophosphorylation, CheA (~15 pmol/sample) is mixed with TEDG + 5 m*M* $MgCl_2$ + 50 m*M* KCl. The reaction is initiated by the addition of [$\gamma$-$^{32}$P]ATP (specific activity typically ~5000 cpm/pmol). With CheA the only protein species present, it is convenient to separate labeled CheA–phosphate from unincorporated [$\gamma$-$^{32}$P]ATP and [$^{32}$P]$P_i$ by trichloroacetic acid precipitation onto filters. Aliquots are spotted onto Whatman GF/C glass fiber filter disks and immediately dropped into ice-cold 10% trichloroacetic acid/1% sodium pyrophosphate. After 30 min, the filters are washed three times for 30 min each in ice-cold 5% trichloroacetic acid/1% sodium pyrophosphate, rinsed for 5 min in 95% ethanol, air dried, and assayed by liquid scintillation counting. Because CheA–phosphate is somewhat acid labile, it is important that all samples are treated identically and that exposure to acid is minimized. This procedure results in hydrolysis of approximately 60–70% of the CheA–phosphate. A single large reaction can be sampled at various times to determine phosphorylation rate, or multiple parallel reactions can be incubated for a constant time to determine the effect of varying some other parameter (e.g., salt or ATP concentration, pH).

CheA autophosphorylation is not observed in the absence of divalent cation.[1,3] Among the divalent cations tested, $Mg^{2+}$ best fulfills this requirement and $Mn^{2+}$ also works.[35] Potassium ion (but not $Na^+$) further stimulates the reaction.[2] Similarly, EnvZ autophosphorylation is stimulated by both $Mg^{2+}$ (Refs. 15 and 17) and $K^+$.[13,15] However, cation requirements do differ somewhat between various sensor proteins. Autophosphorylation of PhoM requires both $K^+$ and $Mg^{2+}$.[19] In contrast, FrzE autophosphorylation is supported by $Mn^{2+}$, but not by $Ca^{2+}$, $Cu^{2+}$, $Fe^{2+}$, $Mg^{2+}$, or $Zn^{2+}$.[23b]

Increasing concentration of ATP results in an increased autophosphorylation rate, with a half-maximal value at ~0.2 m*M* ATP and saturation at ~1 m*M* ATP.[3,35] An ATP concentration below saturation may be used to slow down the reaction for the sake of obtaining multiple time points when measuring the initial reaction rate, or to minimize the amount of radioactivity handled while maintaining high specific activity. Thus, reactions are typically performed at 0.1 to 0.5 m*M* ATP.

The pH optimum for CheA autophosphorylation is 8.5 to 9.0.[2] The pH optima of the other chemotaxis phosphotransfer reactions have not been determined. For the sake of consistency and physiological relevance (the

[35] K. A. Borkovich, personal communication (1990).

internal pH of *E. coli* is 7.5–8.0[36]), autophosphorylation and all other reactions are routinely done in TEDG (pH 7.5).

The nucleotide specificity of the autophosphorylation reaction can be assessed by performing parallel reactions with [$\gamma$-$^{32}$P]ATP or [$\gamma$-$^{32}$P]GTP. Labeling with [$\gamma$-$^{32}$P]GTP has not been observed for the three sensor proteins tested so far (Table I). Similarly, confirming evidence that the labeled protein generated in a reaction containing [$\gamma$-$^{32}$P]ATP results from phosphorylation rather than adenylylation may be obtained by the failure to observe labeled protein in a parallel reaction containing [$\alpha$-$^{32}$P]ATP. This is the case for all sensor proteins tested to date (Table I).

To demonstrate that the phosphorylation observed is an intramolecular event (i.e., autophosphorylation), the initial rate of phosphorylation is measured over a range of protein concentrations. The rate of an intramolecular reaction should be independent of protein concentration.[37] This prediction has been confirmed for CheA,[2] EnvZ,[13,15] and FrzE.[23b]

## Purification of Phosphorylated CheA

The protein product of the autophosphorylation reaction can be isolated and used to study the properties of CheA–phosphate directly, or as starting material for further phosphate transfer reactions (see below). CheA can be phosphorylated with approximately 1 mol of phosphate/mol of CheA monomer by performing a CheA autophosphorylation reaction in the presence of saturating ATP.[2] A typical reaction contains 1.4 nmol CheA, 0.8–1 m*M* [$\gamma$-$^{32}$P]ATP (specific activity of ~6000 cpm/pmol), 5 m*M* $MgCl_2$, and 50 m*M* KCl in TEDG. Following a 15-min incubation, the reaction is terminated by the addition of ammonium sulfate (45% saturation) to precipitate CheA–phosphate. The mixture is placed on ice for 20 min to facilitate precipitation and the precipitate collected by a 10-min centrifugation in a microcentrifuge (15,000 *g*) at 4°. The supernatant, containing most of the radioactivity, is discarded. The precipitated CheA is then resuspended in TEDG, and the remaining free ATP removed by gel filtration on a Superose 12 FPLC column (Pharmacia). The column effluent is collected and the fractions assayed for radioactivity. Two peaks are detected; the first corresponds to CheA–phosphate and the second corresponds to unincorporated ATP. Generally, 70–100% of the CheA protein is phosphorylated. Phosphorylated CheA can be stored at −20° for up to 1 month without significant (less than 10%) hydrolysis of the phosphoryl group.

[36] E. Padan and S. Schuldiner, this series, Vol. 125 [27].

[37] J. A. Todhunter and D. L. Purich, *Biochim. Biophys. Acta* **485,** 87 (1977).

### Identification of CheA Histidine Phosphorylation Site

The first step in determining the site of phosphorylation is to test the stability of the phosphoprotein linkage to various chemical treatments. Phosphoramidates are sensitive to hydrolysis in the presence of acid, hydroxylamine, or pyridine, but are relatively stable to alkali.[25] Results of such tests are consistent with the presence of phosphohistidine in CheA,[3] DegS,[23a] EnvZ,[12,17] NtrB,[11] PhoM,[19] PhoR,[18] and VirA[20] (Table I).

The site of phosphorylation in CheA, histidine-48, was identified by purifying $^{32}$P-labeled peptides generated by proteolytic digestion of purified CheA–phosphate. A seven-amino acid phosphorylated peptide with the sequence RAAHSIK was isolated after three rounds of proteolysis, purification of the labeled peptide by reversed-phase HPLC, and identification of the peptide by amino acid analysis.[7] This peptide was further digested to single amino acids by a cocktail of carboxypeptidases B and Y. The products of the carboxypeptidase digestion were analyzed by two-dimensional thin-layer chromatography on silica gel 60 K6 (Whatman, Hillsboro, OR) with chloroform : methanol : 17% (w/w) ammonia (2 : 2 : 1, v/v) in the first dimension and phenol : water (3 : 1, w/w) in the second dimension.[38] The mobility of radiolabeled spots was compared with synthetic phosphohistidine,[39] phosphoarginine (Sigma, St. Louis, MO), phosphoserine (Sigma), phosphothreonine (Sigma), and phosphotyrosine (Sigma) visualized by ninhydrin staining. The mobility of the radiolabeled spot shifted with the sequential removal of amino acids from the carboxy terminus of the peptide as the digest proceeded; the end product of the digestion comigrated with phosphohistidine.

NtrB,[11] PhoM,[19] and VirA[20] also contain phosphohistidine. When base hydrolysates of $^{32}$P-labeled proteins were subjected to two-dimensional thin-layer chromatography (NtrB, VirA) or ion exchange chromatography (PhoM), radioactivity comigrated with a phosphohistidine standard.

### Phosphorylation of CheY by CheA

CheY may be phosphorylated starting either with CheA and ATP or with purified CheA–phosphate. These two reaction conditions yield different types of information, as will be described here and in the following section. The addition of CheY to a CheA autophosphorylation reaction permits preliminary examination of CheY phosphorylation. It is desirable to examine the radioactivity in the two proteins separately, so reactions are terminated by the addition of SDS sample buffer and the products

[38] A. Niederwieser, this series, Vol. 25 [6].

[39] R. C. Sheridan, J. F. McCullough, and Z. T. Wakefield, *Inorg. Synth.* **13,** 23 (1971).

resolved by electrophoresis on a 12.5% polyacrylamide gel. The gel is treated with Stain for 15 min and Destain for 1 hr. After drying and autoradiography, the labeled bands can be cut out of the gel for direct quantitation by liquid scintillation counting. A disadvantage of this assay method is that there is no way to correct for the inevitable small variations in volumes between samples at each step of the experiment. Thus when comparing the amount of labeled protein observed in different samples, large differences are informative, whereas small differences cannot be determined with confidence. Results from different gels also cannot be compared, due to the variable loss of signal resulting from acid treatment.

Under circumstances where CheY is a poor phosphoacceptor, allowing this reaction to proceed through many cycles can permit the accumulation of product, thus making it a sensitive assay for CheY phosphorylation. For example, in a 2.5-min reaction with a fourfold excess of CheY (15 pmol CheA : 60 pmol CheY), phosphorylation of several mutant CheY proteins not labeled using other assay conditions was easily detected.[40] Furthermore, the reaction conditions were purposely adjusted so that no radioactivity was observed in CheA in the wild-type control lane, whereas CheY was strongly labeled. This was informative, because CheA–phosphate accumulated when many mutant CheY proteins were used as substrates. Such a change reflects an altered balance among the rates for the various reactions in the phosphotransfer pathway and suggests either a defect in phosphotransfer from CheA to CheY, or increased stability of the CheY–phosphate formed. A concurrent increase in CheY–phosphate levels would also be predicted in the latter case.

CheZ can also be included in the reaction, typically in an amount equimolar to CheY. The overall reaction dynamics will be significantly altered by the resulting acceleration of CheY dephosphorylation, and the level of CheA–phosphate observed consequently declines. The addition of CheZ is a relatively easy way to screen CheY mutants for resistance to CheZ-stimulated dephosphorylation or, conversely, to assess the dephosphorylation-stimulating activity of CheZ mutants.

### Phosphotransfer from CheA–Phosphate to CheY

One of the most informative assays of the chemotaxis phosphate transfer reactions is the transfer of the phosphoryl group from purified radiolabeled CheA–phosphate to CheY and finally to $P_i$. In the absence of ATP, CheA cannot cycle back to the phosphorylated form, and thus the flow of phosphate through the chemotaxis signal transduction pathway can be

[40] R. B. Bourret, J. F. Hess, and M. I. Simon, *Proc. Natl. Acad. Sci. U.S.A.* **87,** 41 (1990).

analyzed. In a reaction containing a 5- to 10-fold molar excess of CheA–phosphate with respect to CheY, the dephosphorylation of CheY becomes the rate-limiting step in the production of $P_i$, allowing the detection of radiolabeled phosphate in CheY. The addition of CheZ to this reaction causes a substantial increase in the rate of hydrolysis of the phosphoryl group from CheA due to an increased rate of CheY dephosphorylation.[2]

The products of the phosphotransfer assay are analyzed by both SDS–PAGE, to determine the amount of phosphate present in the individual proteins, and thin-layer chromatography (TLC), to quantitate the amount of $P_i$. A typical reaction contains 40 pmol CheA–phosphate (6000 cpm/pmol), 4 pmol CheY, and 5 m*M* $MgCl_2$ in a volume of 30 $\mu$l. Note that in contrast to CheA autophosphorylation, KCl is neither required nor stimulatory for the phosphotransfer reaction. The reaction is initiated by the addition of CheY, and terminated at the desired times by the removal of 2.5-$\mu$l aliquots to 7.5 $\mu$l SDS sample buffer containing prestained molecular weight standards (Bio-Rad; Bethesda Research Laboratories, Gaithersburg, MD). Time points are taken every 10 sec for the first minute, every 15 sec for the second minute, and finally an end point at 5 min. In a parallel control reaction, CheA–phosphate is incubated for 5 min with $MgCl_2$ in the absence of CheY to determine the stability of CheA–phosphate over the course of the reaction. A 1-$\mu$l aliquot of each time point terminated in SDS sample buffer is removed, spotted onto a Polygram CEL 300 PEI (polyethyleneimine-cellulose) TLC plate (Brinkmann, Westbury, NY) and the plate developed with 1 *M* acetic acid/4 *M* LiCl, 8 : 2 (v/v).[41] The TLC plate is air dried and subjected to autoradiography. The remainder of each sample is loaded onto a 15% polyacrylamide minigel (Mighty Small II minigel apparatus, Hoefer Scientific Instruments, San Francisco, CA). Following SDS–PAGE, the gel is immediately dried and subjected to autoradiography. Three bands of radioactivity are detected on the gel: the CheA–phosphate band at ~71 kDa, the CheY–phosphate band at ~14 kDa, and $P_i$ at the dye front. The amount of CheA–phosphate and CheY–phosphate is determined by excising the corresponding bands (using the prestained molecular weight standards and the autoradiograph as a guide) and counting the gel slices in scintillation fluid. It is impossible to accurately measure $P_i$ since it migrates at the dye front and diffuses laterally out of the lanes; however, $P_i$ can be quantitated by TLC. In the TLC system employed, CheA–phosphate and CheY–phosphate remain at the origin, whereas $P_i$ migrates as a diffuse spot with an $R_f$ of approximately 0.7. These regions are excised and counted.

[41] K. Randerath and E. Randerath, this series, Vol. 12A [40].

*Calculations*

Utilizing the counts obtained from the TLC plate in conjunction with the data from the gel, the amount of phosphate present in each species (CheA, CheY, and inorganic phosphate) is corrected for pipetting errors and for low-level spontaneous decay of CheA–phosphate. First, the percentage of $P_i$ in the control sample containing CheA–phosphate with no CheY is used to correct for spontaneous hydrolysis of the phosphoryl group that occurred during storage. This value, typically 5–10%, is subtracted from the amount of $P_i$ on the TLC plate for each point. (Note that this operation simultaneously corrects for background in the $P_i$ samples.) Second, a piece of the plate is counted to determine background, and this value is subtracted from the counts at the origin of the TLC plate. The adjusted values for the origin and $P_i$ are then summed for each time point, and the protein-bound phosphate and $P_i$ are expressed as percentages of this total. Similarly, a piece of the gel is counted to determine background, and this value is subtracted from the counts for each gel slice. The adjusted values for each time point are summed and the fraction of protein-bound phosphate in CheA and in CheY is calculated. Since the percentage of the total counts in both CheA–phosphate and CheY–phosphate from the gel must be equal to the percentage protein bound counts obtained from the TLC, it is then possible to calculate the percentage of total counts which are in CheA–phosphate and CheY–phosphate, with the remaining counts being $P_i$. This normalization of CheA–phosphate, CheY–phosphate, and $P_i$ permits valid comparison of the level of each species at the different time points in the reaction.

*Assay Modifications*

The CheA to CheY phosphotransfer assay can be modified slightly to estimate the rate of hydrolysis of CheA–phosphate by CheY.[2] The conditions described above are optimized to detect the phosphorylated form of CheY, but the reaction occurs too rapidly to determine a rate accurately. Lowering the amount of CheY such that CheA–phosphate is present in a 100- to 200-fold molar excess slows the reaction, allowing for the collection of several data points in the linear phase of the reaction. CheY is diluted in 0.5 mg/ml BSA to enhance protein stability. In addition, only the TLC assay is utilized for this experiment, since the level of phosphorylated CheY is less than 1% of the total label in the reaction and thus cannot be accurately measured in the gel system.

The addition of EDTA inhibits both the transfer of phosphate from CheA to CheY and the dephosphorylation of CheY to produce $P_i$, indicating that divalent cations are required for both reactions.[5,8,42] The cation requirement is not particularly specific. Phosphotransfer from CheA to CheY occurs rapidly in the presence of $Co^{2+}$, $Mg^{2+}$, $Mn^{2+}$, or $Zn^{2+}$, and slowly in the presence of $Ca^{2+}$ or $Cd^{2+}$.[42] Cadmium ion, $Co^{2+}$, $Mg^{2+}$, $Mn^{2+}$, and $Zn^{2+}$ each support CheY autodephosphorylation, whereas $Ca^{2+}$ does not.[42] CheY contains a single divalent metal ion binding site, with a $K_D$ for $Mg^{2+}$ of about 0.5 m*M* at pH 7–8.[42] As mentioned previously, a divalent metal ion is also necessary for autophosphorylation of CheA. The amount of EDTA required to stop the phosphotransfer from CheA to CheY is less than the amount needed to stop dephosphorylation of the CheY protein. Therefore, high levels of EDTA are necessary to trap CheY in the phosphorylated form. The presence of $Mg^{2+}$ similarly affects phosphate transfer reactions in other two-component regulatory systems; NtrB to NtrC phosphotransfer is accelerated by $Mg^{2+}$ (Ref. 11) and EnvZ to OmpR phosphotransfer requires $Mg^{2+}$ (Ref. 17).

## Purification and Use of CheY–Phosphate

The purification of CheY–phosphate allows the questions of whether CheA is involved in either CheY dephosphorylation or in the acceleration of CheY dephosphorylation by CheZ to be addressed. Isolation of CheY–phosphate from a reaction containing 0.5 nmol purified CheA–phosphate and 1 nmol CheY is accomplished by allowing the reaction to proceed for only 15 sec and then adding 100 m*M* EDTA.[5] The reaction products are separated by gel filtration on a Superose 12 FPLC column (Pharmacia) in TEDG containing 70 m*M* EDTA. Three peaks of radioactivity are detected: one corresponding to CheA–phosphate (~300 kDa), one to CheY–phosphate (~14 kDa), and one to $P_i$ eluting at the included volume of the column. The radioactivity associated with the CheY–phosphate peak indicates that only about 5% of the CheY molecules present are in the phosphorylated form; however, the CheY–phosphate isolated by this procedure is stable and can be stored for several weeks at $-20°$. Dephosphorylation of CheY is initiated by the addition of 100 m*M* $MgCl_2$/100 m*M* Tris, pH 9.0 (the Tris is necessary to buffer against the drop in pH that occurs upon the addition of the $Mg^{2+}$ to the EDTA). Aliquots are removed at 5- to 10-sec intervals for 1 min, terminated in SDS sample buffer, and analyzed on TLC. The dephosphorylation of CheY is not affected by the addition of equimolar amounts of CheA.[5] The addition of equimolar amounts of CheZ accelerates the dephosphorylation of CheY

[42] G. S. Lukat, A. M. Stock, and J. B. Stock, *Biochemistry* **29,** 5436 (1990).

with no requirement for CheA.[5] However, $Mg^{2+}$ is required for maximal stimulation of CheY dephosphorylation by CheZ.[42]

The known phosphatase activities of two-component regulatory systems differ in rate and molecular organization, presumably to meet the particular response time requirements of each system. NtrC has an autophosphatase activity that is somewhat slower than that of CheY, and NtrC–phosphate is also stabilized by EDTA.[10,11] In contrast, FrzE-phosphate,[23b] OmpR–phosphate,[14] and VirG–phosphate[22] are relatively stable (half-life ~1 hr) and lack the autophosphatase activity observed in CheY and NtrC entirely.

In the case of Ntr and Omp (but not Che or Vir), stimulated regulator dephosphorylation is observed in the presence of the sensor. Dephosphorylation of NtrC is accelerated by the combined presence of the NtrB and $P_{II}$ proteins and ATP[9,10]; similarly, dephosphorylation of OmpR is accelerated by the combined presence of EnvZ and ATP.[14,16] The rapid time scale of the chemotaxis reactions conceivably may prevent the detection of a similar role for CheA in CheY dephosphorylation by the experimental design employed, or additional proteins (e.g., CheW, Tar; see [16] in this volume) may be required. Alternatively, CheA may not be involved in CheY dephosphorylation at all. VirA apparently does not affect VirG dephosphorylation, in either the presence or absence of ATP.[22]

## Identification of CheY Aspartate Phosphorylation Site

Acyl phosphates are sensitive to hydrolysis in the presence of acid, base, or hydroxylamine.[25] Results of such tests are consistent with the presence of phosphoaspartate in CheY,[43] DegU,[23a] FrzE,[23b] NtrC,[11] OmpR,[17] PhoB,[18] PhoMORF2,[19] and VirG[22] (Table I).

If stability to various chemical treatments suggests the presence of an acyl phosphate, confirmation depends on replacing the labile phosphate group with a stable modification. Reduction of an aspartylphosphate with $NaBH_4$ generates homoserine.[44] Use of $[^3H]NaBH_4$ thus chemically and radioactively marks the site of phosphorylation. Several possibilities then exist for analyzing smaller pieces of the reduced, labeled protein. The site of CheY phosphorylation was identified by subjecting proteolytic fragments to mass spectrometry.[8] The site of VirG phosphorylation was identified by sequencing radioactive proteolytic fragments by standard Edman degradation.[22] Paper electrophoresis was used to identify homoserine lactone among the acid hydrolysis products of NtrC.[11]

[43] A. M. Stock, D. C. Wylie, J. M. Mottonen, A. N. Lupas, E. G. Nifa, A. J. Ninfa, C. E. Schutt, and J. B. Stock, *Cold Spring Harbor Symp. Quant. Biol.* **53,** 49 (1988).

[44] C. Degani and P. D. Boyer, *J. Biol. Chem.* **248,** 8222 (1973).

## Establishing Biological Relevance of *in Vitro* Assays

The *in vitro* phosphorylation assays described in this chapter provide a powerful method for characterizing a regulated protein phosphorylation cascade. It is absolutely critical, however, to establish a clear link between *in vitro* observations and the biological phenomenon under investigation. A detailed discussion is outside the scope of this chapter, but two general strategies are briefly mentioned. The first is to obtain mutants in the pathway by either random or site-specific mutagenesis, characterize the biochemical (phosphorylation) properties of the mutant proteins, and compare these properties to the phenotype of the mutant organism. This approach has been used successfully in the Che,[7,8,40,45] Omp,[16,17,46] Spo,[24a] and Vir[20,22,46a] systems.

A second strategy is to establish that phosphorylation occurs *in vivo,* and directly monitor the effects of environmental stimuli or signaling pathway mutations on phosphorylation in the organism. The relative stability of the phosphoryl group made this feasible for OmpR.[47] Bacteria were grown in minimal medium containing [$^{32}$P]$P_i$ and lysed. OmpR was immunoprecipitated from soluble extracts and subjected to SDS–PAGE followed by autoradiography and immunoblotting. Another technique that may be more generally applicable has been described.[48] The characteristic mobilities during two-dimensional gel electrophoresis of the phosphorylated and nonphosphorylated proteins are established using controls prepared *in vitro*. The charge difference due to the phosphoryl group provides a means for separation. Both forms of the protein are visualized by [$^{35}$S]methionine labeling in the actual experiment. The signal can also be strengthened by overproducing the protein of interest.

## Acknowledgments

We thank Kathy Borkovich and Kenji Oosawa for useful discussions during the development of these assays, and Kathy Borkovich, Juan Davagnino, and Hong Ma for critical reading of the manuscript. This work was supported by Damon Runyon-Walter Winchell Cancer Fund Fellowship DRG-915 (to J.F.H.), National Research Service Award Fellowship AI107798 (to R.B.B.), and Grant AI19296 from the National Institutes of Health (to M.I.S.).

[45] K. Oosawa, J. F. Hess, and M. I. Simon, *Cell (Cambridge, Mass.)* **53,** 89 (1988).
[46] K. Kanamaru, H. Aiba, S. Mizushima, and T. Mizuno, *J. Biol. Chem.* **264,** 21633 (1990).
[46a] T. Roitsch, H. Wang, S. Jin, and E. W. Nester, *J. Bacteriol.* **172,** 6054 (1990).
[47] S. Forst, J. Delgado, A. Rampersaud, and M. Inouye, *J. Bacteriol.* **172,** 3473 (1990).
[48] O. Amster-Choder and A. Wright, *Science* **249,** 540 (1990).

# [16] Coupling of Receptor Function to Phosphate-Transfer Reactions in Bacterial Chemotaxis

*By* KATHERINE A. BORKOVICH and MELVIN I. SIMON

## Introduction

Cell-surface receptor proteins transduce changes in concentration of specific chemicals in the environment into intracellular signals which enable the cell to respond to the stimulus. In the bacterial chemotaxis system, the binding of specific ligands to transmembrane receptors, such as Tar, results in regulation of swimming behavior.[1–3] One class of ligands, called repellents, causes the bacterium to exhibit increased "tumbling" behavior, which allows the cell to randomly change direction. The other class of ligands (attractants) leads to increased "smooth" swimming behavior. Tumbling and smooth swimming episodes are regulated such that bacteria swim up increasing concentration gradients of attractants and away from repellent stimuli. Change in swimming behavior (excitation) is achieved by changing the direction of rotation of a rotary motor at the base of the flagellum. Four chemotaxis genes, *cheA, cheY, cheW,* and *cheZ,* are required for the excitation signal. It was previously shown by *in vitro* assays (see [15] in this volume) that the cytoplasmic *cheA* gene product is an autophosphorylating histidine kinase that can rapidly transfer phosphate to the CheY protein.[4–7] The CheZ protein accelerates hydrolysis of CheY–phosphate.[5] These results, in combination with other genetic evidence, suggested that these enzymatic reactions are the basis of the signal transduction pathway for bacterial chemotaxis. The system was reconstituted using Tar-containing membranes and purified soluble chemotaxis proteins, thus demonstrating a role for the Tar chemoreceptor and

[1] R. C. Stewart and F. W. Dahlquist, *Chem. Rev.* **87,** 997 (1987).

[2] R. M. Macnab, *in* "*Escherichia coli* and *Salmonella typhimurium*: Molecular and Cellular Biology" (F. C. Neidhardt, ed.), p. 732. Am. Soc. Microbiol., Washington, D.C., 1987.

[3] D. E. Koshland, Jr., *Biochemistry* **27,** 5829 (1988).

[4] J. F. Hess, K. Oosawa, P. Matsumura, and M. I. Simon, *Proc. Natl. Acad. Sci. U.S.A.* **84,** 7600 (1987).

[5] J. F. Hess, K. Oosawa, N. Kaplan, and M. I. Simon, *Cell (Cambridge, Mass.)* **53,** 79 (1988).

[6] D. Wylie, A. M. Stock, C.-Y. Wong, and J. B. Stock, *Biochem. Biophys. Res. Commun.* **151,** 891 (1988).

[7] K. Oosawa, J. F. Hess, and M. I. Simon, *Cell (Cambridge, Mass.)* **53,** 89 (1988).

the CheW protein in modulating the phosphorylation reactions.[8] The procedure that proved successful for these experiments is detailed here. In addition, reference to heterologous systems using chimeras made by fusing portions of Tar to other receptors proteins for testing their effect on phosphate transfer reactions and/or signaling will be presented for comparison.

*Solutions*

LB: 1% Tryptone, 0.5% (w/v) yeast extract, 1% (w/v) NaCl
3-$\beta$-Indoleacrylic acid: 20 mg/ml in 100% ethanol. Make fresh and protect from light
Isopropyl-$\beta$-D-thiogalactoside (IPTG): 100 m*M* in water
Laemmli sample buffer (4×): 260 m*M* Tris-HCl, 20% (v/v) 2-mercaptoethanol, 8% (w/v) sodium dodecyl sulfate (SDS), 40% (v/v) glycerol, pH 6.8.[9]
Buffer A: 50 m*M* Tris-HCl, 1 m*M* phenylmethylsulfonyl fluoride (PMSF), 0.5 m*M* ethylenediaminetetraacetic acid (EDTA), 2 m*M* dithiothreitol (DTT), and 20% (v/v) glycerol, pH 7.5
Buffer B: 50 m*M* Tris-HCl, 1 m*M* PMSF, 0.5 m*M* EDTA, 2 m*M* DTT, and 10% (v/v) glycerol, pH 7.5
Buffer C: 100 m*M* Tris-HCl, 5 m*M* EDTA, 5 m*M* 1,10-phenanthroline (1,10-phe), 2 m*M* PMSF, and 10% (v/v) glycerol, pH 7.5
Buffer D: 50 m*M* Tris-HCl, 2 *M* KCl, 5 m*M* EDTA, 3 m*M* 1,10-phe, 1 m*M* PMSF, and 10% (v/v) glycerol, pH 7.5
Buffer E: 50 m*M* Tris-HCl, 1 m*M* 1,10-phe, 1 m*M* PMSF, and 10% (v/v) glycerol, pH 7.5
TEDG: 50 m*M* Tris-HCl, 0.5 m*M* EDTA, 2 m*M* DTT, and 10% (v/v) glycerol, pH 7.5.[5]
TE: 10 m*M* Tris-HCl, 1 m*M* EDTA, pH 7.5

## Protein Assay

The total protein concentration of each preparation is measured using the Bio-Rad (Richmond, CA) protein reagent concentrate, with bovine serum albumin as a standard.

[8] K. Borkovich, N. Kaplan, J. F. Hess, and M. I. Simon, *Proc. Natl. Acad. Sci. U.S.A.* **86,** 1208 (1989).
[9] U. K. Laemmli, *Nature* (*London*) **227,** 680 (1970).

## SDS-PAGE Analysis

Sodium dodecyl sulfate-polyacrylamide gel electrophoresis (SDS-PAGE) is performed according to Laemmli,[9] using 12.5% slab resolving gels. The position of proteins on the gel is determined by Coomassie staining[10] or Western blot analysis.[11]

## Purification of Soluble Chemotaxis Proteins

CheA and CheY are purified as described in [15] in this volume. CheW is expressed from the plasmid pDV4[12] in strain KO685. Plasmid pDV4 contains the *Escherichia coli cheA* and *cheW* genes under control of the *Serratia marcescens trp* promoter (see [15] in this volume). For purification of CheW, strain KO685 with the pDV4 plasmid is cultivated in LB containing 100 $\mu$g/ml ampicillin at 37° until logarithmic phase ($OD_{600}$ 1.0–1.2). Expression of CheW (and CheA) is induced by addition of 3-$\beta$-indoleacrylic acid to 100 $\mu$g/ml final concentration 1 hr prior to harvesting. The CheW protein is purified using a modification of existing procedures.[8,13] All steps are performed at 4°, unless otherwise noted. Cells are collected by centrifugation at 4000 *g* for 10 min and washed once in buffer A. The resuspended pellets are sonicated using the same buffer and centrifuged at 27,000 *g* for 15 min, followed by a 165,000 *g* centrifugation step (1 hr) to remove cell debris and membranes. CheW is precipitated from the supernatant by bringing it to 42% $(NH_4)_2SO_4$ saturation and centrifugation at 27,000 *g* for 15 min. The pellet is resuspended in and dialyzed against buffer B. The dialyzed solution is applied to an Affi-Gel Blue (Bio-Rad, Richmond, CA) column equilibrated in buffer B, and CheW is eluted in the void volume and wash fractions using buffer B. This step removes most of the contaminating CheA protein, which remains bound to the column. The fractions containing CheW are immediately applied to a DEAE-Trisacryl M (IBF, Biotechnics, Villeneuve-la-Garenne, France) column equilibrated with buffer B. The column is then washed with buffer B, and CheW eluted using a linear gradient of 10–400 m*M* NaCl in buffer B. The fractions containing CheW are identified by subjecting a small aliquot of each fraction to SDS-PAGE analysis, followed by Coomassie staining and checking for the presence of the 18-kDa CheW band. The

[10] K. Weber and M. Osborn, *J. Biol. Chem.* **244,** 4406 (1969).

[11] W. N. Burnette, *Anal. Biochem.* **112,** 195 (1981).

[12] J. Stader, P. Matsumura, D. Vacante, G. E. Dean, and R. M. Macnab, *J. Bacteriol.* **166,** 244 (1986).

[13] A. M. Stock, J. Mottonen, T. Chen, and J. B. Stock, *J. Biol. Chem.* **262,** 535 (1987).

CheW-containing fractions are pooled, concentrated (Amicon, Danvers, MA; ultrafiltration cell) and chromatographed over a Superose 12 FPLC (Pharmacia, Piscataway, NJ) column using TEDG buffer at room temperature. CheW elutes as a single peak corresponding to a dimer of $M_r$ ~35,000–40,000 during this analysis. The fractions containing CheW protein are collected, concentrated, and stored at −20°. The concentrated preparation is stable for at least a year, and can be freeze–thawed without loss of activity. Using this procedure, approximately 2.5 mg of purified CheW protein can be isolated from 2 liters starting culture of cells.

## Preparation of Receptor-Containing Membranes

Tar-containing membranes are isolated from a strain containing a plasmid which overexpresses the protein using a *tac* promotor.[8] Plasmids which express mutant Tar proteins were constructed by replacement of a restriction fragment containing the wild-type sequence with a corresponding mutant fragment.[14] In this way, both "tumble" and "smooth" receptor-encoding vectors were constructed. These mutant Tar proteins proved useful in verification of the specificity of the activation of phosphorylation (see below). All plasmids are maintained in a strain deleted of all four *Escherichia coli* chemoreceptors.[15] This strain is also used as a source of negative control membranes. The four strains are cultivated in LB containing 100 μg/ml ampicillin at 37°. Tar expression is induced by addition of IPTG to 1 m*M* final concentration in late logarithmic phase, followed by another 3 hr of incubation at 37°. Membranes containing Tar are prepared using a modification of the procedure of Bogonez and Koshland.[16] All steps are performed at 4°. Cells are collected by centrifugation at 4000 *g* for 10 min and washed in buffer C. Inclusion of protease inhibitors, especially 1,10-phenanthroline and glycerol, is necessary to prevent proteolysis of the receptor during isolation of membranes.[17] Cell-free lysates are made by sonicating the washed cells in the same buffer, followed by centrifugation twice at 10,000 *g* for 10 min. Membranes are collected by centrifugation at 165,000 *g* for 1 hr. The membranes are washed twice by resuspension in buffer D (use of a Dounce homogenizer expedites this step) and centrifuging at 165,000 *g* for 15 min. This high-salt wash removes peripherally associated membrane proteins from the membrane fraction.[18]

[14] N. Mutoh, K. Oosawa, and M. I. Simon, *J. Bacteriol.* **167,** 992 (1986).

[15] K. Oosawa, N. Mutoh, and M. I. Simon, *J. Bacteriol.* **170,** 2521 (1988).

[16] E. Bogonez and D. E. Koshland, Jr., *Proc. Natl. Acad. Sci. U.S.A.* **82,** 4891 (1985).

[17] D. L. Foster, S. L. Mowbray, B. K. Jap, and D. E. Koshland, Jr., *J. Biol. Chem.* **260,** 11706 (1985).

[18] E. A. Wang and D. E. Koshland, Jr., *Proc. Natl. Acad. Sci. U.S.A.* **77,** 7157 (1980).

Excess KCl is removed by washing the membranes in buffer E. The pellet is resuspended in a small volume of buffer E and stored at $-20°$. Approximately 4 mg of total integral membrane protein is recovered per liter starting culture of cells using this procedure. These membrane preparations are stable and can be repeatedly freeze–thawed with no change in activating activity for the receptor.

The relative amounts of the receptors present per milligram of total protein in each membrane preparation are assessed by subjecting an equivalent amount of total protein from each sample to SDS-PAGE, followed by Coomassie staining or Western blot analysis. Western blot analysis is accomplished using a Tar-specific antiserum[14] and an alkaline phosphatase-conjugated secondary antibody (Promega, Madison, WI) followed by visualization using the Protoblot color development reagent (Promega). The results of such an analysis show that the level of Tar varies per milligram of total membrane protein among the three different Tar-containing preparations.[8] For reactions where the activity of an equal amount of Tar is to be compared from these preparations, the more concentrated sample is diluted with the negative control membranes before addition to the assays. This is necessary because the presence of membranes affected the extent of phosphorylation in the absence of Tar and CheW, possibly due to sequestration of added CheA or CheY proteins or the hydrolytic activity of residual integral membrane proteins in these preparations[8] (K. A. Borkovich and M. I. Simon, unpublished observations, 1989).

## Assay Procedure

Typical coupled assay reactions contain 50 m*M* Tris-HCl (pH 7.5), 50 m*M* KCl, 5 m*M* $MgCl_2$, 1 m*M* [$\gamma$-$^{32}$P]ATP (~1000 cpm/pmol), purified proteins, plasma membranes (approximately 4 $\mu$g total membrane protein), and other components in a final volume of 20 $\mu$l (using water) at room temperature. The purified proteins and membrane fractions are diluted using either buffer B or TE, and added in small volumes (2–3 $\mu$l each) to the Tris-HCl, KCl, $MgCl_2$, and water. After preincubation for 15 min, reactions are initiated by the addition of labeled ATP using a 10-fold concentrated stock. After the appropriate time (between 3 sec and 2.5 min), reactions are terminated by the addition of Laemmli sample buffer to 1 $\times$ final concentration (without heating). The extent of phosphorylation is assessed by subjecting the samples to SDS-PAGE using 12.5% slab gels, followed by a 20-min fixation in 25% (v/v) methanol, 5% (v/v) acetic acid, a 2-min stain in 25% (v/v) methanol, 5% (v/v) acetic acid, 0.1% (w/v) Coomassie Blue, and destaining 30 min in 5% (v/v) acetic acid at room temperature. It is important to limit the exposure of phosphorylated proteins to acid to ensure their stability, for the reasons noted in [15] in this volume. The gel is dried under vacuum and autoradiographed. The bands containing labeled CheA–phosphate or CheY–phosphate can be easily excised and counted for radioactivity in scintillation fluid.

Initial assays contain approximately the same molar amount (40–50 pmol) of the purified chemotaxis proteins and the receptor, with an incubation time of 2.5 min. This allows the identification of an increase in the level of both CheA–phosphate and CheY–phosphate in the reactions containing wild-type receptor-containing membranes, CheA, and CheY.[8] Maximal stimulation is achieved by the inclusion of CheW. Corresponding reactions containing the *tumble* mutant receptor also give increased levels of phosphorylated proteins, while those with negative control or smooth receptor-containing membranes do not. The production of CheY–phosphate is then optimized by varying the ratio of the CheA kinase to the substrate CheY, such that there is no residual CheA–phosphate remaining after a 10-sec reaction. This ratio is found to be approximately 5 pmol CheA:200 pmol CheY. The optimal level of CheW (approximately 80 pmol, using the amount of Tar in 4 $\mu$g total membrane protein) is also measured using the predetermined amounts of CheA and CheY and varying the concentration of CheW. Using the optimized levels of the various components, the sitmulation of CheY–phosphate production in reactions containing the wild-type receptor is 300-fold greater relative to the negative control after 10 sec of reaction (Fig. 1).[8] There is comparable stimulation of CheY–phosphate accumulation using the *tumble* mutant membranes, but not the *smooth* mutant membranes. It is necessary to obtain short reaction times (<10 sec) to be in the linear range of CheY–phosphate production; after this point, the level of phosphorylated protein diminishes (see Fig. 1).

The ligand dependency of the receptor activation of phosphorylation is also explored in this system. In preliminary assays, addition of 1 m$M$ aspartate (adjusted to pH 7.5) to reactions containing wild-type receptor causes loss of activation, while reactions containing *tumble* or *smooth* receptor, or no receptor, are relatively unaffected (Table I). The effect of aspartate on reactions with wild-type Tar was monitored using concentrations ranging from $10^{-6}$ to $10^{-3}$ $M$. Aspartate inhibited activation of phosphate transfer in a dosage-dependent manner. An Eadie–Scatchard plot of the data (Fig. 2) gives a $K_D$ of 3 $\mu M$ for the aspartate effect; this concentration is also in the range which influences chemotaxis strongly *in vivo*,[19] and is identical to the binding constant of solubilized Tar for aspartate.[16] Taken together, these rèsults support a model in which the unliganded receptor and the CheW protein in combination stimulate phos-

[19] S. Clarke and D. E. Koshland, Jr., *J. Biol. Chem.* **254,** 9695 (1979).

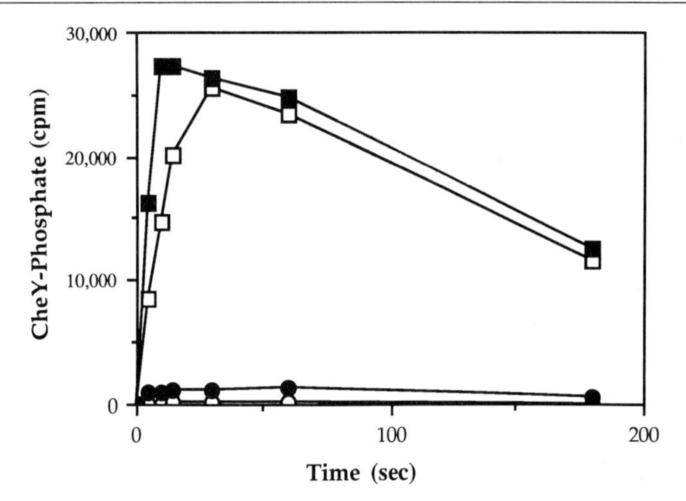


FIG. 1. Kinetics of CheY–phosphate production using the four membrane preparations. Reactions were performed using 5 pmol of CheA, 40 pmol of CheW, 200 pmol of CheY, and approximately 8 $\mu$g of total membrane protein. ○, No receptor; ●, *smooth*; □, *tumble*; ■, wild type. (Taken from Ref. 8.)

TABLE I
DEPENDENCE OF CheY-PHOSPHATE PRODUCTION ON RECEPTOR AND ASPARTATE[a]

| Source of receptor | Aspartate (1 m*M*)[b] | Maximum CheY–phosphate (%) |
|---|---|---|
| Wild type | – | 100 |
| Wild type | + | 6 |
| *Tumble* | – | 48 |
| *Tumble* | + | 43 |
| *Smooth* | – | 5 |
| *Smooth* | + | 1 |
| None | – | 1 |
| None | + | 0 |

[a] Reactions were performed under the conditions described for Fig. 1. Maximum CheY–phosphate produced was 16 pmol. (Taken from Ref. 8.)

[b] +, Aspartate added; –, no aspartate.

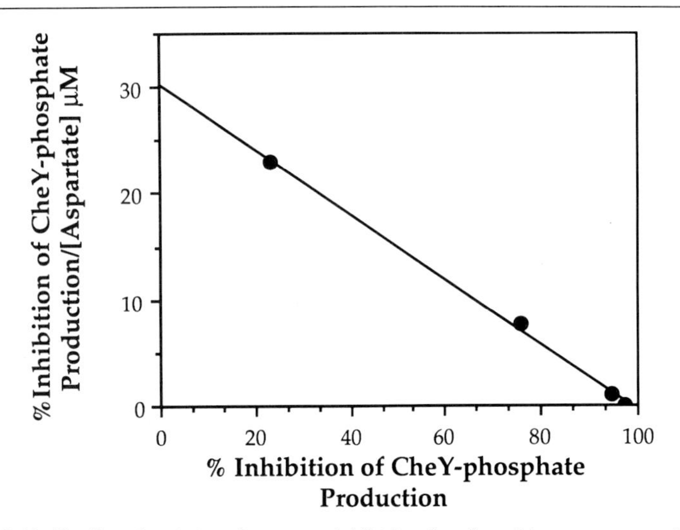


FIG. 2. Eadie–Scatchard plot of aspartate inhibition data for wild-type receptor. Reactions contained the amount of proteins and wild-type receptor-containing membranes described in the legend to Fig. 1 and 0, 1, 10, 100, or 1000 $\mu M$ L-aspartate. Incubation was for 10 sec.

phorylation of CheA and/or phosphate transfer to CheY, resulting in increased levels of CheY–phosphate, the putative tumble signal.

## General Considerations and Limitations

A high-level expression of Tar was necessary in order to identify its activity as an activator of the phosphorylation reactions. When membranes isolated from a strain containing only the chromosomal copy of Tar were used in the assay, no measurable activation was found.

Although Tar from *Salmonella typhimurium* has been solubilized using octylglucoside, and the solubilized protein shown to have activity both in binding aspartate and being a substrate for methylation,[16] it is not possible to perform the phosphorylation assays using solubilized Tar from *E. coli*. This is because at the concentration of octylglucoside required to solubilize Tar (1%), CheA autophosphorylation activity is greatly inhibited. Also, the effect of repellents for Tar, such as $Ni^{2+}$ and $Co^{2+}$, cannot be tested in the coupled assay, as they inhibit the $Mg^{2+}$-requiring CheA kinase.

The disk assay mentioned in the preceding chapter is of limited use in quantitating the levels of phosphorylated proteins produced in the reconstituted assay system. This is because integral membrane proteins that are independent of the chemotaxis system also become phosphorylated during the assay, leading to high background. Also, the membranes tend to trap unincorporated [$^{32}$P]ATP or [$^{32}$P]$P_i$, which contributes to errors in quantitation. Both of these problems are circumvented by using the SDS-PAGE analysis procedure.

## Tar Chimeras

It has been shown that the activity of two different chimeric proteins, which contain the periplasmic ligand-binding domain of Tar fused to the carboxy terminus of a different receptor protein, can be regulated by aspartate. A chimera containing the amino terminus of Tar joined to the C-terminus kinase domain of the bacterial transmembrane sensor protein EnvZ (Tar–EnvZ or Taz) has been constructed by M. Inouye and coworkers.[20] Various truncated EnvZ proteins, containing the C-terminus, have been demonstrated to autophosphorylate and then to transfer phosphate to the OmpR protein.[21–23] One of these truncated proteins, EnvZ115, activates transcription of *ompF* in *in vitro* transcription assays containing OmpR.[21] Normally, regulation of both *ompC* and *ompF* expression *in vivo* is regulated by the osmolarity of the growth medium, which is sensed by the full-length EnvZ protein.[24] However, in the heterologous system, cells which contain the Taz hybrid protein show aspartate-regulated expression of *ompC* transcription *in vivo*, with OmpR being required for the effect.[20] This indicates that the ligand-binding domain of Tar can apparently regulate the activity of a heterologous receptor–kinase domain *in vivo*, although Tar itself does not contain such a carboxy-terminal kinase activity.

Another chimeric protein consisting of the ligand-binding region of Tar joined to the tyrosine kinase domain of the insulin receptor has been studied (aspartate–insulin receptor or AIR).[25] Partially purified AIR from bacteria catalyzed phosphorylation of poly(Glu-Tyr), with aspartate addition increasing the rate 1.5-fold. Additionally, membranes containing AIR showed aspartate-enhanced phosphorylation of several bacterial membrane proteins. The phosphorylation of these bacterial proteins is probably not catalyzed by AIR directly, but instead proceeds through activation of a bacterial serine kinase, as the phosphorylated proteins contained phosphoserine and not phosphotyrosine. However, these results suggest

[20] R. Utsumi, R. E. Brissette, A. Rampersaud, S. A. Forst, K. Oosawa, and M. Inouye, *Science* **245,** 1246 (1989).

[21] M. M. Igo, A. J. Ninfa, and T. J. Silhavy, *Genes Dev.* **3,** 598 (1989).

[22] H. Aiba, T. Mizuno, and S. Mizushima, *J. Biol. Chem.* **264,** 8563 (1989).

[23] S. Forst, J. Delgado, and M. Inouye, *Proc. Natl. Acad. Sci. U.S.A.* **86,** 6052 (1989).

[24] S. Forst and M. Inouye, *Annu. Rev. Cell. Biol.* **4,** 21 (1988).

[25] G. R. Moe, G. E. Bollag, and D. E. Koshland, Jr., *Proc. Natl. Acad. Sci. U.S.A.* **86,** 5683 (1989).

that the ligand-binding domain of Tar can regulate both the phosphate-transfer activity of a eukaryotic receptor kinase domain, and also cause it to behave as an activator of other kinases in a heterologous system.

### Acknowledgments

We thank Elliot Altman, Bob Bourret, Juan Davagnino, Fred Hess, Chris O'Day, and Peggy Saks for critical reading of the manuscript. This work was supported by National Research Service Award Fellowship GM11223 (to K.A.B.) and by Grant AI19296 from the National Institutes of Health (to M.I.S.).

## [17] Analyzing Protein Phosphorylation in Prokaryotes

*By* JEAN-CLAUDE CORTAY, DIDIER NÈGRE, and ALAIN-JEAN COZZONE

In prokaryotes, the interest in protein phosphorylation has taken longer to gather momentum than in eukaryotes. In fact, even the occurrence of this chemical modification in microorganisms has long been a matter of controversy. The first attempt to characterize a protein kinase activity in bacteria was made in 1969 by Kuo and Greengard.[1] The authors reported the presence in *Escherichia coli* extracts of a cyclic AMP-dependent enzyme that could catalyze the phosphorylation by ATP of histones, which are exogenous basic proteins. Soon thereafter, two different protein kinases, regulated in a reciprocal fashion by cyclic AMP, were described in oral streptococci, also phosphorylating histones and protamines,[2] and a few more reports were published on this topic.[3–5] But still no definite conclusion could be drawn on the existence of a protein kinase activity in prokaryotes, namely because of the irreproducibility of certain results, as in the case of ribosomal protein phosphorylation,[3,6] or because of the incomplete chemical characterization of the phosphorylated moiety of the proteins. The latter point was critical, since bacteria are known to contain some kinds of kinases (e.g., polyphosphate kinase) that are quite different

[1] J. F. Kuo and P. Greengard, *J. Biol. Chem.* **244,** 3417 (1969).

[2] R. L. Khandelwal, T. N. Spearman, and I. R. Hamilton, *FEBS Lett.* **31,** 246 (1973).

[3] J. Gordon, *Biochem. Biophys. Res. Commun.* **44,** 579 (1971).

[4] E. Kurek, N. Grankowski, and E. Gasior, *Acta Microbiol. Pol.* **4,** 171 (1972).

[5] D. M. Powers and A. Ginsburg, *in* "Metabolic Interconversions of Enzymes" (E. H. Fischer, E. G. Krebs, and E. R. Stadtman, eds.), p. 131. Springer-Verlag, New York, 1973.

[6] E. Kurek, N. Grankowski, and E. Gasior, *Acta Microbiol. Pol.* **4,** 177 (1972).

from protein kinases and whose *in vitro* activity is stimulated especially by basic proteins,[7,8] thus leading to possible errors in the investigations of protein phosphorylation.[7,9]

The first clear demonstration of a protein kinase in a prokaryotic system came from an analysis of virus-infected bacteria. Rahmsdorf *et al.*[10,11] showed that a cyclic nucleotide-independent protein kinase appears in *E. coli* following infection with bacteriophage T7. The induced enzyme can phosphorylate both endogenous and exogenous proteins, and the products of the reaction have the chemical characteristics of phosphoserine and phosphothreonine. However, the authors simultaneously showed[11] that this protein kinase is, in fact, encoded by a specific viral gene, since its appearance is prevented by ultraviolet irradiation of the phage genome but not by that of the host genome. It was therefore generally assumed for several years that uninfected bacteria did not carry any protein kinase activity and, consequently, that protein phosphorylation was restricted to eukaryotic cells.[9,11]

In the past decade, however, this concept has been reversed by conclusive evidence that bacteria do contain specific protein kinases.[12] At this time, the occurrence of protein phosphorylation has been demonstrated in over 30 different species that belong to the 2 bacterial kingdoms, the eubacteria and archaebacteria, as well as to the line of cyanobacteria, which strongly suggests that this chemical modification is a universal phenomenon in microorganisms.[12] The methods used in the corresponding experiments share a number of common features that will be presented in this chapter.

## Labeling of Phosphoproteins

The detection of phosphorylated proteins necessitates their initial labeling either *in vivo* in a culture medium containing ortho-[$^{32}$P]phosphate or *in vitro* at the expense of [$\gamma$-$^{32}$P]ATP.

[7] H. C. Li and G. G. Brown, *Biochem. Biophys. Res. Commun.* **53,** 875 (1973).

[8] N. Agabian, O. M. Rosen, and L. Shapiro, *Biochem. Biophys. Res. Commun.* **49,** 1690 (1972).

[9] I. Pastan and S. Adhya, *Bacteriol. Rev.* **40,** 527 (1976).

[10] H. J. Rahmsdorf, P. Herrlich, S. H. Pai, M. Schweiger, and H. G. Wittmann, *Mol. Gen. Genet.* **127,** 259 (1973).

[11] H. J. Rahmsdorf, S. H. Pai, H. Ponta, P. Herrlich, and R. Roskoski, *Proc. Natl. Acad. Sci. U.S.A.* **71,** 586 (1974).

[12] A. J. Cozzone, *Annu. Rev. Microbiol.* **42,** 97 (1988).

*Labeling in Vivo*

Cells are grown at 30 or 37° in a low-phosphate medium[13] containing 100 m*M* Tris-HCl, pH 7.5, 37 m*M* $NH_4Cl$, 27 m*M* KCl, 2.5 m*M* $MgCl_2$, 140 $\mu M$ $Na_2SO_4$, 176 $\mu M$ $Na_2HPO_4$, 7.5 $\mu M$ $FeCl_3$, and 150 $\mu M$ $CaCl_2$. This medium is supplemented with either 20 m*M* glucose, 40 m*M* glycerol, 50 m*M* sodium acetate, 50 m*M* sodium citrate, or 50 m*M* sodium succinate as the carbon source. Other low-phosphate media with similar composition can be used as well.[14–17] To activate the growth of cells, casein hydrolysate (1 mg/ml) or yeast extract (0.5 mg/ml) can be added. However, it is wise to purify these compounds before use by eliminating the phosphate[18] that is frequently contained in commercial preparations. The same treatment should be applied to broth preparations whenever used for growing bacteria, although this type of rich medium usually does not favor the detection of phosphorylated proteins. The labeling of cells is achieved in the presence of carrier-free ortho-[$^{32}$P]phosphate (20–100 $\mu$Ci/ml). Double-labeling experiments are performed by adding simultaneously either [$^{35}$S]sulfate (100–200 $\mu$Ci/ml; specific activity 25–40 $\mu$Ci/mg) or [$^{35}$S]methionine (5–10 $\mu$Ci/ml; specific activity 900–1100 Ci/mmol). The labeling time varies from 15 min to a few hours.

*Labeling in Vitro*

A typical incubation mixture[19,20] contains 0.8–1 mg/ml of protein, 25 m*M* Tris-HCl at pH 7.5, 5 m*M* $MgCl_2$, 5 m*M* 2-mercaptoethanol, and 100 $\mu M$ (100 $\mu$Ci/ml) [$\gamma$-$^{32}$P]ATP (specific activity 1000–3000 Ci/mmol). Incubation is carried out for 15–60 min at 30–37°. When required, 20–100 $\mu M$ cyclic AMP or cyclic GMP is added. Some assays are carried out in the presence of 2–10 m*M* sodium fluoride, an inhibitor of certain phosphatases, or in the presence of molybdate or vanadate in the micromolar range.[21,22]

[13] M. Manaï and A. J. Cozzone, *Biochem. Biophys. Res. Commun.* **91,** 819 (1979).
[14] M. Garnak and H. C. Reeves, *Science* **203,** 1111 (1979).
[15] G. Ferro-Luzzi Ames and K. Nikaido, *Eur. J. Biochem.* **115,** 525 (1981).
[16] A. C. Borthwick, W. H. Holms, and H. G. Nimmo, *Biochem. J.* **222,** 797 (1984).
[17] J. Babul and D. G. Fraenkel, *Biochem. Biophys. Res. Commun.* **151,** 1033 (1988).
[18] H. Inouye, S. Michaelis, A. Wright, and J. Beckwith, *J. Bacteriol.* **146,** 668 (1981).
[19] M. Dadssi and A. J. Cozzone, *FEBS Lett.* **186,** 187 (1985).
[20] J. Y. Wang and D. E. Koshland, *J. Biol. Chem.* **256,** 4640 (1981).
[21] T. M. Chiang, J. Reizer, and E. H. Beachey, *J. Biol. Chem.* **264,** 2957 (1989).
[22] D. B. Karr and D. W. Emerich, *J. Bacteriol.* **171,** 3420 (1989).

## Preparation of Phosphoprotein Samples

### *Total Cellular Phosphoproteins*

After labeling *in vivo*, cells are collected by low-speed centrifugation, suspended in a buffer made of 10 m$M$ Tris-HCl, pH 7.4, 5 m$M$ $MgCl_2$, and 50 $\mu$g/ml pancreatic ribonuclease, then opened by alumina grinding or repeated ultrasonic disruption. The cellular extract is incubated for 15 min at 4° in the presence of 50 $\mu$g/ml pancreatic deoxyribonuclease and treated with 0.12 vol of 3% (w/v) sodium dodecyl sulfate (SDS)–10% (v/v) 2-mercaptoethanol. After centrifugation for 20–30 min at 30,000 *g* at 4°C, the supernatant fraction (S30) is collected, proteins are precipitated for 8–12 hr with 5 vol of 95% (v/v) acetone at −20°, then centrifuged and dried under vacuum.[23] Alternatively, proteins are precipitated by 5–10% (w/v) trichloroacetic acid (TCA) or 25 to 80% (w/v) ammonium sulfate.[22,24] This preparation contains total cellular phosphoproteins devoid of contaminating nucleic acids. It can also be obtained, through the same procedure, from a cellular extract incubated *in vitro* with radioactive ATP. Comparative experiments performed in the presence of protease inhibitors throughout the extraction and purification steps indicate that no degradation of phosphoprotein takes place in these experimental conditions.[23,25] Similarly no loss of phosphoryl groups by phosphatase action seems to occur.[22]

### *Phosphoproteins of Subcellular Fractions*

In most cases the analysis of phosphoproteins concerns total cellular preparations. However, experiments have been performed to study phosphoproteins from individual subcellular fractions.[23,26] For this purpose, the conventional techniques of purification can be used. Thus, to prepare cytoplasmic proteins, fraction S30 is collected and further centrifuged for 180 min at 225,000 *g*, the resulting supernatant is removed, and proteins are precipitated as above. The pellet contains the total ribosomes which are suspended in 1 vol of 10 m$M$ Tris-HCl, pH 7.7, 100 m$M$ magnesium acetate, and treated for 60 min at 4° with 2 vol of glacial acetic acid. Ribosomal proteins thus solubilized are dialyzed against 1 $M$ acetic acid and lyophilized.[27] On the other hand, nucleoids can be prepared by the

[23] J. C. Cortay, C. Rieul, B. Duclos, and A. J. Cozzone, *Eur. J. Biochem.* **159,** 227 (1986).
[24] M. Zylicz, J. H. LeBowitz, R. McMacken, and C. Georgopoulos, *Proc. Natl. Acad. Sci. U.S.A.* **80,** 6431 (1983).
[25] A. M. Turner and N. H. Mann, *FEMS Microbiol. Lett.* **57,** 301 (1989).
[26] C. S. Mimura, F. Poy, and G. R. Jacobson, *J. Cell. Biochem.* **33,** 161 (1987).
[27] J. S. Hardy, C. G. Kurland, P. Voynow, and G. Mora, *Biochemistry* **8,** 2897 (1969).

low-salt spermidine technique[28] after conversion of bacterial cells to spheroplasts, proteins are then precipitated with 15% (w/v) trichloroacetic acid, washed with a mixture of acetone–ether (1 : 1), and dried under vacuum. Proteins can also be prepared from membranes purified by detergent treatment of cellular extracts and centrifugation through EDTA–sucrose step gradient.[29]

## Separation and Detection of Phosphoproteins

The techniques more frequently utilized to separate phosphoproteins are polyacrylamide gel electrophoresis and, to a lesser extent, chromatography.

### *One-Dimensional Electrophoresis*

Migration can be performed in nondenaturing conditions by using 7.5% (w/v) polyacrylamide gels at pH 7.4.[30–32] However, phosphoproteins are more often analyzed under denaturing conditions in the presence of 0.5–1% (w/v) sodium dodecyl sulfate by using continuous 7.5–12% gels,[33] or various linear or exponential gradient gels which involve acrylamide concentrations ranging from 4 to 30% (w/v).[34,35] But, except for prepurified phosphoprotein preparations containing a limited number of molecules,[36,37] the resolving power of the one-dimensional electrophoresis technique is generally insufficient to separate all the constituents of a complex protein mixture, e.g., total cellular extract. Two-dimensional analytical systems are then required.

### *Two-Dimensional Electrophoresis*

Protein samples (20–200 $\mu$g) are analyzed mostly by the two-dimensional technique described by O'Farrell.[38] Separation in the first dimension is achieved by either isoelectric focusing to equilibrium in pH 5 to 7

[28] T. Kornberg, A. Lockwood, and A. Worcel, *Proc. Natl. Acad. Sci. U.S.A.* **71,** 3189 (1974).
[29] K. Ito, T. Sato, and T. Yura, *Cell (Cambridge, Mass)* **11,** 551 (1977).
[30] A. C. Peacock and C. W. Dingman, *Biochemistry* **7,** 668 (1968).
[31] G. Antranikian, C. Herzberg, and G. Gottschalk, *Eur. J. Biochem.* **153,** 413 (1985).
[32] D. B. Karr and D. W. Emerich, *J. Bacteriol.* **171,** 3420 (1989).
[33] U. K. Laemmli, *Nature (London)* **227,** 680 (1970).
[34] J. Londesborough, *J. Bacteriol.* **165,** 595 (1986).
[35] G. M. F. Watson and N. H. Mann, *J. Gen. Microbiol.* **134,** 2559 (1988).
[36] J. F. Hess, K. Osawa, P. Matsumura, and M. I. Simon, *Proc. Natl. Acad. Sci. U.S.A.* **84,** 7609 (1987).
[37] J. Reizer, M. J. Novotny, W. Hengstenberg, and M. H. Saier, *J. Bacteriol.* **160,** 333 (1984).
[38] P. H. O'Farrell, *J. Biol. Chem.* **250,** 4007 (1975).

ampholine or nonequilibrium electrophoresis in pH 3 to 10 ampholine.[23] In either case a 4% (w/v) polyacrylamide gel in 9.5 *M* urea is used. Electrophoresis in the second dimension is usually performed in SDS-polyacrylamide gels,[39–41] under conditions similar to those described above for the one-dimensional system. In some instances, migration in the first dimension is done using a nondenaturing gel[42] and separation in the second dimension is done in a 7.5 to 20% (w/v) polyacrylamide gradient gel.[40]

*Detection of Phosphoproteins*

Aside from $^{32}P$ autoradiography or radioactive counting, no specific and sensitive procedure for easily detecting phosphoproteins is currently available. The immunoreactivity with anti-phosphoamino acid antibodies basically affords an efficient means of detection, but its application is often complicated, especially with complex protein mixtures.

After migration of $^{32}P$-labeled proteins, polyacrylamide gels are soaked in 10–16% (w/v) trichloroacetic acid for 30–45 min at 90°.[23,42] This treatment is essential for removing some contaminating phosphorylated molecules, such as polyphosphates or nucleic acids, which may render erroneous the analysis of phosphoproteins. But this also releases the phosphoryl groups from phosphoramidates and acyl phosphates in proteins, and therefore allows only the detection of proteins carrying acid-stable phosphohydroxyamino acids.[43] After hot trichloroacetic acid (TCA) treatment, gels are washed overnight in 5% (w/v) TCA–0.5% (w/v) $Na_2HPO_4$, then incubated for 1–2 hr in 7.5% (v/v) acetic acid–30% (v/v) methanol[23] or in 1% (v/v) glycerol–5% (w/v) TCA–0.5% (w/v) $Na_2HPO_4$–45% (v/v) methanol,[42] and finally dried under vacuum and autoradiographed for 1–4 days. When required, a more sensitive detection of labeled proteins is obtained by fluorography. In this case, after treatment with hot TCA, gels are soaked in glacial acetic acid for 1 hr, then in a 10% (w/v) solution of 2,5-diphenyloxazole in glacial acetic acid. After several washes with water, gels are shrunk to suitable size in 50–70% (v/v) methanol and dried.

In double-labeling experiments where total proteins incorporate [$^{35}S$]sulfur or [$^{35}S$]methionine and phosphoproteins specifically incorporate [$^{32}P$]phosphate, the differential detection of the two radioisotopes can be made.[44] Two films are exposed for each gel: one records directly the $\beta$

[39] B. Averhoff, G. Antranikian, and G. Gottschalk, *FEMS Microbiol. Lett.* **33,** 299 (1986).
[40] M. Wada, K. Sekine, and H. Itikawa, *J. Bacteriol.* **168,** 213 (1986).
[41] M. Enami and A. Ishihama, *J. Biol. Chem.* **259,** 526 (1984).
[42] A. M. Turner and N. H. Mann, *J. Gen. Microbiol.* **132,** 3433 (1986).
[43] B. Duclos, S. Marcandier, and A. J. Cozzone, this series, Volume 201 [2].
[44] P. C. Cooper and A. W. Burgess, *Anal. Biochem.* **126,** 301 (1982).

emissions from $^{35}S$ and the other, shielded by aluminum foil, records scintillation photons from an intensifying screen excited by the $^{32}P$ emissions. Autoradiography is performed at $-80°$ so that there is no detection of $^{35}S$ on the $^{32}P$ film, and only a few percent of the $\beta$ emissions detected on the $^{32}P$ film are found on the $^{35}S$ film, which allows virtually complete discrimination of the two isotopes. Figure 1 shows a typical pattern[23] of the total $^{35}S$-labeled proteins of *E. coli,* and Fig. 2 indicates the location of the 130 $^{32}P$-labeled phosphoproteins present in this bacterium. On the other hand, both nonphosphorylated and phosphorylated proteins can be visualized, in any case, by the standard staining procedures, including Coomassie Blue, PAGE-Blue, and silver nitrate.

### *Chromatography*

For both analytical and preparative purposes, some bacterial phosphoproteins have been purified by chromatographic procedures including ion-exchange and affinity columns, gel filtration and, recently, fast protein liquid chromatography (FPLC). Thus, isocitrate dehydrogenase purification requires DEAE-cellulose, Porcion Red-Sepharose, and Sephadex G-150 columns.[14,16] The phosphoprotein DnaK of *E. coli* is purified by phosphocellulose and DEAE-Sephacel chromatography,[24] and the 61K phosphoprotein of *Streptococcus* is obtained by Sephadex G-100 filtration.[26] The purification of the phosphorylated nitrogen regulator protein is achieved on heparin-Sepharose[45] and that of the chemotaxis phosphoproteins on a Pharmacia Superose 12 FPLC column.[46] A combination of ion-exchange chromatography on DEAE-cellulose and gel filtration on Sephadex G-75 is needed to purify the phosphorylated form of protein HPr from gram-positive bacteria.[37]

## Characterization of Phosphorylated Moiety of Proteins

As in eukaryotes, one important criterion in demonstrating protein phosphorylation in prokaryotes is showing that phosphoryl groups are covalently bound to certain amino acids of protein substrates and characterizing the nature of the linkage involved.

The four major classes of phosphoamino acids (*O*-phosphomonoesters, phosphoramidates, acyl phosphates, and thiophosphates) are all present in bacterial proteins. A number of techniques have been developed to identify these various residues. They are described elsewhere in this vol-

[45] V. Weiss and B. Magasanik, *Proc. Natl. Acad. Sci. U.S.A.* **85,** 8919 (1988).

[46] J. F. Heiss, K. Osawa, N. Kaplan, and M. I. Simon, *Cell (Cambridge, Mass.)* **53,** 79 (1988).

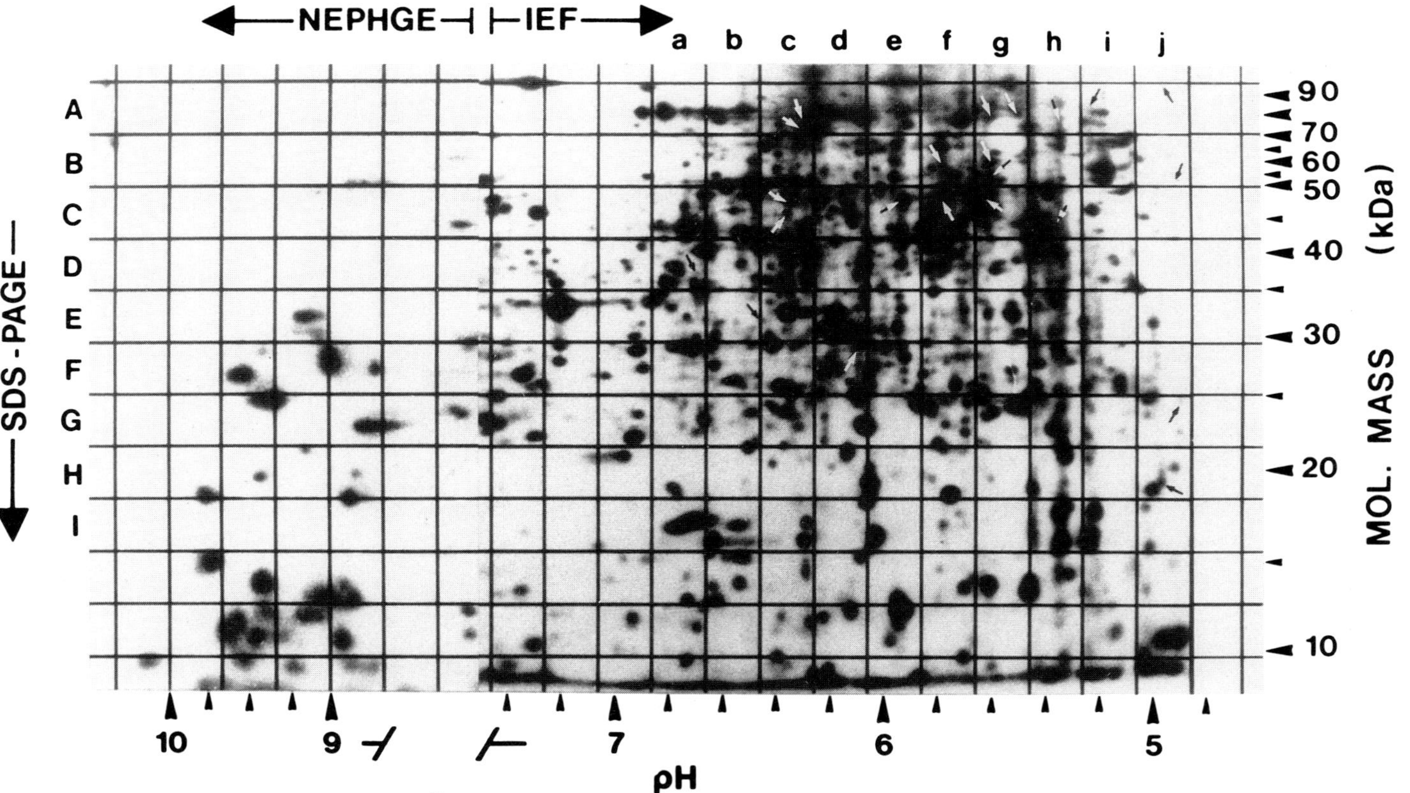


FIG. 1. Two-dimensional separation of the $^{35}$S-labeled total proteins of *E. coli*. Total proteins were extracted from bacteria continuously labeled with radioactive sulfate, and analyzed by the O'Farrell gel technique and autoradiography. Separation in the first dimension was achieved by either isoelectric focusing (IEF) to equilibrium in pH 5 to 7 ampholine, or nonequilibrium gel electrophoresis (NEPHGE) in pH 3 to 10 ampholine, in a 4% acrylamide gel containing 9.5 *M* urea. Migration in the second dimension was performed, in either case, in 12.5% acrylamide and 0.1% SDS. The two corresponding autoradiograms are presented side by side. The arrows indicate the location of the 22 most intensely labeled phosphoproteins described in Fig. 2. (From Ref. 23).

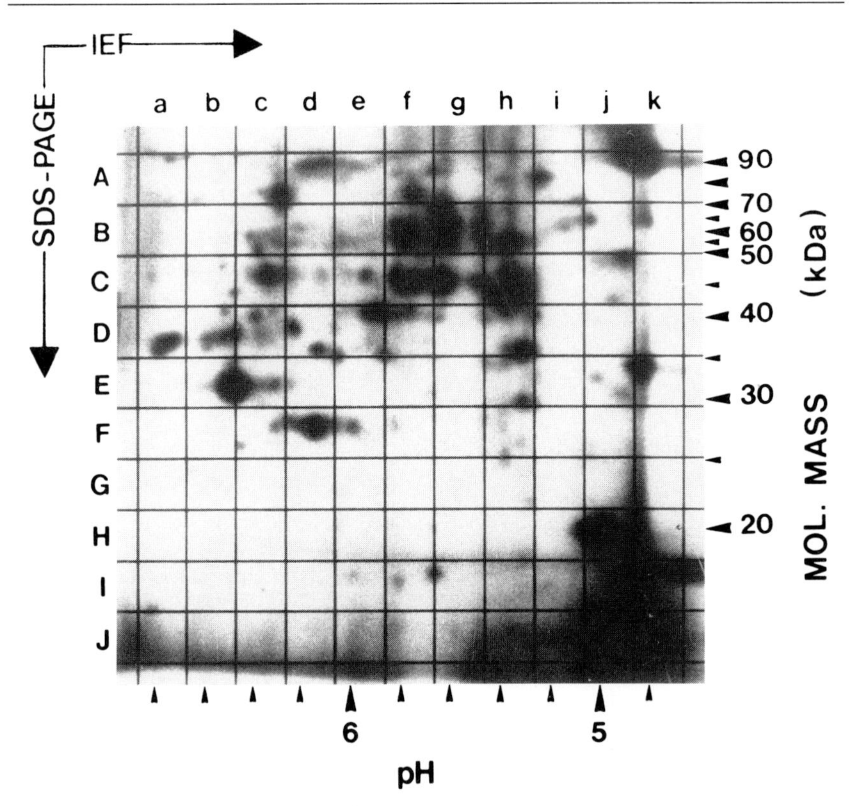


FIG. 2. Autoradiography of the $^{32}$P-labeled phosphoproteins of *E. coli*. Bacteria were cultured on acetate in the presence of radioactive orthophosphate and collected at the beginning of the stationary phase of growth. Proteins were extracted and analyzed as indicated in the legend of Fig. 1. Only the relevant part of the autoradiogram, corresponding to the zone of pH value ranging from 4.9 to 6.8, is presented since all phosphoproteins were detected in this particular area. The even divisions designated by letters a–j in the first dimension refer to decreasing values of isoelectric point, and those designated by A–I in the second dimension refer to decreasing molecular mass. This grid overlay corresponds to that shown in the right-hand part of Fig. 1.

ume.[43] The technique more commonly employed consists of hydrolyzing phosphoproteins either enzymatically or chemically by using acid (6 *N* HCl at 110°) or alkali (3 *N* KOH at 100°), depending on the stability of the phosphoamino acids. The phosphorylated molecules thus released are separated in one-dimensional or two-dimensional chromatographic and/or electrophoretic systems, or else by high-performance liquid chromatography (HPLC), and revealed by staining or autoradiographic procedures.[43]

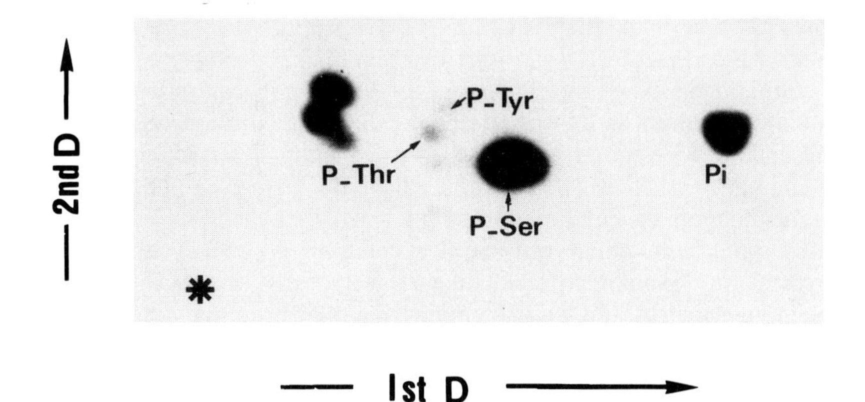


FIG. 3. Nature of the amino acids phosphorylated in the total proteins of *E. coli*. After two-dimensional separation, as in Fig. 2, the phosphoproteins labeled with radioactive orthophosphate were extracted from the gel, pooled, and subjected to acid hydrolysis. Phosphoamino acids were separated by electrophoresis in the first dimension and ascending chromatography in the second dimension, then autoradiographed. Authentic P-Ser, P-Thr, and P-Tyr were run in parallel and revealed by ninhydrin staining. The origin is indicated by an asterisk. The radioactive spots on the left-hand side of diagram likely correspond to unhydrolyzed phosphopeptides.

For example, Fig. 3 shows an autoradiogram of the $^{32}$P-labeled phosphohydroxyamino acids prepared from *E. coli* total proteins and separated by two-dimensional thin-layer electrophoresis/chromatography.[23] In this case, acid hydrolysis of protein samples allows one to eliminate contaminating phosphorylated compounds such as nucleic acids and polyphosphates, but leads to the destruction of acid-labile phosphoamino acids, i.e., phosphoramidates and acyl phosphates.[43] Conversely, alkaline hydrolysis allows isolation of phosphoramidates and phosphotyrosine, but destroys phosphoserine and acyl phosphates.

## Properties and Purification of Enzymes

### *Phosphorylating Enzymes*

The activity of the enzymes catalyzing protein phosphorylation in viruses and bacteria, namely protein kinases, is assayed both *in vivo* and *in vitro*. In the *in vivo* situation, the enzyme activity is detected essentially by $^{32}$P labeling of endogenous proteins, as already mentioned. Bacteria contain two main classes of protein kinases, which phosphorylate different polypeptides: one is in the cytoplasmic fraction and the other is attached to the ribosome/membrane fraction.[5,13] The *in vitro* assays are performed either with endogenous proteins or by making use of exogenous proteins or synthetic peptides. For this type of assay, the composition of the incubation medium is described above, the concentration of substrates being in the micromolar range.[19,47] One difficulty in detecting bacterial protein kinases lies in the fact that they are unable to phosphorylate proteins such as caseins or protamines,[12] which are, on the other hand, readily targeted by eukaryotic kinases. Such high selectivity of bacterial enzymes may explain, at least in part, why early searches employing exogenous substrates failed to demonstrate their presence in cellular extracts.[1,2] However, in some cases, histones seem to be recognized as protein substrates by prokaryotic protein kinases, namely those from *Streptococcus*[21] and *Legionella*[47] strains, and from bacteriophages T7 and T3.[48] Also, phosvitin is phosphorylated in the presence of a cellular extract prepared from *Mycoplasma gallisepticum*,[49] and a good substrate for the bacteriophage T7 enzyme is, uniquely, egg white lysozyme.[48] In addition, protein kinases of *E. coli*,[50] *Rhodospirillum rubrum*,[51] and *Streptococcus pyogenes*[21] can phosphorylate *in vitro* a number of synthetic peptides at serine and/or tyrosine residues.

Another characteristic feature of prokaryotic kinases is their insensitivity to cyclic nucleotides such as cAMP and cGMP. No variation of protein phosphorylation is observed *in vitro* when 20–100 $\mu M$ cAMP or cGMP is present in the incubation mixture.[19,21] Similarly the addition of cAMP at pH 7.0 in the culture medium does not affect the pattern of protein phosphorylation *in vivo*.[19] The independence of bacterial enzymes on cyclic nucleotides is confirmed by the analysis of Δ*cya* mutants, lacking adenylate cyclase activity, which behave like wild-type strains as far as protein phosphorylation is concerned.[15,19]

The techniques employed to purify bacterial and viral protein kinases involve mainly ion-exchange, affinity, and size-exclusion chromatography. Thus, some *E. coli* enzymes are purified, after ammonium sulfate fractionation (35 to 65% saturation), by chromatography on DEAE-Sepha-

[47] A. K. Saha, J. N. Dowling, N. K. Mukhopadhyay, and R. H. Glew, *J. Gen. Microbiol.* **134,** 1275 (1988).

[48] S. H. Pai, H. Ponta, H. J. Rahmsdorf, M. Hirsch-Kauffmann, P. Herrlich, and M. Schweiger, *Eur. J. Biochem.* **55,** 299 (1975).

[49] M. W. Platt, S. Rottem, Y. Milner, M. F. Barile, A. Peterkofsky, and J. Reizer, *Eur. J. Biochem.* **176,** 61 (1988).

[50] M. Dadssi, B. Duclos, and A. J. Cozzone, *Biochem. Biophys. Res. Commun.* **160,** 552 (1989).

[51] R. H. Vallejos, L. Holuigue, H. A. Lucero, and M. Torruella, *Biochem. Biophys. Res. Commun.* **126,** 685 (1985).

dex A-50 and Sepharose 6B columns.[41] In the case of isocitrate dehydrogenase kinase, whose activity modulates the partition of carbon between the Krebs cycle and the glyoxylate bypass,[14,16] purification needs a three-step procedure: ammonium sulfate fractionation, chromatography on DEAE-Sephacel, and affinity chromatography using isocitrate dehydrogenase immobilized on Sepharose 4B.[52] This particular enzyme exhibits an unusual feature[52]: it contains on the same polypeptide chain not only the kinase activity but also the corresponding phosphatase activity, and therefore is a bifunctional enzyme termed isocitrate dehydrogenase kinase/phosphatase. Besides its purification, it has been analyzed from a genetic point of view and the complete nucleotide sequence of its gene (*aceK*) has been determined.[53] Protein DnaK, which possesses an autophosphorylating activity, is purified by passing successively through phosphocellulose, DEAE-Sephacel, and hydroxylapatite columns. The four kinase activities of *Salmonella typhimurium* are obtained by ATP-agarose chromatography and Sephadex G-75 gel filtration.[20] The two forms of *Streptococcus pyogenes* protein kinase activity are resolved by DEAE-cellulose, Sephadex G-200, and affinity chromatography using a poly(Glu : Tyr)-Sepharose 2B column.[21] The separation of the two protein kinases of *Legionella micdadei* is achieved by QAE-Sephadex A-50 chromatography followed by histone affinity chromatography on Sepharose 4B, then HPLC gel filtration.[47] The purification of the T7 bacteriophage kinase takes advantage, on the one hand, of the ATP-accepting capacity of the enzyme by using an affinity Cibacron Blue 3G-A column which permits its binding to the chromophoric group of Dextran Blue. On the other hand, it makes use of the peculiar substrate specificity of the enzyme by employing an affinity egg white lysozyme-Sepharose column.[48] A protein kinase that phosphorylates two periplasmic transport proteins in *E. coli* has been purified to homogeneity by means of DEAE-Sephacel chromatography followed by isoelectric focusing, then chromatofocusing on Mono P and Mono Q columns.[54]

One can distinguish two main classes of phosphorylating activities that differ essentially in the nature and stability of the phosphoamino acid residues that they generate in target proteins, and in the type of molecules that they can use as phosphoryl donors.

The first class of enzymes consists of protein kinases that utilize nucleoside triphosphates, namely ATP, to phosphorylate hydroxyamino acids of proteins, yielding acid-stable phosphoamino acids. Most of the enzymes

[52] D. C. Laporte and D. E. Koshland, *Nature (London)* **300,** 458 (1982).

[53] J. C. Cortay, F. Bleicher, C. Rieul, H. C. Reeves, and A. J. Cozzone, *J. Bacteriol.* **170,** 89 (1988).

[54] C. Urban and R. T. F. Celis, *J. Biol. Chem.* **265,** 1783 (1990).

whose purification is described above belong to this class. Target proteins are generally modified at serine and, to a lesser extent, at threonine residues. In addition, the occurrence of a protein-tyrosine kinase activity has been reported first in *E. coli* and *S. typhimurium,*[55] then in a variety of bacterial species.[51,56,57] In *E. coli,* the majority of the phosphotyrosine recovered from partial acid hydrolysates of proteins seems to derive from glutamine synthetase.[58] It has been suggested that a large proportion of the phosphotyrosine detected in this enzyme would in fact arise from its adenylylation rather than phosphorylation, the nucleotide being linked via a phosphodiester bond to the phenolic hydroxyl of tyrosine.[58] Nevertheless, conclusive evidence has been brought that bacteria, namely *Acinetobacter calcoaceticus,* do contain a protein-tyrosine kinase activity.[59]

The second class of phosphorylating enzymes consists of phosphotransferases that catalyze the transfer of phosphate groups from phosphopolypeptides or metabolites such as phosphoenol pyruvate to acidic and basic amino acids of proteins, namely aspartate and histidine, respectively, yielding acid-labile phosphoamino acids. This type of enzyme is involved, for example, in the regulation of nitrogen assimilation. In this system, protein NtrB ($NR_{II}$) activates protein NtrC ($NR_{I}$) by a phosphorylation reaction which in turn activates the transcription of the *glnA* gene encoding glutamine synthetase.[60] The phosphorylation of NtrC by NtrB occurs in two steps: first, NtrB is modified at a histidine residue using ATP as a phosphate donor, then phospho–NtrB transfers its phosphate group to an aspartate residue of NtrC. Protein NtrB is a protein kinase which can autophosphorylate, whereas NtrC cannot, and phosphorylated NtrC is capable of autodephosphorylation.[61] A number of other "two-component" regulatory systems which also involve a phosphotransfer reaction between a sensor protein and a regulator protein have been described. These include, in particular, chemotaxis (see [15] and [16] in this volume), porin expression, and sporulation.

Interestingly, in the vectorial mechanism that controls the sugar-phosphate accumulation by expulsion of intracellular sugars in gram-positive bacteria, both classes of phosphorylating enzymes are involved.[62] Indeed, protein HPr can be phosphorylated, on the one hand, at a histidine residue

[55] J. C. Cortay, B. Duclos, and A. J. Cozzone, *J. Mol. Biol.* **187,** 305 (1986).
[56] M. Dadssi and A. J. Cozzone, *Int. J. Biochem.* **22,** 493 (1990).
[57] K. Kelly-Wintenberg, T. Anderson, and T. C. Montie, *J. Bacteriol.* **172,** 5135 (1990).
[58] R. Foster, J. Thorner, and G. S. Martin, *J. Bacteriol.* **171,** 272 (1989).
[59] M. Dadssi and A. J. Cozzone, *J. Biol. Chem.* **265,** 20996 (1990).
[60] A. J. Ninfa and B. Magasanik, *Proc. Natl. Acad. Sci. U.S.A.* **83,** 5909 (1986).
[61] J. Keener and S. Kustu, *Proc. Natl. Acad. Sci. U.S.A.* **85,** 4976 (1988).
[62] J. Reizer, J. Deutscher, and M. H. Saier, *Biochimie* **71,** 989 (1989).

by a specific phosphotransferase and, on the other hand, at a serine residue by a protein kinase.

*Dephosphorylating Enzymes*

The phosphoprotein phosphatases are less well defined than the phosphorylating enzymes in terms of their structure, regulation, and substrate specificity. The assay for phosphatase activity is usually performed *in vitro* by measuring, under chase conditions, the loss of $^{32}P$ radioactivity from prelabeled individual or total phosphoproteins. The composition of the reaction medium is similar to that for protein kinase assay[19,49] (see above).

*Salmonella typhimurium* contains at least two soluble protein phosphatase activities which can be purified by Sephadex G-100 filtration.[20] They differentially dephosphorylate specific endogenous phosphoproteins *in vitro* and are both insensitive to 5 m*M* NaF, which indicates that they differ from the nonspecific acid phosphatases harbored by the bacterium (which lacks alkaline phosphatase). Their activity is also unaffected by 100 m*M* iodoacetamide. By contrast, *Mycoplasma gallisepticum* harbors a soluble protein phosphatase which is strongly inhibited by 10 m*M* NaF.[49] In *Clostridium thermohydrosulfuricum* two phosphatases that dephosphorylate distinct substrate proteins are found, one in the cytoplasm and the other in the particulate fraction of the cell.[34] The phosphoryl group of the phosphoserine residue borne by the protein HPr of *Streptococcus faecalis* is hydrolyzed by a soluble protein phosphatase of 75 kDa.[63] The enzyme is purified in a five-step procedure including chromatography on DEAE-cellulose (three times), Sephacryl S-200, and hydroxylapatite. It is stimulated by inorganic phosphate, an inhibitor of the HPr kinase, but is insensitive to fructose 1,6-bisphosphate and other glycolytic intermediates. ATP, but not ADP, severely inhibits this enzyme. EDTA is also a potent inhibitor, but divalent cations such as $Mg^{2+}$, $Mn^{2+}$, and $Co^{2+}$ overcome its inhibitory action.[63] The case of *E. coli* isocitrate dehydrogenase phosphatase is special because, as already indicated, this activity and the isocitrate dehydrogenase kinase activity are, in fact, physically associated with the same protein, which can be purified by the three-step procedure mentioned above.[52]

[63] J. Deutscher, U. Kessler, and W. Hengstenberg, *J. Bacteriol.* **163,** 1203 (1985).

# [18] Expression, Separation, and Assay of Protein Kinase C Subspecies

*By* KOUJI OGITA, YOSHITAKA ONO, USHIO KIKKAWA, and YASUTOMI NISHIZUKA

A large number of extracellular signals which stimulate inositol phospholipid hydrolysis appear to operate through a bifurcating intracellular signal pathway which involves $Ca^{2+}$ mobilization and protein kinase C activation.[1] Molecular cloning analysis has revealed that protein kinase C exists as a family of multiple subspecies.[2-11] Initially, four cDNA clones designated $\alpha$, $\beta$I, $\beta$II, and $\gamma$ were isolated,[12] and these clones were shown to be conserved among different mammalian species such as human, bovine, rat, and rabbit.[2-6] These subspecies have closely related structures with several highly conserved regions in this protein kinase family. This heterogeneity comes from different genes as well as from alternative splicing of a single RNA transcript.[3,13] Later, another group of cDNA clones

[1] Y. Nishizuka, *Nature (London)* **334,** 661 (1988).

[2] P. J. Parker, L. Coussens, N. Totty, L. Rhee, S. Young, E. Chen, S. Stabel, M. D. Waterfield, and A. Ullrich, *Science* **233,** 853 (1986).

[3] L. Coussens, P. J. Parker, L. Rhee, T. L. Yang-Feng, E. Chen, M. D. Waterfield, U. Francke, and A. Ullrich, *Science* **233,** 859 (1986).

[4] Y. Ono, T. Kurokawa, T. Fujii, K. Kawahara, K. Igarashi, U. Kikkawa, K. Ogita, and Y. Nishizuka, *FEBS Lett.* **206,** 347 (1986).

[5] J. L. Knopf, M.-H. Lee, L. A. Sultzman, R. W. Kriz, C. R. Loomis, R. M. Hewick, and R. M. Bell, *Cell (Cambridge, Mass.)* **46,** 491 (1986).

[6] S. Ohno, H. Kawasaki, S. Imajoh, K. Suzuki, M. Inagaki, H. Yokokura, T. Sakoh, and H. Hidaka, *Nature (London)* **325,** 161 (1987).

[7] Y. Ono, T. Fujii, K. Igarashi, U. Kikkawa, K. Ogita, and Y. Nishizuka, *Nucleic Acids Res.* **16,** 5199 (1988).

[8] G. M. Housey, C. A. O'Brian, M. D. Johnson, P. Kirschmeier, and I. B. Weinstein, *Proç. Natl. Acad. Sci. U.S.A.* **84,** 1065 (1987).

[9] Y. Ono, T. Fujii, K. Igarashi, U. Kikkawa, K. Ogita, and Y. Nishizuka, *J. Biol. Chem.* **263,** 6927 (1988).

[10] S. Ohno, Y. Akita, Y. Konno, S. Imajoh, ad K. Suzuki, *Cell (Cambridge, Mass.)* **53,** 731 (1988).

[11] D. Schaap, P. J. Parker, A. Bristol, R. Kriz, and J. Knopf, *FEBS Lett.* **243,** 351 (1989).

[12] The nomenclature of the cDNA clones used in this chapter is as described by Nishizuka.[1] The $\alpha$, $\beta$I, $\beta$II, and $\gamma$ isolated from different species encode 672, 671, 673, and 697 amino acids, respectively.

[13] Y. Ono, U. Kikkawa, K. Ogita, T. Fujii, T. Kurokawa, Y. Asaoka, K. Sekiguchi, K. Ase, K. Igarashi, and Y. Nishizuka, *Science* **236,** 1116 (1987).

was obtained that has structures similar to but distinct from the previous four cDNA clones.[8–11] Differential tissue and cellular expression and different enzymatic activity of these subspecies has been observed.[1] Thus, it is important to isolate each subspecies and study the properties of this enzyme family. This chapter describes a method for the transient expression of protein kinase C subspecies, $\alpha$, $\beta$I, $\beta$II, and $\gamma$, in COS-7 cells using the cDNA clones isolated from rat brain.[4,7,13] The methods for the separation and assay of these subspecies are also described.

## Construction of Expression Plasmids

The expression vector designated pTB701 is employed for the construction of expression plasmids of protein kinase C.[9,13,14] The expression vector is constructed from Okayama and Berg vectors[15] and has a single cloning site of *Eco*RI downstream from the Abelson murine leukemia virus long terminal repeat,[16] simian virus 40 (SV40) origin of DNA replication, and SV40 early promoter regions. The whole insert of the protein kinase C cDNA clones is incorporated into the *Eco*RI site of the pTB701. The resulting expression plasmids for $\alpha$, $\beta$I, $\beta$II, and $\gamma$ contain inserts of 3305 nucleotides from clone $\lambda$CKR$\alpha$5,[7] 3190 nucleotides from clones $\lambda$CKR152 and $\lambda$CKR108,[4,13] 3406 nucleotides from clones $\lambda$CKR152 and $\lambda$CKR107,[4,13] and 3025 nucleotides from clone $\lambda$CKR$\gamma$1,[7] respectively.

## Transfection

COS-7 cells are transfected by the calcium phosphate coprecipitation technique[17] with glycerol shock.[18] Fresh monolayers of COS-7 cells, in 10-cm diameter plates (50 plates) each containing 10 ml of Dulbecco's modified Eagle's medium (DMEM) containing 5% fetal calf serum (FCS), are transfected with 30 $\mu$g each of the plasmid DNA. After a transfection period of 3.5 hr, the cells are shocked with glycerol for 3 min at room temperature, and fresh medium is added to each plate. After incubation for 18 hr at 37°, the medium is replaced by 15 ml of fresh Dulbecco's modified Eagle's medium containing 5% fetal calf serum. Cells are harvested after an additional 48-hr incubation at 37°.

[14] U. Kikkawa, Y. Ono, K. Ogita, T. Fujii, Y. Asaoka, K. Sekiguchi, Y. Kosaka, K. Igarashi, and Y. Nishizuka, *FEBS Lett.* **217,** 227 (1987).

[15] H. Okayama and P. Berg, *Mol. Cell. Biol.* **3,** 280 (1983).

[16] E. P. Reddy, M. J. Smith, and A. Srinivasan, *Proc. Natl. Acad. Sci. U.S.A.* **80,** 3623 (1983).

[17] F. L. Graham and A. J. van der Eb, *Virology* **52,** 456 (1973).

[18] C. Gorman, R. Padmanabhan, and B. H. Howard, *Science* **221,** 551 (1983).

Separation

Protein kinase C expressed in COS-7 cells is separated into each subspecies by hydroxyapatite column chromatography using a fast protein liquid chromatography (FPLC) system (Pharmacia, Piscataway, NJ).[13,14] To obtain good resolution by the hydroxyapatite column, it is necessary to partially purify the protein kinase C from the COS-7 cell extracts before this column chromatography. For this purpose, it is possible to use an ion-exchange column chromatography by Mono Q column (Pharmacia). The following steps are carried out at 0 to 4°. The COS-7 cells (approximately $3 \times 10^8$ cells from 50 plates) are homogenized by sonication (Kontes, Vineland, NJ; sonifier or equivalent) for 1 min in 5 ml of 20 m*M* Tris-HCl at pH 7.5, containing 0.25 *M* sucrose, 10 m*M* ethylene glycol bis($\beta$-aminoethyl ester)-*N,N,N′,N′*-tetraacetic acid (EGTA), 2 m*M* ethylenediaminetetraacetic acid (EDTA), and 20 $\mu$g/ml leupeptin. The homogenate is centrifuged for 60 min at 100,000 *g*, and the supernatant is diluted with 6 vol of 20 m*M* Tris-HCl at pH 7.5, containing 0.5 m*M* EGTA, 0.5 m*M* EDTA, and 10 m*M* 2-mercaptoethanol (buffer A), filtered through a membrane filter (pore size 0.22 $\mu$m, Millex-GS, Millipore, Bedford, MA, or equivalent), and applied to a Mono Q column ($1 \times 10$ cm, Pharmacia HR10/10) that is connected to a Pharmacia FPLC system and equilibrated with buffer A. Protein kinase C is eluted by application of a 164-ml linear concentration gradient of NaCl (0 to 0.6 *M*) in buffer A at a flow rate of 2 ml/min. Fractions (4 ml each) are collected. Protein kinase C activity appears at fractions 8 to 12 at 0.18 *M* NaCl. The protein kinase fractions are pooled and diluted with an equal volume of 20 m*M* potassium phosphate at pH 7.5, containing 0.5 m*M* EGTA, 0.5 m*M* EDTA, 10% (w/v) glycerol, and 10 m*M* 2-mercaptoethanol (buffer B), and applied to a packed hydroxyapatite column ($0.78 \times 10$ cm, type S; Koken, Tokyo, Japan) which is connected to the Pharmacia FPLC system and equilibrated with buffer B. The protein kinase is eluted by application of an 84-ml linear concentration gradient of potassium phosphate (20 to 250 m*M*) in buffer B at a flow rate of 0.4 ml/min, and fractions (1 ml each) are collected. Active fractions are pooled and dialyzed against buffer A containing 10% (w/v) glycerol. This enzyme preparation can be stored at least for several months in the presence of 0.05% (w/v) Triton X-100 at −80°. Rat brain protein kinase C as a control is purified from its cytosol fraction by DEAE-cellulose (DE-52; Whatman, Milford, MA), threonine-Sepharose, and TSK phenyl-5PW (Toyo Soda, Tokyo, Japan) column chromatographies as described,[19] and

[19] U. Kikkawa, M. Go, J. Koumoto, and Y. Nishizuka, *Biochem. Biophys. Res. Commun.* **135,** 636 (1986).

resolved into distinct fractions by the hydroxyapatite column chromatography as described above.

### Assay

Protein kinase C is assayed by measuring the incorporation of $[^{32}P]P_i$ into calf thymus H1 histone from $[\gamma\text{-}^{32}P]ATP$, essentially as described previously.[19] The reaction mixture (0.25 ml) contains 20 m*M* Tris-HCl at pH 7.5, 5 m*M* magnesium acetate, 10 $\mu M$ $[\gamma\text{-}^{32}P]ATP$ (50 to 100 cpm/pmol), 200 $\mu$g/ml of histone H1, 8 $\mu$g/ml of phosphatidylserine, 0.8 $\mu$g/ml of diacylglycerol (1,2-diolein) or 10 ng/ml of 12-*O*-tetradecanoylphorbol 13-acetate (TPA), various concentrations of $CaCl_2$, and the enzyme fraction. Phosphatidylserine and diacylglycerol, stored separately as chloroform solutions at $-20°$, are mixed first in chloroform, dried, suspended in 20 m*M* Tris-HCl at pH 7.5, sonicated, and then added to the reaction mixture as lipid micelles as described.[19] TPA, dissolved in dimethyl sulfoxide, is diluted with distilled water and added directly to the reaction mixture with sonicated phosphatidylserin. The final concentrations of dimethyl sulfoxide should be kept less than 0.01% (v/v), because dimethyl sulfoxide at high concentrations sometimes varies protein kinase activity. Basal activity is measured in the presence of 0.5 m*M* EGTA, instead of $CaCl_2$. It is convenient to prepare a mixed solution of Tris-HCl, magnesium acetate, $[\gamma\text{-}^{32}P]ATP$, and histone H1. The reaction is started by the addition of this solution to the assay tube that contains $CaCl_2$, lipid micelles, and the enzyme fraction. After 3 min at 30°, the reaction is stopped by the addition of 3 ml of 25% trichloroacetic acid (TCA). Acid-precipitable materials are collected on a nitrocellulose membrane filter (pore size 0.45 $\mu$m; Toyo Roshi, Tokyo, Japan or equivalent) in a suction apparatus. The membrane filter is washed three times with 25% TCA, and the radioactivity is quantitated by Cerenkov counting.

### *Comments*

The protein kinase C activity recovered from the transfected cells by Mono Q column chromatography is severalfold higher than that from the control cells. Figure 1 shows the separation of the protein kinase C subspecies from the transfected cells by hydroxyapatite column chromatography. The rat brain protein kinase C was first reported to be resolved into three distinct fractions on hydroxyapatite column chromatography by K.-P. Huang *et al.*[20] These fractions are called Type I, II, and III according

[20] K.-P. Huang, H. Nakabayashi, and F. L. Huang, *Proc. Natl. Acad. Sci. U.S.A.* **83,** 8535 (1986).

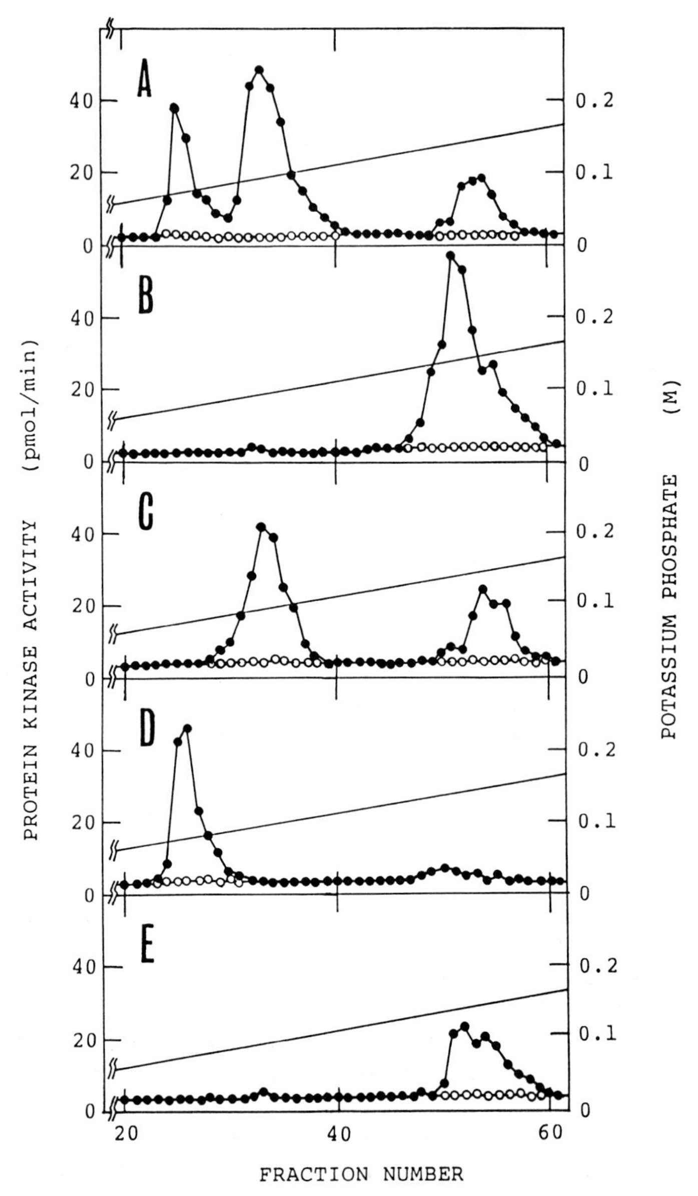


FIG. 1. Separation of protein kinase C subspecies by hydroxyapatite column chromatography. Protein kinase C subspecies expressed in COS-7 cells and the enzyme preparation purified from rat brain were applied to hydroxyapatite column as described in the text. Enzyme activity was assayed in the presence of phosphatidylserine, diacylglycerol, and 0.5 m*M* $CaCl_2$ (●), or in the presence of 0.5 m*M* EGTA instead of phosphatidylserine, diacylglycerol, and $CaCl_2$ (○). Solid line indicates the concentrations of potassium phosphate. (A) Rat brain protein kinase C. (B)–(D) Protein kinase C from COS-7 cells transfected with the expression plasmid of protein kinase C of $\alpha$, $\beta$II, and $\gamma$, respectively. (E) Protein kianse C from control COS-7 cells. (Taken from Ref. 14.)

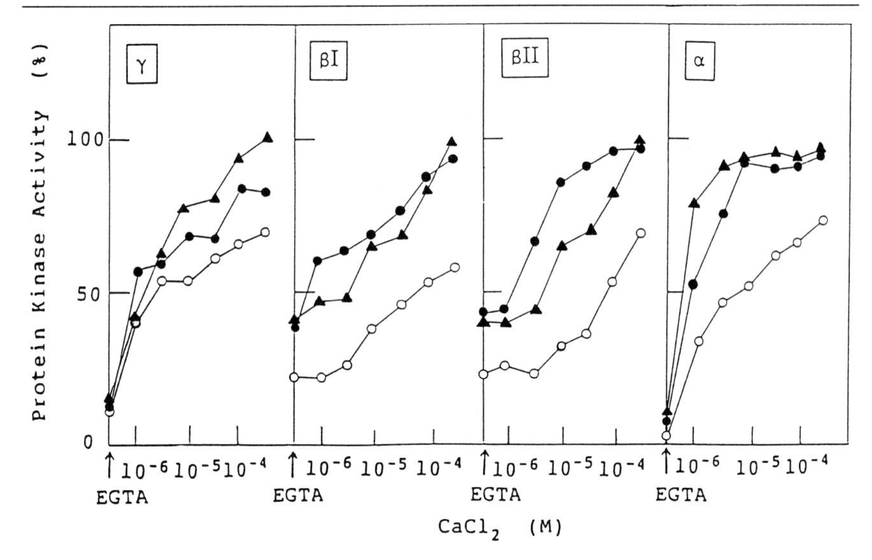


FIG. 2. Activities of protein kinase C subspecies expressed in COS-7 cells. Protein kinase C subspecies expressed in COS-7 cells were separated and assayed at various concentrations of $CaCl_2$ as described in the text in the presence of phosphatidylserine and diacylglycerol (▲), phosphatidylserine and TPA (●), and phosphatidylserine alone (○). The arrows indicate that EGTA (0.5 m*M* final concentrations) was added instead of $CaCl_2$.

to the order of elution from the column. Purified rat brain protein kinase C can be separated into the three fractions reproducibly by a high-performance liquid chromatography hydroxyapatite column connected to the FPLC system. The elution profiles from the hydroxyapatite column indicate that protein kinase C subspecies $\alpha$, $\beta$I and $\beta$II,[21] and $\gamma$ correspond with the enzyme fractions type III, II, and I, respectively.[22] The correspondence is confirmed by immunoblot analysis using subspecies-specific antibodies raised against synthetic polypeptides containing predicted amino acid sequences for each protein kinase C subspecies,[23] and the expressed enzymes are shown to have an approximate molecular weight of 80,000, identical with that of enzymes purified from rat brain on sodium dodecyl sulfate-polyacrylamide gel electrophoresis using immunoblot analysis.[23] Protein kinase C subspecies expressed in COS-7 cells have properties characteristic of this enzyme family, as shown in Fig. 2. Each subspecies

[21] Protein kinase C of $\beta$I subspecies is shown to be eluted from the column at the position identical with $\beta$II subspecies.[13]

[22] The untransfected COS-7 cells have an endogenous protein kinase C of $\alpha$ subspecies, which is recognized by a subspecies-specific antibody.

[23] A. Kishimoto, N. Saito, and K. Ogita, this volume [37].

is activated by diacylglycerol or TPA in the presence of phosphatidylserine and $Ca^{2+}$. However, the requirements for $Ca^{2+}$ are slightly different for each subspecies and properties are similar to those of each native enzyme.[24] The yields of protein kinase C differ among the subspecies as shown in Fig. 1, and depend on the efficiency of the expression in each experiment. However, it is estimated that several micrograms of protein kinase C are recovered from $3 \times 10^8$ transfected cells based on specific activities and staining profiles using immunoblot analysis. The expression and purification procedures can be scaled down by a factor of approximately 10 for 5 culture plates using Mono Q column HR5/5 (0.5 × 5 cm) and the same size hydroxyapatite column under comparable conditions, if only small amounts of the enzymes are needed. Another group of cDNA clones of protein kinase C[8-11] has been studied using the same procedures.

[24] K. Sekiguchi, M. Tsukuda, K. Ase, U. Kikkawa, and Y. Nishizuka, *J. Biochem. (Tokyo)* **103,** 759 (1988).

# [19] Purification of Protein Kinase C Isotypes from Bovine Brain

*By* P. J. PARKER and R. M. MARAIS

## Introduction

The purification of protein kinase C (PKC) isotypes from brains of laboratory animals has been developed very effectively, but there has been little development of protein kinase C purification from larger animals. Nevertheless it became desirable to pursue the large-scale purification of these kinases from larger animal sources for reasons of ethics and cost. Two procedures are described below which were developed specifically for the routine bulk copurification and subsequent separation of PKC-$\alpha$, -$\beta$, and -$\gamma$ isotypes. In order to monitor the efficiency of copurification and separation, it has been essential to employ monospecific antisera.[1] The first procedure is used to prepare milligram amounts of highly purified PKC, containing the $\alpha$, $\beta$, and $\gamma$ isotypes. The second procedure is a

[1] R. M. Marais and P. J. Parker, *Eur. J. Biochem.* **182,** 129 (1989).

scaleup of the first and is used for the routine production of several milligrams of slightly less pure PKC.

## Procedure 1

A single bovine brain (~500 g) is homogenized within 3 min of slaughter into 1 liter of 20 m*M* Tris-HCl, pH 8.0, 10 m*M* EGTA, 10 m*M* EDTA, 10 m*M* benzamidine, 0.3% (v/v) 2-mercaptoethanol, 50 $\mu$g/ml phenylmethylsulfonyl fluoride (homogenization buffer) using a 1-liter Waring blender (3×, 5 sec each, at high-speed setting). All procedures are carried out at 4°. This homogenate is then transported to the laboratory on ice for centrifugation (27,000 *g*, 30 min). The supernatant is adjusted to pH 8.0 with 1 *M* Tris base and diluted 1 : 1 with homogenization buffer. This is then loaded onto a DEAE-cellulose column [DE-52 (Whatman, Clifton, NJ), 6 × 18 cm, 5 ml/min] which is preequilibrated in 20 m*M* Tris-HCl, pH 7.5, 2 m*M* EDTA, 10 m*M* benzamidine, 0.3% (v/v) 2-mercaptoethanol (TEBM buffer). After loading, the column is washed with 600 ml of TEBM buffer. The kinase is eluted stepwise using 0.125 *M* NaCl in TEBM buffer and 16-ml fractions are collected.

The PKC which elutes from the column is detected by determining the histone III-S (Sigma, St. Louis, MO) kinase activity in the presence and absence of phosphatidylserine and 12-*O*-tetradecanoylphorbol 13-acetate (TPA)[1]; the PKC elutes toward the trailing edge of the protein peak. The PKC-containing fractions are pooled (65–150 ml) and solid ammonium sulfate is added to give a final concentration of 18% (w/v). The solid ammonium sulfate must be added slowly, over a period of about 5 min, with constant stirring to avoid local high concentrations of salt which reduce recovery of PKC. The suspension is clarified by centrifugation (5000 *g* for 20 min) and the supernatant is loaded onto a phenyl-Sepharose column (3.5 × 5 cm; 2 ml/min) equilibrated in TEBM buffer containing 1.2 *M* ammonium sulfate. The column is then washed with 100 ml of equilibration buffer and eluted stepwise with TEBM buffer (i.e., without ammonium sulfate). This step serves both to concentrate the DEAE eluate and to remove calcium-activated neutral proteases which remain bound to the phenyl-Sepharose column.

It is not possible to determine accurately the precise position of elution of PKC from the phenyl-Sepharose column due to the high concentration of ammonium sulfate which interferes with the assay. The protein eluted from the phenyl-Sepharose column is therefore pooled without assay and applied to a Sephacryl S-200 column (5 × 90 cm; 2 ml/min) equilibrated in TEBM buffer containing 0.02% sodium azide. This column is developed overnight and 15-ml fractions are collected which are assayed for PKC activity. The PKC-containing fractions are pooled, made 380 m*M* with NaCl (from a 5 *M* NaCl stock), and loaded (the final volume loaded must not exceed 120 ml, see below) onto a threonine-Sepharose column[2] (1.5 × 29 cm; 1 ml/min) equilibrated in TEBM buffer containing 380 m*M* NaCl. The column is then washed with 120 ml of equilibration buffer and eluted with a linear gradient (50 ml) of 0.38 to 1.2 *M* NaCl in TEBM buffer, followed by an isocratic wash at 1.2 *M* NaCl. Fractions containing PKC activity are pooled, made 0.02% in Triton X-100, and dialyzed overnight against 20 m*M* Tris-HCl, pH 7.5, 2 m*M* EDTA, 1 m*M* dithiothreitol (DTT), 0.02% (v/v) Triton X-100, 50% (v/v) glycerol (buffer A). This preparation of PKC is stored at −20°.

The initial column steps in the procedure gives a two- to threefold purification with the major purification being achieved with the threonine-Sepharose column. The threonine-Sepharose column is selective for PKC when loaded under these salt conditions. If the column is loaded in the absence of NaCl, the majority of the proteins in the Sephacryl S-200 sample bind and PKC coelutes with them when the NaCl gradient is applied. Thus less PKC in a lower state of purity is recovered. When the threonine-Sepharose column is loaded in 380 m*M* NaCl, however, the affinity of PKC for the column is reduced and the kinase tends to elute from the column during the wash step. Providing the column dimensions and the volume of the load, wash, and gradient used are as stated, this problem can be overcome. Larger columns can be used but must be optimized to give efficient purification, with attention being paid to the relative length and width of the column (see procedure 2, below).

The order in which the columns are used can be altered and successful purification can still be achieved. For example, a crude extract from a single brain can be made 18% (w/v) with ammonium sulfate and the supernatant loaded directly onto the phenyl-Sepharose column, followed by the Sephacryl S-200 and threonine-Sepharose columns as described above. The PKC which is recovered is about 90% pure as judged by silver-stained SDS–PAGE, but can be taken to homogeneity if passed through a 2-ml DEAE-cellulose column essentially under the conditions described above. When this is done however, the recoveries are lower (~1–2 mg) although the concentration of protein is higher (~1 mg/ml).

The interaction between PKC and the threonine-Sepharose column is not understood but a serine-Sepharose column (made in the same way, but using serine instead of threonine) is as efficient at purifying PKC-$\alpha$, -$\beta$,

[2] U. Kikkawa, M. Go, J. Koumoto, and Y. Nishizuka, *Biochem. Biophys. Res. Commun.* **135,** 636 (1986).

and -$\gamma$ as is the threonine-Sepharose column. Indeed, a serine-Sepharose column has proved useful in the purification of PKC-$\varepsilon$.[3] This suggests that the interaction is a pseudoaffinity step with PKC recognizing its target residues.

The PKC prepared by this procedure is purified about 100- to 120-fold compared to the crude extract and is homogeneous as judged by silver-stained SDS–PAGE, thus indicating the abundance of this enzyme in fresh brain tissues. The sample contains PKC-$\alpha$, -$\beta$, and -$\gamma$, although the bulk of the PKC-$\beta$ is the $\beta_2$ splice product.[4] The yield of the PKC mix is 5–10 mg with a specific activity of 1500 units/mg against histone III-S. There is no detectable histone kinase activity in the absence of phospholipids and in the mixed micelle assay the kinase is absolutely dependent on TPA or diacylglycerols (DG) for activation.[1] General characteristics of these enzymes are described elsewhere.[1]

When stored in buffer A at $-20°$ the kinase activity is stable for at least 9 months. However, after this time there is often a sudden loss (within about 2 weeks) of dependence on phospholipids and TPA for kinase activity, resulting in a constitutively active kinase; this kinase activity then gradually declines over about 2 months. The loss of dependence on phospholipids and TPA can be temporarily reversed by the prompt addition of fresh DTT to 1 m*M*, but the "rescued" PKC is only active for about 2 months after which lipid dependence is once again lost. It is therefore suggested that the kinase be stored in numerous small aliquots at $-20°$ and that only one aliquot be used at a time to avoid excessive exposure to oxidizing conditions.

### *Comments*

This purification scheme was designed to give high recoveries of PKC from bovine brain. The first three steps in the purification are standard column chromatographies and the fourth is a pseudoaffinity chromatography step. The brain should weigh about 500 g; this size gives a smooth homogenate within the protein capacity of the DEAE column. It is essential to remove the brain from the carcass and immerse it in cold homogenization buffer as soon as possible after slaughter, since the half-life of PKC at this stage appears to be about 1 min (we usually achieve this within 3 min). All the equipment (Waring blender, centrifuge bottles, rotors, etc.) and buffers to be used should be precooled to 4° and we prefer to use

[3] D. Schaap and P. J. Parker, *J. Biol. Chem.* **265,** 7301 (1990).

[4] Y. Ono, U. Kikkawa, K. Ogita, T. Fujii, T. Kurokawa, A. Yoshinori, Y. Asaoka, K. Sekiguchi, K. Ase, K. Igarashi, and Y. Nishizuka, *Science* **236,** 1116 (1987).

polypropylene tubes and centrifuge bottles for all steps; 2-mercaptoethanol is added to the buffers immediately before use.

Apart from the Sephacryl S-200 column, all the columns are freshly packed for each preparation and regenerated immediately after use. The DEAE column is regenerated by alternate incubations (1 hr at room temperature) in 2 liters each of 0.5 *M* NaOH, 0.5 *M* HCl, and 0.5 *M* NaOH with extensive water washes between each solution. The phenyl-Sepharose column is regenerated by incubation in 200 ml each of water, ethanol, 1-butanol, ethanol, and finally water. The threonine-Sepharose column is regenerated by incubation in water, 2 *M* NaCl, 1% (v/v) Triton X-100 followed by extensive washing with water. The DEAE, phenyl-Sepharose, and threonine-Sepharose gels are stored at 4° in 0.1 *M* Tris-HCl, pH 8.0, with 0.02% (w/v) sodium azide. The Sephacryl S-200 column is washed by passing 20 liters of TEBM buffer containing 0.02% (w/v) sodium azide through it; the column is stored in this buffer. In our experience older column matrices perform better than new, unused matrices, because the newer matrices appear to have more nonspecific binding sites and therefore give poorer recoveries. New columns are therefore routinely treated with buffer containing 1 mg/ml bovine serum albumin (BSA) followed by one cycle of regeneration before their first use.

This purification procedure must be completed within 2 days of starting in order to achieve high recoveries. Chromatography on the DEAE-cellulose and phenyl-Sepharose columns is therefore done on the same day that the brain is collected (an early start is therefore recommended). The Sephacryl S-200 column is loaded on the evening of the first day and developed overnight. The threonine-Sepharose column is loaded and eluted on the second day; the pool from this column is dialyzed overnight and placed at −20° on the morning of the third day.

### Procedure 2

Due to practical considerations the procedure described above cannot be employed over the same time scale with more starting material. As discussed earlier, the order in which the columns are used can be varied without compromising the purity of the final product. This observation leads to the development of a purification protocol which was designed specifically to generate larger amounts of purified PKC.

Three bovine brains are homogenized within 3 min of slaughter into 4 liters of homogenization buffer (as described in procedure 1) using a 4-liter Waring blender three times for 5 sec each, at a low setting. This homogenate is transported at 4° to the laboratory and then centrifuged (4000 rpm, 30 min). The supernatant is passed through glass wool to remove any fatty particles and adjusted to pH 8.0 with 1 *M* Tris base. The extract is made 18% (w/v) in ammonium sulfate and clarified by centrifugation (4000 rpm, 30 min). The supernatant is made 45% (w/v) final concentration in ammonium sulfate and the precipitate collected by centrifugation (4000 rpm, 30 min). The pellet is resuspended to give a final volume of 800 ml in TEBM buffer (see procedure 1) containing 0.02% (v/v) Triton X-100 (TEBMT). This suspension is gently homogenized (two strokes, loose-fitting 50-ml Dounce homogenizer) and after standing for 20 min clarified by centrifugation (27,000 *g*, 30 min). The supernatant is loaded onto a Sephacryl S-200 HR column (10 × 180 cm, 1 liter/hr) equilibrated in TEBMT buffer and developed overnight (at 1 liter/hr); fractions of 150–200 ml are collected. The PKC is detected by activity and the PKC-containing fractions are pooled. The pool is concentrated by application to a DEAE-cellulose column (3.5 × 10 cm; 2 ml/hr) equilibrated in TEBMT buffer and the PKC is eluted by application of TEBMT buffer containing 0.15 *M* NaCl. The eluted PKC is made 0.4 *M* in NaCl and applied to a threonine-Sepharose column (5 × 20 cm, 2 ml/min) equilibrated in 0.4 *M* NaCl TEBMT buffer which is washed with two column volumes of equilibration buffer and developed with a linear gradient (volume) of 0.4 to 1.2 *M* NaCl in TEBMT. The PKC activity elutes at the top of the gradient and the fractions for pooling are routinely determined by a combination of activity assay and gel electrophoresis; this is most conveniently achieved on a Phast gel system (Pharmacia, Piscataway, NJ). The pool is dialyzed into buffer A containing 10 m*M* benzamidine and stored at −20°.

*Comments*

This scaleup procedure permits a recovery of 10–20 mg of PKC mixture per brain (i.e., 30–60 mg for a standard three-brain preparation). The preparation is 85–90% pure as judged by SDS–gel electrophoresis. As discussed above the dimensions of the threonine-Sepharose column are important for achieving purity.

## Separation of PKC Isotypes

The separation protocol described below is designed to separate the PKC produced from procedure 1 but can also be used to separate the PKC from procedure 2. Both preparations of PKC are well suited to separation on hydroxylapatite columns since they are highly purified and therefore give separated pools which are not contaminated by different polypeptides. Also, the amount of PKC produced means that the recoveries from the hydroxylapatite column are good (~30% overall) and the isotypes are eluted at workable concentrations (~100 $\mu$g/ml).

For the separation of PKC-$\alpha$, -$\beta$, and -$\gamma$, 5 mg of PKC from procedure 1 (in 10–15 ml buffer A) is mixed with an equal volume of 20 m$M$ potassium phosphate, pH 7.5, 0.5 m$M$ EDTA, 0.5 m$M$ EGTA, 0.067% (v/v) 2-mercaptoethanol, 10% (v/v) glycerol (buffer B). This is loaded (using a Superloop) onto a hydroxylapatite column (1 × 15 cm, 10 ml/hr, DNA grade; Bio-Rad, Richmond, CA) which is preequilibrated in buffer B and attached to a fast protein liquid chromatography (FPLC) system (Pharmacia). The column is washed with 10 ml of buffer B and the PKC is eluted with a triphasic phosphate gradient: shallow from 20 to 90 m$M$ (1100 min), steeper from 90 to 150 m$M$ phosphate (50 min), and isocratic at 150 m$M$ phosphate (300 min). Fractions of 1.5 ml are collected into tubes containing 20 $\mu$l of 4% (v/v) Triton X-100. The PKC activity against histone III-S is determined and three peaks of activity are detected. These are pooled, dialyzed against buffer A, and stored at $-20°$.

The three peaks of activity represent separated PKC-$\gamma$, -$\beta$, and -$\alpha$, respectively, in the order of elution. The isotype pools are homogeneous as judged by silver-stained SDS–PAGE and are all dependent on lipids and TPA/DG for activity in the mixed micelle assay.[1] The gradient profile described permits near-baseline resolution of PKC-$\gamma$ and PKC-$\beta$ and also the elution of PKC-$\alpha$ at a reasonable concentration. With linear gradients from 20 to 150 m$M$ phosphate, we find that PKC-$\alpha$ elutes as a broad, dilute peak at the end of the run, or that PKC-$\gamma$ and -$\beta$ are not resolved. Efficient separation is dependent on the amount of PKC loaded, the column dimensions, and the flow rates, so care should be taken to optimize these. A simple linear scaleup of the column can be used for larger amounts of protein without compromising the separation.

*Comments*

The separation of the PKC isotypes is a modification of the method of Huang and colleagues[5] that permits baseline separation of PKC-$\gamma$ and -$\beta$ and the elution of PKC-$\alpha$ at a workable concentration. Incomplete separation frequently occurs if only partially purified PKC is applied to the hydroxylapatite column and it is important to monitor individual peaks with antisera.[1] Under certain conditions we have observed multiple peaks of PKC that in fact each represent the same complex mixture of isotypes.

The fractions are collected into Triton X-100 because this stabilizes the kinase as it elutes (which is at low concentration) and therefore improves recoveries. Attempts to run the hydroxylapatite column in 0.02% Triton X-100 resulted in complete loss of isotype resolution. The high-resolution

[5] K.-P. Huang, H. Nakabayashi, and Y. Yoshida, *Proc. Natl. Acad. Sci. U.S.A.* **83,** 8535 (1986).

separation of the PKC isotypes seen in this procedure can be achieved only with DNA-grade hydroxylapatite. The hydroxylapatite is prepared by swelling the matrix in water and removing fine particles which block the column and cause a loss in resolution. Any unused gel can be stored in water at 4° for up to 2 months. The column is poured (in water) the night before it is to be used and is equilibrated in buffer B overnight. The matrix tends to pack down during this process and also during loading; this should be counteracted by readjusting the flow adaptors so that they are in contact with the top of the column at the start of the wash and gradient cycles.

# [20] Purification and Analysis of Protein Kinase C Isozymes

*By* KUO-PING HUANG and FREESIA L. HUANG

Protein kinase C (PKC) consists of a family of closely related proteins that display lipid-stimulated kinase activity.[1] Two groups of PKC isozymes having different requirements for $Ca^{2+}$ have been identified by molecular cloning. The first group of PKCs, encoded by the $\alpha$, $\beta$, and $\gamma$ genes, exhibits $Ca^{2+}$/phosphatidylserine (PS)/diacylglycerol (DAG)-stimulated kinase activity and the second group, encoded by $\delta$, $\varepsilon$, and $\zeta$ genes, exhibits PS/DAG-stimulated activity. The native PKCs encoded by the $\alpha$, $\beta$, and $\gamma$ genes have been purified to near homogeneity,[2] whereas those encoded by the $\delta$, $\varepsilon$, and $\zeta$ genes have not yet been isolated. The biochemical characteristics of the latter group of PKCs have been investigated with the enzymes transiently expressed in COS cells transfected with each of these cDNAs.[3,4] Transcripts of all these enzymes are highly enriched in the central nervous system, whereas the peripheral tissues appear to express some but not all species of PKC.[1]

PKC as a collection of $Ca^{2+}$/PS/DAG-stimulated kinases has been purified from many sources. The existence of PKC isozymes was first discovered during purification of the enzyme by hydroxylapatite column chromatography.[2] The three $Ca^{2+}$/PS/DAG-stimulated kinases with corresponding $Ca^{2+}$/PS-dependent phorbol ester-binding activity peaks were designated a PKC I, II, and III. The relationships of these enzymes and

[1] Y. Nishizuka, *Nature* (*London*) **334,** 661 (1988).

[2] K.-P. Huang, H. Nakabayashi, and F. L. Huang, *Proc. Natl. Acad. Sci. U.S.A.* **83,** 8535 (1986).

[3] Y. Ono, T. Fujii, K. Ogita, U. Kikkawa, K. Igarashi, and Y. Nishizuka, *J. Biol. Chem.* **263,** 6927 (1988).

[4] D. Schaap, P. J. Parker, A. Bristol, R. Kriz, and J. L. Knopf, *FEBS Lett.* **243,** 351 (1989).

the cloned PKC cDNAs were subsequently established by (1) immunocytochemical localization of these isozymes in neurons where transcripts of specific cDNAs have been identified by *in situ* hybridization histochemistry,[5,6] (2) immunoreactivity of the isozyme-specific antibodies toward the expression products of COS cells transfected with different cDNAs,[5] and (3) hydroxylapatite chromatographic separation of PKCs expressed in COS cells transfected with different cDNAs.[7] It is certain that PKC I, II, and III are derived from $\gamma$, $\beta$, and $\alpha$ genes, respectively. The PKC II fractions from hydroxylapatite column contain both the alternatively spliced $\beta$I and $\beta$II subspecies, which are similar in many respects.[8] The histone kinase activity inherent in PKC I, II, and III accounts for greater than 90% of the total $Ca^{2+}$/PS/DAG-stimulated enzyme in rat brain extracts as determined by immunoprecipitation with a preparation of polyclonal antibody against all three enzymes.[9] PKC I, II, and III are structurally homologous and behave similarly during purification by ion-exchange, gel-filtration, hydrophobic, and affinity column chromatography. Thus, by using hydroxylapatite column chromatography as a last step of purification, these isozymes are separated and each purified in milligram quantity.

### *Materials*

The following materials were obtained from the indicated sources: Histone-IIIS, protamine, protamine sulfate, poly(lysine, serine) (3 : 1) (PLS), poly(arginine, serine) (3 : 1) (PAS), poly(lysine)-agarose, EGTA, pepstatin A, and phenylmethylsulfonyl fluoride from Sigma (St. Louis, MO); leupeptin, chymostatin, histone H1 from Boehringer Mannheim (Mannheim, Germany); phosphatidylserine (PS), lysophosphatidylserine (lyso-PS), phosphatidylcholine (PC), phosphatidylethanolamine (PE), phosphatidylglycerol (PG), phosphatidic acid (PA), cardiolipin (CL), and dioleoylglycerol (DAG) from Avanti Polar Lipids (Birmingham, AL); phorbol 12,13-dibutyrate from LC Services Corporation (Woburn, MA); phenyl-Sepharose, Mono Q column, and Nonidet P-40 (NP-40) from Pharmacia LKB Biotechnology, Inc. (Piscataway, NJ); DE-52 and GF/C glass fiber filters from Whatman (Clifton, NJ); rabbit brain myelin basic protein

[5] F. L. Huang, Y. Yoshida, H. Nakabayashi, J. L. Knopf, W. S. Young, III, and K.-P. Huang, *Biochem. Biophys. Res. Commun.* **149,** 946 (1987).

[6] S. J. Brandt, J. E. Niedel, R. M. Bell, and W. S. Young, III, *Cell (Cambridge, Mass.)* **49,** 57 (1987).

[7] U. Kikkawa, Y. Ono, K. Ogita, T. Fujii, Y. Asaoka, K. Sekiguchi, Y. Kosaka, K. Igarashi, and Y. Nishizuka, *FEBS Lett.* **217,** 227 (1987).

[8] Y. Ono, U. Kikkawa, K. Ogita, T. Fujii, T. Kurukawa, Y. Asaoka, K. Sekiguchi, K. Ase, K. Igarashi, and Y. Nishizuka, *Science* **236,** 1116 (1987).

[9] K.-P. Huang and F. L. Huang, *J. Biol. Chem.* **261,** 14781 (1986).

from Calbiochem (San Diego, CA); hydroxylapatite (BioGel HT) from Bio-Rad (Richmond, CA); and [$\gamma$-$^{32}$P]ATP and [$^3$H]phorbol 12,13-dibutyrate from Du Pont–New England Nuclear.

## Assay of Protein Kinase C

Protein kinase C activity is measured at 30° in 25 $\mu$l of 30 m*M* Tris-HCl buffer, pH 7.5, containing 6 m*M* magnesium acetate, 0.12 m*M* [$\gamma$-$^{32}$P]ATP, 0.25 m*M* EGTA, 0.4 m*M* $CaCl_2$, 1 mg/ml histone IIIS, and 40 $\mu$g/ml PS and 8 $\mu$g/ml DAG or 0.04% Nonidet P-40, 100 $\mu$g/ml PS, and 20 $\mu$g/ml DAG. PKC is added to the reaction mixture to initiate the reaction. Under no circumstances should the enzyme be preincubated with the reaction mixture containing phospholipid, which will cause irreversible inactivation of the enzyme. The mol% values of 100 $\mu$g/ml PS (0.12 m*M*) and 20 $\mu$g/ml DAG (0.03 m*M*) in 0.04% Nonidet P-40 (0.66 m*M*) are approximately 15 and 4, respectively. The $Ca^{2+}$/PS-independent activity is measured in the presence of 2 m*M* EGTA without $Ca^{2+}$ and lipids. Phospholipids and DAG in chloroform are evaporated to dryness under $N_2$ and resuspended in 20 m*M* Tris-HCl buffer, pH 7.5, with sonication and vortexing, whereas CL is dissolved in dimethyl sulfoxide (DMSO). Inclusion of DMSO or ethanol up to 5% does not significantly influence the kinase activity when assayed with PS/Nonidet P-40 mixed micelles. Measurements of $^{32}$P incorporation into protein is done by spotting 20 $\mu$l of the reaction mixture to an ITLC (Gelman instant thin-layer chromatography sheet, type SG) strip (1 × 9.5 cm) 1.5 cm from the bottom previously spotted with 20 $\mu$l of 15% (w/v) trichloroacetic acid (TCA) containing 50 m*M* ATP and followed by chromatography for 6 min in a beaker containing 5% TCA and 0.2 *M* KCl (with protamine as substrate, 30% TCA solution is used). After the strips are air dried, the origin (area approximately 1 cm below and 1.5 cm above the line of origin), which contains the phosphorylated protein, is excised for counting in a scintillation counter. One unit of protein kinase activity is defined as the amount of enzyme catalyzing the incorporation of 1 nmol of phosphate from ATP into protein substrate per minute.

### *Buffers Used in Purification of PKC Isozymes*

Homogenizing buffer: 20 m*M* Tris-HCl, pH 7.5, containing 1 m*M* dithiothreitol (DTT), 5 m*M* EGTA, 2 m*M* EDTA, 0.5% Nonidet P-40, 0.5 m*M* phenylmethylsulfonyl fluoride (PMSF), and 2 $\mu$g/ml each of leupeptin, chymostatin, and pepstatin A.

Buffer A: 20 m*M* Tris-HCl, pH 7.5, containing 1 m*M* dithiothreitol, 0.5 m*M* EGTA, 0.5 m*M* EDTA, and 10% (v/v) glycerol

Buffer B: 20 m$M$ $KPO_4$, pH 7.5, containing 1 m$M$ dithiothreitol, 0.5 m$M$ EGTA, 0.5 m$M$ EDTA, and 10% glycerol

Buffer C: 300 m$M$ $KPO_4$, pH 7.5, containing 1 m$M$ dithiothreitol, 0.5 m$M$ EGTA, 0.5 m$M$ EDTA, and 10% glycerol

*Purification Procedure*

*Preparation of Crude Brain Extracts.* Fresh rat brains (120 g wet wt) from 80 male Sprague-Dawley rats (200–250 g) are homogenized in 600 ml of ice-cold homogenizing buffer using a Polytron at setting 5 with four 15-sec bursts. The homogenate is centrifuged at 34,000 rpm at 4° for 1 hr using Beckman 35 rotors. The supernatant fluid is decanted carefully to avoid the turbid fluffy layer. The combined fluffy layer and the pellet are extracted once again with 400 ml of the homogenizing buffer.

*DEAE-Cellulose Column Chromatography.* The combined high-speed supernatant fluid is adjusted to pH 7.5 by adding solid Tris and applied to a DEAE-cellulose (DE-52) column (4.0 × 16 cm) equilibrated with buffer A. The surface of the column is periodically stirred up gently with a glass rod to avoid clotting of the DEAE-cellulose. The column is washed with 1 liter of buffer A and PKC is eluted with a KCl gradient delivered by an LKB-Ultrograd gradient maker at a flow rate of 120 ml/hr. The gradient is designed in such a way so that the kinase is eluted in a stepwise fashion with 0.15 $M$ KCl in buffer A for 7 hr and then a gradual increase to 0.3 $M$ KCl over 4 hr. PKC eluted with 0.15 $M$ KCl is pooled (700–800 ml) and concentrated by ultrafiltration with three 400-ml Amicon (Danvers, MA) cells fitted with YM10 membranes.

*Phenyl-Sepharose Column Chromatography.* To the concentrated solution (approximately 200 ml), solid KCl is added to a final concentration of 1.5 $M$ and applied to a phenyl-Sepharose column (2.0 × 12 cm) equilibrated with buffer A containing 1.5 $M$ KCl. The column is washed with 100 ml of the equilibration buffer and the kinase is eluted with a 1.5–0 $M$ KCl gradient over 5 hr at a flow rate of 100 ml/hr. For reproducibility and high yield, it is recommended that new phenyl-Sepharose should be used for each experiment.

*Gel-Filtration Chromatography on Sephacryl S-200.* Pooled PKC-containing fractions from phenyl-Sepharose column chromatography are concentrated by ultrafiltration with Amicon cells fitted with YM10 membranes. The concentrated solution (7–10 ml) is applied to a Sephacryl S-200 column (2.5 × 95 cm) equilibrated with buffer A. Fractions of 3.5 ml are collected.

*Poly(lysine)-Agarose Column Chromatography.* PKC from Sephacryl S-200 column chromatography (40–50 ml) is applied to a poly(lysine)-

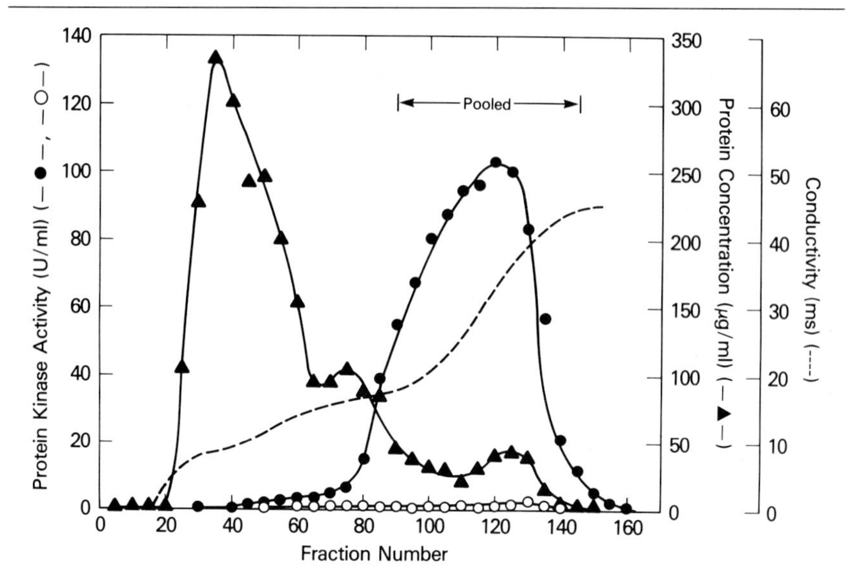


FIG. 1. Elution profile of protein kinase C (PKC) from poly(lysine)-agarose column. PKC purified from Sephacryl S-200 column was applied to a poly(lysine)-agarose column (1.5 × 14 cm) equilibrated with buffer A. Elution of the enzyme was done with a 0–0.8 *M* KCl gradient (– – –) in buffer A delivered by an LKB Ultrograd gradient maker at a flow rate of 100 ml/hr. Fractions of 5 ml were collected for the measurement of PKC activity in the presence of $Ca^{2+}$/phosphatidylserine/diacylglycerol (●) or EGTA (○). Protein concentrations were determined by dye binding and expressed as micrograms per milliliter (▲).

agarose column (1.5 × 14 cm) equilibrated with buffer A. The column is washed with 50 ml of buffer A and eluted with a 0–0.3 *M* KCl gradient over 3.5 hr and 0.3–0.8 *M* KCl over 2 hr at a flow rate of 100 ml/hr. Fractions of 5 ml are collected. The kinase is eluted as a broad peak and the enzyme eluted at the higher end of the KCl gradient (fractions 110–140) is near homogeneous (Fig. 1). This purification step is highly reproducible and the column can be reused many times following washing with buffer A containing 1.5 *M* KCl after each run.

*Hydroxylapatite Column Chromatography.* PKC obtained from poly-(lysine)-agarose column chromatography (200–250 ml) is concentrated in Amicon ultrafiltration cells fitted with YM5 membranes. It should be noted that extensive concentration and mixing of the solution at this stage may cause irreversible aggregation of the enzyme. The protein concentration should be maintained at less than 1 mg/ml. The enzyme solution is exchanged with buffer B in the Amicon cell by repetitive dilution and concentration. The enzyme (10–15 ml) is applied to a hand-packed hydroxylapatite column (1.0 × 7 cm) equilibrated with buffer B at a flow rate of 0.5

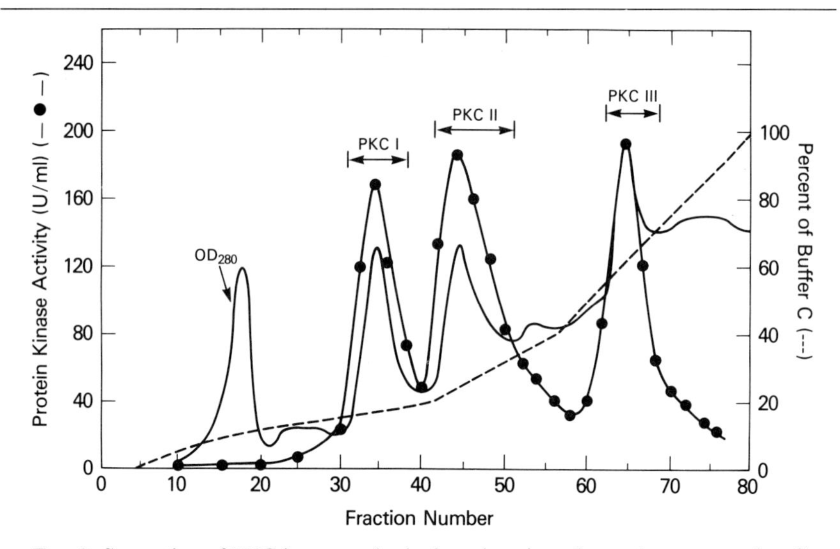


FIG. 2. Separation of PKC isozymes by hydroxylapatite column chromatography. Concentrated and dialyzed PKC (9 ml in buffer B) from poly(lysine)-agarose column was applied to a hand-packed hydroxylapatite column (1.0 × 7 cm) attached to a Pharmacia FPLC unit at a flow rate of 0.5 ml/min. The column was washed with 10 ml of buffer B and the elution was initiated by the program described in the text. Fractions of 2.5 ml were collected for the measurement of PKC activity with $Ca^{2+}$/PS/DAG (●). Absorbance at $OD_{280}$ (full scale, 0.5 arbitrary unit) and percentage of buffer C increment are as indicated (---). Fractions containing PKC I (31–38), II (42–50), and III (62–68) were concentrated by ultrafiltration fitted with Amicon YM5 membrane and dialyzed against buffer A.

ml/min controlled by a Pharmacia fast protein liquid chromatography (FPLC) system. Following washing with 10 ml buffer B, chromatography is carried out by using buffer B and buffer C according to the following program: 0–10 ml, buffer C is maintained at 0%; 10–40 ml, buffer C is increased to 10%; 40–100 ml, buffer C is increased to 20%; 100–140 ml, buffer C is increased to 40%; 140–200 ml, buffer C is increased to 100%; and 200–220 ml, buffer C is kept at 100%. Three kinase activity peaks, PKC I, II, and III, are eluted at 18, 25, and 60% buffer C, respectively (Fig. 2). The performance of the hand-packed column can be improved by adjusting the liquid volume above the packed material to a minimum. The pooled enzymes are concentrated in Amicon ultrafiltration cells fitted with YM5 membranes and dialyzed against buffer A.

*Yield and Purity*

Table I summarizes the results of a typical purification experiment using 120 g of fresh rat brain obtained from 80 rats. The recoveries of PKC I, II, and III range between 5 and 10% with the specific activity of

TABLE I
PURIFICATION OF RAT BRAIN PKC ISOZYMES

| Fraction | Total protein (mg) | Specific activity (units/mg) | Recovery (%) | Relative purification (-fold) |
|---|---|---|---|---|
| Crude extract | 6250 | 7.2 | 100 | 1 |
| DEAE-cellulose | 690 | 60 | 92 | 8 |
| Phenyl-Sepharose | 170 | 180 | 68 | 25 |
| Sephacryl S-200 | 76 | 360 | 60 | 50 |
| Poly(lysine)-agarose | 14 | 1460 | 45 | 202 |
| Hydroxylapatite | | | | |
| PKC I | 1.4 | 2100 | 6.5 | 290 |
| PKC II | 1.6 | 2420 | 8.6 | 336 |
| PKC III | 1.1 | 3200 | 7.8 | 444 |

2000–3000 U/mg. The purified enzymes when stored in the concentrated form (between 0.2 and 1 mg/ml) at −70° are stable for several months without loss of enzymatic activity. Milligram quantities of each isozyme with a purity of greater than 95%, as determined by SDS–PAGE, can be obtained within 1 week. In spite of the fact that a major Coomassie Blue-stained protein band is detected for all these isozymes by SDS–PAGE, multiple bands are observed by isoelectric focusing with a pH gradient of 3–10. All of the heterogeneous protein bands from each isozyme are reactive with each of their respective isozyme-specific antibodies, indicating that these proteins either have undergone posttranslational modification or consist of heterogeneous subspecies such as the alternatively spliced $\beta$I and $\beta$II. The cause of this microheterogeneity for these isozymes has not been worked out; nevertheless, phosphorylation has been implicated.

*Comment on Procedure*

Purification of PKC has been described previously by many groups, including those methods described by Nishizuka and co-workers in three previous volumes of this series.[10–12] The method described in this chapter employs poly(lysine)-agarose column chromatography to achieve a high degree of purity and hydroxylapatite column chromatography to separate the PKC isozymes. The lengthy procedure is designed for the preparation

[10] U. Kikkawa, R. Minakuchi, Y. Takai, and Y. Nishizuka, this series, Vol. 99, p. 288.
[11] T. Kitano, M. Go, U. Kikkawa, and Y. Nishizuka, this series, Vol. 124, p. 349.
[12] M. Go, J. Koumoto, U. Kikkawa, and Y. Nishizuka, this series, Vol. 141, p. 424.

of nearly homogeneous preparation of each isozyme in milligram quantities. We have included 0.5% Nonidet P-40 in the homogenization buffer to extract the cytosolic as well as the particulate fraction-associated PKC. This procedure has also been used successfully for the purification of the monkey brain PKC isozymes and is likely to be applicable for the same enzymes from other sources. For the identification and preparation of small amounts of each isozyme, the cellular extracts are initially fractionated by chromatography on a Mono Q column (HR 5/5, Pharmacia) equilibrated with buffer A and eluted with a 0–0.4 *M* KCl linear gradient. The enzyme obtained from the Mono Q column is concentrated and exchanged with buffer B in an ultrafiltration cell fitted with Amicon YM10 membrane. PKC isozymes can be separated by a small, hand-packed hydroxylapatite column (0.5 × 7 cm) or a Koken (Koken Co., Ltd., Tokyo, Japan) high-performance liquid chromatography (HPLC) hydroxylapatite column. It should be noted that not all the commercially available HPLC hydroxylapatite columns are equally effective for the separation. For the sake of economy and reproducibility, we recommend the hand-packed BioGel HT hydroxylapatite column.

### *Analysis of the Purified PKC Isozymes*

*Differential Sensitivity to Tryptic Proteolysis*. Incubation of the purified enzymes with trypsin (PKC/trypsin = 100) in one-half strength buffer A at 30° results in a rapid degradation of PKC I and II, whereas PKC III is relatively resistant to tryptic digestion.[13] The half-life of PKC I is less than 1 min and that for PKC II is approximately 5 min. This method can be used to selectively destroy PKC I and II in a mixture containing all three isozymes. Degradation of PKC III can be facilitated by the addition of $Ca^{2+}$/PS/DAG.[13] Two major proteolytic fragments of 45–48 kDa, the catalytic domain, and of 33–38 kDa, the regulatory domain, are generated from all three PKCs. For the preparation of the catalytic and regulatory fragments, for instance, from PKC II, the enzyme is digested with trypsin and the reaction is terminated by the addition of soybean trypsin inhibitor [inhibitor/trypsin (w/w) = 500]. The reaction mixture is then applied to a Mono Q column equilibrated with buffer D (buffer A plus 10% ethylene glycol and 0.2% Nonidet P-40). The column is eluted with a 0–0.7 *M* KCl linear gradient in buffer D at a flow rate of 0.5 ml/min over 40 min and fractions of 0.5 ml are collected. The regulatory and catalytic fragments are eluted at 0.2 and 0.4 *M* KCl, respectively (Fig. 3). Since the regulatory fragment is relatively hydrophobic, it is necessary to include detergent in

[13] F. L. Huang, Y. Yoshida, J. R. Cunha-Melo, M. A. Beaven, and K.-P. Huang, *J. Biol. Chem.* **264,** 4238 (1989).

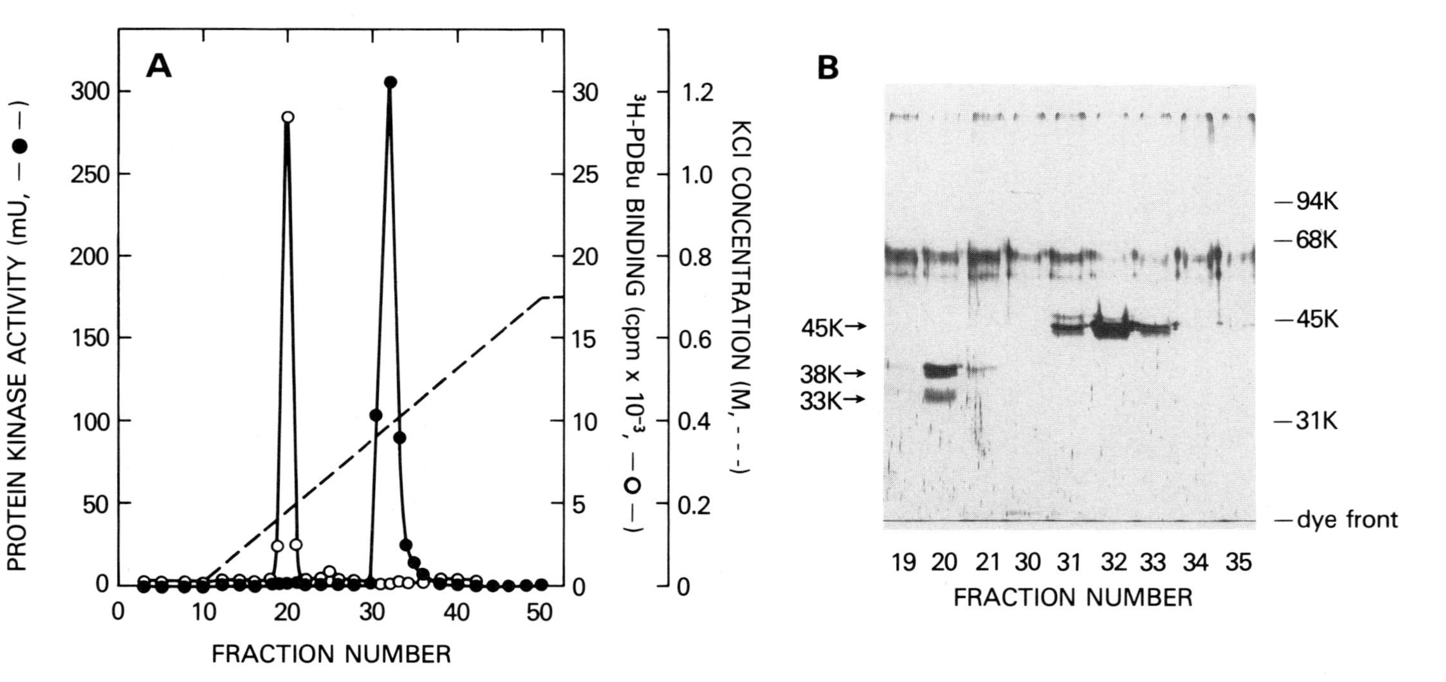


FIG. 3. Separation of the proteolytic fragments of PKC II by chromatography on Mono Q column. PKC II (50 $\mu$g/ml) was incubated with *N*-tosyl-1-phenylalanine chloromethyl ketone (TPCK)-treated trypsin (0.625 $\mu$g/ml) in 10 m*M* Tris-HCl buffer, pH 7.5, containing 0.25 m*M* EDTA, 0.25 m*M* EGTA, 0.5 m*M* dithiothreitol, 5% glycerol, and 0.5 m*M* $CaCl_2$ at 30° for 7 min. The reaction was terminated by the addition of soybean trypsin inhibitor. The reaction mixture was applied to a Mono Q column equilibrated with buffer A containing 10% ethylene glycol and 0.2% Nonidet P-40. (A) Chromatography was carried out by using 0–0.7 *M* KCl linear gradient at a flow rate of 0.5 ml/min and fractions of 0.5 ml were collected for the assay of protein kinase activity (●) in the presence of $Ca^{2+}$/PS/DAG and the binding of [$^3$H]PDBu in the presence of PS (○). The kinase activity in the presence of EGTA was equal to that in the presence of activators. (B) Protein staining (using Gelcode from Upjohn) of fractions containing phorbol ester-binding (lanes 19–21) and protein kinase (lanes 30–35) activities.

the elution buffer to isolate both fragments. In the absence of the detergent, only the catalytic fragment can be eluted from the column.

*Differential Sensitivity of PKC Isozymes to Phospholipid-Induced Inactivation.* Acidic phospholipids, such as PS, PA, PG, and CL, support the activation of PKC by $Ca^{2+}$/DAG. However, incubation of these phospholipids with PKC in the absence of a divalent cation results in the differential inactivation of PKCs.[14] The PS-induced inactivation of PKC is measured in one-half strength buffer A containing 0.25 mg/ml bovine serum albumin, 20–50 $\mu$g/ml of PKC, and 10 $\mu M$ PS vesicles at 30° for 3 min. The kinase activity with protamine sulfate as substrate without $Ca^{2+}$ and lipid cofactor is reduced by greater than 90% for PKC I, whereas PKC II and III are only slightly inactivated. Among the various phospholipids tested, PS appears to be the most potent; the neutral phospholipids, such as PC and PE, are ineffective. Inactivation of PKCs by PS is dose and time dependent and the resulting inactivated enzyme cannot be reactivated by removal of PS by detergent or by digestion of PS with bacterial phospholipase C. Since PKC I is most sensitive to PS-induced inactivation, this method can also be used for the selective destruction of PKC I. These findings also indicate the ill effects of preincubation of PKCs with PS during the kinase activity measurement. The phospholipid-induced inactivation of PKC is most effective with pure PS vesicles but much less with PS/Nonidet P-40 mixed micelles or PS/PC (1 : 4) mixed vesicles. Hence, assay of PKC activity with PS/Nonidet P-40 mixed micelles or PS/PC mixed vesicles is preferable when prolonged incubation is required. The acidic phospholipid-induced inactivation of PKC in the absence of divalent cation is due to a direct interaction of the lipid with the catalytic domain; this interaction can be retarded by divalent cations. The PS vesicles also interact with the regulatory domain either in the presence or absence of divalent cation, however, without affecting the phorbol ester-binding activity.

*Differential Sensitivity of PKC Isozymes to Activation by Phospholipids.* The phospholipid/detergent mixed micelles assay method is used in these experiments.[15] All the phospholipids, except CL, which is dissolved in DMSO, are suspended in 20 m$M$ Tris-HCl buffer, pH 7.5, with sonication and vortexing. The lipid/detergent mixed micelles are prepared by adding Nonidet P-40 to glass test tubes containing DAG or phorbol ester previously evaporated off ethanol and followed by the addition of phospholipid. In the presence of 4 mol% DAG or 1.6 $\mu M$ PDBu, all three kinases

[14] K.-P. Huang and F. L. Huang, *J. Biol. Chem.* **265,** 738 (1990).

[15] K.-P. Huang, F. L. Huang, H. Nakabayashi, and Y. Yoshida, *J. Biol. Chem.* **263,** 14839 (1988).

TABLE II
KINETIC PARAMETERS OF PKC ISOZYMES WITH DIFFERENT PROTEIN SUBSTRATES[a]

| | PKC I | | PKC II | | PKC III | |
|---|---|---|---|---|---|---|
| Protein substrate | $V_{max}$ (units/mg) | $K_m$ ($\mu$g/ml) | $V_{max}$ (units/mg) | $K_m$ ($\mu$g/ml) | $V_{max}$ (units/mg) | $K_m$ ($\mu$g/ml) |
| Histone IIIS | 2,301 ± 126 | 139 ± 20 | 5,238 ± 330 | 176 ± 27 | 10,852 ± 1,247 | 346 ± 42 |
| MBP | 4,229 ± 149 | 75 ± 7 | 9,908 ± 880 | 167 ± 31 | 10,239 ± 329 | 59 ± 5 |
| PLS | 650 ± 22 | 24 ± 2 | 2,321 ± 75 | 63 ± 9 | 2,720 ± 235 | 61 ± 9 |
| Protamine | 464 ± 22 | 32 ± 2 | 1,931 ± 55 | 48 ± 3 | 1,919 ± 51 | 39 ± 2 |

[a] Protein kinase activity was measured with mixed micelles assay containing 15 mol% PS, 4 mol% DAG, 0.01–1.0 mg/ml of histone IIIS, myelin basic protein, or protamine, and 18, 12, and 8 ng of PKC I, II, and III, respectively. Protein kinase activity with PLS as substrate was measured under the same condition without $Ca^{2+}$/PS/DAG but containing 2 m$M$ EGTA. The kinetic parameters were expressed as mean ± SE. PLS, Poly(lysine, serine) (3 : 1); MBP, myelin basic protein.

are maximally stimulated by PS or CL. Other phospholipids, such as PA, PG, lyso-PS, and PI are less effective, and PC and PE are ineffective. With the exception of CL, a similar order of preference for these phospholipids is also observed for [$^3$H]phorbol 12,13 dibutyrate (PDBu) binding with each of the three PKC isozymes. In the presence of 15 mol% CL, binding of [$^3$H]PDBu is less than 10% of that with PS for all three isozymes. The kinase activities of these enzymes can, however, be distinguished by assay in the presence of CL without DAG or PDBu. In the presence of 15 mol% CL alone, PKC I activity is approximately 60% of that with CL and DAG. In comparison, PKC II and III activities are less than 10% of their respective activities with CL and DAG.

*Stimulation of PKC Isozymes by DAG and PDBu.* All three PKCs are stimulated by saturated levels of DAG or PDBu to a similar extent; however, the $A_{1/2}$ values of these activators for these isozymes are different. The $A_{1/2}$ values of PDBu in the presence of 15 mol% PS for PKC I, II, and III are 14.3 ± 1.1, 43.7 ± 4.2, and 52.4 ± 3.8 n$M$, respectively. The $A_{1/2}$ values of DAG for PKC I, II, and III are 0.24 ± 0.01, 0.59 ± 0.08, and 0.41 ± 0.05 mol%, respectively. When the kinase activity is measured in the presence of CL/Nonidet P-40 mixed micelles, the $A_{1/2}$ values of PDBu are 13.6 ± 4.8, 228 ± 21, and 96 ± 14 n$M$ and those of DAG are 0.4 ± 0.16, 8.3 ± 0.6, and 3.2 ± 0.5 mol% for PKC I, II, and III, respectively. The $A_{1/2}$ value of PDBu or DAG for PKC I is not significantly different when assayed with either PS or CL. In comparison, the $A_{1/2}$ values of PDBu or DAG for PKC II and III are significantly higher in the presence of CL than PS. Thus, the $A_{1/2}$ value of DAG or PDBu for PKC I differs most significantly with PKC II and III in the presence of CL/Nonidet P-40 mixed micelles. Overall, PKC I appears to be most sensitive to activation by DAG and PDBu.

*Comparison of Protein Substrate Specificities of PKC Isozymes.* All three PKCs phosphorylate histone IIIS and myelin basic protein dependent on $Ca^{2+}$/PS/DAG. Phosphorylation of PLS, PAS, and protamine sulfate is independent of all the effectors. Peptide mapping analysis of histone H1 and myelin basic protein phosphorylated by the three PKCs revealed the same sites of phosphorylation. Analysis of the kinetic parameters of the PKC isozyme-catalyzed phosphorylation of the various protein substrates shows some differences among them. The $K_m$ values of histone IIIS for PKC I and II are approximately one-half of that for PKC III. The $K_m$ value of myelin basic protein for PKC II is almost twice as much as those for PKC I and III. The $K_m$ values of protamine for all three PKCs are comparable. Among the four protein substrates tested, both PKC I and II have the highest $V_{max}$ values with myelin basic protein, whereas the $V_{max}$ values with histone IIIS and myelin basic protein are almost equivalent for PKC III. The $V_{max}$ values of PLS and protamine for all three PKCs range between 20 and 40% of those with histone IIIS. In general, the $K_m$ values of various protein substrates for PKC I are less than those for PKC II and III (Table II).

# [21] *Xenopus* Ribosomal Protein S6 Kinase II

*By* ELEANOR ERIKSON, JAMES L. MALLER, and R. L. ERIKSON

In stage VI *Xenopus* oocytes, ribosomal protein S6 is almost completely unphosphorylated.[1] In contrast, in oocytes that have progressed through germinal vesicle breakdown and completed maturation or in ovideposited unfertilized eggs, all the S6 is in the maximally phosphorylated form, containing 4–5 mol of phosphate/mol of S6.[1] Extracts of unfertilized eggs display a 10- to 20-fold increase in S6 kinase activity that can be resolved into two peaks by DEAE-Sephacel chromatography.[2] Purification of the S6 kinase activity from the second peak, S6 kinase II, is described here. The purified enzyme has been used to generate peptide sequence information for molecular cloning. The data on the biochemical analysis

[1] P. J. Nielsen, G. Thomas, and J. L. Maller, *Proc. Natl. Acad. Sci. U.S.A.* **79,** 2937 (1982).
[2] E. Erikson and J. L. Maller, *J. Biol. Chem.* **261,** 350 (1986).

and cDNA clones suggest that the gene(s) for this enzyme(s) which we have denoted *rsk*, for ribosomal S6 kinase, may encode a family of related proteins.[3-5]

## Assay

### *Principle*

*Xenopus* S6 kinase II catalyzes the transfer of the $\gamma$ phosphate of ATP to serine residues in the S6 protein of 40S ribosomal subunits. Direct analysis of the phosphorylation sites by sequencing radiolabeled S6 peptides[6] demonstrates that this enzyme phosphorylates the following serine residues (shown as italic characters) in the carboxyl region of S6: 232-RRL*SS*LRA*S*TSK*S*-244.[7,8] The possibility that serine-247 is also phosphorylated by S6KII is still under investigation. The nomenclature and identification of these residues is based on the assumption that the sequence of *Xenopus* S6 is identical to that of rat liver S6 (sequence RRLSSLRA was identified but the remainder of sequence 240–249 was not).

#### *Reagents*

[$\gamma$-$^{32}$P]ATP, 1 m*M*, specific activity 1–5 cpm/fmol
40S ribosomal subunits, ~4 to 5 $\mu M$
Twofold concentrated assay buffer (40 m*M* HEPES, pH 7.0, 20 m*M* $MgCl_2$, 6.6 m*M* 2-mercaptoethanol, 200 $\mu$g of bovine serum albumin/ml, 6.7 $\mu$g of the heat-stable inhibitor of the cAMP-dependent protein kinase/ml)
Fivefold concentrated electrophoresis sample buffer [350 m*M* Tris-HCl, pH 6.8, 10% (v/v) glycerol, 15% (w/v) SDS, 0.01% Bromphenol Blue, 25% (v/v) 2-mercaptoethanol]

[3] S. W. Jones, E. Erikson, J. Blenis, J. L. Maller, and R. L. Erikson, *Proc. Natl. Acad. Sci. U.S.A.* **85,** 3377 (1988).
[4] D. A. Alcorta, C. M. Crews, L. J. Sweet, L. Bankston, S. W. Jones, and R. L. Erikson, *Mol. Cell. Biol.* **9,** 3850 (1989).
[5] L. J. Sweet, D. A. Alcorta, S. W. Jones, E. Erikson, and R. L. Erikson, *Mol. Cell. Biol.* **10,** 2413 (1990).
[6] R. E. H. Wettenhall, R. H. Aebersold, L. E. Hood, and S. H. B. Kent, this series, Vol. 201 [15].
[7] R. E. H. Wettenhall, E. Erikson, and J. L. Maller, in preparation.
[8] Amino acids are identified by the single-letter code.

### *Procedure*

Fifteen microliters of twofold concentrated assay buffer is added to each tube, an appropriate amount of enzyme is added, and the reactions are started by the addition of a mixture of 40S subunits, [$\gamma$-$^{32}$P]ATP, and $H_2O$ to bring the final concentration of ATP and 40S subunits to 100 and 0.4 $\mu M$, respectively, and the final volume to 30 $\mu$l. Reactions are incubated at 30° for 15 min and are stopped by the addition of 8 $\mu$l of fivefold concentrated electrophoresis sample buffer. After incubation at 95° for 2 min the products of the reaction are resolved by electrophoresis through SDS–12.5% polyacrylamide gels according to Laemmli.[9] The gels are stained, destained, dried, and radiolabeled S6 is visualized by autoradiography. Radioactivity in S6 is quantified by liquid scintillation spectrometry of the excised bands.

## Protein Purifications

### *Protein Kinase Inhibitor*

The heat-stable inhibitor protein of the cAMP-dependent protein kinase is prepared by the method of Whitehouse and Walsh.[10]

### *40S Ribosomal Subunits*

The method described here, using *Xenopus laevis* ovary, is a slight modification of that developed for *Artemia salina* cysts.[11] In this and the following section all steps are performed a 0–4°, except as indicated, and the pH of all buffers is adjusted at ambient temperature. The ovary is removed from a mature animal into 10 m$M$ HEPES, 83 m$M$ NaCl, 1 m$M$ $MgCl_2$, 0.5 m$M$ $CaCl_2$, 1 m$M$ KCl, pH 7.9, and minced. Thirty grams of tissue (frozen tissue may be used) is homogenized in 4 vol of 35 m$M$ Tris-HCl, 50 m$M$ KCl, 1.5 m$M$ $MgCl_2$, 250 m$M$ sucrose, pH 7.6, by five passes in a motor-driven Potter–Elvehjem homogenizer. The homogenate is filtered through glass wool and centrifuged at 44,000 $g$ for 15 min. The lipid cap is removed and the supernatant is centrifuged at 220,000 $g_{max}$ for 2 hr. The supernatant is discarded, the tubes and pellets are rinsed with the homogenization buffer, and the tubes are dried. The pellets are overlaid with a total of 10 ml of 50 m$M$ Tris-HCl, 13 m$M$ $MgCl_2$, 725 m$M$ KCl, 3 m$M$ 2-mercaptoethanol, pH 7.6, and kept for 15 hr. The pellets are

[9] U. K. Laemmli, *Nature (London)* **227,** 680 (1970).

[10] S. Whitehouse and D. A. Walsh, this series, Vol. 99, p. 80.

[11] M. Zasloff and S. Ochoa, this series, Vol. 30, p. 197.

resuspended gently and then shaken slowly for 6–8 hr. The preparation is homogenized gently, incubated at 30° for 15 min, and centrifuged at 3000 *g* for 5 min. The supernatant is layered onto six 34-ml 5–20% (w/w) sucrose gradients containing 50 m*M* Tris-HCl, 700 m*M* KCl, 15 m*M* $MgCl_2$, 10 m*M* 2-mercaptoethanol, pH 7.8, and centrifuged at 21,400 rpm in a SW28 Beckman rotor for 15 hr. Fractions are collected and the absorbance at 260 nm is determined. The fractions containing the 40S subunits are pooled and diluted with 8 vol of 50 m*M* Tris-HCl, 10 m*M* $MgCl_2$, 5 m*M* 2-mercaptoethanol, pH 7.8, 0.2 vol of ice-cold ethanol is added with swirling,[12] and after 1 hr the suspension is centrifuged at 10,000 *g* for 15 min. The supernatant is discarded and the precipitate is gently resuspended in approximately 1 ml of 50 m*M* Tris-HCl, 200 m*M* KCl, 10 m*M* magnesium acetate, 0.1 m*M* EDTA, 20 m*M* 2-mercaptoethanol, 250 m*M* sucrose, pH 7.8. The ribosome concentration is calculated according to the formula 17.2 $OD_{260}$ units = 1 nmol of 40S subunits.[13] Aliquots of the suspension are frozen in dry ice–ethanol and stored at −70°. Before use, 1 vol of ethylene glycol is added to the thawed suspension, and thereafter it is stored at −20°. Yield is ~12 nmol/30 g of ovary.

## S6 Kinase II

*Reagents*

Buffer A: 20 m*M* Tris-HCl, 5 m*M* $MgCl_2$, 2 m*M* EGTA, 2 m*M* dithiothreitol, pH 7.1

Buffer B: 55 m*M* β-glycerophosphate, 5 m*M* $MgCl_2$, 5 m*M* EGTA, 100 m*M* NaCl, 2 m*M* dithiothreitol, 0.05% Brij 35, 10% (v/v) ethylene glycol, pH 6.8

Buffer C: 10 m*M* potassium phosphate, 5 m*M* $MgCl_2$, 2 m*M* EGTA, 2 m*M* dithiothreitol, 0.01% Brij 35, 10% (v/v) ethylene glycol, pH 6.8

Buffer D: Buffer A plus 0.01% Brij 35 (Pierce Chemical Co., Rockford, IL) and 10% (v/v) ethylene glycol

Buffer E: 20 m*M* HEPES, 5 m*M* $MgCl_2$, 2 m*M* EGTA, 2 m*M* dithiothreitol, 0.01% Brij 35, 10% (v/v) ethylene glycol, pH 7.0

*Homogenization and Ammonium Sulfate Precipitation.* For production of eggs *X. laevis* females are injected with 75 IU of pregnant mare serum gonadotropin and, 3–4 days later, with 800 IU of human chorionic gonadotropin 12 hr before eggs are desired. Eggs are ovideposited and collected at ambient temperature in tap water containing 0.1 *M* NaCl. Pooled eggs are dejellied in 2% cysteine, pH 7.8 and washed extensively

[12] M. S. Kaulenas, *Anal. Biochem.* **41,** 126 (1971).

[13] C. Darnbrough, S. Legon, T. Hunt, and R. J. Jackson, *J. Mol. Biol.* **76,** 379 (1973).

with 50 m$M$ Tris-HCl, 0.1 $M$ NaCl, pH 7.0. Approximately 80 ml of dejellied eggs is homogenized in 3 vol of 55 m$M$ $\beta$-glycerophosphate, 5 m$M$ $MgCl_2$, 5 m$M$ EGTA, 50 m$M$ NaF, 100 m$M$ $Na_4P_2O_7$, 5 m$M$ phosphotyrosine, 2 m$M$ dithiothreitol, 0.1 m$M$ phenylmethylsulfonyl fluoride, 10 $\mu$g of leupeptin/ml, pH 6.8, by five passes with a hand-operated Dounce homogenizer. The homogenate is centrifuged at 8000 $g$ for 15 min, the lipid layer removed, and the supernatant centrifuged at 44,000 $g$ for 30 min. Solid ammonium sulfate is added slowly with stirring to the supernatant to 20% saturation (114 mg/ml), and after 1 hr the suspension is centrifuged at 17,000 $g$ for 20 min. Solid ammonium sulfate is added slowly with stirring to the supernatant to 40% saturation (an additional 123 mg/ml), and after 1 hr the suspension is centrifuged at 17,000 $g$ for 20 min. The precipitate is suspended in 55 m$M$ $\beta$-glycerophosphate, 5 m$M$ $MgCl_2$, 5 m$M$ EGTA, 2 m$M$ dithiothreitol, 0.1 m$M$ phenylmethylsulfonyl fluoride, pH 6.8 (1/10 vol of the amount of homogenization buffer used) with the aid of a hand-operated homogenizer and dialyzed overnight vs the same buffer (twice, 1 liter each). The dialyzed sample is centrifuged at 12,000 $g$ for 10 min. The precipitate is discarded, and the supernatant can be frozen, if desired. If the preparation is to be frozen ethylene glycol is added to 10% (v/v), the preparation is frozen in dry ice–ethanol, and stored at $-70°$. After thawing, phenylmethylsulfonyl fluoride and leupeptin are added to 0.1 m$M$ and 10 $\mu$g/ml, respectively, and the preparation is centrifuged at 17,000 $g$ for 20 min.

*DEAE-Sephacel Chromatography.* The supernatants from two dialyzed 20–40% ammonium sulfate fractions (~500 mg of protein, representing 160 ml of dejellied eggs) are adjusted to a conductivity of 4.0 mmho with distilled water. This preparation is applied at a flow rate of 40 ml/hr to a column of DEAE-Sephacel (2.5 × 25 cm) previously equilibrated with buffer A containing 25 m$M$ NaCl, and proteins are eluted with a 900-ml linear gradient of NaCl (25 to 700 m$M$) in buffer A at a flow rate of 50 ml/hr. Fractions of 5 ml are collected into tubes containing 0.5 ml 150 m$M$ $\beta$-glycerophosphate, 5 m$M$ dithiothreitol, pH 6.8 and samples of 1 $\mu$l are assayed for S6 kinase activity. Two peaks of activity are detected, eluting between 35 and 70 m$M$ NaCl, and 95 and 120 m$M$ NaCl, as shown in Fig. 1. Fractions are pooled on the basis of activity and the absorbance profile. The pooled fractions can be frozen at this point. Ethylene glycol is added to 10% (v/v), the preparation is frozen in dry ice–ethanol, and stored at $-70°$. Further purification of the S6 kinase in the second peak (S6 kinase II) is described below.

*Sephacryl S-300 Chromatography.* The frozen pools of peak II from two separate DEAE columns are thawed, phenylmethylsulfonyl fluoride and leupeptin are added to 0.1 m$M$ and 10 $\mu$g/ml, respectively, and the

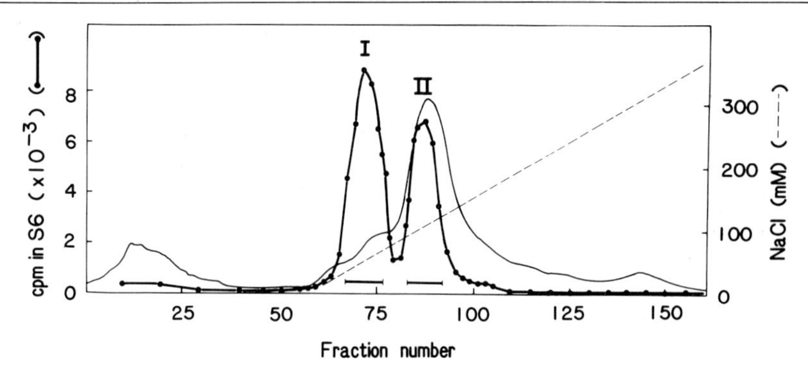


FIG. 1. DEAE-Sephacel chromatography of the dialyzed 20–40% ammonium sulfate fraction. The column was equilibrated and eluted as described in the text. Samples of 1 $\mu$l were assayed for S6 kinase activity and fractions comprising peak I and peak II were pooled as indicated by the solid bars. Total recovery of activity from the column was 35%. $A_{278}$ (——).

preparation is centrifuged at 17,000 *g* for 20 min. One volume of cold, neutral, saturated ammonium sulfate (767 mg/ml) is added to the supernatant slowly with stirring and after 1 hr the suspension is centrifuged at 20,000 *g* for 20 min.The precipitate is gently resuspended in 1 ml of buffer B, and after 20 to 30 min is homogenized gently and centrifuged at 16,000 *g* for 10 min. The clear yellowish brown supernatant is applied at a flow rate of 16 ml/hr to a Sephacryl S-300 column (2.5 × 65 cm) previously equilibrated with buffer B, and the column is eluted with the same buffer at the same flow rate. Fractions of 2.3 ml are collected and 1-$\mu$l samples are assayed. S6 kinase elutes from this column at a position corresponding to an $M_r \sim 150{,}000$ (175–200 ml). Fractions are pooled on the basis of activity and the absorbance profile.

*Mono S Chromatography*. The pooled fractions from the previous step are adjusted to a conductivity of 3.8 mmho with distilled water. One-half of the preparation is applied at a flow rate of 1 ml/min to a Mono S HR 5/5 column of a Pharmacia (Piscataway, NJ) fast protein liquid chromatography (FPLC) system previously equilibrated with buffer C containing 25 m*M* KCl. The column is washed with the same buffer, and proteins are eluted at a flow rate of 1 ml/min with a discontinuous gradient of KCl (25 to 600 m*M*) in buffer C (140 m*M* KCl at 2 ml, 284 m*M* at 27 ml, and 600 m*M* at 32 ml). Fractions of 0.5 ml are collected, and 0.5-$\mu$l (5 $\mu$l of a 1 : 10 dilution) samples are assayed. Activity elutes at approximately 200 m*M* KCl, and fractions are pooled on the basis of activity and the absorbance profile. The remainder of the preparation is fractionated in the same manner, and equivalent fractions are pooled.

TABLE I
PURIFICATION OF S6 KINASE II FROM *Xenopus* UNFERTILIZED EGGS

| Purification step[a] | Protein[b] (mg) | Activity (nmol/min) | Specific activity[c] (nmol/min/mg) | Yield (%) | Purification (-fold) |
|---|---|---|---|---|---|
| DEAE peak II | 167 | 17.5 | 0.10 | 100 | 1 |
| Sephacryl S-300 | 24 | 54.0 | 2.2 | 308 | 22 |
| Mono S | 2.5 | 36.4 | 14.4 | 208 | 144 |
| Mono Q | 0.51 | 27.7 | 53.7 | 158 | 537 |
| Heparin-Sepharose | 0.088 | 10.5 | 119 | 60 | 1190 |

[a] Purification and yield are calculated relative to the combined peak II fractions from two DEAE-Sephacel columns. This represents 320 ml of dejellied unfertilized eggs as starting material.

[b] Measured by the method of M. M. Bradford [*Anal. Biochem.* **72,** 248 (1976)], with bovine serum as a standard.

[c] For these determinations 40S ribosomal subunits were present at a concentration of 0.4 $\mu M$, approximately one-tenth of the $K_m$ of the enzyme for subunits.

*Mono Q Chromatography.* The pooled fractions from the previous step are diluted with 5 vol of 20 m*M* Tris-HCl, 5 m*M* $MgCl_2$, 2 m*M* EGTA, 2 m*M* dithiothreitol, 0.01% Brij 35, pH 7.0, and applied at a flow rate of 1 ml/min to a Mono Q HR 5/5 column previously equilibrated with buffer D containing 20 m*M* NaCl. The column is washed with the same buffer, and proteins are eluted at a flow rate of 1 ml/min with a discontinuous gradient of NaCl (20–600 m*M*) in buffer D (165 m*M* at 2.5 ml, 310 m*M* at 27.5 ml, and 600 m*M* at 32.5 ml). Fractions of 0.5 ml are collected and 0.5-$\mu$l samples are assayed. Activity elutes at approximately 200 m*M* NaCl, and fractions are pooled on the basis of activity and the absorbance profile.

*Heparin-Sepharose Chromatography.* The pooled fractions from the previous step are diluted with 2.5 vol of buffer E and applied at a flow rate of 7 ml/hr to a heparin-Sepharose column (0.7 × 8.5 cm) previously equilibrated with buffer E, and proteins are eluted at a flow rate of 10 ml/hr with a 60-ml linear gradient of NaCl (0 to 600 m*M*) in buffer E. Fractions of 1 ml are collected and 1-$\mu$l samples are assayed. Activity elutes at approximately 470 m*M* NaCl. Fractions are pooled on the basis of activity (there is no absorbance detectable) and dialyzed against 55 m*M* $\beta$-glycerophosphate, 5 m*M* $MgCl_2$, 5 m*M* EGTA, 2 m*M* dithiothreitol, 50% (v/v) ethylene glycol. The enzyme is stored in this buffer at −20°, and has been stable for up to several years.

A summary of the purification is given in Table I. S6 kinase II is purified approximately 1200-fold from the DEAE step, with a yield of 60%. An

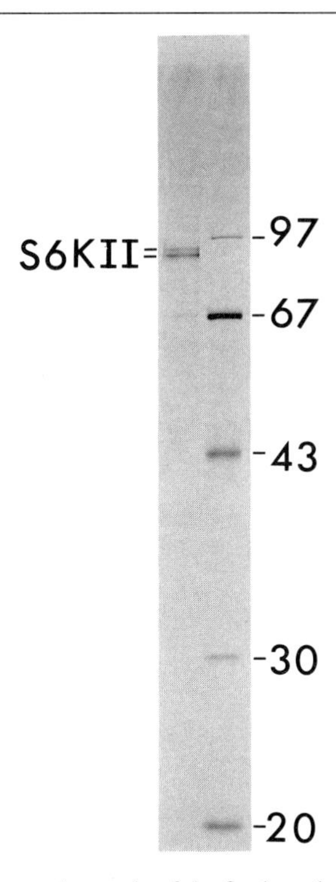


FIG. 2. Purity of the S6 kinase. A sample of the final product was electrophoresed through a SDS–10% polyacrylamide gel. The gel was silver stained as described by B. R. Oakley, D. R. Kirsch, and N. R. Morris [*Anal. Biochem.* **105,** 361 (1980)]. The positions of molecular weight standards ($\times 10^{-3}$) and of S6 kinase II are indicated. Left lane, S6 kinase II; right lane, molecular weight standards.

apparent increase in recovery after S-300 chromatography may reflect removal of an inhibitor. Analysis of the final preparation by SDS–polyacrylamide gel electrophoresis and silver staining is shown in Fig. 2.

### *Comments*

S6 kinase II has been shown to phosphorylate the peptide LRRASLG (kemptide, Sigma Chemical Co., St. Louis, MO) at a significant rate.[2,14] S6 kinase activity and kemptide kinase activity coelute during all fractionations following the DEAE-Sephacel column. Therefore assay of kemptide kinase activity instead of S6 kinase activity can be used to follow purification of S6 kinase II from the Sephacryl S-300 column onward. For this

[14] E. Erikson, D. Stefanovic, J. Blenis, R. L. Erikson, and J. L. Maller, *Mol. Cell. Biol.* **7,** 3147 (1987).

assay the concentration of kemptide is 100 $\mu M$, the specific activity of the [$\gamma$-$^{32}$P]ATP is 100–500 cpm/pmol, and 2-$\mu$l samples of the fractions are assayed. The reaction is stopped by the addition of 15 $\mu$l glacial acetic acid and the phosphorylated peptide is isolated on P81 (Whatman, Inc., Clifton, NJ) phosphocellulose paper.[15] With either assay, from the Sephacryl S-300 column onward, activity elutes as a single symmetrical peak and no activity is detectable in the flow-through fractions.

## Properties

S6 kinase II migrates as a closely spaced doublet with $M_r$ 92,000 on SDS–polyacrylamide gel electrophoresis (Fig. 2). Renaturation of activity from this protein after SDS–polyacrylamide gel electrophoresis verified that the $M_r$ 92,000 protein is an S6 kinase.[14] This enzyme sediments with an apparent $M_r$ 55,000 on glycerol gradient centrifugation, and elutes with an apparent $M_r$ 70,000 to 80,000 from Sephacryl S-200, and with an apparent $M_r$ 150,000 from Sephacryl S-300. The apparent $K_m$ for ATP is 28 $\mu M$ and for 40S subunits is 5 $\mu M$. The optimum $Mg^{2+}$ concentration is 10 to 30 m$M$. GTP does not serve as a substrate. S6 kinase II exhibits a broad pH optimum between 6.25 and 8.25, and incorporation of radiolabel into S6 increases with increasing temperature from 20 to 37°. The $IC_{50}$ values for several compounds are as follows: $\beta$-glycerophosphate, 15 m$M$; NaF, 25 m$M$; *p*-nitrophenylphosphate, 12.5 m$M$; poly(arginine), 25 $\mu$g/ml; heparin, 30 $\mu$g/ml.

S6 kinase II also phosphorylates peptides corresponding to residues 229–239 and 232–239 of S6, AKRRRLSSLRA and RRLSSLRA, respectively. At peptide concentrations of 100 $\mu M$, kemptide and S6(229–239) are phosphorylated at the same rate (1.4 $\mu$mol/min/mg), whereas S6(232–239) is phosphorylated at 30% of this rate. Among protein substrates examined, microtubule-associated protein-2, tau proteins, protamine, and glycogen synthase are phosphorylated at approximately the same rate as S6 in 40S subunits, and bovine cardiac and rabbit skeletal muscle troponin I, bovine and rabbit brain myelin basic protein, and lamin C are phosphorylated at 5–20% of this rate. For these comparisons protein substrates were present at 0.1 mg/ml (approximately 0.6 to 6 $\mu M$), and 40S subunits were present at 0.4 $\mu M$, approximately one-tenth of the $K_m$. Phosphoserine is the only phosphoamino acid detectable in these substrates. Trace amounts of radiolabel can be incorporated into bovine and rabbit skeletal muscle troponin T, H1 histone, and $\alpha$-casein, but

[15] D. B. Glass, R. A. Masaracchia, J. R. Feramisco, and B. E. Kemp, *Anal. Biochem.* **87,** 566 (1978).

no incorporation is detectable in H4 histone, mixed histones, phosvitin, phosphorylase *b*, protein phosphatase inhibitor 1, or rabbit skeletal muscle troponin C. S6 kinase II also undergoes autophosphorylation, with phosphoserine being the only phosphoamino acid detectable. In several substrates the sequence phosphorylated is of the type R-*x*-*y*-*S*-*z*[16]; however, the data on the sites phosphorylated in S6 indicate that serine residues in other sequences can also be phosphorylated. Studies on the regulation of S6 kinase II indicate that it is activated by phosphorylation by other serine/threonine protein kinases.[17,18]

## S6 Kinase I

Purification of the S6 kinase activity from the first DEAE-Sephacel peak, S6 kinase I, has also been achieved. This enzyme migrates with $M_r$ 90,000 on SDS–polyacrylamide gel electrophoresis. Most of its properties are very similar to those of S6 kinase II, but several differences are also evident.[19] Polyclonal antisera raised against purified S6 kinase II do not react with S6 kinase I[14]; however, polyclonal antisera raised against recombinant cDNA-encoded antigen produced in *Escherichia coli* recognize both S6 kinase I and II,[18] indicating that these two enzymes are related.

## Molecular Cloning

### *Peptide Sequences*

Approximately 100 $\mu$g of the enzyme is reduced, carboxymethylated, and gel purified. The protein is digested with 6 $\mu$g of trypsin and the tryptic peptides are resolved on a high-performance liquid chromatography (HPLC) Vydac $C_{18}$ column (218-TP546, 4.6 mm × 25 cm) with a gradient of 0–40% $CH_3CN$ in 0.1% ammonium acetate, pH 6.5. Selected fractions are further resolved by HPLC on a Brownlee $C_8$ column (RP-300, 2.1 mm × 22 cm) with a 2–80% $CH_3CN$ gradient in 0.1% trifluoroacetic acid. Absorbance is monitored at 220 nm and peptide-containing fractions are collected by hand. Selected fractions are applied directly to precycled Polybrene-coated glass filters and sequenced with an Applied Biosystems (Foster City, CA) model 470A gas-phase sequenator with a model 120A on-

[16] E. Erikson and J. L. Maller, *Second Mess. Phosphoprot.* **2,** 135 (1988).
[17] T. W. Sturgill, L. B. Ray, E. Erikson, and J. L. Maller, *Nature* (*London*) **334,** 715 (1988).
[18] E. Erikson and J. L. Maller, *J. Biol. Chem.* **264,** 13711 (1989).
[19] E. Erikson and J. L. Maller, *J. Biol. Chem.* **266,** 5249 (1991).

TABLE II
PEPTIDE SEQUENCES FROM S6 KINASE II[a]

| Peptide | Sequence obtained | |
|---|---|---|
| | Matched | |
| 85-3 | LTDFGLSK | |
| 89-5 | ICDFGFAK | |
| 96-2 | DLKPSNILYVDESGNPESIR | |
| 96-3 | DLKPENILLDEEGHIK | |
| | Not matched | Closest match predicted |
| 84-2 | ISGTDAGQLYAMK | ItppDAnQLYAMK |
| 89-6 | ADPSQFELLK | ADqSdFvLLK |
| 99-6 | ADPSHFEFLK | ADqSdFvlLK |

[a] Peptides from S6 kinase II that yielded unambiguous sequence data are shown. Sequences that are predicted by one or more of the cDNA clones are given, and are boxed in Fig. 3. Peptides 84-2, 89-6, and 99-6 were not predicted. The closest match observed for the *Xenopus* clones is shown, with the lower case letters indicating the unmatched amino acids found in the predicted sequences.

line analyzer. Unambiguous sequences were obtained from seven peptides (Table II).

*Screening Xenopus cDNA Library*

The information for peptide 96-3 is used to design the probe [TT(TGA)AT(GA)TGNCC(TC)TC(TC)TC(GA)TC] and that for peptide 84-2 is used to design the probes [TTCATNGC(GA)TA(TC)AA(TC)TG] and [TTCATNGC(GA)TANAG(TC)TG]. Peptides 96-3 and 96-2 are also used to design the "guess-mers" CTTGATGTGGCCCTCCTCATCCAC-CAGGATGTTCTCAGGCTTCAGGTC and GCGGATGGACTCAGGG-TTGCCAGACTCATCCACATACAGGATGTTGGATGGCTTCAGGTC, respectively.

A *Xenopus* ovarian cDNA library in λgt10 is obtained from D. A. Melton of Harvard University (Cambridge, MA). Hybridization with degenerate olgionucleotides is carried out for 12–14 hr at 45° for the 20-mer and at 37° for 17-mer in 0.05 *M* Tris-HCl, pH 7.1, 1 *M* NaCl, 0.1% (w/v) sodium pyrophosphate, 1% (w/v) SDS, 10% (w/v) dextran sulfate, 0.2% (w/v) polyvinylpyrrolidone, 0.2% (w/v) Ficoll, 0.2% (w/v) bovine serum albumin, 100 μg *E. coli* tRNA/ml. After hybridization, the filters are washed twice with 0.05 *M* Tris-HCl, pH 7.5, 1 *M* NaCl, 0.1% sodium pyrophosphate, 1% SDS at the same temperature used for hybridization. The filters are then washed in 3.2 *M* tetramethylammonium chloride, 1% SDS at 49° for the 20-mer and at 45° for the 17-mer. Hybridization to the unique sequence oligonucleotides is for 10–12 hr at 45° in 0.5 *M* sodium phosphate, pH 7.2, 7% SDS, 1% bovine serum albumin. The filters are washed in 0.15 *M* NaCl, 0.015 *M* sodium citrate, pH 7.0, 0.5% SDS at 45°.

In our laboratory over 200 positive plaques were detected with the degenerate probes after screening $10^6$ phage. The cDNA inserts in the phage were grouped into four classes by restriction endonuclease analysis. One clone with the longest insert was selected from each class for Southern analysis with the two "guess-mers." Two of the four clones hybridized to both probes. Both of these clones hybridized to mRNAs of 3.3–3.4 kilobases (kb) in *Xenopus* ovarian poly(A)$^+$ RNA. These clones were designated A14-1a and A3-1a.

*DNA Sequence Analysis: Predicted Proteins*

The inserts from the selected clones are subcloned into the *Eco*RI site of bluescript KS+ (Stratagene Cloning Systems, La Jolla, CA) and the sequence determined for both strands by the chain termination method of Sanger *et al.*[20] with a combination of nested deletions and specific oligonucleotide primers.[21,22] Sequence analysis of A14-1a shows that it is 3071 nucleotides long and reveals an open reading frame of 2199 nucleotides, predicting a protein of 733 amino acids ($M_r$ 83,000). Clone A3-1a is 2991 nucleotides long with an open reading frame of 1887 nucleotides predicting a protein of 629 amino acids $M_r$ 71,000). Clones A14-1a and A3-1a show 91% sequence identity with each other in the coding region where they overlap. The A3-1a clone has a termination codon resulting in a predicted 629-amino acid protein. If this termination codon were not there, the predicted protein would terminate at the same amino acid as that for clone A14-1a (Fig. 3). There is no information pertaining to the possibility that the termination codon is a cloning artifact, but it should be noted that two nucleotide changes in A14-1a are required to generate it. The predicted $M_r$ of 83,000 for the protein encoded by A14-1a falls short of the size of purified S6 kinase II observed, $M_r$ 92,000. This discrepancy can be explained by a shift in mobility due to phosphorylation on activation of the enzyme in *Xenopus* oocytes.[18]

Peptides 85-3, 89-5, and 96-3 are predicted by the sequence of both A14-1a and A3-1a. Peptide 96-2 is accurately predicted except for position 15, where peptide sequencing shows asparagine but DNA sequencing predicts aspartic acid by both clones. Careful examination of both the peptide and DNA sequence data does not permit a resolution of this

[20] F. Sanger, S. Nicklen, and A. R. Coulson, *Proc. Natl. Acad. Sci. U.S.A.* **74,** 5463 (1977).
[21] S. Henikoff, *Gene* **28,** 351 (1984).
[22] E. C. Strauss, J. A. Kobori, G. Siu, and L. E. Hood, *Anal. Biochem.* **154,** 353 (1986).

```
Xe-alpha   MPLAQLVNLW PEVAVVHEDP ENGHGSPEEG GRH....... .......... .TSKDEVVVK EFPITHHVKE GSEKADQSDF VLLKVLGQGS FGKVFLVRKI
Xe-beta    MPLAQLVDLW PEVELVHEDT ENGHGGPEDR GRH....... .......... .TSKDEVVVK EIPITHHVKE GAEKADQSHF VLLKVLGQGS FGKVFLVRKI
Mu-1       MPLAQLKEPW PLMELVPLDP ENGQTSGEEA GLQPS..... .......... ...KDEAILK EISITHHVKA GSEKADPSQF ELLKVLGQGS FGKVFLVRKV
Mu-2       .......... .......... .......... .......... .......... .......... .......... .......... .......... ..........
AV36       MPLAQLAEPW PNMELVQLDT ENGQAAPEEG GNPPCKAKSD ITWVEKDLVD STDKGEGVVK EINITHHVKE GSEKADPSQF ELLKVLGQGS FGKVFLVRKI
           1                                                                                                        100
Consensus  MPLAQL---W P----V--D- ENG----E-- G--------- ---------- ---K-E---K E--ITHHVK- G-EKAD-S-F -LLKVLGQGS FGKVFLVRK-

Xe-alpha   TPPDANQLYA MKVLKKATLK VRDRVRTKME RDILADVHHP FVVRLHYAFQ TEGKLYLILD FLRGGDLFTR LSKEVMFTEE DVKFYLAELA LGLDHLHSLG
Xe-beta    TPPDANQLYA MKVLKKATLK VRDRVRTKME RDILADVHHP FVVRLHYAFQ TEGKLYLILD FLRGGDLFTR LSKEVMFTEE DVKFYLAELA LGLDHLHSLG
Mu-1       TRPDSGHLYA MKVLKKATLK VRDRVRTKME RDILADVNHP FVVKLHYAFQ TEGKLYLILD FLRGGDLFTR LSKEVMFTEE DVKFYLAELA LGLDHLHSLG
Mu-2       .......... .........K VRDRVRTKME RDILVEVNHP FIVKLHYAFQ TEGKLYLILD FLRGGDLFTR LSKEVMFTEE DVKFYLAELA LALDHLHSLG
AV36       TPPDSNHLYA MKVLKKATLK VRDRVRTKIE RDILADVNHP FVVKLHYAFQ TEGKLYLILD FLRGGDLFTR LSKEVMFTEE DVKFYLAELA LGLDHLHSLG
           101                                                                                                      200
Consensus  T-PD---LYA MKVLKKATLK VRDRVRTK-E RDIL--V-HP F-V-LHYAFQ TEGKLYLILD FLRGGDLFTR LSKEVMFTEE DVKFYLAELA L-LDHLHSLG
                       96-3       85-3
Xe-alpha   IIYRDLKPEN ILLDEEGHIK LTDFGLSKEA IDHEKKAYSF CGTVEYMAPE VVNRQGHSHS ADWWSYGVLM FEMLTGSLPF QGKDRKETMT LILKAKLGMP
Xe-beta    IIYRDLKPEN ILLDEEGHIK LTDFGLSKEA IDHEKKAYSF CGTVEYMAPE VVNRQGHSHG ADWWSYGVLM FEMLTGSLPF QGKDRKETMT LILKAKLGMP
Mu-1       IIYRDLKPEN ILLDEEGHIK LTDFGLSKEA IDHEKKAYSF CGTVEYMAPE VVNRQGHTHS ADWWSYGVLM .......... .GKDRKETMT LILKAKLGMP
Mu-2       IIYRDLKPEN ILLDEEGHIK LTDFGLSKES IDHEKKAYSF CGTVEYMAPE VVNRRGHTQS ADWWSFGVLM FEMLTGTLPF QGKDRKETMT MILKAKLGMP
AV36       IIYRDLKPEN ILLDEEGHIK LTDFGLSKEA IDHEKKAYSF CGTVEYMAPE VVNRQGHSHS ADWWSYGVLM FEMLTGSLPF QGKDRKETMT LILKAKLGMP
           201                                                                                                      300
Consensus  IIYRDLKPEN ILLDEEGHIK LTDFGLSKE- IDHEKKAYSF CGTVEYMAPE VVNR-GH--- ADWWS-GVLM FEMLTG-LPF QGKDRKETMT -ILKAKLGMP

Xe-alpha   QFLSNEAQSL LRALFKRNPT NRLGSAMEGA EEIKRQPFFS TIDWNKLFRR EMSPPFKPAV TQADDTYYFD TEFTSRTPKD SPGIPPSAGA HQLFRGFSFV
Xe-beta    QFLSNEAQSL LRALFKRNAT NRLGSGVEGA EELKRHPFFS TIDWNKLYRR ELSPPFKPSV TQPDDTYYFD TEFTSRTPKD SPGIPPSAGA HQLFRGFSFV
Mu-1       QFLSTEAQSL LRALFKRNPA NRLGSGPDGA EEIKRHIFYS TIDWNKLYRR EIKPPFKPAV AQPDDTFYFD TEFTSRTPRD SPGIPPSAGA HQLFRGFSFV
Mu-2       QFLSPEAQSL LRMLFKRNPA NRLGAGPDGV EEIKRHSFFS TIDWNKLYRR EIHPPFKPAT GRPEDTFYFD PEFTAKTPKD SPGIPPSANA HQLFRGFSFV
AV36       QFLSAEAQSL LRALFKRNPA NRLGSGPDGA EEIKRHPFYS TIDWNKLYRR EIKPPFKPAV GQPDDTFYFD TEFTSRTPKD SPGIPPSAGA HQLFRGFSFV
           301                                                                                                      400
Consensus  QFLS-EAQSL LR-LFKRN-- NRLG----G- EE-KR--F-S TIDWNKL-RR E--PPFKP-- ----DT-YFD -EFT--TP-D SPGIPPSA-A HQLFRGFSFV

Xe-alpha   APALVEEDAK KTSSPPVL.S .VPKTHSKNI LFMDVYTVRE TIGVGSYSVC KRCVHKGTNM EYAVKVIDKT KRDPSEEIEI LRRYGQHPNI IALKDVYKEG
Xe-beta    APVLVEEDAK KTSSPPVL.S .VPKTHSKNV LFTDVYTVRE TIGVGSYSVC KRCVHKGTNM EYAVKVIDKS KRDPSEEIEI LRRYGQHPNI ITLKDVYEEC
Mu-1       ATGLMEDDGK PRTTQAPLHS VVQQLHGKNL VFSDGYVVKE TIGVGSYSVC KRCVHKATNM EYAVKVIDKS KRDPSEEIEI LLRYGQHPNI ITLKDVYDDG
Mu-2       A..ITSDDES QAMQTVGVHS IVQQLHRNSI QFTDGYEVKE DIGVGSYSVC KRCIHKATNM EFAVKIIDKS KRDPTEEIEI LLRYGQHPNI ITLKDVYDDG
AV36       ATGLM.EDSK VKPAQPPLHS VVQQLHGKNI QFSDGYVVKE AIGVGSYSVC KRCIHKTTNM EYAVKVIDKS KRDPSEEIEI LLRYGQHPNI ITLKDVYDDG
           401                                                                                                      500
Consensus  A------D-- ---------S -V---H---- -F-D-Y-V-E -IGVGSYSVC KRC-HK-TNM E-AVK-IDK- KRDP-EEIEI L-RYGQHPNI I-LKDVY---
```

```
                                                                                 96-2         89-5
Xe-alpha   NSIYVVTELM RGGELLDRIL RQKFFSEREA SSVLFTVCKT VENLHSQGVV HRDLKPSNIL YVDESGDPES IRICDFGFAK QLRADNGLLM TPCYTANFVA
Xe-beta    NSIYLVTELM RGGELLDRIL RQKFFSEREA CSVLFTVCKT VEYLHSQGVV HRDLKPSNIL YVDESGDPES IRICDFGFSK QLRAENGLLM TPCYTANFVA
Mu-1       KHVYLVTELM RGGELLDKIL RQKFFSEREA SFVLHTISKT VEYLHSQGVV HRDLKPSNIL YVDESGNPEC LRICDFGFAK QLRAENGLLM TPCYTANFVA
Mu-2       KYVYVVTELM KGGELLDKIL RQKFFSEREA SAVLFTITKT VEYLHAQGVV HRDLKPSNIL YVDESGNPES IRICDFGFAK QLRAENGLLM TPCYTANFVA
AV36       KYVYLVTELM RGGELLDKIL RQKFFSEREA SSVLHTICKT VEYLHSQGVV HRDLKPSNIL YVDESGNPES IRICDFGFAK QLRAENGLLM TPCYTANFVA
           501                                                                                                       600
Consensus  ---Y-VTELM -GGELLD-IL RQKFFSEREA --VL-T--KT VE-LH-QGVV HRDLKPSNIL YVDESG-PE- -RICDFGF-K QLRA-NGLLM TPCYTANFVA

Xe-alpha   PEVLKRQGYD EGCDIWSLGI LLYTMLAGYT PFANGLGDTP EEILARIGSG KFTLRGGNWN TVSAAAKDLV SRMLHVDPHK RLTAKQVLQH EWITKRDALP
Xe-beta    PEVLKRQGYD EGCDIWSLGI LLYTMLAGYT PFANGPGDTP EEILARIGS* KFTLRGGNWN TVSAAAKDLV SRMLHVDPHQ RLNAKQVLQH EWITKRDMLP
Mu-1       PEVLKRQGYD EGCDIWSLGI LLYTMLAGYT PFANGPSDTP EEILTRIGSG KFTLSGGNWN TVSETAKDLV SKMLHVDPHQ RLTAKQVLQH PWITQKDKLP
Mu-2       PEVLKRQGYD AACDIWSLGV LLYTMLTGYT PFANGPDDTP EEILARIGSG KFSLSGGYWN SVSDTAKDLV SKMLHVDPHQ RLTAALVLRH PWIVHWDQLP
AV36       PEVLKRQGYD EGCDIWSLGV LLYTMLAGCT PFANGPSDTP EEILTRIGGG KFSVNGGNWD TISDVAKDLV SKMLHVDPHQ RLTAKQVLQH PWITQKDSLP
           601                                                                                                       700
Consensus  PEVLKRQGYD --CDIWSLG- LLYTML-G-T PFANG--DTP EEIL-RIG-G KF---GG-W- --S--AKDLV S-MLHVDPH- RL-A--VL-H -WI---D-LP

Xe-alpha   QSQLNRQDV. HLVKGAMAAT YSALNSSKPT PLLQPIKSSI LAQRR.VKKL PSTTL*
Xe-beta    QSQLNRQDV. HLVKGAMAAT YSALNSSKPT PQLQPIKSSV LAQRR.VKKL PSITL*
Mu-1       QSQLSHQDL. QLVKGAMAAT YSALNSSKPT PQLKPIESSI LAQRR.VRKL PSTTL*
Mu-2       QYQLNRQDAP HLVKGAMAAT YSALN.RNQS PVLEPVGRST LAQRRGIKKI TSTAL*
AV36       QSQLNYQDV. QLVKGAMAAT YSALNSSKPS PQLKPIESSI LAQRR.VKKL PSTTL*
           701                                                756
Consensus  Q-QL--QD-- -LVKGAMAAT YSALN----- P-L-P---S- LAQRR---K- -S--L-
```

FIG. 3. Comparison of the predicted amino acid sequences of the various *rsk* proteins. The bold characters and bold underlines indicate putative ATP-binding sites. The thin underline indicates a degenerate but potential ATP-binding site. The boxed areas show predicted sequences from cDNA clones that match four of the peptide sequences from purified S6 kinase II. The University of Wisconsin Genetics Computer Group programs were used for sequence assembly analysis and manipulation. The programs GAP and PRETTY were used for protein comparison and alignment. Xe-alpha, *Xenopus* clone A14-1a; Xe-beta, *Xenopus* clone A3-1a; Mu-1, mouse clone 1; Mu-2, mouse clone 2 (partial); AV36, chicken; *, termination codon.

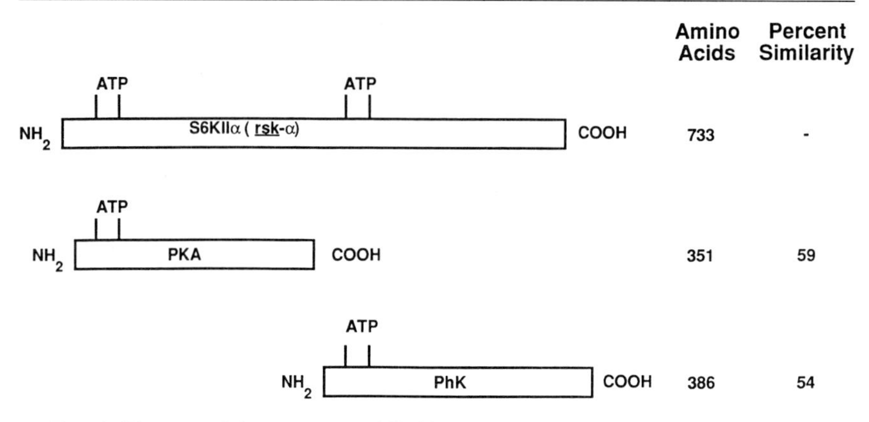


FIG. 4. Diagram of the sequence of S6 kinase II compared with other protein kinases. The amino-terminal region of S6 kinase II also shows similarity with cGMP-dependent protein kinase and bovine protein kinase C.[3] The locations of putative ATP-binding sites are indicated. S6KIIα (*rsk-α*), S6 kinase II; PKA, catalytic subunit of cAMP-dependent protein kinase; PhK, catalytic (γ) subunit of phosphorylase *b* kinase.

discrepancy. The remaining three peptides for which there is unambiguous information are not predicted by the DNA sequences (Table II). The overall sequence of the proteins predicted by the clones is given in Fig. 3. The predicted protein sequence shows strong similarities to two previously sequenced catalytic subunits of protein kinases, the cAMP-dependent enzyme, PKA, and phosphorylase *b* kinase (PhK) (Fig. 4). The similarity to PKA is greatest in the amino terminal domain, whereas PhK homology is greatest in the carboxyl-terminal domain. These similarities also extend to potential ATP-binding sites, raising the possibility that S6 kinase II has two catalytically active domains.

To confirm that these clones encode an S6 kinase, recombinant DNA is used to express the protein in *E. coli* by established techniques. The rabbit polyclonal antisera obtained recognize protein with S6 kinase activity in column fractions containing both S6 kinase I and S6 kinase II[18] and a related enzyme in chicken cells.[5]

*Low-Stringency Screening of Other cDNA Libraries*

The *Xenopus* A14-1a cDNA insert is used to screen chicken, mouse, *Drosophila,* and human cDNA libraries successfully. Only partial sequences are available for the human and *Drosophila* clones. The data on the chicken and two (one partial) mouse clones are presented in Fig. 3. Chicken and mouse cDNA libraries are prepared by standard protocols in λZAP (Stratagene) and screened at 30,000 plaques/150-mm diameter cul-

TABLE III
IDENTITY BETWEEN *rsk* cDNAs AND PROTEINS[a]

| | Nucleic acid sequences | | | | Amino acid sequences | | | |
|---|---|---|---|---|---|---|---|---|
| Source | Xα | Xβ | C | M-1 | Xα | Xβ | C | M-1 |
| Xβ | 93 | — | — | — | 95 | — | — | — |
| C | 75 | 75 | — | — | 86 | 87 | — | — |
| M-1 | 72 | 73 | 80 | — | 85 | 85 | 92 | — |
| M-2 | 71 | 70 | 74 | 71 | 79 | 83 | 85 | 84 |

[a] The percentage identity among the sequences of the cDNA inserts and among the predicted amino acid sequences of the various *rsk* clones is shown. Xα, *Xenopus* A14-1a; Xβ, *Xenopus* A3-1a; C, chicken; M-1, mouse clone 1; M-2, mouse clone 2 (partial).

ture dish using GeneScreen plus nylon membranes (Du Pont, NEN Research Products, Boston, MA). Hybridizations are performed in 30% formamide, 1 *M* NaCl, 50 m*M* sodium phosphate, pH 6.5, 5 m*M* EDTA, 1% SDS at 37°. The filters are washed in 5 m*M* sodium phosphate, pH 6.5, 400 m*M* NaCl, 5 m*M* EDTA, 1% SDS at 37°. Bacteriophage is rescued from λZAP by superinfection with bacteriophage R408 according to the commercial protocols (Stratagene) or subcloned into Bluescript plasmid (Stratagene) for sequencing. Both strands are sequenced using a combination of nested deletions and specific oligonucleotide priming.

The nucleic acid sequences of the chicken and mouse clones predict proteins that are remarkably similar in overall sequence, including the apparent two protein kinase domains (Table III, Fig. 3). The predicted proteins all contain the four peptides predicted by the *Xenopus* clones. Moreover, in contrast to the case with *Xenopus,* position 15 of peptide 96-2 is predicted as asparagine. It is also of interest that peptide 89-6, which is not predicted by either of the two *Xenopus* cDNAs, is predicted by the chicken and the complete mouse cDNAs. As the data in Table III show, the protein sequence identity for the region available is very high for the two mouse clones, yet their nucleic acid sequences are no more closely related than, for example, mouse and *Xenopus*. This suggests that the genes must have diverged very early in evolution.

## Comments

S6 kinase I and S6 kinase II were originally identified as major contributors to S6 phosphorylation in unfertilized *Xenopus* eggs after resolution on DEAE columns.[2] cDNA cloning and antiserum raised against recombinant protein show that homologous enzyme(s) are present in many species and cell types. All cDNA clones predict the unusual two-domain structure. Not all the significant peptide sequences obtained from S6 kinase II (Table II) are predicted by the cDNA clones. The S6 kinase II used for generation of the peptides was highly purified and all peptides from which usable sequence was obtained were present at nearly equal ratios. Taken together, these data suggest that other mRNAs for S6 kinase I or S6 kinase II remain to be cloned. Moreover, it seems likely that S6 kinase I and S6 kinase II are members of a large protein kinase family and that within S6 kinase I and II there may be more than one representative.

## Acknowledgments

We thank David A. Alcorta for helpful discussions and generous contributions in the computer analysis of sequence information. This work was supported by grants from the National Institutes of Health (DK28353 to J.L.M. and CA42580 to R.L.E.).

# [22] Purification and Properties of Mitogen-Activated S6 Kinase from Rat Liver and 3T3 Cells

*By* HEIDI A. LANE and GEORGE THOMAS

## Introduction

When quiescent cells in culture are stimulated to proliferate, one of the earliest mitogenic responses is the multiple phosphorylation of the 40S ribosomal protein S6.[1,2] *In vivo* and *in vitro* studies indicate that this event either triggers or facilitates the two- to threefold increase in protein synthesis required for quiescent cells to reenter the cell cycle.[3,4] The residues phosphorylated on S6 *in vivo* have been identified as five of seven serines located in a 32-amino acid peptide at the carboxyl end of the molecule.[5] Tryptic peptide analysis shows that the five phosphates are added in a specific order, and it is apparently the later sites of phosphorylation which modulate increased rates of protein synthesis.[3,4]

Many protein kinases have been reported to phosphorylate S6 *in vitro*.[3]

[1] G. Thomas, M. Siegmann, and J. Gordon, *Proc. Natl. Acad. Sci. U.S.A.* **76,** 3952 (1979).
[2] S. Decker, *Proc. Natl. Acad. Sci. U.S.A.* **78,** 4112 (1981).
[3] S. C. Kozma, S. Ferrari, and G. Thomas, *Cell Signalling* **1,** 219 (1989).
[4] J. W. B. Hershey, *J. Biol. Chem.* **264,** 20823 (1989).
[5] J. Krieg, J. Hofsteenge, and G. Thomas, *J. Biol. Chem.* **263,** 11473 (1988).

However, based on their modes of activation and their ability to generate S6 phosphopeptide maps equivalent to those observed *in vivo,* only the S6 kinase II from *Xenopus* eggs[6] and the mitogen-activated S6 kinase from mammalian cells[7] appear relevant. The latter enzyme is a rare polypeptide of $M_r$ 70,000 that undergoes autophosphorylation and is activated by a number of mitogens and oncogenes. Initially to detect the activated enzyme, phosphatase inhibitors were found to be necessary,[8] leading to the finding that the S6 kinase itself is activated by serine/threonine phosphorylation.[3,4]

The mitogen-activated S6 kinase was first purified from sodium orthovanadate (vanadate)-treated 3T3 cells.[7,9] When establishing the procedure for purification, two important factors had to be considered: how to maximally activate the enzyme and how to preserve its activity throughout the purification. Vanadate treatment was chosen as the method of activation in 3T3 cells for three reasons. First, it was as efficient at activating the S6 kinase as the most potent mitogens tested, including serum and platelet-derived growth factor (PDGF). Second, activation was persistent rather than transient. Third, vanadate was considerably less expensive to employ for large-scale treatment of cells than mitogens. To maintain the enzyme in its active form phosphatase inhibitors, such as $\beta$-glycerol phosphate, and metal chelators were present during the early stages of the purification.[7,9] Consequently, phosphatase 2A was identified as the major inactivator of the S6 kinase in cell extracts.[10] This led to a strategy in which phosphatases were removed at the first step of purification by means of cation-exchange chromatography.[7,9]

Due to the apparent low abundance of this kinase in 3T3 cells (210 ng of pure enzyme was recovered from 227 mg of 3T3 cell extract protein),[9] a large-scale purification procedure was established from rat liver,[11] in which sufficient amounts of kinase could be produced for protein sequencing and antibody production. A number of methods of activation were investigated, including refeeding of starved animals and intraperitoneal injection of vanadate, insulin, and cycloheximide. In this case, cycloheximide, which is known to mediate many of the early mitogenic responses,

[6] E. Erikson, J. L. Maller, and R. L. Erikson, this volume [21].

[7] P. Jenö, L. M. Ballou, I. Novak-Hofer, and G. Thomas, *Proc. Natl. Acad. Sci. U.S.A.* **85,** 406 (1988).

[8] I. Novak-Hofer and G. Thomas, *J. Biol. Chem.* **259,** 5995 (1984).

[9] P. Jenö, N. Jäggi, H. Luther, M. Siegmann, and G. Thomas, *J. Biol. Chem.* **264,** 1293 (1989).

[10] L. M. Ballou, P. Jenö, and G. Thomas, *J. Biol. Chem.* **263,** 1188 (1988).

[11] S. C. Kozma, H. A. Lane, S. Ferrari, H. Luther, M. Siegmann, and G. Thomas, *EMBO J.* **8,** 4125 (1989).

was found to be the most potent agent.[11] The purification procedure was based on the same principles as those applied to 3T3 cell extracts. This involved moving through the purification as quickly as possible, applying classical chromatographic procedures at the beginning to remove bulk proteins, and using ATP affinity chromatography as a final step of purification. We have been able to improve the procedure for purifying the kinase by including a number of new steps and eliminating less efficient ones.

This chapter will concentrate on the application of this improved procedure to the purification of the mitogen-activated S6 kinase from cycloheximide-treated rat liver and, on an analytical scale, from serum-stimulated 3T3 cells. These procedures permit purification of the protein kinase from either source in 2 to 3 days with higher yields and total activities than previously reported. Unlike the earlier purification protocols, two affinity columns are employed: a threonine column,[12] and a powerful peptide column based on the carboxy-terminal 32 amino acids of rat liver ribosomal protein S6,[5] similar to that described by Price *et al.*[13] Finally, the properties of the S6 kinase will be briefly reviewed and the enzymes obtained from each source compared.

*Solutions*

Assay dilution buffer contains 50 m*M* morpholinopropane sulfonic acid (MOPS), pH 7.0, 1 m*M* dithiothreitol (DTT), 10 m*M* $MgCl_2$, and 0.2% (w/v) Triton X-100. Extraction buffer for liver and 3T3 cells contains 15 m*M* pyrophosphate, pH 6.8, 5 m*M* ethylenediaminetetraacetic acid (EDTA), 1 m*M* DTT, and 1 m*M* benzamidine. Extract dilution buffer contains 1 m*M* $KH_2PO_4/K_2HPO_4$, pH 6.8. FFQ buffer contains 20 m*M* triethanolamine, pH 7.4, 1 m*M* EDTA, 50 m*M* NaF, 1 m*M* DTT, and 1 m*M* benzamidine. Mono Q and threonine buffer are the same as FFQ buffer except that they also contain 0.1% (w/v) Triton X-100. FFQ and Mono Q dilution buffers contain 1 m*M* triethanolamine, pH 7.4. Phenyl buffer contains 20 m*M* Tris-HCl, pH 7.4, 10 m*M* $MgCl_2$, 1 m*M* EDTA, 1 m*M* benzamidine, and 20 m*M* $\beta$-glycerol phosphate. Mono S buffer contains 50 m*M* MOPS, pH 6.8, 1 m*M* EDTA, 10 m*M* pyrophosphate, 1 m*M* DTT, 1 m*M* benzamidine, and 0.1% (w/v) Triton X-100. Peptide buffer contains 20 m*M* diethanolamine, pH 8.5, 1 m*M* EDTA, 1 m*M* benzamidine, 1 m*M* DTT, and 0.1% (w/v) Triton X-100. S-200 buffer is the same as peptide buffer, except that it also contains 0.1 *M* NaCl. The pH of all solutions is adjusted at 4°.

[12] U. Kikkawa, M. Go, J. Koumoto, and Y. Nishizuka, *Biochem. Biophys. Res. Commun.* **135,** 636 (1986).

[13] D. J. Price, R. A. Nemenoff, and J. Avruch, *J. Biol. Chem.* **264,** 13825 (1989).

## General Procedures

### *Protein Determination*

Routine protein assays during enzyme purification are carried out either by measurement of absorbance at 280 nm ($OD_{280}$), by the Lowry assay, as modified by Peterson,[14] or by a fluorometric assay based on *o*-phthaldialdehyde (OPA) derivatization, as previously described.[9] Bovine serum albumin (Bio-Rad, Richmond, CA) is used as a calibration standard.

### *S6 Kinase Assays*

S6 kinase activity is measured using 40S ribosomal subunits prepared from rat liver.[15] Reaction mixtures contain, in a final volume of 10 $\mu$l, 50 m*M* MOPS, pH 7.0, 1 m*M* DTT, 10 m*M* $MgCl_2$, 100 $\mu M$ ATP, 3 $\mu$Ci [$\gamma$-$^{32}$P]ATP (3000 Ci/mmol; Amersham, Amersham, England), 10 m*M* *p*-nitrophenyl phosphate, 20 $\mu$g 40S subunits; 10 units of the heat-stable inhibitor of cAMP-dependent protein kinase (PKI), and 5 $\mu$l of enzyme fractions that have been appropriately diluted in assay dilution buffer (see text). After incubation at 37° for 30 min, reactions are terminated by addition of 5 $\mu$l of electrophoresis sample buffer [0.4 *M* Tris-HCl, pH 6.8, 0.5 *M* DTT; 10% (w/v) SDS, 50% (v/v) glycerol, 0.16 mg/ml Bromophenol Blue]. The samples are boiled for 3 min and analyzed by sodium dodecyl sulfate-polyacrylamide gel electrophoresis (SDS–PAGE) as described below. However, to obtain a signal as rapidly as possible when assaying columns, the ATP concentration is lowered to 5 $\mu M$, the incubation time is shortened to 15 min, and gels are autoradiographed wet. It should also be noted that the efficiency with which the kinase phosphorylates S6 is dependent on the 40S ribosome preparation. Extreme care should, therefore, be taken to ensure that there is no degradation of S6 or "nicking" of the ribosomal RNA during the purification from rat liver.[15] If this occurs, much lower kinase activities are measured. Autophosphorylation reactions are performed as previously described.[7]

### *SDS Gel Electrophoresis*

SDS–PAGE is performed under reducing conditions according to the procedure of Laemmli.[16] The polyacrylamide gel consists of a separating gel [15% (w/v) acrylamide, 0.07% (w/v) bisacrylamide, 0.38 *M* Tris-HCl, pH 8.7, 0.1% (w/v) SDS] and a stacking gel [5% (w/v) acrylamide, 0.13%

[14] G. L. Peterson, *Anal. Biochem.* **83,** 346 (1977).
[15] G. Thomas, J. Gordon, and H. Rogg, *J. Biol. Chem.* **253,** 1101 (1978).
[16] U. K. Laemmli, *Nature (London)* **277,** 680 (1970).

(w/v) bisacrylamide, 0.125 *M* Tris-HCl, pH 6.8; 0.1% (w/v) SDS]. The running buffer is 0.025 *M* Tris, pH 8.3, 0.19 *M* glycine, 0.1% (w/v) SDS. A slab gel system is used (dimensions, 20 × 8 × 0.5 cm) and up to 24 samples are loaded onto each gel. When assaying S6 kinase activity, electrophoresis is performed at a constant current of 100 to 120 mA, requiring between 30 and 20 min, respectively. A large amount of heat is generated under these conditions, which could cause the glass plates to crack. We, therefore, employ a cooling fan (Indola, type TV 250, Italy), blowing air directly at the gels from a distance of 30–50 cm. After electrophoresis, the gels are stained with Coomassie Blue [50% (v/v) methanol; 10% (v/v) acetic acid; 0.1% (w/v) SERVA Blue R (Westbury, NY)] for 5 min and destained [20% (v/v) methanol, 10% (v/v) acetic acid] in an electrophoretic destainer (Biotec, Basel, Switzerland) for 8 min. The stained gels are dried on a slab gel dryer (Hoefer Scientific Instruments, San Francisco, CA) and autoradiographed [Kodak X-OMAT AR film (Eastman Kodak, Rochester, NY)]. The protein band corresponding to ribosomal protein S6 is then excised and the $^{32}$P incorporated is quantitated by liquid scintillation spectroscopy. Autophosphorylated fractions are electrophoresed at 50 mA (constant current) for ~1 hr and silver stained according to the procedure of Wray *et al.*[17] The gels are dried and autoradiographed as above. Molecular weight markers used for SDS–PAGE (Bio-Rad) include phosphorylase *b* (98,000), bovine serum albumin (68,000), ovalbumin (45,000), carbonic anhydrase (31,000), trypsin inhibitor (21,000), and lysozyme (14,400).

### *Peptide Maps*

Aliquots of purified rat liver or Swiss mouse 3T3 S6 kinase are autophosphorylated and separated by SDS–PAGE as stated above. Autoradiography is used to locate the labeled 70,000 protein, which is excised and eluted as previously described.[18] Aliquots of $^{32}$P-labeled protein (~10,000 cpm) are either digested with chymotrypsin (Worthington, Freehold, NJ) or cleaved chemically with cyanogen bromide (Fluka, Ronkonkoma, NY). The samples are then separated by SDS–PAGE as stated above, except that the bisacrylamide concentration is increased to 0.5%. Gels are Coomassie stained, dried, and autoradiographed as above.

## Column Preparation

All columns and gels are obtained from Pharmacia LKB (Piscataway, NJ) and all columns are run on the Pharmacia fast protein liquid chromatography (FPLC) system, unless otherwise stated.

[17] W. Wray, T. Boulikas, V. P. Wray, and R. Hancock, *Anal. Biochem.* **118,** 197 (1981).
[18] J. Martin-Pérez and G. Thomas, this series, Vol. 146, p. 369.

*Fast Flow S (FFS) Sepharose*

FFS Sepharose, available in preswollen form, is washed according to the manufacturer's instructions and 1 liter is packed by gravity into a BioProcess column T113/30 (column dimensions, 11.3 × 10 cm). A Watson-Marlow 501 U pump (Falmouth, England) is connected with the column via 1/4-in. polypropylene tubing using Jaco connectors (Pharmacia). All the connector threads are wrapped with Teflon tape to avoid leakage caused by high back pressures. The column is then equilibrated at 290 ml/min in four column volumes of a solution containing one part rat liver extraction buffer and two parts extract dilution buffer. During this time compression of the gel occurs and the upper plunger must be lowered onto the gel surface before loading. This is best accomplished while running buffer through the column at a speed of approximately 100 ml/min. After use, the column is regenerated with four column volumes of the same buffer mixture containing 1 *M* NaCl. The column is then thoroughly washed with double distilled water and stored in double distilled water containing 0.02% (w/v) sodium azide at 4°.

*Mono S*

An HR5/5 Mono S column (0.5 × 5 cm) is washed according to the manufacturer's instructions and then equilibrated with four column volumes of Mono S buffer. After use, the column is reversed and regenerated with a 2 *M* salt solution, thoroughly washed with double distilled water and stored in 24% (v/v) ethanol at 4°. As this column is exposed to crude extracts (see below) back pressures can rise to unacceptable levels within one or two runs. This problem is usually solved by replacing the upper filter of the column.

*Fast Flow Q (FFQ) Sepharose*

FFQ Sepharose, also available in preswollen form, is washed according to the manufacturer's instructions and 300 ml is packed by gravity into a BioProcess column T113/15 (column dimensions, 11.3 × 3 cm). The column is connected to a Watson-Marlow 501 U pump and equilibrated with four column volumes of FFQ buffer at 275 ml/min. As with the FFS column, the plunger must be lowered during the equilibration process. After use, the column is regenerated with four column volumes of FFQ buffer containing 1 *M* NaCl, thoroughly washed with double distilled $H_2O$, and stored in double distilled $H_2O$ containing 0.02% (w/v) sodium azide at 4°.

*Mono Q*

An HR5/5 Mono Q column (0.5 × 5 cm) is washed according to the manufacturer's instructions and equilibrated with four column volumes of

Mono Q buffer. After use, it is regenerated and stored as the Mono S column.

*Phenyl-TSK*

A 54-ml phenyl-TSK column (21.5 × 15 cm) (Beckman), fitted to the FPLC system via compression screw unions (Pharmacia), is equilibrated with four column volumes of phenyl buffer containing 0.9 *M* ammonium sulfate. After use, the column is regenerated with 20 vol of high-performance liquid chromatography (HPLC)-grade water (Millipore, Bedford, MA), followed by 10 vol of either 10% acetonitrile or 10% methanol, in which it is then stored at 4°.

*Sephacryl S-200*

Sephacryl S-200 is washed with double distilled $H_2O$, degassed, and then 419 ml is packed by gravity into a Bio-Rad Econo column (2.5 × 85.5 cm). The column is calibrated with dextran blue (2,000,000; Pharmacia), bovine serum albumin (68,000; Sigma, St. Louis, MO), ovalbumin (45,000; Serva), and ribonuclease A (13,000; Sigma). The void and internal volumes are 135 and 374 ml, respectively. The column is equilibrated in S-200 buffer (Preciflow pump; Lambda, Naters, Switzerland). Equilibration is started 24 hr before use, at a flow rate of 29 ml/hr. After use, the column is stored in double distilled $H_2O$ containing 0.02% (w/v) sodium azide.

*Threonine-Sepharose*

L-Threonine (Merck, Darmstadt, Germany) is coupled to AH-Sepharose 4B as previously described.[12] The coupled gel (4 ml) is packed by gravity into a column (dimensions, 1.3 × 5 cm) and equilibrated in four column volumes of threonine buffer. Flow rates >1.5 ml/min are avoided as this could cause a collapse of the gel matrix. After use, the column is regenerated by successive washings with 2 *M* NaCl, double distilled $H_2O$, 1% (w/v) Triton X-100, followed by double distilled $H_2O$. The column is stored in threonine buffer containing 1.2 *M* NaCl and 0.02% (w/v) sodium azide at 4°.

*S6 Peptide-Sepharose*

Activated CH-Sepharose 4B (1 ml), previously swollen and washed with ice-cold 1 m*M* HCl, is resuspended in 1 vol of coupling buffer (0.1 *M* $NaHCO_3$, pH 8.0, 0.5 *M* NaCl). The S6 carboxyl peptide (35 mg; sequence: K-E-A-K-E-K-R-Q-E-Q-I-A-K-R-R-R-L-S-S-L-R-A-S-T-S-K-S-E-S-S-Q-K),[5] dissolved in 1 ml of coupling buffer, is added and the coupling reaction allowed to proceed for 1 hr, mixing end over end, at room temperature. Coupling is spontaneous, through the primary $\alpha$-amino group of the peptide, and a stable peptide linkage is formed. When developing this procedure, it was found that the ratio of peptide to active ester groups should not fall below 1.7 $\mu$mol : 1 $\mu$mol, respectively. Any decrease in the concentration of the peptide results in a lower affinity column. This could be due to the $\varepsilon$-amino groups coupling with excess active groups, altering the presentation of the peptide to the kinase.[19] After incubation, the uncoupled peptide is retained for reuse and any remaining active groups are blocked by treatment with 1 ml of 0.1 *M* Tris-HCl, pH 8.0, for 1 hr, mixing end over end, at room temperature. The coupled gel is washed with coupling buffer and then alternately with high- and low-pH buffer solutions (50 m*M* Tris-HCl, pH 8.0; 0.5 NaCl followed by 50 m*M* sodium acetate, pH 4.0; 0.5 *M* NaCl). This is repeated five times and acts to remove excess ligand. The gel is washed with a large volume of double distilled $H_2O$, packed by gravity into an HR5/5 column (0.5 × 5 cm), and equilibrated in four column volumes of peptide buffer. The maximum flow rate is 1.2 ml/min. After use, the column is regenerated with four column volumes of peptide buffer containing 1.5 *M* NaCl, and stored at 4° in peptide buffer containing 0.02% (w/v) sodium azide.

## Purification of S6 Kinase from Rat Liver

### *Liver Extract Preparation*

Adult, male Wistar rats (100), each weighing approximately 400 g, are injected intraperitoneally with 4 ml of cycloheximide (5 mg/ml; Calbiochem, San Diego, CA) in phosphate-buffered saline (PBS), pH 7.4. Before injection the rats are lightly anesthetized with $CO_2$. For this purpose a large glass lyophilizer is used with dry ice in the base and a tight-fitting lid. The rats are sufficiently anesthetized after 15–30 sec. Longer exposure leads to unconsciousness and death. Six rats may be anesthetized and injected at one time. The injections must be performed quickly as the rats fully recover from the anesthesia within 15–30 sec. After 1 hr, the rats are again lightly anesthetized and then sacrificed by decapitation in the order in which they were injected. The livers are quickly removed, washed in cold PBS, weighed, and homogenized in 2.5 vol of ice-cold extraction buffer made 0.1 m*M* in phenylmethylsulfonyl fluoride (PMSF). Homogenization is at full speed for two 30-sec bursts in a Sunbeam XPA blender, followed by centrifugation at 27,500 *g* for 20 min at 4° (Sorvall RC2-B

[19] P. Jenö and G. Thomas, this volume [14].

centrifuge and GSA rotor). All subsequent steps in the purification are carried out at 4°.

*Fast Flow S Cation-Exchange Chromatography*

The centrifuged liver extract is poured through cheesecloth to remove lipids, made 0.1 m*M* in PMSF, and then diluted with 2 vol of extract dilution buffer, so that the final conductivity is <2.5 mS/cm. At higher conductivities the kinase does not bind well to the column. The diluted extract is loaded onto the equilibrated FFS column at 290 ml/min. It is important that the procedures leading from the extraction to the loading of the FFS column be carried out as quickly as possible, to reduce losses from the action of proteases and phosphatases. In general, the extract is collected from approximately 20 rat livers, centrifuged, and is loaded onto the column by the time the next batch is ready to be loaded. In this way, extracts are applied to the column within 30–45 min of preparation. After being loaded, the column is washed at the same flow rate with FFS equilibration buffer until the $OD_{280}$ drops to baseline. The column is developed at 100 ml/min with a 10-liter linear salt gradient from 0 to 0.5 *M* NaCl in FFS equilibration buffer. The $OD_{280}$ and conductivity levels of the eluate are monitored continually. The S6 kinase activity elutes between 0.2 and 0.3 *M* NaCl, after the main peak of protein, at the point where the $OD_{280}$ has returned to near-basal levels, and within a conductivity window of 18 to 30 mS/cm. Using these criteria, the pool is collected directly without fractionating and assaying the column. Based on conductivity and the $OD_{280}$, collection of the pool usually begins after 4.6–4.8 liters of the gradient has passed through the column, with a final pool size of around 3 liters. This method of pooling has been controlled by assaying at specific points during the development of the column (see Fig. 1A). The benefit of this strategy is that the eluate may be loaded immediately onto the next column, thereby preserving kinase activity. It should also be noted that an on-line industrial UV cell is used during column development which can be fitted into the UV recorder of the FPLC and is compatible with the flow rates employed.[20]

*Fast Flow Q Anion-Exchange Chromatography*

The FFS pool is made 0.1 m*M* in PMSF and 20 $\mu M$ in 8-bromoadenosine 3′5′-cyclic monophosphate (8-Br-cAMP; Boehringer, Mannheim, Germany) and incubated on ice for 10 min. The 8-Br-cAMP causes dissociation of the regulatory and catalytic subunits of cAMP-dependent protein

[20] S. Ferrari and G. Thomas, this volume [12].

kinase. In this form the catalytic subunit does not bind to anion-exchange resins and copurification with the S6 kinase can be prevented. After incubation, the FFS pool is diluted with 7 vol of FFQ dilution buffer and applied to the equilibrated FFQ column at a flow rate of 275 ml/min. The difficulty of handling such large volumes has caused problems at this stage. These problems are minimized by sequential dilution of small portions (~500 ml) of the FFS pool just before the required time for loading. This approach also reduces the time the kinase is exposed to buffers containing low amounts of phosphatase inhibitor. The conductivity of the diluted pool should be <4 mS/cm, otherwise some S6 kinase activity is found in the flow through and the portion that binds elutes as a broad peak during development of the column. The column is washed at the same flow rate with FFQ buffer until the $OD_{280}$ drops to baseline, and developed with a 3-liter linear salt gradient from 0 to 0.5 *M* NaCl in FFQ buffer, at 50 ml/min. The kinase elutes at ~0.36 *M* NaCl. Fractions of 50 ml are collected, employing the LKB Superrac fraction collector (rack C). A dilution of 1 : 10 is used for the S6 kinase assay and the fractions indicated in Fig. 1B

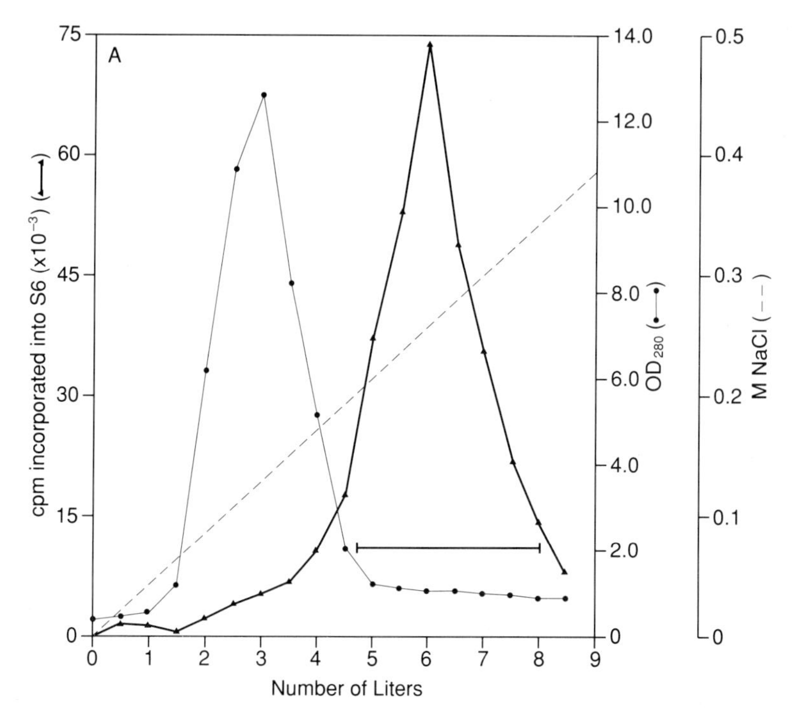


FIG. 1. Elution profile of rat liver S6 kinase on (A) FFS, (B) FFQ, and (C) phenyl-TSK columns. For details see text. Bar indicates pooled fractions.

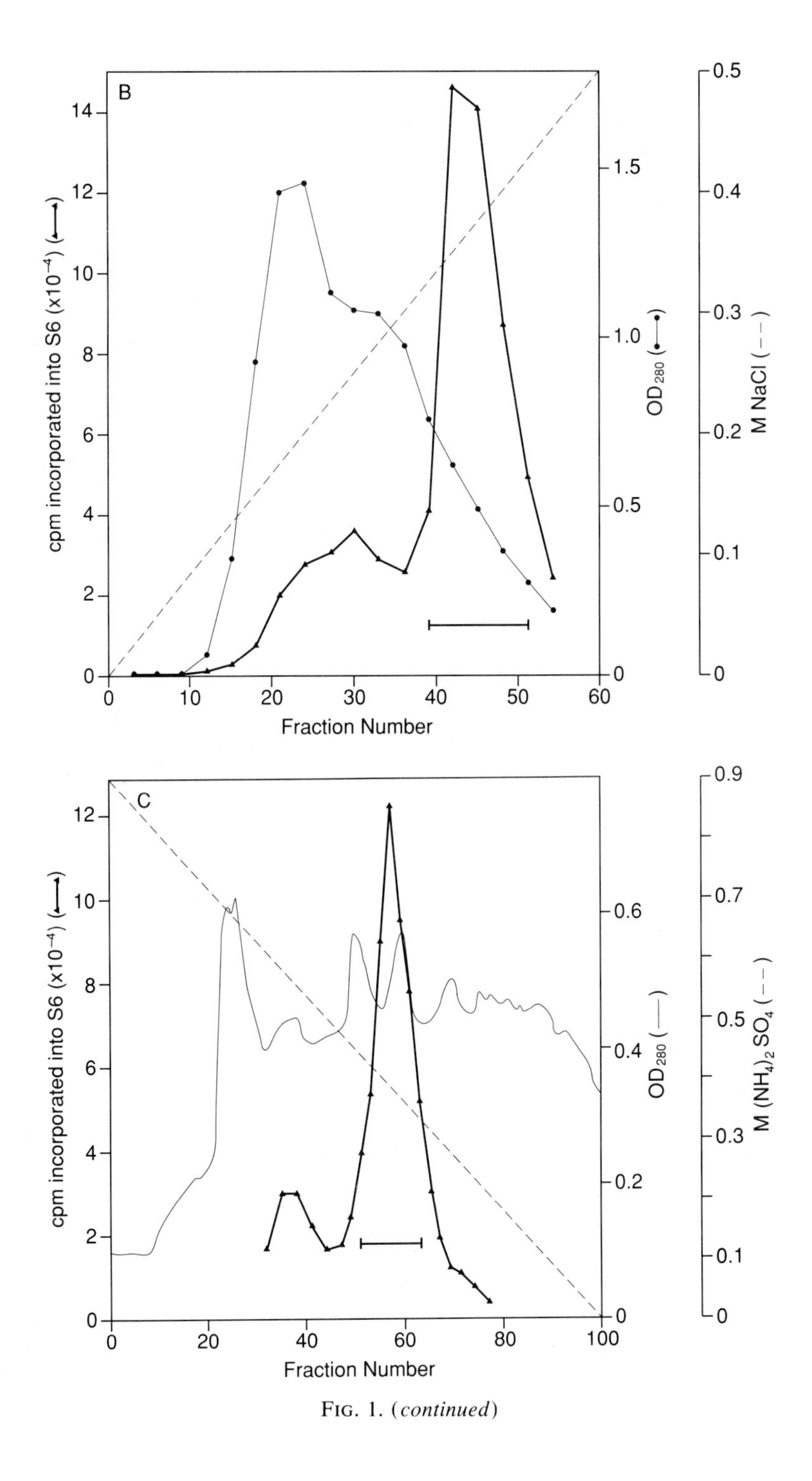


FIG. 1. (*continued*)

are pooled. As with the FFS column it should also be feasible to pool the fractions by conductivity and $OD_{280}$.

*Phenyl-TSK Chromatography*

The FFQ pool is made 0.1 m*M* in PMSF and slowly diluted with a 1/3 vol of 3.6 *M* ammonium sulfate in double distilled $H_2O$. The ammonium sulfate is prefiltered through a 0.22-$\mu$m Nalgene filter unit and must be added dropwise with continual stirring to avoid local protein precipitation. The pool, now 0.9 *M* in ammonium sulfate, is again filtered through a 0.22-$\mu$m Nalgene filter unit to avoid blockage of the filters or the column leading to high back pressure. After filtration, the pool is applied to the equilibrated phenyl column at a flow rate of 6 ml/min. The column is washed with phenyl buffer plus 0.9 *M* ammonium sulfate until the $OD_{280}$ reaches basal levels. A 540-ml linear salt gradient from 0.9 to 0 *M* ammonium sulfate in phenyl buffer is run at 3 ml/min and 5.4-ml fractions are collected. The fractions are assayed at a 1:20 dilution. On this column we generally observe two peaks of activity, a minor peak at ~0.57 *M* ammonium sulfate and a major one at ~0.38 *M* ammonium sulfate. The minor peak, which constitutes 10–15% of the total activity loaded onto the column, has not been analyzed further and is not included in the pooled fractions (see Fig. 1C). The pooled fractions are made 0.1 m*M* in PMSF and dialyzed overnight in threonine buffer [Spectra/Por dialysis membrane; Spectrum (Los Angeles, CA); molecular weight cutoff 12,000–14,000]. In general, we begin the dialysis step at the end of the first day of the purification protocol. However, since the enzyme elutes at relatively low salt, it may be possible to dilute the phenyl pool and load the sample onto the next column slowly overnight.

*Threonine Affinity Chromatography*

When the dialyzed pool has reached a conductivity of ~1.6 mS/cm it is applied, at a flow rate of 1.5 ml/min, onto the equilibrated threonine column. The column is washed with threonine buffer until the $OD_{280}$ reaches baseline. Development is at 1.5 ml/min with a linear salt gradient from 0 to 1.2 *M* NaCl in threonine buffer, and a fraction size of 1 ml. At 0.36 *M* NaCl, however, the gradient is placed on hold, until the $OD_{280}$ reaches baseline, and then continued. This plateau method, used previously in the purification of protein kinase C,[12] enables us to separate a large bulk of proteins from the S6 kinase, which elutes at 0.9 *M* NaCl (see Fig. 2A). The fractions are diluted 1:25 for the kinase assay and those indicated in Fig. 2A are pooled.

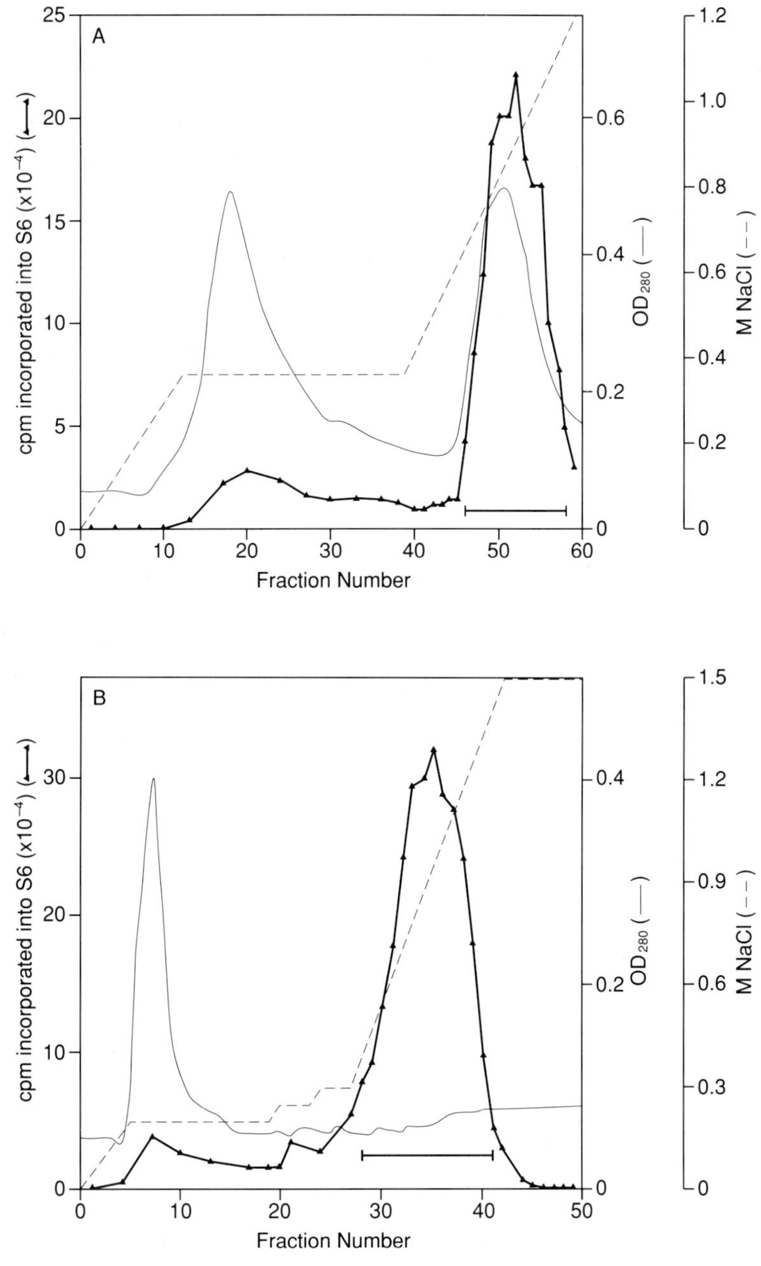


FIG. 2. Elution profile of rat liver S6 kinase on (A) threonine and (B) S6 peptide affinity columns. For details see text. Bar indicates pooled fractions.

*S6 Peptide Affinity Chromatography*

The threonine pool is diluted 1 : 14 with peptide buffer, so that the conductivity is between 5 and 10 mS/cm, and loaded onto the equilibrated peptide column at a flow rate of 1.2 ml/min. The column is washed until the $OD_{280}$ reaches baseline, and developed at 0.5 ml/min with a linear salt gradient from 0 to 1.5 *M* NaCl in peptide buffer, with a fraction size of 1 ml. Here, the gradient is held at three consecutive plateaus—0.2, 0.25, and 0.3 *M* NaCl—and then continued. This again allows the clear separation of a large bulk of proteins from the S6 kinase, which elutes at 0.95 *M* NaCl (see Fig. 2B). At 1.5 *M* NaCl the gradient is held again for eight fractions. This is to ensure that all the kinase activity is eluted from the column. The fractions are collected in siliconized tubes, diluted 1 : 20 for the kinase assay, and those indicated in Fig. 2B are frozen separately in liquid nitrogen.

*Results and Comments*

As shown by SDS–PAGE and silver stain analysis (Fig. 3), the range of proteins present in the active pools is quite large, up to the threonine column pool (Fig. 3, lane F). After the peptide column, however, at the peak of S6 kinase activity, a major band of $M_r$ 70,000 is present (Fig. 3, lane G) and only two minor high-molecular-weight contaminants ($M_r >$ 100,000) are observed. When all the fractions present in the active pool of the peptide column are analyzed, a third contaminant ($M_r$ 34,000) can be detected in fractions 31 and 32 (Fig. 4). If required, these three proteins can be excluded by gel filtration. By visual examination the kinase is 50 to 60% pure at this stage. The 70,000 band migrates at the position previously described for the S6 kinase and has the ability to autophosphorylate (Fig. 3, lane H), a property which directly parallels the S6 kinase activity (not shown). That this band represents the S6 kinase is supported by sequencing of cyanogen bromide peptides derived from the 70,000 protein, which shows that it belongs to the family of serine/threonine kinases (S. Ferrari and N. Totty, personal communication). Furthermore, the sequence data show that it is distinct from the *Xenopus* S6 kinase II.[21,22]

After the peptide column, the enzyme emerges with a specific activity of 0.7 $\mu$mol/min/mg of protein (see Table I). This value is higher than

[21] S. W. Jones, E. Erikson, J. Blenis, J. L. Maller, and R. L. Erikson, *Proc. Natl. Acad. Sci. U.S.A.* **85,** 3377 (1988).

[22] D. A. Alcorta, C. M. Crews, L. J. Sweet, L. Bankston, S. W. Jones, and R. L. Erikson, *Mol. Cell. Biol.* **9,** 3850 (1989).

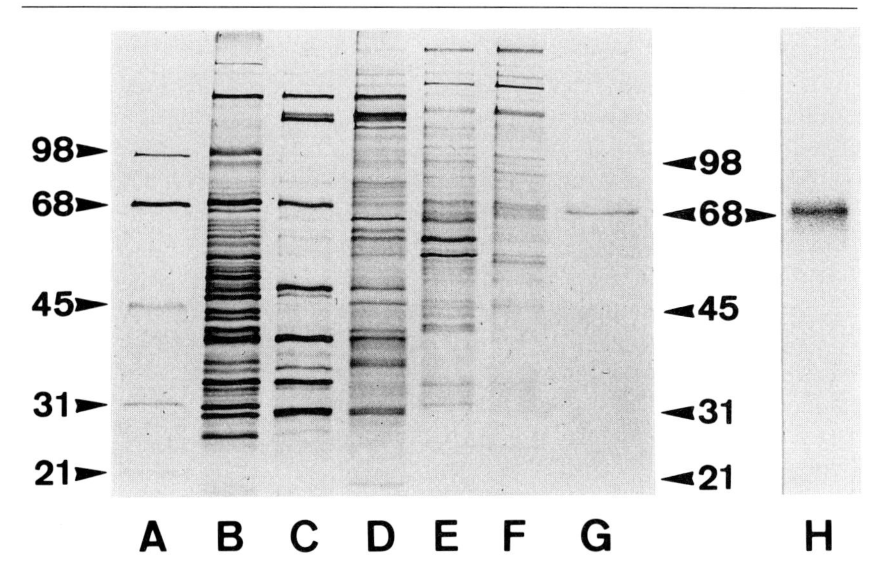


FIG. 3. SDS-PAGE of S6 kinase purification from rat liver at each step of chromatography. Samples containing the indicated amounts of protein were electrophoresed and silver stained as described in General Procedures. Lane A, molecular weight markers ($\times 10^{-3}$; 50 ng/protein); lane B, liver extract (2 $\mu$g); lane C, FFS pool (1 $\mu$g); lane D, FFQ pool (1 $\mu$g); lane E, phenyl pool (1 $\mu$g); lane F, threonine pool (1 $\mu$g); lane G, S6 peptide peak fraction (25 ng); lane H, autoradiogram of the S6 peptide peak fraction incubated alone with [$\gamma$-$^{32}$P]ATP (see General Procedures).

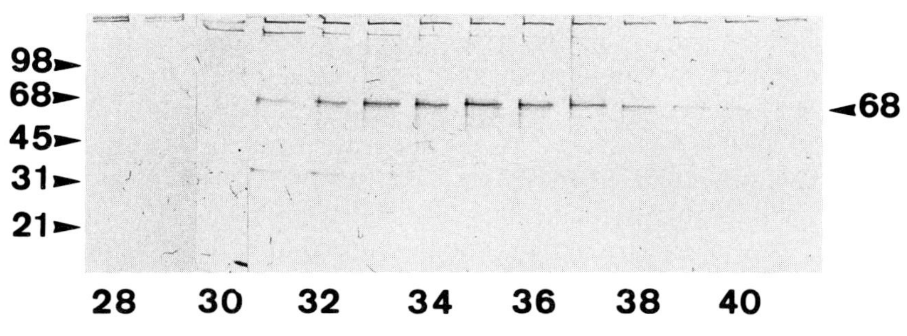


FIG. 4. SDS-PAGE analysis of purified rat liver S6 kinase. A 12-$\mu$l aliquot from each of the peak fractions of S6 kinase activity (28–41), corresponding to the S6 peptide affinity chromatography shown in Fig. 2B, was electrophoresed and silver stained as described in General Procedures. Molecular weight markers ($\times 10^{-3}$) are indicated.

TABLE I
PURIFICATION OF S6 KINASE FROM RAT LIVER

| Purification steps | Volume (ml) | Total protein (mg) | Specific activity (nmol/min/mg) | Total activity (nmol/min) | Recovery (%) | Relative purification |
|---|---|---|---|---|---|---|
| Liver extract | 7,800 | 193,440.00 | 0.01 | 2500 | 100.0 | 1.0 |
| FFS pool | 3,200 | 5,090.00 | 0.18 | 924 | 37.0 | 18.0 |
| FFQ pool | 520 | 676.00 | 0.87 | 590 | 23.6 | 87.0 |
| Phenyl pool | 60 | 28.14 | 8.90 | 249 | 10.0 | 890.0 |
| Threonine pool | 11 | 8.69 | 16.00 | 142 | 5.7 | 1,600.0 |
| S6 peptide pool | 14 | 0.16 | 707.00 | 111 | 4.4 | 70,700.0 |

reported for any S6 kinase purified to date.[9,11,13,23–25] Since the total amount of S6 kinase present in the peptide pool is 70 to 90 μg, purification to homogeneity should result in specific activities of 1 to 2 μmol/min/mg of protein. To preserve activity the first steps of purification are carried out as quickly as possible and, in practice, the whole procedure is performed within 2 days. The recovery, which ranges from 4 to 7%, is higher than we have previously reported,[11] and illustrates the efficiency of this approach. The most powerful step is the S6 peptide affinity column, which gives a 44-fold purification and a 54-fold reduction in protein, resulting in a final purification factor of 70,700 (see Table I). As noted in the introduction, the 32-amino acid peptide used for this column is based on a cyanogen bromide peptide that we previously showed to contain the sites of S6 phosphorylation.[5] A similar peptide was first introduced into the purification of the S6 kinase by Price *et al.*[13] This peptide, however, contains -*G*-*G*-S-Q-K at the carboxyl terminus versus -*E*-*S*-S-Q-K that we reported by direct protein sequencing. The latter sequence has been confirmed by sequence data derived from both the mouse and rat S6 genes.[3] It seems unlikely that this difference would affect the binding of the kinase, since the last five amino acids contain only the last site of S6 phosphorylation. However, care should be taken when constructing the peptide. In our previous reports we relied on affinity chromatography on ATP-Sepharose at this stage of the purification. The S6 peptide column has proved to be more efficient and more stable.

## Purification of S6 Kinase from 3T3 Cells

### *Cell Culture and Extract Preparation*

Fifty 15-cm cell culture plates (Falcon, Lincoln Park, NJ) are seeded with Swiss mouse 3T3 cells, and maintained as previously described.[10] After 7 days, the cells have grown to confluency and are judged to be quiescent since no mitoses are observed. On day 8 the cells are stimulated for 1 hr at 37° by the addition of fetal calf serum (GIBCO, Grand Island, NY) to a final concentration of 10%. The plates are then placed on ice, the medium removed, and the cells washed twice with 10 ml of ice-cold extraction buffer made 0.1 m*M* in PMSF. All subsequent steps of the purification are carried out at 4°. The cells on each plate are scraped with a rubber policeman into ~500 μl of extraction buffer and homogenized

[23] E. Erikson and J. L. Maller, *J. Biol. Chem.* **261,** 350 (1986).

[24] D. Tabarini, A. Garcia de Herreros, J. Heinrich, and O. M. Rosen, *Biochem. Biophys. Res. Commun.* **144,** 891 (1987).

[25] J. S. Gregory, T. G. Boulton, B. Sang, and M. H. Cobb, *J. Biol. Chem.* **264,** 18397 (1989).

with 15 strokes of a Dounce homogenizer (Glaskeller, Basel, Switzerland). The homogenate is centrifuged at 17,000 *g* for 10 min at 4° (Sorvall RC2-B centrifuge and SS-34 rotor) followed by a high-speed centrifugation at 340,000 *g* for 1 hr at 2° (Beckman L8-80 M ultracentrifuge and 70.1 Ti rotor). The high-speed centrifugation step is necessary, as it removes ribosomes and other cell debris, which make it otherwise difficult to filter the sample before loading onto the Mono S column (see below). The supernatants are made 0.1 m*M* in PMSF and frozen in liquid nitrogen until required.

*Mono S Cation-Exchange Chromatography*

Prepared cell extracts (23 mg protein in 30 ml) are thawed, made 0.1 m*M* in PMSF, and diluted 1 : 1 with extract dilution buffer so that the conductivity is ~3 mS/cm. The diluted pool is then filtered through a 0.22-μm Nalgene filter unit and loaded onto the equilibrated Mono S column at 1 ml/min. The column is washed at the same flow rate with Mono S buffer, until the $OD_{280}$ reaches baseline, and developed with a 30-ml linear salt gradient from 0 to 0.5 *M* NaCl in Mono S buffer, at 0.5 ml/min. The S6 kinase elutes at a salt concentration of ~0.2 *M* NaCl. Fractions of 1 ml are collected, assayed for S6 kinase activity at a 1 : 10 dilution, and the fractions indicated in Fig. 5A are pooled.

*Mono Q Anion-Exchange Chromatography*

The Mono S pool (1.0 mg protein in 4 ml) is made 0.1 m*M* in PMSF and 20 μ*M* in 8-Br-cAMP (see FFQ chromatography step of rat liver S6 kinase purification), and incubated for 10 min on ice. After incubation, the pool is diluted with 4 vol of Mono Q buffer so that the conductivity is ~7.0 mS/cm. The pool is filtered through a 0.22-μm Millex-GV filter unit (Millipore) and loaded onto the equilibrated Mono Q column at 1 ml/min. The column is washed at the same flow rate with Mono Q buffer until the $OD_{280}$ reaches baseline. The gradient is a 30-ml linear salt gradient from 0 to 0.5 *M* NaCl in Mono Q buffer, at 0.5 ml/min. The S6 kinase elutes at a salt concentration of ~0.32 *M* NaCl. Fractions of 1 ml are collected and assayed at at 1 : 10 dilution. The fractions indicated in Fig. 5B are pooled.

*Sephacryl S-200 Gel Filtration*

The Mono Q pool (0.18 mg protein in 4 ml) is loaded onto the equilibrated Sephacryl S-200 column at a flow rate of 0.48 ml/min. The column is developed at the same speed and 180 fractions of 1.9 ml are collected, employing the LKB Superrac fraction collector (rack B). S6 kinase activity

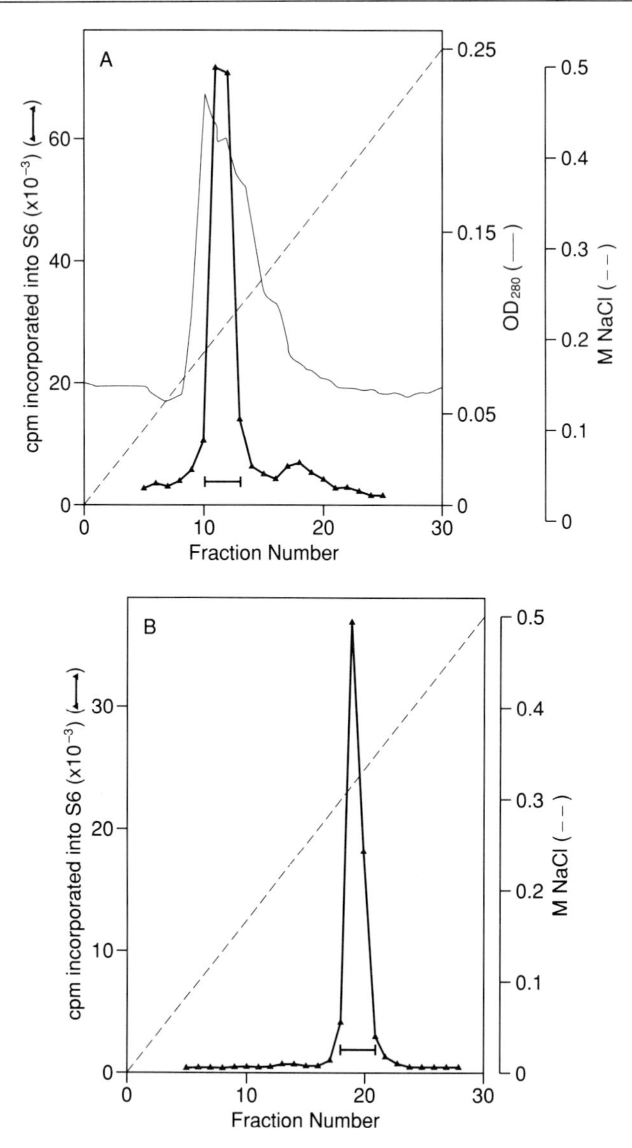


FIG. 5. Elution profile of 3T3 cell S6 kinase on (A) Mono S, (B) Mono Q, (C) S-200 [arrowheads indicate elution position of dextran blue (0), bovine serum albumin (1), and ovalbumin (2)], and (D) S6 peptide columns. For details see text. Bar indicates pooled fractions.

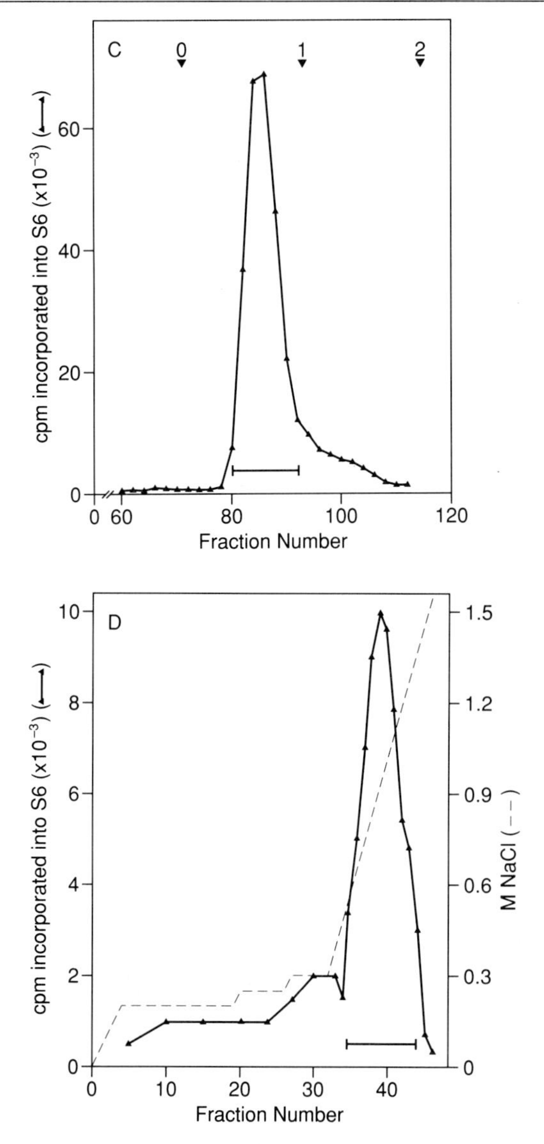


FIG. 5. (*continued*)

TABLE II
PURIFICATION OF S6 KINASE FROM 3T3 CELLS

| Purification steps | Total activity (pmol/min) | Recovery (%) |
|---|---|---|
| 3T3 cell extract | 2530 | 100 |
| Mono S pool | 1810 | 72 |
| Mono Q pool | 1320 | 52 |
| S200 pool | 1010 | 40 |
| S6 peptide pool | 780 | 31 |

elutes at an $M_r$ of ~80,000, before the bovine serum albumin standard. The S6 kinase assay is performed without diluting the samples and the fractions shown in Fig. 5C are pooled.

### *S6 Peptide Affinity Chromatography*

The procedure is as described for the rat liver S6 kinase purification, except there is no dilution of the S-200 pool (24 ml, protein not detectable by OPA) before loading and, as the protein concentration is too low to be detected by $OD_{280}$, the three plateaus at 0.2, 0.25, and 0.3 *M* NaCl are held for 15, 6, and 5 fractions, respectively. The individual fractions are collected in siliconized tubes, assayed at a 1 : 10 dilution, and those shown in Fig. 5D (10 ml, protein not detectable by OPA) are frozen separately in liquid nitrogen or in 50% glycerol at −20°.

### *Results and Comments*

A typical purification procedure is shown in Table II. The total recovery is 31%, which is significantly higher than we previously reported.[9] This is probably due to the smaller scale purification and lower volumes, which enable quick processing of the enzyme through the early stages of the purification. Again, this procedure is completed within 2 days, freezing only after the cell extraction step.

The enzyme elutes from the Sephacryl S-200 column at $M_r$ ~80,000, as previously reported.[9,11] When the preparation from the final stage of the purification is autophosphorylated and analyzed by SDS–PAGE, a single autophosphorylating band is observed at $M_r$ 70,000 (Fig. 6). This property again parallels the S6 kinase activity profile (not shown).

The 3T3 cell has provided a model system in which the function and regulation of the S6 kinase, *in vivo,* may be directly investigated. This purification procedure, although on a small scale, has proved invaluable for such studies.

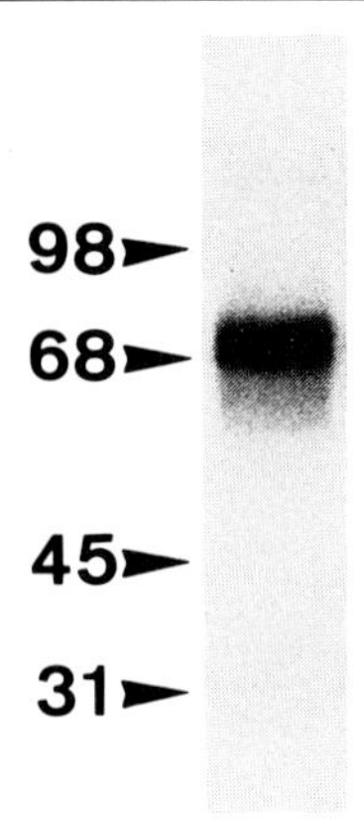


FIG. 6. Autoradiography of autophosphorylated S6 kinase purified from 3T3 cells. A portion (3 $\mu$l) of the S6 peptide pool was incubated alone with [$\gamma$-$^{32}$P]ATP, separated by SDS-PAGE, and autoradiographed, as described in General Procedures. The film has been overexposed to reveal contaminating bands. Molecular weight markers ($\times 10^{-3}$) are indicated.

## Concentration and Storage of Purified S6 Kinase

For some applications it is often necessary to further concentrate the purified S6 kinase. We routinely use a 250-$\mu$l Mono Q anion-exchange column (0.5 $\times$ 1.3 cm). This is stored and equilibrated as previously stated. Before loading, the S6 peptide column pool is diluted with Mono Q dilution buffer to a conductivity of ~7 mS/cm. At this stage the S6 kinase is very pure and, in the case of the 3T3 cell enzyme, in very low amounts. Heat-treated (56° for 30 min) cytochrome *c* (checked for purity by SDS–PAGE electrophoresis and silver staining; see General Procedures) is, therefore, added to the dilution buffer at a concentration of 0.1 mg/ml. The presence of this carrier protein does not interfere with the binding of the S6 kinase to the column. After loading, the column is washed with Mono Q buffer and a 2-ml linear salt gradient from 0 to 0.5 M NaCl in Mono Q buffer is run at 0.2 ml/min. This is followed by a 1.5-ml wash with 0.5 *M* NaCl, to compensate for the internal volume of the column. Fractions of 100 $\mu$l are collected in siliconized tubes and assayed for S6 kinase activity at a dilution of 1/10 (3T3 cell enzyme preparation) or 1/50 (rat liver enzyme preparation). The elution characteristics are as previously shown in Fig. 5B, with the kinase activity peaking at ~0.32 *M* NaCl. Using this technique, the purified S6 kinase can be concentrated to a volume of 0.6–0.8 ml. Activity losses are negligible in the case of the rat liver enzyme preparation, but may be as high as 10% in the case of the 3T3 cell enzyme

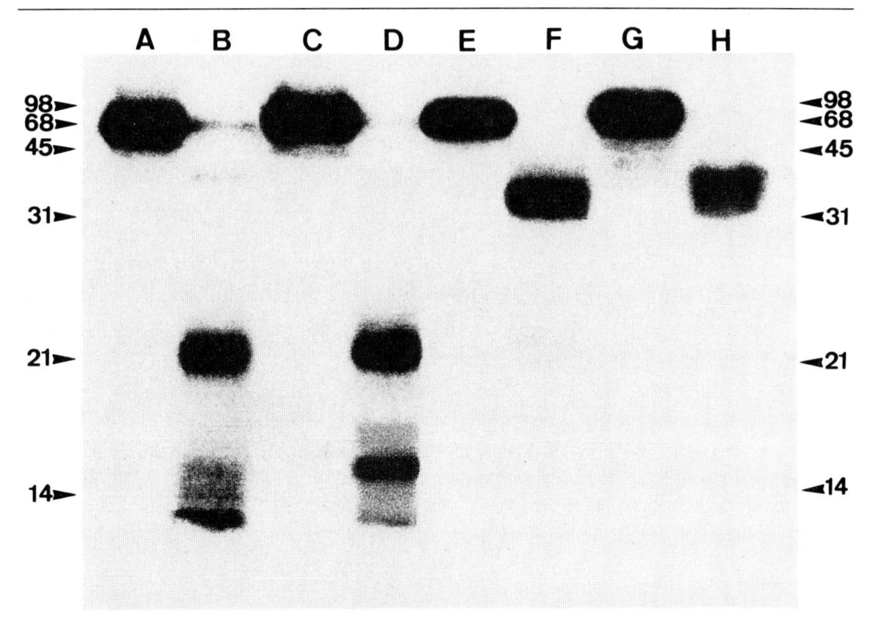


FIG. 7. Peptide maps of autophosphorylated S6 kinase from cycloheximide-treated rat liver (lanes A, B, E, and F) or from serum-stimulated Swiss 3T3 cells (lanes C, D, G, and H). S6 kinase preparations were autophosphorylated followed by electrophoresis (lanes A, C, E, and G) or treated with either chymotrypsin (lanes B and D) or cyanogen bromide (lanes F and H) prior to electrophoresis and autoradiography (see General Procedures). Approximately 10,000 cpm was loaded onto each lane. Molecular weight markers ($\times 10^{-3}$) are indicated. (Reprinted from Kozma *et al.*[11])

preparation. For this reason, we are currently investigating the application of a smaller S6 peptide column in the 3T3 cell S6 kinase purification procedure. This would remove the need for a further concentration step.

For storage of the purified S6 kinase we have found liquid nitrogen to be adequate. Although we have not thoroughly examined long-term stability, in general we have found the enzyme to be stable provided repeated freezing and thawing is avoided. In the case of the 3T3 cell S6 kinase, storage in 50% (v/v) glycerol at $-20°$ is also used. In this state the enzyme retains its stability and is in a convenient form for repeated use without freezing and thawing.

*Properties of S6 Kinase from Rat Liver and 3T3 Cells*

Detailed biochemical analyses have been performed in order to investigate the kinetic and structural characteristics of the S6 kinase. This approach has been used as a tool to compare the enzymes purified from the two different sources. As already mentioned, both purified enzymes have a molecular weight of 70,000 and are capable of autophosphorylation. In terms of kinetic properties, a number of parameters have been investigated. Both purified enzymes are dependent on $Mg^{2+}$ (optimum concentration 5 m*M*) and are inhibited by micromolar quantities of $Mn^{2+}$ (>60% inhibition at 25 $\mu M$) in the presence of 5 m*M* $Mg^{2+}$ (unpublished data). The $K_m$ for ATP of the 3T3 cell enzyme is 28 $\mu M$,[9] and for the rat liver enzyme 25 $\mu M$ (unpublished data). As already stated, the 3T3 cell enzyme is activated by phosphorylation and preferentially dephosphorylated by a type 2A phosphatase.[10,26] It has been shown[11] that both purified enzymes, which are stable in the absence of phosphatase 2A, are rapidly inactivated at equivalent rates by the addition of phosphatase 2A.

In terms of structural relatedness, peptide maps of both enzymes have been compared. For the generation of peptide maps both molecules are incubated with cyanogen bromide or chymotrypsin following autophosphorylation. Analysis of the cleaved labeled fragments by SDS–PAGE shows almost identical patterns (Fig. 7). Based on the peptide maps and the properties listed above, the two enzymes are judged to be equivalent.

## Acknowledgments

We thank S. Ferrari, S. C. Kozma, D. G. Grosskopf, and D. Reddy for their critical reading of the manuscript. We also thank C. Wiedmer for her excellent secretarial assistance.

[26] L. M. Ballou, M. Siegmann, and G. Thomas, *Proc. Natl. Acad. Sci. U.S.A.* **85,** 7154 (1988).

# [23] M Phase-Specific cdc2 Kinase: Preparation from Starfish Oocytes and Properties

*By* JEAN-CLAUDE LABBÉ, JEAN-CLAUDE CAVADORE, and MARCEL DORÉE

The M phase-specific cdc2 kinase (also called M phase-specific H1 histone kinase) is a serine/threonine kinase; it is $Ca^{2+}$, diacylglycerol, and cyclic nucleotide independent, and was first discovered in maturing oocytes of starfish and amphibian oocytes. This kinase demonstrates dramatic changes in its activity during the course of the eukaryotic cell cycle.[1–4] It is identical to MPF (the M phase-promoting factor[5,6]) and is thus sufficient to push $G_2$-arrested cells into M phase[7] or to induce a variety of mitotic events in cell-free systems.[8–11] MPF has no species specificity and is able to induce M phase across phylogenetic boundaries.[12] Here we describe a rapid and efficient procedure to purify the M phase-specific cdc2 kinase from starfish oocytes at first meiotic metaphase. One milligram of the final preparation catalyzes the transfer of about 5 $\mu$mol of phosphate from ATP to H1 histone in 1 min under standard conditions. The kinase is a complex formed by the stoichiometric association of one molecule of cdc2 (a 34-kDa catalytic subunit) with one molecule of cyclin B (a 42-kDa regulatory subunit).[13] Both subunits are homologs to products of cell cycle control genes in yeast.[14] The main step in this procedure is affinity chromatography on the yeast p13$^{suc1}$ protein, which binds to the yeast cdc2/CDC28 proteins and their homologs in higher eukaryotes.[14,15] Any kinase complex containing a cdc2 homolog should be purified using only minor modifications of this procedure, provided the catalytic subunit is free to interact physically with the yeast p13$^{suc1}$ protein.

## Preparing Starfish Oocytes, Homogenates, and High-Speed Supernatants

### *Animals and Oocyte Extracts*

Several starfish species can be obtained in huge quantities from marine biological stations. Females should be captured during their breeding season (which depends on species) and kept in running seawater before use.

[1] A. Picard, G. Peaucellier, F. Le Bouffant, C. Le Peuch, and M. Dorée, *Dev. Biol.* **109,** 311 (1985).
[2] A. Picard, J. C. Labbé, G. Peaucellier, F. Le Bouffant, C. Le Peuch, and M. Dorée, *Dev., Growth Differ.* **29,** 93 (1987).
[3] J. P. Capony, A. Picard, G. Peaucellier, and M. Dorée, *Dev. Biol.* **117,** 1 (1986).
[4] J. C. Labbé, A. Picard, E. Karsenti, and M. Dorée, *Dev. Biol.* **127,** 157 (1988).
[5] M. Lohka, *J. Cell Sci.* **92,** 131 (1989).
[6] M. Dorée, *Curr. Opin. Cell Biol.* **2,** 269 (1990).
[7] J. C. Labbé, A. Picard, G. Peaucellier, J. C. Cavadore, P. Nurse, and M. Dorée, *Cell (Cambridge, Mass.)* **57,** 253 (1989).
[8] M. Lohka, J. Hayes, and J. L. Maller, *Proc. Natl. Acad. Sci. U.S.A.* **85,** 3009 (1988).
[9] T. Tuomikoski, M. A. Felix, M. Dorée, and J. Grunberg, *Nature (London)* **342,** 942 (1989).
[10] F. Verde, J. C. Labbé, M. Dorée, and E. Karsenti, *Nature (London)* **343,** 233 (1990).
[11] M. Peter, J. Nakagawa, M. Dorée, J. C. Labbé, and E. A. Nigg, *Cell (Cambridge, Mass.)* **61,** 591 (1990).
[12] T. Kishimoto, R. Kuriyama, H. Kondo, and H. Kanatani, *Exp. Cell Res.* **137,** 121 (1982).
[13] J. C. Labbé, J. P. Capony, D. Caput, J. C. Cavadore, J. Derancourt, M. Kaghad, J. M. Lelias, A. Picard, and M. Dorée, *EMBO J.* **8,** 3053 (1989).
[14] C. Norbury and P. Nurse, *Biochim. Biophys. Acta* **989,** 85 (1989).
[15] L. Brizuela, G. Draetta, and D. Beach, *EMBO J.* **6,** 3507 (1987).

Marine biological stations like that in Roscoff, France can provide laboratory facilities to prepare starfish extracts. We selected the starfish *Marthasterias glacialis* for studying the kinase. Females of this species contain about $10^6$ transparent oocytes, all similar in size (about 160 $\mu$m in diameter) and arrested with a prominent nucleus (the germinal vesicle, 60 $\mu$m in diameter) at first meiotic prophase. Such oocytes can be induced to enter M phase very synchronously in the laboratory by stimulation with 1-methyladenine, the hormone responsible for meiotic maturation in starfish.[16]

### *High-Speed Extracts*

Oocytes are prepared free of follicle cells and the jelly coat which surrounds the oocytes and then treated with 1-methyladenine. Homogenates are prepared 40 min after hormonal stimulation, when the oocytes have reached first meiotic metaphase.

### *Experimental Protocol*

1. Cut an arm of the starfish with a pair of scissors. Remove the two ovaries it contains (they occupy most of the space within an arm in a ripe female) and place them in a 10-cm Petri dish containing 50 ml of ice-cold artificial calcium-free seawater (10 m*M* KCl, 30 m*M* $MgCl_2$, 17 m*M* $MgSO_4$, 500 m*M* NaCl) adjusted to pH 5.5 with HCl. Tear the ovaries with fine forceps. Remove any fragments by filtration through cheesecloth. The absence of $Ca^{2+}$ releases the oocytes from follicle cells, and the acidic pH solubilizes the jelly coat.
2. Centrifuge the oocytes at low speed (less than 100 *g*) at 4° for 1 min. Resuspend the oocytes (packed volume about 5 ml) in 5–10 vol of ice-cold calcium-free seawater at pH 5.5 and centrifuge again. Repeat the washing procedure once more, then resuspend the oocytes in at least 10 vol artificial calcium-free seawater buffered at pH 8.2 with 2 m*M* Tris-HCl, and maintain at room temperature.
3. Repeat steps 1 and 2 for each arm. Pool the oocytes.
4. Add 1-methyladenine to the final concentration of 1 $\mu M$ (from a 1 m*M* stock solution in distilled water). Germinal vesicle breakdown, easily detected by microscopic observation, should occur about 20 min after 1-methyladenine addition.
5. Wait for 20 min after germinal vesicle breakdown, then collect the oocytes using low-speed centrifugation (<100 *g*) and wash them quickly (less than 30 sec) with 10 vol of an ice-cold buffer containing 144 m*M*

[16] H. Kanatani, H. Shirai, K. Nakanishi, and T. Kurokawa, *Nature (London)* **211,** 273 (1969).

β-glycerophosphate, 34 m$M$ EGTA, 27 m$M$ $MgCl_2$, 1.8 m$M$ dithiothreitol (DTT), 200 m$M$ sucrose, 200 m$M$ KCl at pH 7.8 (buffer A). Pellet the oocytes by low-speed centrifugation (<100 $g$) for 1 min. All the subsequent steps are performed at 0–4°.

6. Add 2 vol (approximately 80 ml) of twofold diluted buffer A. Crush the oocytes in a Potter homogenizer with a Teflon pestle (three strokes). Centrifuge the homogenate at 12,000 $g$ for 15 min in a fixed angle rotor.

7. Collect the supernatant and centrifuge it again for 40 min at 140,000 $g$ in a swinging bucket rotor (TST 28.38, Kontron, Zurich, Switzerland). Separate the supernatant (approximately 60 ml) into 12-ml aliquots (plastic vials) and freeze in liquid nitrogen. They can be kept at −70° for several years without detectable loss of activity.

## Production of Yeast p13$^{suc1}$ Protein

### *Bacterial Strain*

The intronless form of the fission yeast *suc1*$^+$ gene[17,18] is inserted in an ampicillin-resistant plasmid under the control of the gene 10 promoter of the T7 bacteriophage[15] and introduced in *Escherichia coli* strain BL21 (DE3) LysS19, which contains both a chromosomal copy of the *lacUV5* promoter and a chloramphenicol-resistant plasmid.[20]

### *Bacterial Extracts*

Bacteria are grown in the presence of both ampicillin and and chloramphenicol and induced by isopropyl-β-D-thiogalactopyranoside (IPTG). Cells are harvested by centrifugation, resuspended in a buffer containing 50 m$M$ Tris-HCl, pH 8.0,and 2 m$M$ EDTA, and broken with a French press.[21] The soluble proteins are concentrated by 30–50% ammonium sulfate precipitation, the 50% pellet solubilized in 300 ml of buffer B (50 m$M$ Tris-HCl, pH 7.5, 150 m$M$ NaCl), and heated for 3 min at 70° to inactivate proteases. Centrifugation at 5000 $g$ for 15 min removes a large amount of heat-denatured bacterial proteins and yields a clear supernatant containing at least 93% of the accumulated p13$^{suc1}$ protein.

[17] J. Hayles, D. Beach, B. Durkacz, and P. Nurse, *Mol. Gen. Genet.* **202,** 291 (1986).

[18] J. Hindley and G. Phear, *Gene* **31,** 129 (1984).

[19] F. Studier and B. Moffat, *J. Mol. Biol.* **189,** 113 (1986).

[20] The bacterial strain overexpressing the yeast p13$^{suc1}$ protein was constructed in the Paul Nurse laboratory (Cell Cycle Control Group Microbiology Unit, Department of Biochemistry, University of Oxford U.K.).

[21] Large amounts of bacterial extracts can be obtained from any specialized fermentation units, like the Laboratoire de Chimie Bacterienne du CNRS (Marseille, France).

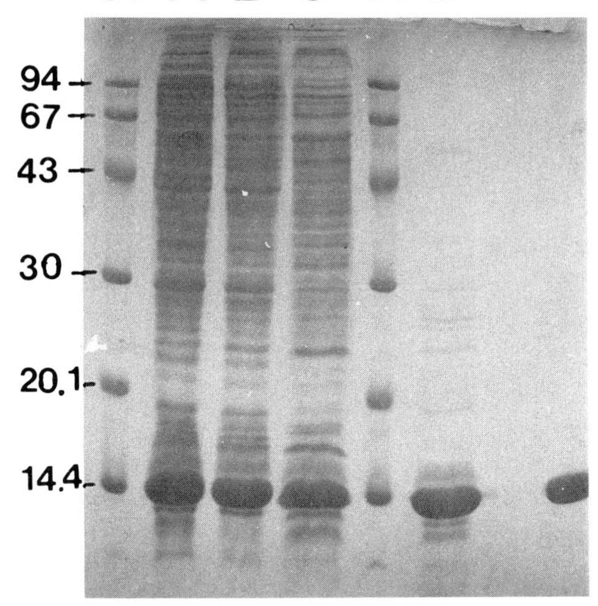


FIG. 1. Purification of the yeast p13*suc1* protein from bacterial extracts. Analysis by SDS–PAGE and Coomassie Blue staining of the successive steps of the purification. *From left to right:* Molecular weight ($\times 10^{-3}$) markers (St), bacterial extract (A), 30–50% ammonium sulfate precipitation pellet (B), heat-step supernatant (C), pooled fractions after AcA 44 gel filtration (D), pooled fractions after ionic exchange on Mono Q (E).

*Purification of Yeast p13*$^{suc1}$ *Protein*

The above supernatant is loaded onto an AcA 44 gel-filtration column (100 × 5 cm; IBF, Villeneuve-la-Garenne, France) equilibrated in buffer B. Elution is performed at 100 ml/hr. Fractions containing p13$^{suc1}$, as determined by SDS–PAGE (15% polyacrylamide), are pooled, dialyzed extensively against water, and lyophilized. A 10-liter batch yields 1.6 g of the yeast protein, approximately 90% pure.

Final purification to apparent homogeneity (Fig. 1) is achieved by ion-exchange fast protein liquid chromatography (FPLC) on a Mono Q column [HR 10/10 (Pharmacia, Uppsala, Sweden), flow rate 3 ml/min] using a linear gradient of 0 to 1 *M* NaCl in 50 m*M* Tris-HCl, pH 7.5. The yeast protein p13$^{suc1}$, eluting at 0.15 *M* NaCl, is extensively dialyzed against distilled water and lyophilized.

The purified p13$^{suc1}$ protein is conjugated to CNBr-activated Sepharose

4B (Pharmacia) at 3.5 mg of protein/ml of gel according to the instructions of the manufacturer.

## Purification of Starfish cdc2 Kinase

All the following steps are performed at 4°.

1. Dialyze the thawed starfish extracts (60 ml) against 2 liters of buffer C (5 m*M* β-glycerophosphate, 1.5 m*M* $MgCl_2$, 1 m*M* DTT adjusted at pH 7.3) for 3 hr with two buffer changes.
2. Clarify the dialyzed extract by centrifugation at 100,000 *g* for 1 hr (TFT 50.38 rotor; Kontron). Ionic strength of the supernatant should be <2.5 mmho. If higher, dilute with distilled water.
3. Load the supernatant onto a 2.5 × 40 cm DEAE-cellulose column (Whatman, Maidstone, England) equilibrated with buffer C containing 35 m*M* NaCl. Wash the column with the same buffer until absorbance at 280 nm has returned to its basal level. Elute proteins in one step with buffer C containing 200 m*M* NaCl. Collect 10-ml fractions and assay them for H1 histone kinase activity (see below, Assay of cdc2 Kinase Activity).
4. Load fractions containing H1 histone kinase activity onto a 5-ml column of $p13^{suc1}$-Sepharose 4B previously equilibrated with buffer C containing 200 m*M* NaCl (load first the fractions of higher activity). Wash the column, first with 5 ml of buffer C containing 200 m*M* NaCl, then with 50 ml of a buffer containing 50 m*M* Tris-HCl, pH 6.8, 1 m*M* EGTA, 1 m*M* DTT (buffer D). Elute cdc2 kinase with 6 ml of buffer D containing 30 mg/ml of $p13^{suc1}$, followed by 6 ml of buffer D.
5. Pool the 12 ml of buffer D containing the cdc2 kinase and $p13^{suc1}$ and load onto a Mono S column (HR 5/5; Pharmacia) equilibrated with buffer D. Elute with a linear gradient of 0 to 0.6 *M* NaCl in buffer D (flow rate 0.8 ml/min). The yeast $p13^{suc1}$ protein does not bind to the matrix and is recovered in the flow through. H1 histone kinase activity is quantitatively recovered as a sharp peak eluting at 0.3 *M* NaCl (Fig. 2). Active fractions contain only two polypeptides of apparent $M_r$ 34,000 and 47,000 associated in a stoichiometric 1 : 1 ratio (Fig. 3). These polypeptides have been identified by direct microsequencing to be cdc2 and cyclin B, the catalytic and the regulatory subunits of the M phase-specific cdc2 kinase.[13]

### *Assay of cdc2 Kinase Activity*

The reaction mixture (40 μl) contains 20 m*M* HEPES–NaOH, pH 7.4, 10 m*M* $MgCl_2$, 0.2 m*M* [$\gamma$-$^{32}$P]ATP (50 cpm/pmol), 1 mg/ml calf thymus H1 histone (Boehringer, Mannheim, Germany). The reaction is initiated with the addition of enzyme and incubation is carried at 30° for 5 min. The

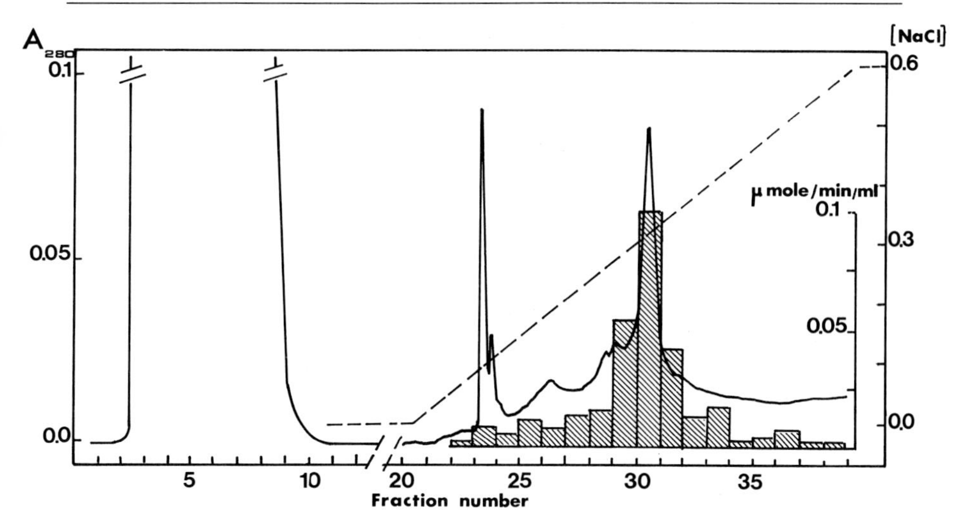


FIG. 2. Typical elution profile of the M phase-specific cdc2 kinase from Mono S column. Absorbance at 280 nm (solid line) was monitored continuously and H1 histone kinase activity (histogram) assayed in each fraction (0.8 ml). No kinase activity was detected in the flow through.

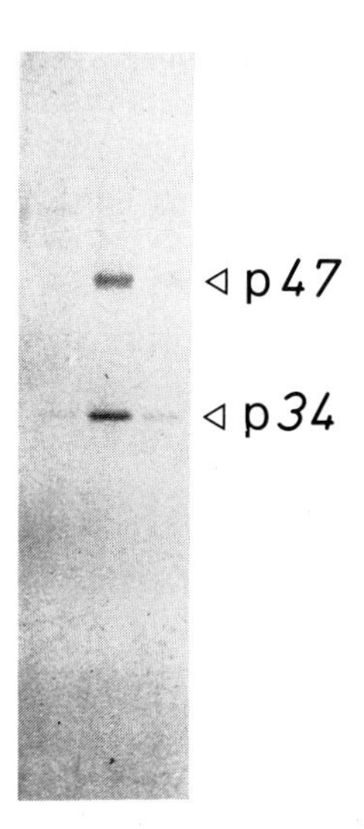


FIG. 3. Coomassie Blue staining of the final preparation of the starfish cdc2 kinase after its analysis by SDS–PAGE (12.5% polyacrylamide). Three successive fractions through the final peak of activity are shown, with the middle fraction containing the highest activity.

reaction is terminated by pipetting aliquots of the mixture onto 1 × 1 cm pieces of P81 ion-exchange paper (Whatman).[22] The papers are immersed in tap water and subjected to four washes for 5 min followed by drying and counting in liquid scintillation solution.

## Properties of Purified cdc2 Kinase

### *Stability*

The kinase can be stored without detectable loss of activity for several months at −70° after freezing it as such after the final step in its purification (0.1 mg/ml of protein). It loses about one-half of its activity after three consecutive freezing–thawing steps under such conditions. No detectable loss of activity is observed when the kinase is kept at 0° for 6 hr.

### *Specificity*

H1 histone is the best *in vitro* substrate of starfish cdc2 kinase known at present. Phosphorylation seems limited to the carboxyl- and amino-terminal domains of H1 histone at sites containing the K-S/T-P-K or K-S/T-P-X-K sequences, which are also phosphorylated at mitosis.[23] The major nucleolar proteins nucleolin and No. 38, as well as type B lamins, are phosphorylated *in vitro* on evolutionary conserved sites that also correspond to those observed during mitosis *in vivo*.[11,24,25] Nucleolin and No. 38 phosphorylation occur on both serine and threonine residues in T/S-P-X-K sequences, while a target site for cdc2 kinase in type B-lamins has been identified as serine in an L-S-P-T-R motif. This suggests that the consensus sequence for phosphorylation by the M phase cdc2 kinase is X-T/S-P-X basic. Nonetheless, the starfish kinase also phosphorylates peptides containing the R-S-P-T-S*-P-S-Y and R-T-P-S-T*-P-S-Y motifs at the indicated serine and threonine residues (asterisks), suggesting that proteins containing simply X-T/S-P-X sequences, not X-T/S-P-X basic, may also be putative substrates. Although it does not contain any X-S-P-X basic motifs, the cyclin B subunit readily undergoes phosphorylation on at least two serine residues in highly purified preparations of the starfish cdc2 kinase.[13]

[22] J. J. Witt and R. Roskowski, *Anal. Biochem.* **66,** 253 (1975).

[23] T. A. Langan, J. Gauthier, M. Lohka, R. Hollingsworth, S. Moreno, P. Nurse, J. Maller, and R. A. Sclafani, *Mol. Cell. Biol.* **9,** 3860 (1989).

[24] P. Belenguer, M. Caizergues-Ferrer, J. C. Labbé, M. Dorée, and F. Amalric, *Mol. Cell. Biol.* **10,** 3607 (1990).

[25] M. Peter, J. Nakagawa, M. Dorée, J. C. Labbé, and E. A. Nigg, *Cell (Cambridge, Mass.)* **60,** 791 (1990).

*Kinetic Properties*

The apparent $K_m$ for ATP and calf thymus H1 histone are 50 and 2 $\mu M$, respectively. GTP can also serve as a phosphate donor. The optimum magnesium concentration is 10–15 m$M$. Starfish cdc2 kinase can also use $Mn^{2+}$ as a divalent cation. The optimum pH is between 7.2 and 7.7 with 20 m$M$ HEPES–NaOH as a test buffer. The kinase is not inhibited by heparin (3 $\mu$g/ml) or by $Ca^{2+}$ (0.3 m$M$). In contrast, pyrophosphate ($IC_{50}$ 2 m$M$) and $Zn^{2+}$ ($IC_{50}$ 1 m$M$) are potent inhibitors.

## Induction of Mitotic Events

Microinjection of the homogeneous M phase-specific cdc2 kinase into oocytes arrested at first meiotic prophase induces nuclear envelope breakdown, chromosome condensation, and spindle formation even in the absence of protein synthesis and is active across phylogenetic boundaries, working so far in all tested species, both vertebrate and invertebrate.[7] Methods for microinjection into starfish and amphibian oocytes are described elsewhere.[26,27] In starfish and *Xenopus* recipient oocytes, the minimal activity of the cdc2 kinase should be 30 and 10 pmol of phosphate to H1 histone/$\mu$l of cytosol, respectively, for successful induction of entry into M phase.[7] The kinase also triggers premature chromosome condensation when injected in $G_1$ cells.[7,28] It induces a variety of mitotic events when added to extracts prepared from *Xenopus* eggs arrested at $G_2$, including a reduced mean elongation rate (from 20 to 4 $\mu$m/min) and an increased turnover of centrosome-nucleated microtubules,[10] and the inhibition of endocytotic vesicle fusion.[9] The M phase-specific cdc2 kinase is also capable of inducing disassembly of nuclear lamina on incubation with isolated nuclei.[11] It has been reported that MPF, partially or extensively purified from amphibian oocytes, fails to induce nuclear envelope breakdown and chromosome condensation in oocytes or in cell-free systems in the absence of $\beta$-glycerophosphate.[29] This is not the case for pure starfish cdc2 kinase, which induces all M phase-specific events in the absence of any added chemical.

## Application to *Xenopus* Eggs

The M phase-specific cdc2 kinase can be prepared from unfertilized *Xenopus* eggs (arrested at second meiotic metaphase) with only minor

[26] T. Kishimoto, *Methods Cell Biol.* **27,** 379 (1986).

[27] J. L. Maller, this series, Vol. 99, p. 219.

[28] N. J. C. Lamb, A. Fernandez, A. Watrin, J. C. Labbé, and J. C. Cavadore, *Cell (Cambridge, Mass.)* **90,** 151 (1990).

[29] E. Erickson and J. L. Maller, *J. Biol. Chem.* **264,** 19577 (1989).

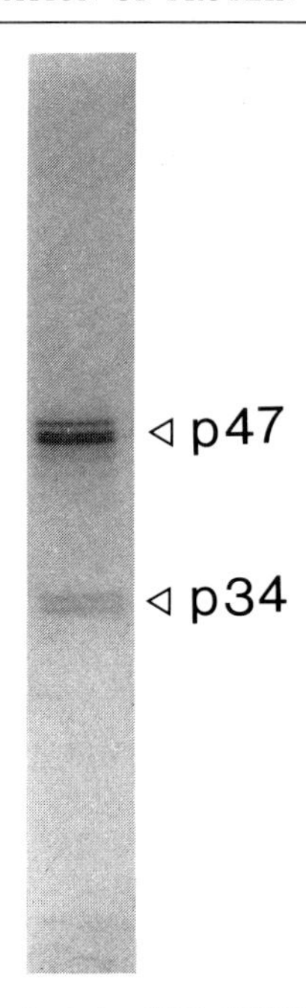


FIG. 4. Autoradiogram after analysis by SDS–PAGE (12.5% polyacrylamide) of the M phase-specific cdc2 kinase prepared from metaphase II-arrested *Xenopus* oocytes labeled with [$^{35}$S]methionine.

modifications of the procedure used for starfish oocytes. Both enzymes have very similar if not identical properties, including specificity.[29]

*Preparation of Homogenates*

Female *Xenopus laevis* are first injected with 125 units of pregnant mare gonadotropin (PMSG; Intervet, Boxmeer, Holland). Three to 5 days later they are injected with 1000 units of human chorionic gonadotropin (HCG, Sigma, St. Louis, MO) and kept in 0.1 *M* NaCl. Eggs laid during the following 12- to 18-hr period are collected and washed several times with gentle manual agitation with 10 vol of 2% (w/v) cysteine hydrochloride (Sigma) adjusted to pH 7.8 with NaOH. The cysteine treatment removes the jelly coat and decreases by about twofold the apparent volume of pelleted eggs. As soon as the jelly has been removed, the eggs are rapidly washed, first with 0.1 *M* NaCl, then with an ice-cold buffer containing 80 m*M* sodium β-glycerophosphate, 15 m*M* sodium EGTA, 10 m*M* $MgCl_2$, 1 m*M* DTT at pH 7.3. Finally, the eggs are homogenized in 1 vol of the same buffer, using a Potter with a Teflon pestle. Homogenates are centrifuged at 100,000 *g* for 40 min at 4° in a swinging bucket rotor. The soluble fraction below the yellow fat layer is collected, frozen in liquid nitrogen, and kept at −70° until use.

*Purification of Xenopus M Phase-Specific cdc2 Kinase*

Frozen extract (100 ml) is thawed, clarified by centrifugation at 100,000 *g* for 30 min, then ammonium sulfate is added to 45% saturation. The pellet of proteins is solubilized in 25 ml of a buffer containing 5 m*M* sodium β-glycerophosphate, 1 m*M* sodium EGTA, 1.5 m*M* $MgCl_2$, 1 m*M* DTT at pH 7.3. Final desalting is achieved by passing this material through Sephadex G-25 disposable columns (a battery of PD-10 columns from Pharmacia, Uppsala, Sweden). After adjusting the ionic strength to 2 mmho, the filtrate is applied to a 150-ml DEAE-cellulose column equilibrated with the above buffer. Elution from the DEAE column, binding to $p13^{suc1}$-Sepharose, and elution with $p13^{suc1}$ in excess, as well as the final Mono S step, are performed exactly as described for the M phase-specific cdc2 kinase from starfish oocytes. The *Xenopus* kinase elutes at 275 m*M* NaCl. Figure 4 shows the final preparation after chromatography on Mono S. Besides cdc2 (a doublet of $M_r$ 34,000 polypeptides), it contains a few polypeptides with an apparent $M_r$ of 47,000. We found that at least one of these components is cyclin $B_1$[30], as determined by direct microsequencing.

[30] J. Minshull, J. Blow, and T. Hunt, *Cell (Cambridge, Mass.)* **56,** 947 (1989).

# [24] Purification of Protein Kinases That Phosphorylate the Repetitive Carboxyl-Terminal Domain of Eukaryotic RNA Polymerase II

*By* LARS J. CISEK and JEFFRY L. CORDEN

RNA polymerase II (Pol II), the enzyme that transcribes protein-coding genes in eukaryotic cells, is a multisubunit complex made up of between 10 and 12 polypeptides with a combined molecular weight of >500K.[1,2]

[1] R. Roeder, *in* "RNA Polymerases" (R. Losick and M. Chamberlin, eds.). Cold Spring Harbor Lab., Cold Spring Harbor, New York, 1976.
[2] A. Sentenec, *CRC Crit. Rev. Biochem.* **1,** 31 (1985).

Analysis of Pol II subunit sequences has revealed extensive homologies among the largest subunits of Pol II and the other multisubunit eukaryotic and prokaryotic RNA polymerases.[3-7] These subunits form a globular catalytic core that is likely to be involved in the fundamental function of the protein—polymerization of ribonucleotides into a complementary RNA. In addition to the catalytic core, Pol II contains an unusual extension at the carboxyl terminus of its largest subunit.[4,8] This carboxyl-terminal domain (CTD) consists of tandem repeats of the consensus sequence Tyr-Ser-Pro-Thr-Ser-Pro-Ser. This repetitive moiety is not found in prokaryotic RNA polymerase or in eukaryotic RNA polymerases I or III, but is found in all Pol II sequenced[9] where it plays an essential,[10,11] albeit unknown, role.

The Pol II largest subunit can exist *in vivo* in two different forms designated IIo and IIa. There two forms display different mobilities on sodium dodecyl sulfate (SDS)-polyacrylamide gels (SDS-PAGE) due to the phosphorylation state of the CTD, with IIo, the lower mobility form, being highly phosphorylated.[12] IIo is detected as the form functioning in the elongation reaction *in vivo*.[12,13] The change from IIa to IIo has been associated with the transition from the initiation to the elongation phase of the transcription reaction,[14] a point at which a $\beta$–$\gamma$ phosphate bond hydrolysis is required.[15,16]

The observation that the CTD of actively elongating Pol II is highly phosphorylated and that this modification could be an obligate part of the

[3] J. M. Ahearn, Jr., M. S. Bartolomei, M. L. West, L. J. Cisek, and J. L. Corden, *J. Biol. Chem.* **262,** 10695 (1987).

[4] L. A. Allison, M. Moyle, M. Shales, and C. J. Ingles, *Cell (Cambridge, Mass.)* **42,** 599 (1985).

[5] J. Biggs, L. L. Searles, and A. L. Greenleaf, *Cell (Cambridge, Mass.)* **42,** 611 (1985).

[6] D. Falkenberg, B. Dworniczak, D. M. Faust, and E. K. F. Bautz, *J. Mol. Biol.* **195,** 929 (1987).

[7] D. Sweetser, M. Nonet, and R. A. Young, *Proc. Natl. Acad. Sci. U.S.A.* **84,** 1192 (1987).

[8] J. L. Corden, D. L. Cadena, J. M. Ahearn, and M. E. Dahmus, *Proc. Natl. Acad. Sci. U.S.A.* **82,** 7934 (1985).

[9] J. L. Corden, *Trends Biochem. Sci.* **15,** 383 (1990).

[10] M. S. Bartolomei, N. F. Halden, C. R. Cullen, and J. L. Corden, *Mol. Cell. Biol.* **8,** 330 (1988).

[11] L. A. Allison, J. K-C. Wong, V. D. Fitzpatrick, M. Moyle, and C. J. Ingles, *Mol. Cell. Biol.* **8,** 321 (1988).

[12] D. L. Cadena and M. E. Dahmus, *J. Biol. Chem.* **262,** 12468 (1987).

[13] B. Bartholomew, M. E. Dahmus, and C. F. Meares, *J. Biol. Chem.* **261,** 14226 (1986).

[14] J. M. Payne, P. J. Laybourn, and M. E. Dahmus, *J. Biol. Chem.* **264,** 19621 (1989).

[15] D. Bunick, R. Zandomeni, S. Ackerman, and R. Weinmann, *Cell (Cambridge, Mass.)* **29,** 877 (1982).

[16] M. Sawadogo and R. G. Roeder, *J. Biol. Chem.* **259,** 5321 (1984).

| | |
|---|---|
| **Consensus** | YSPTSPS |
| **hepta-four** | $(SPTSPSY)_4$ |
| **hepta-five** | $(SPTSPSY)_5$ |
| **Arg-hepta** | $RRR(YSPTSPS)_4$ |
| **hepta-six** | $(SPTSPSY)_6$ |

FIG. 1. Sequences of heptapeptide. The amino acids are represented by their one-letter codes: P, proline; R, arginine; S, serine; T, threonine; Y, tyrosine. The consensus repeat is shown at the top, and the names of synthetic peptides are given next to their sequence on the left.

transcription process, argues that elucidating the function of the CTD and its role in the transcription process would be facilitated by the study of CTD phosphorylation. For this reason we designed an assay to detect CTD kinase and have purified enzymes which catalyze this reaction. We have used the assays described below to isolate two protein kinase complexes, designated CTD kinases, that phosphorylate serine residues in the consensus heptapeptide repeat that composes the CTD. In a previous publication we reported the purification of one of these activities.[17] More recent refinements have led to the discovery of the second CTD kinase. Unexpectedly, both of these complexes contain $p34^{cdc2}$, a gene product required for progression through the cell cycle.

## Preparation of Peptide Substrates

Because the CTD is composed almost entirely of tandem heptapeptide repeats with the consensus sequence Tyr-Ser-Pro-Thr-Ser-Pro-Ser, we reasoned that a few of the heptapeptide repeats would, in isolation, adopt a structure similar to heptapeptides within the CTD. A series of heptapeptides (hepta) were synthesized for use in two assay systems as potential protein kinase substrates (Fig. 1). We have found that peptides with three or fewer repeats are poor as substrates in these systems.

Peptide substrates are prepared on an Applied Biosystems (Foster City, CA) model 430A peptide synthesizer using *tert*-butyloxycarbonyl (*t*-Boc) chemistry. Following synthesis, peptides are cleaved and deblocked using HF, then chromatographed on a preparative Vydac reversed-phase $C_{18}$ (1 × 25 cm) column (The Separations Group, Hesperia, CA) using a gradient of $H_2O$ with 0.1% (v/v) trifluoroacetic acid (TFA) to

[17] L. J. Cisek and J. L. Corden, *Nature* (*London*) **339,** 670 (1989).

acetonitrile/2-propanol (2 : 1) with 0.1% TFA at 4 ml/min as follows: 0 min, 0% organic; 30 min, 52% organic; 40 min, 76% organic; 45 min, 100% organic; 50 min, 100% organic. Appropriate fractions are pooled, lyophilized, and, for Arg-hepta, chromatographed on a Mono S 5/5 column (Pharmacia, Piscataway, NJ) using a multiphasic gradient of 0 to 2.2 *M* ammonium acetate, pH 5.5, as follows: 0 ml, 0 *M* salt; 2 ml, 0 *M* salt; 7 ml, 0.33 *M* salt; 27 ml, 0.75 *M* salt; 35 ml, 2.2 *M* salt. Arg-hepta elutes as a sharp peak at 0.45 *M* salt. Peptides hepta-four, -five, and -six are subject to a second reversed-phase purification step on an analytical Vydac (4.6 × 250 mm) column (The Separations Group) employing the previous solvents at 1.1 ml/min as follows: 0 min, 6% organic; 1 min, 6% organic; 21 min, 30% organic; 28 min, 100% organic; 35 min, 100% organic. The elution times are 13.6, 14.7, and 15.5 min for hepta-four, hepta-five, and hepta-six, respectively. Peptides are sequenced on an Applied Biosystems model 370 peptide sequencer to confirm correct synthesis, neutralized with ammonium bicarbonate, lyophilized, and resuspended in $H_2O$. Peptides are stored as a working stock at −20° or lyophilized at −70°. Peptide concentrations are determined based on $A_{274}$ with $\varepsilon = 5600$ for Arg-hepta and hepta-four, $\varepsilon = 7000$ for hepta-five, and $\varepsilon = 8400$ for hepta-six.

## Carboxyl-Terminal Domain Kinase Filter-Binding Assay

The principal assay used for the purification of CTD kinase employs the peptide Arg-hepta (Fig. 1) as the phosphate acceptor. After incubating the peptide in the presence of [$\gamma$-$^{32}$P]ATP and enzyme, labeled peptide substrate is separated from unincorporated ATP by binding the arginine residues of the peptide to phosphocellulose paper (Whatman, Clifton, NJ; P81), while removing the unbound ATP by washing.[18] This assay can also be used to assess the relative affinity of the enzyme for various heptapeptides (or other substrates) by providing a nonbinding substrate (such as hepta-four) to serve as a competitor in the reaction.

### *Procedure*

Samples are diluted in 60 m*M* KCl, 50 m*M* Tris-HCl, pH 7.9, 10 m*M* $MgCl_2$, 0.1 m*M* dithiothreitol (DTT), and 10% glycerol. A 4-$\mu$l diluted sample is incubated for 30 min at 30° in a 15-$\mu$l total volume using the following reaction conditions: 60 m*M* KCl, 50 m*M* Tris-HCl, pH 7.9, 10 m*M* $MgCl_2$, 0.1 m*M* DTT, 200 $\mu M$ ATP (0.1 Ci/mmol [$\gamma$-$^{32}$P]ATP), and 300 $\mu M$ Arg-hepta. We typically use 1 : 4 dilutions for crude fractions and

[18] E. A. Kuenzel and E. G. Krebs, *Proc. Natl. Acad. Sci. U.S.A.* **82,** 737 (1985).

~1:30 dilutions for purified preparations. The reaction is stopped by addition of an equal volume of 1 *M* acetic acid saturated with $Na_2EDTA$. The contents are then spotted onto 2 × 2 cm square Whatman P81 filters. The filters are briefly air dried, washed four times for 3 min each with 200 ml of 75 m*M* phosphoric acid ($H_3PO_4$), then briefly with a 50:50 mix of 75 m*M* $H_3PO_4$: ethanol, and dried under a heat lamp. Filters are then counted in a scintillation counter, and yield between 10,000 and 100,000 cpm in active fractions with backgrounds of 100–500 cpm. A unit of activity is defined as the quantity of enzyme required to transfer 1 μmol of phosphate/min to the peptide under these conditions.

## Carboxyl-Terminal Domain Kinase Sodium Dodecyl Sulfate Gel Assay

A gel assay has also been employed in the purification of CTD kinases. Heptapeptides consisting of the consensus repeat alone (hepta-four, -five, or -six; Fig. 1) are provided as a substrates for phosphorylation in the presence of [$\gamma$-$^{32}$P]ATP. Labeled products are separated using SDS-PAGE and autoradiographed. This assay has the advantage of utilizing the resolution of SDS-PAGE to separate the exogenous peptide from endogenous substrates which may be present in the extract or fraction. The filter assay, on the other hand, will bind and generate a signal from any strongly positive-charged molecule which is phosphorylated. For instance, histone H1 kinase activity can be measured by providing histone H1 as a substrate instead of heptapeptides in the filter-binding assay. Another potential disadvantage of the filter assay system is that the substrate contains the sequence Arg-Arg-Arg-Tyr-Ser, which will serve (under proper conditions) as a substrate for cAMP-dependent kinase and any other kinase which might recognize this sequence. We have observed the phosphorylation of this site only with the purified catalytic subunit of cAMP-dependent kinase.

### *Procedure*

The gel assay is performed under the same reaction conditions as the filter assay except that hepta-four, -five, or -six is used as the phosphate-accepting substrate. The reaction is stopped by addition of SDS-PAGE loading buffer with EDTA [100 m*M* Tris-HCl, pH 6.8, 1% (w/v) SDS, 1% (v/v) 2-mercaptoethanol, 10% (v/v) glycerol, and 20 m*M* EDTA] and boiling for 3 min. The reaction products are electrophoresed on a 15% (30:0.8 acrylamide : bisacrylamide, v/v) discontinuously buffered SDS-polyacrylamide gel[19] until the dye front is run from the gel. Optimal resolution of

[19] U. K. Laemmli, *Nature (London)* **227,** 680 (1970).

the labeled peptide is obtained when the samples are electrophoresed rapidly, thus minimizing diffusion of the peptide. The gel is fixed for 15 min in methanol : trichloroacetic acid (TCA) : acetic acid : water (3 : 1 : 1 : 5, v/v), dried under vacuum onto Whatman DE81 paper, and autoradiographed. Peptide-containing bands are cut out of the gel and counted in a scintillation counter for quantification. Note that these phosphopeptides display aberrant mobility compared to protein markers, with monophosphorylated hepta-six migrating at ~24K and hepta-four migrating at ~20K on 15% polyacrylamide gels. The apparent molecular weight varies with the percentage of polyacrylamide in the gel, a property which allows for manipulation of the position of the phosphopeptide band when comparing peptide to other substrates.

The filter-binding and gel assays are equivalent in that they detect the same peaks of kinase activity in fractionated extracts. The gel assay is ideally suited for studies in crude systems because the labeled substrate can be resolved from endogenous substrates. It is more cumbersome and less quantitative, however, than the filter-binding assay. The gel system also separates the monophosphorylated and the various polyphosphorylated forms which, while useful, further complicate the quantitation of total phosphate transfer. The two types of substrates (gel and filter binding) provide complementary methods to evaluate CTD kinase activities.

The CTD itself has become available for use as a substrate in a gel assay system. For this assay the isolated whole-mouse CTD, purified from *Escherichia coli* producing a TrpE-CTD fusion protein,[20] is provided at 250 n*M* in place of heptapeptide using the same reaction conditions previously described. The products of this reaction are resolved via SDS-PAGE on an 8% gel and autoradiographed. This assay allows for the observation of phosphate transfer as well as a mobility shift analogous to the IIa to IIo ($CTD_a$ to $CTD_o$) transition (see Fig. 6 for example). This assay, in combination with those described earlier, allows for the study and quantitation of both phosphate transfer and mobility shift associated with phosphorylation of the CTD of Pol II.

## Purification of Carboxyl-Terminal Domain Kinases

### *Buffers*

TED: 10 m*M* Tris-HCl, pH 7.9, 1 m*M* ethylenediaminetetraacetic acid (EDTA), 5 m*M* dithiothreitol (DTT), 0.3 m*M* phenylmethylsulfonyl fluoride (PMSF), 0.5 mg/liter leupeptin, 0.7 mg/liter pepstatin, 0.1 mg/

[20] J. Zhang and J. L. Corden, *J. Biol. Chem.* **266,** 2290 (1991).

liter *N*-tosyl-L-phenylalanine chloromethyl ketone (TPCK), 5 m*M* phosphoserine, 5 m*M* $\beta$-glycerol phosphate, and 1 m*M* levamisole

TMD: 50 m*M* Tris-HCl, pH 7.9, 10 m*M* $MgCl_2$, 2 m*M* DTT, 25% sucrose (w/v), 50% glycerol (v/v), 0.3 m*M* PMSF, 0.5 mg/liter leupeptin, 0.7 mg/liter pepstatin, 0.1 mg/liter TPCK, 5 m*M* phosphoserine, 5 m*M* $\beta$-glycerol phosphate, and 1 m*M* levamisole

Buffer A: 50 m*M* Tris-HCl, pH 7.9, 10 m*M* 2-mercaptoethanol (2-ME), 2 m*M* EDTA, 0.3 m*M* PMSF, 0.5 mg/liter leupeptin, 0.7 mg/liter pepstatin, 0.1 mg/liter TPCK, and 10% glycerol (v/v)

Buffer C: 50 m*M* $KPO_4$, pH 7.0, 2 m*M* 2-ME, 1 m*M* EDTA, and 10% glycerol (v/v)

Buffer D: 50 m*M* HEPES, pH 7.4, 2 m*M* 2-ME, 1 m*M* EDTA, and 10% glycerol (v/v)

Buffer E: 50 m*M* Tris-HCl, pH 7.9, 2 m*M* 2-ME, 1 m*M* EDTA, 0.05% 3-[(3-cholamidopropyl)dimethylammonio]-1-propane sulfonate (CHAPS), and 10% glycerol (v/v)

Buffer F: 20 m*M* Tris-HCl, pH 7.0, 5 m*M* $MgCl_2$, 2 m*M* EGTA, 0.5 m*M* DTT, and 10% glycerol (v/v)

*Preparation of Starting Extract and Ammonium Sulfate Precipitation*

A mouse ascites tumor line [Ehrlich–Lettre ascites carcinoma ATCC (Rockville, MD) CCL 77] demonstrated the highest specific activity of CTD kinase of several cell lines and tissues tested. As these cells are convenient to produce in large quantities, they have been used as the starting source for the purification. The cells can be grown and harvested from the animal, rinsed in phosphate-buffered saline (PBS), pelleted, quick frozen in liquid nitrogen, and stored at $-70°$ until needed. The cells may also be grown under contract by commercial sources (Pel Freeze, Inc., Rogers, AR).

CTD kinase activity does not partition discretely into cytoplasmic or nuclear fractions and thus a whole-cell extract is used as the starting source for purification. The initial extract is prepared in a manner similar to a Manly transcription extract.[21] Typically, 125 ml of frozen packed cells (1 vol) is thawed with agitation in a total of 500 ml TED (4 vol) using multiple small-volume washes to remove the cells as they thaw. The thawed cells are collected and transferred onto ice, and all subsequent steps are performed at 4° or on ice. Cells are Dounce homogenized in 40-ml batches with 15 strokes of a B pestle. Homogenized cells are then added to 500 ml TMD (4 vol) and the solution is stirred gently on a magnetic mixer while

[21] J. L. Manley, A. Fire, A. Cano, P. A. Sharp, and M. L. Gefter, *Proc. Natl. Acad. Sci. U.S.A.* **77,** 3855 (1980).

130 ml of saturated ammonium sulfate (~1 vol) is added over 3 min. The resulting slurry is allowed to stir for 20 min and then centrifuged in Beckman type 50.2 Ti rotors at 43,000 rpm for 3 hr, or Beckman type 45 Ti rotors (Beckman Instruments, Palo Alto, CA) at 45,000 rpm for 4 hr. The supernatant is decanted and brought to 40% saturation in ammonium sulfate by slow addition of the solid (0.183 g/ml) while the solution is gently stirred on a magnetic mixer. The solution is allowed to precipitate for 30 min and then centrifuged in a Sorvall GSA rotor (Dupont Company, Wilmington, DE) at 9000 rpm (13,000 *g*) for 20 min. The supernatant is decanted, and the pellets are resuspended in buffer A. The ammonium sulfate cut is further diluted in buffer A until the conductivity is equivalent to 75 m*M* KCl–buffer A (~2500 ml). This solution is clarified by centrifugation for 15 min at 7000 rpm (8000 *g*) in a GSA rotor.

*DEAE-Sepharose Chromatography*

The clarified protein solution is applied to a 500-ml (5 × 25 cm) DEAE-Sepharose (Pharmacia) column equilibrated to 75 m*M* KCl–buffer A, washed with one-half column volume 75 m*M* KCl–buffer A, one column volume 100 m*M* KCl–buffer A, and eluted with a linear gradient of 1700 ml from 100 to 400 m*M* KCl–buffer A (Fig. 2). The assay profile reveals two peaks of CTD kinase activity eluting between 175 and 240 m*M* salt. The biphasic peak is divided into E1 and E2, referring to the elution order, and further purification is carried out independently. We discard approximately 60–80 ml at the nadir to assure clean separation of these peaks. E2 is the enzyme we have previously described as CTD kinase.[17] Note that the inclusion of protease inhibitors in buffer A is particularly important as a protease which degrades the substrate (at least!) is present in some of the active fractions. The protease inhibitors used, the inclusion of phosphatase inhibitors, and the gradient slope were modified from our earlier work[17] to resolve E1.

## Purification of E1

*Heparin-Sepharose Chromatography of E1*

Fractions composing E1 are pooled and diluted in buffer A until the conductivity is equivalent to 75 m*M* KCl-buffer A. This solution is then applied to a 50 ml (2.5 × 10 cm) heparin-Sepharose (Pharmacia) column equilibrated to 75 m*M* KCl–buffer A. The column is then washed with one-half column volume of 75 m*M* KCl–buffer A, one column volume 100 m*M* KCl–buffer A, and eluted with a linear gradient of 500 ml from 100 to 700 m*M* KCl–buffer A. E1 CTD kinase activity elutes as a single peak between 280 and 330 m*M* salt.

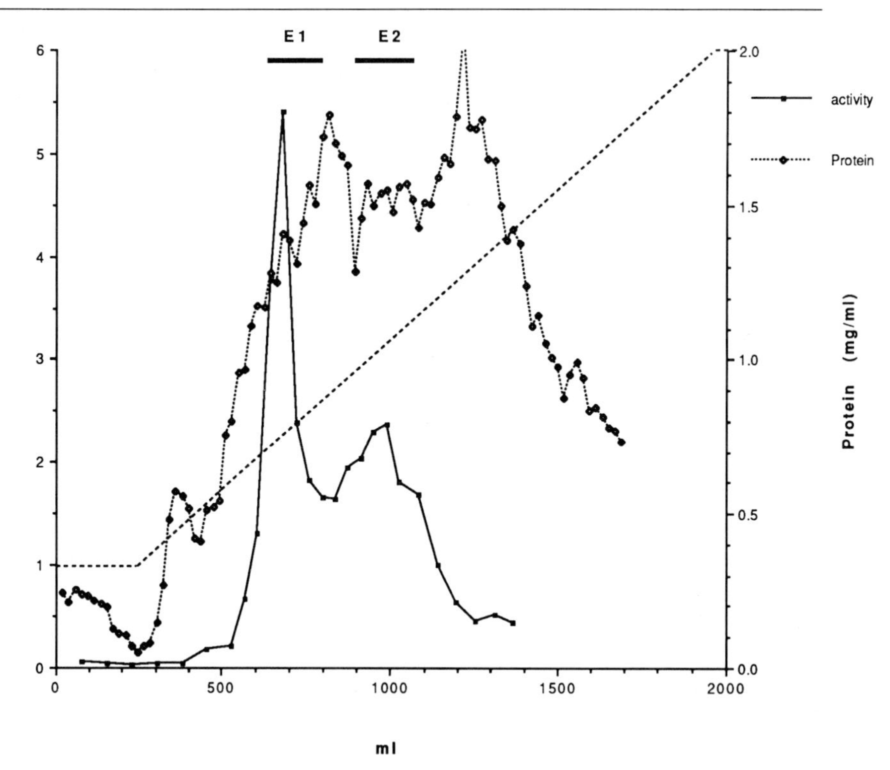


FIG. 2. DEAE-Sepharose chromatography. A whole-cell extract was prepared, ammonium sulfate precipitated, and applied to a 500-ml DEAE-Sepharose column as described. The column was washed and eluted with a gradient of 100 to 400 m*M* KCl–buffer A. Protein, activity, and gradient shape employed are indicated. Protein concentration was determined by Bio-Rad (Richmond, CA) protein assay. CTD kinase activity was assayed on diluted (1 : 8) aliquots of the fractions using the Arg-hepta substrate. The progression of the gradient was interpolated from conductivity measurements. The locations of pools made for subsequent purification of E1 and E2 are indicated by bars above the profile.

## *Mono Q Chromatography of E1*

Active fractions from heparin-Sepharose chromatography are pooled and diluted in buffer D until the conductivity is equivalent to 50 m*M* KCl–buffer D. The solution is then loaded on a Mono Q 10/10 column (Pharmacia) and eluted using a multiphasic gradient of 50 m*M* to 1 *M* KCl–buffer D on an FPLC as follows: 0 ml, 50 m*M* KCl–buffer D; 20 ml, 50 m*M* KCl–buffer D; 180 ml, 350 m*M* KCl–buffer D; 210 ml, 1.0 *M* KCl–buffer D. The flow rate is reduced to 0.2 ml/min from 85 to 145 ml in the program to provide optimal resolution during CTD kinase elution. The CTD kinase activity peak falls between 190 and 240 m*M* salt. Fine (1 ml) fractionation of this region reveals several sharp peaks of kinase activity. These peaks are presumed to represent different phosphate (charge) isomers of the same activity. We have noticed no difference in the enzymatic properties of these peaks. Because the specific activity of the initial fractions are higher (fewer contaminants), the leading portion (typically between 190 and 215 m*M* salt) of the CTD kinase peak is pooled for subsequent purification.

*Phenyl-Superose Chromatography of E1*

The fractions pooled from the Mono Q column are brought to 25% saturation in ammonium sulfate by the addition of 1/3 vol of a saturated solution, and the detergent CHAPS is added to 0.05% from a 10% stock. This solution is clarified by centrifugation at 11,000 rpm in a GSA rotor (20,000 *g*) for 15 min and applied to a phenyl-Superose 5/5 column (Pharmacia) equilibrated in 1 *M* $(NH_4)_2SO_4$ in buffer E. The column is eluted with a multiphasic gradient of 1 to 0 *M* $(NH_4)_2SO_4$–buffer E on an FPLC as follows: 0 ml, 1 *M* $(NH_4)_2SO_4$–buffer E; 2 ml, 1 *M* $(NH_4)_2SO_4$–buffer E; 6 ml, 750 m*M* $(NH_4)_2SO_4$–buffer E; 10 ml, 750 m*M* $(NH_4)_2SO_4$–buffer E; 35 ml, 150 m*M* $(NH_4)_2SO_4$–buffer E; 38 ml, 0 m*M* $(NH_4)_2SO_4$–buffer E. CTD kinase activity elutes sharply between 525 and 450 m*M* salt. Silver staining of the peak fraction reveals p62 and p34 as the only polypeptides whose staining intensities reflect the activity profile. Four additional polypeptides are typically observed at this stage of the purification. We have found this enzyme to be remarkably stable when stored on ice or at 4° in these solutions.

*Mono S Chromatography*

CTD kinase E1 can be further purified and concentrated using a Mono S 5/5 column. Peak fractions from the phenyl-Superose column are pooled and diluted in buffer C until the conductivity is reduced to less than that of a 100 m*M* KCl–buffer C solution. This sample is immediately applied to a Mono S 5/5 column and eluted using a multiphasic gradient from 50 to 1000 m*M* KCl–buffer C on an FPLC as follows: 0 ml, 50 m*M* KCl–buffer C; 2 ml, 50 m*M* KCl–buffer C; 27 ml, 300 m*M* KCl–buffer C; 39 ml, 1 *M* KCl–buffer C. p62 and p34 elute coincident with the CTD kinase activity in ~1 ml at 170 m*M* salt (Fig. 3).

## Purification of E2

*Heparin-Sepharose Chromatography of E2*

Purification of E2 on heparin-Sepharose is performed as for E1. The pooled fractions containing E2 are diluted with buffer A until the conduc-

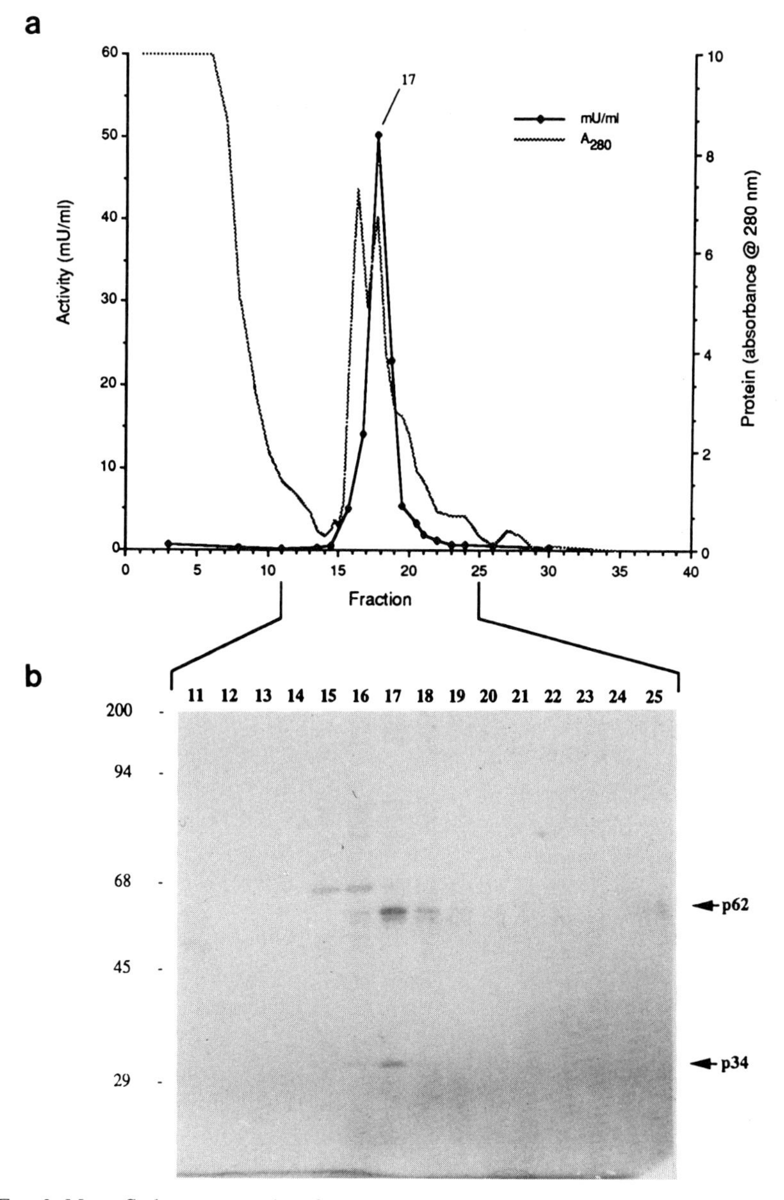


FIG. 3. Mono S chromatography of E1. The phenyl-Superose fractions were diluted and applied to a Mono S 5/5 column under FPLC control. The column was developed using 50 to 1000 m*M* KCl–buffer C as described. (a) Protein, CTD kinase activity, and gradient shape employed are indicated. Protein elution was monitored by in-line UV monitoring at 280 nm (Pharmacia UV-M, FPLC). CTD kinase activity assays were conducted on diluted aliquots (1 : 30) of the fractions using Arg-hepta as substrate. (b) SDS-PAGE of Mono S chromatography fractions. Samples from the Mono S column were resolved by SDS-PAGE on a 12% gel and silver stained. The positions of marker proteins are indicated to the left by their molecular weight ($\times 10^{-3}$). The locations of p34 and p62 are marked on the right.

tivity is less than 75 m*M* KCl–buffer A. The solution is loaded on a 50-ml heparin-Sepharose column and eluted as for E1. CTD kinase activity elutes between 290 and 350 m*M* salt, slightly later in the gradient than E1.

*Mono Q Chromatography of E2*

Active fractions from the heparin column are pooled and diluted as for E1 in buffer D, and applied to a Mono Q 10/10 column. The column is eluted with the same gradient used for E1, and the activity peak is found between 200 and 255 m*M* salt, again slightly later in the gradient than for E1. This column, when finely fractionated, does not reveal discrete multiple peaks as in E1 chromatography.

*Phenyl-Superose Chromatography of E2*

Active fractions from the Mono Q column are pooled and brought to 25% saturation in ammonium sulfate and 0.05% CHAPS as for E1. The pool is applied to a phenyl-Superose 5/5 column equilibrated to 1 *M* $(NH_4)_2SO_4$–buffer E and eluted as for E1. The CTD kinase peak is found between 550 and 475 m*M* salt. Silver staining of these fractions reveals p58 and p34 as the major polypeptides and three additional polypeptides detectable at a lower abundance. p58 and p34 are the only polypeptides whose intensities reflect the activity profile (Fig. 4). There is no p62 detectable by silver staining following SDS-PAGE. This enzyme is also remarkably stable when stored on ice or at 4° as it elutes from the phenyl-Superose column.

*Matrex Green A Chromatography of E2*

The active fractions from phenyl-Superose column may be further purified on a 1-ml (0.5 × 5 cm, HR 5/5, Pharmacia) Matrex Green A (Amicon, Danvers, MA) column. Active fractions from the phenyl-Superose column are pooled and desalted using a Sephadex G-50 column equilibrated in 50 m*M* $NH_4Cl$–buffer F with a bed volume three times that of the input solution. Fractions beyond the void volume are collected and retained for loading until the conductivity of the eluting solution is greater than that of 200 m*M* $NH_4CL$–buffer F. These fractions are pooled and immediately loaded on the Matrex Green A column equilibrated with 100 m*M* $NH_4Cl$–buffer F, and eluted with a multiphasic gradient from 100 to 1000 m*M* $NH_4Cl$ on an FPLC as follows: 0 ml, 100 m*M* $NH_4Cl$; 2 ml, 100 m*M* $NH_4Cl$; 5 ml, 150 m*M* $NH_4Cl$; 40 ml, 550 m*M* $NH_4Cl$; 45 ml, 1 *M* $NH_4Cl$. The enzyme activity profile demonstrates a relatively broad elution, but the peak is clearly well separated from other proteins present.[17]

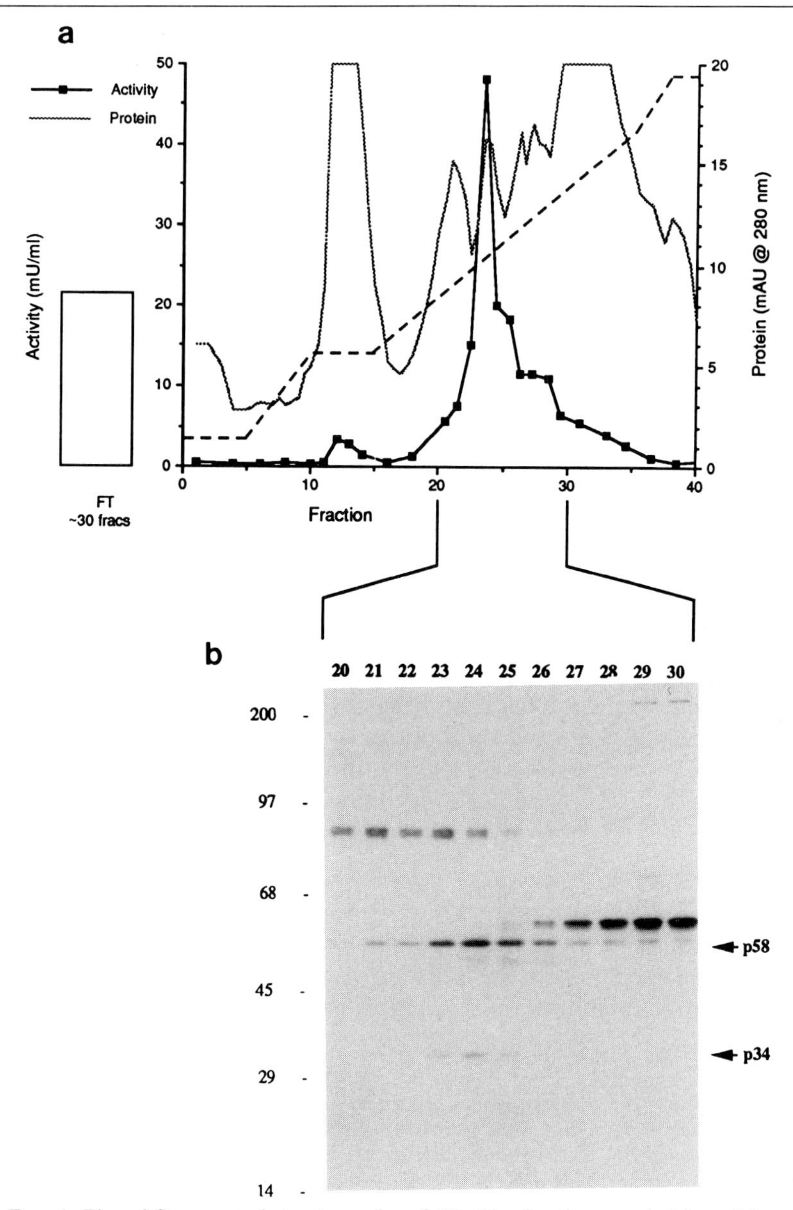


FIG. 4. Phenyl-Superose chromatography of E2. The fractions pooled from Mono Q chromatography are brought to 1 *M* $(NH_4)_2SO_4$, and passed over a phenyl-Superose 5/5 column as described. The column is eluted with a gradient of 1 to 0 *M* $(NH_4)_2SO_4$–buffer E as described. (a) The flow through (FT) is indicated by the bar graph, on the left, and the

TABLE I
PURIFICATION OF MOUSE CTD KINASES

| Step[a] | Volume (ml) | Protein (mg) | Activity (mU) | Specific activity (mU/mg) | Yield (%) |
|---|---|---|---|---|---|
| WCE | 1120 | 4334 | 848 | 0.196 | 100 |
| AmSO PPC | 2400 | 2136 | 2760 | 1.29 | 325 |
| DEAE 1 | 180 | 148 | 808 | 5.46 | 95 |
| DEAE 2 | 215 | 280 | 600 | 2.14 | 71 |
| Heparin 1 | 75 | 66.0 | 472.5 | 7.17 | 56 |
| Heparin 2 | 83 | 103.8 | 630 | 6.08 | 74 |
| Mono Q 1 | 20 | 5.6 | 448 | 80.0 | 53 |
| Mono Q 2 | 28 | 6.7 | 484.4 | 72.1 | 57 |
| Phenyl 1 | 2 | 0.220 | 99.6 | 453 | 12 |
| Phenyl 2 | 2 | 0.096 | 118.8 | 1238 | 14 |

[a] WCE, Whole-cell extract; AmSO PPC, ammonium sulfate cut.

Although the recovery of activity from this column when assayed as the fractions are first collected is ~95%, we typically do not use this column because these fractions have proved unstable in our hands, with a half-life measured in days. It is possible that this instability is due to the low protein concentration of fractions from this column. Our efforts to stabilize CTD kinase purified using this column by adding bovine serum albumin (BSA) or by reintroducing the buffer components from 500 m*M* $(NH_4)_2SO_4$–buffer E have not been successful.

### *Purification Summary*

Table I summarizes a preparation using the above purification protocol. The increase in recovered activity following ammonium sulfate precipitation is explained in part by reduced background, and presumably by elimination of inhibitors or side reactions (phosphatases or proteases).

gradient fractions are depicted in the line graph. The approximate number of fractions represented in the flow through is indicated below the bar graph. Protein, activity, and gradient shape employed are indicated. Protein elution was based on in-line UV monitoring at 280 nm (Pharmacia UV-M, FPLC). CTD kinase activity assays were conducted on diluted aliquots (1 : 30) using Arg-hepta as the substrate. The gradient shape is that programmed for the run. (b) SDS-PAGE analysis of phenyl-Superose chromatography. The samples from the phenyl-Superose column were run on a 12% polyacrylamide gel and silver stained. The positions of the marker proteins are given by their molecular weight ($\times 10^{-3}$) at the left, and the locations of p34 and p58 are indicated. The p34 band stains an orange color on silver stain, which photographs poorly relative to the black color of p58 on gels with high background.

The DEAE column splits CTD kinase activity into two fraction as described, and the yield noted for the E1 and E2 from this initial detection of the individual activities is followed in parentheses. Overall, purification is greater than 2000-fold for E1 and 6000-fold for E2, and we recover 25% of the initial CTD kinase activity present in the whole-cell extract or 8% from the activity in the ammonium sulfate pellet. We typically do not use the Mono S or Matrex Green A columns because we find CTD kinase to be less stable following these steps, and purification of small aliquots from the phenyl-Superose fractions on either Mono S or Matrex Green A have given poor yields.

## Carboxyl-Terminal Domain Kinases Contain $p34^{cdc2}$

In previous work we identified the 34K component of mouse CTD kinase E2 as the product of the mouse *cdc2* gene.[17] This protein kinase catalytic subunit is well conserved among eukaryotes and is thought to play a central role in regulating the cell cycle.[22] First identified in yeast by mutations causing cell cycle-specific arrest at "Start,"[23] cdc2 is required for entry into M phase as well as for the commitment to cell division. Further genetic studies have identified a number of genes whose products interact with $p34^{cdc2}$ to achieve cell cycle control.[22]

Biochemical approaches have also been used to define other components of the cell cycle regulatory pathway. Injection of *Xenopus* egg extract into oocytes induces M-phase and subsequent maturation into eggs.[24] $p34^{cdc2}$ has been shown to be a component of this M-phase-promoting factor (MPF).[25,26] The second component of MPF is cyclin B.[27,28] Cyclins were first identified in sea urchin eggs as proteins that accumulate during the cell cycle and are degraded at M phase.[29–31] Genetic analysis

[22] P. Nurse, *Nature (London)* **344,** 503 (1990).

[23] L. H. Hartwell, *Bacteriol. Rev.* **38,** 164 (1974).

[24] P. S. Sunkara, D. A. Wright, and P. N. Rao, *Proc. Natl. Acad. Sci. U.S.A.* **76,** 2799 (1979).

[25] W. G. Dunphy, L. Brizuela, D. Beach, and J. Newport, *Cell (Cambridge, Mass.)* **54,** 423 (1988).

[26] J. Gautier, C. Norbury, M. Lohka, P. Nurse, and J. Maller, *Cell (Cambridge, Mass.)* **54,** 433 (1988).

[27] G. D. Draetta, F. Luca, J. Westendorf, L. Brizuela, J. Ruderman, and D. Beach, *Cell (Cambridge, Mass.)* **56,** 829 (1989).

[28] L. Meijer, D. Arion, R. Golsteyn, J. Pines, L. Brizuela, T. Hunt, and D. Beach, *EMBO J.* **8,** 2275 (1989).

[29] T. Evans, E. T. Rosethal, J. Younglow, D. Distel, and T. Hunt, *Cell (Cambridge, Mass.)* **33,** 389 (1983).

[30] K. I. Swenson, K. M. Farrell, and J. V. Ruderman, *Cell (Cambridge, Mass.)* **47,** 861 (1986).

[31] N. Standart, J. Minshull, J. Pines, and T. Hunt, *Dev. Biol.* **124,** 248 (1987).

in yeast demonstrated that one class of mutations that suppresses *cdc2* mutations maps to *cdc13*, a gene encoding a cyclin.[32] At present, two members of this family have been identified in human cells,[33,34] more have been identified in amphibians and marine invertebrates,[30,35,36] evidence suggests that a number exist in yeast,[32,37,38] and preliminary polymerase chain reaction (PCR) amplification of cDNA using primers to conserved cyclin protein sequences suggests that at least four cyclin-like molecules exist in mouse.[39] Finally, in experiments where we have followed p34 through our purification procedure with immunoblotting, we have observed additional peaks of 34K cross-reacting material not associated with CTD kinases or cyclin B, suggesting that other complexes may be present.

The current model is that $p34^{cdc2}$ must associate with a cyclin (or cyclin-like molecule) to form an active kinase complex. The activity of this complex is regulated, in a cell cycle-specific fashion, through the availability of the cyclin component and phosphorylation of the $p34^{cdc2}$ and cyclin subunits.[22] The regulatory function of cdc2 is presumably carried out through cell cycle-specific phosphorylation of effectors, although the identity of these critical target proteins is unknown. The presence of $p34^{cdc2}$ in our purified CTD kinases suggests that RNA polymerase may be one of these targets. Histone H1 has also been identified as a potential target of the p62/p34 (cyclin B/cdc2) kinase; indeed, MPF was independently purified as M phase histone H1 kinase.[40,41] Two other potential physiologic substrates in cycling cells have been identified as caldesmon[42] and myosin regulatory light chains,[43,44] suggesting a role for cdc2 in cytokinesis.

We have used antibodies against $p34^{cdc2}$ and cyclins to clarify the relationship between CTD kinases and other $p34^{cdc2}$-containing com-

[32] R. N. Booher and D. Beach, *EMBO J.* **6,** 3441 (1987).

[33] J. Pines and T. Hunter, *Cell (Cambridge, Mass.)* **58,** 833 (1989).

[34] J. Pines and T. Hunter, *Nature (London)* **346,** 760 (1990).

[35] J. Pines and T. Hunt, *EMBO J.* **6,** 2987 (1987).

[36] J. Minshull, J. J. Blow, and T. Hunt, *Cell (Cambridge, Mass.)* **56,** 947 (1989).

[37] J. A. Hadawiger, C. Wittenberg, H. E. Richardson, M. de Barros Lopes, and S. Reed, *Proc. Natl. Acad. Sci. U.S.A.* **86,** 6255 (1989).

[38] H. E. Richardson, C. Wittenberg, F. Cross, and S. I. Reed, *Cell (Cambridge, Mass.)* **59,** 1127 (1989).

[39] C. Nelson and J. L. Corden, unpublished observation (1990).

[40] D. Arion, L. Meijer, L. Brizuela, and D. Beach, *Cell (Cambridge, Mass.)* **55,** 371 (1988).

[41] J. C. Labbe, A. Picard, G. Peaucellier, J. C. Cavadore, P. Nurse, and M. Doree, *Cell (Cambridge, Mass.)* **57,** 253 (1989).

[42] S. Yamashiro, Y. Yamakita, R. Ishikawa, and F. Matsumura, *Nature (London)* **344,** 675 (1990).

[43] T. D. Polard, L. Satterwhite, L. Cisek, J. Corden, M. Sato, and P. Maupin, *Ann. N.Y. Acad. Sci.* **582,** 120 (1990).

[44] L. Satterwhite, L. J. Cisek, J. L. Corden, and T. D. Pollard, submitted for publication.

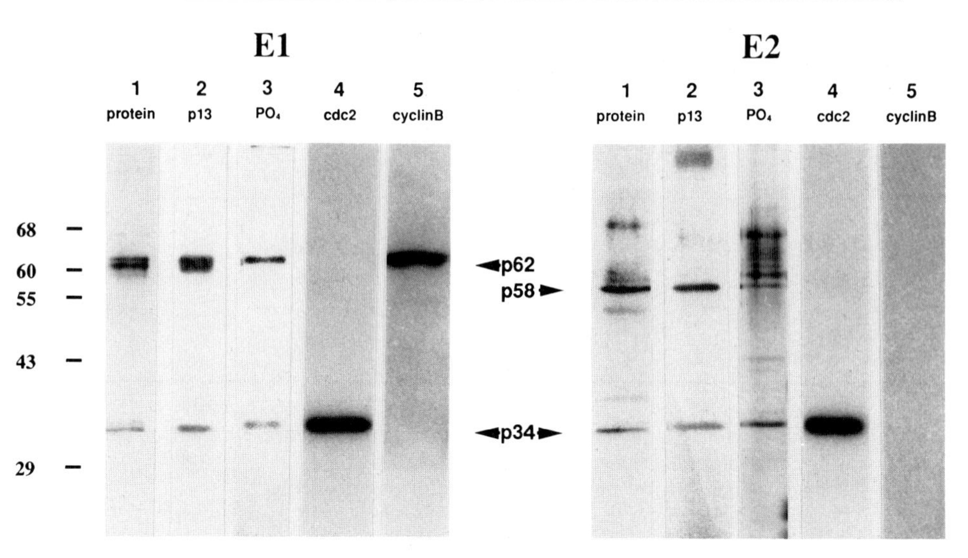


FIG. 5. Analysis of purified CTD kinases. Shown are purified CTD kinases E1 and E2. The lanes represent the appearance of CTD kinases when silver stained (1); precipitated with p13$^{suc1}$-Sepharose[47] and silver stained (2); incubated with $Mg^{2+}$-ATP and silver stained (3); Western blotted using anti-cdc2 sera[45] (4); and Western blotted with anti-cyclin B antisera[33] (5). The positions of the marker proteins are shown to the left ($\times 10^{-3}$) while the locations of p34, p58, and p62 are shown between the panels. The anti-cdc2 serum "PSTAIR"[45] was used at a 1 : 500 dilution, and the anti-cyclin B serum[33] was used at a 1 : 500 dilution.

plexes. In Fig. 5, lanes 1 show the subunits of purified CTD kinases E1 and E2. Both enzymes contain subunits migrating at 34K. We have used antiserum from rabbits exposed to the peptide VEGVPSTAIRELLKE ("PSTAIR"), which represents a sequence unique to cdc2,[45] in the Western blots of the same E1 and E2 samples. Lanes 4 (Fig. 5) show that the smaller subunit of both E1 and E2 react with anti-p34$^{cdc2}$ serum. Similar results are obtained with serum raised against CDNQIKKM, which represents the seven carboxyl-terminal amino acids of human cdc2.[46] Furthermore, Table II shows that this anti-C-terminal serum is able to immunoprecipitate both E1 and E2 CTD kinase activity. Together with peptide sequencing[17] these results show that p34$^{cdc2}$ is a component of CTD kinases E1 and E2.

The presence of p34$^{cdc2}$ and a higher molecular weight polypeptide in purified preparations suggested that these larger components could be cyclins. Anti-cyclin B antibodies[33] react with the 62K band of E1 but not

[45] M. Lee and P. Nurse, *Nature* (*London*) **327,** 31 (1987).

[46] G. Draetta and D. Beach, *Cell* (*Cambridge, Mass.*) **54,** 17 (1988).

TABLE II
PRECIPITATION OF CTD KINASE ACTIVITY[a]

| Kinase | Activity (%) | | | |
|---|---|---|---|---|
| | p13 Sepharose[b] | Control serum[c,d] | Anti-cdc2[c,e] | Anti-cyclin B[c] |
| E1 | | | | |
| Pellet | 89 | 1 | 49 | 58 |
| Supernatant | — | 99 | 51 | 42 |
| E2 | | | | |
| Pellet | 92 | 1 | 51 | 3 |
| Supernatant | — | 99 | 49 | 97 |

[a] Purified E1 or E2 CTD kinase was adsorbed with the indicated reagent.
[b] The activity noted is the percentage of a reaction where enzyme is added to pelleted p13$^{suc1}$-Sepharose but not washed with bead buffer.
[c] The activity given is the percentage of total recovered activity.
[d] Preimmune serum from the rabbit used to raise anti-cyclin B antibodies.[33]
[e] Anti-CDNQIKKM.[46]

with the 58K subunit of E2 (lanes 5, Fig. 5). Thus, E1 appears to correspond to MPF/M phase histone H1 kinase, the kinase activity involved in entry into M phase. The identity of the 58K subunit of E2 is unknown. It fails to react with anti-cyclin B and demonstrates only a weak cross-reactive band at 58K with anti-cyclin A (gift of J. Pines, The Salk Institute, San Diego, CA). This antibody generates a strong signal at 60K in the whole-cell extract. As this serum was raised to a peptide representing the 40 carboxyl-terminal amino acids the possibility that p58 represents a proteolysed cyclin A cannot be excluded from this result. Immunoprecipitation results (Table I) are consistent with the identification of E1 as cyclin B/p34$^{cdc2}$ and E2 as a p34$^{cdc2}$ kinase.

A further property that suggests that the p58 of E2 is not cyclin A is shown in lanes 2 (Fig. 5). These lanes show silver-stained lanes of the purified E1 and E2 CTD kinases after the enzymes have been incubated in the presence of ATP. When autophosphorylated, the mobility of p62 is slightly retarded while p58 is shifted to a series of bands with progressively lower mobility. The shift in mobility correlates with phosphate incorporation into these lower mobility forms, and the protein can be shifted quantitatively to a position corresponding to 65K. This unusual result has not been previously reported for cyclins, suggesting that the p58 subunit is a yet unidentified cyclin, or that it may belong to a different class of p34$^{cdc2}$-interacting proteins. Preliminary results from sequenced tryptic peptides have failed to reveal homologies to cyclins. Also of note is that immunoblots conducted on p58 after phosphorylation to generate the ladder of bands as shown in Fig. 5, lanes 3 show no binding of anti-cyclin A or B antibodies.

A third protein that has been shown to interact with p34$^{cdc2}$ is p13$^{suc1}$. Mutations in the *suc1* gene of yeast suppresses some *cdc2* mutations, and this protein binds to p34 with high affinity.[22] When p13$^{suc1}$ is coupled to Sepharose it can be used as an affinity matrix to bind a number of p34$^{cdc2}$ complexes.[47,48] Figure 5, lanes 2 show that both subunits of E1 and E2 bind to p13 Sepharose beads, and CTD kinase activity is also bound to p13 beads (Table II).

Hydrodynamic data also support the assignment of enzymatic activity to the complexes of p62/p34 and p58/p34. E1 behaves as a heterodimer of p62/p34 with $S_{20,w}$ = 5.2S in isokinetic sucrose gradients and $r_s$ = 42.6 Å in gel-filtration experiments, which predicts a native moelcular weight of 93K. E2 displays $S_{20,w}$ = 5.2S in isokinetic sucrose gradients and $r_s$ = 40.7 Å in gel filtration to give an estimated native molecular weight of 89K. The high affinity of the components for each other is demonstrated by cofiltration in the presence of 2 *M* urea or 900 m*M* KCl and is further suggested by our inability to separate the subunits without loss of enzymatic activity. Together, these results suggest that the CTD kinases are composed of p34$^{cdc2}$ complexed with cyclin B in the case of E1 and with a novel 58K subunit in the case of E2.

## Enzymatic Properties of Carboxyl-Terminal Domain Kinases

Both enzymes are protein serine/threonine kinases with rather broad substrate specificities. They are $Mg^{2+}$ dependent with optima of ~10 m*M*. Manganese ion but not $Ca^{2+}$, $Zn^{2+}$, or $Cu^{2+}$ can replace $Mg^{2+}$ in the reaction. Phosphorylation occurs across a broad pH range (7.0 to 8.2 being optimal) and is inhibited by high-salt concentration (>250 m*M*). Neither enzyme shows dependence on calmodulin, $Ca^{2+}$, phospholipids, or cyclic nucleotides.

Both E1 and E2 will catalyze a γ-phosphate transfer from ATP, GTP, or dATP, but not from dTTP, UTP, CTP, or AMP-PNP. We have estimated the $K_m$ for ATP and GTP at 30 and 180 μ*M*, respectively, for both E1 and E2; the $K_m$ for dATP is likely higher based on a lower rate when dATP is used to shift the CTD on SDS-PAGE. CTD kinases are inhibited by 5,6-dichloro-1-β-D-ribofuranosylbenzimidazole (DRB, a nucleotide an-

[47] L. Brizuela, G. Draetta, and D. Beach, *EMBO J.* **6,** 3507 (1987).
[48] L. Brizuela, G. Draetta, and D. Beach, *EMBO J.* **6,** 3507 (1989).

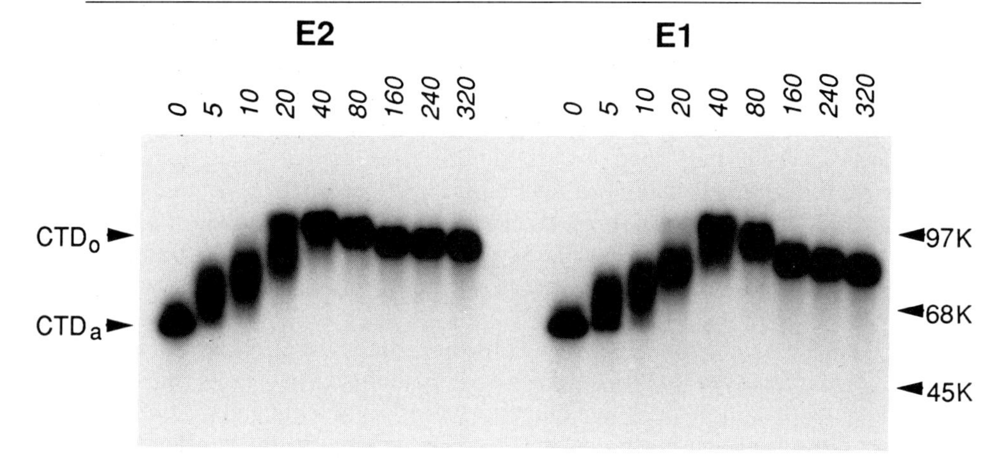


FIG. 6. Labeling and mobility shift of the mouse CTD. The labeled mouse CTD was incubated with ATP and CTD kinase E1 or E2 for the times given (in minutes). The positions of prestained markers ($\times 10^{-3}$) are indicated, as is the location of the shifted ($CTD_o$) and unshifted ($CTD_a$) CTD.[12,20] The mouse CTD was purified from an *E. coli* expression system by cyanogen bromide cleavage, size fractionation, and reversed-phase HPLC.[20] The 50 pmol of CTD was incubated with casein kinase II prepared from Ehrlich cells following the method of Hathaway and Traugh (this series, Vol. 99, p. 317) in the presence of 100 $\mu$Ci of [$\gamma$-$^{32}$P]ATP (3000 Ci/mmol) for 30 min at 30°. Casein kinase II was inactivated by boiling for 5 min, and the labeled CTD was then used as the phosphate acceptor in a reaction containing 250 n*M* CTD, 2 m*M* ATP in standard assay salts. Equal units of CTD kinase activity were added to the E1 and E2 reactions.

alog which has been shown to inhibit transcription[49]) with half-maximal velocity at 35 $\mu M$ under the standard reaction conditions.

The $K_m$ for the phosphate acceptor of E1 has been estimated at 189 and 243 $\mu M$ for hepta-six and Arg-hepta, respectively. The affinity of E2 for peptides substrates has been similarly determined as 200 $\mu M$ for hepta-six and 212 $\mu M$ for Arg-hepta. The entire mouse CTD has been used as substrate. We have observed greater phosphate incorporation using the CTD at 250 n*M*, which provides the heptapeptide repeat at 13 $\mu M$, than with peptide substrates at 300 $\mu M$, with a concentration of individual repeat units of 1.2 m*M*. This observation suggests that the whole CTD serves as a significantly better substrate than heptapeptides.

The ability of CTD kinases to shift the electrophoretic mobility of the CTD in a fashion analogous to the IIa to IIo ($CTD_a$ to $CTD_o$) transition is shown in Fig. 6. Both enzymes shift a CTD substrate to a form displaying

[49] P. Seghal, J. Darnell, and I. Tamm, *Cell (Cambridge, Mass.)* **9,** 473 (1976).

a slower mobility, and the apparent molecular weight of the shifted species is consistent with the mobility of the excised CTD derived from *in vivo*-modified Pol IIo.[12,20] Two phases of the mobility shift are seen in this *in vitro* reaction. Initially a retardation in mobility is seen, presumably due to phosphate groups interfering with SDS binding, and not simply to increased molecular weight. Then a higher mobility phase is seen at longer incubation times, which reflects additional phosphate incorporation as demonstrated by increased signal when [$\gamma$-$^{32}$P]ATP is provided for CTD kinases and a uniformly increasing mobility to a plateau in non-SDS-denaturing gels (not shown). Thus, the E1 and E2 CTD kinases catalyze a IIa to IIo transition consistent with that seen *in vivo*.

The rate of phosphate transfer to Arg-hepta has been determined as 1 nmol phosphate/$\mu$g/min for E1 (Mono S) and 0.32 nmol $PO_4$/$\mu$g/min for E2 (phenyl-Superose) under the standard assay conditions. Phosphoamino acid analysis of purified labeled hepta-six reveals serine to be the sole amino acid acceptor for both E1 and E2. In nonconsensus sequence peptides or the mouse CTD, which contain threonine at positions 2 and/or 5 in some of the repeats (8), phosphothreonine can be detected in addition to phosphoserine. Direct sequencing of consensus peptides phosphorylated by CTD kinase E2 reveal phosphoserines at positions 2 and 5 (Y*S*PT*S*PS) of the repeat.[20]

Both E1 and E2 have been shown to phosphorylate other substrates. Figure 7 shows the results of protein kinase assays done on DEAE fractions using not only the CTD peptide substrate but also casein, histone H1, and the regulatory light chain of chicken myosin. Casein is the only one of these substrates that does not show a peak of activity coincident with the CTD kinase activity. When histone H1 (#1004875; Boehringer Mannheim, Mannheim, Germany) is provided as substrate, we can detect a peak of histone H1 kinase activity that copurifies with CTD kinases E1 and E2 [the lysine-rich histone fraction obtained from Sigma (St. Louis, MO) #H 5505 demonstrated significantly lower incorporation]. One of the myosin regulatory light chain kinase activities also copurifies with CTD kinase activity on the DEAE column.

The histone H1 and myosin regulatory light chain kinase activity in the DEAE pools E1 and E2 copurify with CTD kinase throughout the subsequent steps of the purification. These activities are found to be precipitable with anti-cdc2 antibodies and p13$^{suc1}$-Sepharose in the intermediate and final stages of purification. Furthermore, heptapeptides, histone H1, and myosin regulatory light chains will compete with each other in phosphorylation reactions (not shown). These observations imply that a p34$^{cdc2}$ kinase catalyzes the reaction to all three of these substrates. We

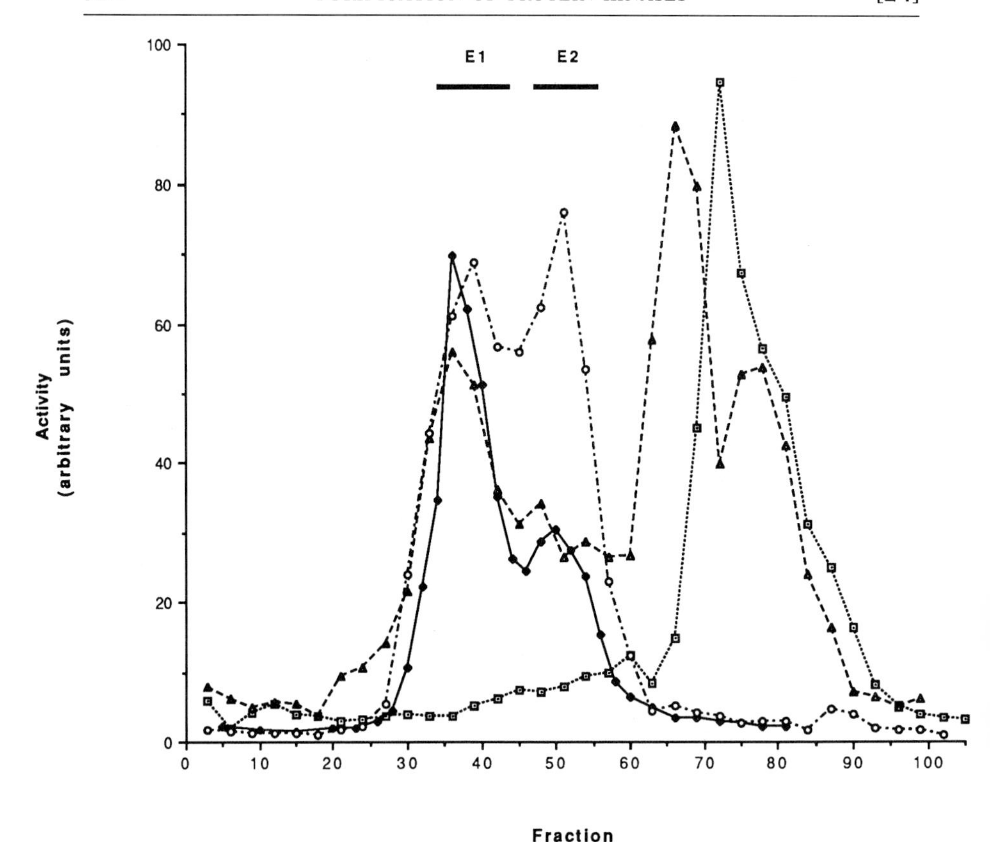


FIG. 7. Additional substrates analyzed on DEAE-Sepharose chromatography. The DEAE-Sepharose column fractions (shown in Fig. 2) were also assayed using casein, histone H1, and myosin regulatory light chains (MLC). The data were scaled for clarity of presentation [the multiplication factors used were as follow: (◆) Arg-hepta, 0.001; (⊡) casein, 0.0013; (▲) MLC, 0.009; and (○) histone H1, 0.003]. The assay conditions were the standard reaction conditions using Arg-hepta at 300 $\mu M$, casein (Sigma) at 2 mg/ml, histone H1 (Boehringer Mannheim Biochemicals) at 1 mg/ml, myosin regulatory light chains at 0.2 mg/ml. The locations of pools made from this preparation for E1 and E2 are indicated.

have determined a relative activity ratio (the activity measured using histone H1 as substrate divided by the activity measured when Arg-hepta is the phosphate acceptor) for E1 and E2 as 2.5 and 0.54, respectively.

## Other Carboxyl-Terminal Domain Kinases

CTD kinases from several other organisms have been reported. Lee and Greenleaf have purified an enzyme from *Saccharomyces cerevisiae* using an assay requiring both shift and phosphorylation of a $\beta$-galactosidase–yeast CTD fusion protein.[50] This enzyme appears to be distinct from the enzymes described here in that it contains three subunits of molecular weight 58K, 38K, and 32K. The p32 subunit does not cross-react with anti-CDC28 antibodies, but the predicted protein sequence of the p58 clone shows strong similarity to p34,[51] suggesting that it may be the catalytic subunit of this yeast CTD kinase. Creation of a null mutation in the p58 gene has shown that this gene is not essential for yeast growth,[51] although the resulting strains are temperature sensitive. A likely explanation for the viability of the null mutant is that there are other CTD kinases in yeast.

Several groups have described the partial purification of human CTD kinases. Stevens and Maupin have used the synthetic peptide Lys(Tyr-Ser-Pro-Thr-Ser-Pro-Ser)$_4$ in a column-binding assay, similar to the filter-binding assay described here, to identify and partially purify a CTD kinase from HeLa cell extracts.[52] This protein kinase has the same nucleotide utilization and reaction condition optima as mouse CTD kinases. This human enzyme is inhibited by 5,6-dichlorobenzimidazole riboside (DRB). Payne *et al.*[14] have characterized an activity capable of shifting Pol IIa to Pol IIo from transcription extracts. This activity is similar to the mouse CTD kinases as it can use ATP, GTP, or dATP to produce a mobility shift on SDS-PAGE. Further purification will be necessary to determine whether either of these enzymes has the same subunit structure as mouse or yeast CTD kinases.

A CTD kinase activity has also been partially purified from wheat germ.[53] This activity has a native molecular weight of ~200K, substantially larger than the mouse CTD kinases. Furthermore, the wheat germ activity will not use GTP as a phosphate donor. These differences may simply reflect a difference between the animal and plant CTD kinases or they may be representative of different members of a set of enzymes present in all species.

## Conclusion and Perspective

The purification of CTD kinases and the identification of p34$^{cdc2}$ as a component of these enzymes raise several interesting questions about the functional role of these enzymes. The first question concerns the bewildering array of *in vitro* substrates—does this diversity reflect the *in*

[50] J. M. Lee and A. L. Greenleaf, *Proc. Natl. Acad. Sci. U.S.A.* **86,** 3624 (1989).
[51] A. Greenleaf, personal communication (1990).
[52] A. Stevens and M. K. Maupin, *Biochem. Biophys. Res. Commun.* **159,** 508 (1989).
[53] T. J. Guilfoyle, *Plant Cell* **1,** 827 (1989).

*vivo* situation? While no direct experimental proof exists for the *in vivo* phosphorylation of any of the proposed $p34^{cdc2}$ substrates, several are known to undergo cell cycle variation in phosphorylation, supporting a possible role for cdc2 *in vivo*. In studies of elutriated and serum-stimulated cells we have noted a cell cycle variation in CTD kinase activity, with maximal activity in $G_2$/M and early $G_1$ and low activity in S and early $G_2$ (unpublished observation). In preliminary experiments addressing the phosphorylation of Pol II we have noted cell cycle variation in the mobility of the largest subunit consistent with the variation in $p34^{cdc2}$ kinases.[54]

The observations that actively transcribing Pol II is highly phosphorylated,[12,13] that $\beta$–$\gamma$ phosphate bond hydrolysis is required in initiation,[15,16] and that the IIa to IIo conversion occurs at or near initiation[14] were among the motivations for these studies. How then does the identification of $p34^{cdc2}$ kinase as a CTD kinase fit the current models of CTD function? One current model suggests that the CTD binds to either DNA or transcription factors during initiation, and phosphorylation releases Pol II from these interactions to proceed with elongation.[9] Suzuki has presented support for a possible heptapeptide–DNA interaction,[55] and DNA band-shift activity has been detected using the mouse CTD.[56] Phosphorylation of the CTD in a cell cycle-regulated fashion could then modify the ability of Pol II to form these postulated interactions with DNA or factors, for example by preventing the establishment of these initiation contacts during $G_2$/M. This role is intriguing given that the cdc2 "consensus" site (S/TPXX, where X is often a basic residue) may be a DNA-binding motif.[57,58] Could a critical role for cdc2 be the regulation the SPXX class of DNA-binding proteins?

Another approach to exploring the *in vivo* specificity is to examine the effects of *cdc2* mutants on the phosphorylation of Pol II. When a temperature-sensitive *CDC28* mutant is grown at nonpermissive temperature no gross change in the level of phosphorylation of Pol II is detected.[59] The observation by Lee and Greenleaf that a null mutant of a yeast CTD kinase subunit is viable[51] suggests the presence of more than one CTD kinase. If multiple CTD kinases are employed by the cell *in vivo* then it will be necessary to examine a combined mutant to demonstrate changes in bulk Pol II phosphorylation. Only through combined genetic and biochemical approaches will the role of the CTD kinases be established.

[54] E. Barron, unpublished observation (1989).

[55] M. Suzuki, *Nature (London)* **344,** 562 (1990).

[56] H. Reinhoff, J. Zhang, and J. L. Corden, unpublished observation (1990).

[57] M. E. A. Churchill and M. Suzuki, *EMBO J.* **8,** 4189 (1989).

[58] M. Suzuki, *EMBO J.* **8,** 797 (1989).

[59] P. A. Kolodziej, N. Woychik, S. M. Liao, and R. A. Young, *Mol. Cell. Biol.* **10,** 1915 (1990).

Acknowledgments

We acknowledge Van Vogel, Brian Byers, and Clark Riley for assistance in protein sequencing and peptide substrate synthesis. We are grateful for the mouse CTD substrate provided by Jie Zhang. We are indebted to Lisa Satterwhite for myosin regulatory light chains, help with assays, photography, and her critical review of this manuscript. We extend our appreciation to Paul Nurse, David Beach, and Jonathan Pines for providing the antibodies used in these studies. We thank Christine Nelson for her technical assistance with DNA cloning and sequencing.

# [25] Purification and Properties of Growth-Associated H1 Histone Kinase

*By* TIMOTHY C. CHAMBERS and THOMAS A. LANGAN

Extensive phosphorylation of H1 histone occurs in eukaryotic cells during growth and division.[1] Beginning in late $G_1$ phase and continuing through S and $G_2$ phases, phosphorylation reaches a peak as cells enter mitosis and falls sharply as cells enter the $G_1$ phase. All H1 molecules are phosphorylated and both serine and threonine residues are modified with maximum levels of six or more phosphates per molecule. Phosphorylation is catalyzed by a chromatin-bound, $Ca^{2+}$- and cyclic nucleotide-independent protein kinase termed growth-associated H1 histone kinase.[2] The activity of the enzyme, which is absent in nongrowing cells,[3] increases sharply as cells proceed from $G_2$ phase to mitosis and falls abruptly as cells enter $G_1$ phase, closely following the pattern of H1 phosphorylation. The enzyme has been detected in a wide variety of proliferating mammalian cells[4] as well as in the slime mold *Physarum polycephalum*[5] and *Xenopus* oocytes.[6] Recently it has been shown that growth-associated H1 kinase is a homolog of yeast $cdc2^+$/CDC28 protein kinases which control entry into mitosis.[7] This finding provides firm experimental evidence that this type of protein kinase and its phosphorylated substrates are crucial

[1] H. R. Matthews and E. M. Bradbury, *in* "Cell Biology of Physarum and Didymium" (H. C. Aldrich and T. W. Daniel, eds.), p. 317. Academic Press, New York, 1982.

[2] H. R. Matthews and V. D. Huebner, *Mol. Cell. Biochem.* **59,** 81 (1985).

[3] T. A. Langan, *Methods Cell Biol.* **19,** 127 (1978).

[4] A. M. Edelman, D. K. Blumenthal, and E. G. Krebs, *Annu. Rev. Biochem.* **56,** 567 (1987).

[5] T. C. Chambers, T. A. Langan, H. R. Matthews, and E. M. Bradbury, *Biochemistry* **22,** 30 (1983).

[6] R. A. Masaracchia, J. L. Maller, and D. A. Walsh, *Arch. Biochem. Biophys.* **194,** 1 (1979).

[7] T. A. Langan, J. Gautier, M. Lohka, R. Hollingsworth, S. Moreno, P. Nurse, J. Maller, and R. A. Sclafani, *Mol. Cell. Biol.* **9,** 3860 (1989).

participants in the control of mitotic entry in eukaryotic cells. The present chapter describes methods for the purification and summarizes the properties of growth-associated H1 histone kinase from Novikoff rat hepatoma cells.

## Assay Method

Reaction mixtures of 0.25 ml contain 50 m*M* Tris-HCl, pH 7.5, 5 m*M* $MgCl_2$, 1 m*M* dithiothreitol (DTT), 1 mg/ml calf thymus H1 histone, 0.5 m*M* [$\gamma$-$^{32}$P]ATP (specific activity 3–30 $\times$ $10^3$ cpm/nmol), and enzyme sufficient to catalyze the transfer of 0.3 to 2.0 nmol of phosphate to H1. Reactions are started by the addition of [$\gamma$-$^{32}$P]ATP, incubated for 10–20 min at 37°, and terminated by the addition of 2.25 ml of 27% trichloroacetic acid. After standing 10 min on ice, insoluble material is collected on HAWP Millipore (Bedford, MA) filters, washed with 35 ml of 25% trichloroacetic acid, and $^{32}$P measured by Cerenkov radiation. Endogenous incorporation of phosphate into enzyme protein, determined in reaction mixtures in which H1 histone is withheld until after termination with trichloroacetic acid, is subtracted. Under the condition of the assay, phosphate incorporation into H1 is linear with time and amount of enzyme. One unit of enzyme activity is defined as that amount which catalyzes the transfer of 1 $\mu$mol of phosphate/hr to H1 histone.

## Purification of Growth-Associated H1 Kinase

### *Buffers*

Buffer A: 20 m*M* Tris-HCl, pH 7.5, 0.2 *M* ammonium sulfate, 1 m*M* DTT, 0.2 m*M* EDTA, 0.05% (w/v) Triton X-100

Buffer B: 20 m*M* Tris-HCl, pH 7.5, 15 m*M* ammonium sulfate, 15 m*M* NaCl, 1 m*M* DTT, 0.2 m*M* EDTA, 0.05% Triton X-100

Buffer C: 50 m*M* Tris-HCl, pH 7.5, 15 m*M* ammonium sulfate, 1 m*M* DTT, 0.2 m*M* EDTA, 0.05% Triton X-100

### *Procedure*

Novikoff rat hepatoma cells are grown in suspension in Swims medium 67 supplemented with 5% newborn calf serum, as described.[8] Typically, eight 2-liter flasks, each containing 1.2 liters of cell culture, are prepared

[8] G. A. Ward and P. G. W. Plagemann, *J. Cell. Physiol.* **73,** 213 (1969).

and cells harvested during exponential growth ($0.8–1.2 \times 10^6$ cells/ml) by centrifugation of 600-ml batches at 500 $g$ for 10 min. Individual cell pellets are resuspended in medium, sedimented in 30-ml polycarbonate tubes, frozen in a mixture of dry ice–ethanol, and stored at −70°. The purification procedure is carried out at 4°. The enzyme fractions are collected and stored in polypropylene tubes whenever possible. Concentration of enzyme fractions is performed in an Amicon (Danvers, MA) ultrafiltration cell using YM10 membranes. Spectra/Por tubing is used for dialysis.

*Step 1. Preparation of Washed Chromatin Extract and Ammonium Sulfate Precipitation.* Sixteen frozen cell pellets derived from a total of 9.6 liters of culture (approximately $10^{10}$ cells) are used in a typical preparation. Each cell pellet is homogenized in 10 ml of 75 m$M$ NaCl, 24 m$M$ EDTA, pH 8.0, for 20 sec with a Brinkmann Polytron type PT10 set at half-maximum speed. Crude chromatin is sedimented (1500 $g$ for 20 min) and individual pellets washed twice by resuspension in 10 ml of 50 m$M$ Tris-HCl, pH 8.0, and recentrifuged. Washed chromatin pellets are combined, the volume adjusted to 40 ml with 50 m$M$ Tris-HCl, pH 8.0, and 4.4 ml of 4 $M$ NaCl added slowly with stirring. After 5 min of stirring, the suspension is centrifuged at 100,000 $g$ for 2 hr. To the supernatant, 1 $M$ potassium phosphate, pH 7.5, is added to a final concentration of 0.2 $M$. Unneutralized saturated ammonium sulfate solution is then added to 17.5% saturation and the precipitate is removed by centrifugation (35,000 $g$, 20 min). The supernatant is collected, the ammonium sulfate concentration increased to 35% saturation, and after stirring for 20 min the precipitate is collected by centrifugation (35,000 $g$, 20 min). The precipitate is suspended in 1.5 ml of 25 m$M$ potassium phosphate, pH 7.5, stirred for 1 hr, and insoluble material removed by centrifugation (35,000 $g$, 30 min).

*Step 2. Sephadex G-200 Chromatography.* The sample from the previous step is applied to a column ($0.9 \times 150$ cm) of Sephadex G-200 equilibrated with buffer A and eluted at a flow rate of 2.8 ml/hr. Fractions of 1.4 ml are collected and 5-$\mu$l aliquots assayed for H1 kinase activity. Enzyme activity elutes as a single peak of apparent $M_r$ 125,000. Active fractions are pooled, concentrated fourfold by ultrafiltration as described above, and dialyzed overnight vs 100 vol of buffer B. Insoluble material is removed by centrifugation.

*Step 3. DEAE-Cellulose Chromatography.* The dialyzed sample is applied to a column ($0.9 \times 18$ cm) of DEAE-cellulose equilibrated with buffer B. The column is washed with 15 ml of buffer B and eluted at a flow rate of 5.3 ml/hr with a linear gradient (total volume 140 ml) of 15 to 150 m$M$ ammonium sulfate in buffer B. Fractions are collected at 30-min intervals and 10-$\mu$l aliquots assayed for H1 kinase activity, which typically elutes as a broad peak at 20–50 m$M$ ammonium sulfate. Active fractions are pooled, concentrated sixfold by ultrafiltration, and dialyzed overnight vs 100 vol of buffer C. Any insoluble material is removed by centrifugation.

*Step 4. Phosphocellulose Chromatography*. Phosphocellulose is precycled as described[9] and a column (0.9 × 18 cm) equilibrated with buffer C. The dialyzed sample is applied, the column washed with 30 ml of buffer C and then eluted at a flow rate of 5.6 ml/hr with a linear gradient (140 ml) of 0 to 500 m$M$ KCl in buffer C. Fractions of 2.8 ml are collected and H1 kinase activity in 10-$\mu$l aliquots determined. The elution pattern of activity typically shows a broad, complex peak eluting at 50–160 m$M$ KCl (pool A) followed by a large peak of higher specific activity eluting at 200 m$M$ KCl (pool B). Fractions are combined into these two separate pools and concentrated by ultrafiltration to a volume of 0.2–0.8 ml.

*Step 5. Sucrose Density Gradient Sedimentation*. Exponential sucrose density gradients are prepared as described previously[10] with 5% (w/v) and 20.4% (w/v) sucrose in the reservoir and mixing chambers, respectively, and with a mixing volume of 9.64 ml. The sucrose solutions contain 0.2 $M$ ammonium sulfate, 20 m$M$ Tris-HCl, pH 7.5, 1 m$M$ DTT, and 0.2 m$M$ EDTA. The phosphocellulose pool B enzyme fraction (0.2–0.3 ml) is layered onto the sucrose density gradient and centrifuged for 65 hr at 40,000 rpm in a Beckman (Palo Alto, CA) SW40 rotor. Fractions of 0.45 ml are collected with an Isco (Lincoln, NE) gradient fractionator and 5-$\mu$l aliquots assayed for H1 kinase activity. As shown in Fig. 1A, the enzyme sediments as a single, symmetrical peak.

### *Subunit Composition*

Sodium dodecyl sulfate (SDS) gel electrophoresis of the sucrose gradient fractions shows that two polypeptides of $M_r$ 33,000 and 60,000 precisely cosediment with the enzyme activity (Fig. 1B). Immunochemical studies with antibody to the yeast cdc2$^+$/CDC28 protein kinases have demonstrated that the 33-kDa component is the protein kinase catalytic subunit.[7] Currently available evidence suggests that the 60-kDa component is related to cyclin.[11] Cyclins are found associated with other homologs of cdc2$^+$/CDC28 protein kinases[12,13] and probably participate in regulation of kinase activity as cells enter and exit metaphase.

[9] V. M. Kish and L. J. Kleinsmith, this series, Vol. 40, p. 198.

[10] K. S. McCarty, Jr., R. T. Vollmer, and K. S. McCarty, *Anal. Biochem.* **61,** 165 (1974).

[11] T. C. Chambers and T. A. Langan, *J. Biol. Chem.* **265,** 16940 (1990).

[12] G. Draetta, F. Luca, J. Westendorf, L. Brizuela, J. Ruderman, and D. Beach, *Cell (Cambridge, Mass.)* **56,** 829 (1989).

[13] J. Gautier, J. Minshull, M. Lohka, M. Glotzer, T. Hunt, and J. L. Maller, *Cell (Cambridge, Mass.)* **60,** 487 (1990).

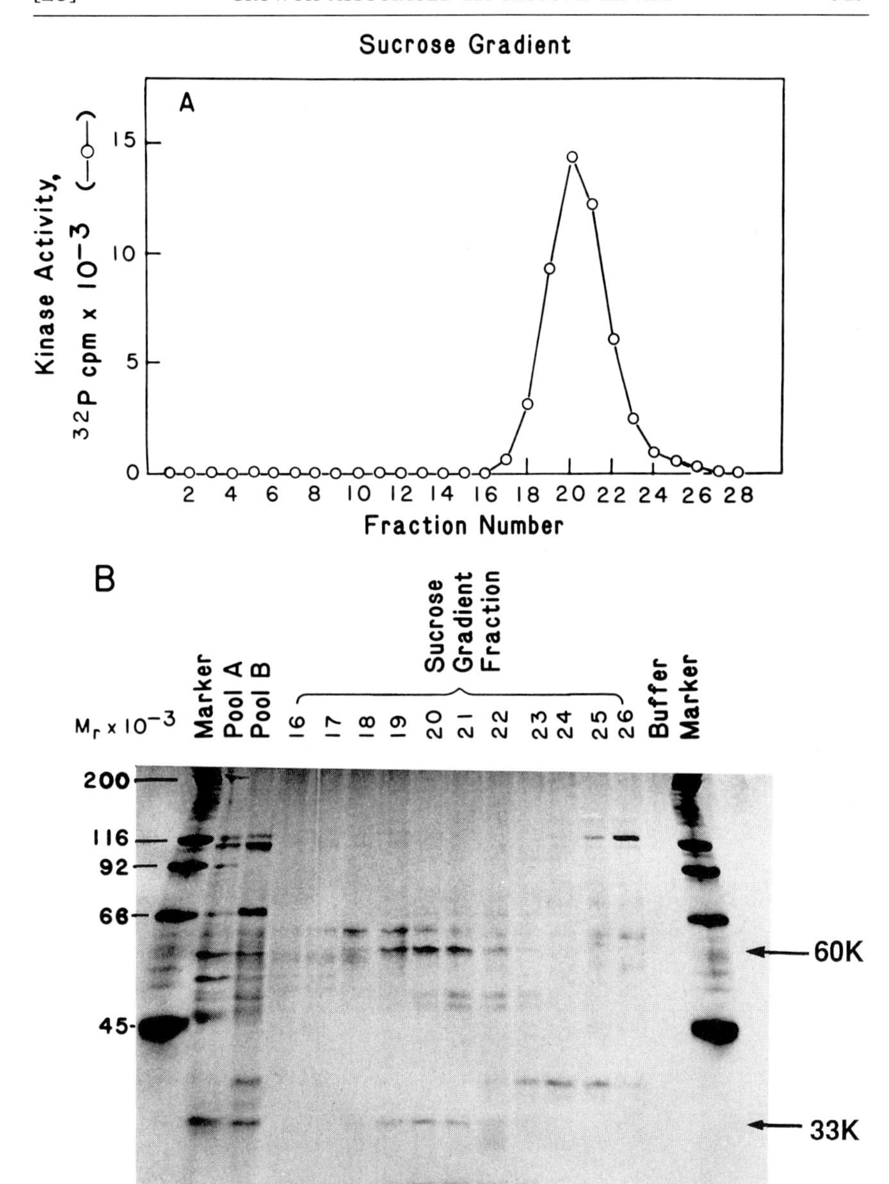


FIG. 1. Sucrose density gradient centrifugation of H1 kinase. A concentrated phosphocellulose pool B fraction (0.275 ml, 2.5 units of activity) was layered on a 5–20.4% exponential sucrose density gradient and sedimented for 65 hr at 40,000 rpm in a Beckman SW40 rotor.

TABLE I
PURIFICATION OF GROWTH-ASSOCIATED H1 KINASE

| Step | Total protein[a] (mg) | Total units | Units/ milligram | Yield (%) | Purification (-fold) |
|---|---|---|---|---|---|
| Chromatin | 237 | 16.3 | 0.07 | 100 | 1 |
| Salt extract | 68 | 14.7 | 0.21 | 90 | 3 |
| $(NH_4)_2SO_4$ precipitate | 21.5 | 20.3 | 0.94 | 124 | 13 |
| Sephadex G-200 | 4.1 | 23.0 | 5.6 | 141 | 80 |
| DEAE-cellulose | 0.65 | 18.4 | 28 | 112 | 404 |
| Phosphocellulose pool A | 0.112 | 6.1 | 54 | 37 | 777 |
| Phosphocellulose pool B | 0.017 | 2.7 | 160 | 17 | 2285 |
| Sucrose gradient | 0.002 | 1.3 | 650 | 8 | 9300 |

[a] Protein was determined for fractions through the $(NH_4)_2SO_4$ step by the method of Lowry *et al.*[14] and subsequently through the phosphocellulose step by that of McKnight.[15] The protein concentration of the sucrose gradient fraction was estimated by comparison of the intensity of staining of polypeptide bands separated by SDS–PAGE with those of standard proteins.

## *Summary of Purification*

A summary of the purification is presented in Table I. Since the enzyme is purified from logarithmically growing cells (mitotic index <3%), the preparation is derived almost entirely from cells in interphase. The enzyme has also been purified through the DEAE-cellulose step from mitotically enriched cell populations (mitotic index >75%) and found to have the same properties as those of the interphase enzyme described below. Growth-associated H1 kinase cannot be measured accurately in the whole-cell homogenate because of the presence of other H1 kinases such as the

[14] O. H. Lowry, N. J. Rosebrough, A. L. Farr, and R. J. Randall, *J. Biol. Chem.* **193,** 265 (1951).

[15] G. S. McKnight, *Anal. Biochem.* **78,** 86 (1977).

Fractions of 0.45 ml were collected. (A) H1 kinase activity in 5-μl aliquots of fractions. The top of the gradient is at the left. (B) SDS–polyacrylamide gel electrophoresis of sucrose gradient fractions. Aliquots (70 μl) of individual fractions were analyzed in adjacent lanes as indicated. The volume analyzed corresponded to 0.1 unit of activity in the peak fraction 20. Other lanes were as follow: Marker, molecular weight standards; Pool A, 0.1 unit of phosphocellulose pool A enzyme; Pool B, 0.1 unit of phosphocellulose pool B enzyme; Buffer, 70 μl of a 1 : 1 mixture of the sucrose solutions used to form the gradient. The gel was stained with silver nitrate. The positions of migration of the 60K and 33K bands corresponding to enzyme activity are indicated by the arrows.

cAMP-dependent protein kinase and histone kinase 2. Therefore, the extent of purification has been calculated relative to washed chromatin, which does not contain these other H1 kinases.[11] Enzyme purified by this procedure has a specific activity of 650 units/mg, which corresponds to the transfer to substrate of ~11 μmol of phosphate/min/mg. The yield of 8% includes a modest but reproducible apparent activation of the enzyme which occurs during purification prior to phosphocellulose chromatography. Despite the extensive purification, the preparation is not completely homogeneous. In addition to the 33- and 60-kDa subunits of the enzyme, the peak fractions from the sucrose gradient contain a small number of minor contaminating polypeptides (Fig. 1B).

## Properties

The properties of the growth-associated H1 kinase were determined using enzyme purified through the DEAE-cellulose step. At this stage of purification, the enzyme is free of detectable protease and protein phosphatase activities (measured, respectively, with H1 and phosphorylated H1 as substrates) and is stable when stored at 4°. Maximum activity is obtained at 5 m*M* $MgCl_2$ and in the pH range 7 to 9. The apparent $K_m$ values for calf thymus H1 histone and ATP are 0.8 and 110 μ*M*, respectively. GTP is utilized inefficiently with maximum reaction velocity occurring at 1.6 m*M* GTP. Activity is not inhibited by the heat-stable inhibitor of cAMP-dependent protein kinase. Identified substrates for the enzyme include H1 histone from numerous sources and related proteins such as the erythrocyte-specific H5 histone, the testis-specific H1t histone and bovine H1°.[11] Stoichiometry ranges from 3 to 6 phosphates per molecule depending on the H1 subtype examined.[11,16] The core histones H2A, H2B, H3, and H4, the high mobility group proteins 1, 2, 14, and 17, ribosomal protein S6, protamine, and casein are not substrates.[11,17] Sequences of the type Lys-Ser/Thr-Pro-Lys and Lys-Ser/Thy-Pro-X-Lys appear to be structural determinants for substrate recognition by growth-associated H1 histone kinase.[3,11]

## Acknowledgment

This work was supported by Grant CA12877 from the National Cancer Institute, United States Department of Health and Human Services.

[16] T. A. Langan, *J. Biol. Chem.* **257,** 14835 (1982).
[17] T. A. Langan, *Methods Cell Biol.* **19,** 143 (1978).

# [26] Purification of Type I$\alpha$ and Type I$\beta$ Isozymes and Proteolyzed Type I$\beta$ Monomeric Enzyme of cGMP-Dependent Protein Kinase from Bovine Aorta

*By* SHARRON H. FRANCIS, LYNN WOLFE, and JACKIE D. CORBIN

Levels of cGMP-dependent protein kinase activity in most mammalian tissues are low when compared with those of cAMP-dependent protein kinase. However, some tissues such as lung, cerebellum, and smooth muscle contain significant amounts of this enzyme activity.[1] The cGMP-dependent protein kinase from bovine lung is a soluble enzyme and has been extensively studied with regard to its mechanism of catalysis, cyclic nucleotide binding, structural organization, and amino acid sequence.[2,3] In contrast to other protein kinases, evidence for isozymic forms of the cGMP-dependent protein kinase has been limited. A membrane-associated cGMP-dependent protein kinase from rabbit aorta has been described,[4,5] and peptide maps of this enzyme on SDS–PAGE are very similar to those of the soluble lung enzyme. deJonge[6] has described a novel cGMP-dependent protein kinase from porcine intestinal brush border membranes. In contrast to the dimeric lung cGMP-dependent protein kinase (subunit $M_r$ 76,000), the enzyme from brush border membranes has a monomeric structure ($M_r$ 86,000), exhibits a lower isoelectric point, and a different phosphopeptide pattern. Due to the apparently distinct physical properties of the soluble form of the lung cGMP-dependent protein kinase and the particulate cGMP-dependent protein kinase from the intestinal brush border, these enzymes have been designated as type I and type II forms of this kinase.[6] Subsequently, bovine aorta has been shown to contain two distinct forms of the soluble cGMP-dependent protein kinase.[7] One form is identical to the well-characterized enzyme from bovine lung and has

[1] T. M. Lincoln and J. D. Corbin, *Adv. Cyclic Nucleotide Res.* **15,** 139 (1983).

[2] K. Takio, R. D. Wade, S. B. Smith, E. G. Krebs, K. A. Walsh, and K. Titani, *Biochemistry* **23,** 4207 (1984).

[3] S. O. Doskeland, O. K. Vintermyr, J. D. Corbin, and D. Ogreid, *J. Biol. Chem.* **262,** 3534 (1987).

[4] J. E. Casnellie, H. E. Ives, J. D. Jamieson, and P. Greengard, *J. Biol. Chem.* **255,** 3770 (1980).

[5] H. E. Ives, J. E. Casnellie, P. Greengard, and J. D. Jamieson, *J. Biol. Chem.* **255,** 3777 (1980).

[6] H. R. deJonge, *Adv. Cyclic Nucleotide Res.* **14,** 315 (1981).

[7] L. Wolfe, J. D. Corbin, and S. H. Francis, *J. Biol. Chem.* **264,** 7734 (1989).

been designated type Iα; the second form of the enzyme is very similar to the type Iα isozyme, but exhibits notable differences in both structural and kinetic properties. The latter isozyme has been designated type Iβ,[7] and it differs from the type Iα in the amino acid sequence of its dimerization and inhibitory domains. The amino acid sequences are identical from serine-90 to the carboxyl-terminal phenylalanine-670 in the type Iα.[7,8] During the course of the purification of the type Iβ, a cGMP-dependent monomeric species of this enzyme is generated by endogenous proteolysis.[9] This chapter describes the purification of bovine aorta type Iα, type Iβ, and type Iβ monomer.

*Materials*

- Bovine aortas (100): Transferred from a local slaughterhouse to the laboratory on ice or purchased frozen from a commercial supply house (Pel-Freez, Rogers, AR) and stored at −70° until use
- Buffer A: 10 m*M* $KH_2PO_4$, 1 m*M* EDTA, 25 m*M* 2-mercaptoethanol, pH 6.8
- DEAE-cellulose, 2.0 liters (Whatman, Clifton, NJ; DE-52)
- DEAE-Sephacel, 7.0 ml (Pharmacia, Piscataway, NJ)
- 8-(6-Aminohexyl)amino-cAMP-agarose, 2.0 ml (P-L Biochemicals) or synthesized by the method of Dills *et al.*[10]
- Heptapeptide (Arg-Lys-Arg-Ser-Arg-Ala-Glu; Peninsula Laboratories, Belmont, CA)
- [$\gamma$-$^{32}$P]ATP (Du Pont–New England Nuclear, Boston, MA)
- Synthetic peptide inhibitor (PKI 5-24) of the cAMP-dependent protein kinase (Peninsula Laboratories)
- Phosphocellulose paper (Whatman P-81)
- Histone IIA mixture (Sigma, St. Louis, MO)
- Nitrocellulose filter paper (Millipore, Bedford, MA; HAWP-0.45 $\mu$m)
- [$^3$H]cGMP, specific activity 15–30 Ci/mmol (Amersham Corporation, Arlington Heights, IL) Saturated solution of $(NH_4)_2SO_4$, 0–4°

## cGMP-Dependent Protein Kinase Assay

The cGMP-dependent protein kinase activity is determined using an assay similar to that described previously.[11] The incorporation of [$^{32}$P]$P_i$ from [$\gamma$-$^{32}$P]ATP into a synthetic heptapeptide (Arg-Lys-Arg-Ser-Arg-Ala-

[8] W. Wernet, V. Flockerzi, and F. Hofmann, *FEBS Lett.* **251,** 191 (1989).

[9] L. Wolfe, S. H. Francis, and J. D. Corbin, *J. Biol. Chem.* **264,** 4157 (1989).

[10] W. L. Dills, J. A. Beavo, P. J. Bechtel, K. R. Myers, L. J. Sakai, and E. G. Krebs, *Biochemistry* **15,** 3724 (1976).

[11] R. Roskoski, this series, Vol. 99, p. 3.

Glu; Peninsula Laboratories, Inc.) is used in the assay since this peptide is a relatively more specific substrate for the cGMP-dependent protein kinase than for the cAMP-dependent protein kinase.[12] The typical assay is conducted using 50 $\mu$l of an assay mixture which contains 20 m*M* Tris, pH 7.4, 200 $\mu M$ ATP, 136 $\mu$g/ml of the heptapeptide noted above, 20 m*M* $MgCl_2$, 100 $\mu M$ 3-isobutyl-1-methylxanthine, 1 $\mu M$ synthetic peptide inhibitor of the cAMP-dependent protein kinase,[13] and 30,000 cpm/$\mu$l [$\gamma$-$^{32}$P]ATP. The reaction is conducted in the absence and presence of 10 $\mu M$ cGMP in order to determine cGMP-dependent kinase activity. A 10-$\mu$l aliquot of enzyme is added to the 50 $\mu$l of assay mixture, and the reaction is allowed to proceed at 30° for 10 min. To terminate the reaction, a 50-$\mu$l aliquot is removed from each reaction tube and then spotted onto individual phosphocellulose papers (Whatman P-81, 2 × 2 cm). The phosphocellulose papers are immediately dropped into 75 m*M* phosphoric acid (10 ml/paper), washed with four changes of phosphoric acid to remove unreacted [$\gamma$-$^{32}$P]ATP, and once with ethanol. The papers are then dried using a hair dryer and placed into vials containing scintillation fluid and counted. One unit of enzyme activity is defined as 1 $\mu$mol of phosphate transferred from ATP to the peptide substrate per minute per milligram of enzyme.

## [$^3$H]cGMP-Binding Assay

The assay for [$^3$H]cGMP binding is conducted using a Millipore filtration assay similar to that developed by Doskeland and Ogreid.[14] Fifty microliters of an assay mixture containing 50 m*M* $KH_2PO_4$, pH 6.8, 1 m*M* EDTA, 0.5 mg/ml histone IIA (Sigma), 1 $\mu M$ cGMP, and 0.3 $\mu M$ [$^3$H]cGMP is combined with 20 $\mu$l of the enzyme fraction, and the reaction is incubated at 30° for 30 min. The binding reaction is terminated by addition of 2 ml of ice-cold saturated $(NH_4)_2SO_4$, and the samples are then filtered through a nitrocellulose Millipore filter (0.45-$\mu$m pore size) under vacuum. The filters are washed with 6 ml of the cold $(NH_4)_2SO_4$ solution (three times, 2 ml each) to remove free cGMP, after which each filter is placed into a counting vial containing 1.5 ml 2% (w/v) aqueous sodium dodecyl sulfate (SDS) and mixed on a Vortex mixer. Ten milliliters of an aqueous scintillant is added, and after vigorous shaking the samples are counted.

[12] D. B. Glass, and E. G. Krebs, *J. Biol. Chem.* **257,** 1196 (1982).

[13] H. C. Cheng, B. E. Kemp, R. B. Pearson, A. J. Smith, L. Misconi, S. M. Van Patten, and D. A. Walsh, *J. Biol. Chem.* **261,** 989 (1986).

[14] S. O. Doskeland and D. Ogreid, this series, Vol. 159, p. 147.

*Step 1. Preparation of Crude Extract*

Either fresh or frozen aortas can be used for the preparation of the isoenzymes of the cGMP-dependent protein kinase. When fresh tissue is used, 100 aortas are purchased from a local slaughterhouse, packed in ice for transporting to the laboratory, and used immediately. When frozen aortas are used in the preparation, they are purchased from a commercial source (Pel-Freez) and are stored at $-70°$. Twenty hours prior to use, the frozen aortas are unpacked, spread evenly on plastic lab bench paper in the cold room, and covered lightly with another sheet of plastic to allow them to thaw slowly. All steps in the purification procedure are conducted at 0–4°. The aortas are cut lengthwise, and the tunica media layer containing most of the smooth muscle is crudely dissected from the aortas (total tissue dissected from 100 aortas was approximately 10 kg). Fine scissors are used to form a cleavage plane approximately halfway through the wall of the aorta, and the two sections of the vessel are pulled apart manually. The innermost sections are immediately ground in a chilled meat grinder (General Slicing, commercial duty, Southeastern Food Equipment, Antioch, TN) using a medium grind plate and weighed into 500-g portions, each of which is combined with 2 liters of buffer A. The mixture is then placed in a Waring blender (commercial, 4 liters) and homogenized on high speed (three times, 30 sec each). The resulting homogenate is centrifuged at 12,000 *g* for 30 min, and the supernatant is filtered through glass wool into a glass or plastic carboy. The volume of the clear filtered supernatant is approximately 20 liters.

*Step 2A. Batch Adsorption to DEAE-Cellulose for Preparation of Type I$\alpha$, Type I$\beta$, and Monomeric Proteolyzed Form of Type I$\beta$*

To the filtered supernatant is added Whatman DE-52 cellulose (100 ml of settled resin per liter of supernatant), and the mixture is incubated for 2 hr with frequent stirring to effectively mix the resin with the extract. The DE-52 cellulose is separated from the supernatant using a glass Büchner funnel with a fritted disk (coarse) and a vacuum pump. The resin is filtered in small batches and is washed with a total of 10 liters of buffer A to remove unbound supernatant proteins and to reduce the volume quickly. The resin is never taken to dryness during the filtration–washing step. The washed DE-52 cellulose in the funnel is then suspended in a small amount of buffer A, and the resulting slurry is poured into a large glass column ($10.5 \times 112$ cm) and allowed to settle. The packed resin is washed with 4 vol of buffer A, and the cGMP-dependent protein kinase is eluted batchwise using 3 liters buffer A containing 300 m*M* NaCl. Six fractions of 500 ml each are collected, and each fraction is assayed for cGMP-dependent protein kinase activity. The fractions containing the peak kinase activity are pooled, and the protein is precipitated with $(NH_4)_2SO_4$ (370 g/liter). The enzyme thus precipitated is pelleted by centrifugation at 12,000 *g* for 30 min at 4°, and the protein pellet is resuspended in 100–150 ml of cold buffer A. The enzyme fraction is dialyzed for 2–3 hr against buffer A and then recentrifuged to remove insoluble material. The $(NH_4)_2SO_4$ precipitation step is used primarily to decrease the volume of the enzyme fraction to be applied to the affinity gel in step 3. Alternatively, the fractions containing the cGMP-dependent protein kinase activity can be applied directly to the affinity matrix in the elution buffer (buffer A containing 0.3 *M* NaCl). If practical, the latter procedure is preferred, since the higher salt concentration appears to facilitate the speed with which the affinity gel can be loaded. In very low salt concentrations, the rate of loading the pool onto the affinity resin can become very slow, apparently due to the formation of a fine precipitate in the fraction.

*Step 2B. Batch Adsorption to DEAE-Sephacel for Preparation of Only Type Iβ Form of cGMP-Dependent Protein Kinase*

The filtered supernatant prepared as described above in step 1 is mixed with DEAE-Sephacel (30 ml packed resin/liter of supernatant) which is preequilibrated in buffer A. The mixture is then stirred frequently for 2–3 hr. The DEAE-Sephacel resin is then separated from the supernatant by filtration of small batches through a glass Büchner funnel with a fritted disk under vacuum, and the resin is washed on the funnel with buffer A containing 0.1 *M* NaCl. The slurry of resin is then transferred to a glass column (5 × 15 cm) and washed overnight with 3 liters of buffer A containing 0.14 *M* NaCl. The type Iβ form of cGMP-dependent protein kinase is eluted from the DEAE-Sephacel with 800 ml of buffer A containing 0.28 *M* NaCl; 200-ml fractions of the eluate are collected and assayed for cGMP-dependent protein kinase activity. All of the enzyme activity is usually located in the second 200-ml fraction. The fraction containing the kinase activity is immediately applied to the 8-(6-aminohexyl)amino-cAMP-agarose affinity gel as described below.

*Step 3. Affinity Gel Chromatography on 8-(6-Aminohexyl)amino-cAMP-agarose*

The cGMP-dependent protein kinase derived from the DEAE-cellulose (step 2A) is applied to a 3.5 ml column of 8-(6-aminohexyl)amino-cAMP-Sepharose 4B which is preequilibrated in buffer A. The enzyme fraction is applied to the column at a rate of one drop/6 sec, and the resin is washed sequentially with 200 ml buffer A containing 2 *M* NaCl, 10 ml buffer A containing 10 m*M* 5′-AMP, and then 10 ml buffer A. The cGMP-dependent protein kinases and the regulatory subunits of the cAMP-dependent protein kinase are eluted from the affinity resin using buffer A containing 10 m*M* cAMP. It may be necessary to readjust the pH of buffer A after dissolving the cAMP. Approximately eight 1-ml fractions are collected over a 1- to 1.5-hr period. Since the affinities of the regulatory subunits of the cAMP-dependent protein kinase for the cAMP affinity matrix are much higher than the affinities of the cGMP-dependent protein kinases for this resin, the rapid rate of elution preferentially releases the cGMP-dependent protein kinases and minimizes the amount of regulatory subunits present in these fractions. Most of the regulatory subunits can be eluted by stopping the column for an overnight period before elution. The fractions are assayed for kinase activity, and the protein patterns on 10% SDS–PAGE stained with Coomassie Blue are used to determine fractions to be pooled for use in the final purification step of the enzymes on DEAE-Sephacel. Fractions containing large amounts of the regulatory subunits are not included in the pooled enzyme.

When the initial purification follows the procedure described in step 2B, the steps used for the affinity column are essentially the same except that a 1.5-ml resin bed is used. After the enzyme pool is loaded onto the column, the resin bed is then washed sequentially with 350 ml buffer A containing 1 *M* NaCl, 10 ml buffer A, and the type Iβ is eluted with buffer A containing 10 m*M* cAMP as described above.

*Step 4. DEAE-Sephacel Column Chromatography to Separate Isozymic Forms and Monomeric Form of cGMP-Dependent Protein Kinases*

The pooled fractions from the cAMP affinity column are applied to a DEAE-Sephacel column (0.9 × 12 cm) which is equilibrated in buffer A, and the column is washed with buffer A containing 50 m*M* NaCl (~400 ml) to remove cAMP remaining from the preceding affinity column. The extensive wash of the column at this step removes free cAMP and decreases the amount of the type Iα associated with cAMP. A linear 300-ml NaCl gradient (50–300 m*M*) in buffer A is used to resolve the isozymic forms of the cGMP-dependent protein kinases from the DEAE-Sephacel (Fig. 1A). Elution with this salt gradient allows the intact native forms of the cGMP-dependent protein kinases to be distinguished from a proteolyzed form of the enzyme (monomer) as well as providing resolution from remaining regulatory subunits of the cAMP-dependent protein kinase. Fractions are assayed for cGMP-dependent protein kinase activity and for [$^3$H]cGMP binding activity. Protein concentration in the fractions is

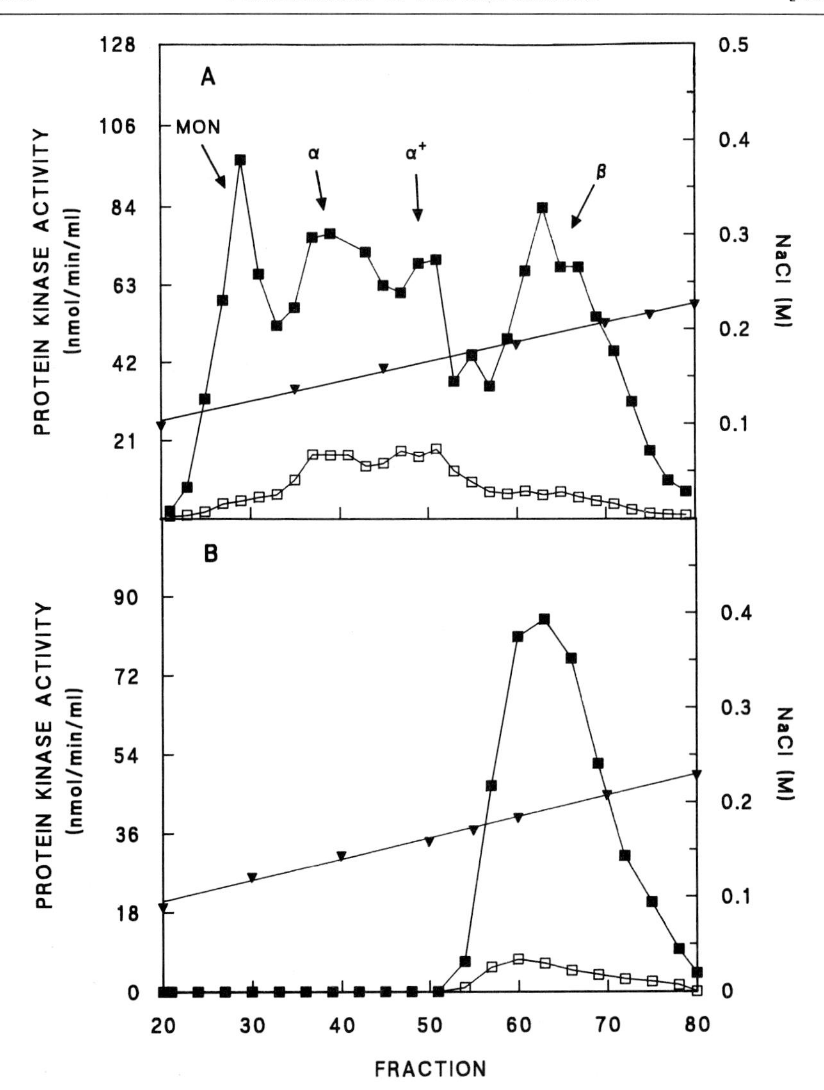


FIG. 1. Separation of the isozymic forms of the cGMP-dependent protein kinase from bovine aorta by DEAE-Sephacel chromatography. Enzyme activities were assayed in the absence (□) and presence (■) of 10 $\mu M$ cGMP as described in the text. (A) Elution of the type I$\beta$ monomer (MON), type I$\alpha$ ($\alpha$), type I$\alpha$ containing bound cyclic nucleotide ($\alpha^+$), and type I$\beta$ ($\beta$). (B) Appearance of only the type I$\beta$ isoenzyme of the cGMP-dependent protein kinase in the elution profile when the preparation is performed according to the procedure described in method 2B. Sodium chloride concentration is indicated (▼).

determined using the method of Bradford[15] with fraction V crystallized bovine serum albumin as the standard.

If step 2A is used in the initial purification step, then the final DEAE-Sephacel column will contain multiple peaks of cGMP-dependent protein kinase activity as shown in Fig. 1A. The first peak in the elution profile is a monomeric species (MON) which is produced during the purification by proteolysis of the dimeric type I$\beta$ isoenzyme by an endogenous protease. The monomeric species of the enzyme is not present in the crude extract. The second peak of activity in the activity profile ($\alpha$) in Fig. 1A corresponds to the elution of the type I$\alpha$ isoenzyme. The type I$\alpha$ is distributed in a broad peak of enzyme activity. The front half of this peak ($\alpha$) contains the type I$\alpha$ with no cyclic nucleotide bound, and the trailing edge of the peak contains type I$\alpha$ with bound cyclic nucleotide ($\alpha^+$). When cyclic nucleotides are bound to the type I$\alpha$ isoenzyme, the elution of the enzyme from DEAE resins is shifted to a more electronegative position. The last peak of cGMP-dependent protein kinase activity ($\beta$) on this column is the type I$\beta$ isoenzyme. There is no evidence of a shifted form of the type I$\beta$ enzyme containing bound cyclic nucleotide in this profile. The difference in the migration of these three forms of the kinase on 10% SDS–polyacrylamide gels is shown in Fig. 2.

If step 2B is used as the initial step in the purification, then the final DEAE-Sephacel column will have only one major peak of cGMP-dependent protein kinase activity, i.e., the type I$\beta$ (Fig. 1B). The type I$\alpha$ isoenzyme is removed when the first DEAE-Sephacel column is washed with buffer A containing 0.14 *M* NaCl. When interest is mainly focused on the type I$\beta$ form of the enzyme, step 2B is the preferred procedure since the elution position of the cyclic nucleotide-bound form of the type I$\alpha$ overlaps significantly with that of the type I$\beta$ form as shown in Fig. 1A. This is true despite extensive prewashing of the final DEAE-Sephacel to remove the cAMP used to elute the enzyme from the cAMP affinity matrix. However, the monomeric species of the type I$\beta$ form of the enzyme is also absent when step 2B is used.

Using either of the above purification procedures, the yield of pure type I$\beta$ cGMP-dependent protein kinase from 100 aortas is approximately 0.7–1.0 mg of protein. When the purification is conducted as described in step 2A, the type I$\alpha$ isoenzyme is also purified, and the yield of type I$\alpha$ from 100 aortas ranges from 1.2 to 2.6 mg. The enzymes from these preparations are at least 98% pure as judged by SDS–PAGE followed by staining with either a silver stain or Coomassie Blue (Fig. 2).

[15] M. M. Bradford, *Anal. Biochem.* **72,** 248 (1976).

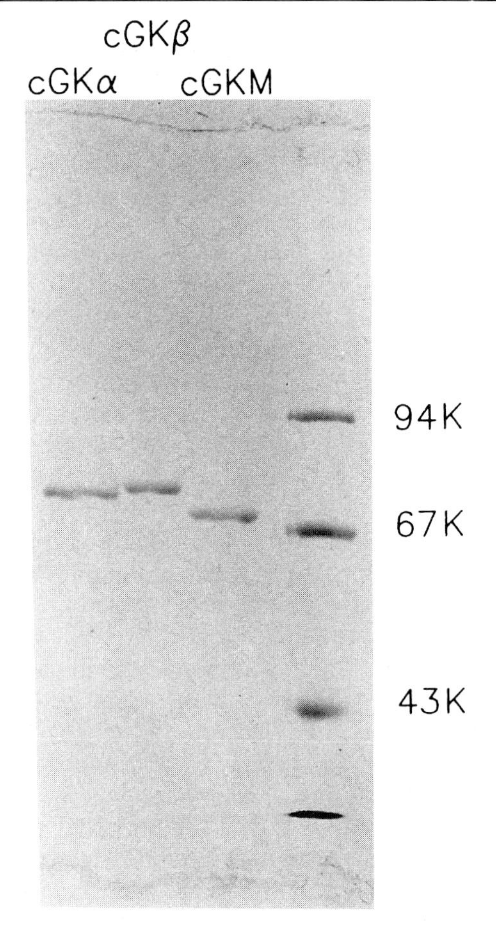


FIG. 2. SDS–polyacrylamide gel electrophoresis of the type Iβ monomer (cGKM), type Iα (cGKα), and type Iβ (cGKβ) on 8% polyacrylamide gels stained with Coomassie Blue.

TABLE I
PHYSICAL PROPERTIES OF TYPE Iα, TYPE Iβ, AND TYPE Iβ MONOMER OF cGMP-DEPENDENT PROTEIN KINASE

| Enzyme | Subunit $M_r$ by SDS–PAGE ($\times 10^{-3}$) | Stokes radius (nm) | Sedimentation coefficient (S) | Calculated native $M_r$ ($\times 10^{-3}$) |
|---|---|---|---|---|
| Type Iα | 78 | 5.3 | 7.5 | 178 |
| Type Iβ | 80 | 5.4 | 7.0 | 170 |
| Type Iβ monomer | 70 | 4.5 | 4.4 | 90 |

TABLE II
FUNCTIONAL PROPERTIES OF TYPE Iα, TYPE Iβ, AND TYPE Iβ MONOMER OF cGMP-DEPENDENT PROTEIN KINASE

| Enzyme | cGMP bound (mol/subunit) | $V_{max}$ ($\mu$mol $P_i$/min/mg) | $K_a$ (n$M$) | | Autophosphorylation sites (residues) |
|---|---|---|---|---|---|
| | | | cGMP | 8-Br-cGMP | |
| Type Iα | 1.63 | 2.6 | 290 | 60 | Ser, Thr |
| Type Iβ | 1.71 | 2.7 | 440 | 580 | Ser |
| Type Iβ monomer | 1.84 | 2.6 | 800 | ND[a] | ND |

[a] ND, not determined.

## General Comments

The physical and functional properties of the purified type Iα, type Iβ, and type Iβ monomer of cGMP-dependent protein kinase are shown in Tables I and II. Types Iα and Iβ coelute on HPLC-DEAE chromatography (Bio-Rad, Richmond, CA; TSK-DEAE-5-PW) but can be readily separated on either DEAE-cellulose (DE-52) or DEAE-Sephacel. The 8-(6-aminohexyl)amino-cAMP-agarose is used as the affinity matrix since the type Iβ isoform does not bind with high affinity to the 6-(2-aminoethyl)amino-cAMP-agarose. The affinity resin which has been used in these preparations is no longer available from P-L Biochemicals, but the derivatized resin can be synthesized according to the method of Dills *et al.*[10] Although 8-(6-aminohexyl)amino-cAMP-agarose is commercially available, variation in the behavior of different batches has presented problems in the purification. The $(NH_4)_2SO_4$ precipitation step following the DEAE-cellulose in step 2A is used only if the volume of the enzyme eluted from the DEAE-cellulose exceeds 1 liter. Proteolytic breakdown of both isoenzymes of the cGMP-dependent protein kinase by endogenous proteases is more likely to occur if the time required for loading the fraction onto the affinity matrix is excessively lengthy (>24 hr). The amount of the monomeric species of the type Iβ enzyme in different preparations is quite variable, and this form may be generated through the action of an endogenous protease at the affinity gel step. The amino acid sequence of the monomer begins at glutamine-62 in type Iβ, which is comparable in position to proline-47 in the sequence of type Iα isoform. In no instance has evidence for a monomeric form of the type Iα isoenzyme been found in the bovine aorta preparations of the cGMP-dependent protein kinases.

## Acknowledgments

The authors would like to thank Bruce Todd, Alfreda Beasley-Leach, and Wayne Price for their technical assistance in this project.

# [27] Purification of Mitogen-Activated Protein Kinase from Epidermal Growth Factor-Treated 3T3-L1 Fibroblasts

*By* THOMAS W. STURGILL, L. BRYAN RAY, NEIL G. ANDERSON, and ALAN K. ERICKSON

Mitogen-activated protein (MAP) kinase is a serine/threonine kinase, $M_r$ 42,000, which is rapidly activated in quiescent cells in response to the addition of growth factors.[1–4] The enzyme was originally named for the *in vitro* substrate, microtubule-associated protein 2, which allowed its identification.[5] Microtubule-associated protein 2 may not be a relevant substrate for MAP kinase *in vivo,* and other *in vitro* substrates have been found (S6 kinase II,[3] myelin basic protein[6]). The acronym, MAP kinase, is less confusing and is recommended as the name of the enzyme. MAP kinase may also stand for "mitogen-activated protein" kinase, consistent with activation of the enzyme by several growth factors.

MAP kinase is of interest primarily for two reasons. First, the enzyme is activated by a mechanism requiring phosphorylation on both tyrosine and threonine residues.[4] Second, MAP kinase is activated prior to activation of S6 kinase,[5] and *in vitro* MAP kinase substantially reactivates *Xenopus* S6 kinase II previously inactivated by protein-serine/threonine phosphatases.[3] These findings suggest that MAP kinase is important in controlling the $G_0$ to $G_1$ transition of the cell cycle.

Here we describe reliable procedures for preparation of activated MAP kinase in amounts useful for analytical, as opposed to structural, studies. The method is modified from that of Ray and Sturgill.[1] Information regarding purification and yield are not yet available, due to the extremely low abundance of the protein and the nonspecific nature of the *in vitro* substrates.

[1] L. B. Ray and T. W. Sturgill, *J. Biol. Chem.* **263,** 12721 (1988).
[2] A. J. Rossomando, D. M. Payne, M. J. Weber, and T. W. Sturgill, *Proc. Natl. Acad. Sci. U.S.A.* **86,** 6940 (1989).
[3] T. W. Sturgill, L. B. Ray, E. Erikson, and J. L. Maller, *Nature (London)* **334,** 715 (1988).
[4] N. G. Anderson, J. L. Maller, N. K. Tonks, and T. W. Sturgill, *Nature (London)* **343,** 651 (1990).
[5] L. B. Ray and T. W. Sturgill, *Proc. Natl. Acad. Sci. U.S.A.* **84,** 1502 (1987).
[6] N. G. Ahn, J. E. Weiel, C. P. Chan, and E. G. Krebs, *J. Cell Biol.* **107,** 695a (1988).

*Buffers*

All water used is equivalent in purity to water prepared by Milli Q (Millipore, Bedford, MA). Buffers A–F are adjusted to pH at 4°. Buffers B–F are shielded from light and kept cold to minimize breakdown of *p*-nitrophenyl phosphate (pNPP) into dinitrophenol and phosphate.

Buffer A: 100 m*M* Tris/HCl, 100 m*M* NaCl, 8 m*M* EGTA, pH 7.0, kept as a stock at 4°

Buffer B: 5.94 g of disodium *p*-nitrophenyl phosphate (pNPP) (Serva, Westbury, NY) and 0.062 g dithiothreitol (Sigma, St. Louis, MO) dissolved in 100 ml of buffer A kept on ice. This buffer is used to prepare buffers C–F, usually on the day preceding the preparation

Buffer C: 20 ml of buffer B plus 60 ml water

Buffer D: 20 ml of buffer B plus 1.61 g NaCl plus 60 ml of water

Buffer E: 25 ml of buffer B plus 1.32 g of NaCl plus 75 ml of water

Buffer F: 25 ml of buffer B plus 60 ml of ethylene glycol (Ameresco, Solon, Ohio, high-purity grade) plus 15 ml of water. (Some preparations of ethylene glycol contain contaminants toxic to enzymes, and therefore use of reagent-grade ethylene glycol or better is suggested.)

Buffers C–F are adjusted to pH 7.5 on ice with 1.0 *N* HCl, filtered through 0.22-$\mu$m Durapore GV (Millipore) membranes, and stored (4°). Larger amounts of these buffers may be prepared and stored in brown glass bottles at −20°. Each contains 25 m*M* Tris/HCl, 40 m*M* pNPP, 2 m*M* EGTA, and 1 m*M* dithiothreitol (DTT), and the following additional components: Buffer C, 25 m*M* NaCl; buffer D, 370 m*M* NaCl; buffer E, 250 m*M* NaCl; buffer F, 25 m*M* NaCl, 60% ethylene glycol (v/v). Although buffers C–F contain Tris, the majority of the buffering capacity is attributable to the added pNPP.

Krebs–Ringer–bicarbonate–HEPES: 120 m*M* NaCl, 4.7 m*M* KCl, 1.2 m*M* $MgSO_4$, 1.2 m*M* $CaCl_2$, 24 m*M* $NaHCO_3$, 10 m*M* HEPES, adjusted to pH 7.5 at 37° with NaOH

Homogenization buffer: 10 ml of buffer C (4°) containing sodium orthovanadate[7] (1 m*M*), phenylmethylsulfonyl fluoride (PMSF) (0.2 m*M*), leupeptin (1 $\mu$g/ml), aprotinin (1 $\mu$g/ml), and pepstatin A (5 $\mu M$). Protease inhibitors and vanadate are added just prior to use from 100× stock solutions

Other reagents: DE-52 cellulose (Whatman, Clifton, NJ) is preequilibrated with 25 m*M* Tris/HCl, 0.2 m*M* EGTA, pH 7.5 (4°) and kept as a stock 1 : 1 slurry at 4° for 4–8 weeks. A stock solution (100 $\mu$g/ml)

[7] J. A. Gordon, this series, Vol. 201 [41].

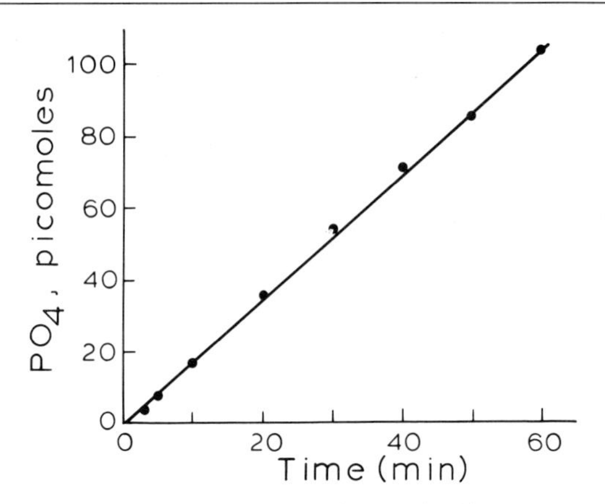


FIG. 1. Phosphorylation of bovine myelin basic protein (Sigma) by MAP kinase, purified by sequential chromatography with DE-52 cellulose (Whatman), phenyl-Superose (Pharmacia-LKB), and Superose 12 (Pharmacia-LKB) from EGF-treated fibroblasts (30 n*M*, 10 min).

of epidermal growth factor (EGF) (Collaborative Research, Bedford, MA) is prepared by dissolving the contents of one vial (Cat. No. 40010, receptor grade) in 1.0 ml of water. Aliquots (0.1 ml) are stored at −20°

## Phosphotransferase Assays

The peak of MAP kinase from phenyl-Superose may be identified by a gel-dependent assay using bovine MAP-2 as substrate.[1] Studies by Ahn *et al.*[6] suggest that myelin basic protein is a substrate for MAP kinase, and we have also demonstrated that myelin basic protein (Sigma) is readily phosphorylated by MAP kinase (Fig. 1). The filter assay using myelin basic protein as substrate is more convenient and reproducible than the gel-based assay with MAP-2 (not commercially available), and is recommended for quantitative studies.

The standard assay procedure, using bovine myelin basic protein (Sigma M-1891) as substrate, is essentially that described by Cicirelli *et al.*,[8] with some modification. Assays (40-$\mu$l total volume) contain 5 $\mu$l of sample to be assayed plus 20 $\mu$l of 2 mg/ml myelin basic protein and 15 $\mu$l of 133 $\mu M$ [$\gamma$-$^{32}$P]ATP (5000 cpm/pmol/27 m*M* magnesium acetate, both in 20 m*M* HEPES, pH 7.5 (20°) [final concentrations: 1 mg/ml myelin basic protein, 50 $\mu M$ [$\gamma$-$^{32}$P]ATP (5000 cpm/pmol), 10 m*M* magnesium acetate, 18 m*M* HEPES, pH 7.5]. Reactions were initiated with ATP/Mg mix and

[8] M. F. Cicirelli, S. L. Pelech, and E. G. Krebs, *J. Biol. Chem.* **263**, 2009 (1988).

incubated at 30°. After 10 min, reactions were stopped by transferring 30-$\mu$l aliquots onto 1-cm squares of P81 (Whatman, Clifton, NJ) phosphocellulose paper, which were dropped into 300 ml of 180 m$M$ phosphoric acid (22°), using the apparatus[9] described by Corbin and Reimann. The papers were washed five times (300 ml/wash) for 5 min in 180 m$M$ phosphoric acid, rinsed with 95% ethanol, dried, and counted in scintillation cocktail [Triton X-100 : toluene (30 : 70) containing 1 gm/liter Omnifluor (New England Nuclear, Boston, MA)].

### *Cells*

Swiss mouse 3T3 or 3T3-L1 fibroblasts[10] are grown to confluence in 100-mm plastic dishes (Falcon) in Dulbecco's modified Eagle's medium containing 10% (v/v) fetal calf serum, and used 2–5 days after becoming quiescent.

### *Stimulation and Preparation of Lysate*

Ten or 20 plates of quiescent fibroblasts are washed three times with 8 ml of Krebs–Ringer–bicarbonate–HEPES (KRB–HEPES) (37°), and incubated in KRB–HEPES (5 ml) for 1 hr (37°).

The cells are stimulated by adding 10 $\mu$l of the EGF stock solution (30 n$M$, final) directly into the medium at the margin of the plate, moving quickly from plate to plate in numerical order, and then mixed by gentle agitation. Timing of the incubation with EGF is begun at the onset of mixing. Stimulation is continued for 10 min at 37°. The plates are then placed on ice, and, in a cold room, the medium is rapidly aspirated and replaced with 8 ml of 0.15 $M$ NaCl (4°). Plates are aspirated in sequence so that the time of exposure of the cells to mixed, 37° buffer containing EGF thus approximates 10 min for each plate. Cells are scraped into 800 $\mu$l/plate (10-plate preparation) or 400 $\mu$l/plate (20-plate preparation) of freshly prepared homogenization buffer. The steps of aspiration and scraping are best done by two people, as speed is important.

The scraped cells are homogenized in a Potter–Elvejhem homogenizer (Thomas, Swedesboro, NJ; No. BB495) with 30 up-and-down movements of a motor-driven Teflon pestle, keeping the homogenizer immersed in ice-water slush. The crude lysate is centrifuged (4°) for 5 min at 30,000 $g$ in a Sorvall RC5B centrifuge, and the supernatant removed and placed on ice.

[9] J. D. Corbin and E. M. Reimann, this series, Vol. 38, p. 287.

[10] H. Green and O. Kehinde, *Cell (Cambridge, Mass.)* **1,** 113 (1974).

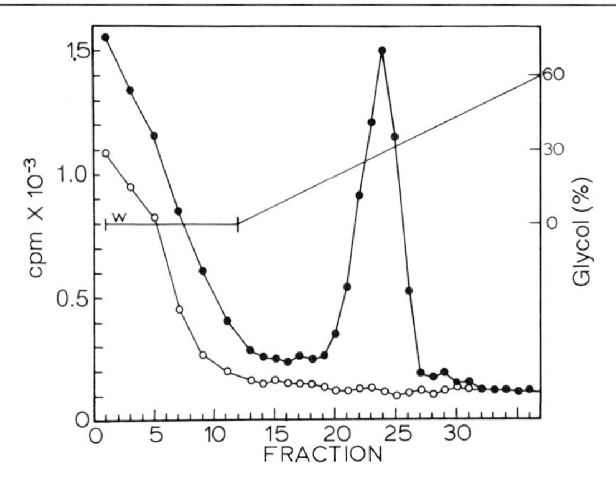


FIG. 2. Elution of MAP kinase from phenyl-Superose using a gradient of increasing ethylene glycol and decreasing sodium chloride. The peak of MAP kinase activity from insulin-treated (80 n*M*, 10 min) 3T3-L1 adipocytes (●) elutes in the gradient at ~37% ethylene glycol and ~100 m*M* NaCl; the profile of kinase activity from control cells is also shown (○). In this experiment, supernatants were filtered through 0.22-$\mu$m Durapore GV membranes and applied directly to phenyl-Superose, omitting the batch DE-52 step. MAP kinase was assayed with bovine microtubule-associated protein 2 as substrate. (From Ray and Sturgill[1] with permission.)

## *DE-52 Cellulose and Phenyl-Superose Chromatography*

All chromatographic steps are conducted in a cold room (4°). A 1-ml (bed volume) column of DE-52 cellulose (Whatman) in a disposable 2-ml Econo column (Bio-Rad, Richmond, CA) is equilibrated with 4 ml of buffer C. Buffers E and F are loaded into the syringe pumps A and B, respectively, of a Pharmacia (Piscataway, NJ) FPLC (fast protein liquid chromatography) system. As buffer F is quite viscous, it is loaded manually, at 4 ml/min. A phenyl-Superose 5.5 HR analytical column (Pharmacia-LKB) is then equilibrated with 5 ml of buffer E (250 m*M* NaCl).

The supernatant (~8 ml, ~1–2 mg/ml total protein) from the EGF-treated cells is applied to the DE-52 cellulose column at ~0.4 ml/min. The column is then washed three times with 2 ml of buffer C (25 m*M* NaCl) at the same flow rate, and then eluted with 3.5 ml of buffer D (370 m*M* NaCl). The eluate is filtered through a 0.22-$\mu$m Durapore GV (Millipore) filter, and applied to the phenyl-Superose column at 0.1 ml/min, using a 10-ml Superloop.

The DE-52 cellulose batch step was used initially to remove lipid, from differentiated 3T3-L1 adipocytes, that might potentially harm the phenyl-Superose column. However, filtered (0.22 $\mu$m) supernatants from these cells may be applied directly to phenyl-Superose (Fig. 2). The DE-52 cellulose step has been retained for other cells because it rapidly concentrates MAP kinase, minimizing the volume to be applied, and it alone affords some purification (approximately threefold, based on total protein[11]). Our routine preparations of MAP kinase now include the batch DE-52 cellulose step.

Once application to phenyl-Superose is complete, the column is washed with 6 ml of buffer E (250 m*M* NaCl) at 0.15 ml/min, and then eluted with a 12-ml linear gradient of 0 to 100% buffer F (60% ethylene glycol) (pump B) at 0.15 ml/min, collecting 0.5-ml fractions in disposable polystyrene tubes (Sarstedt, Princeton, NJ). The flow rate is reduced to 0.1 ml during the last 4 ml of the gradient. Empirically, we have found with these buffers that a back pressure of 1.5 MPa should not be exceeded to avoid compression of the column. The fractions may be stored overnight at −70°.

The fractions are thawed, placed on ice, and assayed for phosphotransferase activity using myelin basic protein. Elution of MAP kinase from phenyl-Superose is illustrated in Fig. 2, in this case using the gel-dependent assay with MAP-2 as substrate to identify the peak of MAP kinase activity. An identical peak is detected with myelin basic protein as substrate. MAP kinase elutes at ~0.1 *M* NaCl and 37% ethylene glycol (v/v). The elution position is highly reproducible with any given column.

### *Concentration and Storage*

The peak three or four fractions of MAP kinase activity are spiked with bovine serum albumin (BSA) (ICN, Lisle, IL; 3× crystallized) to 0.2 mg/ml and concentrated (4°) in a Centricon 10 (Amicon, Danvers, MA), following the manufacturer's recommendations, to a convenient volume, usually 300–400 μl. At 6000 rpm in a Sorvall (Dupont, Wilmington, DE) SM24 rotor (outer row), concentration takes approximately 2 hr. The concentrated phenyl-Superose peak fractions (now containing ~1 mg/ml BSA) may be assayed, aliquoted, and stored at −70°. A preparation of MAP kinase from 10 plates of insulin-stimulated cells was used for the reactivation experiments with S6 kinase II.[3,12] The phenyl-Superose fraction does not appreciably phosphorylate histones, ribosomal protein S6, protamine, or casein, indicating absence of other contaminating kinases.[1] A useful control can be prepared from corresponding fractions from unstimulated cells.[3,13] MAP kinase from phenyl-Superose should be stored

[11] A. J. Rossomano and T. W. Sturgill, unpublished data (1988).

[12] E. Erikson, J. L. Maller, and R. L. Erikson, this volume [21].

[13] L. B. Ray and T. W. Sturgill, *Proc. Natl. Acad. Sci. U.S.A.* **85,** 3753 (1988).

TABLE I
ESTIMATED YIELD OF MAP KINASE ACTIVITY[a]

| Preparation | Stored units (U) | Number of 100-mm plates[b] |
|---|---|---|
| 1 | 1.7 | 20 |
| 2 | 1.0 | 10 |
| 3 | 1.5 | 20 |
| 4 | 2.4 | 20 |
| 5 | 0.7 | 10 |

[a] One unit of MAP kinase transfers 1 nmol of phosphate/min to myelin basic protein in the standard assay (see text). MAP kinase was purified by sequential chromatographies on DE-52 cellulose, phenyl-Superose, and Superose 12.

[b] Number of plates of EGF-treated 3T3-L1 fibroblasts used as the starting material.

in aliquots at $-70°$, and can be frozen and thawed if necessary with little loss of activity.

*Gel Filtration*

The enzyme may be further purified by gel filtration on Superose 12.[1] A Superose 12 column is equilibrated with 20 m$M$ HEPES, 10 m$M$ NaCl, 1 m$M$ DTT, 10% ethylene glycol, pH 7.5 (4°). The stored phenyl-Superose fraction is thawed, applied to the column via a 200-$\mu$l sample loop at 0.2 ml/min, and eluted in 0.3-ml fractions at the same flow rate. MAP kinase elutes as a single symmetrical peak (prior to carbonate dehydratase and after ovalbumin), in approximately five fractions. The peak fractions are pooled and spiked with BSA (0.2 mg/ml, final) and protease inhibitors [phenylmethylsulfonyl fluoride (0.2 m$M$), leupeptin (1 $\mu$g/ml), aprotinin (1 $\mu$g/ml), pepstatin A (5 $\mu M$)]. The spiked fractions are concentrated in a Centricon 10 (see above) to 300–400 $\mu$l, divided into aliquots, and stored at $-70°$. Repeated freezing and thawing of the gel-filtered enzyme should be avoided.

While it has not previously been useful to define a unit of MAP kinase activity, this is now possible after removal of other kinases by phenyl-Superose chromatography, using standard conditions for the myelin basic protein assay (see above). Using this assay, a unit of MAP kinase is defined as that amount of MAP kinase transferring 1 nmol of phosphate/min to myelin basic protein in the standard assay. For purposes of comparison, Table I lists the total number of units stored for several preparations of gel-filtered MAP kinase.

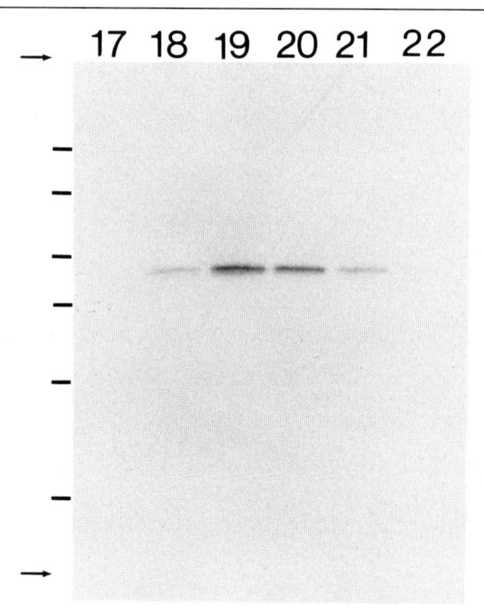


FIG. 3. Analysis, by conventional SDS–gel electrophoresis and autoradiography, of $^{32}P$-labeled proteins eluting in fractions spanning the peak of MAP kinase activity from Superose 12. The preparation was from $^{32}P$-labeled 3T3-L1 adipocytes treated with insulin (80 n$M$, 10 min). The top and bottom of the gel are indicated with arrows. Migration points of marker proteins (see dashes at left) were 97, 66, 43, 36, 31, and 22 kDa (top to bottom). (Reprinted from Ray and Sturgill[1] with permission.)

## Concluding Remarks

Following sequential fractionation on DE-52 cellulose, phenyl-Superose, and Superose 12, MAP kinase is considerably purified and appears as one of the principal silver-stained bands in the preparation not attributable to BSA.[1] If cells are labeled with $^{32}PO_4$ prior to stimulation, several $^{32}P$-labeled proteins are present in the phenyl-Superose peak fractions and of these only MAP kinase is phosphorylated on tyrosine.[12] Following gel filtration, MAP kinase is the only $^{32}P$-labeled protein detectable in the peak fractions (Fig. 3),[1,13] and contains phosphothreonine and phosphotyrosine, but no phosphoserine (Fig. 4).[13] Properties most useful in identification of MAP kinase are given in Table II.

Recovery of MAP kinase activity is dependent on rapid inhibition of and resolution from phosphoprotein phosphatases. Thus, speed, maintenance of low temperature (4°), and inclusion of fresh phosphatase inhibitors are paramount concerns. Sequential chromatographic steps, which

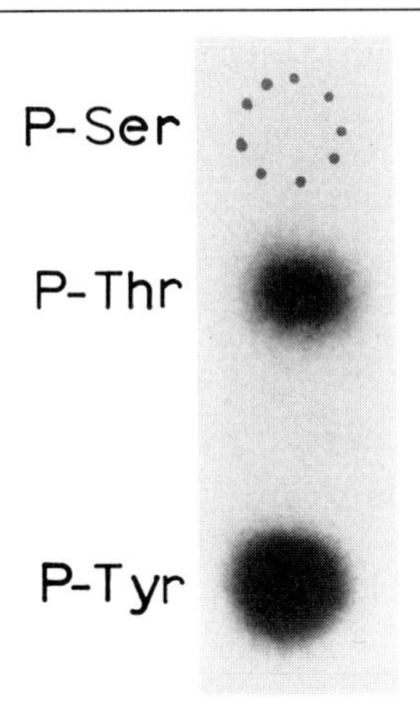


FIG. 4. Phosphoamino acid analysis of $^{32}$P-labeled MAP kinase. (From Ray and Sturgill[13] with permission.)

TABLE II
IDENTIFYING PROPERTIES FOR MAP KINASE

| Property | Ref. |
|---|---|
| Molecular weight: ~40K | *a, b* |
| p*I*: ~6.8 | *b* |
| Binding to phenyl-Superose (Pharmacia-LKB) in low-salt buffers and elution at relatively high concentrations of ethylene glycol | *a* |
| Rapid and transient activation in cells by diverse mitogens and hormones | *a, c* |
| Containing phosphotyrosine and phosphothreonine, but little or no phosphoserine | *d* |
| Inactivation by | |
| Protein-serine phosphatase 2A | *e* |
| Protein-tyrosine phosphatase CD-45 | *f, g* |

[a] L. B. Ray and T. W. Sturgill, *J. Biol. Chem.* **263,** 12721 (1988).

[b] A. J. Rossomando, D. M. Payne, M. J. Weber, and T. W. Sturgill, *Proc. Natl. Acad. Sci. U.S.A.* **86,** 6940 (1989).

[c] C. M. Ely, K. M. Oddie, J. Litz, A. J. Rossomando, S. B. Kanner, T. W. Sturgill, and S. J. Parsons, *J. Cell Biol.* **110,** 731.

[d] L. B. Ray and T. W. Sturgill, *Proc. Natl. Acad. Sci. U.S.A.* **85,** 3753 (1988).

[e] T. W. Sturgill, L. B. Ray, E. Erikson, and J. M. Maller, *Nature (London)* **334,** 715 (1988).

[f] N. G. Anderson, J. L. Maller, N. K. Tonks, and T. W. Sturgill, *Nature (London)* **343,** 651 (1990).

[g] For preparation of these phosphatases, see this series, Vol. 201 [13, 33–38].

are compatible without buffer exchange, are used to avoid dialysis or dilution.

Myelin basic protein and MAP-2 are not specific substrates for MAP kinase. With the procedure described, both substrates are useful and identify the same peak. If the procedure or source is altered, additional peaks of kinase activity phosphorylating one or the other of these substrates may be observed. Reliance on the properties in Table II is then recommended.

Cloning the gene for MAP kinase and producing monoclonal antibodies which specifically bind to this protein will be necessary to address definitively many of the unanswered questions about MAP kinase, both those pertaining to mechanism and those dealing with the relationship of MAP kinase to the pp42 band and to kinase activities[14-16] reported in the literature with similar properties.

### Acknowledgments

These investigations were supported by grants from the National Institutes of Health (RO1-DK41077) and the American Cancer Society (BC546). We thank Anthony Rossomando and D. Michael Payne, two of our colleagues from the Department of Microbiology, University of Virginia, for many helpful suggestions.

[14] C. Hanekom, A. Nel, C. Gittinger, A. Rheeder, and G. Landreth, *Biochem. J.* **262,** 449 (1989).

[15] M. Hoshi, E. Nishida, and H. Sakai, *Eur. J. Biochem.* **184,** 477 (1989).

[16] C. M. Ely, K. M. Oddie, J. Litz, A. J. Rossomando, S. B. Kanner, T. W. Sturgill, and S. J. Parsons, *J. Cell Biol.* **110,** 731 (1990).

# [28] Purification and Characterization of β-Adrenergic Receptor Kinase

*By* JEFFREY L. BENOVIC

Persistent activation of a cell with a given stimulus often results in a blunting of the normal stimulus response. This process, referred to as desensitization, is prevalent in diverse biological systems.The β-adrenergic receptor (βAR)-coupled adenylate cyclase system has proved to be an important model for the study of desensitization phenomena.[1,2] The βAR,

[1] D. R. Sibley and R. J. Lefkowitz, *Nature (London)* **317,** 124 (1985).

[2] J. L. Benovic, M. Bouvier, M. G. Caron, and R. J. Lefkowitz, *Annu. Rev. Cell Biol.* **4,** 405 (1988).

when activated by catecholamines such as epinephrine, stimulates the enzyme adenylate cyclase via its agonist-promoted interaction with the G protein, $G_s$. This process serves to regulate intracellular cAMP levels and thus the activation of the cAMP-dependent protein kinase. The $\beta$-adrenergic receptor is a member of the large family of G protein-coupled receptors, containing many structural features which are common to all the members of this family that have been sequenced to date.[3] These include the presence of seven putative transmembrane $\alpha$-helices, which are thought to serve an important role in ligand binding, an extracellular glycosylated amino terminus, and intracellular domains involved in G protein interaction. An additional structural feature common to many of these receptors is the presence of potential intracellular sites of regulatory phosphorylation. The $\beta$AR, for example, contains two consensus sites for phosphorylation by the cAMP-dependent kinase as well as an acidic Ser/Thr-rich carboxyl terminus, which appears to serve as the locus of phosphorylation by a more specific kinase termed the $\beta$-adrenergic receptor kinase or $\beta$ARK.

Studies have clearly implicated receptor phosphorylation as playing a major role in desensitization. For the $\beta$AR rapid desensitization is mediated by at least two different protein kinases. The cAMP-dependent kinase appears to be involved in phosphorylation and desensitization of the receptor in a heterologous or agonist nonspecific fashion.[1,4] That is, any compound that raises intracellular cAMP, thus activating the cAMP-dependent kinase, will lead to receptor phosphorylation and desensitization. Phosphorylation of the receptor by protein kinase C may also play a role in heterologous desensitization. In contrast, there also appears to be a much more agonist-specific or homologous pattern of desensitization which is often observed. This type of desensitization appears to involve the $\beta$-adrenergic receptor kinase, an enzyme that specifically phosphorylates the agonist-occupied or activated form of the receptor.[5] The focus of this chapter is to present some of the current methodologies used for the study of $\beta$ARK.

## Assays for $\beta$ARK Activity

$\beta$ARK specifically phosphorylates the agonist-occupied form of the $\beta_2$-adrenergic receptor. Thus, this represents the most specific and definitive assay for $\beta$ARK activity. However, this assay is labor intensive and

[3] H. G. Dohlman, M. G. Caron, and R. J. Lefkowitz, *Biochemistry* **26,** 2657 (1987).

[4] J. L. Benovic, L. J. Pike, R. A. Cerione, C. Staniszewski, T. Yoshimasa, J. Codina, M. G. Caron, and R. J. Lefkowitz, *J. Biol. Chem.* **260,** 7094 (1985).

[5] J. L. Benovic, R. Strasser, M. G. Caron, and R. J. Lefkowitz, *Proc. Natl. Acad. Sci. U.S.A.* **83,** 2797 (1986).

expensive. It requires an initial purification of the $\beta$AR followed by reconstitution of the receptor into phospholipid vesicles. The reconstituted receptor can then be used as a substrate for the enzyme with phosphorylation being visualized and/or quantitated after electrophoresis and autoradiography.

A much simpler assay for $\beta$ARK activity utilizes the ability of the enzyme to phosphorylate rhodopsin in a light-dependent fashion.[6] This assay uses urea-treated bovine rod outer segments (ROS) as the substrate. Enough ROS for a few thousand assays can be prepared from bovine retinas in about 12 hr. The procedure described below details the preparation of urea-treated ROS and the phosphorylation procedure using either crude homogenates or purified enzyme preparations. Assays using purified and reconstituted $\beta$AR as the substrate will not be detailed here. However, sufficient detail for purification, reconstitution, and phosphorylation of the $\beta$AR is presented elsewhere.[5,7] It should be emphasized, however, that if one is attempting to purify $\beta$ARK or a related kinase from an untested source (i.e., a different tissue or species) it is important to validate the specificity of the kinase preparation using the appropriate substrates.

*Preparation of Urea-Treated Bovine Rod Outer Segments*

Urea-treated rod outer segments are prepared using the procedures of Shichi and Somers[8] and Wilden and Kuhn.[9] The procedure is described for 50 bovine retinas but can be easily scaled up or down. All steps in this procedure should be carried out at 4° in a darkroom using a safelight.

1. Thaw 50 frozen dark-adapted bovine retinas (George Hormel Co., Austin, MN) in 50 ml of ice-cold homogenization buffer (dissolve 68 g sucrose in 132 ml 65 m*M* NaCl, 2 m*M* $MgCl_2$, 10 m*M* Tris-acetate, pH 7.4) in a 200-ml Erlenmeyer flask. Shake vigorously for 4 min and then centrifuge at 4000 rpm for 4 min in an SS34 rotor.

2. Add 100 ml of 10 m*M* Tris-acetate, pH 7.4 buffer to the supernatant. Centrifuge at 4000 rpm for 4 min and resuspend the pellets in a minimal volume of 0.77 *M* sucrose buffer (in 10 m*M* Tris-acetate, pH 7.4, 1 m*M* $MgCl_2$). Pool the tubes and adjust the volume to ~30 ml with 0.77 *M* sucrose buffer.

3. Dounce homogenize the sample (40 ml Wheaton homogenizer, A pestle, Millville, NJ) with 15 vigorous strokes. Draw up the homogenate

[6] J. L. Benovic, F. Mayor, Jr., R. L. Somers, M. G. Caron, and R. J. Lefkowitz, *Nature (London)* **321,** 869 (1986).

[7] J. L. Benovic, R. G. L. Shorr, M. G. Caron, and R. J. Lefkowitz, *Biochemistry* **23,** 4510 (1984).

[8] H. Shichi and R. L. Somers, *J. Biol. Chem.* **253,** 7040 (1978).

[9] U. Wilden and H. Kuhn, *Biochemistry* **21,** 3014 (1982).

through a 20.5-gauge needle followed by an additional Dounce homogenization with 5 strokes. The sample can then be layered onto a stepwise sucrose gradient composed of 5 ml of 1.14 *M* sucrose, 5 ml of 1.00 *M* sucrose, and 5 ml of 0.84 *M* sucrose.The sucrose buffers are all dissolved in 10 m*M* Tris-acetate, pH 7.4, 1 m*M* $MgCl_2$.

4. Centrifuge the tubes at 26,000 rpm for 30 min in an SW27 rotor (Sorvall AH-627, 36-ml swinging bucket). Remove the red-orange band at the 0.84–1.00 sucrose interface with a sucrose needle. The total volume should be ~25 ml. Dilute the sample with 25 ml of 0.77 *M* sucrose buffer, vortex, and centrifuge at 18,000 rpm for 20 min in an SS34 rotor. If desired, the pellets can be stored at −20° at this point.

5. Resuspend the pellets in a total volume of 25 ml of 50 m*M* Tris-HCl, pH 7.4, 5 m*M* EDTA, 5 *M* urea. Sonicate the sample for 4 min in a tube immersed in ice. This step denatures the endogenous kinase activity without harm to the rhodopsin.

6. Dilute the sample with 100 ml of 50 m*M* Tris-HCl, pH 7.4 and spin in an ultracentrifuge at 30,000 rpm for 45 min. Resuspend the pellets in Tris buffer and recentrifuge a total of three times. The final pellet should be resuspended in a total volume of ~5 ml of Tris buffer. Drawing the sample through a 20.5-gauge needle helps the resuspension. The urea-treated ROS can be aliquoted and stored at −80° until needed. Tubes should be wrapped in aluminum foil and uncovered immediately before use.

### *Phosphorylation of Urea-Treated Rod Outer Segments*

The phosphorylation of urea-treated ROS by crude kinase preparations requires the separation of the phosphorylated rhodopsin from other endogenous phosphorylated proteins. The separation of rhodopsin is readily accomplished by quenching the reaction with sodium dodecyl sulfate (SDS) sample buffer followed by electrophoresis on a 10% polyacrylamide gel and autoradiography. In addition, if one is assaying a soluble kinase preparation it helps to pellet the phosphorylated ROS in a tabletop ultracentrifuge or microcentrifuge. The pellet is then resuspended in SDS sample buffer before electrophoresis. This procedure significantly reduces the background phosphorylation due to other phosphorylated proteins and free ATP. The resultant bands can then be cut and counted if quantitative results are needed.

When assaying more purified kinase preparations it is possible to use filtration methods to separate the phosphorylated rhodopsin from endogenous phosphorylated proteins and free ATP. This is possible because the ROS are membranes which will be trapped on glass fiber filters (e.g.,

GF/C; Whatman, Clifton, NJ). It should be noted that βARK is very sensitive to inhibition by salt[10] (e.g., 0.1 *M* NaCl inhibits βARK 90%). Thus, the ionic strength in the reaction should be kept as low as possible. The procedure for assaying βARK activity using urea-treated ROS is detailed below.

1. Prepare an assay mixture containing (per 10 assays):

   20/2 buffer (20 m*M* Tris-HCl, pH 7.5, 2 m*M* EDTA), 300 μl
   $MgCl_2$ (1 *M*), 2 μl
   ATP (10 m*M*), 3 μl
   [$\gamma$-$^{32}$P]ATP (NEG-002A from NEN Du Pont, Boston, MA), ~10 μCi

2. Assays are set up containing 30 μl of assay mixture (from step 1), ~2 μl of urea-treated ROS, and ~5 μl of the kinase preparation. The samples are incubated for 20 min at 30° in room light. It is important that the ROS be kept in the dark until just before use or else the amount of phosphate incorporated into rhodopsin will be significantly reduced.

3. Following the incubation period the reactions can be quenched several different ways. If the kinase preparation is crude the reactions can be stopped by the addition of 1 ml of cold 100 m*M* sodium phosphate, pH 7, 5 m*M* EDTA buffer followed by centrifugation in a microcentrifuge. The pellets can be resuspended in 100 μl of SDS sample buffer (sonication helps with the resuspension) followed by electrophoresis on a homogeneous 10% (w/v) polyacrylamide gel. The centrifugation step is not absolutely necessary, but it does significantly improve the signal-to-noise ratio on the autoradiogram. With more purified kinase preparations it is possible to stop the reactions by filtration on GF/C glass fiber filters. The filters are then washed four to five times with 4 ml of cold phosphate/EDTA buffer. Obviously, one important control is the demonstration that the phosphorylation reaction is light dependent.

## βARK Purification

### *Step 1. Tissue Source*

An important consideration in the purification of a low-abundance protein is the tissue source. While βARK is a ubiquitous protein it does vary significantly in concentration from tissue to tissue.[10] Bovine cerebral cortex has provided the best source for βARK; it is readily available, inexpensive, and has the highest level of βARK activity of any bovine

[10] J. L. Benovic, F. Mayor, Jr., C. Staniszewski, R. J. Lefkowitz, and M. G. Caron, *J. Biol. Chem.* **262,** 9026 (1987).

tissue examined. To inhibit proteolysis the tissue should either be used fresh or immediately frozen in liquid $N_2$ and stored at $-135°$ until use.

*Step 2. Extraction*

All procedures are carried out at 4°. Tissue (250 g) is thawed and minced in 1250 ml of homogenization buffer [20 m*M* Tris-HCl, pH 7.5, 10 m*M* EGTA, 10 μg/ml leupeptin, 10 μg/ml benzamidine, 5 μg/ml pepstatin, 0.1 m*M* phenylmethylsulfonyl fluoride (PMSF)]. The sample is homogenized using two 30-sec bursts with a tissue disrupter (Polytron equipped with a low-foaming PTA 20TS probe; Brinkmann, Westbury, NY). The homogenate is then centrifuged for 30 min at 13,000 rpm in a GSA rotor.

*Step 3. Ammonium Sulfate Precipitation*

The volume of the supernatant is measured and 150 g of ammonium sulfate is added/1000 ml of supernatant. The sample is stirred for 5 min before centrifugation for 30 min at 20,000 rpm in a TZ-28 rotor (the GSA rotor can be used if a TZ-28 rotor is not available). The supernatant is removed and additional ammonium sulfate is added (same amount as above). The sample is stirred for 5 min and centrifuged for 30 min at 20,000 rpm as described above. The supernatant is discarded while the pellet is resuspended in ~150 ml of column buffer (5 m*M* Tris–HCl, pH 7.5, 2 m*M* EDTA, 5 μg/ml leupeptin, 5 μg/ml benzamidine, 5 μg/ml pepstatin, 0.1 m*M* PMSF) using the Polytron homogenizer on low speed. The sample is centrifuged for 15 min at 19,000 rpm (SS34 rotor) and the supernatant is then recentrifuged at 48,000 rpm for 60 min (50.38 rotor or equivalent).

*Step 4. Chromatography on Ultrogel AcA 34*

The supernatant is carefully loaded onto an Ultrogel AcA 34 column (4.4 × 200 cm; Amicon, Danvers, MA) which has been equilibrated for at least 3 days with column buffer. The flow rate on the column should not exceed 50–55 ml/hr. βARK activity elutes after ~2400 ml of buffer has passed through the column. The activity should be assessed using urea-treated ROS or reconstituted βAR as the substrate. Following SDS-polyacrylamide gel electrophoresis the products can be visualized by autoradiography (Fig. 1, lanes 2 and 5). In addition, the agonist-dependent nature of the phosphorylation reaction should also be assessed (Fig. 1, lanes 1 and 4).

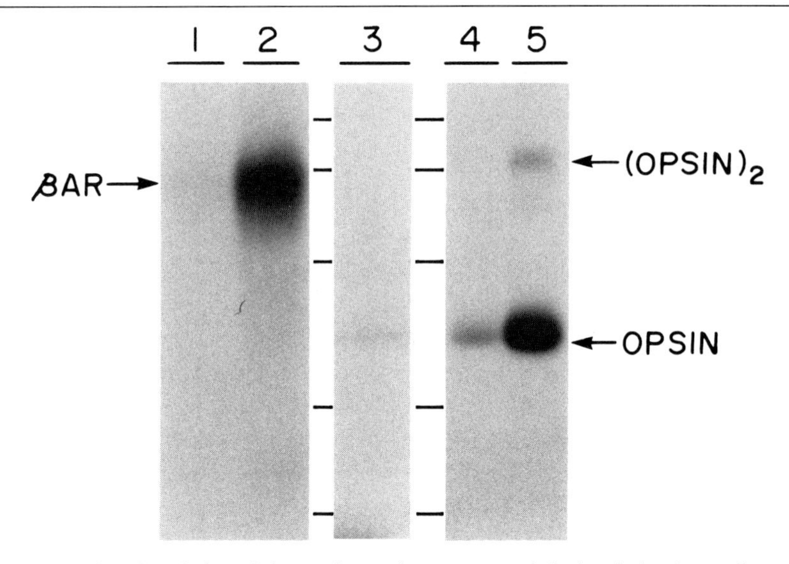


FIG. 1. Phosphorylation of the $\beta$-adrenergic receptor and rhodopsin by the $\beta$-adrenergic receptor kinase. Purified, reconstituted hamster lung $\beta$-adrenergic receptor or urea-treated ROS were incubated for 60 min at 30° with 20 m*M* Tris-HCl, pH 7.5, 20 m*M* NaCl, 5 m*M* sodium phosphate, 2 m*M* EDTA, 5 m*M* $MgCl_2$, 5 m*M* NaF, 0.05 m*M* [$\gamma$-$^{32}$P]ATP (2000 cpm/pmol), and 0.3 $\mu$g $\beta$ARK in a total volume of 50 $\mu$l. Lane 1 contains 0.6 pmol $\beta$AR; lane 2, 0.6 pmol $\beta$AR and 20 $\mu M$ (−)-isoproterenol; lane 3, 0.6 pmol rhodopsin in the light; lane 4, 250 pmol rhodopsin in the dark; lane 5, 250 pmol rhodopsin in the light. The samples were run on SDS-polyacrylamide gel electrophoresis before autoradiography. Relative molecular weight standards (94K, 67K, 45K, 30K, and 20.1K) are indicated by marks between lanes 2 and 3 and between lanes 3 and 4.

### *Step 5. Chromatography on DEAE-Sephacel*

The peak fractions from the AcA 34 column are pooled (~80 ml) and diluted with an equal volume of column buffer containing 0.04% Triton X-100. The sample is then applied to a 35-ml (2.2 × 9 cm) DEAE-Sephacel column equilibrated with column buffer containing 0.02% Triton X-100. The column is washed with buffer containing Triton until the absorbance comes down to baseline (1–2 hr). The column is then eluted with a 300-ml linear gradient from 0 to 40 m*M* NaCl in column buffer with Triton. A flow rate of 30 ml/hr is used and ~5-ml fractions are collected in 12 × 75 mm polypropylene tubes. The fractions are then assayed for $\beta$ARK activity using urea-treated ROS or $\beta$AR. While some of the $\beta$ARK activity does not bind to the column we typically recover ~50% of the activity in a peak eluting between 10 and 20 m*M* NaCl.

*Step 6. Chromatography on CM-Fractogel*

The peak fractions from the DEAE-Sephacel column are pooled and applied to a 1.2-ml (1 × 1.5 cm) CM-Fractogel column equilibrated with column buffer containing 0.02% Triton X-100. The column is washed with Triton buffer containing 20 m*M* NaCl and is then eluted with an 80-ml linear gradient from 20 to 100 m*M* NaCl in Triton buffer. A flow rate of ~20 ml/hr is used and ~3-ml fractions are collected in polypropylene tubes. Fractions are assayed for βARK activity and should be stored at 4° (do not freeze). βARK typically elutes at ~60 m*M* NaCl. Using this purification procedure an ~20,000-fold purification is obtained with an overall recovery of 10–20%. Purity is typically >50% although peak fractions are often >90% pure (Fig. 2). The inclusion of Triton X-100 in the last two steps of purification greatly enhances the recovery and stability of the enzyme. These preparations lose little activity even after 1 year at 4°.

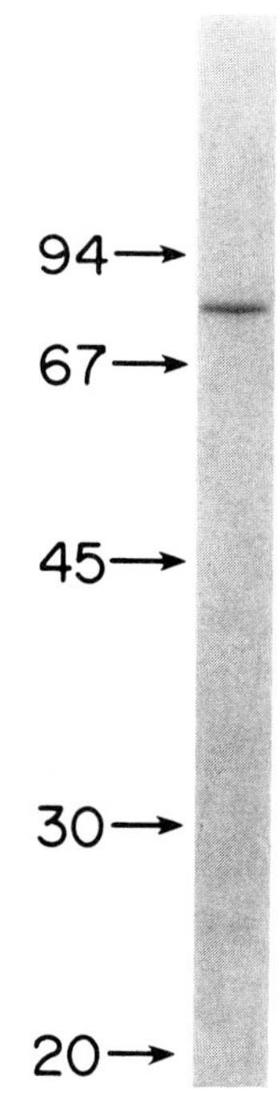


FIG. 2. Purification of bovine brain βARK. βARK was purified from bovine brain by successive chromatography on Ultrogel AcA 34, DEAE-Sephacel, and CM-Fractogel. Sample (~0.5 μg) was run on a 10% SDS-polyacrylamide gel, which was then stained with Coomassie Blue. Relative molecular weight standards (× $10^{-3}$) are indicated on the left.

Substrate Specificity of βARK

Several lines of investigation have attempted to address the substrate specificity of βARK. These studies have focused on (1) assessing the agonist-dependent nature of the phosphorylation reaction, (2) determining the sites of phosphorylation on the receptor, and (3) determining the receptor specificity for phosphorylation by βARK. As discussed above, βARK has a strong preference for phosphorylating the agonist-occupied or activated form of the receptor. The agonist-dependent nature of this reaction has been characterized by studying the ability of full and partial agonists to promote receptor phosphorylation by βARK.[11] Partial agonists appear to promote reduced receptor phosphorylation when compared with full agonists even when the receptor is fully occupied. In addition, the kinetics of the phosphorylation reactions suggest that the major difference between phosphorylation of agonist vs partial agonist vs unoccupied receptor is in the rate or extent ($V_{max}$) at which the receptor is phosphorylated.

The sites of βARK phosphorylation on the mammalian $\beta_2$AR remain somewhat obscure although several lines of evidence suggest they are localized to the carboxyl-terminal tail of the receptor. It has been demonstrated that most, if not all, of the sites on the βARK-phosphorylated $\beta_2$AR are lost after carboxypeptidase treatment, which removes the carboxyl tail of the receptor.[12] In addition, mutagenesis studies also target the carboxyl tail as a major phosphorylation domain. Using cells expressing a mutant $\beta_2$AR lacking 50 amino acids at the carboxyl terminus, it has been demonstrated that the mutant receptor does not undergo the agonist-promoted phosphorylation observed for the wild-type receptor.[13] Finally, synthetic peptides derived from the carboxyl terminus of the hamster $\beta_2$AR appear to serve as specific, albeit poor, substrates for βARK.[14]

Although the βAR-coupled adenylate cyclase system has proved to be a useful model for studying agonist-specific desensitization, many other G protein-coupled receptors also undergo desensitization when activated. Thus, it is not unreasonable to speculate that one mechanism by which these other receptors undergo desensitization might also involve an agonist-induced receptor phosphorylation. The notion that each receptor has its own specific protein kinase, however, seems unlikely. A more plausible

[11] J. L. Benovic, C. Staniszewski, F. Mayor, Jr., M. G. Caron, and R. J. Lefkowitz, *J. Biol. Chem.* **263,** 8856 (1988).

[12] H. G. Dohlman, M. Bouvier, J. L. Benovic, M. G. Caron, and R. J. Lefkowitz, *J. Biol. Chem.* **262,** 14282 (1987).

[13] M. Bouvier, W. P. Hausdorff, A. DeBlasi, B. F. O'Dowd, B. K. Kobilka, M. G. Caron, and R. J. Lefkowitz, *Nature (London)* **333,** 370 (1988).

[14] J. L. Benovic, J. Onorato, M. J. Lohse, H. G. Dohlman, C. Staniszewski, M. G. Caron, and R. J. Lefkowitz, *Br. J. Clin. Pharmacol.* **30,** 3S (1990).

hypothesis is that one or more general receptor kinases act to phosphorylate multiple agonist-occupied receptors. Several lines of evidence suggest that $\beta$ARK may in fact be a general receptor kinase. Stimulation of S49 lymphoma cells with a $\beta$-agonist leads to the translocation of up to 80% of the $\beta$ARK activity from the cytosol to the plasma membrane.[15] Similar levels of $\beta$ARK translocation are also observed when these cells are activated by prostaglandin $E_1$,[15] which activates adenylate cyclase, or by somatostatin,[16] which inhibits cyclase. This suggests that the $PGE_1$ and somatostatin receptors might also be regulated by $\beta$ARK phosphorylation. Additional studies have directly demonstrated that purified $\beta$ARK is able to phosphorylate the agonist-occupied human platelet $\alpha_2$-adrenergic receptor[17] and the chick cardiac muscarinic acetylcholine receptor[18] to high stoichiometries. Taken together these results strongly suggest that $\beta$ARK is a general adenylate cyclase-coupled receptor kinase.

## Inhibition of $\beta$ARK

A number of compounds which inhibit the $\beta$-adrenergic receptor kinase have been identified.[19] These compounds include polyanions, such as heparin and dextran sulfate, which have $IC_{50}$ values of $\sim 0.15\ \mu M$ (Table I). Heparin is the most potent inhibitor of $\beta$ARK identified to date and has been useful for correlating $\beta$AR phosphorylation and desensitization in permeabilized A431 cells.[20] A number of other anions such as poly(aspartic acid) and poly(glutamic acid) as well as inositol hexasulfate are also good inhibitors of $\beta$ARK. In contrast, inositol hexaphosphate, pyridoxal phosphate and 2,3-diphosphoglycerol are poor inhibitors of the enzyme. The ability of polyanions to inhibit $\beta$ARK largely parallels results seen with casein kinase II, where heparin is a potent inhibitor.[21] Conversely, heparin

[15] R. H. Strasser, J. L. Benovic, M. G. Caron, and R. J. Lefkowitz, *Proc. Natl. Acad. Sci. U.S.A.* **83,** 6362 (1986).

[16] F. Mayor, Jr., J. L. Benovic, M. G. Caron, and R. J. Lefkowitz, *J. Biol. Chem.* **262,** 6468 (1987).

[17] J. L. Benovic, J. W. Regan, H. Matsui, F. Mayor, Jr., S. Cotecchia, L. M. F. Leeb-Lundberg, M. G. Caron, and R. J. Lefkowitz, *J. Biol. Chem.* **262,** 17251 (1987).

[18] M. M. Kwatra, J. L. Benovic, M. G. Caron, R. J. Lefkowitz, and M. M. Hosey, *Biochemistry* **28,** 4543 (1989).

[19] J. L. Benovic, W. C. Stone, M. G. Caron, and R. J. Lefkowitz, *J. Biol. Chem.* **264,** 235 (1989).

[20] M. J. Lohse, R. J. Lefkowitz, M. G. Caron, and J. L. Benovic, *Proc. Natl. Acad. Sci. U.S.A.* **86,** 3011 (1989).

[21] G. M. Hathaway and J. A. Traugh, *Curr. Top. Cell. Regul.* **21,** 101 (1982).

TABLE I
INHIBITORS OF $\beta$ARK

| Compound | Class | $IC_{50}$ ($\mu M$)[a] |
|---|---|---|
| Heparin | Anion | 0.15 |
| Dextran sulfate | | 0.15 |
| Polyaspartic acid | | 1.3 |
| Inositol hexasulfate | | 13.5 |
| Inositol hexaphosphate | | 3600 |
| Pyridoxal phosphate | | 900 |
| Poly(lysine) | Cation | 69 |
| Spermine | | 1600 |
| Chlorpromazine | Phospholipid-interacting | 43 |
| Tamoxifen | | 40 |
| D-Sphingosine | | 27 |
| Digitonin | | 50 |
| Triton X-100 | | 54 |
| Tween 20 | | 27 |
| Trifluoperazine | Calmodulin antagonist | 35 |
| Sangivamycin | Nucleoside analog | 67 |
| H7 | | 250 |

[a] $IC_{50}$ is the concentration of compound which inhibited $\beta$ARK phosphorylation of rhodopsin by 50%.

is a poor inhibitor of rhodopsin kinase,[22] the enzyme involved in phosphorylating the light-activated form of rhodopsin. This demonstrates that heparin is not a general inhibitor of all agonist-dependent protein kinases.

Cations such as poly(lysine) and spermine also inhibit $\beta$ARK activity albeit much more weakly than do the polyanions. While polycations are able to partially relieve polyanion inhibition they do not activate $\beta$ARK as is observed for casein kinase II[21] and rhodopsin kinase.[22] A variety of compounds which interact with phospholipids also inhibit $\beta$ARK with $IC_{50}$ values from 27 to 54 $\mu M$. These compounds include chlorpromazine, tamoxifen, and sphingosine as well as numerous detergents. It is not clear whether these compounds directly inhibit $\beta$ARK or whether they interfere with the ability of $\beta$ARK to interact with its substrate (agonist-activated rhodopsin or $\beta$AR in phospholipid). Several other compounds tested also inhibited $\beta$ARK, including trifluoperazine, a calmodulin antagonist, and sangivamycin and H7, nucleoside analogs.

[22] K. Palczewski, A. Arendt, J. H. McDowell, and P. A. Hargrave, *Biochemistry* **28,** 8764 (1989).

Structure of $\beta$ARK

The structure of $\beta$ARK has been elucidated by isolation of a cDNA encoding this enzyme.[23] This was accomplished by initially treating purified $\beta$ARK with cyanogen bromide, resolving and sequencing the resultant peptides, and then screening a cDNA library with synthetic oligonucleotide probes derived from the peptide sequences. Using this strategy a clone which contained an open reading frame of 2067 base pairs (bp) was isolated. This clone encodes a protein of 689 amino acids (79.7 kDa) whose topology suggests an amino-terminal domain of ~197 amino acids, a central catalytic domain of ~239 amino acids, and a carboxyl-terminal domain of ~253 amino acids. The catalytic domain of $\beta$ARK has highest homology with the cAMP-dependent protein kinase (33.1% identity) and with protein kinase C (33.7%). When the $\beta$ARK cDNA was inserted into a mammalian expression vector, a protein which specifically phosphorylated the agonist-occupied $\beta_2$AR was expressed, providing further evidence that this cDNA encodes $\beta$ARK.

Several lines of evidence suggest that $\beta$ARK may be a member of a multigene family. When genomic DNA is probed with a fragment of the $\beta$ARK cDNA, multiple hybridizing species are observed. Moreover, low-stringency hybridization studies have enabled the isolation of a second class of cDNA clones. This cDNA encodes a protein of 688 amino acids (79.6 kDa) with an overall homology of 84.2% with $\beta$ARK. Current efforts are focused on isolating additional members of this gene family as well as characterizing the substrate specificity and functional role of these novel kinases.

[23] J. L. Benovic, A. DeBlasi, W. C. Stone, M. G. Caron, and R. J. Lefkowitz, *Science* **246,** 235 (1989).

# [29] Adenosine Monophosphate-Activated Protein Kinase: Hydroxymethylglutaryl-CoA Reductase Kinase

*By* DAVID CARLING, PAUL R. CLARKE, and D. GRAHAME HARDIE

The AMP-activated protein kinase is a recently defined activity which is regulated by both AMP and by phosphorylation,[1] and which modulates the activity of key enzymes of lipid metabolism, i.e., acetyl-CoA carboxyl-

[1] D. G. Hardie, D. Carling, and A. T. R. Sim, *Trends Biochem. Sci.* **14,** 20 (1989).

ase,[2,3] hydroxymethylglutaryl (HMG)-CoA reductase,[2,3] and hormone-sensitive lipase/cholesterol esterase.[4] Although purified in our laboratory based on its ability to inactivate acetyl-CoA carboxylase, we have obtained good evidence that it is identical with the HMG-CoA reductase kinase(s) described previously in this series[5] and elsewhere.[6-9] Thus the acetyl-CoA carboxylase kinase and HMG-CoA reductase kinase activities copurify to near homogeneity (4800-fold) from rat liver.[3] In addition, the two activities of the purified preparation show an excellent correlation during studies of stimulation by 5′-AMP, inactivation by protein phosphatases, and inactivation by the reactive ATP analog, 5′-fluorosulfonylbenzoyladenosine.[3] Since there is now good evidence that this kinase regulates acetyl-CoA carboxylase[10] and hormone-sensitive lipase/cholesterol esterase,[4] as well as HMG-CoA reductase, *in vivo,* the name HMG-CoA reductase kinase is no longer appropriate. We have adopted the precedent set in nomenclature of many other protein kinases and named the activity for its allosteric activator, 5′-AMP.

HMG-CoA reductase kinase has been previously assayed by its ability to inactivate HMG-CoA reductase in crude rat liver microsomes,[5-9] using a radioisotopic assay for HMG-CoA reductase. However, this assay is insensitive and time consuming; it uses expensive radioisotopes and is subject to interference by HMG-CoA lyase and mevalonate kinase.[11] We have improved the procedure by using partially purified, solubilized HMG-CoA reductase, which can be assayed spectrophotometrically. This assay is quicker, less expensive, and subject to less interference; but obtaining precise results is still difficult because the assay is quantitative only in the range where small decreases in HMG-CoA reductase activity occur. AMP-activated protein kinase is assayed much more conveniently and accurately by measuring the incorporation of radioactivity from [$\gamma$-$^{32}$P]ATP into purified acetyl-CoA carboxylase.[2,3] However, one requires large amounts of the substrate protein, which must be purified on a regular basis,

[2] D. Carling, V. A. Zammit, and D. G. Hardie, *FEBS Lett.* **223,** 217 (1987).

[3] D. Carling, P. R. Clarke, V. A. Zammit, and D. G. Hardie, *Eur. J. Biochem.* **186,** 129 (1989).

[4] A. J. Garton, D. G. Campbell, D. Carling, D. G. Hardie, R. J. Colbran, and S. J. Yeaman, *Eur. J. Biochem.* **179,** 249 (1989).

[5] T. S. Ingebritsen and D. M. Gibson, this series, Vol. 71, p. 486.

[6] Z. H. Beg, J. A. Stonik, and H. B. Brewer, *Proc. Natl. Acad. Sci. U.S.A.* **76,** 4375 (1979).

[7] T. S. Ingebritsen, H. Lee, R. A. Parker, and D. M. Gibson, *Biochem. Biophys. Res. Commun.* **81,** 1268 (1978).

[8] H. J. Harwood, K. G. Brandt, and V. W. Rodwell, *J. Biol. Chem.* **259,** 2810 (1984).

[9] A. Ferrer and F. G. Hegardt, *Arch. Biochem. Biophys.* **230,** 227 (1984).

[10] A. T. R. Sim and D. G. Hardie, *FEBS Lett.* **233,** 294 (1988).

[11] G. C. Ness, C. D. Spindler, and G. A. Benton, *J. Biol. Chem.* **255,** 9013 (1980).

and there is some variability between preparations in both phosphorylation state and contamination with endogenous kinase(s), as well as occasional problems of proteolysis. This assay is also not specific because acetyl-CoA carboxylase can be phosphorylated by at least six other protein kinases.[1] We have developed a completely specific and more convenient assay[12] involving phosphorylation of a synthetic peptide (SAMS peptide) with the sequence HMRSAMSGLHLVKRR.[13a] This peptide is based on the sequence from His-73 to Lys-85 in rat acetyl-CoA carboxylase,[13,14] except that the fifth residue is alanine rather than serine, to abolish a site (corresponding to Ser-77) phosphorylated by cyclic AMP-dependent protein kinase.[13] In addition, two arginines were added at the C terminus to make the peptide bind tightly to phosphocellulose paper. The AMP-activated protein kinase phosphorylates the peptide on the seventh residue,[12] corresponding to Ser-79 on acetyl-CoA carboxylase, the site which is phosphorylated most rapidly by the kinase and which appears to regulate acetyl-CoA carboxylase activity.[13]

## Assay Methods

### *Assay of HMG-CoA Reductase Kinase Activity*

*Principle.* HMG-CoA reductase is solubilized and partially purified from rat liver microsomes using nonionic detergent. The preparation is incubated with MgATP and the kinase, and the loss of HMG-CoA reductase activity is monitored by assaying that enzyme spectrophotometrically, following the oxidation of NADPH at 340 nm. The spectrophotometric assay cannot be used with microsomal HMG-CoA reductase because of interference from turbidity and from other NADPH-oxidizing enzymes.

*Materials.* HMG-CoA reductase is solubilized and partially purified from rat liver microsomes by a modification of the method of Kennelly *et al.*[15] as described previously.[3]

[12] S. P. Davies, D. Carling, and D. G. Hardie, *Eur. J. Biochem.* **186,** 123 (1989).

[13] M. R. Munday, D. G. Campbell, D. Carling, and D. G. Hardie, *Eur. J. Biochem.* **175,** 331 (1988).

[13a] A, Alanine; G, glycine; H, histidine; K, lysine; L, leucine; M, methionine; R, arginine; S, serine; V, valine.

[14] F. Lopez-Casillas, D. H. Bai, X. Luo, I. S. Kong, M. A. Hermodson, and K. H. Kim, *Proc. Natl. Acad. Sci. U.S.A.* **85,** 5784 (1988).

[15] P. J. Kennelly, K. G. Brandt, and V. W. Rodwell, *Biochemistry* **22,** 2784 (1983).

*Reagents*

Buffer A (20 m*M* Tris/HCl, pH 7.4, 1 m*M* dithiothreitol (DTT), 50 $\mu M$ leupeptin, 1 m*M* phenylmethylsulfonyl fluoride (PMSF), 0.01% Brij W1)
Buffer B (200 m*M* potassium phosphate, pH 7.0)
NaF (500 m*M*)
Dithiothreitol (100 m*M*)
AMP (10 m*M*)
$MgCl_2$ (25 m*M*) plus ATP (1 m*M*)
EDTA (100 m*M*)
NADPH (2 m*M*)
HMG-CoA (3 m*M*)

*Procedure*. Twenty-seven microliters of partially purified HMG-CoA reductase [50 units (U)/ml] in buffer A is added to 2 $\mu$l dithiothreitol, 1 $\mu$l AMP, and 10 $\mu$l kinase in buffer A. The reaction is started by addition of 10 $\mu$l MgATP, and, after incubation for 10 min at 30°, is stopped by adding 10 $\mu$l of EDTA. HMG-CoA reductase is then assayed by adding 50 $\mu$l of the mixture to 250 $\mu$l buffer B, 50 $\mu$l NADPH, 20 $\mu$l dithiothreitol, 50 $\mu$l NaF, and 30 $\mu$l water to a cuvette at 37°. After recording the change in absorbance at 340 nm during incubation for several minutes, the reaction is started by adding 50 $\mu$l HMG-CoA. The HMG-CoA reductase activity is defined as the rate of NADPH oxidation in the presence of HMG-CoA minus that in its absence. Control incubations are normally carried out in the absence of kinase fraction: with crude preparations it may also be necessary to carry out controls in the absence of MgATP, and to correct for MgATP-independent inactivation by the kinase fraction.

*Units*. One unit of HMG-CoA reductase reduces 1 $\mu$mol of HMG-CoA (2 $\mu$mol NADPH oxidized)/min at 37°. One unit of HMG-CoA reductase kinase inactivates 1 unit of HMG-CoA reductase/min at 30°.

### *Assay of Acetyl-CoA Carboxylase and Peptide Kinase Activity*

*Principle*. AMP-activated protein kinase is measured by its ability to incorporate radioactivity from [$\gamma$-$^{32}$P]ATP into acetyl-CoA carboxylase or into the synthetic SAMS peptide. Acetyl-CoA carboxylase is separated from unreacted ATP by trichloroacetic acid precipitation, while in the case of the peptide this is achieved by binding of the peptide to phosphocellulose paper.

*Materials*. Acetyl-CoA carboxylase is purified from mammary glands of lactating rats using the avidin-Sepharose procedure.[13] The SAMS peptide is synthesized and purified as described previously.[12]

*Reagents*

HEPES buffer [50 m$M$ sodium HEPES, pH 7.0, 1 m$M$ EDTA, 1 m$M$ dithiothreitol, 50 m$M$ NaF, 10% (w/v) glycerol]
Acetyl-CoA carboxylase [1.2 mg/ml (= 5 $\mu M$) in HEPES buffer]
SAMS peptide (500 $\mu M$ in HEPES buffer)
AMP (1 m$M$ in HEPES buffer)
[$\gamma$-$^{32}$P]ATP (1 m$M$, 50–300 cpm/pmol) plus $MgCl_2$ (25 m$M$)
Bovine serum albumin (BSA) (10 mg/ml)
Trichloroacetic acid (TCA) [25% (w/v)]
Orthophosphoric acid [1% (v/v)]
Whatman P81 phosphocellulose paper (cut into 1.5-cm squares)

*Procedure.* Assays are carried out in 1.5-ml microcentrifuge tubes. Ten microliters of substrate (acetyl-CoA carboxylase or peptide) is mixed with AMP (5 $\mu$l) and kinase fraction (5 $\mu$l, diluted in HEPES buffer). The reaction is started by adding 5 $\mu$l of MgATP, and incubation is for 10 min at 30°. For acetyl-CoA carboxylase kinase assays, the reaction is stopped by adding 1 ml of trichloroacetic acid and 30 $\mu$l of serum albumin. The precipitated protein is recovered by centrifugation in a microcentrifuge (5 min). The pellet is washed three times with 1 ml of trichloroacetic acid, and the radioactivity in the washed pellet determined by Cerenkov counting of the microcentrifuge tube. A 20-$\mu$l sample of MgATP is counted at the same time to determine the specific radioactivity. For peptide kinase assays, the reaction is stopped by pipetting 15 $\mu$l of the mixture onto a P81 paper square (Whatman, Clifton, NJ) and dropping it into 500 ml of phosphoric acid. The squares are stirred gently for 20 min with 3 changes of phosphoric acid (up to 100 squares can be washed together), rinsed in acetone, air dried, and counted in 10 ml of toluene-based scintillation fluid. Vials of scintillation fluid can be reused after the squares are removed. A 10-$\mu$l sample of the MgATP is pipetted onto a paper square, dried, and counted in the same way to determine the specific radioactivity.

For both assays, two reaction blanks should be carried out, one without substrate and one without kinase. For the peptide assay, the blank without kinase is not significantly different from a blank containing buffer only, but with some preparations of acetyl-CoA carboxylase there can be significant phosphorylation in the absence of added kinase.

*Units.* One unit of acetyl-CoA carboxylase kinase catalyzes the phosphorylation of 1 $\mu$mol of phosphate into acetyl-CoA carboxylase/min at 30°. One unit of peptide kinase catalyzes the phosphorylation of 1 $\mu$mol of phosphate into the SAMS peptide/min at 30°. Activities are ~2.5-fold higher using the peptide assay, due to the higher $V_{max}$ value obtained with that substrate.[12] The number of units of activity obtained using the peptide kinase assay are six- to sevenfold higher than those using the HMG-CoA reductase kinase assay.

*Purification Procedure*

*Reagents*

Buffer A [0.05 *M* Tris/HCl (pH 8.4 at 4°), 0.25 *M* mannitol, 50 m*M* NaF, 5 m*M* sodium pyrophosphate, 1 m*M* EDTA, 1 m*M* EGTA, 1 m*M* dithiothreitol, 0.1 m*M* phenylmethylsulfonyl fluoride, 1 $\mu$g/ml soybean trypsin inhibitor, 1 m*M* benzamidine]

Buffer B [buffer A, except pH 7.5 (4°) without mannitol, plus 10% (w/v) glycerol, 0.02% (w/v) Brij 35]

Buffer C [buffer B, except 50% (w/v) glycerol]

Buffer D [20 m*M* ethanolamine/HCl, pH 9.0, 50 m*M* NaF, 10% (w/v) glycerol, 5 m*M* sodium pyrophosphate, 1 m*M* EDTA, 1 m*M* EGTA, 1 m*M* dithiothreitol, 0.02% (w/v) Brij 35]

*Procedure.* Male Wistar rats (180–250 g) are maintained on a standard chow diet and are sacrificed by stunning and cervical dislocation. Livers are rapidly removed and homogenized in 2 vol of buffer A using three 30-sec bursts in a blender, with cooling in ice between each burst. All subsequent steps are performed at 4°, in buffer B unless stated otherwise. The homogenate is centrifuged (3000 *g*, 10 min) and 25% (w/v) polyethylene glycol 6000 added to a final concentration of 2.5%. After centrifugation (30,000 *g*, 15 min), 25% (w/v) polyethylene glycol 6000 is added to a final concentration of 6%. The precipitate is collected by centrifugation (30,000 *g*, 15 min), resuspended, and applied to a DEAE-Sepharose Fast Flow column (6 × 15 cm). The column is washed until the $A_{280}$ is <0.05, and then eluted with buffer B plus 0.2 *M* NaCl. Fractions containing AMP-activated protein kinase are dialyzed to remove NaCl and centrifuged (50,000 *g*, 15 min); the supernatant is applied to a column (2 × 10 cm) of Blue-Sepharose. AMP-activated protein kinase is eluted with a linear gradient (500 ml) from 0 to 1 *M* NaCl. Active fractions are made 40% saturated with ammonium sulfate, kept on ice a few minutes, and the precipitate collected by centrifugation (10,000 *g*, 10 min). After resuspension and dialysis to remove ammonium sulfate, the preparation is applied to a Mono Q (HR 5/5) column (Pharmacia) at 1 ml/min, the column washed until the $A_{280}$ is <0.05, and the kinase eluted with a 20 ml gradient from 0 to 0.4 *M* NaCl. Active fractions are concentrated to ~0.5 ml in a Centricon 30 concentrator (Amicon, Danvers, MA) and applied to a Superose 12 gel-filtration column at 0.25 ml/min. The peak fractions are dialyzed into buffer C and stored at −20°. Under these conditions the kinase is stable for at least 2 months.

Further purification on a small scale (aliquots containing <100 μg protein) can be achieved by dialyzing the enzyme into buffer D and applying it to a Mono Q (HR 5/5) column equilibrated in the same buffer. The column is washed until the $A_{280}$ is <0.05, and the kinase eluted with a 20-ml gradient of buffer containing NaCl from 0.15 to 0.23 *M*.

## Purity and Stability of Purified Enzyme

Table I shows that the acetyl-CoA carboxylase kinase and HMG-CoA reductase kinase activities of the purified preparation remain at a constant ratio of 2.8 ± 0.3 through the five steps up to the Superose 12 column. The ratio of peptide kinase : acetyl-CoA carboxylase kinase also remains constant at 2.4 ± 0.3. All three activities also copurify through the final Mono Q (pH 9) column. Using the peptide kinase assay, the final preparation has a specific activity of 3.0 μmol/min/mg; this is comparable with the specific activities obtained for other, well-characterized protein kinases which have been purified to homogeneity (see [6], [8], [26], [27], and [30–36] in this series, Vol. 99). The final preparation contains several polypeptides detectable by Coomassie Blue Staining after SDS-polyacrylamide gel electrophoresis. However, only one polypeptide of apparent molecular weight 63K is labeled using 5′-fluorosulfonylbenzoyl[$^{14}$C]adenosine, which abolishes both acetyl-CoA carboxylase kinase and HMG-CoA reductase kinase activities.[3] Since this reagent reacts at both the allosteric (AMP) and catalytic (ATP) sites,[3,16] both sites must be on the 63K subunit. Additional evidence that this is the catalytic subunit includes its autophosphorylation using [γ-$^{32}$P]ATP,[3] and a mobility shift on SDS-polyacrylamide gel electrophoresis after protein phosphatase treatment.[17]

The enzyme from the Superose 12 step can be stored unfrozen at −20° in buffer C for at least 2 months without significant loss of activity. This preparation is not pure but contains no other kinases which phosphorylate acetyl-CoA carboxylase or HMG-CoA reductase. The Mono Q (pH 9) preparation appears to be less stable, but this has not been studied systematically.

## Properties

The purified kinase is stimulated three- to sevenfold by AMP with, under standard assay conditions, a half-maximal effect at 1.5 *μM*.[3] Variability in the degree of stimulation may be due to trace contamination of

[16] A. Ferrer, C. Caelles, N. Massot, and F. G. Hegardt, *J. Biol. Chem.* **262,** 13507 (1987).
[17] D. Carling and D. G. Hardie, unpublished observations (1989).

TABLE I

COPURIFICATION OF ACETYL-CoA CARBOXYLASE (ACC) AND HMG-CoA REDUCTASE (HMGR) KINASE ACTIVITIES FROM RAT LIVER[a]

| Fraction | Protein (mg) | ACC kinase activity: Total activity (mUnits) | ACC kinase activity: Specific activity (mUnits/mg) | ACC kinase activity: Purification (-fold) | ACC kinase activity: Yield (%) | HMGR kinase activity: Total activity (mUnits) | HMGR kinase activity: Specific activity (mUnits/mg) | HMGR kinase activity: Purification (-fold) | HMGR kinase activity: Yield (%) | Activity ratio (ACC/HMGR) |
|---|---|---|---|---|---|---|---|---|---|---|
| 1. Postmitochondrial supernatant | 12,897 | 3,417 | 0.26 | 1.0 | 100 | 1,146 | 0.089 | 1.0 | 100 | 3.0 |
| 2. 6% PEG pellet | 2,381 | 3,333 | 1.4 | 5.0 | 98 | 1,055 | 0.44 | 5.0 | 92 | 3.2 |
| 6% PEG supernatant | 7,885 | 0 | 0 | — | — | 315 | 0.004 | 0.4 | 27 | — |
| 3. DEAE-Sepharose | 260 | 3,016 | 11.6 | 45 | 88 | 1,304 | 5.0 | 56 | 114 | 2.3 |
| 4. Blue-Sepharose | 60 | 1,398 | 23.3 | 90 | 41 | 508 | 8.5 | 95 | 44 | 2.8 |
| 5. Mono Q | 6.6 | 381 | 57.7 | 222 | 11 | 134 | 20 | 228 | 11.7 | 2.8 |
| 6. Superose 12 | 1.02 | 188 | 184 | 708 | 5.5 | 70.5 | 69 | 779 | 6.2 | 2.7 |

[a] The results are taken from a preparation of eight 250-g male rats.[3]

the kinase with AMP (contamination of ATP solutions can also be a problem). For the same reason, stimulation by AMP is rarely evident during purification until after the DEAE-Sepharose step. Activation by 5′-AMP is very specific, with a number of analogs, including ADP and cyclic 3′,5′-AMP, being ineffective.[3] However, ATP and 8-bromo-AMP antagonize activation by AMP, whereas many nucleotides and nucleosides, including ADP and adenosine, inhibit the basal catalytic activity at high concentrations.[3]

If the enzyme is purified in the absence of protein phosphatase inhibitors (NaF and pyrophosphate), it rapidly loses activity due to dephosphorylation. The inactive form can be reactivated in the presence of MgATP, and this is stimulated by low concentrations of palmityl-CoA.[2,3] Reactivation occurs only in partially purified enzyme (DEAE-Sepharose step), apparently because the kinase kinase responsible is lost on further purification. The kinase kinase is very poorly characterized, but it is known that it is not cyclic AMP-dependent protein kinase.[12] Kinase purified in the presence of protein phosphatase inhibitors as far as the second Mono Q (pH 9) step, which appears to be in the fully phosphorylated form,[17] is inactivated by 70–80% by treatment with the catalytic subunits of protein phosphatase-1 or -2a.[2,3]

## Comparison with Other HMG-CoA Reductase/Acetyl-CoA Carboxylase Kinase Preparations

Two groups have reported purification of HMG-CoA reductase kinase from rat liver microsomes, and subunit molecular weights of 58K (Ref. 7) and 105K (Ref. 9) were claimed. In neither case was evidence presented that the polypeptide identified corresponded to the kinase and, in the one case where it was quoted,[9] the specific activity was very low. While we cannot rule out the possibility that there is a distinct microsomal isoenzyme, our data suggest that it could be only a minor component.[3] Ferrer and Hegardt[16] have described purification of HMG-CoA reductase kinase from the soluble fraction of rat liver. No details of specific activity were given, but 5′-fluorosulfonylbenzoyl[$^{14}$C]adenosine labeled a prominent polypeptide at 70K. This appears to be different from our 63K polypeptide, because on SDS-polyacrylamide gel electrophoresis it migrated just behind a bovine serum albumin marker,[16] whereas the 63K polypeptide migrates just in front.[3] The preparation from Ferrer and Hegardt did in fact contain a minor polypeptide which labeled with 5′-fluorosulfonyl[$^{14}$C]benzoyl-adenosine and which migrated in the approximate position of the 63K polypeptide. This component may represent the kinase.

One group[18] has purified an acetyl-CoA carboxylase kinase from rat liver, and since the first two steps in the purification were identical to ours it is likely that they purified a form of the AMP-activated protein kinase. A subunit molecular weight of 170K was claimed,[18] but once again no evidence was presented that the polypeptide identified corresponding to the kinase, and the specific activity was extremely low.

Acknowledgments

Studies in this laboratory were supported by the British Heart Foundation, and by a studentship from the Science and Engineering Research Council (to P.R.C.).

[18] B. Lent and K. H. Kim, *J. Biol. Chem.* **257,** 1897 (1982).

# [30] Purification of Platelet-Derived Growth Factor $\beta$ Receptor from Porcine Uterus

*By* LARS RÖNNSTRAND and CARL-HENRIK HELDIN

Platelet-derived growth factor (PDGF), a major mitogen for connective tissue cells, is structurally a 30-kDa dimeric molecule composed of disulfide-bonded A and B polypeptide chains.[1] The three different isoforms, PDGF-AA, -AB, and -BB, bind to two distinct receptor types with different specificities and affinities; the PDGF $\alpha$ receptor (also called A-type receptor) binds all three isoforms with high affinity, whereas the $\beta$ receptor (also called B-type receptor) binds only PDGF-BB with high affinity, PDGF-AB with a 10-fold lower affinity, and appears not to bind PDGF-AA. The $\alpha$ and $\beta$ receptors are structurally related molecules which both have ligand-stimulatable protein tyrosine kinase activities; their cDNAs have been cloned.[2–4] In this chapter, the purification of the PDGF $\beta$ receptor from porcine uterus is described.

[1] C.-H. Heldin and B. Westermark, *Cell. Regul.* **1,** 555 (1990).

[2] Y. Yarden, J. A. Escobedo, W.-J. Kuang, T. L. Yang-Feng, T. O. Daniel, P. M. Tremble, E. Y. Chen, M. E. Ando, R. N. Harkins, U. Francke, V. A. Friend, A. Ullrich, and L. T. Williams, *Nature (London)* **323,** 226 (1986).

[3] L. Claesson-Welsh, A. Eriksson, B. Westermark, and C.-H. Heldin, *Proc. Natl. Acad. Sci. U.S.A.* **86,** 4917 (1989).

[4] T. Matsui, M. Heidaran, M. Toru, N. Popescu, W. LaRochelle, M. Kraus, J. Pierce, and S. A. Aaronson, *Science* **243,** 800 (1989).

## Assay for PDGF Receptor

Whereas binding of $^{125}$I-labeled PDGF to cultured cells can easily be determined, attempts to construct a useful assay for binding of $^{125}$I-labeled PDGF to membrane preparations or detergent-solubilized fractions have been futile. This is probably due to "stickiness" of the basic PDGF molecule and to the fact that the affinity of the receptor decreases after solubilization with detergents. We have therefore employed the ligand-stimulated autophosphorylation of the PDGF $\beta$ receptor as an assay in the purification.[5]

Briefly, the receptor-containing fraction is incubated for 10 min at 0° with 100 ng of PDGF purified from human platelets (containing PDGF-AB and PDGF-BB[6]) or recombinant PDGF-BB[7]; [$\gamma$-$^{32}$P]ATP is then added and the incubation is continued for another 10 min at 0°. The incubation mixture (40 $\mu$l) contains 0.1% Triton X-100, 5% glycerol, 0.15 *M* NaCl, 0.02 *M* HEPES, pH 7.4, 0.5 m*M* EGTA, 0.5 m*M* dithiothreitol, 0.7 mg/ml bovine serum albumin, and 15 $\mu$*M* [$\gamma$-$^{32}$P]ATP containing $5 \times 10^6$ cpm of radioactivity. The phosphorylation is stopped by heating at 95° for 3 min in reducing SDS-sample buffer,[8] followed by alkylation with iodoacetamide. The samples are then analyzed by SDS gel electrophoresis,[8] using 5–10% polyacrylamide gradient gels, and autoradiography. Occasionally, receptor autophosphorylation is quantified by determining the $^{32}$P radioactivity in excised receptor bands by Cerenkov radiation. Theoretically, this assay detects both $\alpha$ receptors (about 170 kDa) and $\beta$ receptors (170–190 kDa, depending on the source).

### *Source of PDGF Receptor*

Potential sources of PDGF receptors include cultured cells and tissues. Since there are no cell lines available that overexpress PDGF receptors, it is necessary to grow very large quantities of cells (thousands of roller bottles) to obtain nanomole amounts of pure receptor.[2] We therefore explored the possibility of using a tissue as starting material in receptor purification. Membranes were prepared (see below) from about 20 different porcine organs and subjected to the PDGF receptor autophosphorylation

[5] L. Rönnstrand, P. M. Beckmann, B. Faulders, A. Östman, B. Ek, and C.-H. Heldin, *J. Biol. Chem.* **262,** 2929 (1987).

[6] A. Hammacher, U. Hellman, A. Johnson, K. Gunnarsson, A. Östman, B. Westermark, Å. Wasteson, and C.-H. Heldin, *J. Biol. Chem.* **263,** 16493 (1988).

[7] A. Östman, G. Bäckström, N. Fong, C. Betsholtz, C. Wernstedt, U. Hellman, B. Westermark, P. Valenzuela, and C.-H. Heldin, *Growth Factors* **1,** 271 (1989).

[8] G. Blobel and B. Dobberstein, *J. Cell Biol.* **67,** 835 (1975).

assay; the uterus was found to be the richest source of PDGF receptor, and was chosen as starting material for purification.[5]

*Preparation of Membranes from Porcine Uterus*

Membranes are prepared from pig uteri obtained from a local slaughterhouse. The uteri are used fresh or stored at $-20°$ prior to use. About 5 kg is thawed and processed at a time; all subsequent procedures are performed at 0–4°. The material is first ground in a meat grinder and then homogenized (Ultra-Turrax, Janke and Kunkel Ginbtl, IKA Werk, Stafven, Switzerland; maximal speed for 3 min in $N_2$ atmosphere) in four times its weight of 0.15 *M* NaCl, 10 m*M* sodium phosphate buffer, pH 7.4, 5 m*M* EDTA, 1 m*M* phenylmethylsulfonyl fluoride (PMSF), 1 m*M* dithiothreitol (DTT).

The homogenate is cleared from debris by centrifugation for 10 min at 1300 *g* and then filtered through two layers of cheesecloth and recentrifuged at 5800 *g* for 15 min. The pellet is discarded and a crude membrane pellet is obtained by centrifugation at 13,700 *g* for 45 min using Beckman high-speed centrifuges (Beckman Instruments, Palo Alto, CA) equipped with JA-10 rotors. The pellets are resuspended in the same buffer, combined, and washed twice by repeated centrifugations. Approximately 1.5 g of membrane proteins is obtained. Membranes are stored at $-70°$.

During membrane preparation, it is essential to include in the homogenization buffer phenylmethylsulfonyl fluoride and EDTA to prevent proteolytic degradation of the PDGF $\beta$ receptor. In particular, the receptor is readily degraded to a 130-kDa form by a $Ca^{2+}$-dependent protease.[9] It is also important to prevent oxidation of the receptor. The homogenization is therefore performed under an atmosphere of nitrogen and 1 m*M* dithiothreitol is included in all buffers throughout the purification.

*Chromatography on Wheat Germ Agglutinin (WGA)-Sepharose*

About 400 mg of WGA (purchased from Separationscentralen, Biomedical Center, Uppsala, Sweden) is coupled to CNBr-activated Sepharose 4B (20 ml; Pharmacia Biochemicals, Uppsala, Sweden) according to the description from the manufacturer.

Approximately 1.5 g of uterus membrane proteins is solubilized in 400 ml of 2.5% (v/v) Triton X-100, 10% (v/v) glycerol, 20 m*M* HEPES, pH 7.4, 4 m*M* EGTA, 1 m*M* dithiothreitol. After 30 min of incubation on ice, the solubilized material is centrifuged at 100,000 *g* for 30 min. The resulting supernatant is pumped through the WGA column at a flow rate of approxi-

[9] B. Ek and C.-H. Heldin, *Eur. J. Biochem.* **155**, 409 (1986).

mately 30 ml/hr. The column is washed with 100 ml 0.2% Triton X-100, 10% glycerol, 0.15 *M* NaCl, 20 m*M* HEPES, 1 m*M* EGTA, 1 m*M* dithiothreitol, pH 7.4, at a flow rate of 50 ml/hr. After a second wash in the same buffer, but lacking NaCl, bound material is eluted with 50 ml of 0.3 *M* *N*-acetylglucosamine, 0.2% Triton X-100, 10% glycerol, 20 m*M* HEPES, pH 7.4, 1 m*M* EGTA, 1 m*M* dithiothreitol, at a flow rate of 20 ml/hr.

### *Chromatography on an FPLC Mono Q Column*

The eluate of the lectin column is applied directly onto a fast protein liquid chromatography (FPLC) Mono Q column (bed volume 2 ml, Pharmacia P-L Biochemicals) and eluted with a gradient from 0 to 0.3 *M* NaCl in 0.2% Triton X-100, 10% glycerol, 20 m*M* HEPES, pH 7.4, 1 m*M* EGTA, 1 m*M* dithiothreitol, at a flow rate of 2 ml/min. Two-ml fractions are collected. Individual fractions are analyzed for PDGF receptor content by the autophosphorylation assay. A 175-kDa component, which is phosphorylated after PDGF stimulation, is observed in three or four fractions eluting at about 0.25 *M* NaCl.

### *Chromatography on Anti-Phosphotyrosine-Sepharose*

In order to make an anti-phosphotyrosine immunoglobulin column, the immunoglobulin fraction of a rabbit antiserum against phosphotyrosine[10] is affinity purified on an L-phosphotyrosine-Sepharose column [5 ml, made by coupling 10 mg of L-phosphotyrosine (Sigma, St. Louis, MO) to activated CH-Sepharose 4B (Pharmacia Biochemicals)]. After washing with 50 ml of 0.15 *M* NaCl, 0.01 *M* phosphate buffer, pH 7.4, the antibodies are eluted with 25 ml 40 m*M* phenylphosphate, 0.15 *M* NaCl, 0.01 *M* phosphate buffer, pH 7.4 and run directly onto a 2-ml protein A-Sepharose column. The column is washed with 20 ml 0.15 *M* NaCl, 0.01 *M* phosphate buffer, pH 7.4, and eluted with 50 m*M* sodium citrate buffer, pH 3.0. Two-milliliter fractions are collected in tubes containing 200 $\mu$l of 1 *M* Tris, pH 7.4. The purified antibodies are dialyzed extensively against 0.15 *M* NaCl, 0.01 *M* phosphate, pH 7.4. The affinity-purified antibodies (4 mg is obtained from 100 ml of immune serum) are then coupled to CNBr-activated Sepharose 4B (2 mg immunoglobulin/ml of gel).

The active fractions from the Mono Q chromatogram are pooled and incubated overnight on ice in the presence of 3 m*M* $MnCl_2$ and 15 $\mu M$ ATP. During this long incubation time, the background activity of the

[10] B. Ek and C.-H. Heldin, *J. Biol. Chem.* **259,** 11145 (1984).

receptor kinase leads to autophosphorylation on tyrosine residues of the receptor, even in the absence of ligand. [$\gamma$-$^{32}$P]ATP (100 $\times$ $10^6$ cpm) is also included to allow the determination of the stoichiometry of phosphorylation and to monitor the purification. There are two major advantages of using the background kinase activity for the autophosphorylation of the receptor, rather than stimulating autophosphorylation by the ligand: it does not consume large quantities of ligand and the purified receptor will not contain bound ligand.

The phosphorylated Mono Q fractions are applied to a 2-ml anti-phosphotyrosine column at a flow rate of 5 ml/hr. The column is then washed with 10 ml of 0.2% Triton X-100, 10% glycerol, 0.5 *M* NaCl, 20 m*M* HEPES, pH 7.4, 1 m*M* EGTA, 1 m*M* dithiothreitol, and eluted with 10 ml of the same buffer supplemented with 40 m*M* phenylphosphate.

### *Dephosphorylation of Phosphorylated Receptor by Alkaline Phosphatase*

The eluate from the immunoaffinity column is dialyzed against 0.2% Triton X-100, 10% glycerol, 0.15 *M* NaCl, 5 m*M* benzamidine, 1 m*M* $MgCl_2$, 1 m*M* EGTA, 20 m*M* HEPES, pH 7.4, 1 m*M* dithiothreitol, at 4° for 6 hr with one change of dialysis buffer. Thereafter, 500 units (U) of alkaline phosphatase (from calf intestine, grade I, Catalog No. 108146; Boehringer Mannheim, Mannheim, Germany) is added, and the dialysis is continued for 12 hr against the same dialysis buffer. In order to separate the dephosphorylated receptor from the phosphatase, the dialysate is applied to a 200-$\mu$l DEAE-Sepharose CL-4B column, whereafter the column is washed with 5 ml of the dialysis buffer. The phosphatase does not bind to the column under these conditions, and the receptor is eluted with 2 ml of 0.2% Triton X-100, 10% glycerol, 0.5 *M* NaCl, 20 m*M* HEPES, pH 7.4, 1 m*M* EGTA, 1 m*M* dithiothreitol. The recovery of receptor during this step is approximately 70%.

Starting from 1.5 g of membrane proteins, about 160 $\mu$g of receptor is obtained (Table I), which on analysis by SDS gel electrophoresis and silver staining gives one homogeneous band of 170K[5] (Fig. 1). The receptor undergoes autophosphorylation after stimulation with PDGF-BB (the phosphorylated receptor migrates as 175K), and to a lesser extent after stimulation with PDGF-AB, but not after stimulation with PDGF-AA,[11] indicating that it represents the PDGF $\beta$ receptor (Fig. 1).

[11] C.-H. Heldin, A. Ernlund, C. Rorsman, and L. Rönnstrand, *J. Biol. Chem.* **264,** 8905 (1989).

TABLE I
PURIFICATION OF PDGF β RECEPTOR FROM PORCINE UTERUS MEMBRANES[a]

| Step | Total protein (mg) | Specific activity[b] (pmol phosphate incorporated/min/mg) | Purification (-fold) | Yield (%) |
|---|---|---|---|---|
| Membranes solubilized with Triton X-100 | 1280 | 0.023 | 1 | 100 |
| Eluate from WGA-Sepharose | 14.2 | 2.0 | 90 | 99 |
| Pool from Mono-Q chromatogram | 2.5 | 5.3 | 233 | 46 |
| Eluate from anti-phosphotyrosine-Sepharose after dephosphorylation | 0.16 | 14.3 | 630 | 8 |

[a] Modified from Ref. 5.
[b] Activity of the receptor preparations was estimated as the amount of [$^{32}$P]phosphate that was incorporated per minute into the 170-kDa receptor; this component was cut out from an SDS gel electropherogram and its content of $^{32}$P radioactivity was determined as Cerenkov radiation.

## General Comments

The assay used to follow the purification would recognize PDGF α as well as β receptors, since PDGF-BB binds to and activates both receptor types. However, the receptor obtained seemed to consist entirely of β receptors. Analysis of the various fractions in the purification procedure by PDGF-AA-stimulated phosphorylation followed by immunoprecipitation with antisera recognizing the β receptor only, or both receptor types,[12] revealed no or very low amounts of α receptors. It is thus possible that porcine uterus does not contain α receptors, or contains only low amounts. Notably, immunohistochemical stainings using β receptor-specific monoclonal antibodies revealed that cells of most normal tissues do not contain PDGF β receptors, despite the fact that the corresponding cells cultured *in vitro* possess β receptors.[13] The staining in the uterus was localized to the endometrium;[13] this tissue is thus one of the few normal tissues that contains PDGF β receptor-expressing cells.

An alternative procedure for the purification of PDGF receptors would be to use immobilized specific antibodies. A major disadvantage with

[12] L. Claesson-Welsh, A. Hammacher, B. Westermark, C.-H. Heldin, and M. Nistér, *J. Biol. Chem.* **264,** 1742 (1989).
[13] L. Terracio, L. Rönnstrand, A. Tingström, K. Rubin, L. Claesson-Welsh, K. Funa, and C.-H. Heldin, *J. Cell Biol.* **107,** 1947 (1988).

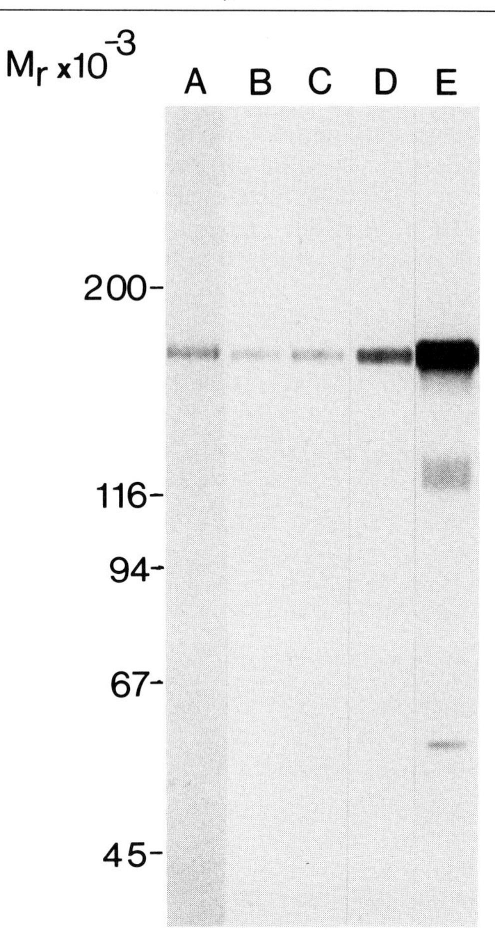


FIG. 1. Characterization of the purified PDGF β receptor. An aliquot of the purified receptor was analyzed by SDS gel electrophoresis and silver staining (A). The receptor was also subjected to the autophosphorylation assay in the absence of ligand (B) or after stimulation with 100 ng of either PDGF-AA (C), PDGF-AB (D), or PDGF-BB (E). Samples were analyzed by SDS gel electrophoresis; an autoradiogram of the gel is shown (lanes B–E). The 130K component visible in lane E is a proteolytic degradation product of the receptor.[9] [Taken from C.-H. Heldin and L. Rönnstrand, *in* "The Receptors" (G. Litwack, ed.), p. 303. Humana Press, Clifton, New Jersey, 1990.

such a procedure is, however, that it is difficult to elute the receptor from these columns with retained activity, since the kinase activity of the receptor is susceptible to low pH and to chaotropic agents.

This chapter describes the purification to homogeneity of nanomole quantities of functionally active PDGF $\beta$ receptors, from a readily available source. The method should be particularly useful for the preparation of receptor for kinetic experiments and for experiments aimed at characterizing the substrate specificity of the receptor kinase.

Acknowledgments

We thank Ingegärd Schiller for valuable help in the preparation of this manuscript.

# [31] Affinity Purification of Active Epidermal Growth Factor Receptor Using Monoclonal Antibodies

*By* JUSTIN HSUAN and PNINA YAISH

## Introduction

### *Epidermal Growth Factor Receptor*

The cell surface receptor for epidermal growth factor (EGF) is a transmembrane glycoprotein of apparent molecular weight commonly between 150,000 and 170,000 as estimated by SDS-PAGE. In fact this receptor is able to bind other ligands competitively with EGF, notably including transforming growth factor $\alpha$, with similar affinity.

The receptor is derived from a single gene but structural heterogeneity is derived from differential glycosylation and phosphorylation of the extracellular and intracellular regions, respectively. The former is thought to comprise at least four domains by examination of the primary structure: two homologous, globular (L) domains that may be directly involved in binding ligand and two homologous, cysteine-rich (S) domains, arranged as L1, S1, L2, S2. This region is followed by a single transmembrane helix. The intracellular region comprises a regulatory juxtamembrane domain, a protein-tyrosine kinase domain, and a putatively flexible autophosphorylation domain at the C terminus.[1]

The EGF receptor is expressed in a wide variety of mammalian cell

[1] J. J. Hsuan, G. Panayotou, and M. D. Waterfield, *Prog. Growth Factor Res.* **1,** 23 (1989).

types, with the important exception of hemopoietic lineages, and structurally similar proteins have been found in avian and insect species.

### *Purification Schemes*

Although much has been learned from the many independent studies that have addressed key issues, many important questions still remain unanswered regarding the structure, regulation, and activity of the EGF receptor, such as the tertiary structure of the molecule, the definition of the ligand-binding site, the mechanism of receptor activation by ligand, and the identity of important cellular substrates. As one approach to achieving these ends, a source of pure and active receptor has been established using monoclonal antibody affinity to purify receptor from the A431 human epidermal carcinoma cell line.[2]

This particular approach has the great advantage over those purification schemes that use ligand affinity, conventional chromatography, or certain other types of antibody affinity in that it allows a relatively rapid isolation of receptor using extremely mild conditions. The yield is sufficient for many types of functional analysis and we are currently exploring the possibility of scaling the preparation up to a level that allows structural studies.

This type of purification has been achieved using an anti-carbohydrate antibody (9A) that is directed against the blood group A antigen, present on the EGF receptor of A431 cells. The receptor can be eluted from this monoclonal antibody (MAb) using competition with a simple, dialyzable monosaccharide. In contrast, while the use of anti-polypeptide antibodies or ligand affinity frequently gives higher yields, more severe physical conditions are required for the recovery of receptor. Anti-phosphotyrosine antibody affinity has been widely used for the analysis and purification of tyrosine kinases under mild conditions, but this method may be inefficient if little tyrosine phosphosphorylation is present on the receptor or if many other proteins are phosphorylated on tyrosine residues. It may also preferentially select highly phosphorylated receptor subpopulations.

## Assay

### *Principle*

Saturation ligand binding is routinely exploited to assay EGF receptor in membranes and in detergent solution. In the latter case a high-affinity monoclonal antibody is used to separate bound from free radiolabeled

[2] P. J. Parker, S. Young, W. J. Gullick, E. L. V. Mayes, P. Bennet, and M. D. Waterfield, *J. Biol. Chem.* **259,** 9906 (1984).

ligand. This procedure has the disadvantage that it does not determine the ligand-dependent kinase activity of the receptor. This is an important point as the kinase activity is first a distinct and second a more labile property relative to ligand-binding ability. Accordingly, an assay for peptide phosphorylation activity is also described here.

In contrast to the peptide kinase assay (below), a receptor phosphorylation assay is relatively specific to EGF receptor kinase activity. While this is often used with SDS-PAGE for the detection of active receptor, autophosphorylation does not easily lend itself to quantitative kinetic analysis. Polypeptide phosphorylation is an important complementary assay to peptide phosphorylation. Common substrates include synthetic poly(glutamic acid, tyrosine) polymers and various glycolytic enzymes, which are generally separated from receptor by SDS-PAGE. Conditions for this assay have been described elsewhere.[3]

Greater specificity of the peptide kinase assay can be obtained using receptor immunoprecipitated on monoclonal antibody EGFR1 and protein A-Sepharose after the incubation with EGF, essentially as described below.

*Procedures*

*$^{125}$I-Labeled Epidermal Growth Factor Radioimmunoassay*.[4] Duplicate aliquots for assay of between 10 and 40 $\mu$l are incubated with a saturating concentration of 0.2 $\mu$M $^{125}$I-labeled EGF (Amersham, Aylesbury, UK) of known specific activity and 67 n*M* monoclonal antibody EGFR1[5] (2 $\mu$g/sample; Amersham) and made up to a final volume of 200 $\mu$l with 10 m*M* sodium phosphate, pH 7.4, containing 0.2% (v/v) Triton X-100 and 150 m*M* NaCl (Triton buffer). After tumbling for 1 hr at room temperature, 20 $\mu$l of a 50% (v/v) mixture of protein A-Sepharose (Pharmacia, Piscataway, NJ) in Triton buffer is added. After a further 30-min tumbling at room temperature, the pellet is washed with three 1-ml aliquots of Triton buffer at 4° and then counted in a $\gamma$ counter. Background, nonspecific binding is estimated by the use of normal mouse serum in place of EGFR1.

*Peptide Kinase Assay*. Samples containing EGF receptor are incubated in 25 m*M* HEPES, pH 7.4, containing 10 m*M* $MgCl_2$, 0.1 m*M* $Na_3VO_4$, 0.5 m*M* EGTA either with or without 150 n*M* EGF for 30 min at 20° in a final volume of about 100 $\mu$l. Samples are transferred to 30° and [$\gamma$-$^{32}$P]ATP

[3] N. Reiss, H. Kanety, and J. Schlessinger, *Biochem. J.* **239**, 691 (1986).

[4] W. J. Gullick, D. J. H. Downward, J. J. Marsden, and M. D. Waterfield, *Anal. Biochem.* **141,** 253 (1984).

[5] M. D. Waterfield, E. L. Mayes, P. Stroobant, P. L. P. Bennet, S. Young, P. N. Goodfellow, G. S. Banting, and B. Ozanne, *J. Cell. Biochem.* **20,** 149 (1982).

(specific activity routinely 5000–10,000 cpm/pmol; Amersham) and peptide RR-src (RRLIEDAEYAARG, synthesized on an Applied Biosystems 430A automated peptide synthesizer using Fmoc chemistry, or available from Sigma, St. Louis, MO) are then added to final concentrations of 100 $\mu M$ and 1 m$M$, respectively. Samples of 10 $\mu$l are removed at 2-min intervals, rapidly spotted onto ~2-cm$^2$ pieces of P81 cellulose filters (Whatman, Clifton, NJ), and immediately quenched in 30% (v/v) acetic acid, 0.5% (v/v) phosphoric acid with gentle stirring. After two further acid washes of 10 min each the filters are washed in ethanol for 10 min and dried before scintillation counting. Samples are assayed in duplicate and the background is estimated by omission of the peptide substrate. It is possible to substitute other peptide substrates for RR-src, such as angiotensin II or peptides derived from the autophosphorylation sites of the EGF receptor.[6] This method has been successfully used with immunoprecipitates (it is important to perform the incubation with ligand prior to immunoprecipitation), purified receptor in the presence of 0.1% (v/v) Triton X-100, and intact membrane preparations.

*Units.* The EGF receptor is thought to stoichiometrically bind a single molecule of EGF, hence the results of ligand-binding assays are expressed as either moles or moles/liter. The unit of kinase activity is commonly expressed as picomoles of phosphate incorporated per minute (pmol/min), and specific activity is expressed as units per picomole of receptor (U/pmol or pmol/min/pmol) measured at 30°.

## *Purification*

*Antibody Preparation.* The original generation of anti-receptor antibodies used intact A431 cells injected into BALB/c mice. Fusion was with X-63 myeloma cells and screening was initially by binding to cultured A431 cells and a panel of 10 monoclonal hybridomas was isolated. Antibody 9A ($IgG_3$) was chosen for the purification scheme as it can be readily purified from ascites fluid using protein A affinity chromatography and has a relatively high binding affinity.[2]

*Affinity Matrix Preparation.* We routinely couple 30 mg 9A to 15 ml Affi-Gel 10 (Bio-Rad, Richmond, CA) in 0.1 $M$ HEPES buffer, pH 7.5. The antibody is dialyzed against pH 7.5 buffer and the matrix is prepared as described by the manufacturer. After tumbling overnight and standing for a further 12 hr at 4°, free reactive groups are blocked by treatment with 15 ml 0.1 $M$ ethanolamine/HCl, pH 8.0 for 1 hr. The product is washed in phosphate-buffered saline (PBSA) and stored in PBSA containing 0.02% azide at 4°. Coupling efficiency is monitored by the Bradford assay.

[6] J. Downward, M. D. Waterfield, and P. J. Parker, *J. Biol. Chem.* **260,** 14538 (1985).

*Cell Culture.* A431 cells (clone 7; Imperial Cancer Research Fund, London, England) are cultured in Dulbecco's modified Eagle's medium containing 10% (v/v) fetal calf serum, penicillin, and streptomycin (GIBCO-BRL, Uxbridge, UK) at 37° in an atmosphere of saturating humidity and 10% (v/v) carbon dioxide. Stock cultures of cells are passaged 1 : 10 or 1 : 20 at confluence (about every 4 days). Cells should not be allowed to overgrow as they become difficult to detach and less viable. Fresh stocks are used every 2 months. We routinely grow 20 roller bottles for each preparation, 2 of which are used to seed the subsequent culture.

Under these conditions cells grow to a density of $\sim 3.5 \times 10^5/cm^2$ and the doubling time is ~24 hr. Using lower concentrations of newborn calf serum or calf serum causes a decrease in the cell density to below $2 \times 10^5/cm^2$ and an increase in the doubling time by up to 100%. Although there is no apparent change in receptor numbers per cell in different sera as revealed by fluorescence-activated cell sorting (FACS) analysis, large-scale roller bottle cultures are best maintained in 10% (v/v) fetal calf serum. Initial attempts to grow A431 cells on microcarriers have met with only limited success due to extensive aggregation of stirred cultures.

*Lysate Preparation.* Confluent A431 cells ($1-2 \times 10^8$ cells/roller bottle) are aspirated and washed twice with calcium- and magnesium-free PBSA containing 5 m*M* EDTA. Cells are then harvested by incubation at 37° in PBSA containing 5 m*M* EDTA, pelleted by centrifugation at 200 *g* for 10 min at 4°, washed once with PBSA at 4°, and lysed in 50 m*M* HEPES, pH 7.5 containing 150 m*M* NaCl, 1 m*M* EDTA, 1 m*M* EGTA, 0.2 m*M* PMSF, 25 m*M* benzamidine, 10% (v/v) glycerol, and 1% Triton X-100 (lysis buffer, 25 ml/roller bottle) with tumbling for 20 min at 4°. The pH is immediately adjusted to 8.5 with 1 *M* NaOH in order to limit proteolysis. The lysate is cleared by centrifugation first at 3000 *g* for 10 min at 4°, and then at 100,000 *g* for 1 hr at 4°. At this stage the supernatant can be assayed for EGF receptor and stored at −70° if necessary for several months with no apparent loss of activity.

*Monoclonal Antibody Binding.* The lysate is tumbled with the 9A matrix at 4° for 2 hr, followed by washing in a filtration unit (Nalgene, Rochester, NY) with 50 ml PBSA containing 0.5 *M* NaCl, 1 m*M* EDTA, 0.1% Triton X-100 (buffer A), then 50 ml buffer A containing 0.25 *M* D-glucose, and finally 500 ml buffer A, all at 4°. Elution is performed by tumbling the matrix with 15 ml 50 m*M* HEPES, pH 7.5, containing 0.3 *M* *N*-acetylgalactosamine, 0.15 *M* NaCl, 1 m*M* EDTA, 0.05% (v/v) Triton X-100 (elution buffer) for 30 min at 4°. The eluted material is collected by filtration as above. A second volume of elution buffer is added and the procedure repeated. The eluates are pooled, immediately dialyzed against 50 m*M* HEPES, pH 7.5, containing 50% (v/v) glycerol, 1 m*M* dithiothreitol

(DTT), 1 m*M* EDTA and stored at $-20°$. Under these conditions the EGF receptor can be stored for several months with no significant loss of EGF-dependent autophosphorylation. The matrix is washed and stored in PBSA containing 0.05% azide at 4°.

*Mono Q*. At this point the preparation should be ~70% pure as determined by SDS-PAGE and contain ~30 μg receptor/roller bottle used. Further purification and/or concentration may be required for many types of study and we have accordingly developed a simple fast protein liquid chromatography (FPLC) (Pharmacia, Uppsala, Sweden) ion-exchange step that is performed throughout at 4°.

The EGF receptor preparation is diluted with 3 vol of 50 m*M* Tris/HCl, pH 7.4 (4°) containing 1 m*M* 2-mercaptoethanol and 0.1% Triton X-100 (buffer A). This is applied to a Mono Q column (HR5/5) equilibrated with buffer A at 0.1 ml/min. After washing at 0.2 ml/min to constant absorbance with buffer A, receptor is eluted using a single step to 2 *M* NaCl in buffer A. Poor recoveries were obtained using 0.4 *M* NaCl washing and using gradient elution, probably due to charge heterogeneity of the receptor preparation. The elution is monitered by measuring the absorbance at 280 nm using buffer A in the reference cell to offset the high background absorbance. A clearer analysis can be obtained using Bradford reagent or by spiking the preparation with $^{32}$P-labeled receptor, which shows greater than 60% recovery in the peak fractions (Fig. 1).

Fractions containing receptor are finally pooled and immediately dialyzed twice against 50 m*M* Tris/HCl, pH 7.4 (4°) containing 1 m*M* 2-mercaptoethanol, 0.01% Triton X-100, and 50% (v/v) glycerol for 1 hr each. This further concentrates the preparation, decreases the salt concentration, and allows storage at $-20°$ without freezing.

## Comments on Purification

We have found that the yield of receptor decreases with the number of preparations, probably due to loss of 9A viability. In order to improve the efficiency of the antibody affinity step, we have decreased the amount of reducing agent in the buffers and initial experiments have investigated the use of Protein A-Sepharose CL-4B (Pharmacia) in place of Affi-Gel 10. In this case antibody immobilization is only via the Fc region and the antibody can readily be removed and the matrix regenerated.

The efficiency of this method is sensitive to lysis volume, EGF receptor concentration, and antibody concentration.[7] Optimal parameters have been determined only on relatively small cultures (150 cm$^2$) from which

[7] P. Yaish, J. J. Hsuan, and M. D. Waterfield, unpublished results (1989).

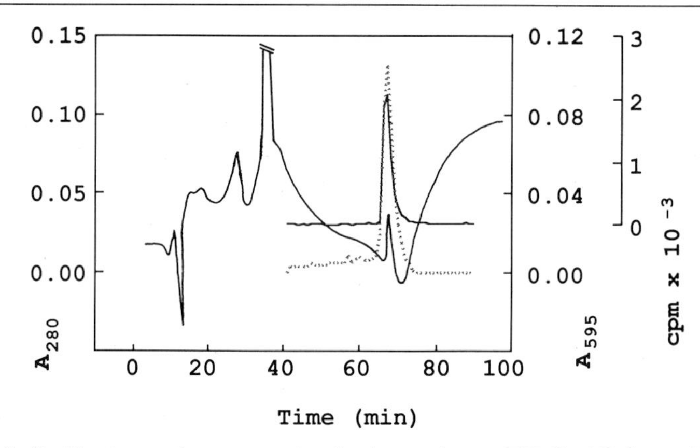


FIG. 1. Purification and concentration by ion-exchange FPLC. Affinity-purified EGF receptor (0.5 ml) in 50% glycerol buffer was diluted to 2 ml with buffer A (see text) and spiked with $^{32}$P-labeled EGF receptor. This was loaded onto a Mono Q FPLC column (HR5/5) at 0.1 ml/min. After 20 min the column was washed in buffer A for 20 min at 0.2 ml/min and then eluted with 2 *M* NaCl in buffer A. Fractions (1 min) were collected from 40 min and counted for Cerenkov radiation (solid line) and aliquots (50 $\mu$l) were analyzed for protein using the Bradford assay (dotted line). Peak fractions were essentially pure and active EGF receptor as determined by SDS-PAGE and EGF-dependent autophosphorylation.

400 $\mu$g MAb 9A can extract at least 70% of the total EGF receptor in 2 ml lysis buffer. The most significant factors that lower the efficiency are decreasing the lysis volume and decreasing the antibody concentration. This may be due to recognition of other factors containing the 9A epitope, such as glycolipids.

Not only is the 9A epitope expressed by other cell components in addition to the EGF receptor, but there appears to be considerable variation in both the glycosylation[8] and expression of the receptor itself.[9]

A relatively rapid method for selecting 9A-positive cells is to use FACS, in which cultured cells are suspended, incubated with 9A, and then with fluorescent-labeled second antibody before automated sorting according to fluorescence intensity. A similar method can be used to select against low EGF receptor-expressing mutants using sorting with, for example, MAb EGFR1[5] or fluoresceinated EGF.[9]

[8] L. H. K. Defize, D. J. Arndt-Jovin, T. M. Jovin, J. Boonstra, J. Meisenhelder, T. Hunter, H. T. de Hey, and S. W. de Laat, *J. Cell Biol.* **107,** 939 (1988).

[9] R. C. Chatelier, R. G. Ashcroft, C. J. Lloyd, E. C. Nice, R. H. Whitehead, W. H. Sawyer, and A. W. Burgess, *EMBO J.* **5,** 1181 (1986).

Properties

The $K_m$ value of the EGF receptor for ATP measured at 0° is lower for autophosphorylation than for substrate phosphorylation: the former has been estimated at 0.21–0.25 $\mu M$,[10,11] although values up to 3 $\mu M$ have been reported,[12] while the latter has been estimated at 2–7 $\mu M$.[12,13] The variation in these data may be a consequence of factors such as different assay buffer compositions and receptor states. Numerous factors are known to affect receptor kinase activity and these are briefly summarized below.

The maximum rate of ligand-stimulated autophosphorylation is similar in the presence of either $Mn^{2+}$ or $Mg^{2+}$, but the basal rate of autophosphorylation is two- to threefold higher in the presence of $Mn^{2+}$. Furthermore $Mn^{2+}$ stimulates peptide phosphorylation in the presence of MgATP, which has been interpreted in terms of two metal ion-binding sites.[14]

The dissociation constants for the binding of both EGF and TGF$\alpha$ to the cellular receptor are approximately 0.1–1.0 n$M$ for high-affinity sites and 2–10 n$M$ for low-affinity sites with a 10- to 20-fold difference between the two classes, except for the avian EGF receptor, which binds murine EGF with 100-fold lower affinity. Following solubilization in the nonionic detergents Nonidet P-40 (NP-40) or Triton X-100 the affinity of both classes decreases by an order of magnitude (see below). Commonly low-affinity sites represent 90% of the total ligand binding, but this fraction is increased by protein kinase C activity. This down modulation of ligand binding appears not to be mediated by phosphorylation at either Thr-654 or Thr-669 in the juxtamembrane domain, as was previously suggested. Thapsigargin, a non-12-*O*-tetradecanoyl phorbol-13-acetate(TPA)-type tumor promoter, also inhibits high-affinity binding, while sphingosine is reported to increase the affinity of low-affinity sites (see below). The mechanism by which these agents act still remains to be defined.

The action of ligand on the kinase activity is primarily to increase the $V_{max}$ value with little effect on the substrate $K_m$ value. The maximum velocity of peptide phosphorylation at 30° in the presence of EGF is extremely sensitive to the state of the receptor and readily decreases during solubilization and purification steps. For example, the $V_{max}$ value

[10] P. J. Bertics, W. S. Chen, L. H. Hubler, C. S. Lazar, M. G. Rosenfeld, and G. N. Gill, *J. Biol. Chem.* **263,** 3610 (1988).

[11] W. Weber, P. J. Bertics, and G. N. Gill, *J. Biol. Chem.* **259,** 14631 (1984).

[12] A. Honnegger, T. J. Dull, D. Szapary, A. Komoriya, R. Kris, A. Ullrich, and J. Schlessinger, *EMBO J.* **7,** 3053 (1988).

[13] C. Erneaux, S. Cohen, and D. L. Garbers, *J. Biol. Chem.* **258,** 4137 (1983).

[14] J. G. Koland and R. A. Cerione, *J. Biol. Chem.* **262,** 2230 (1988).

of angiotensin II phosphorylation has been estimated at 253 pmol/min/pmol receptor for a plasma membrane preparation,[10] 19.5 pmol/min/pmol for lysates,[11] and 11 pmol/min/pmol for purified receptor.[11] The effect of changing receptor concentration may also contribute to changes in specific activity (see below). Prior receptor autophosphorylation does not appear to affect the $V_{max}$ value, but in the absence of ligand the $V_{max}$ value decreases by ~two- to sixfold.[6]

The $K_m$ value for angiotensin II is 0.8–1.1 m*M*,[10–12] 0.2–1.0 m*M* for v-src peptide,[5,10,12] and 0.16–0.22 m*M* for a peptide derived from the P1 autophosphorylation site of the receptor.[5] The effect of prior phosphorylation or the presence of ligand on the $K_m$ value of an exogenous substrate remains unclear, but there is evidence to suggest that the C-terminal autophosphorylation sites of the receptor can competitively inhibit exogenous substrate phosphorylation.[10,12]

Four major autophosphorylation sites termed P1 to P4 have been identified on the EGF receptor, all of which are found in the C-terminal domain. These are at tyrosine residues 1173, 1148, 1068, and 1086, respectively.[6,15] P1 is the major site of tyrosine phosphorylation in response to stimulation by EGF both in solution and in intact cells, probably as a result of its low $K_m$ value relative to the other sites.[6]

## Inhibitors and Activators

As described above, the EGF receptor is sensitive to magnesium and manganese cations and metal ion-chelating agents such as EDTA are able to totally inhibit kinase activity.

Calcium ions do not directly affect the receptor, but the autophosphorylation domain is particularly sensitive to the calcium-activated neutral proteases or calpains. These act rapidly to generate a 150K fragment following cell lysis, which must therefore be performed in the presence of EGTA. Furthermore, there is evidence that C-terminal proteolysis increases protein kinase activity of the receptor by decreasing steric hinderance.[16]

The effects of changes in ionic strength on kinase activity are poorly studied, but there is a clear sensitivity to ammonium sulfate in particular. In this case 0.25 *M* salt has been proposed to stabilize a native, ligand-dependent conformation of the receptor that exhibits far greater stimulation with EGF and manganese ions than in the absence of salt.[14]

Both sphingosine and thapsigargin (a non-TPA-type protein kinase C-

[15] J. J. Hsuan, N. Totty, and M. D. Waterfield, *Biochem. J.* **262,** 659 (1989).
[16] S. Cohen, H. Ushiro, C. Stoscheck, and M. Chinkers, *J. Biol. Chem.* **257,** 1523 (1982).

independent tumor promoter) are able to indirectly modulate the ligand-binding affinity of the EGF receptor. Sphingosine appears to increase ligand affinity and stimulate kinase activity,[17] while thapsigargin down modulates receptor activity by decreasing the binding affinity and the ligand-dependent kinase activity.[18] Both may act by altering the activity of serine/threonine kinases toward the EGF receptor.

Solubilization of the EGF receptor in nonionic detergents increases basal kinase activity and decreases ligand-binding affinities to a single class. The dissociation constant is approximately an order of magnitude below that of the low-affinity membrane sites. The structural basis for these changes remains unclear. The probable basis for the specific subclass of high-affinity sites observed on binding EGF to intact cells is the dimerization of receptor molecules and recently a similar subclass has been identified in highly concentrated solutions of receptor. The dimer form of the receptor is thought to exist in equilibrium with monomers, show high-affinity binding, and elevated kinase activity in the absence of ligand. According to this model the binding of ligand stabilizes dimers relative to monomers.[19]

A corollary of the dimerization model is that certain multivalent ligands may enhance kinase activity independent of ligand and indeed this has been demonstrated using bivalent antibodies. Furthermore, prior immobilization of receptors using affinity matrices has been shown to inhibit the ability of ligand to stimulate kinase activity.

The role of receptor tyrosine phosphorylation is not clear, but the C-terminal domain is thought to act as a weak competitive inhibitor of exogenous substrate phosphorylation that can be relieved by autophosphorylation, proteolysis, or genetic deletion. Further phosphorylation sites may, however, play a clear regulatory role. These include Thr-654, a major site for protein kinase C, which appears to be involved in down regulating receptor kinase activity. Several other sites have been identified by metabolic labeling studies. The precise function of these sites has yet to be determined, but they may be involved in the regulation of ligand affinity and receptor internalization.

Dimethyl sulfoxide is able to directly stimulate EGF receptor kinase activity independent of ligand in either detergent solution or intact membranes.[20] The mechanism of this activation is not known.

Unlike the insulin receptor, the EGF receptor kinase activity is not

[17] R. J. Davis, N. Girones, and M. Faucher, *J. Biol. Chem.* **2263,** 5373 (1988).

[18] K. Takishima, B. Friedman, H. Fujiki, and M. R. Rosner, *Biochem. Biophys. Res. Commun.* **157,** 740 (1988).

[19] J. Schlessinger, *Biochemistry* **27,** 3119 (1988).

[20] R. A. Rubin and H. S. Earp, *J. Biol. Chem.* **258,** 5177 (1983).

stimulated by disulfide reduction using dithiothreitol. The intracellular region is, however, sensitive to oxidation and kinase activity is lost in the absence of reducing agents.

## Concluding Remarks

It seems clear that a large number of parameters can affect the enzymatic properties of the EGF receptor such that comparisons of receptor activity between different cell types, using different preparation procedures, and different assay conditions can give misleading and even contradictory results. The protocol described here is rapid, uses extremely mild conditions, and produces a highly purified and active preparation from A431 cells. No other source of EGF receptor has yet been found to carry the 9A determinant.

Attempts to overproduce active full-length EGF receptor or fragments in bacteria and in yeast (*Schizosaccharomyces pombe*) have been unsuccessful. The baculovirus expression system has been shown to give an active, full-length receptor, but with no improvement in yield over A431 cells.[21] Active fragments have proved more successful in this system: fragments comprising the extracellular and intracellular regions have both been expressed and purified, the latter using monoclonal anti-phosphotyrosine affinity chromatography.[22,23]

[21] M. D. Waterfield and C. Greenfield, this volume [52].

[22] C. Greenfield, I. Hiles, M. D. Waterfield, M. Federwisch, A. Wollmer, T. L. Blundell, and N. McDonald, *EMBO J.* **8,** 4115 (1989).

[23] P. B. Wedergaertner and G. N. Gill, *J. Biol. Chem.* **264,** 11346 (1989).

# [32] Identification of Phosphohistidine in Proteins and Purification of Protein-Histidine Kinases

*By* YING-FEI WEI and HARRY R. MATTHEWS

This chapter contains a potpourri of methods for working with phosphohistidine and protein-histidine kinase. The phosphate group in phosphohistidine is attached to a side-chain nitrogen, either at the 1- or 3-position or both. The nitrogen–phosphate bond is acid labile and the first part of the chapter covers methods for overcoming this and exploiting the alkali stability of phosphohistidine. Because phosphohistidines are not commercially available, they must be synthesized in the laboratory. This is straightforward, either by the traditional method or by a new method, starting with polyhistidine, that is more suited to most biochemistry or molecular biology laboratories. Phospholysine can also be synthesized by this method, starting with polylysine. Detection of proteins containing phosphohistidine in polyacrylamide gels requires neutral pH fixing and staining, which is described. Although histidine phosphorylation in proteins can be inferred by various criteria, the most direct demonstration of its presence is through phosphoamino acid analysis. Phosphoamino acids may be generated by alkaline hydrolysis or enzyme digestion of phosphoproteins; the phosphoamino acid must then be identified. High-performance column chromatographic methods (HPLC) are described using either an ion-exchange column (Mono Q) or a $C_{18}$ reversed-phase column, with detection of the fluorescent derivatives of phosphoamino acids, including the phosphohistidines. These methods can be used to analyze phosphoamino acid content or to prepare pure phosphoamino acids for further study. The second part of the chapter is concerned with the purification of protein-histidine kinase. The purification relies on a new assay, which is described, but otherwise uses conventional fast protein liquid chromatography (FPLC) with Q-Sepharose, Mono Q, and Mono S columns. Further purification and molecular weight determination are carried out on a Sepharose S-100 column.

## Introduction

Protein-histidine kinase is an enzyme that transfers the $\gamma$ phosphate group of ATP to a histidine residue in a protein substrate. Phosphohistidine has been found in several cellular proteins, particularly in (1) bacterial cells, where it is involved in the sugar phosphotransferase system,[1] (2) the chemotactic response of bacteria,[2] and (3) the regulation of *nif* gene expression in *Escherichia coli*.[3] However, in the first case the source of the phosphate group is phosphoenolpyruvate[1] and in the other cases the phosphorylatable histidine residue is on the kinase itself.[4] Smith[5,6] drew attention to nuclear protein-histidine kinases with work on mammalian tissues in the 1970s and a nuclear histidine kinase was reported in the slime

[1] J. Reizer, M. H. Saier, Jr., J. Deutscher, F. Grenier, J. Thompson, and W. Hengstenberg, *CRC Crit. Rev. Microbiol.* **15,** 297 (1988).

[2] J. F. Hess, R. B. Bourret, and M. I. Simon, *Nature* (*London*) **336,** 139 (1988).

[3] V. Weiss and B. Magasanik, *Proc. Natl. Acad. Sci. U.S.A.* **85,** 8919 (1988).

[4] J. B. Stock, A. M. Stock, and J. M. Mottonen, *Nature* (*London*) **344,** 395 (1990).

[5] J. M. Fujitaki, G. Fung, E.Y. Oh, and R. A. Smith, *Biochemistry* **20,** 3658 (1981).

[6] D. L. Smith, B. B. Bruegger, R. M. Halpern, and R. A. Smith, *Nature* (*London*) **246,** 103 (1973).

mold *Physarum polycephalum*.[7] More recently, a protein-histidine kinase was discovered in *Saccharomyces cerevisiae*[8] and purified.[8a]

Phosphohistidine is very unstable under the low pH conditions used in most kinase assays and in many protein purification and analysis techniques.[9,10] Specific methods for working with proteins containing phosphohistidine and for assaying protein-histidine kinase are necessary and are described below.

Amino acid analysis of the phosphohistone H4 produced by the reaction catalyzed by the purified form of the protein-histidine kinase shows the presence of phosphothreonine as well as phosphohistidine (unpublished, 1989). This could be either a direct product of the same kinase or a contaminating kinase, or it could be a secondary transfer (possibly intramolecular) of phosphate from phosphohistidine to threonine. The latter possibility is favored since substrate treated with diethyl pyrocarbonate to specifically modify histidine is not phosphorylated at all by protein-histidine kinase.[8a] Partial amino acid sequence data obtained by Dr. R. Aebersold of the University of Vancouver (unpublished, 1990) indicate that the protein-histidine kinase whose isolation is described here is absent from the National Biomedical Research Foundation protein sequence data base.

### *Phosphohistidine Synthesis*

Phosphohistidine is required as a standard in phosphoamino acid analysis of the products of protein-histidine kinase activity. Phosphohistidine has a short half-life and is not available commercially. It can be synthesized in the laboratory by either the phosphoramidate method[11] or the polyhistidine method. There are three forms of phosphohistidine, $N^1$-phosphohistidine, $N^3$-phosphohistidine, and $N^1$,$N^3$-diphosphohistidine. Both the above methods give a mixture of these compounds but the conditions for the phosphoramidate method can be readily adjusted to maximize the production of one or the other of the phosphohistidines.

*Phosphoramidate Method.* This method consists of two steps: the synthesis of potassium phosphoramidate and the reaction of phosphoramidate with histidine to produce phosphohistidine. The following method is based on that described by Sheridan *et al.*[11]

[7] V. D. Huebner and H. R. Matthews, *J. Biol. Chem.* **260,** 16106 (1985).

[8] Y. F. Wei, J. E. Morgan, and H. R. Matthews, *Arch. Biochem. Biophys.* **268,** 546 (1989).

[8a] J. Huang, Y. F. Wei, Y. Kim, L. Osterberg, and H. R. Matthews, *J. Biol. Chem.* **266** (in press).

[9] D. E. Hultquist, R. W. Moyer, and P. D. Boyer, *Biochemistry* **5,** 322 (1966).

[10] D. E. Hultquist, *Biochim. Biophys. Acta* **153,** 329 (1968).

[11] R. C. Sheridan, J. F. McCullough, and Z. T. Wakefield, *Inorg. Synth.* **13,** 23 (1971).

*Potassium phosphoramidate synthesis*

*Reagents*

Phosphoryl chloride, 30 ml
Ammonium hydroxide (10%, v/v), 500 ml
Acetone, 1500 ml
Acetic acid, 30 ml
Ethanol, 1050 ml
Ethyl ether, 50 ml
Potassium hydroxide (50%, w/v), 20 ml

1. Add 9.15 ml (0.1 mol) of reagent-grade phosphoryl chloride (phosphorus oxychloride, $POCl_3$) dropwise into 150 ml of ice-cold 10% ammonium hydroxide (0.75 mol $NH_3$) in a fume hood. Stir the solution vigorously on ice for 15 min. There is some fuming and evolution of heat, after which a clear solution is obtained.

2. Add 500 ml of acetone to this solution. Transfer the mixture to a 1-liter separation funnel and let it sit at room temperature for a few minutes until two layers form. Collect the bottom layer.

3. Neutralize the bottom layer to approximately pH 6 with about 5 ml of acetic acid.

4. Cool the solution to 5–10° for about 20 min to induce crystallization of ammonium hydrogen phosphoramidate.

5. Filter the ammonium hydrogen phosphoramidate product through Whatman (Clifton, NJ) No. 1 filter paper. Wash the precipitate on the filter successively with 100% ethanol and then ether, about 15 ml each. Air dry for several hours or overnight in the hood. About 40% of the phosphorus in the phosphoryl chloride is converted to ammonium phosphoramidate.

6. Repeat steps 1 through 5 two more times to give a total of about 13 g of ammonium phosphoramidate.

7. Dissolve 11.4 g (0.1 mol) of ammonium hydrogen phosphoramidate in 20 ml of 50% potassium hydroxide solution. Incubate at 50–60° for 10 min to expel ammonia in a fume hood.

8. Cool the solution to 5–10° and neutralize it to pH 6 with 7–8 ml of acetic acid.

9. Add 1 liter of ethanol and let the solution sit at room temperature for at least 30 min to precipitate the potassium salt of phosphoramidate. Collect the potassium phosphoramidate precipitate by filtration through Whatman No. 1 filter paper. Wash the precipitate on the filter successively with 100% ethanol and ether, about 15 ml each. Air dry for several hours or overnight in the hood. The yield is about 84%.

10. Store the product at room temperature in a desiccator. It is reason-

ably stable for several months. Phosphohistidine is made as required using the following reaction.

*Phosphohistidine synthesis:* Dissolve 15 mg of L-histidine (Sigma, St. Louis, MO; H-800) in 1 ml of water; if labeled product is required, add 1 $\mu$Ci/ml [$^{14}$C]histidine. Add 100 mg of potassium phosphoramidate. Incubate the reaction mixture at room temperature for 10 min to 24 hr, depending on the major product required (see Fig. 1). Add 5 ml water and store the diluted reaction mixture at 4°. The product is stable for 1 or 2 days.

*Incubation time:* The reaction of phosphoramidate with histidine shows complex kinetics, with the production of $N^1$-phosphohistidine initially, followed by $N^1$,$N^3$-diphosphohistidine and then $N^3$-phosphohistidine.[9,10] Using the above conditions, a 30-min reaction usually gives comparable amounts of the three forms of phosphohistidine. As the reaction continues, the $N^1$-phosphohistidine peak decreases, the $N^3$-phosphohistidine and $N^1$,$N^3$-diphosphohistidine peaks increase. After 11 hr of incubation, $N^3$-phosphohistidine and $N^1$,$N^3$-diphosphohistidine peaks dominate the products and after 23 hr the major product is $N^3$-phosphohistidine. A time course of the reaction is shown in Fig. 1.

The phosphohistidines can be individually purified from the reaction mixture by ion-exchange chromatography as described below (Fig. 2C).

*Polyhistidine Method.* The potassium phosphoramidate method (above) for synthesizing phosphohistidine involves the fairly extensive use of organic solvents, including ether, for the initial synthesis of phosphoramidate. An alternative method has been developed which is more easily undertaken in a typical biochemistry or molecular biology laboratory.

In the polyhistidine method, histidine is phosphorylated directly with phosphoryl chloride. The disadvantage of phosphorylating the monomer amino acid directly would be that the $\alpha$-amino group is phosphorylated as well as the side chain nitrogens. We avoid this by phosphorylating polyhistidine, which is commercially available. Poly(histidine) is of high enough molecular weight that the concentration of $\alpha$-amino groups is negligible. For some applications, e.g., generation or screening of antibodies to phosphohistidine, polyphosphohistidine can be used directly. When the phosphohistidine monomer is required, it is generated by alkaline hydrolysis of polyphosphohistidine, using the protocol developed for complete alkaline hydrolysis of proteins.

As in the case of synthesis using phosphoramidate, a mixture of phosphohistidines is obtained. The mixture can be fractionated by ion-exchange chromatography as described below and shown in Fig. 2. Under the conditions described, the polyhistidine method yields mainly $N^3$-phosphohistidine plus some $N^1$-phosphohistidine; no diphosphohistidine is produced.

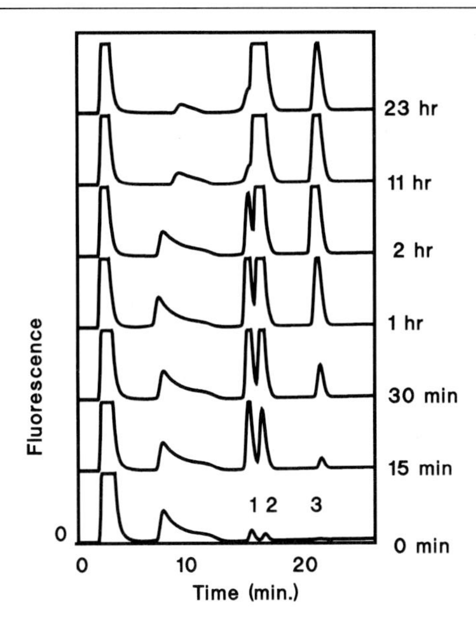


FIG. 1. Time course of phosphohistidine synthesis by the phosphoramidate method. A phosphohistidine synthesis reaction mixture containing 15 mg/ml of L-histidine and 100 mg/ml of potassium phosphoramidate was incubated at room temperature. At the time points shown on the right, 10 $\mu$l of the reaction mixture was taken and diluted five times with water. Fifty microliters of the five-times diluted reaction was loaded onto the Mono Q HR5/5 column, which had been equilibrated with 0.1 *M* *N*-methyldiethanolamine, pH 9.0. The column was eluted with a 20-ml linear gradient from 0.1 to 1.4 *M* *N*-methyldiethanolamine, pH 9.0, at 1 ml/min. The eluant was derivatized on line with *o*-phthalaldehyde (OPA) and detected by fluorescence. The fluorescence profile at each time point is shown, displaced progressively upward, for clarity. Peaks 1 through 3 are $N^1$-phosphohistidine, $N^3$-phosphohistidine, and $N^1$,$N^3$-diphosphohistidine, respectively, in each profile.

The yield of phosphohistidine after alkaline hydrolysis is approximately 21% of the initial polyhistidine, comprising 3% $N^1$-phosphohistidine and 18% $N^3$-phosphohistidine (Fig. 2A). The remaining residues are assumed to be unmodified histidines.

The principal disadvantage of the polyhistidine method is that poly-[$^{14}$C]histidine is not available and hence the method is not suitable for making phospho[$^{14}$C]histidines.

In the case of phospholysine, which can also be synthesized by this method[12] (C. Holman and H. R. Matthews, unpublished, 1988), the diluted alkaline hydrolysate is stable at $-20°$ for at least a year. The stability of phosphohistidine in the diluted alkaline hydrolysate is currently being tested.

[12] O. Zetterqvist and L. Engstrom, *Biochim. Biophys. Acta* **141,** 523 (1967).

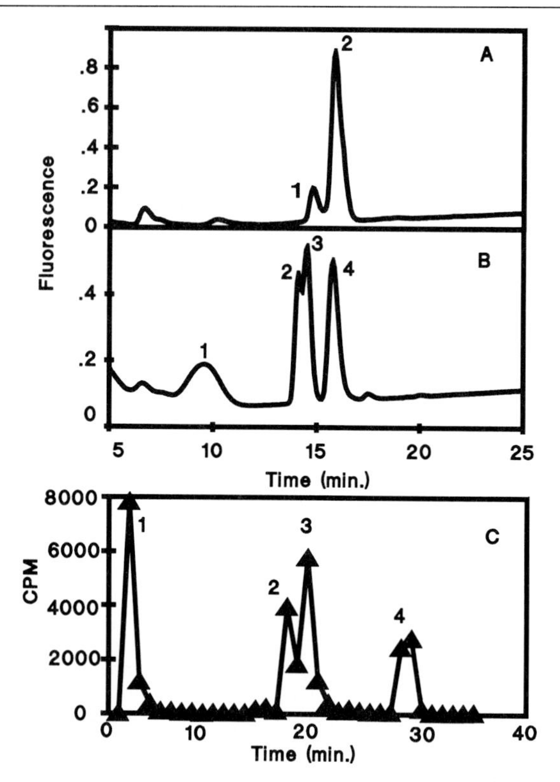


FIG. 2. Separation of phosphoamino acid standards on Mono Q column with *N*-methyldiethanolamine system. (A) Phosphohistidine synthesized from polyhistidine. Two hundred and fifty microliters of 20-times diluted phosphohistidine base hydrolysate (see the text for the details) was loaded onto the Mono Q HR5/5 column. The column was eluted as described in the legend of Fig. 1. Peaks 1 and 2 are $N^1$-phosphohistidine and $N^3$-phosphohistidine, respectively. (B) Other phosphoamino acid standards. Phosphoserine (0.18 $\mu$mol), 0.18 $\mu$mol phosphothreonine, 20 $\mu$l saturated phosphotyrosine, and 100 $\mu$l 30-times diluted phospholysine base hydrolysate in a final volume of 200 $\mu$l were loaded onto the Mono Q HR5/5 column, which had been equilibrated with 0.1 *M* *N*-methyldiethanolamine, pH 9.0. The column was eluted as described in the legend of Fig. 1. Peaks 1 through 4 are phospholysine, phosphothreonine, phosphoserine, and phosphotyrosine, respectively. (C) Phospho[$^{14}$C]histidine: Phospho[$^{14}$C]histidine was synthesized by the phosphoramidate method as described in the text with 1 $\mu$Ci/ml of L-[$^{14}$C]histidine and an incubation time of 30 min. An aliquot of the reaction mixture was diluted five times with water and 400 $\mu$l of this diluted reaction mixture was loaded onto the Mono Q HR10/10 column. The column was eluted with 40 ml of a linear gradient from 0.1 to 1.4 *M* *N*-methyldiethanolamine, pH 9.0, at a flow rate of 4 ml/min. Fractions of 4 ml were collected and subjected to liquid scintillation counting. Peaks 1 through 4 are histidine, $N^1$-phosphohistidine, $N^3$-phosphohistidine, and $N^1$,$N^3$-diphosphohistidine, respectively.

*Procedure*

*Reagents*

Polyhistidine (HCl salt), 50 mg in 1.5 ml water
Triethylamine, 0.1 ml
Phosphoryl chloride, 0.11 ml
NaOH, 5 *M*
Water
NaOH (0.1 *N*), 1 liter
Solid KOH

1. Dissolve 50 mg of polyhistidine (from Sigma Chemical Co., St. Louis, MO; HCl salt, $M_r$ 15,000 to 50,000) in 1.5 ml of water. This will eventually yield approximately 10 mg of phosphohistidines.
2. Saturate the solution with triethylamine by adding 10-$\mu$l aliquots of triethylamine until the solution becomes cloudy. A total of about 50 $\mu$l of triethylamine is required.
3. Add 110 $\mu$l of phosphoryl chloride in 10-$\mu$l increments. The reaction is carried out with constant stirring on ice and the pH is held within the range pH 9.5 to 12.5 by adding 5 *M* NaOH as required. After addition of all the phosphoryl chloride, keep the solution on ice for an additional 30 min.
4. Dilute the solution with water to a final volume of 3 ml. Dialyze it against 0.1 *N* NaOH overnight.
5. Measure the volume of the solution after dialysis and add solid KOH to make the final KOH concentration 3 *N*. Incubate the solution at 105° for 5 hr to hydrolyze the polyphosphohistidine.
6. Dilute the hydrolysate with 10 vol of water and store at 4° for immediate use or frozen at −20° for longer term storage.

*Phospholysine Synthesis*

Phospholysine can be prepared from polylysine as described above for the production of phosphohistidine, except that a larger volume of triethylamine (400 $\mu$l in 50-$\mu$l increments) is required. Phosphoarginine, phosphotyrosine, phosphothreonine, and phosphoserine are available commercially.

*Phosphohistidine Separation and Purification*

Column chromatography methods for analyzing samples containing phosphohistidine are described. These methods require the use of an HPLC or FPLC solvent delivery system capable of delivering linear gradients at moderate pressure (Mono Q ion-exchange column) or high pressure (silica $C_{18}$ reversed-phase column). Modern prepacked columns provide much faster analysis times than open columns.[10,13]

*Fluorescent Derivatives.* The simplest sensitive method for detecting amino acids is to label them with a fluorescent or UV-absorbing group.[14-16] Phosphoamino acid analysis, like unmodified amino acid analysis, can be carried out using either pre- or postcolumn derivatized amino acids. In our laboratory, precolumn derivatizing provides the fastest and most sensitive method for identifying phosphoamino acids. Precolumn methods do not require additional equipment, except for the fluorescence detector. However, the instability of the derivatives and the inability to recover underivatized amino acids are important disadvantages. Postcolumn derivatizing procedures are more complex, slower, and less sensitive than precolumn procedures, but offer the opportunity to recover unmodified amino acids. Both methods are described below.

Both methods require the use of an HPLC system equipped with a fluorescence detector, available from several manufacturers of HPLC equipment. The derivatizing reagent is *o*-phthalaldehyde (OPA) in each case and so the fluorescence excitation and emission wavelengths are as follows: excitation filter passes 370 nm; emission filter passes 418 to 700 nm. For a discussion of derivatization by OPA see Alvarez-Coque *et al.*[17]

For the precolumn method, the column outlet is simply led to the fluorescence monitor and then to a fraction collector, if required. Radioactivity in the fractions can be determined by liquid scintillation counting in the usual way.

The postcolumn method requires the addition of a pump to deliver the derivatizing solution. Well-designed systems for postcolumn derivatization are available commercially, but a satisfactory system can be assembled from off-the-shelf components as follows. A low-pressure metering pump capable of pumping at 1.0 to 1.5 ml/min at pressures up to 500 psi is suitable. If not included with the pump, a simple device to smooth the flow should be provided. This equipment is available from the larger laboratory supply houses.

The output from the pump is connected to the outlet from the column through a T connector and the combined stream is led through a coil of tubing to the fluorescence detector. The volume of the tubing between the

[13] L. Carlomagno, V. D. Huebner, and H. R. Matthews, *Anal. Biochem.* **149,** 344 (1985).

[14] J. Y. Chang, *J. Chromatogr.* **295,** 193 (1984).

[15] M. O. Fleury and D. V. Ashley, *Anal. Biochem.* **133,** 330 (1983).

[16] S. A. Cohen and D. J. Strydom, *Anal. Biochem.* **174,** 1 (1988).

[17] M. C. G. Alvarez-Coque, M. J. M. Hernandez, R. M. V. Camanas, and C. M. Fernandez, *Anal. Biochem.* **178,** 1 (1989).

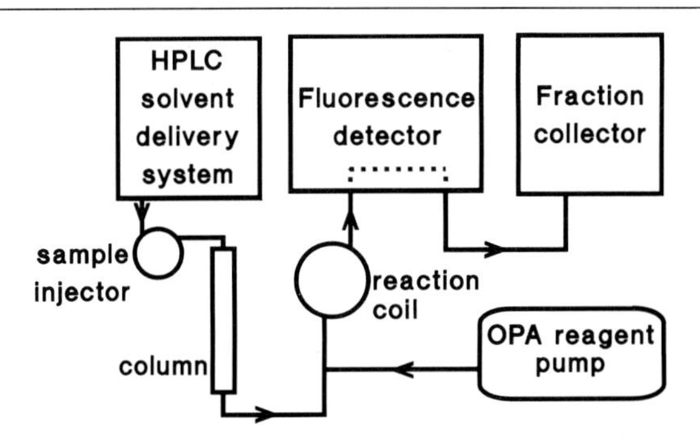


FIG. 3. Postcolumn derivatization. Liquid connections are shown. The column eluant flows through the column, mixes with derivatizing solution, and then passes through the fluorescence monitor. The derivatizing solution is simply pumped, through a pulse dampener if necessary, to the T connection with the column eluant.

T connector and the fluorescence detector determines the reaction time and a coil of internal volume 1 ml has been found to work well. These connections are shown in Fig. 3.

The postcolumn derivatizing solution is stable in the dark at 4° for 1 week. It is made up as follows:

Postcolumn derivatizing solution: 50 g boric acid, 44 g KOH, 3 ml Brij, water to a final volume of 1 liter (the above components can be mixed in larger quantities and stored, if desired), 2 ml 2-mercaptoethanol, *o*-phthalaldehyde (400 mg dissolved in 5 ml methanol)

*Anion-Exchange Chromatography on Mono Q HR5/5.* The three forms of phosphohistidine, $N^1$-phosphohistidine, $N^3$-phosphohistidine, and $N^1$,$N^3$-diphosphohistidine, can be separated in the underivatized state by anion-exchange chromatography. Dowex resins can be used in open columns[9] but in the methods below a prepacked Mono Q column (HR5/5 from Pharmacia, Piscataway, NJ) is used to speed up the chromatography. The highest resolution of phosphohistidine isomers is obtained by elution with *N*-methyldiethanolamine, but $N^1$-phosphohistidine is eluted very close to phosphoserine and $N^3$-phosphohistidine is eluted very close to phosphotyrosine (Fig. 2A and B). Elution with $KHCO_3$ gives better separation of phosphohistidines from the other phosphoamino acids (Fig. 4). Finally, when amino acids are to be recovered from the column a volatile eluant, $NH_4HCO_3$, is needed, although it gives poor resolution (Fig. 5). Hence, for high-resolution preparative work, the amino acid is first separated using a high-resolution eluant and then rechromatographed with $NH_4HCO_3$.

All eluants for these chromatography methods were filtered through a

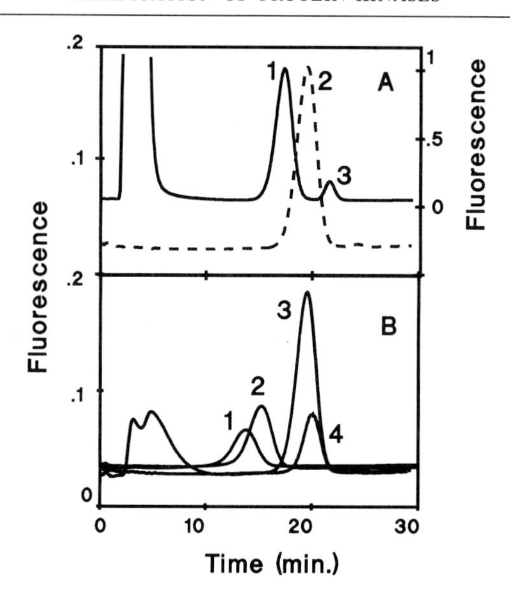


FIG. 4. Separation of phosphoamino acids on Mono Q HR5/5 with a gradient of $KHCO_3$. (A) Phosphohistidine and phosphotyrosine. Phosphohistidine was synthesized from phosphoramidate as described in the text. An aliquot of the reaction mixture was diluted 10 times with water and loaded onto the Mono Q HR5/5 column. The column was eluted with a 15-ml linear gradient from 0.05 to 0.5 *M* $KHCO_3$, pH 8.5, followed by a 5-ml linear gradient from 0.5 to 1 *M* $KHCO_3$ at 1 ml/min. The eluant was detected by OPA derivatization as described in the text. Peak 1 is a mixture of $N^1$-phosphohistidine and $N^3$-phosphohistidine; peaks 2 and 3 are phosphotyrosine and $N^1$,$N^3$-diphosphohistidine, respectively. (B) Other phosphoamino acid standards. Phosphoserine (50 nmol), 50 nmol phosphothreonine, 4 $\mu$l saturated phosphotyrosine, and 20 $\mu$l of 30-times diluted phospholysine base hydrolysate (as described in the text) were loaded onto the Mono Q HR5/5 column in separate runs. The column was eluted as described for (A). Peaks 1 through 4 are phosphothreonine, phosphoserine, phosphotyrosine, and phospholysine, respectively.

0.45- or 0.2-$\mu$m filter and degassed. Samples were clarified by similar filtration, or by centrifugation at room temperature (5 min in a microcentrifuge at approximately 12,000 *g*).

*N-Methyldiethanolamine system:* In this system the three forms of phosphohistidine can be well resolved from each other and from inorganic phosphate but ATP coelutes with diphosphohistidine. For routine analytical work, use a Mono Q 5/5 column (Fig. 2A and B) and the conditions described below. For preparative purposes, a larger column such as a 10/10 column may be used with increased flow rate and gradient volume (Fig. 2C).

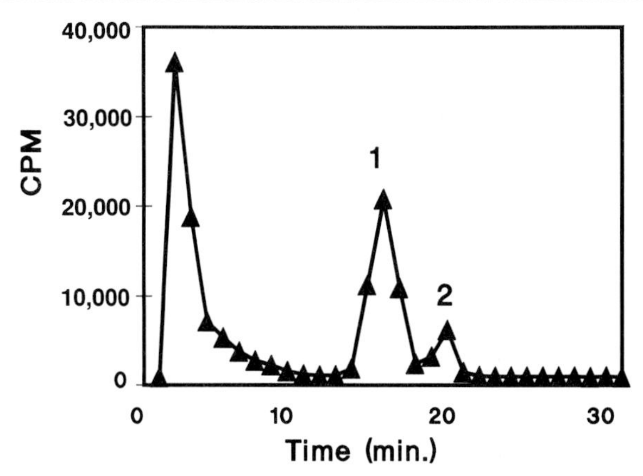


FIG. 5. Separation of phospho[$^{14}$C]histidine on Mono Q HR5/5 with a gradient of $NH_4HCO_3$. A phosphohistidine synthesis reaction containing 1 $\mu$Ci L-[$^{14}$C]histidine, 15 mg/ml L-histidine, and 100 mg/ml of potassium phosphoramidate was carried out. At 30 min after the reaction started, 80 $\mu$l of the reaction mixture was taken and diluted five times with water. Four hundred microliters of the five-times diluted reaction mixture was loaded onto the column, which was equilibrated with 0.1 *M* $NH_4HCO_3$, pH 8.25. The column was then eluted with a 20-ml linear gradient from 0.1 to 1.5 *M* $NH_4HCO_3$ at a flow rate of 1 ml/min. Fractions of 1 ml were collected, mixed with 4 ml of the Bio-Safe II liquid scintillation solution, and subjected to scintillation counting. Peak 1 is a mixture of $N^1$-phosphohistidine and $N^3$-phosphohistidine; peak 2 is $N^1$,$N^3$-diphosphohistidine.

*Reagents*

Water
*N*-Methyldiethanolamine (0.1 *M*), 1 liter
Phosphoamino acid standards (step 2 below)
*N*-Methyldiethanolamine (1.4 *M*), 200 ml
On-line derivatizing solution (see Fluorescent Derivatives, above)

1. Wash the Mono Q HR5/5 column with water and equilibrate the column with three column volumes of 0.1 *M* *N*-methyldiethanolamine, pH 9.0, at a flow rate of 1 ml/min.

2. Inject the sample. For a phosphohistidine standard run, inject 50 $\mu$l of the diluted phosphohistidine reaction mixture (see Phosphohistidine Synthesis, above) onto the column. Other phosphoamino acid standards dissolved in water may be useful (Fig. 2B). Suitable amounts include the following: 180 nmol of phosphoserine and phosphothreonine, 13 $\mu$l of a saturated solution of phosphotyrosine, and 100 $\mu$l of 30 times diluted phospholysine hydrolysate. Phosphoarginine does not bind to the column.

3. Elute the column with a linear gradient (total volume 20 ml) from 0.1 to 1.4 *M* *N*-methyldiethanolamine, pH 9.0, at 1 ml/min. For fluorescence detection, the eluant is derivatized on line with OPA as described above and detected by fluorescence. When the phosphoamino acids are radiolabeled, the eluant is collected with a fraction collector and radioactivity determined by liquid scintillation counting.

4. Elution of several phosphoamino acids is shown in Fig. 2B. The elution volumes of inorganic phosphate and ATP are 10.5 and 20.5 ml, respectively.

*$KHCO_3$ system*

*Eluants*

$KHCO_3$ (0.05 *M*), pH 8.5
$KHCO_3$ (0.5 *M*), pH 8.5
$KHCO_3$ (1 *M*), pH 8.5

Elution with a $KHCO_3$ gradient is an alternative and complementary way to separate and identify the phosphoamino acids. Phosphohistidines are separated from the other phosphoamino acids in this system, but $N^1$- and $N^3$-phosphohistidines coelute and phospholysine elutes close to phosphotyrosine (Fig. 4). Data from both the $KHCO_3$ system and the *N*-methyldiethanolamine system are required for a complete analysis.

The procedure is as described for elution with *N*-methyldiethanolamine, except the column is equilibrated with 0.05 *M* $KHCO_3$, pH 8.5, and eluted with a 15-ml linear gradient from 0.05 to 0.5 *M* $KHCO_3$, pH 8.5, followed by a 5-ml linear gradient from 0.5 to 1 *M* $KHCO_3$, pH 8.5. The elution of phosphohistidine standards is shown in Fig. 4A and other phosphoamino acids in Fig. 4B. The elution volumes of inorganic phosphate and ATP are 4 and 16 ml, respectively.

*$NH_4HCO_3$ system*

*Eluants*

$NH_4HCO_3$ (0.1 *M*), pH 8.25
$NH_4HCO_3$ (1.5 *M*), pH 8.25

Neither of the above eluants is volatile. A volatile eluant is required if the method is to be a useful preparative method for phosphohistidines. Ammonium bicarbonate can be used as an eluant, but it has two disadvantages: lower resolution and interference with postcolumn derivatization by OPA. The column is thus calibrated with phospho[$^{14}C$]histidine (Fig. 5). Elution times are very reproducible so it is not necessary to spike the preparative sample with phospho[$^{14}C$]histidine. If it is necessary to

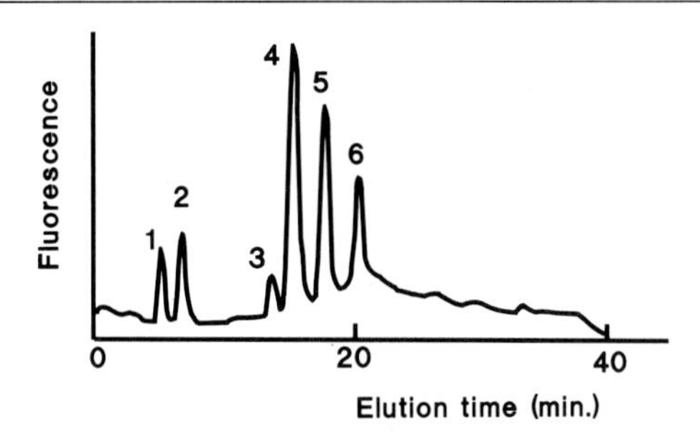


FIG. 6. Reversed-phase column chromatography of phosphoamino acids. A mixture of phosphoamino acids was loaded onto an analytical 25-cm $C_{18}$ reversed-phase column and eluted as described in the text. The peaks were identified by chromatographing each amino acid alone. Peaks 1 through 6 are $N^1$-phosphohistidine, phosphoserine, $N^3$-phosphohistidine, phosphoarginine, phosphothreonine, and phosphotyrosine, respectively.

separate the monophosphohistidine isomers then they should be first separated by elution with *N*-methyldiethanolamine and then each separate compound diluted, injected onto the column again, and eluted with $NH_4HCO_3$ to remove the *N*-methyldiethanolamine.

The procedure is as described for the *N*-methyldiethanolamine system, except that the postcolumn derivatization cannot be used and the column is equilibrated with 0.1 *M* $NH_4HCO_3$, pH 8.25, and eluted with a 20-ml linear gradient of 0.1 to 1.5 *M* $NH_4HCO_3$, pH 8.25. Elution of phospho-[$^{14}C$]histidines is shown in Fig. 5.

*Reversed-Phase HPLC.* Several methods are available for derivatizing amino acids before chromatography so that the derivatives separate well on a reversed-phase $C_{18}$ HPLC column. These derivatives are either strong absorbers of ultraviolet light or they are fluorescent, or both. One of these methods has been adapted to the analysis of phosphoamino acids, including the phosphohistidines. Some, or all, of the other methods may also be adaptable to analysis of phosphohistidines but they have not been tested. The OPA method described below was selected for its simplicity and sensitivity.

The HPLC system required is a standard gradient system with the addition of a fluorescence monitor. The data shown in Fig. 6 were obtained with an ISCO gradient programmer, HPLC pump and injector, and a fluorescence monitor from LDC Milton Roy (Riviera Beach, FL) using the standard lamp and filters as described above.

The column is a standard analytical silica-based 25 cm $C_{18}$ column (e.g., Waters, Milford, MA; $\mu$Bondapak $C_{18}$). A guard column is recommended. The pH of the eluants is chosen to be a compromise between that required for stability of the phosphohistidines and the stability of the silica column. A polymer-based column has also been used[13] at higher pH values.

*Reagents*

Derivatizing solution: 10 mg *o*-phthalaldehyde, 25 $\mu$l ethanethiol, 5.3 ml saturated boric acid, pH 9.5, 14.9 ml methanol (make fresh each day, keep in the dark)

Gradient buffer A: 100 ml 0.3 $M$ $NaH_2PO_4$, 25 ml tetrahydrofuran, 1875 ml water

Gradient buffer B: 45 ml 0.3 $M$ $NaH_2PO_4$, 1100 ml acetonitrile, 855 ml water

1. Equilibrate the column with 70% buffer A + 30% buffer B at 1.5 ml/min.
2. Thoroughly mix equal volumes of sample and derivatizing solution to make the sample mixture. Note the time when mixing began.
3. Place the sample mixture in the injection loop of the HPLC. Wait until 3 min after the time when mixing began and then inject the sample. Small variations in time (± 20 sec) do not affect the chromatography greatly but 3 min is the recommended derivatizing time at room temperature.
4. Elute the column with a gradient from 70% buffer A to 20% buffer A with buffer B at 1.5 ml/min. The gradient time should be 30 min.
5. Phosphoamino acids are eluted as shown in Fig. 6.

### *Detection of Proteins Containing Phosphohistidine*

Ultimately, the presence of phosphohistidine needs to be demonstrated by phosphoamino acid analysis as described above. Phosphoproteins can be digested to amino acids either by alkaline hydrolysis[7] or by proteolytic digestion.[8a] Specific antibodies to phosphohistidine may provide an alternative approach when such antibodies become available. However, for screening purposes, less specific methods can be used. The simplest is to treat the protein sample by mild alkaline hydrolysis, sufficient to remove serine and threonine phosphate, and then study the remaining phosphorylated proteins, for example, by SDS gel electrophoresis. The use of SDS gel electrophoresis as a screen, followed by confirmation using column chromatography and phosphoamino acid analysis, is illustrated in experiments showing the presence of proteins containing phosphohistidine in nuclei of the slime mold *Physarum polycephalum*.[7] Note that acid fixation of SDS gels will deplete the gels of any phosphate label in phosphohistidine. A neutral pH fixing procedure is described below.[7]

Acid-labile phosphate can also be determined by acid washing of gels and comparison of autoradiographs taken before and after the acid wash. This has the advantage that phosphotyrosine will not interfere, but the disadvantage that loss of protein might be interpreted as evidence for acid-labile phosphorylation. Such a difference method is also inherently less sensitive than the positive signal given by the alkali-stable approach. In our laboratory, it has also been difficult to achieve quantitative removal of acid-labile phosphate from proteins in gels by this method, possibly due to chemical transfer of the phosphate to other acceptors.

*SDS Gel Electrophoresis of Acid-Labile Phosphoproteins*

*Reagents*

SDS gel system
Loading buffer, 5× [e.g., 0.5% (w/v) SDS, 50% (v/v) glycerol, 50 m*M* Tris, pH 7.5, 0.005% (w/v) Bromphenol Blue]
NaOH, 3 *N*
Formaldehyde (18.5%, w/v), 50 m*M* $NaPO_4$, pH 8
2-Propanol (25%, v/v), 0.2% (w/v) Coomassie Blue G250, 50 m*M* $NaPO_4$, pH 8
2-Propanol (10%, v/v), 50 m*M* $NaPO_4$, pH 8

1. The protein sample is prepared for SDS gel electrophoresis, avoiding the use of low pH. The sample may be prepared from cells incubated in the presence of [$^{32}$P]phosphate (guanidine hydrochloride is a useful alternative to acid-extraction methods[7]), or proteins may be phosphorylated *in vitro*.

2. Mild alkaline hydrolysis is used to deplete the sample of serine and/or threonine phosphate as follows. Add 3 *N* NaOH to the sample to give a final concentration of 0.5 *N* NaOH and then incubate the mixture at 60° for 30 min.

3. Sample preparation depends on the resolution required. If modest resolution is acceptable, the sample in 0.5 *N* NaOH can be added to one-fifth volume of 5× loading buffer. This works for histone samples run on 17.5% gels[18] but may not give good results in general due to the high pH and ionic strength of the sample. For better resolution, dialyze the sample extensively against 1× loading buffer (diluted 5× loading buffer) without glycerol, then add glycerol to a final concentration of 10%. In either case, incubate the sample at 60° for 5 min and load it onto the SDS gel.

4. If the gel is to be stained, it should first be fixed as described below.

[18] M. H. Schreier, B. Erni, and T. Staehelin, *J. Mol. Biol.* **116**, 727 (1977).

TABLE I
PURIFICATION OF YEAST HISTIDINE KINASE[a]

| Column step | Protein loaded for one column run (mg) | Protein recovered (mg) | Yield of kinase activity (%) | Purification (-fold) | Final specific activity (nmol/15 min/mg) |
|---|---|---|---|---|---|
| S-100 extract | ND | ND | ND | ND | 1.1 |
| Q-Sepharose | 1,320 | 70 | 120 | 22 | 25 |
| Mono Q HR 10/10 | 180 | 8 | 144 | 32 | 800 |
| Mono S HR 5/5 | 10 | 0.1 | 110 | 95 | 77,000 |

[a] Data obtained with individual column runs. In practice, several runs of each step were pooled before proceeding with the following step. ND, Not determined.

For autoradiography without staining, the gel may be simply dried and subjected to autoradiography directly.

Fix the gel by shaking it in 18.5% formaldehyde, 50 m*M* $NaPO_4$, pH 8, for 1.5 hr.

Stain the gel in 25% 2-propanol, 0.2% Coomassie Blue G250, 50 m*M* $NaPO_4$, pH 8 overnight and destain it in 10% 2-propanol, 50 m*M* $NaPO_4$, pH 8.

5. Autoradiography can be carried out by normal procedures, as described.[19]

*Yeast Histidine Kinase Purification*

The purification depends on a new assay and conventional column chromatography. The enzyme is purified through a Q-Sepharose column, a Mono Q column, and a Mono S column. Additional purification can be achieved with a Sepharose S-100 column. Note the different behavior on the Q-Sepharose compared with the Mono Q column, probably due to the loss of a binding factor or domain of the kinase on the first column. The purification is summarized in Table I.

*Assay*. While the existence of phosphohistidine in proteins, and of protein histidine kinases has been known for some time,[5,9,20–22] no extensive purification of a protein histidine kinase has been reported. The major

[19] V. D. Huebner and H. R. Matthews, *in* "Methods in Molecular Biology. Nucleic Acids" (J. M. Walker, ed.), pp. 51–53. Humana Press, Clifton, New Jersey, 1984.

[20] O. Zetterqvist and L. Engström, *Br. J. Haematol.* **12,** 520 (1966).

[21] O. Walinder, O. Zetterqvist, and L. Engström, *J. Biol. Chem.* **243,** 2793 (1968).

[22] Z. B. Rose, *Arch. Biochem. Biophys.* **140,** 508 (1970).

reason has been the lack of a rapid and reproducible assay for the protein histidine kinase activity in the presence of other protein kinases. The most widely used simple assay for protein kinase activity involves precipitation and washing of the phosphoprotein product in 5% or higher concentration of trichloroacetic acid (TCA). The instability of phosphohistidine under these conditions rules out this assay for protein histidine kinase. Huebner and Matthews[7] avoided the TCA precipitation by adsorbing the phosphoprotein product to phosphocellulose paper, following the method of Witt and Roskoski.[23] While this method was satisfactory in some experiments, in others very high backgrounds were experienced. Reproducible control of the background has not been achieved. More recently, Wei introduced the use of Nytran paper (a nylon membrane available from Schleicher and Schuell, Keene, NH). Proteins appear to bind irreversibly to Nytran paper, even at high pH, and reproducibly low backgrounds are achieved. The method is described below (see Nytran Paper Assay).

In these various membrane assays, phosphorylation due to serine/threonine kinases is depleted by mild alkaline hydrolysis of the phosphoprotein product. Hence, the Nytran paper assay measures alkali-stable phosphorylation in general and is not specific for protein-histidine kinase activity. In particular, protein-tyrosine kinase and protein-lysine kinase activities would be measured as efficiently as protein-histidine kinase activity. It is also possible that some serine/threonine phosphate might survive the hydrolysis, depending on the local amino acid sequence.[24] Phosphoprotein analysis (above) is necessary to distinguish these possibilities. Nevertheless, the Nytran paper assay remains the crucial advance that has led to the purification of protein-histidine kinase.

The conditions of alkaline hydrolysis described below are quite harsh and will lead to some degradation of the protein substrate. Reduction of the KOH concentration from 0.5 to 0.1 *N* makes the assay a little less specific, but may be more practical for less stable proteins than histone H4.[25]

ATP binds nonenzymatically to paper, to histone H4, and to proteins in the enzyme sample. The effects of nonenzymatic binding are reduced by pretreatment of the Nytran paper and by washing the loaded papers in solutions containing ATP. The use of appropriate controls is important, and backgrounds of 600–800 cpm can be expected from zero-time controls. Zero-substrate controls may be misleading if there are potential substrates in the enzyme sample, as occurs in crude samples but not purified samples.

[23] J. J. Witt and R. Roskoski, Jr., *Anal. Biochem.* **66,** 253 (1975).

[24] B. E. Kemp, *J. Biol. Chem.* **255,** 2914 (1980).

[25] V. D. Huebner, Ph.D. Thesis, University of California, Davis (1985).

Also, zero-enzyme controls can give misleading background values if ATP binds nonenzymatically to substances in the enzyme sample. Considerable caution is necessary when assaying crude samples and it is recommended that reaction mixtures be examined by SDS gel electrophoresis if there is question about the Nytran paper assay. In spite of these reservations, the Nytran paper assay has been found to be very useful.

*Nytran paper assay*[26]

*Reagents*

Reaction buffer, 10× (150 m*M* $MgCl_2$, 500 m*M* Tris, pH 7.5)
Calf thymus histone H4, 3 mg/ml
ATP, 2 m*M*
Enzyme solution
Water
NaOH, 3 *N*
[$\gamma$-$^{32}$P]ATP
Liquid scintillation fluid

1. Prepare the appropriate substrate as follows. Histone H4 may be purchased from Boehringer Mannheim (Mannheim, Germany) or purified from total calf thymus histone by chromatography on a long (at least 100 cm) column of BioGel P-10 (Bio-Rad, Richmond, CA; 200–400 mesh) eluted with 10 m*M* HCl; the loading sample should be made up in 8 *M* urea at 20 mg/ml total histone and allowed to sit at 4° overnight before loading. Histone H4 elutes just before the urea peak, close to the total volume of the column and after the other histones. Its position should be confirmed by gel electrophoresis. Histone H4 is lyophilized and dissolved in water to make a 3 mg/ml stock solution that is stored at 4° for up to 6 months.

2. Pretreat the Nytran paper to reduce the background as follows. Cut the Nytran paper (from Schleicher and Schuell; 0.45 $\mu$m) into 2 × 2 cm pieces and soak them in 1 m*M* ATP, pH 9, for a few hours or overnight with slow stirring at room temperature. Air dry the papers and store them at room temperature for up to 3 weeks.

3. Prepare the enzyme reaction mixture as follows.

Enzyme reaction mixture: 5 $\mu$l 10× reaction buffer (150 m*M* $MgCl_2$, 500 m*M* Tris, pH 7.5), 10 $\mu$l of 3 mg/ml calf thymus H4, 5 $\mu$l of 2 m*M* [$\gamma$-$^{32}$P]ATP (350 Ci/mol), enzyme, and water to give a final volume of 50 $\mu$l

[26] Y. Wei and H. R. Matthews, *Anal. Biochem.* **190,** 188 (1990).

Incubate the reaction mixture at 30° for 15 min.

4. Add 10 μl of 3 *N* NaOH to the reaction mixture and incubate it at 60° for 30 min.

5. Transfer the reaction mixture onto two pretreated Nytran papers, 25 μl to each paper. Wash the papers with 1 m*M* ATP, pH 9 four times, 30 min each time, at room temperature.

6. Dry the paper under a drying lamp. Put each paper into an 8-ml scintillation vial and add an appropriate liquid scintillation solution, e.g., 6 ml of Bio-Safe II (RPI Corp., Mt. Prospect, IL). Determine the radioactivity using a liquid scintillation counter.

*Yeast S-100 Extract Preparation. Saccharomyces cerevisiae* extract is prepared as described by Manley *et al.*[27] and modified by Swanson and Holland,[28] as follows.

*Reagents*

Baker's yeast (*Saccharomyces cerevisiae*)
Growth medium [2% (w/v) Bacto-peptone, 1% (w/v) yeast extract, 2% (w/v) glucose]
Suspension buffer [50 m*M* Tris, pH 7.5, 5 m*M* $MgCl_2$, 1 m*M* dithiothreitol, 0.2 m*M* EDTA, 20% glycerol, 2.5 m*M* phenylmethylsulfonyl fluoride (PMSF) (added immediately before use from a stock solution of 25 m*M* PMSF in dimethyl sulfoxide)]
Extraction buffer [50 m*M* Tris, pH 7.5, 10 m*M* $MgCl_2$, 2 m*M* dithiothreitol, 25% (w/v) sucrose, 20% (v/v) glycerol]
Saturated $(NH_4)_2SO_4$
Solid $(NH_4)_2SO_4$
Storage buffer [50 m*M* Tris, pH 7.5, 6 m*M* $MgCl_2$, 0.2 m*M* EDTA, 1 m*M* dithiothreitol, 15% (v/v) glycerol]

1. Grow *S. cerevisiae*, strain D273-10B, to its midlog phase (apparent absorbance at 600 nm is between 1 and 5) in a liquid medium containing 2% Bacto-peptone, 1% yeast extract, and 2% glucose.

2. Harvest cells from a total of 6 liters of culture by filtration through a Millipore (Bedford, MA) AP filter.

3. Suspend the cells with suspension buffer. Use 15 ml of suspension buffer for every 2 liters of original cell culture. Add the resulting cell suspension dropwise into liquid nitrogen and store it in liquid nitrogen. All the following steps are carried out at 4° or as indicated.

4. Break the frozen cells by passage through an Eaton press (SLM

[27] J. L. Manley, A. Fire, A. Cano, P. A. Sharp, and M. L. Gefter, *Proc. Natl. Acad. Sci. U.S.A.* **77,** 3855 (1980).

[28] M. E. Swanson and M. J. Holland, *J. Biol. Chem.* **258,** 3242 (1983).

Aminco, Springfield, IL) at 9000 psi and suspend in 40 ml of extraction buffer on ice. Measure the final volume.

5. Add 10% sample volume of saturated $(NH_4)_2SO_4$ into the extraction mixture dropwise with stirring and stir for an additional 20 min at 4°.

6. Centrifuge the suspension at 50,000 rpm for 3 hr in a Beckman Ti 60 rotor at 4°.

7. To the supernatant, add solid $(NH_4)_2SO_4$ (0.4 g/ml of supernatant). Neutralize the solution with dropwise addition of 1 *M* NaOH to pH 7.5 and stir slowly at 4° for another 20 min.

8. Collect the precipitate by centrifugation at 50,000 rpm for 30 min in a Beckman Ti 60 rotor at 4°.

9. Resuspend the precipitate with 10 ml of storage buffer and dialyze it against the same buffer. Store the extract in liquid nitrogen. This is the S-100 extract.

10. Repeat the above extraction about seven times and combine all the extracts to give about 1.5 g of protein.

*Q-Sepharose Chromatography*. Column chromatography on a column of Q-Sepharose fast flow resin (Pharmacia), an anion-exchange resin, is used as the first chromatography step in purification of protein-histidine kinase. Small amounts of extract can be chromatographed on an HR10 column (1-cm diameter) but for normal preparative purposes a K50 column (5-cm diameter) is used. The bed height is approximately 20 cm. The column is packed using the packing reservoir and instructions provided by Pharmacia. In our laboratory, the column is operated using an FPLC apparatus (Pharmacia). The chromatography is performed as follows.

*Reagents*

Buffer A (20 m*M* Tris, pH 7.5, 0.2 m*M* EDTA, 6 m*M* $MgCl_2$, 1 m*M* dithiothreitol)
Equilibrium buffer (75 m*M* NaCl in buffer A)
NaCl (400 m*M*) in buffer A
NaCl (1 *M*) in buffer A
Ethanol, 20%
Solid polyethylene glycol
Storage buffer [50 m*M* Tris, pH 7.5, 6 m*M* $MgCl_2$, 0.2 m*M* EDTA, 1 m*M* dithiothreitol, 15% (v/v) glycerol]
Liquid nitrogen

1. Equilibrate the column (5 × 19 cm) with equilibration buffer at a flow rate of 8 ml/min for at least three column volumes.

2. Dialyze the combined S-100 extracts against 5 liters of equilibration buffer for 2 days with three changes of the buffer (total volume of dialysis buffer, 20 liters).

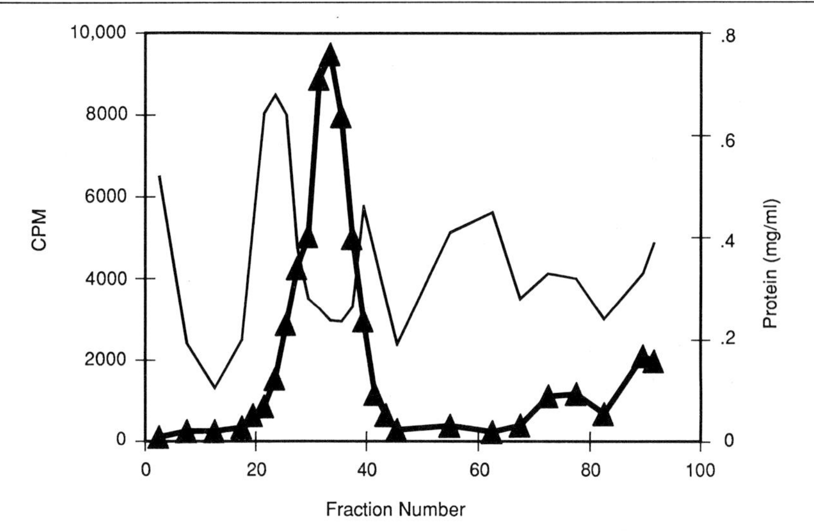


FIG. 7. Chromatography of histidine kinase on Q-Sepharose. The column was equilibrated with three column volumes of 75 m*M* NaCl in buffer A as described in the text. Yeast S-100 extract was dialyzed against buffer A and 150 ml of this dialyzed extract, containing 1.3 g protein, was loaded onto the column. The column was washed with the same buffer until the absorbance at 280 nm returned to baseline and then eluted with an 1840-ml linear gradient from 75 to 400 m*M* NaCl in buffer A at a flow rate of 8 ml/min. Fractions of 20 ml were collected and assayed for histidine kinase activity and protein content. The thicker line with data points represents the kinase activity; the thinner line without data points represents the protein concentration.

3. Centrifuge the dialyzed sample at 10,000 *g* for 15 min at 4°. Collect the supernatant and filter it through a 0.45-$\mu$m filter.

4. Load the filtrate, now 150 ml containing 1.3 g of protein, onto the column at 8 ml/min. Collect the eluant (flow-through fractions) and keep washing the column until the absorbance at 280 nm returns to baseline.

5. Elute the column with a 1840-ml linear gradient from 75 to 400 m*M* NaCl, all in buffer A, at a flow rate of 8 ml/min. Collect 20-ml fractions. Keep all the fractions on ice.

6. Wash the column with one column volume of 1 *M* NaCl in buffer A and three column volumes of water. Store the column in 20% ethanol.

7. Assay the fractions of the column eluant by the Nytran paper assay. The major enzyme activity peak is found at 200–300 m*M* NaCl (Fig. 7).

8. Pool the fractions with peak enzyme activity and concentrate them with polyethylene glycol as follows. Put the enzyme sample into a dialysis bag and cover the bag with precooled solid polyethylene glycol at 4°. Change the wet polyethylene glycol with dry every few hours until the

sample volume decreases to the desired volume (approximately fivefold concentration). The speed of this concentration step depends on the sample volume, the frequency of changing the polyethylene glycol, and the surface area of the dialysis bag. Dialyze the enzyme against storage buffer and store in liquid nitrogen or at −80°. Determine the kinase activity with the Nytran paper assay (see above) and protein concentration[29] in the concentrated sample.

The yield of protein histidine kinase activity from the Q-Sepharose column is typically 120%. The specific activity after this column is typically 25 pmol/15 min/mg and the purification is typically 22-fold.

*Mono Q Chromatography.* The second chromatographic step is also anion-exchange chromatography. The protein-histidine kinase elutes at a lower salt concentration from the second column than it did from the first. Q-Sepharose may be used for the second column with equivalent results, but Mono Q is preferred for its higher resolution. A Pharmacia HR10/10 semipreparative column is routinely used but smaller amounts may be chromatographed on an HR5/5 analytical column using scaled down flow rate and gradient. The HR10/10 Mono Q column is used with an FPLC apparatus as follows.

*Reagents*

Buffer A (20 m*M* Tris, pH 7.5, 0.2 m*M* EDTA, 6 m*M* $MgCl_2$, 1 m*M* dithiothreitol)
NaCl (400 m*M*) in buffer A
NaCl (1 *M*) in buffer A
Water
Ethanol, 24%
Solid polyethylene glycol
Liquid nitrogen
Storage buffer [50 m*M* Tris, pH 7.5, 6 m*M* $MgCl_2$, 0.2 m*M* EDTA, 1 m*M* dithiothreitol, 15% (v/v) glycerol]

1. Wash the column with water at a flow rate of 4 ml/min. Equilibrate the column with at least three column volumes of buffer A.
2. Dialyze the concentrated activity peak from the Q-Sepharose column against 100 vol of buffer A overnight at 4°.
3. Filter the dialyzed sample through a 0.45-$\mu$m filter.
4. Load filtrate containing 180 mg protein onto the column and elute

[29] M. M. Bradford, *Anal. Biochem.* **72,** 248 (1976).

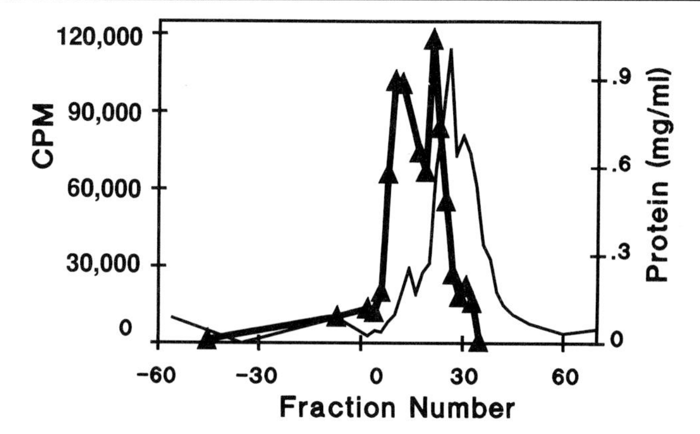


FIG. 8. Chromatography of histidine kinase on an HR10/10 Mono Q column. The column was equilibrated with three column volumes of buffer A at a flow rate of 4 ml/min. The concentrated Q-Sepharose-purified histidine kinase was dialyzed against buffer A and 180 mg of protein was loaded onto the column. The column was eluted with a 160-ml linear gradient from 0 to 400 m*M* NaCl in buffer A at 4 ml/min. Fractions of 2 ml were collected and assayed for histidine kinase activity and protein content. The thicker line with data points represents the kinase activity; the thinner line without data points represents the protein concentration.

the column with a 160-ml linear gradient of buffer A containing 0–400 m*M* NaCl at a flow rate of 4 ml/min.

5. Collect 2-ml fractions and store them on ice.

6. Wash the column with one column volume of 1 *M* NaCl and three column volumes of water. Store the column in 24% ethanol. After several runs, the column should be more extensively cleaned as described by the supplier.

7. Assay the fractions of the column eluant by the Nytran paper assay. The major enzyme activity peak elutes at 25–175 m*M* NaCl (Fig. 8).

8. Pool the enzyme activity peak and concentrate it with polyethylene glycol as described above. Dialyze the enzyme against storage buffer. Store the enzyme in liquid nitrogen or −80°.

The yield of the Mono Q column is up to 144%. The final specific activity is typically 800 pmol/15 min/mg with a purification for this step of 32-fold.

*Mono S Chromatography.* The third chromatographic step is cation-exchange chromatography on a Mono S column. Up to 20 mg of protein may be run on an HR5/5 column as described below (Fig. 9A). The procedure has also been successfully scaled up and run on an HR10/10 column (Fig. 9B).

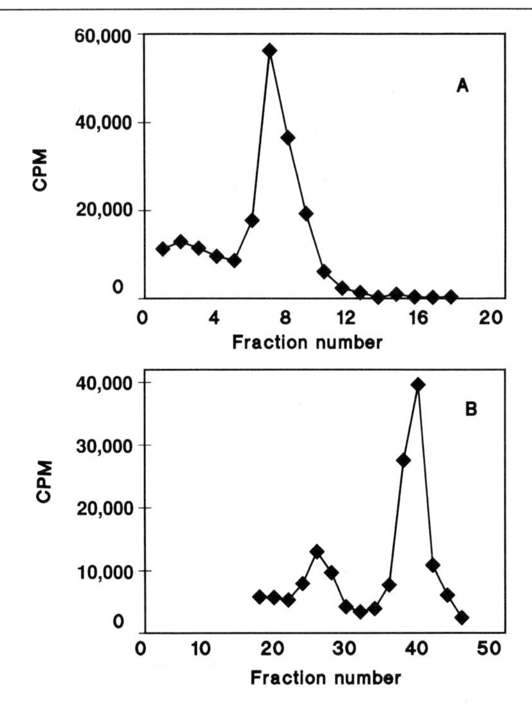


FIG. 9. Chromatography of histidine kinase on Mono S columns. (A) Mono S HR5/5. The column was equilibrated with three column volumes of 70 m*M* NaCl in buffer B at a flow rate of 1 ml/min. The concentrated Mono Q-purified histidine kinase was dialyzed against the same buffer and loaded onto the Mono S column. The column was eluted with a 17.5-ml linear gradient from 75 to 450 m*M* NaCl in buffer B followed by another linear gradient of 4.5 ml from 450 to 1000 m*M* NaCl in buffer B at the same flow rate. Fractions of 1 ml were collected and assayed for the histidine kinase activity. (B) HR10/10. The column was equilibrated with three column volumes of buffer B at a flow rate of 4 ml/min. The concentrated Mono Q-purified histidine kinase was dialyzed against the same buffer and loaded onto the column. The column was eluted with a 140-ml linear gradient from 0 to 450 m*M* NaCl in buffer B at the same flow rate. Fractions of 2 ml/min were collected and assayed for the histidine kinase activity.

*Reagents*

Buffer B (50 m*M* $NaPO_4$, pH 7.5, 6 m*M* $MgCl_2$, 1 m*M* dithiothreitol, 0.2 m*M* EDTA)
NaCl (75 m*M*) in buffer B
NaCl (450 m*M*) in buffer B
NaCl (1 *M*) in buffer B
Water

Ethanol, 20%
Liquid nitrogen
Storage buffer [50 m*M* Tris, pH 7.5, 6 m*M* $MgCl_2$, 0.2 m*M* EDTA, 1 m*M* dithiothreitol, 15% (v/v) glycerol]

1. Wash the column with water and equilibrate the column with 75 m*M* NaCl in buffer B at 1 ml/min.
2. Dialyze the concentrated Mono Q-purified enzyme against 100 vol of 75 m*M* NaCl in buffer B overnight.
3. Filter the dialyzed solution through a 0.45-$\mu$m filter.
4. Load the filtered solution onto the column at 1 ml/min. Elute the column with a 17.5-ml linear gradient of 75 to 450 m*M* NaCl in buffer B followed by another linear gradient of 4.5 ml of 450 to 1000 m*M* NaCl in buffer B.
5. Collect 1-ml fractions and store them on ice.
6. Wash the column with one column volume of 1 *M* NaCl and three column volumes of water and store the column in 20% ethanol.
7. Assay the fractions by the Nytran paper assay. The major protein-histidine kinase peak elutes from the column at 133–200 m*M* NaCl (Fig. 9A).
8. Pool the fractions containing enzyme, dialyze them against storage buffer, and store them in liquid nitrogen.

The yield of the Mono S column is typically 110%. The specific activity of the purified enzyme is typically 77 nmol/15 min/mg and the purification is about 95-fold.

A similar procedure may also be used with a semipreparative Mono S column, HR 10/10. The buffer flow rate is increased to 4 ml/min and the gradient volume is increased to 140 ml. The gradient is changed to start with buffer B with 0 *M* NaCl and increase linearly to 450 m*M* NaCl in buffer B. As shown in Fig. 9B the main kinase peak elutes between 220 and 270 m*M* NaCl.

*Sephacryl S-100 Chromatography*. The final chromatography step uses a high-resolution gel-filtration column.[8a] An HR10 column (Pharmacia) is packed with Sephacryl S-100 HR using a packing reservoir and the instructions provided by the supplier. The column size is 1 cm in diameter by 31 cm long and has a void volume of 8.7 ml determined with Blue dextran. For analytical purposes, small amounts of enzyme may be loaded and recovered if 0.1 mg/ml bovine serum albumin (BSA) is included in the elution buffers. For preparative purposes, larger amounts of enzyme protein are loaded and the BSA is not necessary. The preparative procedure is as follows.

*Reagents*

NaCl (0.15 *M*), 50 m*M* $NaPO_4$, pH 7.5, 6 m*M* $MgCl_2$, 1 m*M* dithiothreitol, 0.2 m*M* EDTA

1. Equilibrate the column with 0.15 *M* NaCl, 50 m*M* $NaPO_4$, pH 7.5, 6 m*M* $MgCl_2$, 1 m*M* dithiothreitol, 0.2 m*M* EDTA at 1 ml/min.
2. Load 200 $\mu$l of the Mono S-purified histidine kinase (2 mg/ml) onto the column and continue to elute the column with the same buffer and flow rate.

In the case of standard molecular weight protein markers, 0.15 ml of 2 mg/ml ovalbumin, 200 $\mu$l of 1 mg/ml carbonate dehydratase, 200 $\mu$l of 2 mg/ml BSA, and 200 $\mu$l of 2 mg/ml RNase are loaded onto the column separately and eluted with the same buffer at the same flow rate.

3. Collect 1-ml fractions and store them on ice.
4. Assay the fractions with the Nytran paper assay. The elution volume of the histidine kinase is 13 ml, which corresponds to a molecular weight of 31,000.
5. Pool the active fractions and store them in liquid nitrogen. Yields up to 110% have been obtained.

## Acknowledgments

We are particularly grateful to Dr. Verena Huebner, Matthew Lewis, Chris Holman, and Sherri Newmeyer, who helped develop several of the methods described here, and to our colleagues, especially Dr. Donal A. Walsh, Jianmin Huang, and Younhee Kim for their help and advice. Much of this work was supported by the American Cancer Society.

# Section IV

# Renaturation of Protein Kinases

# [33] Renaturation and Assay of Protein Kinases after Electrophoresis in Sodium Dodecyl Sulfate–Polyacrylamide Gels

*By* JILL E. HUTCHCROFT, MICHAEL ANOSTARIO, JR., MARIETTA L. HARRISON, and ROBERT L. GEAHLEN

An increasing appreciation of the importance of protein kinases in metabolic regulation has generated considerable interest in methods for their separation, detection, and characterization.[1] A powerful method for the resolution of complex mixtures of proteins containing kinases is sodium dodecyl sulfate–polyacrylamide gel electrophoresis (SDS–PAGE). Although SDS–enzyme complexes are typically devoid of enzyme activity, it has proved possible in many cases to recover activity after removal of the detergent. Thus, procedures can be developed that couple the high resolving power of SDS–polyacrylamide gel electrophoresis with "on the spot" detection of specific enzyme activities. An extensive review of techniques used to localize functionally active enzymes following gel electrophoresis has been previously published in this series.[2] We have adapted and extended such procedures to develop a simple and convenient method for the *in situ* (in the gel) renaturation and assay of protein kinases.[3]

Our gel-renaturation technique is summarized in Fig. 1. Briefly, proteins are separated by electrophoresis through SDS–polyacrylamide gels, renatured by removal of the SDS, and incubated with buffer containing [$\gamma$-$^{32}$P]ATP. The phosphorylation of substrates polymerized into the gel or the autophosphorylation of individual kinases is detected by autoradiography. Individual protein bands may be excised from the gel for quantification of radioactivity, analysis of phosphoamino acid content, or use in subsequent electrophoretic or chromatographic steps. The procedure can be applied both to the study of kinases present in crude protein mixtures and to the characterization of purified or partially purified enzymes.

[1] This work was supported by Grant CA37372 awarded by the National Cancer Institute, National Institutes of Health.

[2] M. J. Heeb and O. Gabriel, this series, Vol. 104, p. 416.

[3] R. L. Geahlen, M. Anostario, Jr., P. S. Low, and M. L. Harrison, *Anal. Biochem.* **153,** 151 (1986).

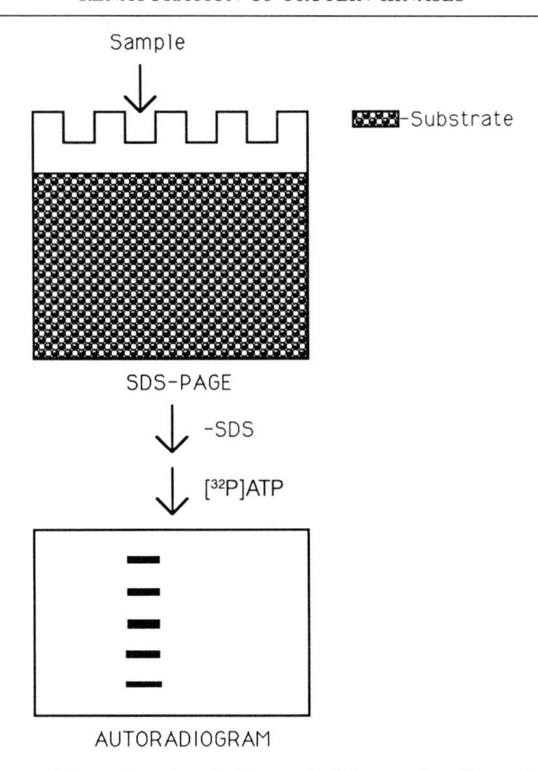


FIG. 1. Diagram of the major steps in the method for renaturation and assay of protein kinase activity following electrophoresis in SDS–polyacrylamide gels.

## General Procedure

1. Protein samples are solubilized in an equal volume of SDS–sample buffer [2.5% (w/v) SDS, 25% (w/v) sucrose, 25 m*M* Tris-HCl, pH 8.0, 2.5 m*M* EDTA, 15 m*M* 2-mercaptoethanol, 2.5 mg% pyronin Y). The sample mixture is then heated in a boiling water bath for 5 min prior to loading on the gel. The boiling step does not inhibit the subsequent renaturation of protein kinase activity. Proteins are separated by electrophoresis on an SDS–PAGE slab gel consisting of a 9% (w/v) acrylamide running gel and a 4% (w/v) stacking gel using the buffer system described by Laemmli.[4] If desired, 1 mg/ml of a protein kinase substrate may be added to the running gel solution prior to polymerization. Electrophoresis is performed overnight at 7 mA/gel until the tracking dye reaches the bottom of the gel.

2. The stacking gel is cut away from the separating gel and discarded.

[4] U. K. Laemmli, *Nature* (*London*) **277,** 680 (1970).

The remaining polyacrylamide gel is carefully placed into a clean plastic dish with a tight lid for the remaining steps. All of the incubations are at room temperature on an orbital shaker.

3. To remove the SDS, the polyacrylamide gel is washed six times with 200 ml of 40 m*M* HEPES, pH 7.4, over a period of 4–6 hr.

4. The gel is then incubated with 50 to 100 ml of phosphorylation buffer [25 m*M* HEPES, pH 7.4, 10 m*M* $MnCl_2$, 250–500 $\mu$Ci [$\gamma$-$^{32}$P]ATP (>7000 Ci/mmol; New England Nuclear, Boston, MA] for 3 hr. From this step on, care should be taken to use the appropriate Plexiglas shielding around the radioactive gel.

5. To remove excess [$\gamma$-$^{32}$P]ATP, the gel is rinsed briefly with distilled water and incubated for 4–12 hr in 600 ml of 40 m*M* HEPES, pH 7.4. To the buffer is added 20 g of Dowex 2X8-50 anion-exchange resin packed into dialysis tubing, which acts to bind the free [$\gamma$-$^{32}$P]ATP.

6. Substrate-containing gels are washed for an additional 3 hr in 40 m*M* HEPES, pH 7.4, containing 1% (w/v) sodium pyrophosphate.

7. The proteins in the gel are fixed by a 1-hr incubation in 250 ml of 10% 2-propanol, 5% acetic acid, 1% sodium pyrophosphate; stained for 2 hr in 250 ml 0.1% Coomassie Brilliant Blue R in 25% 2-propanol, 10% acetic acid; and then destained in 250 ml of 10% 2-propanol, 5% acetic acid. The destaining time can be shortened by adding pieces of packaging foam to the buffer.

8. The gel is vacuum dried and exposed to Kodak (Rochester, NY) X-Omat AR film to detect labeled bands.

## Technical Comments and Applications

The *in situ* renaturation procedure can be used to identify and characterize protein kinase activity in a number of different systems. Some of the parameters that can or have been modified for specific applications are discussed below.

### *Strategies for SDS–PAGE*

The *in situ* renaturation technique is generally applied to the detection of protein kinases either by autophosphorylation or by the phosphorylation of specific substrates included in the gel matrix. The experimental design determines the amount of protein that needs to be loaded onto the gel. Small quantities of purified proteins (0.01 to 2 $\mu$g/gel lane) are sufficient for renaturation of activity in gels containing protein substrates. When detecting protein kinase activities in crude cell extracts by autophosphorylation, the best results are obtained when large amounts of protein are

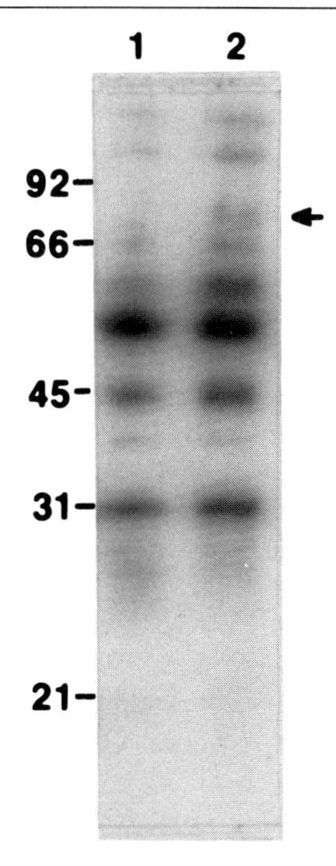


FIG. 2. Detection of protein kinase activity in particulate fractions from mouse spleen T lymphocytes. Postnuclear particulate fractions were prepared as described previously [M. L. Harrison, P. S. Low, and R. L. Geahlen, *J. Biol. Chem.* **259,** 9348 (1984)] from untreated T lymphocytes (lane 1) or from cells treated for 10 min with 100 ng/ml TPA (lane 2). Proteins were separated by electrophoresis using a Hoeffer minigel apparatus. Protein kinases were detected by autophosphorylation following *in situ* renaturation. Radiolabeled proteins were detected by autoradiography. The arrow indicates the migration position of an 80K protein. Molecular weight ($\times 10^{-3}$) given on left-hand side.

loaded (0.25–1.0 mg/gel lane for 12 × 14 × 0.15 cm separating gels). For this purpose, it is not necessary to also include a substrate protein in the separating gel. By using a minigel apparatus, up to 10-fold less protein can be loaded and incubation times can be decreased considerably. An example renaturation experiment using the minigel format is shown in Fig. 2.

Since many protein kinases catalyze autophosphorylation reactions, it is possible to identify several different enzymes in relatively crude protein mixtures. This is illustrated in Fig. 2 for particulate fraction proteins prepared from murine spleen T lymphocytes. The distribution of specific kinases can then be examined by comparing protein fractions prepared from different cells or different subcellular compartments. Changes in the distribution of kinases in response to various stimuli can also be examined using this approach. For example, pretreatment of T lymphocytes with 100 ng/ml 12-*O*-tetradecanoylphorbol 13-acetate (TPA) results in the appearance of an 80-kDa kinase in the membrane preparation (Fig. 2). Such studies can also lead to the identification of new enzymes.[5]

Different protein substrates can be incorporated into the gel matrix. From a technical point of view, the substrate needs only to be soluble under the SDS–PAGE conditions and not so small that it diffuses from the gel. When the substrate is included in the gel solution prior to initiation of polymerization, much of the protein becomes trapped in the gel matrix and does not electrophorese from the gel. Recovery of the protein from the polymerized gel is difficult, suggesting that some of it may even be covalently attached to the acrylamide polymers. Substrate specificity studies may be performed to characterize purified kinases. For example, the active subunits of multisubunit protein kinases such as the $\alpha$ subunit of casein kinase II[3] or the $\gamma$ subunit of phosphorylase kinase[6] have been identified in this fashion. Also, the enzymes capable of phosphorylating specific substrates can be identified in crude protein mixtures, as illustrated by Paudel and Carlson[6] for the phosphorylation of phosphorylase *b* by extracts of rabbit skeletal muscle. The technique should also be useful for the correlation of protein kinase activity with specific protein bands on polyacrylamide gels to verify the identity of purified or partially purified enzymes.

The concentration of acrylamide in the SDS–polyacrylamide gels can be varied to optimize the resolution of kinases of interest. We have performed renaturation assays using concentrations of acrylamide ranging from 9 to 15% in the separating gel mixtures. The renaturation of enzymes separated on two-dimensional gels has been reported.[7] The addition of large amounts (1 mg/ml) of protein substrates to the polyacrylamide separating gel precludes the use of Coomassie Blue or other protein stains because of the high level of background staining. The use of prestained molecular weight markers provides a convenient way of determining the relative sizes of specific protein kinases in such gels. They may also be used to avoid the fixing, staining, and destaining steps.

### *Renaturation Procedure*

In some cases the inclusion of 25% (v/v) 2-propanol in the renaturation washes helps to remove the SDS and renature the enzymes. It has been suggested that 2-propanol may help remove renaturation-inhibiting con-

[5] R. L. Geahlen and M. L. Harrison, *Biochem. Biophys. Res. Commun.* **134,** 963 (1986).
[6] H. K. Paudel and G. M. Carlson, *Arch. Biochem. Biophys.* **264,** 641 (1988).
[7] J. H. Keen, M. H. Chestnut, and K. A. Beck, *J. Biol. Chem.* **262,** 3864 (1987).

taminants present in certain batches of SDS.[8,9] We have had consistently good results using electrophoresis-grade SDS purchased from Bio-Rad (Richmond, CA), without the inclusion of 2-propanol in our buffers. After removal of SDS, denaturants such as 6 *M* guanidine hydrochloride or 8 *M* urea can be used to completely denature the proteins. Removal of the denaturants by subsequent washes allows the gradual refolding of the kinases into their enzymatically active conformations.[10] This has been reported to increase the sensitivity for the detection of $Ca^{2+}$/calmodulin-dependent protein kinase II.[11]

Renaturation times reported in the literature vary widely. The incubations should be long enough to ensure the complete removal of SDS, yet not so long as to cause a significant diffusion of proteins.

*Assay Conditions*

Specific effectors of kinase activity such as $Ca^{2+}$ or $Ca^{2+}$ and calmodulin can be added during the assay to detect kinases that are stimulated by these agents. In fact, the procedure can be utilized to decipher the cofactor requirements of certain enzymes. Harmon *et al.*[12] used the renaturation method to demonstrate that a $Ca^{2+}$-dependent protein kinase from soybean was regulated by $Ca^{2+}$, but did not require calmodulin. Similar results were obtained by Gunderson and Nelson[13] for a $Ca^{2+}$-dependent kinase from *Paramecium*.[14] $Ca^{2+}$/calmodulin dependent kinases from a variety of sources can be detected by including the activators in the assay mixture.[11,14,15]

The phosphorylation buffer may be modified depending on the characteristics of a given protein kinase. For example, the catalytic subunit of cAMP-dependent protein kinase shows a clear preference for $Mg^{2+}$ over $Mn^{2+}$ in the presence of high concentrations of ATP, both in solution and when renatured in gels.[11,16] However, at the low ATP concentrations described in our procedure above, $Mn^{2+}$ is much more effective than $Mg^{2+}$ at supporting the kinase reaction.[3]

It is important to verify that the presumptive kinase activity detected under a given set of experimental conditions is due to covalent binding of

[8] S. A. Lacks, S. S. Springhorn, and A. L. Rosenthal, *Anal. Biochem.* **100,** 357 (1979).
[9] A. Blank, R. H. Sugiyama, and C. A. Dekker, *Anal. Biochem.* **120,** 267 (1982).
[10] K. Weber and D. J. Kuter, *J. Biol. Chem.* **246,** 4504 (1971).
[11] I. Kameshita and H. Fujisawa, *Anal. Biochem.* **183,** 139 (1989).
[12] A. C. Harmon, C. Putnam-Evans, and M. J. Cormier, *Plant Physiol.* **83,** 830 (1987).
[13] R. E. Gundersen and D. L. Nelson, *J. Biol. Chem.* **262,** 4602 (1987).
[14] D. P. Blowers and A. J. Trewavas, *Biochem. Biophys. Res. Commun.* **143,** 691 (1987).
[15] T. Miyakawa, Y. Oka, E. Tsuchiya, and S. Fukui, *J. Bacteriol.* **171,** 1417 (1989).
[16] E. M. Reimann, D. A. Walsh, and E. G. Krebs, *J. Biol. Chem.* **246,** 1986 (1971).

the radiolabeled phosphate rather than nonspecific binding or entrapment of [$\gamma$-$^{32}$P]ATP. This can be done by performing a number of control experiments including the following: (1) the addition of excess unlabeled ATP to the phosphorylation buffer to compete with the [$\gamma$-$^{32}$P]ATP, (2) the performance of labeling studies with [$\alpha$-$^{32}$P]ATP,[11] (3) the reelectrophoresis of an excised $^{32}$P-labeled protein band on a second gel, or (4) analysis of the phosphoamino acid content of excised bands.

In gels containing immobilized substrate, it is possible to differentiate between autophosphorylation of a protein kinase and substrate phosphorylation. This is easily done by reelectrophoresis of the $^{32}$P-labeled protein band in a second SDS–polyacrylamide gel and analysis of the molecular weight of the labeled protein. Recovery of the radiolabeled protein is improved if the proteins are not fixed prior to drying the initial gel.

To be successful, the above techniques depend on the ability of a denatured protein kinase to refold to regain enzymatic activity. It is likely that some enzymes will not renature under these conditions. While many protein-serine/threonine kinases have been renatured from SDS–PAGE gels, relatively few protein-tyrosine kinases have been successfully detected. For example, we have not been successful at renaturing the protein-tyrosine kinase p56$^{lck}$ in gels containing proteins obtained from membranes of LSTRA, a cell line that overexpresses the enzyme.[17,18] Thus, the ability to detect a particular enzyme of interest must be empirically determined. However, for those enzymes that can be detected, *in situ* renaturation can be a simple and useful method for their further characterization.

[17] A. F. Voronova and B. M. Sefton, *Nature (London)* **319,** 682 (1986).

[18] J. D. Marth, R. Peet, E. G. Krebs, and R. M. Perlmutter, *Cell (Cambridge, Mass.)* **43,** 393 (1985).

# [34] Renaturation of Protein Kinase Activity on Protein Blots

*By* JOHN L. CELENZA and MARIAN CARLSON

We describe here a method for renaturing proteins bound to a nitrocellulose membrane and then assaying for protein kinase activity *in situ* on the protein blot. We first developed this assay to prove that a gene with sequence homology to protein kinases (the *SNF1* gene of *Saccharomyces cerevisiae*) in fact encodes a protein with kinase activity (Fig. 1).[1] The

[1] J. L. Celenza and M. Carlson, *Science* **233,** 1175 (1986).

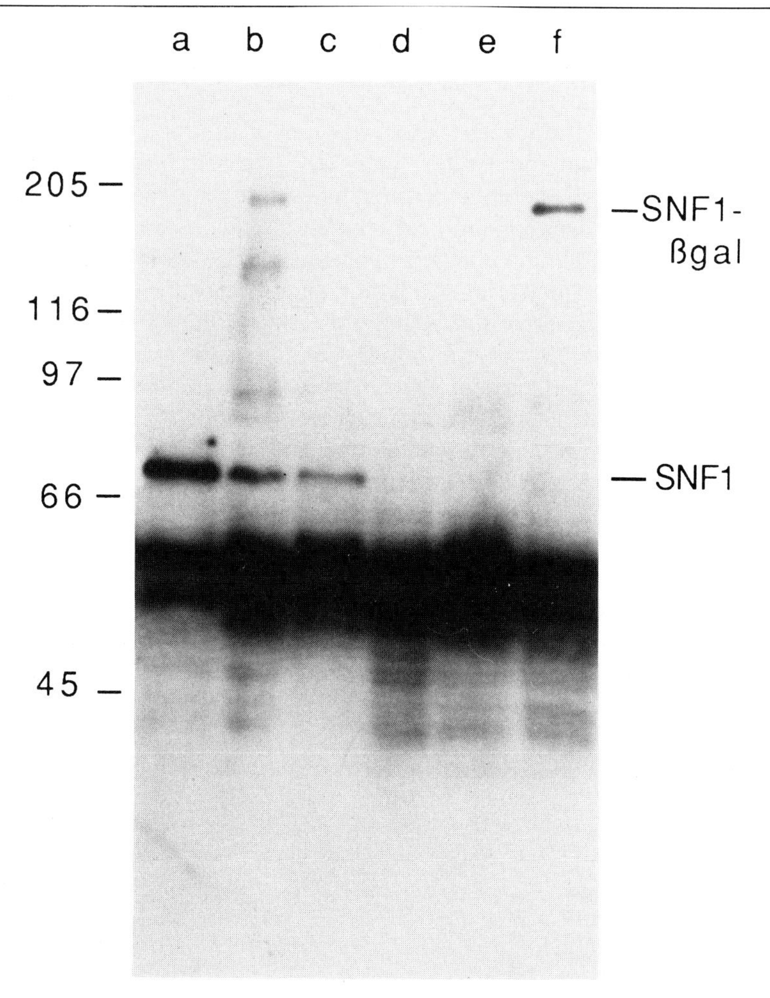


FIG. 1. Assay of the SNF1 protein kinase of *S. cerevisiae* on a protein blot. Proteins were prepared from cultures (glucose repressed, except for lane c) of yeast strains with the following genotypes: (a) wild type (*SNF1*$^+$) carrying the *SNF1* gene on a multicopy plasmid, (b) wild type, (c) wild type, glucose derepressed, (d) *snf1*-$\Delta 3$ (deletion mutation), (e) *snf1::HIS3* (insertion mutation), (f) *snf1::HIS3* carrying a bifunctional *SNF1–lacZ* gene fusion on a plasmid. Proteins (50 $\mu$g) were separated by electrophoresis in 7.5% SDS–polyacrylamide and transferred to nitrocellulose. Protein kinase activity was assayed on the protein blot as described here, with only minor differences. An autoradiograph is shown. The positions of the SNF1 and SNF1-$\beta$-galactosidase (SNF1-$\beta$gal) fusion proteins are marked. The identification of these proteins was confirmed by immunoblot analysis with anti-SNF1 antibody (not shown). Faint bands above the SNF1 band were detected frequently, but not reproducibly (see lane b). Protein size standards are indicated in kilodaltons. [Adapted from Ref. 1 (copyright 1986 by the AAAS).]

method is generally applicable for assaying protein kinase activity and does not require antibody. It is particularly useful for demonstrating that a cloned putative protein kinase gene or cDNA encodes a protein kinase. Other applications are also discussed.

In this method, a protein blot is prepared by transferring electrophoretically fractionated proteins to a nitrocellulose membrane. The proteins bound to the filter are exposed to the denaturant guanidine hydrochloride and then are allowed to renature in buffer. An autophosphorylation reaction is carried out by incubating the filter-bound proteins in buffer containing [$\gamma$-$^{32}$P]ATP. The labeled products are detected by autoradiography. Genetic or immunological methods can be applied to identify the protein kinase activity of interest.

## Method

### *Preparing Protein Blot*

Separate proteins by SDS–polyacrylamide gel electrophoresis[2] and electrophoretically transfer to a nitrocellulose membrane[3] in cold 25 m*M* Tris, 192 m*M* glycine (pH 8.3), without methanol. Carry out the transfer in the cold room and submerge the electroblotter in an ice bath. Apply a voltage gradient of 12–14 V/cm for 90 min.

### *Blocking Filter*

Block the filter by incubation in 30 m*M* HEPES (*N*-2-hydroxyethylpiperazine-*N*′-2-ethane sulfonate), pH 7.5, containing 5% (w/v) nonfat dry milk (Carnation), for 30 min at 25°. This and subsequent incubations and washes are carried out on a rotating platform.

*Comments.* Nonfat dry milk is commonly used as a blocking agent. We have substituted 3% (w/v) gelatin or 100 $\mu$g/ml phenol-extracted salmon sperm DNA as the blocking agent and then omitted nonfat dry milk from renaturation and denaturation buffers. These changes did not affect the array of phosphorylated proteins detected. Omitting any blocking agent reduced labeling severalfold and increased the background. Roussou *et al.*[4] substituted 1% (w/v) bovine serum albumin or 1% (w/v) total histones.

Varying the blocking agent provides a control to determine whether proteins in the blocking agent are serving as a substrate for the kinase of

[2] U. K. Laemmli, *Nature (London)* **227,** 680 (1970).
[3] H. Towbin, T. Staehelin, and J. Gordon, *Proc. Natl. Acad. Sci. U.S.A.* **76,** 4350 (1979).
[4] I. Roussou, G. Thireos, and B. M. Hauge, *Mol. Cell. Biol.* **8,** 2132 (1988).

interest or in some way affecting kinase activity on the blot. The blocking agent can also be chosen deliberately to provide a substrate; for example, the yeast GCN2 protein kinase is not autophosphorylated in this assay but is able to phosphorylate histones coating the filter.[4]

*Denaturation and Renaturation of Filter-Bound Proteins*

Incubate the filter in 100 ml of denaturation buffer for 1 hr at 25°. Wash the filter several times in cold renaturation buffer and incubate in 500 ml of renaturation buffer for 16 hr at 4°. (Suggested volumes are suitable for filters up to 12 × 14 cm.)

Denaturation buffer: 7 *M* guanidine hydrochloride, 50 m*M* Tris-HCl, pH 8.3, 50 m*M* dithiothreitol (DTT), 2 m*M* EDTA, 0.25% (w/v) nonfat dry milk

Renaturation buffer: 50 m*M* Tris-HCl, pH 7.5, 100 m*M* NaCl, 2 m*M* DTT, 2 m*M* EDTA, 0.1% (v/v) Nonidet P-40 (NP-40), 0.25% (w/v) nonfat dry milk

*Comments.* This denaturation and renaturation procedure is adapted[5] from the method of Hager and Burgess.[6] Omission of the denaturation step reduces phosphorylation of SNF1 and other proteins; however, this might not be the case for all protein kinases. This and similar denaturation and renaturation procedures have been effective for renaturing a variety of activities on protein blots: phosphorylation of histones by the yeast GCN2 protein kinase[4]; DNA binding by retroviral gene products required for integration,[7] by a human enhancer-binding protein,[8] and by yeast transcription factor IIIA[9]; and erythrocyte binding by a *Giardia* lectin.[10] Vinson *et al.*[11] described a related method involving graded removal of the denaturant that allowed *in situ* detection of DNA binding activity for a nitrocellulose-bound protein expressed from a recombinant bacteriophage.

[5] M. J. Roth, personal communication (1986).

[6] D. A. Hager and R. R. Burgess, *Anal. Biochem.* **109,** 76 (1980).

[7] M. J. Roth, N. Tanese, and S. P. Goff, *J. Mol. Biol.* **203,** 131 (1988).

[8] H. Singh, J. H. LeBowitz, A. S. Baldwin, Jr., and P. A. Sharp, *Cell (Cambridge, Mass.)* **52,** 415 (1988).

[9] C. K. Wang and P. A. Weil, *J. Biol. Chem.* **264,** 1092 (1989).

[10] H. D. Ward, B. I. Lev, A. V. Kane, G. T. Keusch, and M. E. A. Pereira, *Biochemistry* **26,** 8669 (1987).

[11] C. R. Vinson, K. L. LaMarco, P. F. Johnson, W. H. Landschulz, and S. L. McKnight, *Genes Dev.* **2,** 801 (1988).

*Autophosphorylation Reaction*

Rinse the filter several times in 30 m*M* HEPES, pH 7.5, at 25°. Place the filter in a sealable plastic bag with 5 ml of kinase buffer containing [$\gamma$-$^{32}$P]ATP. Incubate 20–30 min at 25°.

Kinase buffer: 30 m*M* HEPES, pH 7.5, 10 m*M* $MgCl_2$, 2 m*M* $MnCl_2$, 0.1 $\mu M$ ATP, 0.03 $\mu M$ [$\gamma$-$^{32}$P]ATP (3000 Ci/mmol, New England Nuclear, Boston, MA)

*Comments.* The kinase buffer should be adjusted according to the requirements of particular protein kinases. The reaction can also be optimized with respect to ATP concentration. The reported $K_m$ values for most protein kinases are in the range of 10 to 100 $\mu M$.[12] To assay the SNF1 kinase, we originally used only 0.03 $\mu M$ [$\gamma$-$^{32}$P]ATP. Addition of 0.1 $\mu M$ cold ATP did not change the incorporation of label into the SNF1 protein and reduced the background. Addition of 1 $\mu M$ cold ATP reduced the labeling of SNF1 about twofold, and addition of 10 $\mu M$ cold ATP reduced the signal substantially.

*Washing Filter and Autoradiography*

Wash the filter in 30 m*M* HEPES, pH 7.5, for 2 hr at 25° several times, until the radioactivity bound to the filter no longer decreases with continued washing. Dry the filter in air until slightly damp; a damp filter can be successfully washed more extensively, if necessary. Expose to film for 1 to 10 hr at −70° with an intensifying screen for autoradiography. Longer exposures may result in unacceptable background.

*Comments.* If the washing procedure described above does not reduce the background to an acceptable level, the filter can be washed again in 1 *N* HCl for 1 hr at 25°. Washing the filter in 1 *N* HCl for 1 hr at 70° greatly reduces background, but also reduces the signal; we found that the amount of SNF1 protein detectable by subsequent immunoblot assay was also reduced. Roussou *et al.*[4] included 1 m*M* ATP in the washing buffer.

*Identifying Activity of Interest*

*Genetic Methods.* In genetic systems, the protein kinase activity of interest can be identified by examining mutants lacking the protein, overexpressing the protein, or expressing a derivative with different electrophoretic mobility (Fig. 1). Similar methods apply for proteins expressed

[12] P. J. Roach, *in* "Methods in Enzymology" (F. Wold and K. Moldave, eds.), Vol. 107, p. 81. Academic Press, Orlando, Florida, 1984.

from cloned sequences in bacteria; in this case identifying the activity of interest should be trivial.

*Immunological Methods.* If antibody specific to the protein kinase of interest is available, the filter can be used for an immunoblot assay after a suitable autoradiograph is obtained.

### *Analysis of Phosphoprotein*

To demonstrate that the labeling of a protein results from phosphorylation of an amino acid residue, the phosphoprotein can be recovered from the filter by the method of Parekh *et al.*[13] and subjected to phosphoamino acid analysis.[14]

Estimate the amount of protein required to yield sufficient labeled material and, if necessary, run a preparative gel. After carrying out the assay, excise the region of the filter containing the phosphoprotein, taking care to minimize the excised region. Incubate the filter strip in 0.5 ml of 40% (v/v) acetonitrile, 0.1 *M* ammonium acetate, pH 8.9, for 3 hr at 37°. About 30–60% of the radioactive label should be recovered. Lyophilize the eluate. Subject the recovered material to partial acid hydrolysis and separate phosphoamino acids by two-dimensional thin-layer electrophoresis as described by Cooper *et al.*[14] As a control, carry out the same procedure using an equal-sized strip of filter that does not have labeled protein bound to it.

## Applications

This is a rapid method for demonstrating biochemically that a gene that is homologous to protein kinase genes does in fact encode a functional protein kinase. This method does not require either biochemical characterization of the protein or specific antibody. Cloned genes or cDNAs can be expressed in *Escherichia coli,* and the protein produced can be assayed directly. In genetic systems, mutants can be employed to show that a gene encodes a protein kinase. In systems where altered genes can be transformed into the organism or into cultured cells, the gene can be manipulated *in vitro* to change the size and level of expression of the gene product, thereby facilitating proof that the phosphorylated protein is encoded by the gene in question.

If antibody specific to the protein is available, this assay can be used

[13] B. S. Parekh, H. B. Mehta, M. D. West, and R. C. Montelaro, *Anal. Biochem.* **148,** 87 (1985).

[14] J. A. Cooper, B. M. Sefton, and T. Hunter, *in* "Methods in Enzymology" (J. D. Corbin and J. D. Hardman, eds.), Vol. 99, p. 387. Academic Press, New York, 1983.

as an alternate to, or in conjunction with, immune complex assays. This assay offers the advantage that the kinase activity is associated with a particular polypeptide. Thus, it can be used to confirm that the activity detected in an immune complex assay is due to the protein against which the antibody is directed, rather than to a minor protein kinase contaminating the immune complex.

This assay may also be useful during the purification of a protein kinase to assign the catalytic activity to a particular polypeptide.

The activity of a protein kinase is in many cases regulated or modulated in some way. This assay can be useful in distinguishing between possible regulatory mechanisms. For example, if the kinase activity is altered by a mechanism involving covalent modification, differences in activity may be apparent in this assay, presuming that the modification is stable. In contrast, if the kinase activity is modulated by physical association with another protein, then the electrophoretic separation of the two proteins in this procedure should preclude detection of regulatory effects.

## Limitations and Possible Problems

The method is not suitable for detection of any protein kinase that requires two different subunits for activity because the different polypeptides would be separated electrophoretically. A possible remedy would be electrophoresis on a nondenaturing gel so that the polypeptides remain associated.

Some proteins may not regain activity in response to a denaturation and renaturation regimen. It is possible that some proteins may retain activity or renature more efficiently following electrophoresis under nondenaturing conditions.

The activity of a protein kinase may be too low, relative to background, for detection by this assay. If the cloned gene is available, overexpression of the protein would remedy this problem.

Another possible problem is that the protein kinase may comigrate with another kinase so that the activity of interest is obscured. For example, assays of yeast proteins revealed many highly phosphorylated proteins of 45 to 65 kDa (Fig. 1), and it would therefore be difficult to identify a minor activity in this size range. For proteins expressed in *E. coli,* comigration of two activities is less likely to pose a problem as there are few phosphorylated species detected by this assay. If the cloned gene or cDNA encoding the protein kinase is available, the gene can usually be manipulated *in vitro* to shift the activity to a more convenient region of the protein blot. It is often possible to construct a fusion to another protein, such as $\beta$-galactosidase, that is larger than the native protein and yet retains protein kinase activity (Fig. 1). Construction of fusion proteins, or other derivatives, also provides a valuable control. Demonstration that two different-sized derivatives both exhibit kinase activity confirms that the protein of interest is not merely serving as a substrate for another, comigrating kinase.

Related methods are described in [33] and [35] in this volume.

Acknowledgments

We thank Monica Roth for suggesting the method. This work was supported by Public Health Service Grant GM34095 from the N.I.H. and an American Cancer Society Faculty Research Award to M.C.

# [35] Assessing Activities of Blotted Protein Kinases

*By* JAMES E. FERRELL, JR., and G. STEVEN MARTIN

Celenza and Carlson[1,2] have described a method for assessing the activities of protein kinases bound to nitrocellulose-blotting membranes. The method exploits the ability of SDS-denatured enzymes to regain activity after treatment with guanidine and nonionic detergent. The resulting kinase activities might arise from disinhibition of partially denatured enzymes or from renaturation of denatured enzymes.

Here we present a modification of the Celenza and Carlson assay. In this assay, poly(vinylidene difluoride) (PVDF) membranes are used in place of nitrocellulose. The main advantage of PVDF here is that it can be washed with strong bases, which significantly lowers the background of unincorporated radiolabel without diminishing the kinase signal. In addition, milk is not used in the protocol. At least some batches of nonfat dry milk possess detectable levels of kinase activity (G. Schieven and J. E. Ferrell, unpublished results), which means that blotted proteins that appear to possess kinase activity could actually represent preferred substrates for the milk kinase. This precaution may be of a more theoretical than practical significance; we have not found any kinase bands that are detectable when milk is used as a blocking agent that are not when albumin or no blocking agent is used.

This protocol is the most sensitive and reliable of many variations we

[1] J. L. Celenza and M. Carlson, this volume [34].

[2] J. L. Celenza and M. Carlson, *Science* **233,** 1175 (1986).

have tried. Further information about the method and its application to specific cell types can be found elsewhere.[3,4]

## Renaturation of Blotted Kinases

### *Lysis*

To a 35-mm dish of fibroblasts add 0.25 ml of sample buffer preheated to 100°:

Sample buffer; 10% (w/v) glycerol, 5% (w/v) 2-mercaptoethanol, 2.3% (w/v) sodium dodecyl sulfate (SDS) (BDH, Hoefer Scientific, San Francisco, CA; "specially pure"), 62.5 m*M* Tris, adjusted to pH 6.8 with HCl, 10 m*M* EDTA

Scrape the cells into a microcentrifuge tube, vortex, and immerse the tube in boiling water for 3 min. Vortex, pellet for 5 min in a microcentrifuge, and load onto a polyacrylamide gel. Removal of nucleic acids by DNase/RNase treatment and/or shearing is not essential.

*Alternative*. Lyse the cells in ice-cold 1% (w/v) Nonidet P-40, 1% (w/v) sodium deoxycholate, 0.1% (w/v) SDS, 100 m*M* NaCl, 50 m*M* Tris (pH 7.4), 10 m*M* EDTA (pH 8.0), 1% (v/v) aprotinin solution (Sigma, St. Louis, MO), 100 μg leupeptin/ml, 200 μ*M* phenylmethylsulfonyl fluoride (PMSF), 1 m*M* sodium orthovanadate (pH 10.0). Scrape cells into a microcentrifuge tube, pellet nuclei for 5 min in a microfuge, take supernatant, add 0.25 vol 5× sample buffer, boil for 3 min, and load.

### *Electrophoresis*

For crude lysates, load 100 μg protein/lane on a discontinuous slab gel with a 4.5% (w/v) stacking gel and 7.5, 8.5, or 10% (w/v) separating gel. For simpler protein mixtures, load as much as can be loaded without producing grossly distorted bands. Thioglycolate (0.002%, w/v) can be added to the running buffer as a precaution against protein oxidation, although in our experience it does not improve kinase renaturability.

### *Blotting*

Wet the PVDF membrane (Immobilon P, 0.45-μm pore size, Millipore, Bedford, MA) briefly with methanol, then rinse with water and transfer buffer (192 m*M* glycine base, 25 m*M* Tris base). Do not equilibrate gel with transfer buffer. Lay gel on membrane and carry out transfer at 4° for

[3] J. E. Ferrell, Jr. and G. S. Martin, *J. Biol. Chem.* **264,** 20723 (1989).
[4] J. E. Ferrell, Jr. and G. S. Martin, *Mol. Cell. Biol.* **10,** 3020 (1990).

90 min at 5 V/cm [50 V for a Hoefer (San Francisco, CA) tank blotter]. Despite the absence of SDS and methanol, virtually all stainable proteins between molecular weight 30K and 200K are transferred and retained under these conditions. For lower molecular weight proteins, decrease the transfer voltage.

*Denaturation*

Incubate blot for 1 hr at room temperature with gentle rocking in 100 ml denaturation buffer:

Denaturation buffer: 7 *M* guanidine hydrochloride (Sigma, grade I), 50 m*M* Tris base, 50 m*M* dithiothreitol (DTT), 2 m*M* EDTA; adjust pH to 8.3

The guanidine renders the blotting membrane translucent. All of the kinases we have examined will renature even without guanidine treatment, but their activities are generally better with it.

*Renaturation*

Pour off guanidinium solution. Rinse blot in Tris-buffered saline, pH 7.4, briefly. The membrane will become opaque again. Add renaturation buffer:

Renaturation buffer: 140 m*M* NaCl, 10 m*M* Tris-HCl, pH 7.4, 2 m*M* DTT, 2 m*M* EDTA, 1% (w/v) bovine serum albumin (BSA) (ICN, Costa Mesa, CA), 0.1% (w/v) Nonidet P-40 (NP-40)

Incubate with gentle rocking overnight at 4°. This step is essential for kinase renaturation.

*Blocking*

Pour off renaturation buffer. Add the following:

Tris-HCl (pH 7.4), 30 m*M*
Bovine serum albumin, 5% (w/v)

Incubate at room temperature for 1 hr.

*In Situ Phosphorylation*

Pour off blocking solution. Cover the blot with freshly prepared reaction buffer (15 ml for a 12 × 12 cm blot):

Reaction buffer: 30 m*M* Tris-HCl, pH 7.4, 10 m*M* $MgCl_2$, 2 m*M* $MnCl_2$, 100 $\mu$Ci/ml [$\gamma$-$^{32}$P]ATP, 3000–6000 Ci/mmol

Rubbermaid brand plastic containders (Servin' Savers, 470-ml size) are particularly useful for this incubation, as they are relatively easy to decontaminate. Plastic bags can also be used, although the scraping that inevitably occurs when the blot is inserted into and removed from the bag can lead to high backgrounds. Incubate with rocking at room temperature for 30 min.

*Washing*

Pour off the reaction buffer. Add 50 ml of 30 m*M* Tris-HCl, pH 7.4, rinse briefly, and transfer blot to a larger plastic container. Wash eight times, 10 min/wash, as follows:

2× with 250 ml 30 m*M* Tris-HCl (pH 7.4)
1× with 250 ml 30 m*M* Tris-HCl (pH 7.4) + 0.05% Nonidet P-40
2× with 250 ml 30 m*M* Tris-HCl (pH 7.4)
1× with 250 ml 1 *M* KOH
2× with 250 ml 30 m*M* Tris-HCl (pH 7.4)

Let blot dry and carry out autoradiography. Typical exposure times are 12–14 hr at −70° with an intensifying screen to detect the more prominent kinases, longer for minor bands. Typical results are shown in Fig. 1.

Notes

*Sensitivity*

Generally 10–30 electrophoretically distinct kinases can be detected in whole-cell lysates from various sources (human platelets, HeLa cells, mouse and chicken fibroblasts, *Xenopus laevis* oocytes and eggs). We do not know what proportion of the total protein kinases this number represents. There is considerable variation in the amount of kinase necessary to yield a detectable band after renaturation. For example, 50 ng of protein kinase C and 2 ng of p42 (a tyrosine-phosphorylated protein-serine/threonine kinase[4]) can be detected, whereas 200 ng of $pp60^{c\text{-}src}$ cannot.

*Specificity*

To ensure that radiolabeled bands indicate the presence of kinases rather than other ATP-binding proteins, bands of interest should be subjected to phosphoamino acid analysis. Bands are excised and hydrolyzed without elution in 500 μl of 6 *M* HCl at 100° for 1 hr (unlike nitrocellulose, PVDF is impervious to strong acids[5]). Supernatants are lyophilized and

[5] M. Kamps, this series, Vol. 201, p. 21.

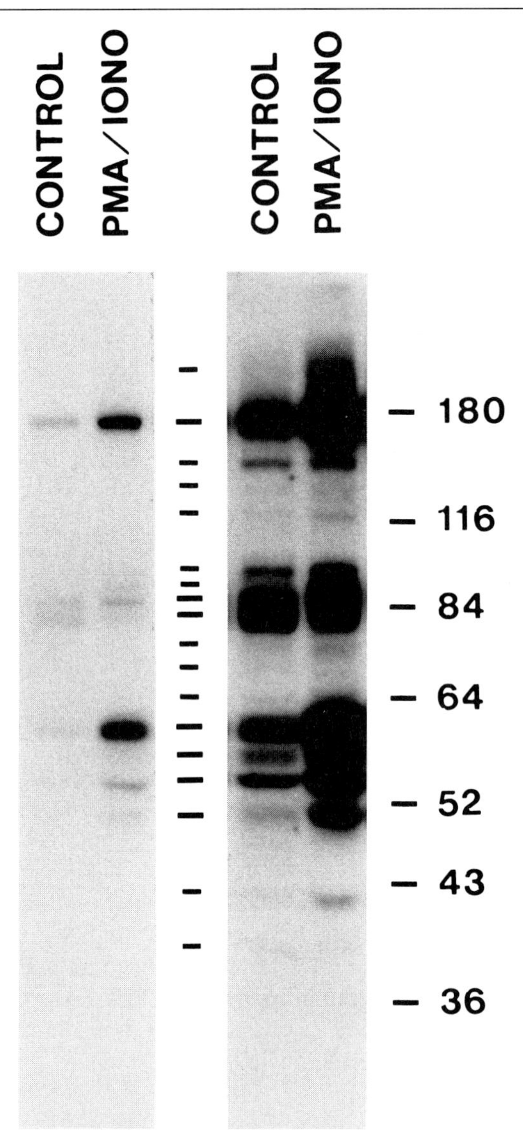


FIG. 1. Assessment of protein kinase activity after transfer to PVDF membranes. Lysates were prepared from resting human platelets (CONTROL) and from platelets incubated for 5 min at 37° with 100 n*M* ionomycin, 1 m*M* calcium, and 50 ng phorbol myristate acetate/ml (PMA/IONO). Approximately 75 $\mu$g protein was loaded per lane. Proteins were separated on an 8.5% polyacrylamide gel, transferred to PVDF, treated with guanidine, and allowed to renature overnight. The renatured proteins were overlaid with 15 ml reaction buffer containing 50 $\mu$Ci/ml [$\gamma$-$^{32}$P]ATP. Kodak (Rochester, NY) X-OMAT AR film was exposed at room temperature with one screen for 45 min (left) or 16 hr (right). Tick marks denote the renaturable kinase bands.

subjected to two-dimensional electrophoresis on thin-layer cellulose plates.[6] Thus far all of the bands we have examined have proved to be *bona fide* kinases, and all have phosphorylated only serine and threonine residues.[3,4]

### *Analysis of Reaction Products*

A variety of proteins are phosphorylated in this assay, including the kinases themselves (autophosphorylation), proteins present in the blocking solution, and proteins comigrating with the renaturable kinases.[3] These products can be analyzed by eluting the reaction products with 2% (w/v) SDS, 1% (w/v) Nonidet P-40, 30 m*M* sodium phosphate (pH 8.6).[7] Blots should not be allowed to dry prior to protein elution. Albumin can be omitted from all of the buffers, which generally results in some decrease in kinase activity, or can be replaced with substrates of particular interest.

### *Quantitative Assessment of Kinase Activity*

For the kinases we have examined, $^{32}P$ incorporation increases linearly with respect to time and protein loading under the assay conditions described here.[3] This linearity allows the assay to be used to assess changes in *in vitro* kinase activity that may arise by covalent modification of the kinases *in vivo*. Figure 1 shows that many kinases increase in activity when intact human platelets are incubated with a calcium ionophore and phorbol myristate acetate.

### *Immunoblotting*

Kinase blots can be rewetted, reblocked, and probed with antibodies and alkaline phosphatase to determine the positions of known protein kinases, even after alkali treatment.[8]

## Acknowledgments

This work was supported by National Research Service Award GM11507 and a Leukemia Society of America Special Fellowship (to J.E.F.) and by Public Health Service Grant CA-17542 and Council for Tobacco Research Grant 2452 (to G.S.M.).

[6] J.-C. Cortay, D. Nègre, and A.-J. Cozzone, this volume [17].
[7] B. Szewczyk and D. F. Summers, *Anal. Biochem.* **168,** 48 (1988).
[8] M. Y. Lim, personal communication.

# [36] High-Performance Liquid Chromatographic Separation and Renaturation of Protein Kinase Subunits: Application to Catalytic Subunit of Phosphorylase Kinase

*By* SCOTT M. KEE, CHIUN-JYE YUAN, and DONALD J. GRAVES

Any attempt to relate the catalytic and regulatory properties of an oligomeric enzyme to its structure involves an understanding of the roles of the different subunits and how they interact with one another. Perhaps the most efficient means of arriving at such understanding is the characterization of isolated subunits in their native states. For some enzymes, however, dissociation and isolation of subunits under nondenaturing conditions are extremely difficult because of strong intersubunit bonding. Such is the case with phosphorylase kinase.[1] This enzyme plays an important role in the regulation of glycogenolysis and consists of four different subunits, $\alpha$, $\beta$, $\gamma$, and $\delta$, with a stoichiometry of $\alpha_4\beta_4\gamma_4\delta_4$.[2–4] The $\alpha$ and $\beta$ subunits are regulatory subunits.[2,5] The $\delta$ subunit is shown to be identical to the calcium-binding protein, calmodulin (CaM), by its primary structure and ability to activate CaM-dependent enzymes.[4] The catalytic role of $\gamma$ was proved by Skuster *et al.*[6] based on the treatment of the holoenzyme with LiBr, a denaturant. Dissociation of the enzyme into its subunits occurred and an active fraction containing the $\gamma$ subunit was obtained by gel filtration and chromatography. Subsequently, catalytically active subunit complexes of $\alpha\gamma\delta$ and $\gamma\delta$ were found to retain the full catalytic activity of the holoenzyme.[7] Amino acid sequencing provided proof that the $\gamma$ subunit contains structural elements similar to those found in the active region of other protein kinases.[8] The catalytic properties of the active $\gamma$ subunit have been described.[9]

Methods are described here for the separation of phosphorylase kinase

[1] G. M. Carlson, P. J. Bechtel, and D. J. Graves, *Adv. Enzymol.* **50,** 41 (1979).
[2] P. Cohen, *Eur. J. Biochem.* **34,** 1 (1973).
[3] T. Hayakawa, J. P. Perkins, D. A. Walsh, and E. G. Krebs, *Biochemistry* **12,** 567 (1973).
[4] P. Cohen, A. Burchell, J. G. Foulkes, P. T. W. Cohen, T. C. Vanaman, and A. C. Nairn, *FEBS Lett.* **92,** 287 (1978).
[5] T. Hayakawa, J. P. Perkins, and E. G. Krebs, *Biochemistry* **12,** 574 (1973).
[6] J. R. Skuster, K.-F. J. Chan, and D. J. Graves, *J. Biol. Chem.* **255,** 2203 (1980).
[7] K.-F. J. Chan and D. J. Graves, *J. Biol. Chem.* **257,** 5948 (1982).
[8] E. M. Reimann, K. Titani, L. H. Ericsson, R. D. Wade, E. H. Fischer, and D. A. Walsh, *Biochemistry* **23,** 4185 (1984).
[9] S. M. Kee and D. J. Graves, *J. Biol. Chem.* **262,** 9448 (1987).

subunits under denaturing conditions and subsequent reactivation of the $\gamma$ subunit[10,11] and for renaturation of the bacterial expressed form of the $\gamma$ subunit of phosphorylase kinase.[12,13]

## Materials and Methods

Phosphorylase kinase can be prepared from rabbit skeletal muscle as described by Hayakawa *et al.*[3] Further purification is obtained by using DEAE-cellulose chromatography.[2] Phosphorylase *b* is prepared as described by Fischer and Krebs.[14]

The kinase activity may be assayed as described by Kee and Graves.[9] The standard assay mixture contains 62 m$M$ HEPES, 42 m$M$ Tris, 2.5 m$M$ dithiothreitol, 0.1 m$M$ $CaCl_2$, 10 m$M$ $MgCl_2$, 3 m$M$ ATP, and 100 $\mu M$ phosphorylase. The final pH of the reaction mixture is 8.2. The reactions are performed at 30° and initiated by adding MgATP. The reaction is terminated by dilution into ice-cold buffer followed by phosphorylase assay. Phosphorylase activity can be determined directly by measuring the amount of inorganic phosphate released from glucose 1-phosphate in the reaction of glycogen synthesis.[15]

### *High-Performance Liquid Chromatographic Separation of Phosphorylase Kinase Subunits*

The separation of phosphorylase kinase subunits on reversed-phase high-performance liquid chromatography (HPLC) was investigated by Crabb and Heilmeyer.[16] They were successful in completely separating the four subunits using programmed gradients that started with 0.1% (v/v) trifluoroacetic acid (TFA) in water and finished with 0.1% TFA in 62% (v/v) acetonitrile. The subunits are eluted according to size, with the $\delta$ subunit eluting first between 43 and 50% acetonitrile and $\alpha$ subunit eluting last at about 58% acetonitrile. They found that optimal resolution and recovery was dependent on the amount of protein applied. With a 5-$\mu$m Vydac $C_{18}$ column (25 × 0.46 cm i.d.), for example, phosphorylase kinase

[10] S. M. Kee and D. J. Graves, *J. Biol. Chem.* **261,** 4732 (1986).

[11] C.-J. Yuan and D. J. Graves, *Arch. Biochem. Biophys.* **274,** 317 (1989).

[12] L.-R. Chen, C.-J. Yuan, G. Somasekhar, P. Wejksnora, J. E. Peterson, A. M. Myers, L. Graves, P. T. W. Cohen, E. F. da Cruz e Silva, and D. J. Graves, *Biochem. Biophys. Res. Commun.* **161,** 746 (1989).

[13] L.-R. Chen, C.-J. Yuan, G. Somasekhar, P. Wejksnora, J. E. Peterson, A. M. Myers, and D. J. Graves, *in* "Information transduction and processing systems from cell to whole body" (O. Hatase and J. E. Wang, eds.), p. 19. Elsevier, New York, 1989.

[14] E. H. Fischer and E. G. Krebs, *J. Biol. Chem.* **231,** 65 (1958).

[15] B. Illingworth and G. T. Coli, *Biochem. Prep.* **3,** 1 (1953).

[16] J. W. Crabb and L. M. G. Heilmeyer, Jr., *J. Chromatogr.* **296,** 129 (1984).

applications of greater than 0.3 mg result in decreased recovery of all subunits and partial contamination of the $\beta$ subunit with the $\alpha$ subunit. When 0.25 mg sample is applied, individual subunit recoveries ranged from 66 to 88%. A much shorter column (3–6 cm) has little effect on resolution and it is found that a 6 × 1 cm column allowed higher yields with higher loads while still maintaining excellent resolution.

*Procedure.* Separation of phosphorylase kinase for the purpose of recovering an isolated, catalytically active $\gamma$ subunit is achieved using a 5-$\mu$m Vydac $C_{18}$ column (25 × 0.46 cm i.d.) and a gradient program and other conditions as described by Crabb and Heilmeyer.[17] Phosphorylase kinase (1–2 mg/ml in 50 m*M* glycerol 1-phosphate, pH 7.0, 2 m*M* EDTA, and 1 m*M* dithiothreitol) is injected directly to the column previously equilibrated in 0.1% TFA. The $\gamma$ subunit is eluted at approximately 50% acetonitrile and 0.09% TFA, with a pH of 2.5. Fractions containing the $\gamma$ subunit can be stored in plastic tubes on ice for 1 to 2 weeks. The $\gamma$ subunit prepared in this way is pure, judged by SDS–PAGE.[10]

### *Renaturation of High-Performance Liquid Chromatography-Isolated $\gamma$ Subunit of Phosphorylase Kinase*

Initially, the HPLC-recovered $\gamma$ subunit (about 100 $\mu$g/ml in HPLC solvent) has no activity when assayed according to the usual conditions, with no dilution or after a 20-fold dilution in 40 m*M* HEPES, pH 6.8, and 5 m*M* dithiothreitol (buffer A) before the assay. Presumably, the $\gamma$ subunit is denatured after undergoing exposure to the low pH and high organic solvent concentrations of the HPLC solvents. A small amount of activity (about 2% of the molar specific activity of the holoenzyme) is observed, however, when the $\gamma$ subunit is assayed in the presence of $Ca^{2+}$ and the $\delta$ subunit of phosphorylase kinase or its equivalent, CaM. The activity is significantly enhanced if $\gamma$ is preincubated with calmodulin and $Ca^{2+}$. Because CaM interacts strongly with the $\gamma$ subunit the presence of native CaM in the renaturation medium is thought to affect the renaturation process by binding to the $\gamma$ subunit and influencing its folding or by trapping the $\gamma$ in its native state, facilitating further conversion of denatured $\gamma$ to native $\gamma$.

*Procedure.* The HPLC-purified $\gamma$ subunit is diluted to 1–5 $\mu$g/ml in an ice-cold solution containing one part assay buffer (0.25 *M* HEPES, 0.25 *M* Tris, 0.6 m*M* $CaCl_2$, pH 8.6) and four parts buffer A plus 60–100 $\mu$g/ml calmodulin and 1.5 mg/ml phosphorylase *b* or bovine serum albumin. This results in a final pH of 8.2 and final concentrations of no more than 5%

[17] J. W. Crabb and L. M. G. Heilmeyer, Jr., *J. Biol. Chem.* **259,** 6346 (1984).

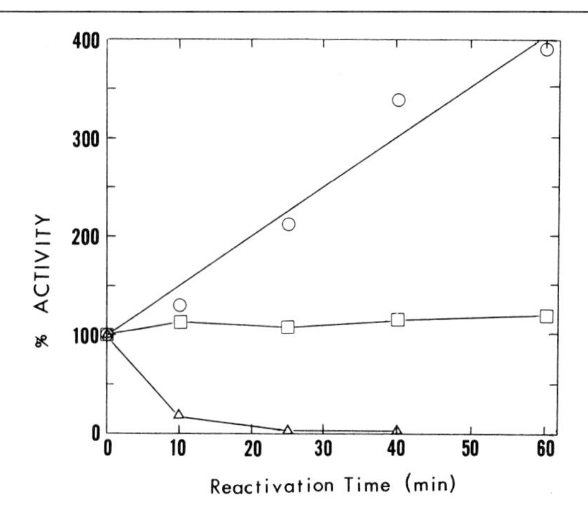


FIG. 1. Temperature effects on reactivation of $\gamma$ subunit.

acetonitrile and 0.01% TFA. The reactivation mixture is kept on ice for several days to attain optimum catalytic activity. With these conditions, the molar specific activity of the $\gamma$ subunit reaches approximately 70% of the molar specific activity of nonactivated phosphorylase kinase. Figure 1 illustrates how different temperatures affect reactivation of the $\gamma$ subunit. The initial activity drops 100-fold within about 25 min of incubation at 30°, while at 22° activity stays constant for at least 1 hr. In contrast, there is a marked increase in activity when incubated on ice. After only 1 hr the phosphorylase kinase activity of the $\gamma$ subunit increases fourfold.

Figure 2 shows the effect other factors have on reactivation. As mentioned above, $Ca^{2+}$ and calmodulin were used for reactivation. It is clear that only the calcium-bound conformation of calmodulin facilitates the reactivation in this case because if $Ca^{2+}$ is absent from the reactivation mixture no reactivation occurs. This is true even though another protein, phosphorylase *b*, is present. Thus, it is clear that reactivation does not occur in a nonspecific manner requiring only the presence of another protein for success. Reactivation also is pH dependent. Reactivation is greatest at pH 8.2 and only 20% as good at pH 6.8 (result not shown). Finally, as shown in Figs. 1 and 2, reactivation is time dependent and under optimal conditions maximum activity requires 1–2 days. When stored on ice, the reactivated $\gamma$ subunit loses no activity for at least 1 month.

In the presence of a low concentration of CaM, phosphorylase *b* or bovine serum albumin does enhance reactivation of the $\gamma$ subunit. If high amounts of CaM (0.5 mg/ml) and 1 m*M* $Ca^{2+}$ are used, however, the

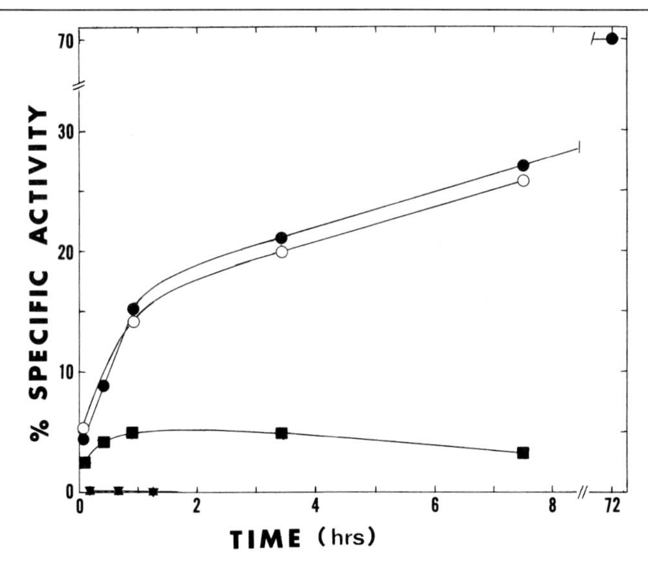


FIG. 2. Effect of other factors on reactivation.

HPLC-purified $\gamma$ subunit can be reactivated even better (Fig. 3). A second procedure for renaturing the HPLC-purified $\gamma$ subunit is to dialyze the enzyme against 8 *M* urea followed by dilution and assay as described.[11] With this procedure CaM is not required for renaturation of the $\gamma$ subunit. HPLC-purified $\gamma$ is flushed with $N_2$ to remove acetonitrile and followed by dialysis against 8 *M* urea, pH 6.8, containing 100 m*M* HEPES, and 1 m*M* dithiothreitol, at 4° for 12 hr. The $\gamma$ subunit is reactivated by a dilution of 8 *M* urea 20-fold with pH 6.8 HEPES buffer or pH 8.2 Tris buffer containing 1 m*M* dithiothreitol and 10% (v/v) glycerol on ice for several hours to reach a maximum activity. By this method the renaturation of the $\gamma$ subunit at pH 8.2 is also more favorable than the renaturation at pH 6.8. Although a total of 4 hr of incubation on ice is needed for HPLC-purified $\gamma$ to reach a maximum reactivation, the $\gamma$ subunit/8 *M* urea is reactivated as soon as 8 *M* urea is diluted with buffers.

### *Isolation of Reactivated $\gamma$ Subunit from CaM Using CaM-Agarose*

To obtain the isolated, reactivated $\gamma$ subunit free of CaM, agarose-bound CaM is used in place of free CaM or $\delta$ in the first reactivation protocol above. Specifically, the reactivation mixture is the same as described in the above section, except that it contained 20% CaM-agarose (Sigma, St. Louis, MO or Bio-Rad, Richmond, CA) and no free CaM. During reactivation the CaM-agarose resin is suspended often. Also, the

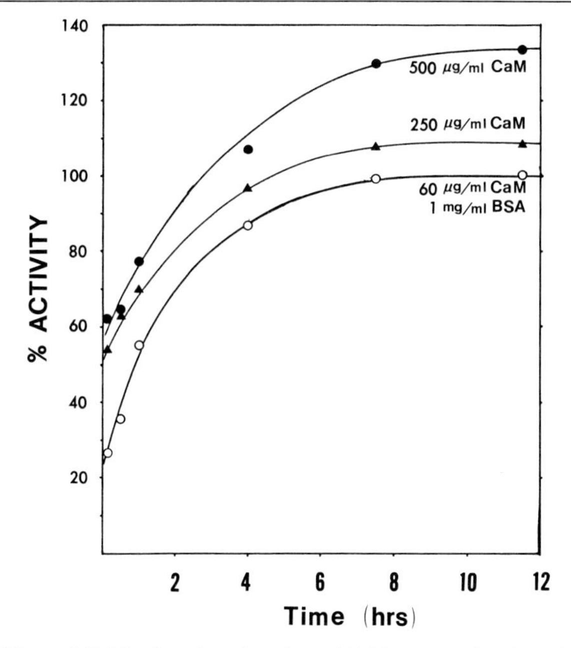


FIG. 3. Effect of CaM, phosphorylase $b$, and BSA on reactivation of $\gamma$ subunit.

final yield can be increased by removing the supernatant from the resin after 1 or 2 days and replacing it with an identical solution containing fresh HPLC-isolated $\gamma$. When reactivation is complete, and the mixture is poured into a small column, the column is washed well with buffer A at pH 8.2. This removes most of the phosphorylase but not the phosphorylase kinase activity. The $\gamma$–$\delta$ bond is known to be extremely strong in the holoenzyme and the LiBr-isolated $\gamma\delta$ complex. Thus, it is not surprising that elution is difficult and is not achieved by a number of reagents, including 10 m$M$ EGTA and 1 m$M$ EGTA plus either 1 $M$ NaCl, 2 $M$ LiBr, 5 mg/ml heparin, or 0.1 $M$ $NaCO_3$ (pH 10). Partial elution occurs with either 1% (v/v) Triton X-100 or an incubation of several hours in the presence of 1 $M$ Tris, pH 7.0, 5 m$M$ dithiothreitol, and 1 m$M$ EGTA. With a combined solution of 1% (v/v) Triton X-100, 1 $M$ Tris, pH 7.0, 1 m$M$ EGTA, and 5 m$M$ dithiothreitol all activity is immediately eluted. This activity decreases significantly (70% in 1 day) if left in the elution buffer. Thus, the Triton X-100 is removed by passing the eluted $\gamma$ subunit activity through a Bio-Beads SM-2 column, but this step results in about 60% loss of activity. The active fractions can then be concentrated, dialyzed against 50% glycerol, 0.1 $M$ HEPES, pH 7.0, 5 m$M$ dithiothreitol, and 0.1 m$M$ $CaCl_2$, and stored at $-20°$ for at least 3 months with little loss of activity.

The resulting γ subunit solution is shown to be free of CaM,[10] but contaminated with residual phosphorylase *b* from the reactivation mixture.

*Renaturation of Bacterial Expressed Chloramphenicol Acetyl Transferase–γ*

The fusion chloramphenicol acetyl transferase (CAT)–γ subunit in *Escherichia coli* is prepared as described by Chen *et al.*[12,13] A fusion plasmid containing a p*tac* promoter, the DNA coding for N-terminal 73 amino acids of bacterial CAT protein, and the cDNA coding for the entire γ subunit of phosphorylase kinase is first constructed. *Escherichia coli* D1210 containing *lacI*$^q$ is, then, transformed with this CAT–kinase plasmid. The plasmid-containing cells are grown at 37° to an OD of 0.2 at 600 nm. At this stage, cells are induced with 2 m*M* isopropyl-β-D-thiogalactoside (IPTG) and allowed to grow for another 4 hr. The CAT–γ fusion protein is synthesized and large inclusion bodies are formed. Essentially, 20% of the cell protein is the fusion protein. Cells with inclusion bodies are then pelleted and fusion protein is isolated as described by Nagai and Thogersen.[18] The fusion CAT–γ subunit, however, is insoluble and is found in the inclusion bodies. It can be dissolved in 6 *M* guanidine hydrochloride and partially dissolved in 8 *M* urea solution. The renaturation of the fusion protein (around 0.6 mg/ml in guanidine hydrochloride) can be achieved by the dilution of 6 *M* guanidine hydrochloride 60-fold or more with either pH 8.2 Tris buffer or pH 6.8 HEPES buffer containing CaM. When incubated with CaM (0.5 mg/ml), 1 m*M* $Ca^{2+}$, and 600 m*M* glycine on ice for 24 hr around 10–15% molecular specific activity of holoenzyme of CAT–γ is found (C.-Y. Huang and D. J. Graves, unpublished result, 1990). Two to 3% of the molecular specific activity of holoenzyme of CAT–γ can also be observed when the high concentration of denaturant, guanidine or urea, is simply diluted with pH 8.2 or 6.8 buffer. The CAT–γ can be further purified by using reversed-phase HPLC or a CaM-Sepharose column as essentially described by Kee and Graves.[10]

## Conclusion

The general applicability of these methods is quite promising. Reversed-phase HPLC is used to separate the subunits of rat liver branched chain α-ketoacid dehydrogenase 2-oxoisovalerate dehydrogenase,[19] and

[18] K. Nagai and H. C. Thogersen, *in* "Methods in Enzymology" (R. Wu and L. Grossman, eds.), Vol. 153, p. 461. Academic Press, San Diego, California, 1987.

[19] B. Zhang, M. J. Kuntz, G. W. Goodwin, R. A. Harris, and D. W. Crabb, *J. Biol. Chem.* **262,** 15220 (1987).

CaM treatment results in reactivation of calcineurin after immobilization on nitrocellulose.[20] Other proteins which strongly interact with other protein subunits might be renatured by dilution of the denaturant in the presence of these subunits. Alternative methods have been designed by Carlson's laboratory (U. of Mississippi Medical Center, Jackson, Mississippi) for renaturation. In one case, a high concentration of denaturant, 8 *M* urea, is used for the separation of the subunits of phosphorylase kinase[21] by gel filtration, and subsequent dilution of the denaturant achieved enzyme activity of the $\gamma$ subunit. A mixture of native $\alpha/\beta$ subunits also can be obtained by this method.

## Acknowledgment

This work was supported by the National Institutes of Health (Grant No. GM-09587).

[20] M. J. Hubbard and C. B. Klee, *J. Biol. Chem.* **262,** 15062 (1987).
[21] H. K. Paudel and G. M. Carlson, *J. Biol. Chem.* **262,** 11912 (1987).

# Section V

# Antibodies against Protein Kinases

# [37] Preparation and Use of Protein Kinase C Subspecies-Specific Anti-peptide Antibodies for Immunostaining[1]

*By* AKIRA KISHIMOTO, NAOAKI SAITO, and KOUJI OGITA

Protein kinase C (PKC) is a multifunctional protein-serine/threonine kinase and generally accepted to be involved in a wide variety of cellular signal transductions.[2,3] Biochemical fractionation of PKC as well as sequence analysis of its cDNA clones have revealed the existence of multiple subspecies of this enzyme with obvious tissue-specific expression. Figure 1 shows schematically the structure of the PKC subspecies identified to date. They are all composed of a single polypeptide, with the $\alpha$, $\beta$I, $\beta$II, and $\gamma$ subspecies each having four conserved ($C_1$–$C_4$) and five variable ($V_1$–$V_5$) regions.[4] Although the second group of $\delta$, $\varepsilon$, and $\zeta$ subspecies lack the region $C_2$, the molecular map of the enzyme molecule is, nevertheless, similar.[5] The conserved region $C_1$ contains a cysteine-rich sequence resembling the consensus sequence of a so-called "zinc finger." This region seems to be essential for phorbol ester binding.[6] Thus, the N-terminal half of the enzyme molecule is a regulatory domain. The C-terminal half, containing the regions $C_3$ and $C_4$, is the protein kinase domain. The conserved region $C_3$ has an ATP-binding sequence, Gly-X-Gly-X-X-Gly---Lys. The regulatory and protein kinase domains are cleaved by limited proteolysis catalyzed by the $Ca^{2+}$-dependent neutral protease, calpain, at

[1] The authors are grateful for the critical reading of the manuscript by, and the discussion with, Drs. Yasutomi Nishizuka and Chikako Tanaka. The skillful technical and secretarial assistance of Y. Yamaguchi, S. Nishiyama, and Y. Kimura is appreciated. This investigation was supported in part by research grants from the Special Research Fund of the Ministry of Education, Science and Culture, Japan; the Science and Technology Agency, Japan; and Sankyo Foundation of Life Science.

[2] Y. Nishizuka, *Nature (London)* **334,** 661 (1988).

[3] U. Kikkawa, A. Kishimoto, and Y. Nishizuka, *Annu. Rev. Biochem.* **58,** 31 (1989).

[4] U. Kikkawa, K. Ogita, Y. Ono, Y. Asaoka, M. S. Shearman, T. Fujii, K. Ase, K. Sekiguchi, K. Igarashi, and Y. Nishizuka, *FEBS Lett.* **223,** 212 (1987).

[5] Y. Ono, T. Fujii, K. Ogita, U. Kikkawa, K. Igarashi, and Y. Nishizuka, *FEBS Lett.* **226,** 125 (1987).

[6] Y. Ono, T. Fujii, K. Igarashi, T. Kuno, C. Tanaka, U. Kikkawa, and Y. Nishizuka, *Proc. Natl. Acad. Sci. U.S.A.* **86,** 4868 (1989).

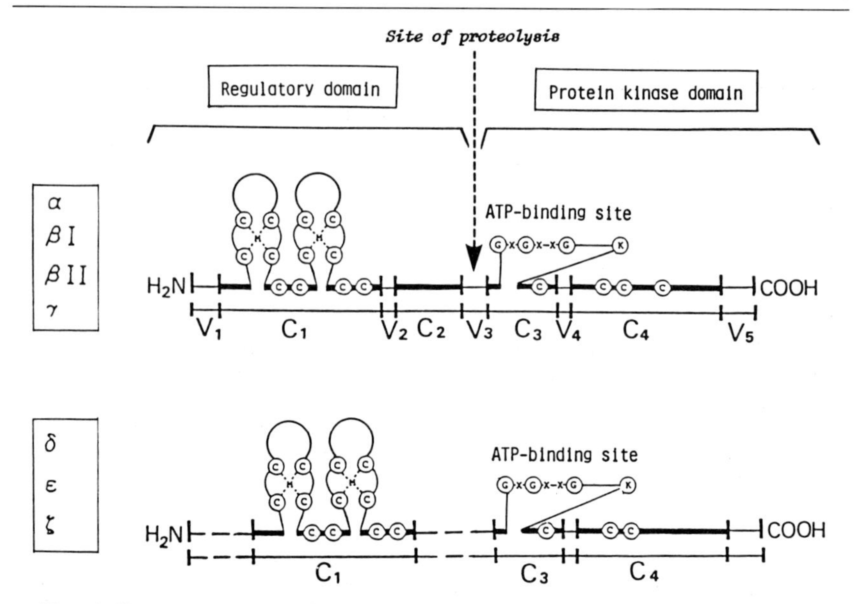


FIG. 1. Common structure of PKC subspecies. C, G, K, X, and M represent cysteine, glycine, lysine, any amino acid, and metal, respectively. Four conserved ($C_1$–$C_4$) and five variable ($V_1$–$V_5$) regions of the larger group are indicated. Other details are described in the text.

one or two specific sites in the variable region $V_3$.[7] Since these specific structures are conserved among various mammals, such as bovine, rat, rabbit, and human, antibodies prepared against the sequence-specific oligopeptides can be employed for immunochemical studies.[8–11] The present chapter will briefly describe the preparation of PKC subspecies-specific anti-peptide antibodies and their application to histochemical studies.

[7] A. Kishimoto, K. Mikawa, K. Hashimoto, I. Yasuda, S. Tanaka, M. Tominaga, T. Kuroda, and Y. Nishizuka, *J. Biol. Chem.* **264,** 4088 (1989).

[8] K. Ase, N. Saito, M. S. Shearman, U. Kikkawa, Y. Ono, K. Igarashi, C. Tanaka, and Y. Nishizuka, *J. Neurosci.* **8,** 3850 (1988).

[9] K. Hosoda, N. Saito, A. Kose, A. Ito, T. Tsujino, K. Ogita, U. Kikkawa, Y. Ono, K. Igarashi, Y. Nishizuka, and C. Tanaka, *Proc. Natl. Acad. Sci. U.S.A.* **86,** 1393 (1989).

[10] N. Saito, A. Kose, A. Ito, K. Hosoda, M. Mori, M. Hirata, K. Ogita, U. Kikkawa, Y. Ono, K. Igarashi, Y. Nishizuka, and C. Tanaka, *Proc. Natl. Acad. Sci. U.S.A.* **86,** 3409 (1989).

[11] A. Ito, N. Saito, M. Hirata, A. Kose, T. Tsujino, C. Yoshihara, K. Ogita, A. Kishimoto, Y. Nishizuka, and C. Tanaka, *Proc. Natl. Acad. Sci. U.S.A.* **87,** 3195 (1990).

TABLE I
RABBIT POLYCLONAL ANTIBODIES AGAINST PKC SUBSPECIES

| Antibodies | Peptide used as antigen: Sequence[a] | Region | Specificity |
|---|---|---|---|
| CKpC$_1\beta$-a | FARKGALRQKNVHEVKNHKF | $\beta:C_1$ | Common |
| CKpV$_3\alpha$-a | LGPAGNKVISPSEDR | $\alpha:V_3$ | $\alpha$ |
| CKpV$_5\alpha$-a | QFVHPILQSAV | $\alpha:V_5$ | $\alpha$ |
| CKpV$_1\beta$-a | PAAGPPPSEGEESTVR | $\beta:V_1$ | $\beta$I and $\beta$II |
| CKpV$_5\beta$I-a[b] | SYTNPEFVINV | $\beta$I : $V_5$ | $\beta$I |
| CKpV$_5\beta$II-a[b] | SFVNSEFLKPEVKS | $\beta$II : $V_5$ | $\beta$II |
| CKpV$_3\gamma$-a | SPIPSPSPSPTDSK | $\gamma:V_3$ | $\gamma$ |
| CKpV$_5\gamma$-a | DARSPTSPVPVPV | $\gamma:V_5$ | $\gamma$ |

[a] Amino acid sequences are represented using one-letter abbreviations. A, Alanine; D, aspartic acid; E, glutamic acid; F, phenylalanine; G, glycine; H, histidine; I, isoleucine; K, lysine; L, leucine; N, asparagine; P, proline; Q, glutamine; R, arginine; S, serine; T, threonine; V, valine.

[b] The sequences of the peptides used contain seven amino acids of the C-terminal end of the $C_4$ region and share some similarities with other PKC subspecies. Thus, lower concentrations of antibodies (less than 0.5 $\mu$g/ml) should be used for immunoblotting and histochemical analyses to avoid the cross-reactivity to other PKC subspecies. The antibodies employed in the previous studies[8-10] were generated using these peptides coupled to bovine thyroglobulin with glutaraldehyde or 1-ethyl-3-(3-dimethylaminopropyl)carbodiimide hydrochloride.

## Preparation of Anti-Peptide Antibodies

### *Principle*

Variable regions of PKC subspecies, mainly C-terminal $V_5$ region and hinge $V_3$ region, are chosen to be antigenic epitopes because of their specific structures (Table I). The peptides are chemically synthesized and coupled to keyhole limpet hemocyanin (KLH) or bovine serum albumin (BSA) using *m*-maleimidobenzoic acid *N*-hydroxysuccinimide ester (MBS)[12] or glutaraldehyde.[13] Then these complexes are employed for the immunization of rabbit with Freund's complete or incomplete adjuvant. The method using MBS is a modification of a technique from the laboratory of M. S. Brown and J. L. Goldstein.[14]

[12] F.-T. Liu, M. Zinnecker, T. Hamaoka, and D. H. Katz, *Biochemistry* **18,** 690 (1979).

[13] M. Reichlin, this series, Vol. 70, p. 159.

[14] J. Sambrook, E. F. Fritsch, and T. Maniatis, *in* "Molecular Cloning, a Laboratory Manual" (2nd ed.), Vol. 3, p. 18.7. Cold Spring Harbor Laboratory, Cold Spring Harbor, 1989.

*Reagents*

MBS (Sigma, St. Louis, MO; Cat. No. M 8759), 30 mg/ml in dimethyl sulfoxide (DMSO)
Glutaraldehyde (Sigma; grade I, Cat. No. G 5882), 3% (v/v) in 0.1 *M* phosphate buffer (pH 7.0)
KLH (Calbiochem, San Diego, CA; Cat. No. 374817), 10 mg/ml in phosphate-buffered salts solution (PBS) (Flow Laboratories, McLean, VA; Cat. No. 28-103-05)
BSA (Sigma; fraction V, Cat. No. A 3350)
Freund's complete adjuvant (Difco, Detroit, MI; Cat. No. 0638-59)
Freund's incomplete adjuvant (Difco, Detroit, MI; Cat. No. 0639-59)
Goat anti-rabbit IgG (ICN, Costa Mesa, CA; Cat. No. 270336)

*Oligopeptides*

The oligopeptides, corresponding to the sequence of each PKC subspecies,[4] are synthesized with a solid-phase peptide synthesizer (Applied Biosystems, Foster City, CA; model 430A) under the manufacturer's standard procedure using *t*-Boc (*t*-butoxycarbonyl)amino acid. Purification is initially made by gel filtration on a Sephadex G-15 column (0.7 × 50 cm), eluting with 2 *M* acetic acid. After lyophilization, the peptides are taken up with 2 ml of 0.1% (v/v) trifluoroacetic acid, and further purified by high-performance liquid chromatography (HPLC) using $C_8$ or $C_{18}$ reversed-phase column chromatography.

*Columns for Purification of Antibodies*

Either goat anti-rabbit IgG (20 mg) or the oligopeptide used for immunization, which is conjugated to a protein other than that used for preparation of antigen (equivalent to 20 mg of protein), is coupled to 2 g of CNBr-activated Sepharose CL-4B (Pharmacia LKB Biotechnology, Piscataway, NJ; Cat. No. 17-0430-01) according to the manufacturer's standard procedure. Resins thus obtained are washed extensively with 0.1 *M* phosphate buffer at pH 8.0, containing 0.02% sodium azide, and stored at 4° until use.

*Preparation of Oligopeptide–Antigens*

To 10 mg KLH in 1 ml PBS, 1.5 mg of MBS in DMSO is added with vigorous vortexing. After stirring for 30 min at room temperature, the solution is gel filtered on a PD-10 column (Pharmacia LKB Biotechnology; Cat. No. 17-0851-01) equilibrated with PBS, then a grayish cloudy void fraction (2 ml) is collected (KLH–MBS solution). The peptide (5 mg) is dissolved in a minimum volume of PBS. If it is not dissolved, 5 *M* NaOH

(5–25 $\mu$l) is added to the suspension until the solution becomes clear. Finally the peptide solution is made up to 500 $\mu$l with PBS and mixed with KLH–MBS solution under vigorous vortexing, then stirred for 1 hr at room temperature. Under the same conditions, the peptides are conjugated with BSA. When glutaraldehyde is used, the peptide (2 mg) and BSA (2 mg) are mixed in 1.5 ml of 0.1 *M* phosphate buffer (pH 7.0) containing 3% (v/v) glutaraldehyde followed by stirring overnight at 4°. Then, the conjugate is dialyzed against a large volume of 0.1 *M* phosphate buffer (pH 7.0) overnight. The peptides conjugated with proteins are stored at −20° until use.

*Preparation of Antibodies*

The conjugate (equivalent to 300 $\mu$g of peptide) is made up to 500 $\mu$l with PBS, emulsified with 500 $\mu$l of Freund's complete adjuvant, and administered to a female New Zealand White rabbit by multiple intracutaneous injections. The same amount of the conjugate emulsified with Freund's incomplete adjuvant is given to the rabbit 2 weeks after the first injection, followed by repeated boosts at intervals of 1 week. The rabbit is bled 4–6 days after each booster administration and the antisera are examined by enzyme-linked immunosorbent assay (ELISA). The titer is increased after the second booster shot and gradually reaches a maximum after the fourth boost. Then, the antisera after the fourth boost are applied to a goat anti-rabbit IgG-conjugated Sepharose CL-4B column (9 × 1 cm) equilibrated with 0.1 *M* phosphate buffer at pH 8.0. After washing the column with 0.1 *M* phosphate buffer at pH. 8.0, the antibody is eluted from the column with 1 *M* acetic acid containing 0.1 *M* glycine, and then dialyzed against 10 m*M* phosphate buffer at pH 7.4 containing 50 m*M* NaCl. When antibodies are subjected to the immunostaining, further purification of antibodies on an oligopeptide-conjugated column (9 × 1 cm) may be suggested to minimize the nonspecific staining. The antibody is eluted from the column as described for goat antibody-conjugated column. The purified antibodies are stored at −80° until use. The antibodies prepared in this study are summarized in Table I. By immunoblotting analysis using these standard materials of PKC subspecies, the antibodies are shown to be specific for the respective enzymes. Typical experimental results obtained with CKpV$_5\alpha$-a are shown in Fig. 2.

## Application of Antibodies to Histochemical Analysis

*Principle*

Various tissues are fixed by an appropriate fixative that maintains morphology of the cell, but does not change the antigenicity of the enzyme.

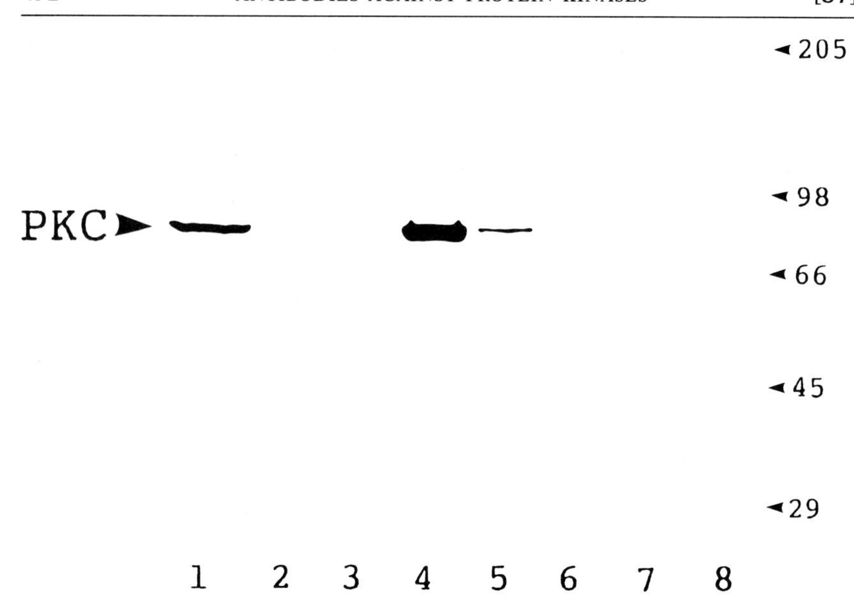


FIG. 2. Immunoblot analysis of the antibody, CKpV$_5\alpha$-a. Samples of PKC subspecies are fractionated by sodium dodecyl sulfate–polyacrylamide gel electrophoresis. After transfer of the proteins to nitrocellulose paper (pore size, 0.22 $\mu$m), the paper is incubated with the antibody, CKpV$_5\alpha$-a (2 $\mu$g/ml), and visualized by the peroxidase–anti-peroxidase method under conditions similar to those for a histochemical analysis (see the section, Application of Antibodies to Histochemical Analysis). Lanes 1–8, crude soluble fraction of rat brain, type I PKC, type II PKC, type III PKC, $\alpha$-PKC, $\beta$I-PKC, $\beta$II-PKC, and $\gamma$-PKC, respectively. The three subtypes of rat brain PKC [type I ($\gamma$), II ($\beta$I + $\beta$II), and III ($\alpha$)] can be purified to homogeneity [Sekiguchi *et al.*, *J. Biochem.* (*Tokyo*) **103,** 759 (1988)]. The four subspecies of PKC ($\alpha$, $\beta$I, $\beta$II, and $\gamma$) can also be obtained from the COS7 cells that are transfected with the cDNAs as described ([18] in this volume). Molecular weights ($\times$ $10^{-3}$) of standard proteins are indicated at right.

When the antibody is raised against oligopeptide conjugated to the carrier protein with glutaraldehyde, Zamboni fixative[15] containing glutaraldehyde is used for fixation. Otherwise periodate–lysine–paraformaldehyde (PLP) fixative[16] is used. Then the sections, which are treated with PKC subspecies-specific antibody, are treated with goat anti-rabbit IgG and rabbit antiperoxidase–peroxidase complex followed by development with diaminobenzidine.[17] For analysis by electron microscopy, the developed sections are further incubated with osmic acid. Then ultrathin sections are cut and

[15] L. Zamboni, *J. Cell Biol.* **35,** 148A (1967).

[16] I. W. McLean and P. K. Nakane, *J. Histochem. Cytochem.* **22,** 1077 (1974).

[17] L. A. Sternberger, P. H. Hardy, Jr., J. J. Cuculis, and H. G. Meyer, *J. Histochem. Cytochem.* **18,** 315 (1970).

stained with uranyl acetate. The method described below is for the analysis of rat central nervous system.

*Reagents*

Zamboni fixative containing glutaraldehyde: 0.5–1% (w/v) glutaraldehyde, 4% (w/v) paraformaldehyde, and 0.2% (w/v) picric acid in 0.1 *M* phosphate buffer (pH 7.4)
PLP fixative: 0.01 *M* sodium *m*-periodate, 0.075 *M* lysine, and 4% (w/v) paraformaldehyde in 0.0375 *M* phosphate buffer (pH 7.4)
TPBS: 0.3% Triton X-100 in PBS
Normal goat serum (Rockland, Gilbertsville, PA): 0.3% (v/v) in either PBS or TPBS
Goat anti-rabbit IgG: 10 $\mu$g/ml in either PBS or TPBS
Rabbit anti-peroxidase/peroxidase complex (ICN; Cat. No. 280231): 1 : 1000 diluted in either PBS or TPBS
3,3′-Diaminobenzidine hydrochloride (Sigma; Cat. No. D 5637): 0.2 mg/ml in 50 m*M* Tris/HCl (pH 7.4)
Entellan (Merck, Darmstadt, Germany; Cat. No. 1866)
Osmium tetroxide (Merck; Cat. No. 24505), 1% (v/v) in PBS
Quetol 812 (Nissin EM, Tokyo, Japan; Cat. No. 340)
Uranyl acetate (Wako Chemical, Osaka, Japan; Cat. No. 219-00692), 1% (w/v) in ethanol [50% (v/v) in $H_2O$]

*Preparation of Brain Sections for Light Microscopy*

Anesthetized rat is perfused from the left ventricle with 0.9% (w/v) NaCl followed by 200 ml of an appropriate fixative at the rate of 15 ml/min. All procedures described below are carried out at 4° unless otherwise indicated. The brain is dissected, immersed in the fixative for 48 hr, washed with 30% sucrose (w/v) in 0.1 *M* phosphate buffer (pH 7.4) for an additional 48 hr, and then cut on a cryostat into frontal sections (20 $\mu$m) which are dipped directly in TPBS. Subsequently, the sections are treated with TPBS for at least 4 days followed by the incubation with 0.3% $H_2O_2$ in TPBS to inhibit endogenous peroxidase activity. After blocking the nonspecific binding sites with 0.3% (v/v) normal goat serum, the sections are incubated with the primary antibody (0.5–5 $\mu$g/ml in TPBS) for 18–36 hr, washed with TPBS, incubated with goat anti-rabbit IgG (10 $\mu$g/ml in TPBS) for 4 hr, washed with TPBS, and incubated with rabbit anti-peroxidase/peroxidase complex solution in TPBS for 90 min. After rinsing three times with TPBS, the sections are developed with 0.2 mg/ml 3,3′-diaminobenzidine hydrochloride containing 0.005% (v/v) $H_2O_2$. For control staining, the primary PKC-specific antibody is routinely replaced by nonimmune rabbit serum. Otherwise, the PKC-specific antibody is first mixed with the oligopeptide used for immunization, which is conjugated to the protein other than that used for preparation of antigen, then employed for the control staining. The sections are, then, dehydrated and mounted in Entellan for light microscopic observation.

### *Preparation of Sections for Electron Microscopy*

Rat brain is fixed and washed with 30% sucrose as described above. The brain is immediately frozen in liquid nitrogen followed by thawing in 30% sucrose. Then, 40-$\mu$m thick frontal sections are cut on a Vibratome (Oxford, Bedford, MA). The sections are treated with the antibody followed by the conjugation of peroxidase and the development as described above except that PBS is used instead of TPBS. Finally, the developed sections are postfixed for 1 hr in 1% osmium tetroxide in PBS, dehydrated in graded alcohol, and then flat embedded on siliconized slides in resin (Quetol 812). After polymerization at 60° for 48 hr, the sections are observed under a photomicroscope and the selected areas are cut off and attached to Epon supports for further sectioning on a Reichert Jung (Wien, Austria) Ultracut E ultramicrotome. Ultrathin sections (60–90 nm thick) are collected on 200-mesh bare grids (Maxtaform, Kent, England, HR24), stained with 1% uranyl acetate in 50% ethanol, and examined with a JEM100SX (Nihon Denshi, Tokyo, Japan) electron microscope at 80 kV.

# [38] Preparation, Characterization, and Use of Isozyme-Specific Anti-protein Kinase C Antibodies

*By* FREESIA L. HUANG, YASUYOSHI YOSHIDA, and KUO-PING HUANG

Multiple forms of protein kinase C (PKC) have been implicated in the regulation of a variety of cellular functions.[1,2] This enzyme family is widely distributed in animal tissues. Members of the PKC gene family are homologous; however, they display distinctive biochemical characteristics and have unique cellular and subcellular localizations.[3,4] It has been predicted that each PKC isozyme may be involved in regulating different cellular

[1] Y. Nishizuka, *Nature (London)* **334,** 661 (1988).

[2] K.-P. Huang, *Trends Neurosci.* **12,** 425 (1989).

[3] K.-P. Huang, F. L. Huang, H. Nakabayashi, and Y. Yoshida, *J. Biol. Chem.* **263,** 14839 (1988).

[4] F. L. Huang, Y. Yoshida, H. Nakabayashi, W. S. Young, III, and K.-P. Huang, *J. Neurosci.* **8,** 4734 (1988).

functions. PKC I, II, and III separated from the hydroxylapatite column chromatography account for the majority of the $Ca^{2+}$/phospholipid/diacylglycerol-dependent kinase activity in mammalian brain extracts.[5,6] PKC I, II, and III have been shown to be coded by the $\gamma$, $\beta$, and $\alpha$ genes, respectively,[7,8] whereas other isozymes, encoded by the $\delta$, $\varepsilon$, and $\zeta$ genes, have not been identified. PKC II fractions from hydroxylapatite column contain the kinases derived from two alternatively spliced $\beta$ cDNA subspecies, $\beta$I and $\beta$II.[9] The biochemical characteristics of the PKCs expressed in COS cells transfected with these latter two cDNAs are similar. Investigation of the role of each PKC isozyme in the regulation of cellular function has become a focal interest. These studies are greatly facilitated by immunological methods to identify, quantify, and localize these enzymes in mammalian tissues and cell cultures. Polyclonal, monoclonal, and specific peptide antibodies have been prepared from many laboratories. In this chapter we will describe the preparation of PKC isozyme-specific antibodies from a preparation of polyclonal antibodies against a mixture of isozymes. These antibodies have been shown to be specific and fairly versatile for use in immunoprecipitation, immunoblotting, and immunostaining.

## Materials

Affinity-purified rabbit anti-goat IgG and biotin-conjugated rabbit anti-goat IgG are from Cappel Biochemical; normal goat and rabbit sera are from Miles; crystallized bovine plasma albumin is from Armour Pharmaceutical (Kankakee, IL); PBS, pH 7.4, is from GIBCO (Grand Island, NY); nitrocellulose membranes are from Schleicher and Schuell (Keene, NH); Pansorbin is from Calbiochem (San Diego, CA); and $^{125}$I-labeled protein A is from Du Pont–New England Nuclear (Boston, MA).

## General Methods

Immunoprecipitation is performed by incubating PKC (50 ng) in buffer A [20 m*M* Tris-HCl buffer, pH 7.5, containing 1 m*M* dithiothreitol (DTT), 0.5 m*M* EDTA, 0.5 m*M* EGTA, and 10% (v/v) glycerol] with antibody in

[5] K.-P. Huang, H. Nakabayashi, and F. L. Huang, *Proc. Natl. Acad. Sci. U.S.A.* **83,** 8535 (1986).

[6] K.-P. Huang and F. L. Huang, *J. Biol. Chem.* **261,** 14781 (1986).

[7] F. L. Huang, Y. Yoshida, H. Nakabayashi, J. L. Knopf, W. S. Young, III, and K.-P. Huang, *Biochem. Biophys. Res. Commun.* **149,** 946 (1987).

[8] U. Kikkawa, Y. Ono, K. Ogita, T. Fujii, Y. Asaoka, K. Sekiguchi, K. Ase, K. Igarashi, and Y. Nishizuka, *FEBS Lett.* **217,** 227 (1987).

[9] Y. Ono, U. Kikkawa, K. Ogita, T. Fujii, T. Kurukawa, Y. Asaoka, K. Sekiguchi, K. Ase, K. Igarashi, and Y. Nishizuka, *Science* **236,** 1116 (1987).

a total volume of 30 μl. After incubation at room temperature for 30 min, 25 μl of 35% Pansorbin suspended in buffer A is added and further incubated at 4° for 30 min. The antigen–antibody complex, which is bound to Pansorbin, is removed by centrifugation at 0°. Aliquots of the resulting supernatant fluid are taken for the measurement of PKC activity.

Immunoblot analysis is carried out according to the method of Towbin *et al.*[10] using reagents from Bio-Rad (Richmond, CA). Samples are separated by SDS–polyacrylamide gel (10%) electrophoresis followed by electrophoretic transfer to nitrocellulose membrane in 4 liters of solution containing 12 g of Tris, 57.6 g of glycine, and 800 ml of methanol (pH 8.3) at a current setting of 0.4 A for 16 hr at 4°. The solution is changed after four or five transfers. The membrane is incubated successively with 3% (w/v) gelatin in TBS (20 m*M* Tris-HCl buffer, pH 7.5, containing 500 m*M* NaCl) for 30 min to block nonspecific binding sites, antibodies in TBS containing 1% gelatin, affinity purified rabbit anti-goat IgG (2000-fold dilution) in TBS containing 1% gelatin for 1 hr, and $^{125}$I-labeled protein A (0.3–0.5 μg containing 2–5 μCi) in TBS containing 1% gelatin for 1 hr. In between incubations, the membrane is washed with TTBS [TBS containing 0.05% (w/v) Tween 20] twice for 10 min each. At the end the membrane is washed with TTBS four times and at least once with water. The membrane is then air dried or dried under a heating lamp and wrapped with a piece of cellophane with Whatman (Clifton, NJ) No. 1 filter paper as a support. The immunoreactive bands are detected by autoradiography. The positions of the molecular weight marker proteins are localized by brief staining with 0.1% (w/v) Amido Black in 25% (w/v) 2-propanol containing 10% (w/v) acetic acid and destained in the same solution without dye.

Indirect immunofluorescence staining of the rat brain slices is used to locate the PKC isozymes in the various neurons. Frozen brains of adult Sprague-Dawley rats previously perfused with cold 4% (w/v) paraformaldehyde in 120 m*M* phosphate buffer, pH 7.4 are used. Serial brain sections (12 μm thick) at various levels are warmed to room temperature and thoroughly rehydrated with PBS. The slices are incubated with goat anti-rat brain PKC antibodies at 4° for 40 hr, PBS two times for 10 min each, biotin-conjugated rabbit anti-goat IgG for 1 hr, PBS two times for 10 min each, Texas Red-conjugated avidin (7 μg/ml avidin, Texas Red/avidin = 3) at room temperature for 1 hr, PBS four times for 10 min each. The stained sections are rinsed with deionized water, air dried, and mounted with buffered poly(vinyl alcohol) for fluorescent microscopy and photography. All the antibodies are diluted in PBS containing 2 mg/ml bovine plasma albumin.

[10] H. Towbin, T. Staehelin, and J. Gordon, *Proc. Natl. Acad. Sci. U.S.A.* **76,** 4350 (1979).

*Preparation of Polyclonal Antibody*

Immunization of a female goat was performed by subcutaneous injection of 2 mg of purified unfractionated PKC (obtained from polylysine column chromatography) in Freund's complete adjuvant at a site near the neck. Booster injections were carried out at 3-week intervals using 1 mg of PKC, also in complete adjuvant. After the third booster injection, test bleeds were made to determine the generation of antibody by immunoprecipitation. Antibodies of high titer were obtained 2 weeks after the third booster injection. A large preparation of goat plasma (1 liter/week) was allowed to coagulate at 4° overnight following the addition of 7 m*M* of $CaCl_2$. The antiserum was obtained by filtering through four layers of cheesecloth. The PKC isozyme-specific antibodies are subsequently purified from it by immunoabsorption. Each purified PKC isozyme is repetitively applied to a 10% (w/v) SDS gel at 20-min intervals with 0.2–0.5 $\mu$g of PKC/slot in a 15-slot gel four or five times. After electrophoresis (4.5 hr after first application), the enzyme is transferred to nitrocellulose sheets and the area containing PKC, as determined by staining of a portion of the side lane, is excised in a strip across the membrane. The membrane strip is cut into three segments, each containing five original slots. A collection of nitrocellulose pieces from two gels totaling 30–60 $\mu$g (based on the initial application) of each PKC isozyme is used for the absorption of the corresponding antibody from the aforementioned polyclonal antibody that recognizes all three isozymes. These membranes are incubated successively with 3% gelatin for 30 min, TBS for 10 min twice, goat anti-rat brain PKC antibody (partially purified by precipitation with 50% ammonium sulfate and extensively dialyzed against buffer A) for 30 min, TTBS for 10 min three times, and TBS for 5 min once. The specific antibody is eluted from the membrane with 0.1 *M* acetic acid, pH 2.8 and neutralized with 2 *M* Tris. The membrane can be reused for several times without significant loss of the enzyme from the membrane. The pooled antibody is concentrated and dialyzed against buffer A in an Amicon (Danvers, MA) ultrafiltration cell fitted with YM10 membrane. The concentrated antibodies are stored at −70° after addition of bovine serum albumin to 1 mg/ml.

*Characterization of PKC Isozyme-Specific Antibodies*

The specificity of the purified antibodies is tested by immunoblotting with the isozyme fractions eluted from the hydroxylapatite column and by the cross-reactivity with the cell extracts from COS cells transfected with $\alpha$, $\beta$, and $\gamma$ cDNAs. The polyclonal antibody cross-reacted with all the PKC fractions eluted from the hydroxylapatite column (Fig. 1D). Purified PKC I (Fig. 1A)-, II (Fig. 1B)-, and III (Fig. 1C)-specific antibodies recog-

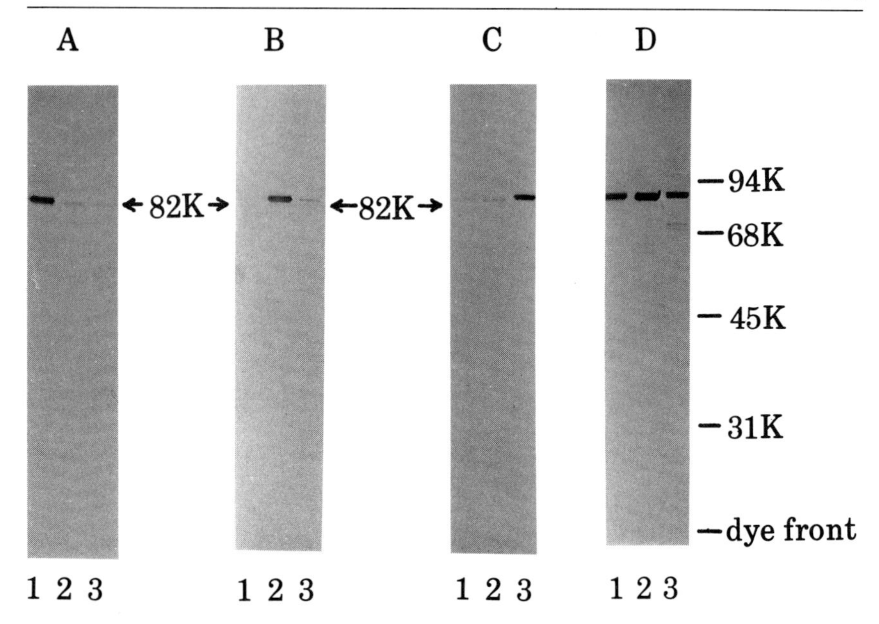


FIG. 1. Characterization of monospecific antibodies against PKC I, II, and III. Approximately 0.1 μg each of the purified PKC I (lane 1), II (lane 2), and III (lane 3) was electrophoretically transferred to nitrocellulose membrane after SDS–PAGE. Each piece of membrane was incubated with monospecific antibodies against PKC I (A), PKC II (B), and PKC III (C), or unfractionated antibodies (D). The immunoreactive bands were detected by 4-chloro-1-naphthol after incubation with horseradish peroxidase-conjugated rabbit anti-goat IgG.

nized those fractions eluted at peak 1, 2, and 3 from the hydroxylapatite column, respectively. PKC I-, II-, and III-specific antibodies also recognized the PKCs transiently expressed in the COS cell transfected with $\gamma$, $\beta$, and $\alpha$ cDNA, respectively.[7] The PKC II-specific antibody cross-reacted with both of the enzymes expressed in COS cells transfected with $\beta$I and $\beta$II cDNAs.[11] Occasionally, the purified antibodies exhibited minor cross-reactivity between different isozymes. These minor contaminants can be eliminated by incubation of the antibody with the nitrocellulose membrane containing the specific contaminant.

### *Application of Isozyme-Specific Antibodies for Quantification and Immunostaining*

Identification of a PKC isozyme in a tissue or cell type is greatly facilitated by the use of the isozyme-specific antibodies. These antibodies can also be used for the quantification of each isozyme in a sample by

[11] K.-P. Huang, F. L. Huang, H. Nakabayashi, and Y. Yoshida, *Acta Endocrinol. (Copenhagen)* **121,** 307 (1989).

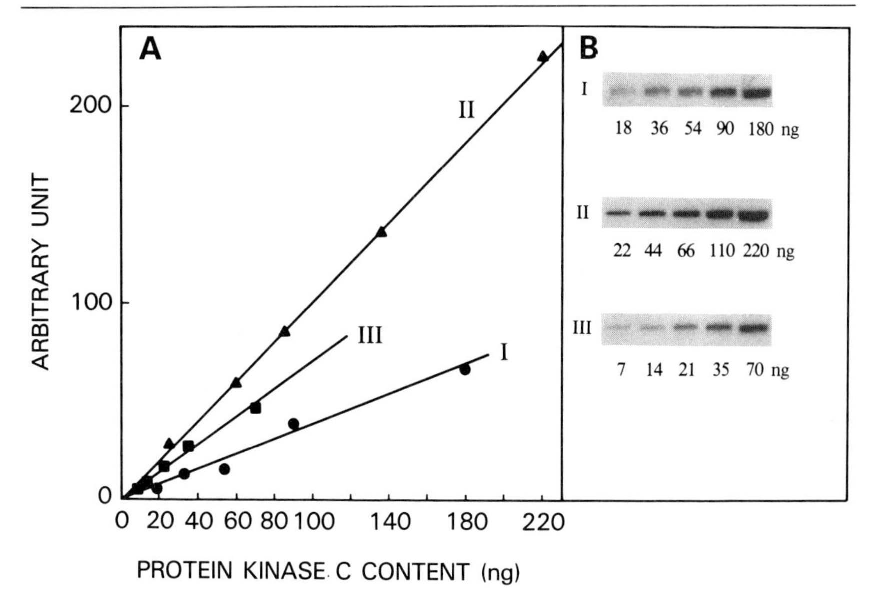


FIG. 2. Quantification of PKC isozymes by immunoblotting. PKC isozymes (20–200 ng) were separated by SDS–PAGE (10% gel) and electrophoretically transferred to nitrocellulose membranes. Each sheet of the membrane was incubated with PKC I-, II-, and III-specific antibodies and $^{125}$I-labeled protein A. The immunoreactive signals were determined by densitometric tracing of the autoradiograms using a scanner equipped with an integrator. (A) The optical densities of the immunoreactive bands as a function of the enzyme contents for PKC I (●), II (▲), and III (■). (B) The autoradiograms containing different quantities of the three isozymes.

using $^{125}$I-labeled protein A binding to the immune complex.[12] A standard curve is constructed by applying known amounts of purified PKC isozymes (0.01–0.2 $\mu$g) to an SDS gel, followed by electrophoretic transfer to nitrocellulose membrane and immunoblotting. The immunoreactive signal determined by densitometric tracings of the autoradiograms is proportional to the amount of PKC applied (Fig. 2). For the determination of the unknown samples, at least three different amounts of the purified enzymes along with the unknown samples are included in the same gel to serve as an internal standard. Exposure of the autoradiogram must be carefully controlled so that a linear relationship for the standard is ensured. For the accuracy of the estimation of the unknown it is essential to apply at least three different amounts of the unknown samples.

These isozyme-specific antibodies have also been successfully used

[12] Y. Yoshida, F. L. Huang, H. Nakabayashi, and K.-P. Huang, *J. Biol. Chem.* **263,** 9868 (1988).

for the cellular localization of these enzymes in rat brain.[4,7,13] Immunostaining of the brain slices with the isozyme-specific antibodies has yielded many interesting findings. For example, indirect immunofluorescent staining of rat cerebellum with PKC I-specific antibody reveals the presence of this enzyme in Purkinje cells (Fig. 3). Examination at a higher magnification shows that both the cell bodies and dendrites of the Purkinje cells are positively stained. Staining with the PKC II-specific antibody revealed the presence of this enzyme in the cell bodies of the granule cells. Staining of the cerebellum with the PKC III-specific antibody shows that both the Purkinje cells and granule cells contain PKC III. Even though both PKC I- and PKC III-specific antibodies stain Purkinje cells, it seems that only the PKC I-specific antibody stains the dendrites of this cell type. Another interesting feature of the localization of the PKC isozymes in the various cell types is the coexistence of multiple forms in a single cell. This is most prominent in the rat hippocampal pyramidal cells, in which all three PKCs are present in high levels. Table I summarizes the presence of the various PKC isozymes in a variety of cell types.

## Comments on Procedures

Polyclonal antibodies for a variety of proteins have been successfully prepared by immunization of the rabbit. However, we were unable to generate antibodies of good titer against rat brain PKCs by using rabbit. The failure may be due to the close relationship between rat and rabbit PKCs. The antibodies we raised in goat appear to have a high potency against all these enzymes. Approximately 10 $\mu$l of the antiserum is sufficient to immunoprecipitate completely 0.3–0.5 $\mu$g of each isozyme. For immunoblotting, a 2000-fold diluted antiserum can detect as low as 10 ng of each isozyme. Analysis of the titration curves of each isozyme by immunoprecipitation with the antiserum reveals that the antibody has different potency against each isozyme.[5] We found that the antibody titer against PKC I or II was higher than that against PKC III in spite of the nearly equivalent amount of each isozyme used in the immunization. We noticed that the titer of the antiserum against PKC I or II was also higher than that against PKC III by immunization of mice. It appears that PKC III is the least antigenic, perhaps because of its widespread presence in the different tissues and its extensively conserved structure during evolution.

The method used for the preparation of isozyme-specific antibodies is

[13] F. L. Huang, W. S. Young, III, Y. Yoshida, and K.-P. Huang, *Dev. Brain Res.* **52,** 121 (1990).

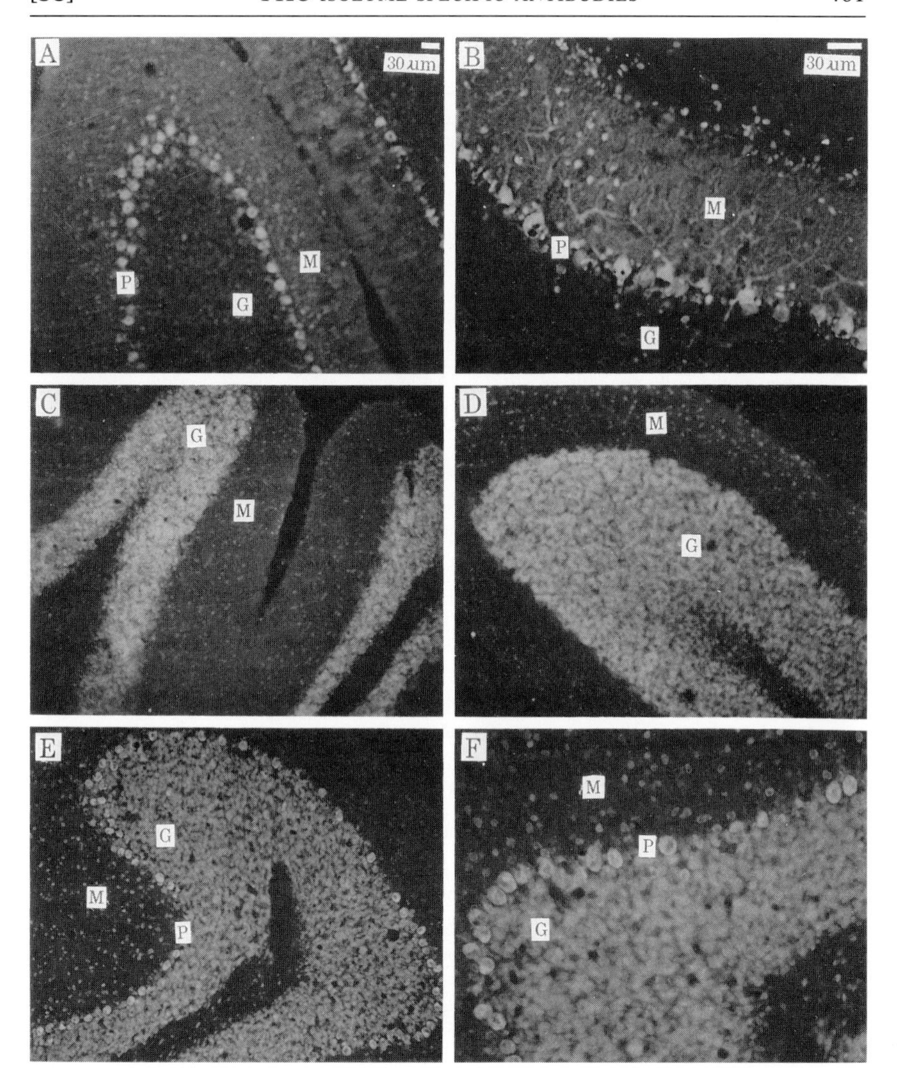


FIG. 3. Indirect immunofluorescent staining of rat cerebellum. Frozen cerebellum sections were stained with monospecific antibodies against PKC I (A and B), II (C and D), and III (E and F). P, Purkinje cells; M, molecular layer; G, granular layer. Bar in (A) is also for (C) and (E), and that in (B) is also for (D) and (F). Control experiments with normal goat serum or antibodies preabsorbed with corresponding PKC isozymes did not result in a positive staining (data not shown).

TABLE I
IMMUNOCHEMICAL IDENTIFICATION OF PKC ISOZYMES IN DIFFERENT CELL TYPES[a]

| Source | PKC I ($\gamma$) | PKC II ($\beta$I + II) | PKC III ($\alpha$) |
|---|---|---|---|
| Purkinje cell (rat cerebellum) | + | | + |
| Granule cell (rat cerebellum) | | + | + |
| Pyramidal cell (rat hippocampus) | + | + | + |
| Granule cell (rat dentate gyrus) | + | + | + |
| Pyramidal cell (rat cerebral cortex) | + | + | + |
| EL-4 mouse thymoma cell | | + | + |
| HL-60 human promyelocytic leukemia cell | | + | + |
| 2H3 rat basophilic leukemia cell | | + | + |
| 3T3-L1 fibroblast | | | + |
| WEHI-231 murine B cell lymphoma | | + | + |
| Granulosa cell (rat ovary) | | + | |
| PC-12 pheochromocytoma | | + | + |

[a] The presence of various PKC isozymes in rat brain was identified by immunofluorescent staining of the brain sections and that of the other cell types was determined by immunoblotting with 100 $\mu$g protein of each cellular extract. The positive signal in the immunoblot analysis represents a PKC content greater than 10 ng equivalent of homogeneous rat brain PKC isozymes.

a modification of that originally developed by Olmsted,[14] who used the derivatized membrane to form covalent linkage between protein and the membrane. We noted that the unmodified nitrocellulose membrane used for routine immunoblotting is quite satisfactory in this application. The loss in the membrane-bound enzymes during the elution cycle appears to be minimal and the membrane can be reused for many times. In spite of the fact that PKC I, II, and III are highly homologous, the purified antibodies do not extensively cross-react. It is likely that the epitopes of these isozymes recognized by the isozyme-specific antibodies are located at the variable sequence regions. Based on the tests of the immunoreactivities of these isozymes during limited proteolysis, we found that these antibodies recognized both the regulatory and catalytic domains of the enzymes. It seems that these antibodies recognize the PKC molecules at multiple sites and, therefore, are excellent reagents for the detection of both the intact and degraded forms of the enzymes.

Purification of the isozyme-specific antibodies from a preparation of polyclonal antibodies against a mixture of isozymes bypasses the expenses

[14] J. B. Olmsted, *J. Biol. Chem.* **256,** 11955 (1981).

required for the preparation of monoclonal antibodies. The starting material, goat immune serum, is plentiful and only the purified isozymes are required for the isolation of the specific antibodies. We found that these purified antibodies have a broader range of applications than the monoclonal antibodies we prepared.

# [39] Generation and Use of Anti-peptide Antibodies Directed against Catalytic Domain of Protein Kinases

*By* GAIL M. CLINTON and NANCY A. BROWN

The molecular architecture of the protein kinases includes regulatory regions which are diverse in their amino acid sequence and a catalytic domain which is composed of sequences that are conserved because they are required for catalysis. The 250 to 300 amino acid residues of the catalytic domain can be divided into highly conserved subdomains that are interspersed with relatively divergent sequences.[1] The highly conserved subdomains contain amino acid residues which are likely to participate directly in catalysis or to provide structural constraints that are necessary to form the active site pocket. Although it has been shown that an invariant lysine residue in the catalytic domain is involved in ATP binding[2-4] (Lys-72 in the cAMP-dependent protein kinase[1]) the functions of amino acid residues in other conserved subdomains are unknown. It remains to be determined which sequences within the catalytic domain form the active site pocket, which may be blocked or buried by regulatory sequences and thus are inaccessible to substrates, and which subdomains may be altered in conformation by binding of ligands and other effectors.

Antibodies against selected amino acid sequences in proteins have been found to be valuable reagents for examining the location of these sequences within a native polypeptide. Immunoprecipitation with anti-peptide antibodies has been used to determine whether the cognate peptide sequence is exposed or buried, and whether its conformation changes following binding of an effector. Ligand-activated autophosphorylation of the platelet-derived growth factor (PDGF) receptor results in enhanced accessibility of sites in the C-terminal autophosphorylation domain to

[1] S. K. Hanks, A. M. Quinn, and T. Hunter, *Science* **241,** 42 (1988).
[2] M. J. Zoller, N. C. Nelson, and S. S. Taylor, *J. Biol. Chem.* **256,** 10837 (1981).
[3] M. D. Kamps, S. S. Taylor, and B. M. Sefton, *Nature (London)* **310,** 589 (1984).
[4] M. W. Russo, T. J. Lukes, S. Cohen, and J. V. Staros, *J. Biol. Chem.* **260,** 5205 (1985).

binding with anti-peptide antibodies.[5,6] Autophosphorylation of the insulin receptor exposes the previously inaccesible autophosphorylation sites in the catalytic domain and allows immunoprecipitation with anti-peptide antibodies.[7,8] The catalytic domain sequence, HRDLAARN [residues 811–818[8a] in the human epidermal growth factor (EGF) receptor], binds anti-peptide antibody only when the protein structure has been opened by an elevation in temperature or by mild denaturation,[9] indicating that this sequence is buried in the native EGF receptor.

Anti-peptide antibodies have also supplied information on the involvement in catalysis of selected sequences within protein kinases. If the antibody binds to a site directly involved in catalysis, it may block interaction of that site with substrates. Antibody directed against a v-*abl* tyrosine kinase sequence which includes the Gly at the C-terminal end of the nucleotide binding motif, Gly-X-Gly-X-X-Gly, inhibits both autophosphorylation and phosphorylation of exogenous substrates.[10] This inhibition is likely due to interference with ATP binding. The kinase activities of the EGF receptor, the insulin receptor, and the IGF-I receptor are neutralized when site-directed antibodies bind to autophosphorylation sites (corresponding to Tyr-416 in Src) within the catalytic domain.[11] A site near the C terminus of the catalytic domain, which contains an invariant Arg (residue 280 in the cAMP kinase), also appears critical for catalysis since anti-peptide antibodies against a sequence within this site (residues 498–512 in Src) neutralize the kinase activity of Src.[12] Kinase activity is not always neutralized, however, by binding of antibodies to sites in the catalytic domain. For example, the EGF receptor retains autophosphorylation activity when complexed to antibodies against either of two highly conserved sequences with the catalytic domain (residues 852–861 and 871–878).[9] Furthermore, there is no apparent effect on kinase activity when anti-

[5] M. T. Keating, J. A. Escobedo, and L. T. Williams, *J. Biol. Chem.* **263,** 12805 (1988).

[6] S. Bishayee, S. Majumdar, C. D. Scher, and S. Khan, *Mol. Cell. Biol.* **8,** 3696 (1988).

[7] R. Perlman, D. P. Bottaro, M. F. White, and R. Kahn, *J. Biol. Chem.* **264,** 8946 (1989).

[8] R. Herrera and O. M. Rosen, *J. Biol. Chem.* **261,** 1980.

[8a] Single-letter abbreviations are used for amino acids: A, alanine; D, aspartic acid; H, histidine; L, leucine; N, asparagine; R, arginine.

[9] N. Brown, L. A. Compton, and G. M. Clinton, unpublished observations (1990).

[10] J. B. Konopka, R. L. Davis, S. M. Watanabe, A. S. Ponticelli, L. Schiff-Maker, N. Rosenberg, and O. N. Witte, *J. Virol.* **51,** 223 (1984).

[11] T. Izume, S. Yoshiyuki, Y. Akanuma, J. Takaku, and M. Kasuga, *J. Biol. Chem.* **263,** 10386 (1988).

[12] L. E. Gentry, L. R. Rohrschneider, J. E. Casnellie, and E. G. Krebs, *J. Biol. Chem.* **258,** 11219 (1983).

peptide antibodies are complexed to sites in the C-terminal domain of Src,[12] the PDGF receptor,[6] insulin receptor,[11,13] and EGF receptor.[9,11,14]

The interaction of antibodies with specific sites outside of the catalytic domain may be used to examine potential regulatory roles of these sites. When an antibody interacts with one site outside the insulin-binding site in the extracellular domain of the insulin receptor $\beta$ subunit, autophosphorylation and kinase activity are inhibited.[15] Similarly, binding of an anti-peptide antibody to a region in the insulin receptor on the cytoplasmic side of and directly adjacent to the transmembrane domain inhibits kinase activity.[13] Binding of an anti-peptide antibody to the pseudosubstrate sequence in the regulatory domain of protein kinase C has been found to activate this kinase in the absence of calcium and phospholipids, possibly by relieving the inhibitory effects of this sequence.[16]

## Generation of Anti-Peptide Antibodies to Sites in Catalytic Domain

There have been several publications describing the generation and use of anti-peptide antibodies.[17–19] The procedures described here are targeted toward the generation of antibodies against conserved sequences in the catalytic domain of protein kinases and the special considerations that may be required in using these antibodies. The methods presented are synthesized from our experience with anti-peptide antibodies directed against four highly conserved sites within the catalytic domain and two nonconserved regions from outside the catalytic domain of the EGF receptor (EGFR).

### *Choosing Amino Acid Sequences from Kinase Domain for Synthesis of Peptide Antigen*

When selecting the amino acid sequence from the catalytic domain to which the anti-peptide antibody will be directed, it is important to consider whether the epitopes will be collinear with the primary sequence. In the absence of information on the three-dimensional structure of the catalytic

[13] R. Herrera, L. Petruzzelli, N. Thomas, H. N. Bramson, E. T. Kaiser, and O. M. Rosen, *Proc. Natl. Acad. Sci. U.S.A.* **82,** 7899 (1985).

[14] W. J. Gullick, J. Downward, and M. D. Waterfield, *EMBO J.* **4,** 2869 (1985).

[15] R. Gherzi, G. Sesti, G. Andraghetti, R. De Pirro, R. Lauro, L. Adezanti, and R. Condera, *J. Biol. Chem.* **264,** 8627 (1989).

[16] M. Makowske and O. M. Rosen, *J. Biol. Chem.* **264,** 16155 (1989).

[17] R. A. Lerner, *Nature (London)* **299,** 592 (1982).

[18] G. Walter and R. F. Doolittle, *Genet. Eng.* **5,** 61 (1983).

[19] G. Walter, *J. Immunol. Methods* **88,** 149 (1986).

domain, whether the sequence is conserved may be considered. Patterns of amino acid sequence conservation can most easily be assessed by consulting the publication of Hanks, Quinn, and Hunter,[1] in which these authors aligned the sequences within the catalytic domains of 65 different protein kinases. Attention should be directed not only to the 11 conserved subdomains (numbered I through XI) but also to the patterns of conservation within and outside of these subdomains. For example, within subdomain VI, the sequence HRDLRAAN (HRD) for the Src family (residues 384–391) is one of the most highly conserved segments in the tyrosine kinase family and may, therefore, represent a site that is structurally or functionally important. Other considerations when choosing the amino acid sequence, including the length and the chemical properties of the amino acid side chains, have been discussed.[17–20]

### *Conjugation of Synthetic Peptide*

There are several procedures for conjugating the peptide to a carrier protein. We have found that cross-linking synthetic peptides with carbodiimide by the following procedure has yielded antibodies which are reactive with highly conserved sequences in the interior of the EGFR.

1. Dissolve 2.5 mg of the synthetic peptide in 1 ml water. Determine absorbance at 214 nm or at the wavelength of maximum absorbance for the particular peptide.
2. Add 7.5 mg hemocyanin (thyroglobulin works also) in 100 $\mu$l of water containing 0.5 mg/ml cross-linker 1-ethyl-(3-dimethylaminopropyl)carbodiimide hydrochloride (EDC) (Sigma, St. Louis, MO). Vortex and let stand at room temperature for 30 min.
3. Dilute to 3 ml with water and put into dialysis tubing with a pore size that will allow free peptide to diffuse out. Dialyze against 100 ml of water overnight.
4. Determine absorbance of dialysate and calculate the amount of peptide in the dialysate to determine the efficiency of conjugation. Alternatively, spot an aliquot of dialysate onto paper and react with ninhydrin to estimate amount of free peptide. We have found that 40 to 50% of the peptide is usually conjugated.

We have also assessed the immunogenicity of peptide conjugated to hemocyanin using glutaraldehyde as the cross-linking reagent. Glutaraldehyde efficiently conjugated the peptide to the carrier protein and the conjugated peptide generated antibody reactive to the immobilized peptide in an enzyme-linked immunosorbent assay (ELISA). However, we have

[20] T. P. Hopp and K. R. Woods, *Proc. Natl. Acad. Sci. U.S.A.* **78,** 3824 (1981).

found that these antibodies did not immunoprecipitate the EGFR if the cognate sequence was in the interior of the protein. The conformation of the peptide that is cross-linked with glutaraldehyde may be different than the conformation of the same sequence when it is located at an internal site in the protein. Anti-peptide antibodies to sites in the flexible C-terminal domain, on the other hand, bound to the EGFR whether glutaraldehyde or carbodiimide was used to prepare the immunogen.

The other procedure we have evaluated for producing immunogen is the synthesis of a heptalysine core to which peptide monomers were attached according to the method of Tam.[21,22] Peptides attached to the polylysine core have been found to be immunogenic, thereby eliminating the need for a carrier protein. However, because the synthesis of the peptide on the polylysine core was more difficult and there did not appear to be improvement in the titer or in the proportion of immunized rabbits that produced antibodies, there seems little reason to use this procedure.

*Immunization*

New Zealand White rabbits of 2.5–3.0 kg are bled for preimmune sera and then injected subcutaneously with conjugated peptide emulsified at a ratio of 1 to 1.2 with Freund's complete adjuvant for the first and last injections, and Freund's incomplete adjuvant for all other injections. The rabbits are given subcutaneous booster injections at 2-week intervals and weekly test bleeds are begun 2–3 weeks after the initial injection. The amount of peptide used for immunization varies from 50 to 500 $\mu$g/injection. Antibodies of highest titer are produced in animals immunized with 50 to 100 $\mu$g peptide conjugated to hemocyanin using EDC. Our experience is that the titer and the time of production of antibody are more affected by differences in individual rabbits rather than by the injection scheme used.

*Monitoring Anti-Peptide Antibody Production*

Production of polyclonal antisera which bind to conserved, internal sites in the catalytic domain needs to be monitored by immunoprecipitation or by Western blotting rather than by ELISA using immobilized, synthetic peptide. A discrepancy between the titer of the antibody that binds to the peptide versus the antibody that binds to the cognate sequence in the whole protein has previously been observed[7,10] and may be caused by differences in how the antigenic determinants are displayed. In several

[21] D. N. Posnett, H. McGrath, and J. P. Tam, *J. Biol. Chem.* **263,** 1719 (1988).

[22] J. P. Tam, *Proc. Natl. Acad. Sci. U.S.A.* **85,** 5409 (1988).

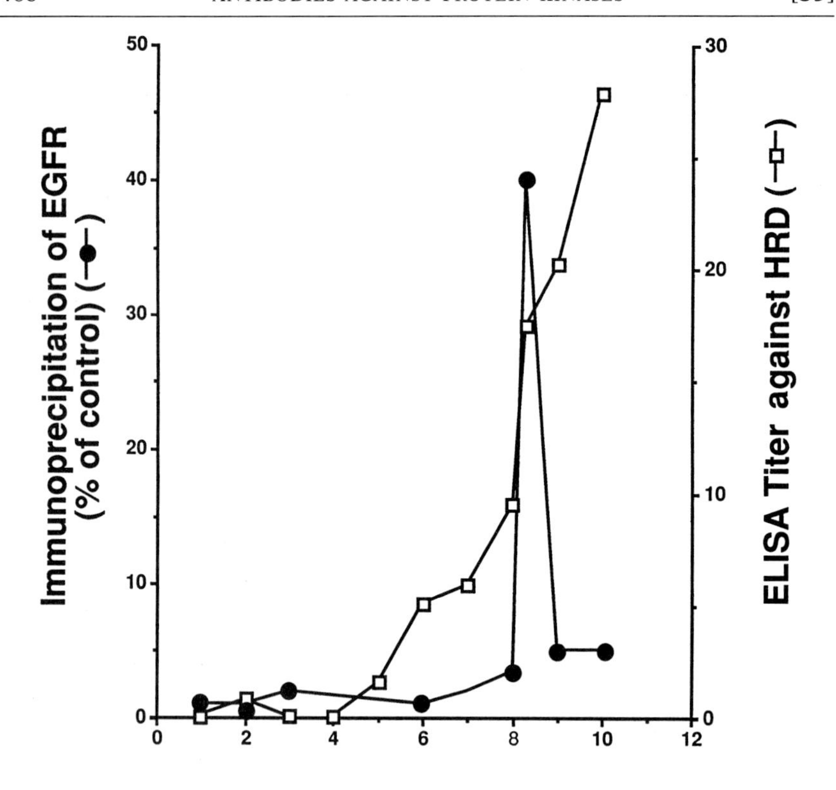


FIG. 1. Production of anti-peptide antibodies to a conserved sequence from the catalytic domain of the EGF receptor (EGFR). Aliquots (50 $\mu$l) of sera from rabbits immunized with HRDLAARN conjugated to hemocyanin were tested for the presence of antibody by immunoprecipitation of $^{32}$P-labeled EGF receptor. Parallel immunoprecipitations were conducted with a control polyclonal antiserum to the extracellular domain of the EGF receptor under conditions of antibody excess where 95% of the EGF receptor was immunoprecipitated. The serum from each test bleed was also tested by ELISA using the peptide immobilized on microtiter plates. The ELISA results are expressed relative to an ELISA using a control anti-peptide antibody against a peptide containing a sequence from the carboxy terminus of the EGF receptor.

cases, we have also found that the antibody reactive with the entire protein was produced transiently and early compared to antibody that reacted with the isolated peptide. In fact, when maximum ELISA titer was attained, the antibody that immunoprecipitated the entire protein was often no longer detectable. Eight different rabbits immunized with conjugated HRDLAARN (residues 811–818 of the EGFR) have displayed this pattern and Fig. 1 illustrates the profile of antibody production in one of these

TABLE I
BUFFERS USED IN IMMUNOPRECIPITATION[a]

| Buffer | NaCl (m$M$) | Concentration (%) | | | Kinase activity[b] |
|---|---|---|---|---|---|
| | | Deoxycholate | SDS | Glycerol | |
| A | — | — | — | 10 | + |
| B | 150 | — | 0.05 | 10 | + |
| C | 150 | 1 | 0.1 | — | − |

[a] All buffers contain 20 m$M$ Tris, pH 7.6, 1% Triton X-100, 1% aprotinin, 1 m$M$ PMSF, 2 m$M$ sodium vanadate.

[b] The autophosphorylation activity of the EGF receptor was measured in each of the buffers.

rabbits. Thus for rabbits immunized with peptides containing highly conserved sequences from the catalytic domain of protein kinases, we recommend that antibody production be monitored on a weekly basis by immunoprecipitation of the target protein kinase and that monitoring begin following the first booster injection.

The most convenient and sensitive method to monitor production of antibodies that immunoprecipitate a protein kinase capable of autophosphorylation is to radiolabel the protein with [$\gamma$-$^{32}$P]ATP in an *in vitro* kinase reaction. Because the antibodies to highly conserved sites may recognize several protein kinases, it may be necessary to have a partially purified preparation or a concentrated source of the specific target protein kinase. For a source of concentrated EGFR, we use membrane vesicles prepared from A431 cells by a previously described procedure.[23] The immunoprecipitation protocol follows:

1. The protein kinase is labeled by autophosphorylation in a reaction containing [$\gamma$-$^{32}$P]ATP. The components in the reaction and their concentration depend on the particular protein kinase to be investigated.
2. The $^{32}$P-labeled protein kinase is denatured in buffer C (Table I) to expose sites that may react with the antibody.
3. The denatured protein kinase is mixed with 50 $\mu$l antiserum from a test bleed and the volume is adjusted to 200 $\mu$l with buffer C (50 $\mu$l antiserum is the maximum amount that can be used before overloading the SDS gel with IgG).
4. The sera from the test bleed and the radiolabeled protein kinase are incubated for 1 hr at 4° with continuous rotation.

[23] S. Cohen, *in* "Methods in Enzymology" (J. D. Corbin and J. D. Hardman, eds.), Vol. 99, p. 379. Academic Press, New York, 1983.

5. Protein A–Sepharose is added at a ratio of one bed volume to one volume of sera and the mixture is incubated with shaking at 4° for 30 min.

6. The immune complex is washed four times with 1 ml of buffer C.

7. The immune complex is suspended into 100 μl of sample buffer containing 63 m*M* Tris-HCl, pH 6.8, 10% (v/v) glycerol, 2% (w/v) SDS, 5% (v/v) 2-mercaptoethanol, and 0.002% (w/v) Bromphenol Blue and incubated in a boiling water bath for 2 min. The released proteins are analyzed by electrophoresis in an SDS-containing gel of 1.5 mm thickness with a well size that holds 200 μl of sample volume.

### *Tests of Specificity of Anti-Peptide Antibodies*

Binding of the anti-peptide antibody to a specific sequence in the target protein is best demonstrated by showing that immunoprecipitation is blocked by preincubation of the antibody with its cognate peptide. To conduct peptide-blocking experiments, the antiserum is incubated without or with different concentrations of peptide for 1 hr at 4° with continuous shaking prior to addition of the radiolabeled antigen. The remainder of the immunoprecipitation protocol is conducted as described above. We find that the concentration of peptide is sometimes important in determining the extent to which a specific antibody is blocked by its cognate peptide. For antibodies prepared to the HRDLAARN sequence, we have found that a peptide concentration of about 15 μg/ml is most effective at blocking immunoprecipitation of the EGFR by 250 ng of peptide affinity purified antibody.

### *Antibody Storage*

We find that anti-peptide antisera raised against some sequences from the catalytic domain in the EGFR, such as HRDLAARN, are unstable when stored outside of the rabbit. This is true only for the antibody that reacts with the entire EGFR, not for that which binds the immobilized peptide in an ELISA. The sera are therefore stored at 4° or at room temperature with the addition of 0.02% (w/v) sodium azide to prevent bacterial growth. Under these conditions, the activity is retained for 1–3 weeks. In contrast, antisera against nonconserved sequences within the EGFR are stable when stored at −70°.

Unstable antibodies such as anti-HRD retain their activity best when bound to a peptide affinity column. The column is prepared using a 20-ml bed volume of Affi-Gel 10 (Bio-Rad, Richmond, CA) coupled to 15 mg of peptide using the procedure recommended by the manufacturer. We bind antibodies to their respective peptide affinity columns using the following procedure:

1. Precipitate the IgG-containing fraction from 15 ml of antiserum by adding a saturated ammonium sulfate solution to a final concentration of 40% of saturation.

2. Centrifuge at 10,000 *g* for 30 min at 4° and resuspend the pellet in 3 ml of 20 m*M* HEPES, 150 m*M* NaCl, pH 7.5. Dialyze against the same buffer overnight at 4°.

3. Slowly add the dialyzed IgG to the peptide affinity column and incubate at 4° for 1 hr.

4. Wash the column with equilibration buffer until the absorbance at 280 nm of the eluate is reduced to background levels.

5. Store the column at 4° in buffer containing 0.02% sodium azide.

Antibodies appear to be stable while bound to the peptide affinity column and can be used after elution from the column. However, they lose activity with time following elution from the column. Elution is accomplished as follows:

1. Add 60 ml of 0.1 *M* sodium acetate, 0.5 *M* NaCl, pH 2 to the column. Collect 0.5-ml fractions directly into 0.5 ml of 0.1 *M* glycine, pH 12, to neutralize each fraction.

2. Monitor the protein content of the fractions by absorbance at 280 nm.

3. The fractions with maximum absorbance may be used directly. Otherwise, combine these fractions and precipitate the IgG using 40% (v/v) ammonium sulfate.

4. Centrifuge and resuspend the IgG pellet in water. Dialyze against 20 m*M* HEPES, pH 7.5 at 4°.

The protease inhibitors aprotinin (0.5%), leupeptin (4 $\mu M$), and phenylmethylsulfonyl fluoride (PMSF) (2 m*M*) (all from Sigma, St. Louis, MO) are added to the purified antibody if it is not used immediately.

## Use of Anti-Peptide Antibodies in Studies of Topology and Function of Sites in Catalytic Domain

### *Determination of Topology*

To determine the topology of a specific sequence in the native protein kinase, immunoprecipitation of the radiolabeled protein by the anti-peptide antibody is conducted in buffer A (Table I) at 4° under conditions of antibody excess. Buffer A is relatively nondenaturing and, in the case of EGFR, stabilizes the kinase activity. Subsequent steps in the immunoprecipitation are carried out as described above. It is critical that a control antibody which is known to immunoprecipitate the protein kinase in its native state be used here as a standard. In addition, the proportion of the protein kinase immunoprecipitated under the nondenaturing conditions of buffer A should be compared with the amount immunoprecipitated in parallel under the denaturing conditions of buffer C. Such a comparison addresses the question of whether conformation has an effect on the efficiency of antibody binding to its cognate sequence. The radiolabeled protein kinase in the immune complex can be most accurately quantitated by SDS gel electrophoresis and scintillation counting of the radiolabeled bands excised from the gel.

It may also be of interest to determine whether partial "opening" of the protein structure is sufficient to expose a particular site in the protein or whether the protein needs to be more thoroughly denatured. This can best be determined by immunoprecipitation of the radiolabeled protein kinase in buffer B (Table I). When buffer B with 0.05% SDS is used for immunoprecipitation the protein kinase activity in the immune complex is about 80% of that found when buffer A is used. We have further found that sites that are inaccessible to antibody in buffer A may be exposed in buffer B. Another way to achieve partial opening of the protein structure is to incubate the radiolabeled protein kinase together with the anti-peptide antibody in either buffer A or buffer B at 34° for 5 min before incubation at 4° for 1 hr.

The effects of modulators of kinase activity on the topology of specific sites in the catalytic domain of protein kinases can be examined by binding of anti-peptide antibodies. For example, one can ask whether the exposure of the sites in the kinase domain is altered by ligand binding or by autophosphorylation of the protein kinase. This approach has been described in more detail for the PDGF receptor[5,6] and the insulin receptor[7,8] tyrosine kinases.

*Identification of Sites Critical for Catalysis*

The following protocol can be used to determine whether binding of anti-peptide antibody to specific sites in the catalytic domain neutralizes kinase activity. It is critical that a control antibody that efficiently immunoprecipitates the protein kinase but does not inhibit kinase activity be included in order to determine the amount of kinase activity that is lost during the immunoprecipitation procedure.

1. Mix 1 vol of anti-peptide antisera with one bed volume of protein A-Sepharose and incubate at 4° for 30 min on a shaking platform.

2. Wash the IgG–protein A complex three times with 10 to 20 vol of buffer A.

3. Mix the protein kinase with the anti-peptide IgG that is bound to protein A-Sepharose in buffer A or buffer B at 4° for 1 hr with shaking.

4. Wash the antigen–antibody complex with about 1 ml of buffer B once and buffer A three times. Care should be taken to work quickly and keep the protein kinase on ice to preserve kinase activity.

5. To test for autophosphorylation suspend the immune complex in about 100 $\mu$l of a kinase reaction mixture containing [$\gamma$-$^{32}$P]ATP. Incubate under conditions that achieve maximum levels of autophosphorylation for the particular protein kinase. (Because autophosphorylation occurs very quickly, it is not feasible to analyze the kinetics of autophosphorylation. Thus the overall level rather than the rate of autophosphorylation is determined.)

6. Wash the immune complex three times with buffer C to remove the unincorporated $^{32}$P and analyze the proteins in the immune complex by SDS gel electrophoresis.

To test for kinase activity toward exogenous protein substrate:

1. Suspend the washed immune complex from step (4) above in a kinase reaction mixture containing [$\gamma$-$^{32}$P]ATP and an exogenous substrate such as enolase or angiotensin for tyrosine kinases such as the EGFR. Incubate the kinase reaction at a temperature where a linear rate of substrate phosphorylation can be maintained.

2. Remove aliquots at different times of incubation, plunge into a boiling water bath to stop the kinase reaction, and separate the soluble substrate from the immune complex by centrifugation.

3. Phosphorylation of the exogenous substrate is determined following separation of the substrate from the other reaction components by techniques such as SDS gel electrophoresis. The rate of phosphorylation of the exogenous substrate can then be determined by quantitation of the $^{32}$P incorporated with time. The rate of substrate phosphorylation will reveal whether the kinase activity is inhibited by binding of anti-peptide antibody to a specific sequence.

In some cases it may be desirable to determine the effects of the antibody on the kinase activity in a soluble reaction where the kinase has not been immunoprecipitated. The advantage is that the effects of antibody on the entire population of kinase molecules rather than on those that are immunoprecipitated can be evaluated. This approach, however, requires that the antibody be of high titer, in excess of the protein kinase, and be highly purified, for example by peptide affinity chromatography.

To test for antibody inhibition of kinase activity in a soluble reaction, add the purified anti-peptide IgG to the protein kinase and incubate the mixture at 4° with shaking for 1 hr. Add the components of the kinase reaction, including [$\gamma$-$^{32}$P]ATP. Remove aliquots at increasing times to determine the level of substrate phosphorylation by the amount of $^{32}$P that is incorporated. Use peptide blocking (discussed above) of the purified IgG to establish that inhibition of kinase activity is caused by antibody binding to a specific sequence.

# Section VI

# Affinity Labeling, Chemical Probing, and Crystallization of Protein Kinases

# [40] Nucleotide Photoaffinity Labeling of Protein Kinase Subunits

*By* BOYD E. HALEY

## Introduction

This chapter presents a series of experimental considerations which may be used to resolve problems faced when using nucleotide photoaffinity probes to detect the subunits involved in the catalytic activity of protein kinases. The ideas presented are general in nature, since each different nucleotide-binding protein will have its own characteristics which must be considered when designing the initial photolabeling experiments. The most impressive use of nucleotide photoaffinity probes has been the identification in crude homogenates of the particular subunit or peptide involved in binding a specific nucleotide. This is particularly true for the photoprobes that contain the azide directly attached to the base of the nucleotide. These probes have the following unique advantages: (1) The chemical species which cross-links is generated from a relatively small azide group and can be defined as being within the active site, since these probes are usually substrates or competitive inhibitors of the enzymes being tested; (2) the reactive species generated has been shown to react with several different amino acid residues and is therefore relatively indiscriminate with regard to the residues available within the active site; (3) these probes can be made with extremely high specific activity (all three phosphates may be $^{32}P$ labeled), which allows the detection of binding proteins even if the efficiency of photoinsertion is much less than 1%; (4) the reactive species generated, ostensibly a nitrene, is of such short half-life that nonspecific labeling, even in crude homogenates, is minimal; (5) they leave the phosphate and ribose moieties open for secondary modifications, which greatly increases the use of these compounds for the enzymatic synthesis of other photoactive molecules of RNA, DNA, $NAD^+$, etc.; (6) the location of the azido group may be placed at the 8-position, which allows for excellent discrimination between adenosine and guanosine nucleotide-binding proteins.

However, there is one important fact that must be understood before designing experiments using the available nucleotide photoaffinity probes. Chemically, all azides are not alike! Conventional wisdom implies that all azides when photoactivated react through a nitrene intermediate. This is obviously not true. Some insert into proteins through short half-life intermediates, whereas others produce long-lived intermediates that behave like classical chemical probes and react only with certain amino acid side chains. An example of this is the observation that most phenylazide-based probes may be first photolyzed and the enzyme added afterward and labeling of the protein will still occur. It has been shown that phenylazides, when photolyzed, rearrange into dehydroazepines, which have long half-lives of reactivity.[1,2] Many scientific opinions about the selectivity of "nitrene"-based photoaffinity probes have been developed based on results obtained using phenylazide analogs. Because dehydroazepines behave like classic chemical intermediates, i.e., pseudophotoaffinity probes, it is not valid to assume that all azido-containing compounds will behave like phenylazides. A contrast in behavior is seen with probes with azides attached to a heterocyclic ring system. These are represented by 8-azidoadenosine-, 2-azidoadenosine-, 8-azidoguanosine-, and 5-azidouridine-containing probes.[3] These analogs seem to work through very short half-life intermediates, since photolysis before addition of protein has not led to labeling in any of the many proteins that we have worked with. Also, photolysis in the presence of several proteins, i.e., a homogenate, results in selective photolabeling of only a few of the available proteins.[4-6] While this indicates that the reactive intermediate is very short lived it is not proof that the intermediate is a nitrene. To date we have very little absolute proof of the nature of the reactive intermediate that is generated on photolysis of most photoaffinity probes in a biological medium. However, by careful experimental design we can test whether or not the probe and procedure being used is giving specific labeling of the active site of the kinase being studied.

## Design of Initial Photolabeling Experiments

The following is presented assuming the investigator knows the binding affinity of the native nucleotide for the enzyme or regulatory protein. The first experiment is done under "best guess" conditions assuming that the photoprobe will bind with approximately the same affinity as the native nucleotide (this is usually the case, but exceptions do occur) and that the enzyme will probably catalytically modify the photoprobe as it would the

[1] E. Leyva and M. S. Platz, *Tetrahedron Lett.* **26,** 2147 (1985).

[2] E. Leyva, M. Young, and M. S. Platz, *J. Am. Chem. Soc.* **108,** 8307 (1986).

[3] R. Potter and B. E. Haley, in "Methods in Enzymology" (C. H. W. Hirs and S. N. Timasheff, eds.), Vol. 91, p. 613. Academic Press, New York, 1983.

[4] R. W. Atherton, S. Khatoon, P. K. Schoff, and B. E. Haley, *Biol. Reprod.* **32,** 155 (1985).

[5] S. Khatoon, S. R. Campbell, B. E. Haley, and J. T. Slevin, *Ann. Neurol.* **26,** 210 (1989).

[6] J. N. Dholakia, B. R. Francis, B. E. Haley, and A. Wahba, *J. Biol. Chem.* **264,** 20638 (1989).

native nucleotide (e.g., if ATP autophosphorylates the enzyme then 8-$N_3$ATP will do the same). It is important to realize that photoaffinity labeling is very dependent on binding affinity and that proteins with $K_d$ values in the millimolar range are less likely to be good candidates for photolabeling than those with micromolar $K_d$ values. Also, it is important to consider the turnover number of the nucleotide-hydrolyzing enzymes since it is obviously difficult to photolabel an enzyme that is hydrolyzing the probe faster than one can finish the photolysis. Likewise, enzymes that autophosphorylate rapidly can be difficult to photolabel with $\gamma$-$^{32}$P-labeled probes under conditions where the enzyme is active. With such enzymes the use of the $\alpha$-$^{32}$P-labeled probe is preferred unless autophosphorylation can be slowed to a minimum by decreasing the temperature or changing metal ions. Consideration for the metal ion requirements is a must. Usually, if the catalytic activity requires $Mg^{2+}$, then the binding and subsequent photoinsertion will also require the presence of a divalent metal ion. However, substitution of another metal ion for $Mg^{2+}$ will sometimes result in photoprobe binding but deter autophosphorylation. This allows photoinsertion into the catalytic site without interference from autophosphorylation events (see Section "Procedures for Increasing Binding Affinity of Probes for Enzymes").

If the entire goal of the experiment is to determine which protein of a complex mixture is binding a particular nucleotide this may be accomplished by addition of the photoprobe to concentrations that saturate the binding site as approximated by the predetermined $K_d$ of the enzyme for its native substrate. This is done in a reaction mix which is normally used to assay the enzyme. The probe is mixed with the enzyme for 10 to 20 sec, then immediately exposed to a low-intensity (1000 $\mu$W or less) UV lamp for about 2 min, and the reaction is immediately quenched with a solution containing sodium dodecyl sulfate (SDS) (to stop enzymatic reactions) and dithiothreitol (DTT) (to reduce the remaining azide) and subjected to SDS–PAGE and autoradiography.[7,8] The success rate of this experiment can be enhanced by considering the general observations presented in Section "Observations on Nucleotide Photoaffinity Labeling."

If the objective of the research is to develop procedures to detect the binding protein at very low levels in crude homogenates or to maximize photoinsertion in order to obtain active site peptides for sequencing, then the photoinsertion must be optimized. For protein kinases it is important to test the photoprobe as a substrate. In the majority of the cases that we have tested the 8-azido- and 2-azidoadenosine probes are substrates for small molecule kinases as well as protein kinases. Using a normal assay,

[7] P. Hoyer, J. R. Owens, and B. E. Haley, *Ann. N. Y. Acad. Sci.* **346,** 280 (1980).
[8] B. E. Haley, *Fed. Proc., Fed. Am. Soc. Exp. Biol.* **42,** 2831 (1983).

minus reagents like DTT which reduce azides, a $K_m$ and an apparent $K_d$ can be determined for the photoprobe. For optimal labeling probe must be present at saturating concentrations. It is important to remember that with a photoaffinity probe it is the concentration of analog that determines the amount of radioactivity incorporated specifically within the binding site, not the total amount of radioactivity added. Increasing concentrations above saturating conditions will increase undesirable, nonspecific labeling but will not substantially increase specific labeling.

The major weakness of photoaffinity probes in general is that they do not give stoichiometric labeling and may modify only a small percentage of the binding sites (values between 0.01 and 0.95 mol to mol have been reported). Two major sets of factors must be considered to increase selective photoinsertion. The first are those factors that increase photoinsertion merely by increasing the affinity of the enzyme for the probe, which increases the residence time within the binding site. An example would be lowering the temperature or ionic strength of the photolysis solution. The second are those factors that increase the efficiency of photoinsertion of the probe without affecting affinity. An example of the latter would be the choice of lamp intensity, lamp wavelength, and length of photolysis. Examples of these factors are given in more detail below.

## Observations on Nucleotide Photoaffinity Labeling

### *General Considerations Involving Effects of Light*

Photolabeling experiments with 8- and 2-azidoadenosine probes can be done in normal room light. Because these analogs are activated only by UV light, which is essentially screened out by glass, plastic, etc., a dark room is not needed.

It is critical to determine the effects of prior photolysis of the probe on its reactivity. This ensures that the probe is not behaving as a pseudophotoaffinity probe. Prephotolysis of the photoprobe followed by immediate addition of enzyme should not result in incorporation of radiolabel, even if the system is exposed to additional photolysis.

The effects of UV light on the stability and nucleotide-binding properties of the enzyme should be determined. This requires that the enzyme be prephotolyzed at varying times without photoprobe present, followed by the addition of photoprobe and photolysis to effect covalent insertion. The level of insertion can be compared to that obtained when enzyme is not prephotolyzed. If the enzyme is very sensitive to UV light a measurable drop in photoinsertion will be observed in the prephotolyzed enzyme. If UV sensitivity is too great it may be possible to change the wavelength of light used to activate the probe by using appropriate screens or lamp sources.

The optimum time of photolysis and the optimum intensity of light used must be determined for each biological system. It takes more time to maximally photoinsert probe into proteins in a homogenate than it does for a purified enzyme due primarily to UV-screening effects of endogenous compounds. Also, each lamp will have different spectral characteristics which will affect the optimum time of photolysis. Lamps with filters represent a special problem. These filters solarize with increased use and the resulting intensity of emitted light of specific wavelengths decreases. If filters are to be left on these lamps their intensity of irradiation should be monitored frequently.

Selecting the wavelength of light used to activate the photoprobe is very important. Because of the high extinction coefficient of the azide-containing chromophore of 8-azidopurine nucleotide analogs it is usually very easy to photolyze them totally with a low-intensity 254-nm UV light with very little deleterious effect on the enzyme being studied. However, certain enzymes are very sensitive to light in this region and longer wavelengths of light in the 320-nm range may work better. One must remember that to photoactivate a probe it must absorb the light you are using. Numerous reports that 8-$N_3$ATP or 8-$N_3$cAMP was photolyzed using 360-nm light sources should be looked at skeptically since these probes do not absorb photons in this region.

Avoid high-intensity UV lamps. The most common mistake made using these probes is the use of very high-intensity lamps and/or too extensive a photolysis time. Using a low-intensity, hand-held UV lamp, designed for viewing thin-layer chromatography (TLC) plates, total photoconversion of micromolar concentrations of these probes in a 1-cm path length is usually accomplished in less than 1 min. The time of photolysis for optimal photoinsertion must be determined for every lamp used and every different reaction mixture. Extensive photolysis past the point that maximum photoinsertion is observed, or photolysis with high-intensity lamps, usually decreases photoinsertion due to multiple photon fragmentation. This decreases photoinsertion through the reactive species generated through the azide group and increases photoinsertion by what is called "direct photoaffinity" labeling (i.e., photolabeling that can be obtained using the unmodified nucleotide). Direct photoaffinity labeling is probably due to free radical formation on aromatic amino acid residues which then react with the nucleotide. The problem with this approach is that it takes extensive photolysis times and these insertion reactions can take place long after the UV light has destroyed the binding integrity of the enzyme. It is important to remember that if one is to claim photoinsertion is due to conversion of the azide to a reactive species this photoinsertion must be seen within the first minute of photolysis or at least coincide with the very rapid destruction of the azide form. It cannot be occurring 10 min later. If this is happening then a pseudophotoaffinity intermediate is being produced or free radical formation of amino acid residues is probably involved.

*Reduction of the Azide of Nucleotide Photoaffinity Probes*

DTT should not be present, as it reduces purine azides very rapidly in a pH-dependent process. If the DTT cannot be removed, try to adjust the pH to 6.5 or less. This will greatly reduce the rate of azide reduction and if it does not reduce active enzyme levels it will result in increased photoinsertion. However, if possible it is best to replace the DTT with a monothiol reagent like mercaptoethanol at low millimolar concentrations. The effects of DTT on 8-$N_3$ATP UV spectra are quite different from the effects of UV light. UV light abolishes absorbance in the 280-nm range, leaving no noticeable maxima below 240 nm. DTT changes the absorption maximum to 274 nm while producing a compound, 8-$NH_2$ATP, that is not photoactive.

When using analogs containing thiophosphates, be careful. Many commercial preparations of these nucleotides also contain DTT to keep the thiol groups from oxidizing. Addition of these analogs appears to give protection when in actuality they are reducing the azide on the photoprobe. We have observed an actual case where ATP$\gamma$S gave protection against 8-[$^{32}$P]$N_3$GTP photoinsertion into a specific GTP-binding protein whereas ATP would not do so. Removal of the DTT from the commercial ATP$\gamma$S eliminated this observed protection.

Many commercial enzyme preparations contain DTT and it may not be listed on the label. If one attempts to photolabel these enzymes they likely will not be successful. Also, if they are used with 8-$N_3$ATP to produce another photoprobe enzymatically, e.g., 8-$N_3$[$\gamma$-$^{32}$P]ATP, 8-$N_3$-*S*-adenosylmethionine, or 8-$N_3$-adenosine-containing RNA, the generated probe will not be an 8-azido analog but will be an 8-amino derivative.

*Other Caveats to Consider When Using Photoaffinity Probes*

Bovine serum albumin (BSA) or similar proteins should not be present in the preparation being photolabeled. These proteins act as "sponges" for the 8-$N_3$-purine nucleotides, removing them from solution. BSA photolabels quite well with these photoprobes but saturation has never been obtained, indicating that the binding is not specific but it is of high capacity.

If the binding protein being photolabeled needs other protein present for stability try to find a substitute for BSA.

The selection of buffer can be critical since certain buffers can absorb UV light, act as potent scavengers, or decrease the affinity of the enzyme for nucleotide. Excellent results have been obtained with Tris and phosphate buffers although others may work as well. Try to avoid buffers with conjugated ring systems or high extinction coefficients.

*Procedures for Increasing Binding Affinity of Probes for Enzymes*

Drop the temperature to 0° or less. This increases the apparent affinity of most kinases for the nucleotide probes and decreases the phosphoryl transfer and hydrolysis rates. It is also possible to photolabel in a frozen matrix. However, this process has given mixed results. Sometimes it improves photoinsertion and sometimes it totally prevents it. It is important to compare photoinsertion at 0° against that at higher temperatures since conformational changes induced by low temperatures can also decrease photoinsertion efficiency.

Keep the ionic strength as low as the enzyme will tolerate. Nucleotide affinity in general is increased by decreasing ionic strength. However, some enzymes will need a certain level of cations (mono- or divalent) for optimal binding.

The presence of divalent metal ions is a critical factor that may be used to great advantage when designing a photolabeling experiment. It is thought that all nucleotide triphosphates exist in cells complexed to $Mg^{2+}$ and that this complex is the native substrate form and is required for binding. With rare exceptions, this has been found to be the case. However, $Mn^{2+}$ may sometimes be substituted for $Mg^{2+}$, producing a complex that binds tighter and yet remains catalytically active. With certain nucleotide-binding proteins photolabeling is increased by changing $Mg^{2+}$ to $Mn^{2+}$.

Calcium ion can also bind to nucleotide triphosphates with an affinity close to that of $Mg^{2+}$. However, the resulting complex is usually not a substrate but a very potent competitive inhibitor of $Mg^{2+}$-nucleotide complex binding. This is probably the reason that high concentrations of $Ca^{2+}$ inhibit most $Mg^{2+}$-dependent ATPases. This observation is used in our laboratory to photolabel enzymes that exhibit very high turnover numbers for $Mg^{2+}$-ATP or enzymes that rapidly autophosphorylate. It is sometimes possible to photolabel autophosphorylating enzymes using the $Ca^{2+}$–8-$N_3$ATP complex without the interference of background autophosphorylation. This is especially true if $Ca^{2+}$-ATP forms a dead-end complex inhibition form with the enzyme. Also the use of $CA^{2+}$–8-$N_3$ATP in homogenates is useful in that the rate of $N_3ATP$ hydrolysis by enzymes present is dramatically reduced and those enzymes that bind the $Ca^{2+}$-chelated form may be more effectively photolabeled. With these experiments it is best that the $Ca^{2+}$ concentration be approximately equal to the nucleotide concentration and *not* be excessive due to other deleterious effects of free $Ca^{2+}$.

Photolabeling of proteins with slow on rates for nucleotide binding presents special difficulties. For example, photolabeling of G regulatory proteins in membranes and homogenates is a problem since the on rate of GTP and GTP analogs is very slow and the added 8-[$^{32}$P]$N_3$GTP is hydrolyzed to the diphosphate before it has time to be bound. This problem can sometimes be surmounted by photolabeling with low micromolar levels of 8-$N_3$[$^{32}$P]GTP in the presence of millimolar concentrations of ATP. This greatly decreases the rate of 8-$N_3$[$^{32}$P]GTP hydrolysis by nonspecific enzymes and allows binding to proteins that are very selective for binding GTP.[9] This problem may also be circumvented by using azidopurine nucleotides with nonhydrolyzable phosphates.

Usually, photoinsertion of nucleotide photoprobes will be maximized at the pH that gives maximum enzyme activity. However, this is not always the case and a determination of the pH optimum for photoinsertion should be made. It is quite possible that changing the pH can decrease the phosphoryl transfer or hydrolytic activity at the active site without changing the apparent $K_d$ of the nucleotide probe. This will increase the residence time of the intact probe within the active site and enhance photoinsertion. It is also possible that changing the ionic species of certain amino acid residues within the enzyme may change the geographical orientation of the photoprobe with regard to active site residues involved in binding, which will influence the efficiency of photoinsertion.

Consideration of allosteric effectors of the enzyme is of prime importance. Any factor that is known to enhance nucleotide binding will probably enhance the binding of the mimicking photoprobe. For example, both GTP and glutarate increase the affinity of glutamate dehydrogenase for $NAD^+$, whereas ADP decreases the affinity. Using the photoprobe 2-$N_3NAD^+$ the effects of these allosteric regulators were consistent with GTP and glutarate increasing and ADP decreasing photoinsertion.[10] They can also have a major effect on the efficiency of photoinsertion due to

[9] B. Francis, J. Overmeyer, W. John, E. Marshall, and B. Haley, *Mol. Carcinog.* **2,** 168 (1989).

[10] H. Kim and B. Haley, *J. Biol. Chem.* **265,** 3636 (1990).

minor changes in the positions of the active site residues involved in binding.

## Validation of Active Site Photolabeling

There are four major items that are absolutely needed to demonstrate the validity of specific photolabeling of the nucleotide-binding subunit or protein. The first is the observation that only a specific protein(s) in a complex or homogenate is photolabeled with the photoprobe being used.[11] This eliminates the possibility that the photoinsertion is due to protein scavenging of the reactive intermediate and nonspecific. It also shows that the photogenerated reactive species is short lived and is not labeling proteins which do not bind it.

The second is the demonstration that photoinsertion saturates and does so at concentrations that are in agreement with the known reversible binding constants of the probe determined in the absence of UV light. It is important here to remember that $K_m$ values do not represent $K_d$ values, although they may be close. However, if a kinase has a $K_m$ of 5 $\mu M$ for 8-$N_3$ATP it is unlikely that saturation of photoinsertion will require 200 $\mu M$ 8-$N_3$ATP.

The third is the observed selective protection against photoinsertion by native compounds that are known to bind to the protein being studied and the inability of other similar compounds, which do not reversibly interact with this protein, to elicit this protection. In the case of most protein kinases this would be ATP. However, with casein kinase II this would be ATP and GTP. Understanding the kinetics of the enzyme being studied is very important. One should not expect enzymes that operate through ordered Bi–Bi mechanisms to behave like those which obey simple Michaelis–Menten kinetics with regard to substrate and product protection against photoinsertion. It is also very important to show that the protective compounds elicit their protective effects at concentrations that are in agreement with their reported binding constants and that the protection is not due to UV-screening effects. This may be accomplished in control experiments by substituting the appropriate nucleoside for the nucleotide which provides protection in a manner that keeps the total absorbance profile identical at all data points collected.

The fourth is a demonstration that the photolabeling is absolutely dependent on activating light. To show dependence on photolysis requires a simple control where photoprobe is mixed with the proteins being tested and treated identically as the photolyzed samples with the exception that

[11] J. R. Owens and B. E. Haley, *J. Biol. Chem.* **259,** 14843 (1984).

it is totally protected from exposure to activating light. This procedure will identify any incorporation of label by enzymatic processes like phosphorylation ($\gamma$-$^{32}$P-labeled analogs) and adenylation ($\alpha$-$^{32}$P-labeled analogs).

## Considerations on Stability of Photoinserted Nucleotide Analog

The major disadvantage of photoaffinity labeling is the usually less than stoichiometric insertion that is observed after separation of the photolabeled protein from the noncovalently attached photoprobe by SDS–PAGE or other techniques. Undoubtedly, much of the problem lies in the lack of efficient photoincorporation in many cases. However, in other cases the instability of the photoinserted product to subsequent experimental work up leads to loss of label and makes it appear as if the efficiency of photomodification is lower than it actually is. An example of this is phosphoenolpyruvate (PEP)-carboxykinase, which is specifically inhibited by one photolysis with 8-$N_3$GTP by over 60% but where the apparent covalent modification is much less.[12] In this situation the PEP-carboxykinase can be reactivated by the addition of DTT, indicating that the amino acid residue being modified by photolysis is a sulfhydryl and that a disulfide is being rapidly formed which locks the enzyme in an inhibited form but releases the photoinserted radiolabel.[12] Also, from our first-hand experience, and from communications from several sources, it has been observed that nucleotide-photolabeled peptides (either from direct or nitrene-generated systems) lose most of the photoinserted radiolabel on high-performance liquid chromatography (HPLC). It appears that the bond breakage that results in the loss of radiolabel occurs as the peptide binds to the resin since the radiolabel comes off in the void volume. It does appear that photoinsertion into tyrosine produces the most HPLC-stable covalent insertion. This has led many to believe that the azide-based probes interact only with tyrosines since their successful isolation by HPLC is most favorable. However, other amino acid residues are modified and these can be isolated if procedures other than HPLC are used to isolate the photolabeled peptide.[10,13]

To determine more accurately the efficiency of photomodification of enzymes by photoaffinity probes we have developed a procedure that, while indirect, works quite well, especially if the binding protein does not have a known or easily measurable enzymatic activity. First, the stability

[12] C. T. Lewis, B. E. Haley, and G. M. Carlson, *Biochemistry* **28,** 9248.

[13] S. King, H. Kim, and B. E. Haley, *in* "Methods in Enzymology" (R. Vallee, ed.), Vol. 196, p. 449. Academic Press, San Diego, California, 1991.

of the protein to photolysis conditions is determined as given above. The protein is then photolyzed in the presence of saturating photoprobe under conditions that give maximum photoinsertion. An aliquot is removed and saved for quantification of photoinsertion, usually by SDS–PAGE, autoradiography, and liquid scintillation counting.[3] Additional photoprobe is added to the remaining sample to saturating concentrations and the sample is rephotolyzed and another aliquot removed for quantification. This can be repeated several times, depending on the stability of the protein to photolysis. If the protein is stable to photolysis and photoinsertion is of low efficiency multiple photolysis will lead to additional photoinsertion at each photolysis step. The amount of additional increase at each step will be dependent on the amount of unphotolabeled protein left after the previous step. If most of the protein is photolabeled on the first photolysis, subsequent photolysis steps will not lead to markedly increased photoinsertion. Using this procedure it has been observed that most proteins are much more covalently modified by photolysis with azidopurine analogs than is calculated by the amount of radiolabel that survives SDS–PAGE or acid precipitations. In the case of interleukin 2 (IL-2) we observed about 10% photoinsertion after one photolysis and SDS–PAGE purification.[14] However, additional photolabeling steps increased the total photoincorporation only slightly (S. Campbell and B. Haley, 1990, unpublished results). IL-2 appears to be quite stable to the photolysis conditions used. This type of experiment is necessary if the objective of the photolabeling experiment is to specifically inhibit a nucleotide-dependent activity, e.g., a specific kinase.

[14] S. Campbell, H. Kim, M. Doukas, and B. Haley, *Proc. Natl. Acad. Sci. U.S.A.* **87,** 1243 (1990).

# [41] Carbodiimides as Probes for Protein Kinase Structure and Function

*By* JOSEPH A. BUECHLER, JEAN A. TONER-WEBB, and SUSAN S. TAYLOR

Structure–function relationships in proteins can be established by several different techniques. Ideally, the roles of particular amino acids can be inferred from the tertiary structure of the enzyme, determined by X-ray crystallography or nuclear magnetic resonance spectroscopy. The proposed function of an amino acid can then be checked by site-directed mutagenesis. Ultimately, the structures of the native and mutant proteins must be compared to ensure that the structure of the mutant protein has not been changed at sites distant from the mutated residue. An alternative method for identifying amino acids that may be near the active site of an enzyme is chemical modification. This approach is particularly valuable because there are many proteins for which no structural data exist. In addition, even when a high-resolution crystal structure is known, the protein is often not cocrystallized with substrates and active site geometry is simply deduced. Thus, it is advantageous to have chemical information in order to interpret the structure accurately and to correlate structure with function.

Ideally, the reagent used for chemical modification should modify one amino acid stoichiometrically, and the reaction should not change the overall folding of the polypeptide chain. Affinity labeling perhaps comes the closest to fulfilling these criteria.[1-3] Unfortunately, affinity-labeling experiments are limited by the availability of suitable affinity-labeling probes. An alternative approach is to modify the protein with a chemical reagent that is specific for a single functional group, such as $\varepsilon$-amino groups or free sulfhydryls.[4] Using this approach, it is often not possible to limit the reaction to a single residue; however, by using differential labeling under a variety of conditions, the modification of specific amino acids can often be correlated with the loss of a particular property of the enzyme.

## Protein Kinases

The protein kinases are a family of enzymes whose three-dimensional crystal structure is not yet available. Although this family of proteins is quite diverse, each eukaryotic member has a homologous catalytic core.[5] Residues essential for common functions shared by all of these enzymes, such as the binding of MgATP or catalysis, might be expected to be conserved in every protein kinase, while amino acids that are important for regulation or substrate specificity would be expected to vary.[6-9]

[1] V. Chowdhry and F. H. Westheimer, *Annu. Rev. Biochem.* **48,** 293 (1979).

[2] S. J. Singer, *Adv. Protein Chem.* **22,** 1 (1967).

[3] R. F. Coleman, *Annu. Rev. Biochem.* **52,** 67 (1983).

[4] R. L. Lundblad and C. M. Noyes, "Chemical Reagents for Protein Modification," Vols. 1 and 2. CRC Press, Boca Raton, Florida, 1984.

[5] S. K. Hanks, A. M. Quinn, and T. Hunter, *Science* **241,** 420 (1988).

[6] S. S. Taylor, J. Bubis, J. Toner-Webb, L. D. Saraswat, E. A. First, J. A. Buechler, D. R. Knighton, and J. Sowadski, *FASEB J.* **2,** 2677 (1988).

[7] T. Hunter, *Cell (Cambridge, Mass.)* **50,** 823 (1987).

[8] S. S. Taylor, J. A. Buechler, and W. Yonemoto, *Annu. Rev. Biochem.* **59,** 971 (1990).

[9] S. S. Taylor, J. A. Buechler, and D. Knighton, *CRC Crit. Rev. Biochem.* 1 (1990).

The catalytic subunit of cAMP-dependent protein kinase is perhaps the simplest protein kinase, because the major regulatory element is part of a separate, dissociable subunit. The inactive holoenzyme consists of a tetramer of two catalytic subunits and two regulatory subunits. When the regulatory subunit binds cAMP, the holoenzyme complex dissociates into two active catalytic subunits and a regulatory subunit dimer.[10] The catalytic subunit phosphorylates serine or threonine residues in protein substrates. The site of phosphorylation is typically flanked on the N-terminal side by basic residues, which are usually arginines. Studies with peptide substrates have suggested that the optimal recognition sequence for the catalytic subunit is Arg-Arg-X-Ser/Thr.[11,12] The catalytic subunit is a good candidate for structural analysis by chemical modification because large quantities of purified protein can be obtained, and because, unlike other protein kinases, the major regulatory component can be easily separated and removed.

In the catalytic subunit, there are several potential functional roles for carboxyl groups, as summarized in Fig. 1.[13] The pH dependence of the activity of the catalytic subunit suggests that a general base with a p$K_a$ of 6.2 is important for removing the hydrogen from the hydroxyl of the serine or threonine residue to be phosphorylated in the protein substrate.[14] In addition, chiral analogs of ATP showed that the reaction occurs by inversion of the configuration of the phosphate group,[15] providing further support for the role of a general base in the catalytic mechanism. This general base is possibly an aspartic or glutamic acid. Several small molecule kinases are predicted to have an aspartic acid that functions as a general base in the active site,[16,17] and they may therefore serve as a precedent for the protein kinases. Additional carboxyl groups at the active site could serve as ligands for the magnesium ion in the MgATP complex. Finally, carboxyl groups in the catalytic subunit may be important for interacting with the two basic residues preceding the phosphorylated amino acid in the substrates of the catalytic subunit.

[10] D. A. Flockhart and J. K. Corbin, *CRC Crit. Rev. Biochem.* **12,** 133 (1982).

[11] J. R. Feramisco, D. B. Glass, and E. G. Krebs, *J. Biol. Chem.* **255,** 4240 (1980).

[12] H. N. Bramson, E. T. Kaiser, and A. S. Mildvan, *CRC Crit. Rev. Biochem.* **15,** 93 (1984).

[13] J. Granot, A. S. Mildvan, H. N. Bramson, N. Thomas, and E. T. Kaiser, *Biochemistry* **20,** 602 (1981).

[14] M. Y. Yoon and P. F. Cook, *Biochemistry* **26,** 4118 (1987).

[15] M. F. Ho, N. H. Bramson, D. E. Hansen, J. R. Knowles, and E. T. Kaiser, *J. Am. Chem. Soc.* **110,** 2680 (1988).

[16] C. M. Anderson, R. E. Stenkamp, R. C. McDonald, and T. A. Steitz, *J. Mol. Biol.* **123,** 207 (1978).

[17] P. R. Evans, G. W. Farrants, and P. J. Hudson, *Philos. Trans. R. Soc. London, Ser. B* **293,** 53 (1981).

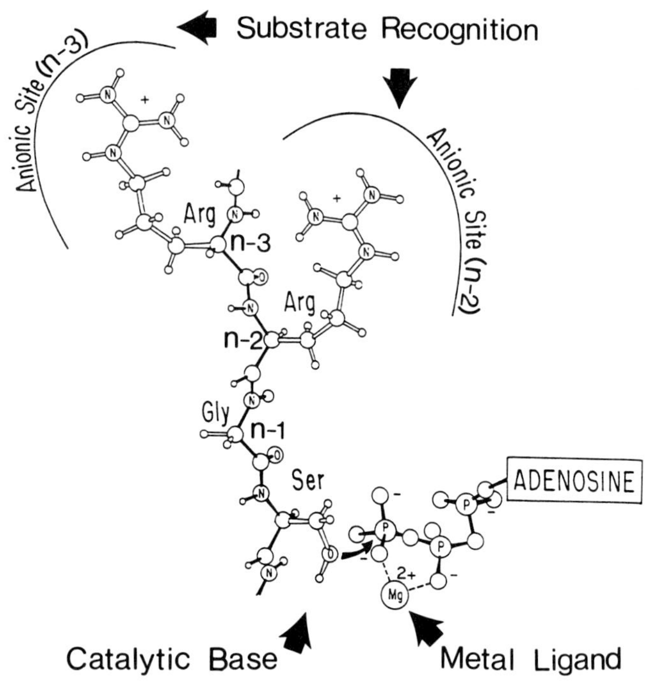


FIG. 1. Potential roles for carboxyl groups in the active site of the catalytic subunit of cAMP-dependant protein kinase. The proposed structure for the bound substrate peptides is based on the NMR data of Granot *et al.*[13]

## General Reaction of Carbodiimides with Proteins

In order to identify specific carboxyl groups in the catalytic subunit that participate in these functions, the protein can be modified with two different carbodiimides, dicyclohexylcarbodiimide (DCCD) and 1-ethyl-3-(3-dimethylaminopropyl)carbodiimide (EDC), the former being hydrophobic and the latter hydrophilic. Although carbodiimides are useful reagents for modifying proteins, the multiple possible reaction pathways must always be kept in mind. Figure 2 shows the reaction mechanism and the various reaction pathways that the activated carboxyl group generated in the reaction can follow. The protonated carbodiimide first reacts with the deprotonated carboxylic acid to form an *O*-acylisourea intermediate. This intermediate is susceptible to nucleophilic attack via reaction pathway **IV,** or the intermediate can rearrange **(III),** or it can hydrolyze **(II).** Two methods for determining the specific sites of modification can be used: (1) the use of radiolabeled carbodiimide may lead to the formation of a labeled *N*-acylurea product **(III),** or (2) a radiolabeled nucleophile, such as aniline

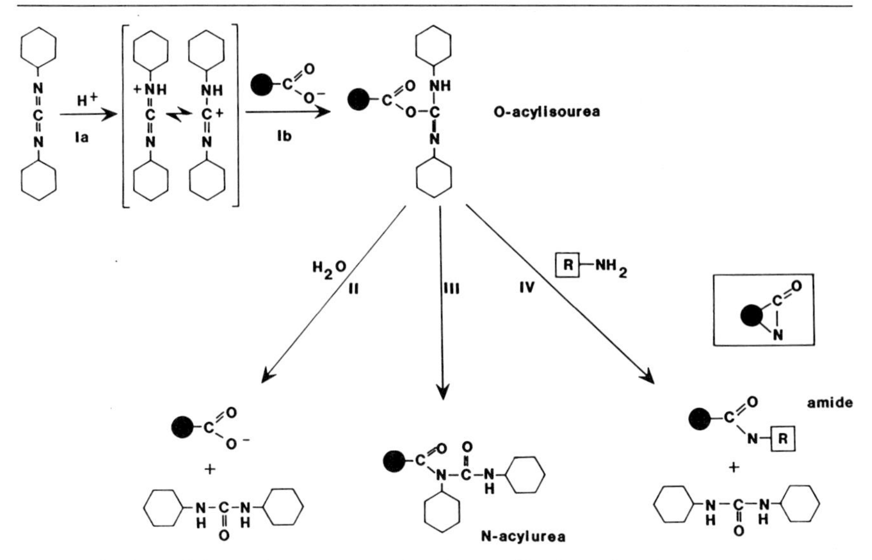


FIG. 2. Potential reaction pathways for the modification of carboxyl groups by carbodiimides. The protein molecule is indicated by the solid black circle. The carbodiimide shown is dicyclohexylcarbodiimide; however, EDC reacts with the same mechanism. (Reprinted with permission from *Biochemistry*.[22])

or glycine ethyl ester, can be used with unlabeled carbodiimide to cause the formation of a labeled amide **(IV)**. Alternatively, the presence of a lysine close to the carboxyl group that is being modified can lead to the formation of an intramolecular cross-link, in which case no radiolabel becomes incorporated into the protein.

### *Materials*

Reagents are obtained from the following sources: dicyclohexyl[$^{14}$C]-carbodiimide (50 mCi/mmol) (Research Products International, Mount Prospect, IL); [$^{3}$H]acetic anhydride (100 mCi/mmol) (ICN, Costa Mesa, CA); acetic anhydride and acetonitrile (HPLC grade) (Fisher Scientific, Tustin, CA); dicyclohexylcarbodiimide (DCCD; Aldrich, Milwaukee, WI; L-1-tosylamido-2-phenyl ethyl chloromethyl ketone (TPCK)-treated trypsin and $\alpha$-chymotrypsin (Sigma, St. Louis, MO); trifluoroacetic acid (TFA, HPLC grade) and 1-ethyl-3-[3-(dimethylamino)propyl]carbodiimide (EDC; Pierce, Rockford, IL); [$^{14}$C]glycine ethyl ester (50 mCi/mmol) (DuPont Co./New England Nuclear Research Products, Boston, MA); 2-(*N*-morpholino)ethanesulfonic acid (MES) and glycine ethyl ester (United States Biochemical Corp., Cleveland, OH).

The peptide substrate, L-R-R-N-S-I,[17a] and the peptide inhibitor, L-R-R-N-A-I, are synthesized at the Peptide and Oligonucleotide Facility at the University of California, San Diego.

The catalytic subunit is purified from porcine heart according to Nelson and Taylor.[18] The recombinant type I regulatory subunit is purified from *Escherichia coli* according to Saraswat *et al.*[19]

### *Chemical Modification with DCCD*

DCCD is a hydrophobic carbodiimide that is only sparingly soluble in water. DCCD has been shown to react with carboxyl groups in hydrophobic regions of proteins, such as membrane-spanning segments and nucleotide binding sites,[20] presumably because of its hydrophobic character. The labeling of the catalytic subunit with DCCD could potentially identify carboxyl groups close to the nucleotide-binding site that function as a ligand of $Mg^{2+}$ or as the general base in the catalytic mechanism.

In order to establish whether DCCD is a potential reagent for selectively modifying an enzyme, it is essential to first characterize the kinetic effects of the reagent. One must achieve first order, time-dependent inhibition and then establish that substrates of the enzyme are capable of protecting against inactivation. The activity of the catalytic subunit is initially measured as a function of DCCD concentration. Enzyme concentrations of 0.8 $\mu M$ at pH 5.2 to 8.0 in 50 m$M$ MES or potassium phosphate buffers containing 5% glycerol are used to establish the kinetics of inhibition. After initiating the reaction by adding an aliquot of DCCD dissolved in absolute ethanol, the enzyme solution is incubated at 37°. Less than 3% total volume of ethanol is added to ensure full activity of the catalytic subunit. At intervals, enzyme activity is determined using a coupled spectrophotometric assay.[21] The catalytic subunit is irreversibly inhibited by DCCD in a linear, time-dependent manner. When the reciprocal of the observed rate of inhibition is plotted against the reciprocal of DCCD concentration, the $K_i$ for inhibition is calculated to be 60 $\mu M$ (Fig. 3). The first order inhibition suggests that inactivation correlates with the covalent modification of a single amino acid, although this can be difficult to prove sometimes when two or more amino acids are labeled at the same rate.

[17a] Single-letter abbreviations are used for amino acids: A, alanine; I, isoleucine; N, asparagine; R, arginine; S, serine.

[18] N. Nelson and S. S. Taylor, *J. Biol. Chem.* **256,** 3743 (1981).

[19] L. D. Saraswat, M. Filutowicz, and S. S. Taylor, *J. Biol. Chem.* **261,** 11091 (1986).

[20] M. Solioz, *Trends Biochem. Sci.* (*Personal Ed.*) **103,** 309 (1984).

[21] P. F. Cook, M. E. Neville, Jr., K. E. Vrana, F. T. Hartl, and R. Roskoski, Jr., *Biochemistry* **21,** 5794 (1982).

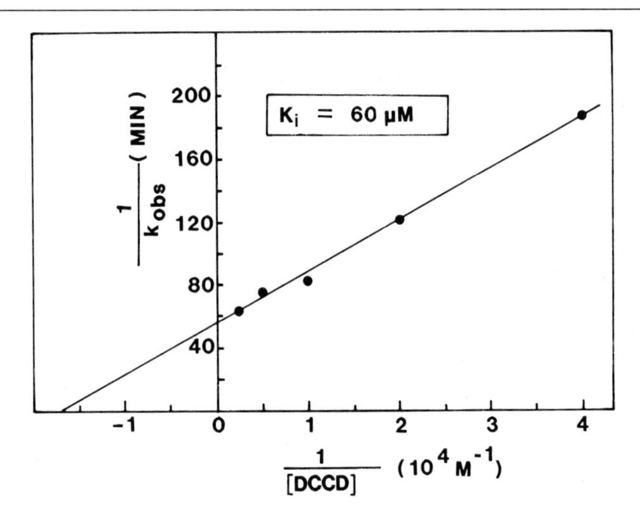


FIG. 3. Effect of DCCD concentration on the rate of inactivation of C-subunit. C-subunit (2.2 $\mu M$) was incubated with DCCD ranging in concentration from 25.0 to 400.0 $\mu M$. The data are presented in the form of a double-reciprocal plot of $1/K_{obs}$ vs. 1/[DCCD]. At each concentration of DCCD, the $K_{obs}$ was determined from the slope of ln $E/E_0$ versus time, where $E_0$ is the activity in the absence of DCCD.

The inactivation of the catalytic subunit also is dependent on pH. DCCD is more reactive at lower pH values since the carbodiimide must be protonated before it can react with carboxyl groups. Hence the lower the pH, the greater the rate of inactivation. It is important to remember, however, that the trapping of the electrophilic intermediate (Fig. 2) with a radiolabeled nucleophile and the rearrangement of radiolabeled carbodiimide to the *N*-acylurea product are both more favorable at higher pH values. When the DCCD reaction is carried out in the presence of MgATP, the catalytic subunit is completely protected from inactivation. In addition, the type I regulatory subunit blocks the effects of DCCD when it is bound to the catalytic subunit as part of the holoenzyme complex. Thus, at least one carboxyl group is modified in the catalytic subunit by DCCD, and modification of this residue(s) results in the loss of enzymatic activity.[22]

Having established that DCCD is an effective inhibitor of the catalytic subunit and that MgATP can protect against inactivation,[22] the next step is to identify the reactive carboxyl group(s). As discussed above, there are two different reaction pathways through which the modification of a carboxyl group by a carbodiimide would generate a radiolabeled amino acid (Fig. 2). Initially, it is not possible to radiolabel the catalytic subunit

[22] J. Toner-Webb and S. S. Taylor, *Biochemistry* **26,** 7371 (1987).

either in a direct manner with [$^{14}$C]DCCD (**III,** Fig. 2), or by trapping the reactive carboxyl groups with an excess of [$^{3}$H]aniline or [$^{14}$C]glycine ethyl ester (**IV,** Fig. 2). It is assumed, therefore, that a nucleophile in the protein reacts with the electrophilic intermediate formed by the reaction of DCCD with a carboxyl group. Lysine or cysteine residues are the most likely nucleophiles.[22] In order to circumvent this cross-linking, the catalytic subunit (26 $\mu M$, 2.5 mg protein/reaction) is treated with 3 m$M$ acetic anhydride (100 m$M$ stock in acetonitrile) prior to modification with DCCD. The reaction is initiated by adding acetic anhydride to the catalytic subunit in 100 m$M$ Tris-HCl (pH 8.0) and 5% (v/v) glycerol at room temperature, and it is terminated after 2 min by adding an excess of 2-mercaptoethanol. Under these conditions, acetic anhydride reacts irreversibly only with lysine residues. The covalent modification of the catalytic subunit by acetic anhydride has been characterized previously.[23] After treatment with acetic anhydride, the catalytic subunit is dialyzed against 50 m$M$ MES (pH 6.2) and 5% glycerol. [$^{14}$C]Glycine ethyl ester (3.5 mCi/mmol) is added to a concentration of 2 m$M$, and the reaction is initiated by adding DCCD (100 m$M$ in acetonitrile) to a final concentration of 0.4 m$M$ and incubating the samples at 37°. After 90 min, the samples are placed on ice and then dialyzed exhaustively against 50 m$M$ $NH_4HCO_3$ (pH 8.1) to remove excess radiolabel. The modified protein is then digested at 37° with TPCK-treated trypsin at a 1 : 50 (w/w) ratio of protease to catalytic subunit. The resulting tryptic peptides are separated by high-performance liquid chromatography (HPLC), and radioactivity profiles such as those shown in Fig. 4 are obtained. The pretreatment of the catalytic subunit with acetic anhydride drastically increases the incorporation of radiolabel into the protein (Fig. 4a and b). When the radiolabeled peptides are purified by HPLC and sequenced, the two sites of modification are identified. The major site corresponds to Asp-184 and the minor site to Glu-91. As shown in Fig. 4c, neither of those carboxyl groups are labeled when MgATP is incubated with the catalytic subunit prior to the modification with DCCD and [$^{14}$C]glycine ethyl ester, indicating that the ATP-binding site is still functional even though acetic anhydride-treated catalytic subunit is inactive.[24]

Because pretreatment with acetic anhydride is required to allow labeling of the protein with DCCD and [$^{14}$C]glycine ethyl ester, and [$^{14}$C]DCCD alone does not lead to radiolabeling of the catalytic subunit,[23] a crosslinking reaction between a lysine residue and a carboxyl group is apparently occurring. The next step is to identify the lysine residue that is becoming cross-linked to Asp-184. To do this, catalytic subunit (29.4 $\mu M$, 2.5 mg) is

[23] J. A. Buechler, T. Vedvick, and S. S. Taylor, *Biochemistry* **28,** 3018 (1989).
[24] J. A. Buechler and S. S. Taylor, *Biochemistry* **27,** 7356 (1988).

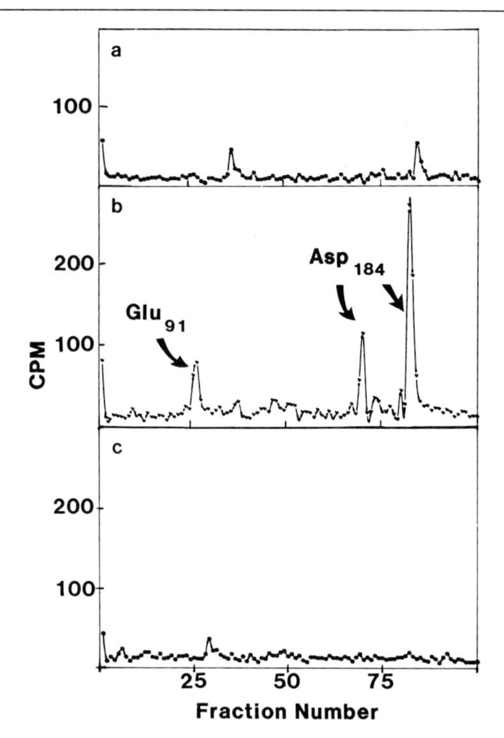


FIG. 4. Effect of modifying lysine residues on the DCCD-mediated incorporation of [$^{14}$C]glycine ethyl ester into the catalytic subunit. Catalytic subunit was treated with (b, c) or without (a) acetic anhydride followed by modification with DCCD in the presence of 2 m$M$ $MgCl_2$ (a, b) or 10 m$M$ ATP and 20 m$M$ $MgCl_2$ (c). After digestion with TPCK-treated trypsin, the resulting peptides were resolved by HPLC using an Altex 3200 system with a Vydac $C_4$ column (0.46 × 25 cm). The buffers employed were 10 m$M$ sodium phosphate, pH 6.9, and $CH_3CN$. Absorbance was monitored at 219 nm with a 100-30 Hitachi spectrophotometer (Hitachi, Los Angeles, CA) equipped with a flow-through cell, and at 280 nm with an SF 769 Kratos spectrophotometer (ABI Analytical/Kratos Division, San Jose, CA) equipped with a flow-through cell. The peptides were eluted with a 136-min linear gradient from 6 to 40% $CH_3CN$ followed by a 10-min linear gradient from 40 to 50% $CH_3CN$. The radioactivity was determined by counting 10% of each 1-ml fraction in Cytoscint.

first modified with 1.3 m$M$ DCCD (100 m$M$ in acetonitrile). Reactions are carried out in 50 m$M$ MES (pH 6.2) and 5% glycerol at 37° for 90 min. The final concentration of acetonitrile is 1.2%. After 90 min, the samples are cooled on ice and then dialyzed against 0.1 $M$ Tris-HCl (pH 8.3). Both samples are then denatured by adding solid urea to a final concentration of 8 $M$. Lysine residues are stoichiometrically modified by incubation with aliquots (75 μl) of [$^3$H]acetic anhydride (1 mCi/mmol, 100 m$M$ in acetonitrile) at room temperature over a 40-min period. All of the lysines should be radiolabeled except for the one involved in the cross-linking reaction. The modified catalytic subunit is dialyzed extensively against 50 m$M$ $NH_4HCO_3$ (pH 8.1) to remove excess reagent. Proteolysis is then carried out at 37° with a 1 : 50 (w/w) ratio of TPCK-treated trypsin to catalytic subunit. The tryptic peptides are separated by HPLC. The peptide containing Asp-184 is lost from the sample that has been treated with DCCD. A new peptide is then isolated and sequenced from the DCCD-treated sample. The amino acids that are cross-linked by DCCD are thus identified as Asp-184 and Lys-72.[25]

*Chemical Modification with EDC*

Unlike DCCD, EDC is very soluble in water. The two carbodiimides react with completely different carboxyl groups in proteins.[20] Presumably EDC modifies amino acids that are in more aqueous environments of a protein. Groups important for peptide recognition, such as the carboxyl groups that may interact with the basic amino acids flanking the phosphorylation site in the protein substrates of the catalytic subunit, should lie near the surface of the enzyme and thus be accessible to labeling by a water-soluble carbodiimide.

The catalytic subunit is modified with EDC under essentially the same conditions as described for DCCD. EDC irreversibly inhibits the catalytic subunit in the absence of substrates, and this inactivation is dependent on the amount of carbodiimide used (data not shown). Inhibition is also a first order process. These reactions are all carried out at 37° in 50 m$M$ MES (pH 6.2) and 5% glycerol. The effect of substrates on the inactivation of the catalytic subunit is determined. Catalytic subunit (22.4 $\mu M$) is incubated on ice for 30 min with MgATP (2 m$M$ ATP and 4 m$M$ $MgCl_2$), MgATP and an inhibitor peptide (2 m$M$ ATP, 2 m$M$ L-R-R-N-A-I, and 4 m$M$ $MgCl_2$), or inhibitor peptide alone (2 m$M$). Glycine ethyl ester (1 $M$ in water) is added to each sample to a final concentration of 20 m$M$. The reactions are initiated by adding EDC (100 m$M$ in water) to a final concentration of 4 m$M$ and incubating the samples at 37°. An aliquot of each tube is checked for activity at 15-min intervals. In contrast to DCCD, which does not inhibit the catalytic subunit in the presence of MgATP, substantial but not complete inactivation is seen with EDC in the presence of nucleotide (data not shown). EDC reacts with MgATP.[26] Thus, the results obtained in the presence of MgATP must be viewed with caution. The inhibition of the catalytic subunit with EDC is dependent on the concentration of MgATP. However, since the initial rate of inactivation is not affected by the concentration of MgATP, the decreased inhibition seen in the presence of high

[25] J. A. Buechler and S. S. Taylor, *Biochemistry* **28,** 2065 (1989).
[26] M. A. Gilles, A. Q. Hudson, and C. L. Borders, Jr., *Anal. Biochem.* (1989).

levels of MgATP is almost certainly due to breakdown of the reagent, EDC. The sample containing MgATP and an inhibitor peptide (L-R-R-N-A-I) is completely protected from inactivation by EDC, whereas peptide alone affords no protection (data not shown). Finally, the type I regulatory subunit, when bound to the catalytic subunit as part of the holoenzyme complex, also blocks the inactivation by EDC, even in the absence of nucleotide (data not shown).[27] The regulatory subunit contains a substrate-like sequence that occupies the substrate-binding site in the holoenzyme complex.[8,10]

In order to identify any specific carboxyl groups whose modification might account for the inhibition of activity, the catalytic subunit (31 $\mu M$, 2.7 mg) is modified with EDC and [$^{14}$C]glycine ethyl ester under three different conditions: (1) in the absence of substrates, (2) in the presence of MgATP, and (3) in the presence of both MgATP and inhibitor peptide. Each sample is incubated with substrates on ice for 30 min. [$^{14}$C]Glycine ethyl ester (3.3 mCi/mmol) is present at a final concentration of 2.8 m$M$. The reaction is initiated by adding EDC (100 m$M$ in water) to a concentration of 0.8 m$M$ and placing the samples in a 37° water bath. After 90 min, the protein solutions are placed on ice and dialyzed against 50 m$M$ $NH_4HCO_3$ (pH 8.1). Each sample is digested with TPCK-treated trypsin, and the tryptic peptides are separated in a manner similar to the procedure described earlier with DCCD (Fig. 5). The sites of modification in the apoenzyme are Glu-107, Glu-170, Asp-241, Asp-328, Asp-329, Glu-331, Glu-332, and Glu-333. Of those amino acids, Glu-170, Asp-328, Glu-332, and Glu-333 are labeled in the apoenzyme and in the presence of MgATP but are completely protected from modification in the presence of MgATP and the inhibitor peptide (Fig. 6).[27]

## Comparison of DCCD and EDC Labeling

Two carbodiimides having different solubility properties in water are used to identify carboxyl groups that might play important roles in the function of the catalytic subunit of cAMP-dependent protein kinase. The carboxyl groups modified by DCCD are completely different from those labeled by EDC (Fig. 7). As might be predicted on the basis of its hydrophobic nature, DCCD shows the most selectivity toward carboxyl groups in the catalytic subunit. Only one major site of labeling, Asp-184, is identified, whereas Glu-91 is modified to a much smaller extent. A third carboxyl group, Glu-230, also reacts with DCCD in the apoenzyme to a minor extent, but its reactivity is not altered by the presence of MgATP. Although

[27] J. A. Buechler and S. S. Taylor, *Biochemistry* **29,** 1937 (1990).

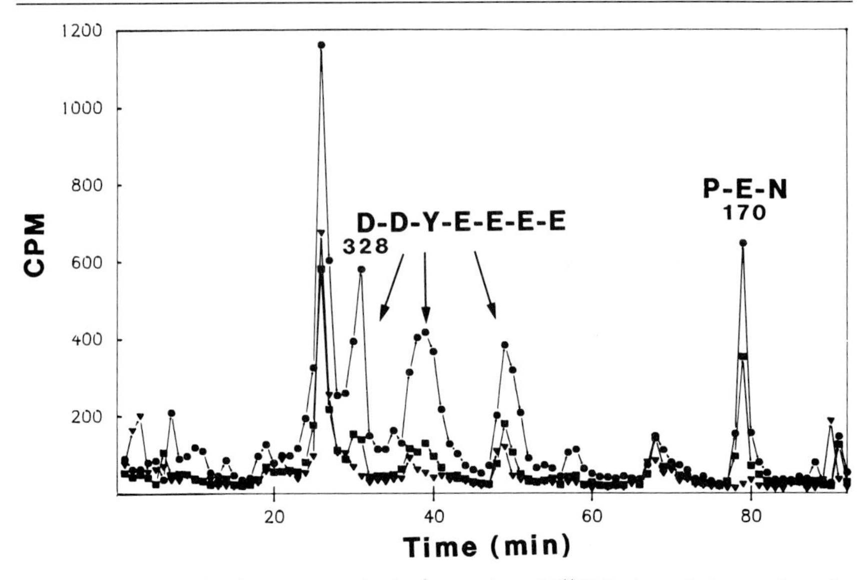


FIG. 5. Effect of substrates on the incorporation of [$^{14}$C]glycine ethyl ester into the catalytic subunit. Catalytic subunit was incubated on ice for 30 min with the following substrates prior to the addition of EDC: (a) 2 m*M* $MgCl_2$ (●), (b) 4 m*M* $MgCl_2$ and 2 m*M* ATP (■), (c) 4 m*M* $MgCl_2$, 2 m*M* ATP, and 35 $\mu M$ inhibitor peptide (▼). After digestion with TPCK-treated trypsin, the resulting peptides were separated by HPLC as described in Fig. 4. The tryptic peptides were eluted with a 100-min linear gradient from 6 to 31% $CH_3CN$ followed by a 15-min linear gradient from 31 to 50% $CH_3CN$ (not shown). Radioactivity was determined by counting 10% of each 1-ml fraction in Cytoscint. The sequence of the peptides in the indicated peaks are shown. (Reprinted with permission from *Biochemistry*.[25])

the intramolecular cross-linking of Asp-184 and Lys-72 mediated by DCCD makes identification of the modified amino acids somewhat more difficult, the ultimate confirmation of the cross-link provides additional evidence that Asp-184 is at the active site which had already been shown for Lys-72 based on earlier affinity labeling.[28] Lys-72, Asp-184, and Glu-91 are all conserved in every eukaryotic protein kinase,[5] suggesting that they will each have an invariant function in the protein kinase family.[6,7] Two roles that potentially would be conserved in all protein kinases are a general base catalyst and a ligand for $Mg^{2+}$ in the MgATP complex, suggesting that Asp-184 and Glu-91 may function in either of those capacities.

EDC most likely reacts with amino acids that are exposed to more hydrophilic environments than those modified by DCCD. The carboxyl groups labeled by EDC and protected from modification by MgATP and

[28] M. J. Zoller, N. C. Nelson, and S. S. Taylor, *J. Biol. Chem.* **256,** 10837 (1981).

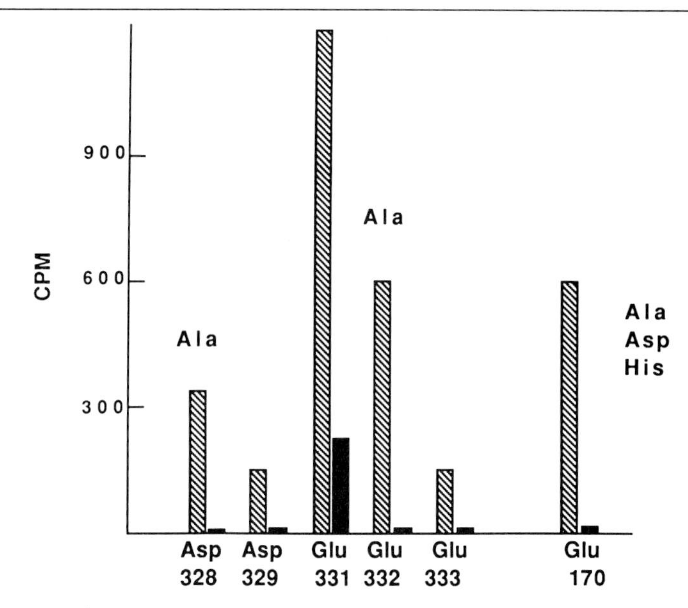


FIG. 6. Relative amounts of radioactivity incorporated into six carboxyl groups in the absence of substrates (hatched bars) and in the presence of MgATP and inhibitor peptide (solid bars).

inhibitor peptide, Glu-170, Asp-328, and Glu-332, may contribute to the recognition of arginine residues flanking the site of phosphorylation in the protein substrate. Glu-170 is located in a highly conserved region of the catalytic core shared by all the protein kinases. A carboxyl group is found at the equivalent position in other protein kinases recognizing basic amino acids in their substrates, and residues with a positive charge at the equiva-

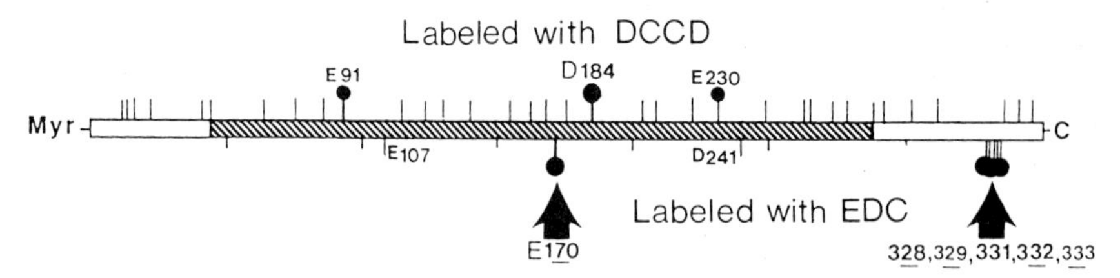


FIG. 7. Amino acids in the catalytic subunit that are modified by EDC and DCCD. The residues modified with DCCD are marked on top, while the carboxyl groups labeled with EDC are shown below the bar. The carboxylic acid residues that are not modified by either carbodiimide are represented by a small line extending out of the bar. The shaded region of the bar represents the conserved catalytic core in the catalytic subunit that is shared between the protein kinase.[5] (Reprinted with permission from *Biochemistry*.[27])

lent position are found in protein kinases with a specificity for acidic groups.[27] The cluster of carboxyl groups at the carboxy terminus of the catalytic subunit is located outside the conserved catalytic core, although Asp-328 and Glu-332 are invariant in all of the known catalytic subunit amino acid sequences. The amino acids in the catalytic subunit that interact with the arginines of the protein substrates would not be expected to be conserved in other protein kinases since the substrate specificity of the various kinases is different.

Several amino acids in the catalytic subunit are labeled by the carbodiimides, DCCD and EDC (Fig. 7). The importance of these amino acids eventually will be verified by site-directed mutagenesis and/or X-ray crystallography. In the interim, however, the chemical modification of the catalytic subunit with carbodiimides provides many insights into several structure–function relationships in the protein kinase family.

Acknowledgment

This work was supported in part by USPHS Grant GM19301 to S.S.T. J.A.B. was supported in part by U.S. Public Health Service Training Grant AM07233.

# [42] Peptide-Based Affinity Labeling of Adenosine Cyclic Monophosphate-Dependent Protein Kinase

*By* W. TODD MILLER

Affinity labeling has proved to be a useful method for probing the structures of protein kinases.[1] Peptide-based affinity-labeling experiments can be used to identify residues and domains in protein kinases which are involved in peptide substrate recognition. The catalytic subunit of the cAMP-dependent protein kinase presents a good model system for the study of kinase active sites; the enzyme is readily purified in milligram quantities from bovine heart,[2] and a synthetic peptide substrate is available (kemptide; Leu-Arg-Arg-Ala-Ser-Leu-Gly) which is phosphorylated as efficiently as natural protein substrates.[3] In addition, nuclear magnetic resonance (NMR) experiments have defined the conformation of kemptide

[1] S. S. Taylor, J. A. Buechler, and W. Yonemoto, *Ann. Rev. Biochem.* **59,** 971 (1990).

[2] J. G. Demaille, K. A. Peters, and E. H. Fischer, *Biochemistry* **16,** 3080 (1977).

[3] B. E. Kemp, D. J. Graves, E. Benjamini, and E. G. Krebs, *J. Biol. Chem.* **252,** 4888 (1977).

bound in the active site of the cAMP-dependent protein kinase.[4,5] Using solid-phase peptide synthesis, chemically or photochemically active amino acid analogs can be incorporated at different positions along the chain of kemptide. The results of peptide-based affinity-labeling experiments can then be correlated with the known conformation of bound substrate. Using this methodology the residues Cys-199 and Thr-197 have been assigned to the active site of the enzyme.[6,7] An advantage of photoaffinity labeling over chemical affinity labeling is that the reactive species is not targeted only to nucleophilic residues. As described below, labels containing photoactivated diaryl ketones are capable of insertion into amino acid side chains which are normally unreactive.

The general experimental strategy for peptide-based affinity labels and photoaffinity labels consists of three stages. In order to measure the extent of labeling, an inactivation assay is carried out. Free catalytic subunit is incubated with various concentrations of the label; for the photoactive peptides, the samples are photolyzed at an appropriate wavelength. Aliquots are withdrawn and assayed for phosphotransferase activity using the spectrophotometric assay described below. In order to distinguish specific from nonspecific affinity labeling, the experiments are repeated in the presence of substrates or substrate analogs. If the inactivation is directed to the active site, the catalytic subunit should be protected by the addition of MgATP, Mg-$\beta,\gamma$-methylene-ATP, and kemptide. Next, radioactive affinity label is prepared and used to determine the stoichiometry of binding of peptide to enzyme. Finally, proteolysis of the enzyme–peptide complex is carried out and high-performance liquid chromatography (HPLC) is used to isolate the radiolabeled fragments. The alkylated sites on the enzyme are identified by peptide sequencing.

## Synthesis and Characterization of Photoactive Peptides

The photoactive amino acid *p*-benzoyl-Phe (Fig. 1) used in these studies is synthesized from *p*-aminobenzophenone via the Meerwein reaction.[8,9] It can also be synthesized using 4-methyl-benzophenone as starting mate-

[4] P. R. Rosevear, D. C. Fry, A. S. Mildvan, M. Doughty, C. O'Brian, and E. T. Kaiser, *Biochemistry* **23,** 3161 (1984).

[5] H. N. Bramson, N. E. Thomas, W. T. Miller, D. C. Fry, A. S. Mildvan, and E. T. Kaiser, *Biochemistry* **26,** 4466 (1987).

[6] H. N. Bramson, N. Thomas, R. Matsueda, N. C. Nelson, S. S. Taylor, and E. T. Kaiser, *J. Biol. Chem.* **257,** 10575 (1982).

[7] S. Mobashery and E. T. Kaiser, *Biochemistry* **27,** 3691 (1988).

[8] W. T. Miller and E. T. Kaiser, *Fed. Proc.* **45,** 1548 (1986).

[9] W. T. Miller and E. T. Kaiser, *Proc. Natl. Acad. Sci. U.S.A.* **85,** 5429 (1988).

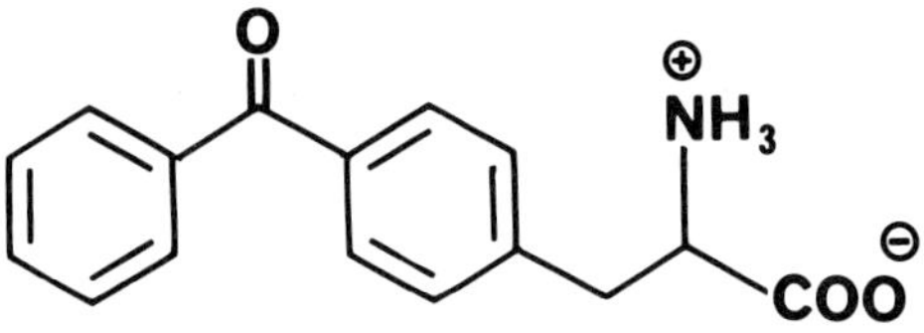

FIG. 1. The structure of *p*-benzoyl-Phe.

rial.[10] In order to incorporate *p*-benzoyl-Phe into kemptide analogs, the butyloxycarbonyl (Boc)-protected amino acid derivative is used in-stepwise solid-phase synthesis on Merrifield resin.[11] Because *p*-benzoyl-Phe is incorporated as a mixture of enantiomers, the diastereomeric peptides are subsequently separated by HPLC using a $C_{18}$ column.[9] In the experiments described here, peptides were synthesized in which residues Leu-1, Ala-4, and Ser-5 of kemptide were replaced by the photoactive amino acid:

Peptide L-1: Leu-Arg-Arg-Ala-(*p*-benzoyl-L-Phe)-Leu-Gly
Peptide D-1: Leu-Arg-Arg-Ala-(*p*-benzoyl-D-Phe)-Leu-Gly
Peptide L-2: (*p*-benzoyl-L-Phe)-Arg-Arg-Ala-Ser-Leu-Gly
Peptide L-3: Leu-Arg-Arg-(*p*-benzoyl-L-Phe)-Ser-Leu-Gly

The structures of the peptides have been confirmed by amino acid analysis, mass spectroscopy, and NMR. The photolability of these peptides is measured by photolysis in 50 m*M* 4-morpholinopropanesulfonic acid (MOPS), pH 7.5. Solutions are irradiated at 25° in a Rayonet RMR-500 photochemical reactor fitted with RMR-3500-Å lamps (Southern New England Ultraviolet Co., Hamden, CT). Samples are clamped 3 cm from the lamps in borosilicate glass tubes. The rates of photolysis of the peptides follow apparent first order kinetics (half-time approximately 70 sec).

The peptides are tested as competitive inhibitors of the catalytic subunit of the cAMP-dependent protein kinase in order to assess their affinities for the enzyme. A spectrophotometric assay is used which takes advantage of the change in absorbance at 430 nm upon phosphorylation of the serine residue in the peptide Leu-Arg-Arg-(*o*-nitro)Tyr-Ser-Leu-Gly.[12] The assays are carried out at pH 7.5 and typically contain 12.3 n*M* enzyme, 50

[10] J. C. Kauer, S. Erickson-Viitanen, H. R. Wolfe, Jr., and W. F. DeGrado, *J. Biol. Chem.* **261,** 10695 (1986).

[11] G. Barany and R. B. Merrifield, *in* "The Peptides" (E. Gross and J. Meienhofer, eds.), Vol. 2, p. 1. Academic Press, New York, 1979.

[12] H. N. Bramson, N. Thomas, W. F. DeGrado, and E. T. Kaiser, *J. Am. Chem. Soc.* **102,** 7156 (1980).

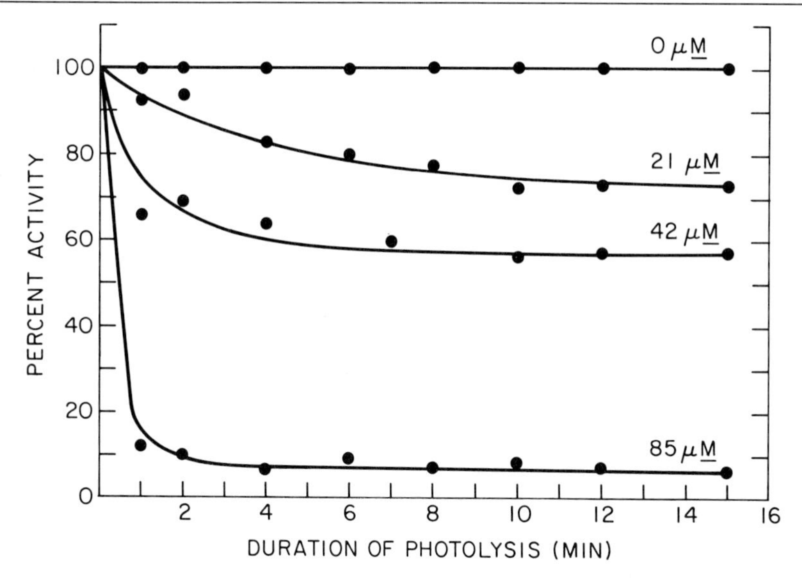


FIG. 2. Photoinactivation of the catalytic subunit (100 n$M$) by peptide L-1. The concentrations of the inactivator are given above the curves. The experiments were performed in 50 m$M$ MOPS buffer, pH 7.5, at 25°. One hundred percent activity was defined as the activity of the enzyme–peptide mixture before photolysis. (Reproduced from Miller and Kaiser.[9])

m$M$ MOPS, 10 m$M$ $MgCl_2$, 150 m$M$ KCl, 2.0 m$M$ ATP, 0.2 mg/ml bovine serum albumin (BSA), and 0.2 m$M$ dithiothreitol (DTT). Lineweaver-Burk plots are constructed and the data are replotted as described by Dixon.[13] The $K_i$ values for the peptides described here fall in the range of 100–200 $\mu M$, indicating that the kinase is tolerant to substitution at positions 1, 4, and 5 of kemptide.

## Peptide-Based Photoaffinity labeling

A typical photoaffinity-labeling experiment is shown in Fig. 2. Catalytic subunit (100 n$M$) in 50 m$M$ MOPS, pH 7.5, is incubated with various concentrations of peptide L-1, the kemptide analog in which the phosphorylatable serine residue is replaced by *p*-benzoyl-L-Phe. The mixture is photolyzed for up to 15 min, and aliquots are withdrawn and assayed for enzyme activity. The enzyme is photoinactivated in a time- and concentration-dependent fashion. As controls for the photoinactivation, the catalytic subunit is photolyzed in the presence of (1) *p*-benzoyl-phenylalanine,

[13] M. Dixon, *Biochem. J.* **55,** 170 (1953).

(2) prephotolyzed peptide L-1, and (3) peptide D-1, the kemptide analog containing *p*-benzoyl-D-Phe in place of serine. No photoinactivation is observed in any of these cases, indicating that tight binding and proper orientation at the enzyme active site is required for the effect. To strengthen this argument, protection experiments are carried out with other active site-directed reagents. The catalytic subunit is protected about 70% when preincubated with saturating concentrations of $MgCl_2$ and 2.0 m*M* ATP or 2.6 m*M* $\beta,\gamma$-methylene-ATP before photolysis in the presence of 100 $\mu M$ peptide L-1. When 270 $\mu M$ kemptide is added to this mixture, 95% protection is observed, consistent with a specific interaction between peptide L-1 and the active site of the catalytic subunit.

To prepare radioactive peptide, the peptide is dissolved in 300 m*M* potassium borate, pH 8.0. The solution is cooled to 0° and [1-$^3$H]acetic anhydride (20 Eq, 500 mCi/mmol; New England Nuclear, Boston, MA) is added. The reaction is typically complete after 10 min as judged by thin-layer chromatography (*n*-butanol/$H_2O$/$CH_3COOH$, 4 : 1 : 1; ninhydrin staining). The reaction mixture is diluted to 50 m*M* borate and loaded onto a CM-Sephadex C-25 column equilibrated in 25 m*M* ammonium acetate. The radioactive acetylated peptide is eluted from the column with a linear gradient to 250 m*M* ammonium acetate.

Radioactive acetylated peptide [100,000 disintegrations/min (dpm)/nmol] is used to determine the stoichiometry of binding of peptide to enzyme. The photolysis is carried out as before, with varying concentrations of peptide and enzyme. Aliquots are removed at various time points and assayed for enzyme activity. Extra aliquots (250 $\mu$l) are also removed to quantitate incorporation of label into the catalytic subunit. To these aliquots is added 100 $\mu$l of 10 mg/ml BSA together with 500 $\mu$l of 50 m*M* MOPS, pH 7.5. The protein in this solution is precipitated by addition of 100 $\mu$l of 70% (w/v) $HClO_4$. The precipitate is collected on a Millipore (Bedford, MA) HA filter and washed extensively with 7% $HClO_4$. The filters are dried, suspended in Hydrofluor (National Diagnostics, Manville, NJ), and counted in a Beckman (Fullerton, CA) LS/7000 liquid scintillation counter. A linear relationship is established between photoinactivation and covalent incorporation of label. The stoichiometry of the inactivation by peptide L-1 is approximately 1 : 1.[9]

## Identification of Modified Residues

To identify the residues on the protein which are modified by peptide inactivators, cleavage of the inactivated enzyme with CNBr and V8 protease is carried out to produce fragments suitable for HPLC analysis. Catalytic subunit (2 mg) is irradiated together with [1-$^3$H]acetyl-peptide (100,000 dpm/nmol) until more than 90% loss of activity results. The excess photolabel is removed by filtration through a Centricon-30 ultrafiltration unit (Amicon, Beverly, MA; molecular weight cutoff, 30K).[7] The modified enzyme is dissolved in 70% formic acid and treated with a 50-fold excess of CNBr (24 hr, 25°). The resultant solution is passed over a Centricon-10 filter (molecular weight cutoff, 10K). For peptide L-1, approximately 60% of the original radioactivity is recovered in the filtrate, indicating that the label is present in peptidic fragments from the CNBr cleavage reaction that are of molecular weight less than 10K. This low-molecular-weight fraction is analyzed directly by HPLC. The fraction containing the remaining 40% of the radioactivity is recovered and digested with *Staphylococcus aureus* V8 protease (1:50, w/w, V8:catalytic subunit; 50 m*M* $NH_4HCO_3$, pH 7.8; 12 hr at 37°).

Peptide fragments from the CNBr cleavage and the V8 digest are then analyzed by HPLC using a reversed-phase $C_{18}$ column (Altex, San Ramon, CA; 4.6 mm × 25 cm). The column is equilibrated in 200 m*M* $NaClO_4$, 25 m*M* $NaH_2PO_4$, pH 2.5, and the radioactive fragments are purified using a linear gradient from 0 to 50% $CH_3CN$ in the same buffer.[14] In the case of peptide L-1, the major peak of radioactivity for the small ($M_r$ < 10K) CNBr fragments elutes at 46 min (Fig. 3B). Major peaks of radioactivity for the larger CNBr fragments elute at 36 and 45 min (Fig. 3C). For both samples, fractions are collected, concentrated, and further purified by HPLC on a $C_{18}$ column using a 60-min gradient, 0 to 50% $CH_3CN$, in 0.1% trifluoroacetic acid (TFA). Radioactive fractions from the second round of HPLC are applied to a gas-phase sequenator and carried through the Edman degradation. The peptide obtained from the small CNBr fragment has the following sequence: Glu-Tyr-Val-Pro-Gly-X, where X is an unidentified phenylthiohydantoin. This sequence corresponds to the amino acids between positions 120 and 124 of the catalytic subunit,[15] with Gly-125 being the site of attachment. The following peptide sequence is obtained from the V8 digest of the larger CNBr fragment: Tyr-Val-Pro-Gly-Gly-Glu-X. In this case, the sequence corresponds to residues 121 through 126 of the enzyme, with sequence termination and observation of an unknown phenylthiohydantoin at the next round of sequencing. These results indicate modification at Met-127. Modification of Met-127 by peptide L-1 apparently blocks the CNBr reaction at that position, generating the fragment of molecular weight greater than 10K. Residues Gly-125 and Met-127 are contained in the CNBr peptide beginning at Met-119 and ending at Met-127 of the catalytic subunit. As a reference, unmodified catalytic

[14] J. L. Meek and Z. L. Rossetti, *J. Chromatogr.* **211,** 15 (1981).

[15] S. Shoji, D. C. Parmelee, R. D. Wade, S. Kumar, L. H. Ericsson, K. A. Walsh, H. Neurath, G. L. Long, J. G. Demaille, E. H. Fischer, and K. Titani, *Proc. Natl. Acad. Sci. U.S.A.* **78,** 848 (1981).

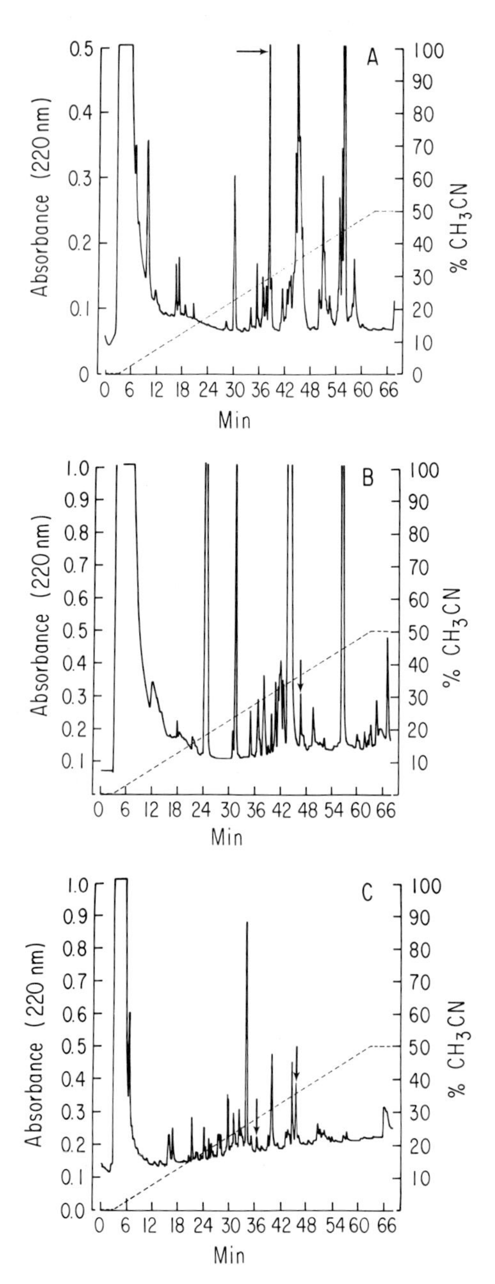


FIG. 3. HPLC purification of the peptide fragments. The solvent system is described in the text. (A) Elution profile of the unmodified catalytic subunit after cleavage with CNBr. The arrow indicates the peptide from Met-119 to Met-127, which was identified by peptide sequencing. (B) Elution profile of the small ($M_r < 10$ K) CNBr fragments of the enzyme after photoinactivation by peptide L-1. The arrow indicates the peak of radioactivity. (C) Elution profile of the larger ($M_r > 10$K) CNBr fragments, after digestion with V8 protease. The arrows indicate the two peaks of radioactivity. (Reproduced from Miller and Kaiser.[9])

subunit (1 mg) is carried through the CNBr cleavage and HPLC analysis. The major HPLC peaks are collected, rechromatographed, and sequenced. The unmodified CNBr peptide from Met-119 to Met-127 elutes at 39 min in the initial gradient (Fig. 3A). By comparison with Fig. 3B, it can be seen that this peak is missing or reduced in the HPLC profile of the modified catalytic subunit.

## Other Peptide-Based Photoaffinity Labels

Analogous experiments to those described above have been carried out with peptide L-2, the peptide in which Leu-1 of kemptide is replaced by *p*-benzoyl-L-Phe. Photolysis of this peptide (740 $\mu M$) together with enzyme (170 n$M$) for 15 min leads to 55% photoinactivation of the catalytic subunit. To demonstrate that the inactivation is active site directed, protection experiments with MgATP are carried out. The catalytic subunit is preincubated with $MgCl_2$ (18.5 m$M$) and ATP (2.7 m$M$) for 5 min before photolysis with peptide L-2. In these experiments, 89% protection from inactivation is observed (vs a control with no MgATP). Radioactive peptide L-2 (prepared by acetylation on the N-terminus with [1-$^3$H]acetic anhydride) is used to determine the stoichiometry of inactivation. A value of 0.3 mol peptide/mol enzyme is obtained in this experiment under conditions in which the catalytic subunit is 50% inactivated. To identify the region of the catalytic subunit modified by this photoaffinity label, the inactivated enzyme (2 mg) is digested with CNBr and V8 protease and analyzed as described above for peptide L-1. Edman degradation of the $^3$H-labeled peptide fragment after HPLC purification gives the following sequence: X-Leu-Lys-Lys-Gln-Glu-Asn-Pro-Ala-Gln-Asn-Thr-X, where residues marked X are unidentified phenylthiohydantoins. This corresponds to the sequence Leu-26 through Thr-36 of the catalytic subunit. The probable site of attachment of the photolabel is Ala-37 because of the drop in yield and sequence termination at that point.[16]

Peptide L-3, in which Ala-4 of kemptide is replaced by *p*-benzoyl-L-Phe, has also been tested as a photoaffinity label of the catalytic subunit of the cAMP-dependent protein kinase. When photolyzed together with the enzyme under the conditions described above, it is an efficient photoinactivator (60% inactivation using 1.4 m$M$ peptide L-3). Surprisingly, radioactive peptide L-3 does not incorporate into the catalytic subunit during photoinactivation (only 0.05 mol/mol), suggesting that a different mechanism of inactivation takes place with peptide L-3 than with peptides L-1 and L-2. Possibilities for such mechanisms have been discussed in other

[16] W. T. Miller and E. T. Kaiser, 1988, unpubished observations.

cases in which photoinactivation does not lead to covalent modification.[17] In any case, the data suggest a model in which the side chain of Ala-4 in kemptide (replaced in peptide L-3) points out of the peptide-binding site, in contrast to the side chains of Leu-1 and Ser-5. The data are consistent with NMR experiments, in which the side chain of Ser-5 points toward the metal–nucleotide complex and the side chain of Ala-4 points away.[5]

These experiments illustrate both the strengths and the limitations of the affinity-labeling approach. The methodology allows the identification of enzyme residues which are involved in substrate recognition. One region important in peptide binding in the catalytic subunit of the cAMP-dependent protein kinase contains the amino acids Gly-125 and Met-127 and has homology to tyrosine kinases. Ala-37 in the N-terminal domain of the kinase apparently also plays a role in substrate recognition. In order to definitively assess the functional contributions of these residues, however, it is necessary to carry out site-directed mutagenesis. Comparisons of the properties of the wild-type and mutant enzymes allow a clearer understanding of the functions of the modified residues.

### Acknowledgments

The author wishes to thank Drs. Neal Bramson, Michael Doughty, and Shahriar Mobashery for their contributions to this project, and to acknowledge especially the late Professor E. T. Kaiser, in whose laboratory at The Rockefeller University this work was carried out.

[17] H. Bayley, "Photogenerated Reagents in Biochemistry and Molecular Biology." North-Holland/Elsevier, Amsterdam, 1983.

# [43] Crystallization of Catalytic Subunit of Adenosine Cyclic Monophosphate-Dependent Protein Kinase

*By* JIANHUA ZHENG, DANIEL R. KNIGHTON, JOSEPH PARELLO, SUSAN S. TAYLOR, and JANUSZ M. SOWADSKI

Progress in fast acquisition of crystallographic data, due to the use of powerful X-ray sources (synchrotron radiation) and the development of area detectors,[1] has increased interest in the process of protein crystalliza-

[1] N. H. Xuong, D. Sullivan, C. Nielsen, and R. Hamlin, *Acta Crystallogr., Sect. B* **B41,** 267 (1985).

tion. The successful crystallization of a protein remains one of the most difficult and elusive tasks in X-ray crystallography. Obtaining high-quality crystals, however, is a crucial prerequisite for determining protein structure. Advances in molecular biology have provided the crystallographer with an array of new techniques to help in the task of crystallizing proteins. First, highly efficient vectors offer the possibility of obtaining sufficient amounts of protein for extensive crystallization trials. Second, expressing many eukaryotic proteins in *Escherichia coli* has eliminated posttranslational modifications which often complicate crystallization.[2] Third, cloning and expressing proteins allows for the generation of mutant proteins that can serve as effective alternatives to wild-type proteins in crystallization experiments.[3] Fourth, site-directed mutagenesis can be used to provide the amino acids necessary for heavy atom derivatization.[4] In addition, expressing protein in selenomethionine-rich media has enabled the direct cotranslational incorporation of the heavy atom derivative as the protein is synthesized.[5]

Nucleation and crystal growth can be monitored by several physical methods. A study of crystal growth by freeze-etch electron microscopy at nearly molecular resolution, reported for the tetragonal form of lysozyme, establishes that crystal growth is dependent on supersaturation conditions.[6] While biochemical methods are well adapted for detecting impurities or sequence heterogeneities in biological macromolecules, the characterization of the physical parameters of a protein solution relevant to the crystallization process is possible only by physiochemical methods. Scattering techniques such as light scattering[7] or small-angle neutron scattering[8] are the most useful for defining conformational heterogeneities or aggregation states of the protein in relation to the conditions leading to crystal nucleation and growth. Whether monodispersity in the protein solution up to supersaturation conditions is a prerequisite for nucleation and crystal growth is an open question. Aggregation and/or nonspecific interactions are likely to prevent a protein from making the appropriate contacts to form a crystal and lead to the formation of disordered amor-

[2] F. A. O. Marston, "DNA Cloning: A Practical Approach" (D. M. Glover, ed.), Vol. 3, p. 59. IRL Press, Oxford, 1987.

[3] G. Kapedia, J. Reizer, S. L. Sutrina, M. H. Saier, P. Reddy, and O. Herzberg, *J. Mol. Biol.* **212,** 1 (1990).

[4] G. F. Hatfull, M. R. Sanderson, P. S. Freemont, P. R. Raccuia, N. D. Grindly, and T. A. Steitz, *J. Mol. Biol.* **208,** 661 (1989).

[5] W. A. Hendrickson, J. R. Horton, and D. M. Le Master, *EMBO J.* **9,** 1665 (1990).

[6] S. D. Durbin and G. Feher, *J. Mol. Biol.* **212,** 763 (1990).

[7] V. Mikol, J.-L. Rondeau, E. Hirsch, and R. Giegé, *J. Cryst. Growth* **99,** 1017 (1990).

[8] R. Geigé, B. Lorber, J.-P. Ebel, D. Morres, J. C. Thierry, B. Jacrot, and G. Zaccaï, *Biochimie* **54,** 357 (1982).

phous precipitates. However, it was suggested on the basis of light scattering that the formation of tetragonal lysozyme nuclei involves the addition of aggregates preformed in the bulk solution and not of lysozyme monomers.[9] In this respect, it would be important to know the exact nature of these putative preformed aggregates in terms of their stoichiometry and self-organization. It is likely that the use of scattering techniques with shorter wavelengths (X rays, neutrons) will be well suited for monitoring the state of a protein solution with regard to its tendency to aggregate under the varied conditions used for its crystallization. Carefully measuring a radius of gyration may help in finding conditions suitable for crystallization. We have addressed this question by small-angle neutron scattering (SANS) with solutions of the catalytic subunit of cAMP-dependent protein kinase (cAPK) and found that the conformation of the enzyme is altered by the presence of a high-affinity, small-sized peptide inhibitor.[10]

The value of such techniques to the crystallographer, however, as well as their potential as interpretive tools that can be used in tandem with crystallographic work, depends largely on the soundness of the crystallographic results. The main hurdle still is to obtain good crystals. The pursuit of this task may be significantly aided by other techniques that may result in observations that extend the scope of the crystallographic results.

Extensive crystallographic work on the catalytic subunit of cAPK has established reproducible crystallization conditions that generate a set of different crystal forms. These different crystal forms almost certainly represent different conformational states of the enzyme and provide a comparative framework that significantly enlarges the scope of the crystallographic work.

The crystallographic work described here was carried out on the catalytic subunit purified from porcine heart,[11] as well as on the catalytic subunit expressed in and purified from *E. coli*.[12] Through systematic adjustments of crystallization conditions, crystals representing three forms of the protein—the apoenzyme, a binary complex with a specific peptide inhibitor, and a ternary complex with a specific peptide inhibitor and MgATP—were obtained in four different space groups.[13–15] Each of these crystals is described separately. Three factors influencing the reproducibil-

[9] M. L. Pusey, *Proc. Int. Conf. Crystall. Biol. Macromol., 3rd, 1989* (1989).

[10] J. Parello, P. A. Timmins, J. M. Sowadski, and S. S. Taylor, *J. Mol. Biol. Chem.* (submitted for publication).

[11] N. C. Nelson and S. S. Taylor, *J. Biol. Chem.* **256,** 3743 (1981).

[12] L. W. Slice and S. S. Taylor, *J. Biol. Chem.* **264,** 20940 (1989).

[13] J. M. Sowadski, N. H. Xuong, D. Anderson, and S. S. Taylor, *J. Mol. Biol.* **182,** 617 (1985).

[14] D. R. Knighton, N. H. Xuong, S. S. Taylor, and J. M. Sowadski, *J. Mol. Biol.* in press.

TABLE I
DETERGENTS USED DURING CRYSTALLIZATION OF MONOCLINIC CRYSTAL ($P2_1$) OF CATALYTIC SUBUNIT

| Detergent[a] | Concentration (%) |
|---|---|
| β-OG | 0.50 |
| MEGA-8 | 1.00 |
| MEGA-10 | 0.17 |

[a] β-OG, Octyl-β-D-glucopyranoside; MEGA-8, octanoyl-*N*-methylglucamide; MEGA-10, decanoyl-*N*-methylglucamide.

ity and control of crystallization conditions are discussed. Crystallographic results correlate with data from the SANS experiments, and indicate that the enzyme is flexible and that crystals in different space groups very likely represent different conformational states of the enzyme. The best crystals are formed by the binary and ternary complexes of the recombinant catalytic subunit and thus emphasize the importance of being able to take advantage of recombinant techniques.

## Materials

Reagents are obtained from the following sources: *threo*-1,4-dimercapto-2,3-butanediol (DTT, dithiothreitol); (Aldrich, Milwaukee, WI); *N,N*-bis(2-hydroxyethyl)glycine (Bicine); (Aldrich); methanol (Fisher Scientific, Pittsburgh, PA); ammonium acetate (Aldrich); polyethylene glycol (Dow).

The peptide inhibitors Ala-kemptide and $IP_{20}$ are synthesized at the La Jolla Cancer Research Foundation and modified in our laboratory.

The catalytic subunit is purified from porcine heart according to Nelson and Taylor.[11] The recombinant catalytic subunit is purified from *E. coli* according to Slice and Taylor.[12]

### *Monoclinic Crystals (Space Group $P2_1$) of Catalytic Subunit Purified from Porcine Heart*

Crystallization of monoclinic crystals was reported previously by Sowadski *et al.*[13] Variations of the crystal growth techniques, usually dialysis and a combination of vapor diffusion and dialysis, have not resulted in significant improvement in the quality of this crystal form. In addition, several detergents (as described in Table I) were used in varying amounts.

[15] J. Zheng, D. R. Knighton, N. H. Xuong, S. S. Taylor, and J. M. Sowadski, *J. Mol. Biol.* (submitted for publication).

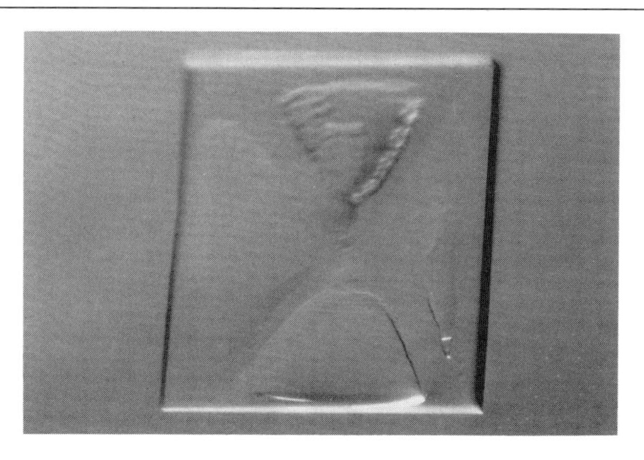

FIG. 1. Monoclinic crystal of the apoenzyme of the catalytic subunit from porcine heart. Space group: $P2_1$; $a$ = 64.24 Å; $b$ = 143.58 Å; $c$ = 48.40 Å; $\alpha = \gamma = 90°$; $\beta$ = 106.9°. Unit cell volume: 427,110.0 $Å^3$. Two molecules per asymmetric unit (asu), assuming 2.7 $Å^3$/Da.

This led to a marked improvement in the overall morphology of crystals and in significantly less twinning; however, the net result was negligible and diffraction of X rays by the crystals still did not exceed 3.5-Å resolution (Fig. 1).

### *Cubic Crystals (Space Group $P4_132$) of Catalytic Subunit Purified from Porcine Heart*

The single most important modification in the crystallization protocol that led to the formation of crystals in a different space group was the careful selection of polyethylene glycol in combination with various low-molecular-weight alcohols. This work is described in detail by Knighton *et al.*[14] Only the main methodological observations, which may be useful in crystallization of other proteins, will be discussed here.

Commercially available polyethylene glycols contain various contaminants that may cause problems in the achievement of stable and reproducible crystallization conditions. Since it is the presence of these contaminants that may have been responsible for the relatively moderate diffraction of the monoclinic crystal described above, all commercially available polyethylene glycols (PEG) were examined with the aim of detecting the presence of ionic species. The results of conductivity measurements are summarized in Table II.

The lowest level of ionic contaminants was detected in PEG manufactured by Dow. It is this PEG that was selected for further crystallization experiments. PEG from other sources appeared to be generally more

TABLE II
CONDUCTIVITY MEASUREMENTS FOR VARIOUS POLYETHYLENE GLYCOLS USED IN CRYSTALLIZATION OF CATALYTIC SUBUNIT[a]

| Polyethylene glycol source | Molecular weight | Conductivity (μS/cm) |
|---|---|---|
| Sigma | 400 | 6, 30 |
| | 1450 | 33.6 |
| | 8000 | 13.8 |
| Carbowax | 400 | 5.2 |
| | 1450 | 13.8 |
| | 8000 | 14.3 |
| BASF | 400 | 4.5 |
| | 1450 | 37.0 |
| | 8000 | 19.1 |
| Dow | 400 | 3.9 |
| | 1450 | 3.7 |
| | 8000 | 14.2 |

[a] In each measurement the concentration of any given polyethylene glycol was constant and equal to 4% (w/v) in water with a constant conductivity of 1.5 μS/cm.

contaminated and also exhibited large differences in contamination between batches. In our experiments, several molecular weights of PEG were used along with several low-molecular-weight alcohols. The results are presented in Table III.

The use of PEG 400 and methanol resulted in the best crystals in the cubic space group. These crystals, along with their diffraction data, are

TABLE III
POLYETHYLENE GLYCOLS DIFFERING BY MOLECULAR WEIGHTS AND ALCOHOLS USED IN CRYSTALLIZATION OF CUBIC CRYSTAL FORM ($P4_132$) OF CATALYTIC SUBUNIT

| | Alcohol[a] | | | |
|---|---|---|---|---|
| Molecular weight | Methanol | Ethanol | Propanol | *n*-Butanol |
| 6000 | Δ | — | — | — |
| 3350 | Δ | — | — | — |
| 1450 | Δ | — | — | — |
| 400 | Δ | Δ | — | — |

[a] Δ denotes successful crystallization.

FIG. 2. Cubic crystal of the apoenzyme of catalytic subunit from porcine heart. Space group: *P4₁32*; $a = b = c = 169.24$ Å. Unit cell volume: 4,847,419.24 Å$^3$. Two molecules per asu assuming 2.7 Å$^3$/Da.

described in Fig. 2. The diffraction of these crystals was no better than that of the monoclinic crystals; hence, although diffraction was observable at 3.5 Å, useful data do not extend beyond 4 Å.

*Hexagonal Crystals (Space Group $P6_122$) of Catalytic Subunit Purified from Porcine Heart*

Both the monoclinic ($P2_1$) and cubic ($P4_132$) crystal forms of the catalytic subunit were formed using the apo form of the enzyme. Since the diffraction of both crystal forms was only moderate, efforts were made to improve the quality of the crystals by cocrystallizing the enzyme with a peptide inhibitor and MgATP under otherwise identical crystallization conditions.

Inhibition studies with the catalytic subunit identified several inhibitors with differing affinities toward the catalytic subunit.[11] The most potent of the inhibitors is the 20-amino acid peptide inhibitor, $IP_{20}$, which includes residues 5–24 of the thermostable protein kinase inhibitor (PKI) originally described by Walsh and colleagues.[16] This 20-amino acid segment that lies near the N terminus of PKI exhibits nearly the same inhibiting potency as the intact 75-residue PKI. All of the inhibitors used for cocrystallization with the catalytic subunit share the consensus recognition sequence, Arg-Arg-X-Ala, but differ in length and in sequence. A modified form of $IP_{20}$, where Tyr-7 was diiodinated in order to provide a heavy atom

[16] H.-C. Cheng, B. E. Kemp, R. B. Pearson, A. J. Smith, L. Misconi, S. M. Van Patten, and D. A. Walsh, *J. Biol. Chem.* **261,** 989 (1986).

TABLE IV
PEPTIDE INHIBITORS USED FOR COCRYSTALLIZATION WITH CATALYTIC SUBUNIT

| Complex | Molar ratio: Protein | $Mg^{2+}$ | ATP | Peptide | Space group |
|---|---|---|---|---|---|
| Apo | 1 | 0 | 0 | 0 | *P4₁32* |
| Ternary complex | | | | | |
| With MgATP and Ala-kemptide | 1 | 5 | 20 | 3 | $P4_132$ |
| With MgATP and $IP_{20}$(I) | 1 | 5 | 20 | 3, 6 | $P4_132$ |
| With MgATP and $IP_{20}$ | 1 | 5 | 20 | 1 | $P6_122$ |

| Inhibitor | Sequence | $K_i$ (n*M*) |
|---|---|---|
| Ala-kemptide | L**RR**AALG[a] | 348,000 |
| $IP_{20}$(I) | TTY*ADFIASGRTG**RR**NAIHD[b] | Unknown |
| $IP_{20}$ | TTYADFIASGRTG**RR**NAIHD | 2.3 |

[a] Consensus recognition sequence indicated in boldface type.
[b] I, Iodinated; Y*, Diiodinated tyrosine.

derivative, was also utilized. While the exact binding affinity of this peptide was not measured, the phosphorylation of Tyr-7 in PKI by the EGF receptor kinase resulted in a 10-fold decrease in binding affinity.[17] The different peptide inhibitors used in the cocrystallization of the catalytic subunit are presented in Table IV.

As Table IV shows, the addition of $IP_{20}$ and MgATP caused the catalytic subunit to crystallize in a hexagonal space group, $P6_122$. Interestingly, cocrystallization of the catalytic subunit with MgATP and both $IP_{20}$ (diiodinated) and the Ala-kemptide inhibitor resulted in crystals having the cubic space group, just as in the case of the free catalytic subunit. The new hexagonal crystal showed significantly better diffraction characteristics than the crystals of the apo form and useful data extends to 3.0 Å. The crystal is shown in Fig. 3. In comparison, the crystals in the cubic space group diffracted to 3.2 and 4.0 Å, respectively.

The fact that in our experiments the catalytic subunit crystallized in the hexagonal space group on the introduction of $IP_{20}$ and MgATP, whereas in its apo form it crystallized in the cubic space group using otherwise identical crystallization conditions, indicates that the hexagonal crystal may arise as a result of a different conformational state of the enzyme. This observation is corroborated by the significantly better quality of the hexagonal crystal.

[17] S. M. Van Patten, G. J. Heisermann, H.-C. Cheng, and D. A. Walsh, *J. Biol. Chem.* **262,** 3398 (1987).

FIG. 3. Hexagonal crystal of ternary complex of porcine catalytic subunit with $IP_{20}$ and MgATP. Space group: $P6_122$; $a = b = 80.16$ Å; $c = 288.07$ Å; $\alpha = \beta = 90°$; $\gamma = 120°$. Unit cell volume: 1,603,083.70 Å$^3$. One molecule per asu, assuming 2.7 Å$^3$/Da.

### *Orthorhombic Crystals (Space Group $P2_12_12_1$) of the Recombinant Murine Catalytic Subunit*

One of the most promising directions for combining crystallographic methods with those of molecular biology is the development of highly effective vectors for expressing large amounts of protein for crystallization. Expression of proteins in *E. coli* also provides a mechanism for eliminating posttranslational modifications which may hinder crystallization and in addition enables one to carry out structure–function studies on mutant forms of the protein, such as ones containing introduced amino acids known to be easily modified by heavy atom derivatives.

The recombinant murine catalytic subunit, whose expression and purification was described previously,[12] is devoid of myristic acid at the N terminus and differs by nine amino acids from the porcine heart catalytic subunit used in the earlier crystallizations. This recombinant enzyme was crystallized under conditions described by Zheng *et al.*[15] Interestingly, the ternary complex of recombinant catalytic subunit with $IP_{20}$ and MgATP crystallized not in the hexagonal space group, as did the ternary complex of the porcine heart enzyme, (Fig. 3) but in an orthorhombic space group, $P2_12_12_1$. This crystal exhibited the best quality yet, and diffracts to 2.7-Å resolution. The crystal, along with its diffraction data, is presented in Fig. 4.

When the enzyme was subsequently cocrystallized with $IP_{20}$ but without MgATP, the resulting crystal was again in the orthorhombic space group and exhibited similarly good diffraction characteristics.

FIG. 4. Orthorhombic crystal of ternary complex of murine recombinant catalytic subunit with $IP_{20}$ and MgATP. Crystals of the binary complex of murine recombinant catalytic subunit with $IP_{20}$ were isomorphous with the ternary crystals. Space group: $P2_12_12_1$; $a$ = 73.70 Å; $b$ = 76.26 Å; $c$ = 80.74 Å. Unit cell volume: 453,790.61 $Å^3$. One molecule per asu, assuming 2.7 $Å^3$/Da. (It has been observed that the crystals of the ternary complex exhibit variability in unit cell dimensions depending on enzyme preparation.)

### *Small-Angle Neutron Scattering Studies in Solution*

Small-angle neutron scattering established that in the absence of inhibitor and MgATP the enzyme adopts a more expanded conformation than that adopted by the enzyme in the binary complex of the enzyme and the peptide inhibitor $IP_{20}$ or the ternary complex of the enzyme, $IP_{20}$ and MgATP.[10] Whereas the radius of gyration ($R_g$) of the apoenzyme is 19.5 ± 0.5 Å, this is decreased to 18.7 ± 0.5 Å in the case of the binary and ternary complexes. These results with the recombinant catalytic subunit of cAPK are to be compared with previous results established by small-angle X-ray scattering (SAXS) in solution with yeast hexokinase, a kinase acting on the small-sized substrate glucose.[18] A decrease of the radius of gyration of 0.9 Å was observed upon the binding of glucose to hexokinase. Such a decrease in the $R_g$ of the enzyme in solution has been interpreted to result from a glucose-induced conformational change. Such a conformational change was deduced to occur previously based on the crystal structures of a yeast hexokinase isoenzyme determined in the presence and in the absence of glucose.[19] It therefore appears that the large atomic displacements inferred from crystallographic studies, which occur in the hexokinase molecule upon substrate binding, reflect a change in the radius

[18] R. C. McDonald, T. A. Steitz, and D. M. Engelman, *Biochemistry* **18,** 338 (1979).

[19] C. M. Anderson, F. H. Zucker, and T. A. Steitz, *Science* **204,** 375 (1979).

of gyration of the enzyme of less than 1 Å, consistent with the Rg's established by scattering in solution. A major conformational event, the closing of the catalytic cleft upon substrate binding, is clearly related therefore to a rather small decrease of the radius of gyration of hexokinase. In this light, an $R_g$ decrease of 0.8 Å upon binding of the $IP_{20}$ peptide inhibitor to the cAPK catalytic subunit could reflect a major conformational change of the enzyme molecule. Since both complexes, the binary and the ternary, of the recombinant catalytic subunit of cAPK display practically the same value for their radii of gyration in solution as determined by SANS,[10] it is concluded that the enzyme adopts the same overall compact conformation in the binary and the ternary complex. In agreement with such a conclusion, we observe that the two complexes crystallize in the same (orthorhombic) space group with identical cell parameters. In the case of the catalytic subunit of cAPK, it is likely that it is the similarities between the conformations adopted by the enzyme in both the binary and ternary complexes, as a consequence of the strong binding of the $IP_{20}$ peptide inhibitor, which are responsible for the identical molecular packing in the crystal. Work is now progressing toward the elucidation of the crystal structure of both enzyme complexes at atomic resolution and we will therefore be able to calculate the corresponding radii of gyration of both complexes and compare these values with the experimental ones measured in the solution state by SANS.[10]

During our SANS studies we noticed that the recombinant apoenzyme has a tendency to aggregate, although only moderately under the conditions used,[10] whereas the ternary complex of the recombinant enzyme and $IP_{20}$ does not aggregate under the same conditions. As shown in Figs. 3 and 4, the binary complex containing either the mammalian or the recombinant enzyme crystallizes with one molecule per asymmetric unit, whereas the apoenzyme from mammalian origin crystallizes with two molecules per asymmetric unit in a different space group (see Figs. 1 and 2). It would be interesting to investigate whether the observed differences in the packing of both types of crystals are related to the aggregation state of the protein solutions. A possibility is that the aggregation, observed by SANS with the recombinant apoenzyme, does not reflect nonspecific intermolecular contacts, but rather the preformation of molecular assemblies, possibly dimers, with a given degree of organization. This might explain why the apoenzyme crystallizes with two molecules per asymmetric unit. The ternary complex with the recombinant enzyme also, however, has a tendency to aggregate in solution, as shown by SANS,[10] but forms crystals in the same space group with identical cell parameters (Fig. 4) as the binary complex, which does not aggregate detectably.

It must be noted that our present studies on crystal growth and scattering from solutions of cAPK catalytic subunits are not directly comparable, since the scattering measurements were not carried out under the supersaturation conditions used for crystal initiation. We are presently planning to monitor the state of the enzyme solutions by X-ray scattering under the conditions used for crystallization. It is hoped that the use of a short wavelength in the angstrom range will afford an accurate description of the intermolecular effects which characterize the protein solution under such conditions.

## Conclusions

Figure 5 summarizes all of the crystal forms of the catalytic subunit of cAPK that have been obtained thus far. All of the crystal forms of the different complexes of the catalytic subunit, with the exception only of the monoclinic crystal of the apoenzyme, were obtained under identical crystallization conditions as described by Knighton *et al.*[14] and Zheng *et al.*[15] The different space groups, therefore, result very likely from different conformational states of the enzyme. Crystals of both the binary and ternary complexes with $IP_{20}$ exhibited better diffraction characteristics than crystals of the apoenzyme. This suggests that one way to obtain crystals of better quality is to manipulate the conformational state of the enzyme by cocrystallization with specific inhibitors.

Conversely, control of the crystallization conditions may allow crystallographic results to corroborate, extend, and allow interpretation of observations derived with other techniques. For instance, the SANS studies showed that a major conformational change in the enzyme accompanied only the binding of $IP_{20}$; no change was detected due to the binding of MgATP itself. Similarly, our results show that both the binary and ternary enzyme complexes crystallize in the same (orthorhombic) space group. The correlations between the results of low-angle scattering studies and crystallographic data are presented in Fig. 6.

Our results also indicate that the ternary complex of murine catalytic subunit expressed in *E. coli* produced a crystal of better quality than did the ternary complex of the catalytic subunit purified from porcine heart. It is difficult to conclude whether this is due to the absence of myristic acid, the amino acid differences between the two forms, microheterogeneity in the mammalian enzyme, or a combination of these factors. It may suggest, however, that another way to improve the quality of crystals is to mutate the protein and to crystallize the mutant protein if crystallization of the wild type protein fails.

Three factors are important for reproducible crystallization. First, the salt of the eluting buffer of the last column must be chosen carefully. Second, the purity of the protein must be verified with isoelectric focusing gels. The protein must not contain typical additives, such as glycerol, and

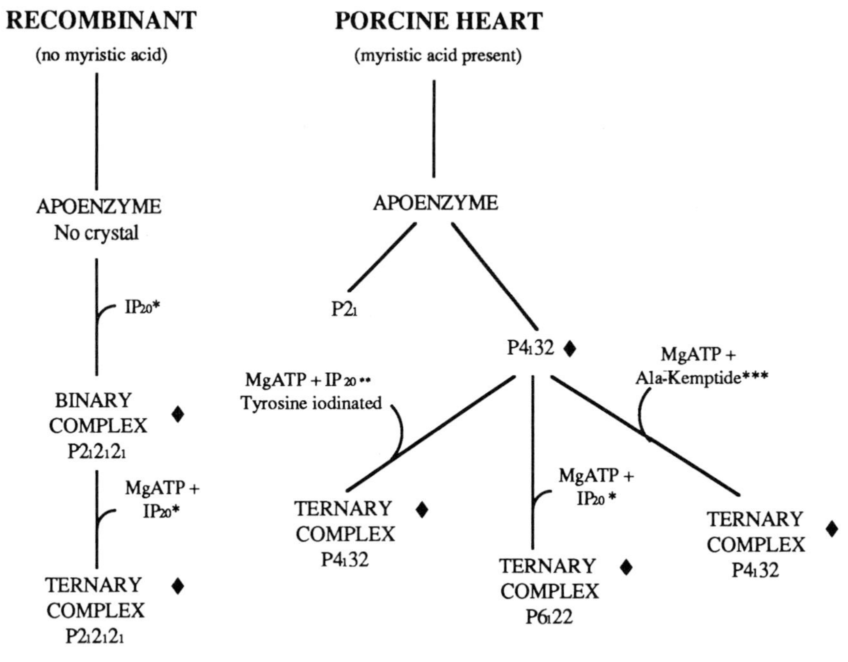


FIG. 5. Summary of all crystal forms obtained during the crystallization study of catalytic subunit of cAMP-dependent protein kinase. ◆, Conditions identical to those presented by Knighton *et al.*[14] and Zheng *et al.*[15] The porcine heart or murine recombinant catalytic subunit was brought to a final concentration of 10 mg/ml and dialyzed against 50 m*M* Bicine buffer (pH 8.0–8.3) and 150 m*M* ammonium acetate. Crystals were grown at 4° by hanging drop vapor diffusion after mixing equal volumes of protein solution, reservoir solution, and 10 m*M* DTT. The reservoir contained 10 m*M* DTT and 8% (w/v) polyethylene glycol 400 (Dow). Before sealing the coverslip, methanol was added to the reservoir to a concentration of 15%. The peptide inhibitors and MgATP were introduced by mixing equal volumes of the protein, the reservoir solution, and 10 m*M* DTT with MgATP and the given inhibitor. A, Alanine; D, aspartic acid; F, phenylalanine; G, glycine; H, histidine; I, isoleucine; L, leucine; R, arginine; S, serine; T, threonine; Y, tyrosine. *, $IP_{20}$ inhibitor: TTYADFIASGRTGRRNAIHD. **, $IP_{20}$ inhibitor, tyrosine iodinated: TTY*ADFIASGRTGRRNAIHD. ***, Ala-kemptide: LRRAALG.

should not be frozen prior to crystallization. Third, all reagents used for crystallization must be of the highest degree of purity. If all of these conditions are met, it is possible to obtain, in identical crystallizations, three different crystal forms representing two different conformational states of the enzyme. Some of those crystals, such as those of the ternary complex with $IP_{20}$, are of much better quality than the other crystals. One may, perhaps, extend this observation to other protein kinases where inhibitors may also be useful for cocrystallization.

In general, our results show that it is possible to generate high-quality

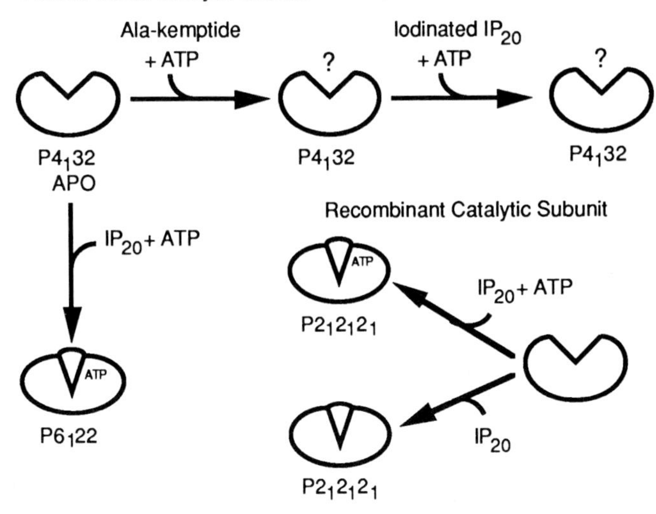


FIG. 6. Proposed correlation between small angle neutron scattering data, space group in which catalytic subunit crystallizes and the conformational state of the enzyme.

crystals, with the help from molecular biology, and it is possible to use the results obtained with one technique to support, extend, and interpret the results obtained with another, provided that crystallization methods are under optimal control.

## Acknowledgments

This work was supported in part by PHS Grant GM19301 to S. Taylor, a CRCC grant to J. M. Sowadski, and by the L. Markey Charitable Trust award. J. Parello is supported by CNRS (France). D. Knighton was supported in part by a USPHS training grants. We credit undergraduate student, Sean M. Bell, for his technical assistance in some of the crystallization studies.

# Section VII

# Cloning of Genes Encoding Protein Kinases

# [44] Use of Degenerate Oligonucleotide Probes to Identify Clones That Encode Protein Kinases

*By* STEVEN K. HANKS and RICHARD A. LINDBERG

Protein kinases play key roles in signal transduction pathways and it is likely that a great many members of this large enzyme family remain undiscovered. The identification and characterization of novel protein kinases should lead to new insights into how cells regulate their activities.

The homologous nature of eukaryotic protein kinase catalytic domains presents investigators with an opportunity to use DNA hybridization to screen recombinant DNA libraries for novel family members. The first successful applications of this approach utilized protein kinase gene fragments as probes under low-stringency hybridization conditions. Almost all novel protein kinases identified in this manner have been very closely related to the protein kinase from which the probe was derived. A description of this low-stringency screening method is presented elsewhere in this volume.[1]

Here we describe an alternative homology cloning approach that uses degenerate oligodeoxynucleotides ("oligos") as hybridization probes. The oligo probes are designed to recognize target sequences that encode short stretches of six to nine highly conserved amino acid residues found within the catalytic domains. Such probes are ideally suited for the purpose of identifying a wide variety of novel protein kinases from a single library screen. Virtually all of the codon possibilities for a conserved stretch can be included in the probe mixture, thereby assuring that many different protein kinase genes (cDNAs) will be recognized. False positives are reduced by targeting two or three conserved stretches and selecting for further characterization only those clones that are identified by all of the probe mixtures. This approach has been successful in identifying both novel protein-serine/threonine kinases[2–4] and novel protein-tyrosine kinases.[5] Degenerate oligodeoxynucleotides can also be used to prime the

[1] M. H. Kraus and S. A. Aaronson, this volume [46].

[2] S. K. Hanks, *Proc. Natl. Acad. Sci. U.S.A.* **84,** 388 (1987).

[3] S. K. Hanks, *Mol. Endocrinol.* **3,** 110 (1989).

[4] M. A. Lawton, R. T. Yamamoto, S. K. Hanks, and C. J. Lamb, *Proc. Natl. Acad. Sci. U.S.A.* **86,** 3140 (1989).

[5] T. Hunter, P. Angel, W. J. Boyle, R. Chiu, E. Freed, K. L. Gould, C. M. Isacke, M. Karin, R. A. Lindberg, and P. van der Geer, *Cold Spring Harbor Symp. Quant. Biol.* **53,** 131 (1988).

polymerase chain reaction and thus identify novel protein kinases through selective gene amplification.[6]

## Selecting Conserved Regions for Probe Targeting

Highly conserved protein kinase amino acid residues have been identified from catalytic domain sequence alignments.[7,8] Three separate regions within the catalytic domains appear to be the best targets for degenerate oligo probes (Table I). These regions are characterized by stretches of six to nine contiguous positions of high conservation. It is notable that slightly different consensus sequences are obtained for these potential probe targets when only subsets of the family are considered. Thus, the search for novel protein kinases can be directed by designing probes that will specifically recognize certain classes or subfamilies. This is also important for reducing probe degeneracy (see the section, Designing Probes, below).

Table I shows separate consensus sequences for receptor protein-tyrosine kinases, *src*-family protein-tyrosine kinases, and protein-serine/threonine kinases. An indication of the extent of conservation is also given in Table I as a consensus score. The protein-tyrosine kinases are a much more conserved group than are the protein-serine/threonine kinases and this is reflected in their consensus scores. The consensus sequences we have shown for the much broader protein-serine/threonine kinase class are most representative of three large subfamilies: the cyclic nucleotide-dependent subfamily, the protein kinase C subfamily, and the $Ca^{2+}$/calmodulin-dependent subfamily. Slightly different and more highly conserved consensus sequences would be obtained if any of these, or other, protein-serine/threonine kinase subfamilies were considered separately.

## Designing Probes

Degenerate oligo probes are designed that will recognize many or all possible codons for the chosen target consensus; however, there is an upper limit to the level of degeneracy that is acceptable. If the degeneracy of the probe is too great, one can expect to encounter problems relating to excessive background hybridization and numerous false positives, not to mention the impracticability of including each oligo type in the hybridization solution at a concentration high enough to allow for annealing to occur at reasonable rate. When using the hybridization and wash condi-

[6] A. F. Wilks, this volume [45].

[7] S. K. Hanks, A. M. Quinn, and T. Hunter, *Science* **241,** 42 (1988).

[8] S. K. Hanks and A. M. Quinn, this volume [2].

TABLE I

CONSENSUS SEQUENCES AND DEGENERATE PROBES FOR RECEPTOR PROTEIN-TYROSINE KINASES (rPTKs), *src*-FAMILY PROTEIN-TYROSINE KINASES (sPTKs), AND PROTEIN-SERINE/THREONINE KINASES (PSKs)

| Kinase | Subdomain[a] | VI | VII | IX |
|---|---|---|---|---|
| rPTKs | Consensus[b] | D - L - A - A - R - N | P - $^{I}_{V}$ - $^{K}_{R}$ - W - $^{T}_{M}$ - A - P - E | D - V - W - $^{A}_{S}$ - $^{F}_{Y}$ - G |
| | Score[c] | (26) (24) (26) (23) (26) (26) | (26) (25) (26) (26) (24) (20) (18) (26) | (26) (24) (25) (26) (25) (26) |
| | Probe | GA$^{C}_{T}$-CTX-GC$^{\substack{A\\C}}_{T}$-GC$^{\substack{A\\C}}_{T}$-$^{A}_{C}$G$^{\substack{A\\C}}_{G}$-AA | CCC-$^{A}_{G}$T$^{C}_{G}$-A$^{AA}_{GG}$-TGG-A$^{C}_{T}$X-GC$^{C}_{T}$-CC$^{C}_{T}$-GA | GA$^{C}_{T}$-GTX-TGG-$^{G}_{T}$CX-T$^{AC}_{TT}$-GG |
| | Degeneracy | (432) | (512) | (256) |
| sPTKs | Consensus | D - L - R - A - A - N | P - I - K - W - T - A - P - E | D - V - W - S - F - G |
| | Score | (8) (8) (8) (7) (8) (8) | (8) (8) (8) (8) (8) (8) (8) (8) | (8) (8) (8) (8) (8) (8) |
| | Probe | GA$^{C}_{T}$-CTX-$^{A}_{C}$G$^{\substack{A\\C}}_{G}$-GC$^{\substack{A\\C}}_{T}$-GC$^{\substack{A\\C}}_{T}$-AA | CCX-AT$^{C}_{T}$-AA$^{A}_{G}$-TGG-AC$^{\substack{A\\C}}_{T}$-GC$^{\substack{A\\C}}_{T}$-CCX-GA | GA$^{C}_{T}$-GTX-TGG-$^{G}_{T}$CX-TT$^{C}_{T}$-GG |
| | Degeneracy | (432) | (576) | (128) |
| PSKs | Consensus | D - L - K - $^{P}_{L}$ - $^{E}_{D}$ - N | G - T - P - $^{D}_{E}$ - Y - $^{I}_{L}$ - A - P - E | D - $^{V}_{W}$ - W - $^{A}_{S}$ - $^{L}_{Y}$ - G |
| | Score | (70) (58) (69) (62) (44) (70) | (55) (53) (33) (28) (54) (31) (54) (69) (69) | (70) (37) (61) (67) (49) (69) |
| | Probe | GA$^{C}_{T}$-CTX-AA$^{A}_{G}$-C$^{C}_{T}$X-GAX-AA | GGC-ACC-CCC-GAX-TA$^{C}_{T}$-$^{A}_{C}$TX-GC$^{\substack{A\\C}}_{T}$-CC$^{\substack{A\\C}}_{T}$-GA | GA$^{C}_{T}$-$^{GT}_{TG}$G-TGG-$^{G}_{T}$C$^{\substack{A\\C}}_{T}$-$^{\substack{C\\CTG}}_{TAT}$-GG |
| | Degeneracy | (512) | (576) | (576) |

[a] Subdomains within the protein kinase catalytic domain are defined in Hanks *et al.*[7]

[b] Consensus is based on an alignment of 26 rPTKs, 8 vertebrate sPTKs, and 70 PSKs.

[c] Consensus score indicates the number of kinases in the subgroup that actually contain the consensus residue.

tions we describe here complete degeneracy is not required because a perfect match is not necessary for annealing. We suggest, therefore, that degeneracy not exceed about 500-fold. If degeneracy levels obtained from the initial back translation are much greater than this, steps should be taken to reduce degeneracy to acceptable levels. This can readily be achieved through a combination of three strategies: (1) targeting particular protein kinase subfamilies, (2) selecting only some of the possible codons at degenerate positions (consult codon-usage tables for the species in question and also consider codon usage in known protein kinase genes), and (3) shortening the probe. Our probes have been as small as 17-mers (6 codons minus 1), but 14-mers (5 codons minus 1) may also be useful.

In Table I we suggest degenerate oligos that can be used as probes for specific recognition of clones that encode either receptor or *src*-family protein-tyrosine kinases, or protein-serine/threonine kinases. Where codon selection was used, the basis was usage frequency in mammalian protein-coding seqeunces.[9] Table I gives only 5′→3′ "sense" oligos, but the complementary "antisense" oligos can also be used for library screening and have the advantage of being able to also hybridize to RNA, which may be useful for preliminary Northern analysis of potential probe targets.

### Oligo Synthesis

Most DNA synthesizers now utilize phosphoramidite chemistry and these machines can be programmed to incorporate several different nucleotides at degenerate positions to an equal extent. Many laboratories now support their own DNA synthesizer, but the oligos can also be obtained through commercial sources. In either case, one should be aware of the potential for probe bias resulting from unequal incorporation of nucleotides at degenerate positions. This problem can be minimized by using a synthesizer that has adjusted flow rates to ensure uniform delivery of the phosphoramidites to the growing chains. Also, fresh phosphoramidites should be used as they degrade at different rates.

### Labeling Oligos

The oligo mixtures are labeled with $^{32}P$ using [$\gamma$-$^{32}P$]ATP and bacteriophage T4 polynucleotide kinase. Enough probe should be prepared so that each oligo in the degenerate mixture will be present in the hybridization solution at a concentration of 0.025 pmol/ml. Thus, for a typical library

[9] S.-I. Aota, T. Gojobori, F. Ishibashi, T. Maruyama, and T. Ikemura, *Nucleic Acids Res.* **16** (Suppl.), r315 (1988).

hybridization volume of 20 ml, 250 pmol of a 500-fold degenerate probe mixture should be included. Set up the kinasing reaction as follows[10]:

250 pmol of 500-fold degenerate oligo mix (1.7 $\mu$g for a 17-mer)
5 $\mu$l 10× kinase buffer [0.5 *M* Tris-HCl (pH 7.6), 0.1 *M* $MgCl_2$, 50 m*M* dithiothreitol (DTT), 1 m*M* spermidine hydrochloride, 1 m*M* EDTA]
100 pmol (300 $\mu$Ci) [$\gamma$-$^{32}$P]ATP (>7000 Ci/mmol)
20 U T4 polynucleotide kinase
$H_2O$ to 50 $\mu$l

Incubate the reaction mixture at 37° for 1 hr. Most of the $^{32}$P will be incorporated into the oligos. The remaining unincorporated isotope does not seem to significantly contribute to background in the hybridization conditions we use (see below) so that purification of labeled oligos is not necessary. The efficiency of transfer of $^{32}$P to oligos can be checked by running 0.5 $\mu$l from the reaction mixture on a 12% acrylamide/urea (sequencing) gel and exposing a sheet of X-ray film to the gel for 1 min. The labeled mixture can be stored at −20° until ready for use in library screening.

## Preparing Library for Screening

We have successfully screened a variety of cDNA libraries for novel protein kinases using this degenerate oligo strategy. An important consideration is the quality, in terms of insert size, of the cDNA library that is used. Thus, if oligo(dT) was used to prime the reverse transcriptase reaction, the inserts must be long enough to extend through 3′ untranslated regions and well into the coding region. And if random priming was used, the inserts must be long enough to include all of the multiple probe target regions. Differential cDNA libraries may be useful for identifying protein kinases expressed specifically in certain cell types or tissues. It should also be possible to screen genomic libraries for protein kinase genes using this approach, particularly in organisms such as *Saccharomyces cerevisiae,* where introns are rare.

Standard procedures[11] are used for plating the library, transferring colonies or phage onto nitrocellulose or nylon filters, and liberating and binding DNA to the filters. For plasmid libraries, the ideal density for screening is ~25,000 colonies/135-mm filter. For phage libraries, the den-

[10] J. Sambrook, E. F. Fritsch, and T. Maniatis, "Molecular Cloning: A Laboratory Manual," 2nd ed. Cold Spring Harbor Lab., Cold Spring Harbor, New York, 1989.

[11] T. Maniatis, E. F. Fritsch, and J. Sambrook, "Molecular Cloning: A Laboratory Manual," Cold Spring Harbor Lab., Cold Spring Harbor, New York, 1982.

sity can be increased to ~50,000 plaques/135-mm filter. Enough replicate filters must be prepared to allow for duplicate filters to be hybridized with each different oligo probe mixture.

The filters are processed for hybridization as follows[11]:

1. Prepare three trays or large dishes containing Whatman 3MM paper or equivalent saturated in either (1) 10% SDS, (2) 0.5 *M* NaOH, 1.5 *M* NaCl, or (3) 1.5 *M* NaCl, 0.5 *M* Tris-HCl (pH 8.0).
2. Sequentially place the filters onto the three saturated papers and allow them to stand for 5–10 min each. Between each step place the filters on dry Whatman 3MM paper to absorb excess solution.
3. Rinse the filters in 2× SSPE with shaking for 5 min.
4. Place the filters on Whatman 3MM paper and allow to dry for 30 min at room temperature.
5. Bake filters in a vacuum oven for 1–2 hr at 80°.
6. Wash in 3× SSPE, 0.1% SDS solution for 3 hr at 65°. Rub the filters with a gloved finger to help remove any remaining bacterial debris. Rinse in 5× SSPE and transfer the filters into freezer bags for hybridization. Prepare a separate bag for each different oligo probe mixture.
7. Prehybridize for at least 1 hr at 37° in a solution consisting of:

5× SSPE
5× Denhardt's [0.1% Ficoll, 0.1% polyvinylpyrrolidone (PVP), 0.1% bovine serum albumin (BSA)]
100 $\mu$g/ml hydrolyzed yeast RNA
0.05% sodium pyrophosphate

Use 1 ml prehybridization solution per filter.

### Hybridization and Washing

Following prehybridization, the labeled oligo mixture is added to the bag and hybridization is allowed to occur during an overnight incubation. Hybridization temperature will vary depending on probe length and composition and ideally should be empirically determined for each probe mixture used. After hybridization the filters are washed with 6× SSC containing 0.05% sodium pyrophosphate. Again, temperature will depend on probe length and composition. We suggest a series of sequential washes, from lower to higher temperature, with autoradiographic exposures in between. This will assure that weak positives, due to nonperfect matches or A/T richness, are not missed. The first wash should be at the hybridization temperature and subsequent washes are then increased in temperature by 5° increments until background levels are judged acceptable. Autoradiography is usually for 2–4 hr at −70° using preflashed X-ray film and an intensifying screen.

Guidelines for hybridization and maximum washing temperatures have been suggested[12] as tabulated below:

| Probe length | Temperature (°C) | |
|---|---|---|
| | Hybridization | Wash maximum |
| 14-mer | 20 | 41 |
| 17-mer | 37 | 53 |
| 20-mer | 42 | 63 |
| 23-mer | 48 | 70 |

An alternative approach for hybridizing and washing the filters utilizes quaternary alkylammonium salts, either tetramethylammonium chloride or tetraethylammonium chloride.[13,14] In these salts, hybridization and dissociation temperatures are independent of base composition and one can therefore wash at higher stringencies without eliminating true positives that are A/T rich. This approach may be chosen in cases where the investigator is confident that all possible codons are included in the probe mixture and thus wants to eliminate any nonperfect matches.

## Picking Positives and Rescreening

Deciding which colonies or phage to isolate for further characterization will depend, to some extent, upon the goals of the individual investigator. It is desirable, of course, to choose only those most likely to actually encode a member of the protein kinase family and to eliminate those most likely to represent false positives. The best protein kinase candidates will be clones that give strong hybridization signals with two or more probe mixtures, even at the highest wash stringencies. Clones that give very strong hybridization signals with only one probe can also be selected if this group is small. We have found that clones positive with more than one probe, but only at low wash stringencies, are usually false positives.

After selecting a reasonable number of positives to pursue further, they

[12] F. M. Ausubel, R. Brent, R. E. Kingston, D. D. Moore, J. G. Seidman, J. A. Smith, and K. Struhl, "Current Protocols in Molecular Biology," Vol. 2. Wiley, New York, 1987.

[13] W. I. Wood, J. Gitschier, L. A. Laskey, and R. M. Lawn, *Proc. Natl. Acad. Sci. U.S.A.* **82,** 1585 (1985).

[14] K. A. Jacobs, R. Rudersdorf, S. D. Neill, J. P. Dougherty, E. L. Brown, and E. F. Fritsch, *Nucleic Acids Res.* **16,** 4637 (1988).

are each subjected to several rounds of rescreening until they can be picked cleanly without contamination by neighboring colonies or plaques. The probe solutions used in the primary library screen can be saved and reused for subsequent rounds of screening.

## Confirming Candidate Clones as Members of Protein Kinase Family

To show that a clone identified by the oligo probes does, indeed, represent a member of the protein kinase family, one must eventually obtain nucleotide sequence data for a portion of the catalytic domain-encoding region and then compare the deduced amino acid sequence with other members of the family. We have used two methods that will give the required sequence information quickly. First, we have used the same probe mixtures used for the library screening to probe Southern blots to identify small restriction fragments [100–300 base pairs (bp)] that contain the probe targets. These small fragments are then subcloned into M13 for dideoxy sequencing and the amino acid sequence is deduced. The second approach involves the use of the degenerate oligos to directly prime double-stranded sequencing reactions from plasmids or phage DNA isolated from positive colonies or plaques. This obviates subcloning but is not as reliable.

Once the amino acid sequence has been deduced, a clone is confirmed as a member of the protein kinase family (a putative protein kinase) on the basis of inferred homology, i.e., by the presence of highly conserved catalytic domain residues. To determine if the clone represents a novel family member or if it has been previously described, one must compare the deduced amino acid sequence with those from all other members of the family. To aid in this analysis, a database of protein kinase catalytic domain sequences has been established.[8]

## Acknowledgments

S.K.H. was supported by USPHS Grant GM-38793. We thank Tony Hunter and Robert W. Holley for providing additional support and encouragement.

# [45] Cloning Members of Protein-Tyrosine Kinase Family Using Polymerase Chain Reaction

*By* ANDREW F. WILKS

## Introduction

A functional protein domain, such as the catalytic domain of a protein kinase, has a well-defined three-dimensional structure, so that key functional elements may be brought together to constitute an active site. In the protein kinase family, evolutionary processes have honed the crucial components of the conserved catalytic domain into a discontinuous pattern of identity. Accordingly, amino acid sequence comparison of the catalytic domains of 27 putative PTKs demonstrated the existence of 11 highly conserved motifs[1] (Fig. 1). These "patches" of homology are excellent candidates for the generation of degenerate oligonucleotide primers for use in the polymerase chain reaction (PCR).[2-5] In order that the exponential amplification of target sequences takes place, the PCR-based approach described herein requires both the presence of annealing sites for the oligonucleotide primers, and the further constraint of their proximity and correct orientation with respect to each other. It is thus possible to trade the gains in specificity derived from these additional constraints for considerable leeway in the degeneracy of the PCR primers employed. This PCR-based cloning protocol has already proved highly effective in revealing previously unknown members of the protein-tyrosine kinase family. The applicability of the approach to other protein families will not be covered, although application of the basic principles described herein would probably prove effective in revealing new members of most protein families.

## Strategies for Selection of Oligonucleotide Primer Sequences

The first step in the selection of primer sequences is the generation of an amino acid sequence alignment of the feature conserved in the protein

[1] S. K. Hanks, A. M. Quinn, and T. Hunter, *Science* **241,** 42 (1988).

[2] K. Mullis, F. Faloona, S. Scharf, R. Saiki, G. Horn, and H. A. Erlich, *Cold Spring Harbor Symp. Quant. Biol.* **51,** 263 (1986).

[3] R. K. Saiki, S. Scharf, F. Faloona, K. B. Mullis, G. T. Horn, H. A. Erlich, and N. Arnheim, *Science* **230,** 1350 (1985).

[4] R. K. Saiki, D. H. Gelfand, S. Stoffel, S. Scharf, R. Higuchi, G. T. Horn, K. B. Mullis, and H. A. Erlich, *Science* **239,** 487 (1988).

[5] A. F. Wilks, *Proc. Natl. Acad. Sci. U.S.A.* **86,** 1603 (1989).

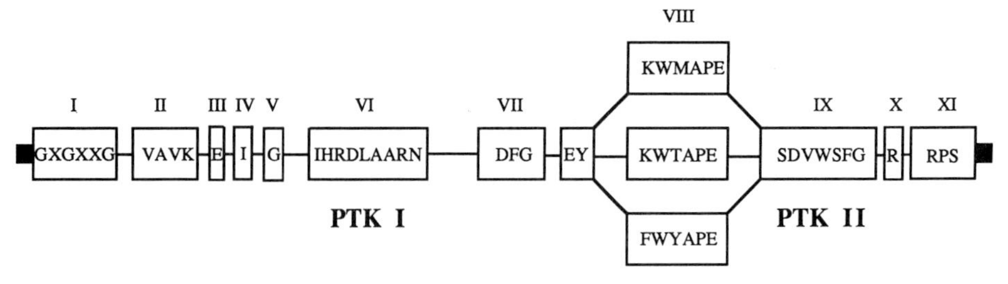


FIG. 1. Catalytic domain of the protein-tyrosine kinase family. The conserved sequence motifs, shown in one-letter code, of the PTK family are shown within homology boxes in a stylized catalytic domain. Thick lines represent extracatalytic domain sequences, thin lines joining boxes represent nonconserved catalytic domain sequences lying between homology boxes. Each conserved motif is additionally labeled by the nomenclature of Hanks *et al.*[1] (motifs I–XI). The locations of the two conserved motifs which have been successfully employed in the application of the polymerase chain reaction (PCR) to the cloning of members of this protein family are shown by the location of PTKI and PTKII. The KWMAPE/KWTAPE/FWYAPE motif appears to be PTK subgroup specific.

family of interest, in this case the catalytic domain of the protein-tyrosine kinase family. In this laboratory a current database of PTK sequences, as well as other protooncogene sequences, is maintained and regularly updated as new sequences are generated, or found in the literature. The original alignments were generated using Staden-based programs on a Vax VMS4.7 computer (Digital Equipment Corp.).

*Sequence Selection*

Obviously the best amino acid motifs for generation of PCR primers are those most highly conserved among the known protein family members. When the amino acid sequences of the catalytic domains of the protein tyrosine kinases are aligned these conserved sequence motifs become easily apparent. Eleven highly conserved domains have been delineated by Hanks *et al.*[1] and their nomenclature has been adopted here (Fig. 1).[5a] As with the selection of any degenerate oligonucleotide sequence for library screening, sequences with minimal codon degeneracy should be selected, although it is possible (indeed usual) that the sequence motifs available for primer generation do not afford this luxury. For example, in selecting the motifs upon which the *PTK*1 and *PTK*2 primers were based

[5a] Single-letter abbreviations are used for amino acids. A, Alanine; R, arginine; N, asparagine; D, aspartic acid; C, cysteine; Q, glutamine; E, glutamic acid; G, glycine; H, histidine; I, isoleucine; L, leucine; K, lysine; M, methionine; F, phenylalanine; P, proline; S, serine; T, threonine; W, tryptophan; Y, tyrosine; V, valine; X, unknown or "other."

(Fig. 2),[6–37] the complexity of the oligonucleotides was kept as low as possible by choosing to avoid the -**R**/**A**- codons in PTKI (subdomain VI, -IHRDL**R**/**A**-) and one of the serine residues (with six codons) in PTKII (subdomain IX, -**S**DVWSFG-). Nonetheless, the presence of a serine in *PTK*2 was ultimately unavoidable.

[6] Y. Ben-Neriah and A. R. Bauskin, *Nature (London)* **333,** 672 (1988).

[7] Q-L. Hao, N. Heisterkamp, and J. Groffen, *Mol. Cell. Biol.* **9,** 1587 (1989).

[8] A. F. Wilks and R. R. Kurban, *Oncogene* **3,** 289 (1988).

[9] E. Shtivelman, B. Lifshitz, R. P. Gale, B. A. Roe, and E. Canaani, *Cell (Cambridge, Mass.)* **47,** 277 (1986).

[10] G. D. Kruh, C. R. King, M. H. Kraus, N. C. Popescu, S. C. Amsbaugh, W. O. McBride, and S. A. Aaronson, *Science* **234,** 1545 (1986).

[11] J. Sukegawa, K. Semba, Y. Yamanishi, M. Nishizawa, N. Miyajima, T. Yamamoto, and K. Toyoshima, *Mol. Cell. Biol.* **7,** 41 (1987).

[12] S. Katamine, V. Notario, C. D. Rao, T. Miki, M. S. C. Cheah, S. R. Tronick, and K. C. Robbins, *Mol. Cell. Biol.* **8,** 259 (1988).

[13] Y. Yamanishi, S.-I. Fukushige, K. Semba, J. Sukegawa, N. Miyajima, K.-I. Matsubara, T. Yamamoto, and K. Toyoshima, *Mol. Cell. Biol.* **7,** 237 (1987).

[14] D. Holtzman, W. D. Cook, and A. R. Dunn, *Proc. Natl. Acad. Sci. U.S.A.* **84,** 8325 (1987).

[15] J. D. Marth, R. Peet, E. G. Krebs, and R. M. Perlmutter, *Cell (Cambridge, Mass.)* **43,** 393 (1985).

[16] K. Strebhart, J. I. Mullins, C. Bruck, and H. Rübsamen-Waigmann, *Proc. Natl. Acad. Sci. U.S.A.* **84,** 8778 (1988).

[17] R. J. Gregory, K. L. Kammermeyer, W. S. Vincent, III, and S. G. Wadsworth, *Mol. Cell. Biol.* **7,** 2119 (1987).

[18] A. Ullrich, A. Gray, A. W. Tam, T. L. Yang-Feng, M. Tsubkawa, C. Collins, W. Henzel, T. LeBon, S. Katthuria, E. Chen, S. Jacobs, U. Franke, J. Ramachandran, and Y. Fujita-Yamaguchi, *EMBO J.* **5,** 2503 (1986).

[19] Y. Ebina, L. Ellis, K. Jarnagin, M. Edery, L. Graf, E. Clauser, J.-H. Ou, F. Masiarz, Y. W. Kan, I. D. Godfine, R. A. Roth, and W. J. Rutter, *Cell (Cambridge, Mass.)* **40,** 747 (1985).

[20] H. Matsushime, L.-H. Wang, and M. Shibuya, *Mol. Cell. Biol.* **6,** 3000 (1986).

[21] D. D. L. Bowtell, M. A. Simon, and G. M. Rubin, *Genes Dev.* **1,** 620 (1988).

[22] M. Takahashi, Y. Buma, T. Iwamoto, Y. Inaguma, H. Ikeda, and H. Hiai, *Oncogene* **3,** 571 (1988).

[23] V. M. Rothwell and L. R. Rohrschneider, *Oncogene Res.* **1,** 311 (1987).

[24] L. Claesson-Welsh, A. Ericksson, B. Westermark, and C.-H. Heldin, *Proc. Natl. Acad. Sci. U.S.A.* **86,** 4917 (1989).

[25] Y. Yarden, J. A. Escobedo, W.-J. Kuang, T. L. Yang-Feng, T. O. Daniel, P. M. Tremble, E. Y. Chen, M. E. Ando, R. N. Harkins, U. Franke, V. A. Fried, A. Ullrich, and L. T. Williams, *Nature (London)* **323,** 226 (1986).

[26] Y. Yarden, W.-J. Kuang, T. Yang-Feng, L. Coussens, S. Munemitsu, T. J. Dull, E. Chen, J. Schlessinger, U. Francke, and A. Ullrich, *EMBO J.* **6,** 3341 (1987).

[27] A. Ullrich, L. Coussens, J. S. Hayflick, T. J. Dull, A. Gray, A. W. Tam, J. Lee, Y. Yarden, T. A. Liberman, J. Schlessinger, J. Downward, E. L. V. Mayes, N. Whittle, M. D. Waterfield, and P. H. Seeburg, *Nature (London)* **309,** 418 (1984).

[28] L. Coussens, T. L. Yang-Feng, E. Chen, A. Gray, J. McGrath, T. A. Liberman, J. Schlessinger, U. Franke, A. Levinson, and A. Ullrich, *Science* **230,** 1132 (1985).

```
                 PTKI                                                 PTKII

                 I  H  R  D  L R/A A A/R N                            S  D  V  W  S  F  G I/V

LTK          ATTCACAGAGACATTGCTGCCCGTAAC         -----        ACAGACTCCTGGTCTTTTGGGGTC
FER          ATACACAGGGACCTTGCTGCAAGAAAC         -----        AGTGACGTGTGGAGCTTTGGCATC
c-fes        ATCCACAGGGACCTGGCTGCTCGGAAC         -----        AGTGATGTGTGGAGCTTTGGCATT
c-abl        ATCCACAGAGATCTTGCTGCCCGAAAC         -----        TCCGACGTCTGGGCATTTGGAGTA
arg          ATCCATAGAGATCTTGCAGCTCGTAAC         -----        TCTGACGTCTGGG
c-yes        ATTCACCGAGATCTTCGGGCTGCTAAT         -----        TCTGATGTCTGGTCATTTGGAATT
c-fgr        ATTCACCGCGACCTGAGGGCAGCCAAC         -----        TCAGACGTGTGGTCCTTTGGGATC
c-lyn        ATTCACCGGGACCTGCGAGCAGCTAAT         -----        TCTGATGTGTGGTCCTTTGGAATC
HCK          ATCCACCGAGACCTCCGAGCTGCCAAC         -----        TCAGACGTCTGGTCCTTTGGTATC
LCK          ATCCATCGGGACCTGCGCGCCGCCAAC         -----        TCAGACGTGTGGTCCTTCGGGATC
c-tkl        ATCCACCGTGACCTGCGGGCTGCCAAC         -----        TCTGACGTCTGGTCCTTTGGCATC
Dsrc28C      ATTCACCGGGATCTGGCTGCCCGCAAC         -----        AGCGATGTGTGGGCATACGGTGTG
IGFIR        GTCCACAGAGACCTTGCTGCCCGGAAT         -----        TCGGACGTCTGGTCCTTCGGGGTC
IR           GTGCATCGGGACCTGGCAGCGAGAAAC         -----        TCTGACATGTGGTCCTTTGGCGTG
c-ros        ATTCACAGGGATCTGGCAGCTAGAAAT         -----        TCTGATGTATGGTCTTTTGGAATT
7-less       GTGCATCGCGACCTGGCCTGCCGGAAC         -----        TCGGATGTGTGGGCCTTTGGAGTG
ret          GTTCATCGGGACTTGGCAGCCAGAAAC         -----        AGTGATGTATGGTCTTTTGGTGTC
CSFIR        ATCCACCGGGACGTAGCAGCTCGAAAC         -----        AGTGATGTGTGGTCCTACGGCATC
PDGF-RA      GTCCACCGTGATCTGGCTGCTCGCAAC         -----        AGTGATGTCTGGTCTTATGGCATT
PDGF-RB      GTTCACCGAGACTTGGCGGCCAGGAAT         -----        AGTGATGTCTGGTCTTTTGGGATC
c-kit        ATTCACAGAGATTTGGCAGCCAGGAAT         -----        AGTGATGTCTGGTCCTATGGGATT
EGFR         GTGCACCGCGACCTGGCAGCCAGGAAC         -----        AGTGATGTCTGGAGCTACGGGGTG
NEU          GTACACAGGGACTTGGCCGCTCGGAAC         -----        AGTGATGTGTGGAGTTATGGTGTG
flg          ATACACCGAGACCTGGCAGCCAGGAAT         -----        AGTGATGTGTGGTCTTTCGGGGTG
bek          ATCCATCGAGATTTGGCTGCCAGAAAC         -----        AGCGATGTCTGGTCCTTCGGGGTG
MET          GTCCACAGAGACTTGGCTGCAAGAAAC         -----        TCAGATGTGTGGTCCTTTGGCGTC
trk          GTGCACCGGGACCTGGCCACACGCAAC         -----        AGCGACGTGTGGAGCTTCGGCGTG
EPH          GTCCACCGGGACCTGGCTGCCAGAAAC         -----        AGCGATGTGTGGAGCTTTGGGATT
ELK          GTGCACCGGGACCTGGCTGCTAGGAAC         -----        AGCGATGTCTGGAGCTACGGGATT
SEA          GTGCACCGGGACCTGGCAGCCAGGAAT         -----        TCAGACGTGTGGTCCTTTGGGGTG
torso        GTGCATAGGGATCTGGCTGCCCGGAAT         -----        AGCGATGTTTGGTCCTTTGGTGTG
JAK1         ATCCACAGGGACCTGGCAACAAGGAAC         -----        TCAGATGTGTGGAGCTTTGGAGTG
JAK2         GTTCACCGGGACCTTGGCAGAAAGAAT         -----        TCTGACGTCTGGTCTTTTGGAGTC

      5'CGG ATCCACAGNGACCT 3'   PTK1                      GATGTCTGGTCCTTTGG
       BamHI        C    TT                            3'CTGCAGACCAGGAAACC TTAAGG 5'   PTK2
                                                          A  C       T      EcoRI

   5'CGGATCC ATNCACAGNGATCTNGCNGCNAGNAA   PTK3      3' AGNCTGCANACCAGNAAAGGNTAN TTAAGG 5'  PTK4
             G    TC     CT         C                  TC   A      T    T   C
```

*Reducing Degeneracy*

Precisely what the limits of primer degeneracy are in this application of the polymerase chain reaction are unknown. Nonetheless, it is probably worth the effort of purging the sequences selected of unnecessary degeneracy. One approach found to be useful in this respect is to reanalyze the motifs selected for primer generation at the nucleotide level. A nucleotide sequence comparison of most of the known *PTK* sequences in the region of the PTKI (subdomain VI) and PTKII (subdomain IX) motifs is shown in Fig. 2, alongside the *PTK*1, *PTK*2, *PTK*3, and *PTK*4 primer sequences derived from this region. These oligonucleotide primers were selected on the basis of the DNA sequences shown, rather than on the protein sequences that they encode. Selection for codon usage is thus an empirical process, and no prejudgment is required based on the codon usage tables available. Where this sort of analysis is not possible, codon usage tables may be employed, although their use may not be necessary.

[29] M. Ruta, R. Howk, G. Ricca, W. Drohan, M. Zabelshansky, G. Laureys, D. E. Barton, U. Francke, J. Schlessinger, and D. Givol, *Oncogene* **3,** 9 (1988).

[30] S. Kornbluth, K. E. Paulson, and H. Hanafusa, *Mol. Cell. Biol.* **8,** 5541 (1988).

[31] M. Park, M. Dean, K. Kaul, M. J. Braun, M. A. Gonda, and G. Vande Woude, *Proc. Natl. Acad. Sci. U.S.A.* **84,** 6379 (1987).

[32] S. C. Kozma, S. M. S. Redford, F. Xioao-Chang, S. M. Saurer, B. Groner, and N. E. Hynes, *EMBO J.* **7,** 147 (1988).

[33] H. Hirai, Y. Maru, K. Hagiwara, J. Nishida, and F. Takaku, *Science* **238,** 1717 (1982).

[34] K. Letwin, S. P. Yee, and T. Pawson, *Oncogene* **3,** 621 (1988).

[35] D. R. Smith, P. K. Vogt, and M. J. Hayman, *Proc. Natl. Acad. Sci. U.S.A.* **86,** 5291 (1989).

[36] F. Sprenger, L. M. Stevens, and C. Nüsslein-Volhard, *Nature (London)* **338,** 478 (1989).

[37] A. F. Wilks, A. Harpur, and R. R. Kurban, unpublished data.

FIG. 2. Sequences used in the derivation of the *PTK*1, *PTK*2, *PTK*3, and *PTK*4 oligonucleotide primers. The nucleotide sequences of 33 *PTK*s are aligned in the region of the PTKI and PTKII motifs. *PTK*1 and *PTK*3 are derived from the PTKI motif (motif VI[1]), *PTK*2 and *PTK*4 are derived from the PTKII motif (motif IX[1]). Sequences were taken from the following sources: LTK,[6] FER,[7] c-*fes*,[8] c-*abl*,[9] *arg*,[10] c-*yes*,[11] c-*fgr*,[12] c-*lyn*,[13] HCK,[14] LCK,[15] c-*tkl*,[16] D*src*28C,[17] insulin-like growth factor receptor (IGFI-R),[18] insulin receptor,[19] c-*ros*,[20] sevenless,[21] *ret*,[22] colony-stimulating factor-I receptor (CSFI-R),[23] platelet-derived growth factor-A receptor (PDGF-RA),[24] platelet-derived growth factor-B receptor (PDGF-RB),[25] c-*kit*,[26] epidermal growth factor receptor (EGF-R),[27] NEU,[28] *flg*,[29] *bek*,[30] MET,[31] *trk*,[32] EPH,[33] ELK,[34] SEA,[35] torso,[36] JAK1 and JAK2.[37] Those *PTK*s shown in bold have been successfully isolated from cDNA using *PTK*1 and *PTK*2. The nucleotides shown in bold type are those that are identical to sequences in the *PTK*1 and *PTK*2 primers. *PTK*3 and *PTK*4 have been designed to include all of the degeneracy possible at any particular codon in the conserved motifs PTKI and PTKII. The symbol **N** represents positions where all four nucleotides are present in an oligonucleotide primer.

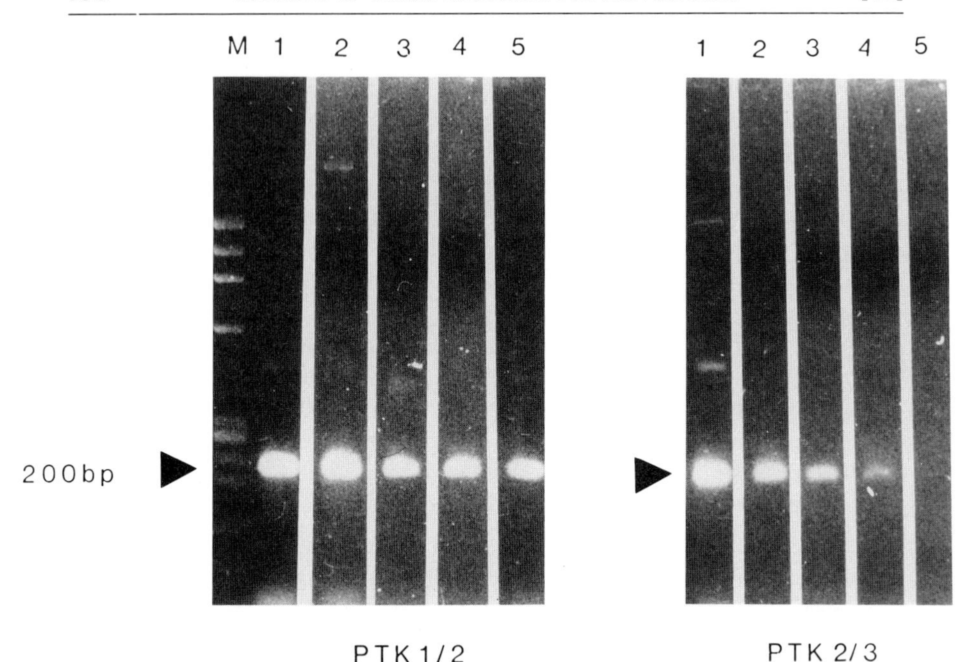


FIG. 3. Effect of temperature on the utility of the *PTK* series oligonucleotides as PCR primers. A 5-μl aliquot of a murine bone marrow cDNA library, in λgt11 (titer $10^{10}$ pfu/ml) was used as a template for each of the PCRs. The primers were employed in pairs: *PTK*1 and *PTK*2 (left-hand panel), or *PTK*3 and *PTK*2 (right-hand panel). Amplification reactions were performed as described in the text (denaturation at 95°; extension at 65°), except that the annealing phase of the cycle was performed at the following temperatures: track 1, 35°; track 2, 40°; track 3, 45°; track 4, 50°; track 5, 55°. The production of the expected ~200-bp *PTK* fragment (arrowed) was detected by ethidium bromide staining in one-tenth aliquot of the reaction mix run on a 2.5% (w/v) agarose gel; the DNA size standards were φX174 *Hae*III digestion products.

### *Length of Oligonucleotide Primers*

Two types of oligonucleotide primers work in the PCR amplification of *PTK*-related sequences. These are short (14–20 bases long), moderately degenerate (8 to 128-fold degenerate) primers,[5] and long (20–35 bases long), highly degenerate (~$10^5$-fold degenerate) primers.[38] In our hands shorter primers, with moderate degeneracy, have provided the best results. Those reactions, primed with the *PTK*4 oligonucleotide, in combination with either *PTK*1 or *PTK*3, failed to amplify whereas the *PTK*3 oligonucleotide was an efficient primer only at the low annealing temperatures (see Fig. 3). There is good evidence that the PCR amplification is

[38] A. F. Wilks, R. R. Kurban, C. M. Hovens, and S. J. Ralph, *Gene* **85,** 67 (1989).

able to sustain at least one pair of mismatched bases in the priming reaction (see below, Wilks *et al.*[38]). Thus the degeneracy of the oligonucleotides used may be reduced still further than a simple decoding of the amino acid sequence might permit.

*Subclass Targeting*

Many protein families have diversified into clearly demarcated subfamilies. Based on a comparison of the catalytic domains and other structural features of the PTK family, there is good evidence for the existence of five to six distinct classes of growth factor receptor PTK, and five to six classes of nonreceptor/cytoplasmic PTK.[1] One important strategic decision must be made in selecting the motifs from which to generate the degenerate oligonucleotide primers. It must be decided whether to target an entire protein family, and suffer the concomitant increase in the degeneracy of the oligonucleotide primers generated, or to target particular subfamilies, which would be expected to demonstrate a somewhat higher degree of sequence identity. This improvement in the level of sequence identity manifests itself on two levels. First, amino acid motifs may be subgroup specific. For example, in the case of the *src*-like kinases the motif labeled as PTKI (subdomain VI) in Fig. 1 is invariably -HRDLRAAN-, whereas all of the growth factor receptor subgroups, and the rest of the nonreceptor/cytoplasmic PTKs, bear the motif -HRDLAARN-. In this case the position of the arginine residue in essentially the same motif in each of the different subclasses of PTK can be used to generate specific subgroup-specific oligonucleotides, but must be taken into consideration when contemplating the generation of a "pan" PTK primer.

The second aspect of the sequence specificity of the different PTK families is exemplified in the use of the different arginine, serine, and leucine codons. Each of these amino acids is encoded by two different subgroups of codons. The utilization of one particular codon type, if demonstrated to be present at a given position in a conserved motif, will usually be retained in the identical position of other members of the same protein subfamily. In order to alter the codon usage at this location, two mutations, each of them possibly deleterious, would have to occur, so there is considerable evolutionary constraint upon their retention. Where subfamilies are sufficiently distinct, however, these evolutionary barriers may have been overcome, and the alternative codon root for the same amino acid has been selected. This is not an esoteric observation, since in Fig. 2 precisely this situation has arisen in the sequence from which the original oligonucleotide primer *PTK*2 has been derived. Of the 33 *PTK*s shown in Fig. 2, only 8 of them (c-*fes*, *FER*, *EGF-R*, *NEU*, *EPH*, *ELK*,

*TRK, AND JAK*1) use the AGC/T root of the serine codon; the remainder use TCX, where X is any nucleotide. The *PTK*2 primer was constructed with TCX in this location, and thus favors the priming of *PTK*s containing this type of serine codon. It was surprising, however, that a proportion of the *PTK*s isolated contained, upon examination of their cDNA sequences, an AGC/T in this location (e.g., c-*fes* and *JAK*1).

### *Toleration of Mismatches in Priming Reaction*

It has been possible to explore the toleration of mismatch in the priming reaction by sequencing multiple isolates of the same amplified *PTK* in the region of the oligonucleotide primers. An assessment of the primer utilization in this application of PCR can thus be made. In two series of experiments, the only mismatches tolerated in the priming reaction of *PTK*1 were at the most highly degenerate base location at position 12. The degeneracy available at the other three potential sites of mismatch (positions 10, 15, and 16) was never exploited. From this information it can be inferred that the priming reaction with this short oligonucleotide is able to tolerate a single mismatch, but that the position of this mismatch with respect to the rest of the oligonucleotide sequence may be important. Similar analysis of the products of amplification at low temperature with the *PTK*3 primer demonstrated that the utilization of the *PTK*3 primer was, not surprisingly, highly variable. Individual clones showed considerable sequence variation in the region of the *PTK*3 primer when compared with the sequence of the mRNA from which they were derived (data not shown). Between 5 and 10 mismatches were detected in each clone. These mismatches are the product of the last priming event to have occurred on the particular DNA molecule cloned into M13, and the data therefore reflect the ability of highly mismatched primers to prime DNA synthesis at the low temperature employed in this experiment. It remains to be determined whether the number of mismatches detected were in some part those accumulated in previous cycles, rendering the target a more suitable template for the mismatched primer to prime on, or whether all of the mismatches were introduced during the last cycle.

## Precautions to Be Taken in Applying Polymerase Chain Reaction Method

The polymerase chain reaction is an extraordinarily sensitive technique, and is quite capable of taking even miniscule quantities of a contaminating *PTK* DNA sequence and amplifying it to predominance even in the presence of a broad spectrum of DNAs encoding other cellular kinases. It is therefore prudent, particularly in laboratories where DNA clones of *PTK*s are handled, to be careful in the preparation of PCR mixes in order to ensure contamination problems are eliminated. The guidelines outlined by Kwok and Higuchi[39] should be followed. Briefly stated, they include the following suggestions: Always wear gloves; use positive displacement pipettes; run negative controls, always; be aware of the possibility of DNA carryover on items of equipment such as transilluminators, gel combs, gel electrophoresis apparatus. These precautions should be followed in *all* steps of the protocol outlined below.

## Template Preparation

Template selection is generally determined by the nature of the project envisaged. DNA amplification can be performed on the single-stranded or double-stranded cDNA produced from between 0.1 and 1.0 $\mu$g of poly(A)$^+$ mRNA. Using cDNA synthesis kits greatly facilitates the generation of this material, and while there are no data to suggest that double-stranded material is better as a template than single-stranded cDNA, we have routinely completed the second strand synthesis prior to performing the PCR step. Alternatively, DNA isolated from a 5-$\mu$l aliquot of a high titer [$10^{10}$ plaque-forming units (pfu)/ml] $\lambda$ phage stock can be used as a template. We have never used genomic DNA in this particular application of the PCR technique, although potentially the range of target sequences should be wider here than in the case of an mRNA population. It is possible that the presence of intron sequences at, or in between, the primer sites may reduce the efficiency of the amplification.

1. When mRNA is the source of the material to be amplified, poly(A)$^+$ mRNA from the cell line or tissue of choice is used to generate oligo(dT)- or random hexamer-primed double-stranded cDNA using an Amersham cDNA synthesis kit, and the conditions suggested by the suppliers. Other cDNA synthesis kits may well work as efficiently, but have not been tried in this capacity. There is no real requirement to synthesize both strands, although the extra step is trivial and may be performed in the same tube as the first strand synthesis.

2. cDNA generated in the above manner, or the DNA isolated from a portion of a $\lambda$-based cDNA library (5 $\mu$l of a high-titer $10^{10}$ pfu/ml phage stock, diluted to 100 $\mu$l), is prepared for amplification by phenol/chloroform extraction (2×), followed by ethanol precipitation without carrier.[40]

[39] S. Kwok and R. Higuchi, *Nature (London)* **339,** 237 (1989).

[40] T. Maniatis, E. F. Fritsch, and J. Sambrook, "Molecular Cloning: A Laboratory Manual." Cold Spring Harbor Lab., Cold Spring Harbor, New York, 1982.

The cDNA is resuspended in 100 μl doubly distilled water, and aliquots of 10–50 μl employed in the amplification reaction.

3. When applying the method to λ-based cDNA libraries, the whole of the λ DNA preparation is used in a single reaction. In several cases we have been able to generate amplified material from the addition of 5 μl of phage stock directly to the amplification reaction, without prior extraction of the DNA.

## DNA Amplification

1. Once prepared for amplification the DNA to be amplified is first diluted to 70 μl and boiled for 2–3 min before being reconstituted into PCR buffer [1× reaction buffer: 50 m*M* KCl, 10 m*M* Tris-HCl, pH 8.3, 1.5 m*M* $MgCl_2$, 0.01% (w/v) gelatin, 200 μ*M* of each dNTP, 1 μ*M* of each oligonucleotide primer, 2.5 units TAQ polymerase], in a final volume of 100 μl. The reaction mix is then overlaid with 100 μl of liquid paraffin, in order to prevent evaporation.

2. The PCR cycle should be determined empirically for each pair of oligonucleotides. In addition, optimal conditions can be expected to vary from thermocycler to thermocycler (in our hands the efficiency of amplification varies from machine to machine; manual transfer from heated water bath to heated water bath remains one of the most efficient ways of cycling samples, but is obviously impracticable for most purposes). In the case of the short, moderately degenerate, *PTK* oligonucleotides described elsewhere[5] the conditions were 1.5 min at 95° (denaturation), 2 min at 37° (annealing), and 3 min at 63° (elongation). Surprisingly, these primers can still efficiently prime the amplification of *PTK*-related sequences at annealing temperatures of 55° (Fig. 3). This is in contrast to the *PTK*3 oligonucleotide, which is only effective at low (e.g., 35–40°) annealing temperatures.

3. Some difficulties have been encountered in cloning material amplified in the above manner. In order to preclude these problems the following strategy for purifying amplified DNA is suggested. First, only 30 cycles of amplification are performed, the polymerase is thus more likely to remain active and efficient at filling the ends of the amplified DNA. The extension phase of the final cycle should be two to three times longer than the normal extension phase.

4. The reaction tube should be removed from the thermocycler as soon as the reaction is complete. The opportunity for exonuclease digestion of the amplified fragments is thus minimized.

5. After chloroform (2 vol) extraction of the reaction, it is diluted to 1 ml and purified using the manufacturer's suggested conditions on an Elutip

(Schleicher and Schuell, Keene, NH) or similar ion-exchange support. For the Elutip, the sample is applied to the tip in 0.2 *M* NaCl, 10 m*M* EDTA, 10 m*M* Tris-HCl, pH 7.5, washed twice with 1 ml of the same buffer, and eluted with 1.5 *M* NaCl, 10 m*M* EDTA, 10 m*M* Tris-HCl, pH 7.5. The eluate is then precipitated with 2 vol of ethanol without carrier.[40]

## Construction of PCR Libraries

The use of primers with different built-in restriction sites enables the directional cloning of the amplified fragment. Almost any vector, either plasmid or bacteriophage, can be used, although, since DNA sequencing is most often going to be the method by which the library os screened, we have used M13mp18 as the vector of choice. If a large PCR library is required, a λ-based vector coupled to an efficient phage packaging extract would provide a good alternative.

1. In order to clone the amplified DNA, the ethanol precipitate is air dried and resuspended in 20 μl of doubly distilled water before being codigested in *Eco*RI buffer (50 m*M* NaCl, 100 m*M* Tris-HCl, pH 7.5, 5 m*M* $MgCl_2$, 100 μg/ml bovine serum albumin) with 20 units each of *Eco*RI and *Bam*HI, for 4 hr at 37°. The DNA is then repurified on an Elutip (Schleicher and Schuell), as before, prior to ligation into *Eco*RI- and *Bam*HI-cleaved M13 mp19.
2. White plaques may be picked at random for DNA sequencing.[41]

## Screening of PCR Libraries

The PCR libraries generated in this laboratory have all been screened by DNA sequencing.[41] Once a reasonable number (20–50) have been processed it is useful to screen the remainder by T tracking. A streamlined version of this protocol is shown below.

### *T Tracking of M13 PCR Libraries*

1. *Annealing:* Mix

30 μl M13 universal primer
60 μl 5× sequencing buffer (1× buffer: 200 m*M* Tris-HCl, pH 7.5, 100 m*M* $MgCl_2$, 250 m*M* NaCl)

For each clone, take 0.75 μl of above mix and add 1.75 μl M13 DNA. Heat to 65°, slow cool to below 35°.

2. *Labeling reaction:* Mix

[41] F. Sanger, S. Nicklen, and A. R. Coulson, *Proc. Natl. Acad. Sci. U.S.A.* **74,** 5463 (1977).

28 μl 0.1 *M* DTT
56 μl labeling mix diluted 1 : 5 (1 × labeling mix: 7.5 μ*M* each of dGTP, dCTP, TTP)
14 μl [$^{35}$S]dATP

3. Just before use, add Sequenase (diluted 1 : 8 in 10 m*M* Tris-HCl, pH 7.5, 1 m*M* EDTA) to the amount of labeling mix you are about to use; e.g., for 50 reactions,

49 μl labeling reaction mix
28 μl diluted Sequenase (United States Biochemical Corporation)

4. Add 1.4 μl of this mixture to each annealing mix (total volume 3.9 μl).
5. Leave at room temperature for 5 min.
6. Add 2.5 μl TTP termination mix (80 μ*M* dATP, dCTP, dGTP, TTP; 8 μ*M* dideoxy-TTP; 50 m*M* NaCl), leave at 37° for 5 min.
7. Add 4 μl of stop solution [95% (w/v) formamide, 20 m*M* EDTA, 0.05% (w/v) Bromphenol Blue, 0.05% (w/v) xylene cyanol FF].
8. Load onto conventional sequencing gel.
9. Those T tracks which appear to contain sequences which are novel can be resequenced using conventional methods and the data processed as described below.

*Establishment of Sequence Database*

The incorporation of two different restriction enzyme sites into the *PTK*1 (*Bam*HI) and *PTK*2 (*Eco*RI) PCR primers means that the amplified material will inevitably clone in a single orientation into the cloning vector. Thus DNA sequences derived from the same sequencing primer will be readily comparable without further treatment. In the work described elsewhere,[5] the bulk of the data was compiled as a sequencing project in a Staden-based program, each new sequence being entered separately. When sequence comparisons were required, a simple sequence comparison program was employed to generate the alignment of two separate entries, and the data routinely printed as a hard copy. As more sequences are generated, it becomes possible to perform the sequence comparisons with representative clones only, so that the overall work load is reduced. Manual comparison methods are feasible, but it is possible to generate many hundreds of individual sequences using this method, and the problems of data management could become overwhelming without careful organization. Table I shows the profile of *PTK* clones isolated from a typical series of PCR libraries.

TABLE I
SCREENING OF PCR LIBRARIES BY DNA SEQUENCING

| Library | FDC-P1[a] | WEHI 3B $D^{+}$ [a] | M$\phi$[b] | Brain[b] |
|---|---|---|---|---|
| Number of clones | 200 | 200 | 24 | 22 |
| Number of M13[c] | 54 | 67 | 6 | 8 |
| Number of primers[c] | 12 | 17 | 0 | 2 |
| c-*fes* | 49 | 7 | 11 | 0 |
| JAK1 | 17 | 4 | 2 | 1 |
| JAK2 | 45 | 94 | 3 | 7 |
| IGFI-R | 15 | 2 | 0 | 0 |
| MET | 4 | 0 | 0 | 1 |
| D*src*28C | 2 | 6 | 0 | 0 |
| LYN | 2 | 1 | 0 | 0 |
| HCK | 0 | 2 | 2 | 0 |

[a] The polymerase chain reaction (PCR) libraries for FDC-P1 and WEHI 3B $D^{+}$ (see Wilks *et al.*[5,6]) were constructed from newly synthesized cDNA.

[b] The macrophage (M$\phi$) and brain PCR libraries were generated from preexisting λgt11 cDNA libraries.

[c] In each case the number of clones without inserts, and those with polymerized primer sequences, is noted.

## Potential Improvements to Method

### *Use of Inosine in Highly Degenerate PCR Primers*

Inosine is a nucleotide base analog which can be considered as "neutral" with respect to its contribution to the melting temperature of DNA duplexes. Thus its inclusion in oligonucleotide probes and primers at positions of most degeneracy avoids the destabilizing effects a mismatch would have on the hybrids formed between target and probe. This property has been exploited in the use of this analog for genomic library screening[42] and in PCR applications.[43,44] A logical extension of this idea would be to circumvent some of the limitations of the methods described herein, particularly with regard to avoiding conserved motifs with amino acids encoded by highly degenerate codons. It can be anticipated that the use of inosine in PCR primers of this nature will extend its utility considerably.

[42] Y. Takahashi, K. Kato, Y. Hayashizaki, T. Wakabayashi, E. Ohtsuka, S. Matsuki, M. Ikehara, and K. Matsubara, *Proc. Natl. Acad. Sci. U.S.A.* **82,** 1931 (1985).

[43] K. Knoth, S. Roberds, C. Poteet, and M. Tamkun, *Nucleic Acids Res.* **16,** 10932 (1988).

[44] F. Libert, M. Parmentier, A. Lefort, C. Dinsart, J. Van Sande, C. Maenhaut, M.-J. Simons, J. E. Dumont, and G. Vassart, *Science* **249,** 569 (1989).

*Plus/Minus Screening*

The screening of PCR libraries generated with the oligonucleotides *PTK*1 and *PTK*2 has been based on a random accumulation of sequence data, followed by subsequent selection of sequences of interest. While in the early phase of this work a high proportion of clones contained new *PTK*-related sequences, later experiments proved to be somewhat less fruitful. Moreover, the selection of clones for sequencing was a purely random process and was uninformative with regard to other potentially important selection criteria, such as expression pattern. An improvement in this respect has been the opportunity to screen PCR libraries for *PTK*-related sequences which show a tissue-specific or developmentally regulated pattern of expression. The PCR library generated from the mRNA source of interest is screened, in duplicate, with $^{32}P$-labeled PCR product from which the library was constructed. The filters are then stripped and reprobed with a PCR probe generated in an identical fashion from an mRNA source other than that used to construct the PCR library. In this way we have been able to generate differentially expressed *PTK*-related clones expressed in epithelial cells but not fibroblasts (A. Ziemiecki and A. F. W., unpublished data). Although this procedure has been employed successfully we have not tested the limits of its sensitivity, nor its utility in other situations, such as, for example, a differentiation system.

# [46] Detection and Isolation of Novel Protein-Tyrosine Kinase Genes Employing Reduced Stringency Hybridization

*By* M. H. KRAUS and S. A. AARONSON

## Introduction

Protein-tyrosine kinases (PTKs) are encoded by a conserved family of ancestrally related genes that have segregated during evolution within the larger family of protein kinases.[1] They share a catalytic domain encoding 250–300 amino acid residues that represents the region of most extensive structural conservation and harbors the intrinsic enzymatic activity. Individual members in distinct subfamilies of both receptor-like and cytoplasmic tyrosine kinases share closer structural and functional homology with each other than with other protein kinases. The higher degree of structural homology among functionally more closely related PTKs pro-

[1] S. K. Hanks, A. M. Quinn, and T. Hunter, *Science* **241,** 42 (1988).

vides a basis for the identification of novel members with predictable functional properties due to their association with individual PTK subfamilies. Based upon concurrent nucleotide sequence conservation, the search for novel *PTK* genes can be directed toward a specific subfamily without detection of more distantly related *PTK* genes, due to cross-hybridization of genomic exon sequences of novel members under quantitatively reduced hybridization stringencies with *PTK* domain probes of known subfamily members. Analyzing normal genomic DNA representing *PTK* genes at single-copy level permitted detection of exon sequences encoding closely related novel PTKs independent of their expression pattern.

Novel genomic restriction fragments are cloned and the most conserved region is mapped by Southern blot analysis. Thus, exons with highest homology to the probe are identified and their structure is determined by nucleotide sequence analysis, establishing the degree of structural relatedness and exon organization of a novel related tyrosine kinase gene. Furthermore, exon-containing probes can be derived for expression analysis as well as the isolation of cDNA clones of novel *PTK*s. Employing probes from the TK domain of v-*erbB*, v-*fms*, and v-*abl* we have isolated with *erbB-2*[2] and *erbB-3*[3] two novel receptor-like tyrosine kinases in the *erbB*/EGF-R family, a second PDGF-R[4] in the *fms*/CSF1-R family, as well as a gene closely related to *abl*,[5] respectively in the human genome.

## Genomic Southern Blot Analysis

The reliable detection of novel *PTK* genes at reduced stringency requires a high sensitivity and specificity of genomic Southern blot analysis.[6] Using a slightly modified procedure we detect human single-copy genomic DNA fragments with a completely matching probe at high stringency after 2–4 hr of film exposure at $-70°$ in the presence of intensifier screens.

1. Ten micrograms genomic DNA is restricted in a final volume of 200–400 $\mu$l. To monitor restriction, bacteriophage $\lambda$ DNA is incubated with a small aliquot of the main reaction at identical DNA/enzyme unit ratio.
2. After complete restriction, DNA samples are purified with an equal

[2] C. R. King, M. H. Kraus, and S. A. Aaronson, *Science* **229,** 974 (1985).
[3] M. H. Kraus, W. Issing, T. Miki, N. C. Popescu, and S. A. Aaronson, *Proc. Natl. Acad. Sci. U.S.A.* **86,** 9193 (1989).
[4] T. Matsui, M. Heidaran, T. Miki, N. Popescu, W. LaRochelle, M. Kraus, J. Pierce, and S. Aaronson, *Science* **243,** 800 (1989).
[5] G. D. Kruh, C. R. King, M. H. Kraus, N. C. Popescu, S. C. Amsbaugh, W. O. McBride, and S. A. Aaronson, *Science* **234,** 1545 (1986).
[6] E. M. Southern, *J. Mol. Biol.* **98,** 503 (1975).

volume of buffered phenol/CIA (chloroform/isoamyl alcohol, 24 : 1) and precipitated from 1 *M* ammonium acetate with 2.5 vol cold ethanol on dry ice for 20 min. The pelleted DNA is carefully washed with cold 75% (v/v) ethanol, lyophilized, and following solubilization in 30 $\mu$l 1× sample buffer loaded on a 0.8–1.2% (w/v) horizontal agarose gel (20 × 25 cm; i.e., BRL, Gaithersburg, MD). Electrophoresis proceeds overnight at 35 V in a Tris-acetate gel/buffer system (1× E) containing 0.5 $\mu$g/ml ethidium bromide with buffer recirculation.

40× E buffer: 1.6 *M* Tris, 0.8 *M* sodium acetate, 40 m*M* EDTA; adjust pH to 7.2 with acetic acid
10× sample buffer: 0.2 *M* Tris-acetate, pH 7.5, 0.02 *M* EDTA, 1% (w/v) SDS, 50% (v/v) glycerol, 0.3% (w/v) Bromphenol Blue

3. Following gel photography, the gel is irradiated for 2 min with 302-nm UV light to improve transfer of larger DNA fragments, and trimmed to 20 × 20 cm. The DNA is denatured by two consecutive gel treatments of 15 min in 0.5 *M* NaOH/1.5 *M* NaCl and equilibrated with two treatments of 15 min each in 1 *M* ammonium acetate/0.02 *M* NaOH. Capillary transfer in this solution to standard 0.45-$\mu$m nitrocellulose proceeds overnight with the bottom of the gel facing the membrane. After transfer, the filter is baked for 2 hr at 80° under vacuum.

4. Standard high-stringency hybridization is conducted in 5× SSC (1× SSC = 0.15 *M* NaCl, 0.015 *M* sodium citrate, pH 7) and 50% formamide at 42°, which establishes conditions 20–25° below the $T_m$ (melting point) of a completely matched DNA hybrid. For hybridization the membrane is placed in a sealing bag, wetted in two-thirds of the final volume (0.075 ml/$cm^2$ filter area), and probe is added in the remaining third of hybridization solution. The bag is sealed and after thorough mixing of the solutions placed between two glass plates in a 42° water bath for 8–16 hr.

Hybridization solution (1×): 5× SSC, 10% (w/v) dextran sulfate, 2.5× Denhardt's solution, 10 m*M* Tris pH 7.4, 50 $\mu$g/ml sheared and boiled salmon sperm DNA, 50% (v/v) formamide, 2–5 × $10^6$ cpm/ml at a DNA concentration of <5 ng/ml

Hybridization buffer containing SSC, dextran sulfate, Denhardt's, and Tris is prepared as 2× stock solution by dissolving dextran sulfate powder in SSC while stirring prior to adding the other components. Nonradioactive and radioactive hybridization solutions are prepared by adding salmon sperm DNA to the formamide, and purified probe is added to the latter solution. Both are then incubated 5 min at 50°, thoroughly mixed, and hybridization buffer is added to 1× final concentration. This step will ensure sufficient denaturation of both carrier and labeled probe DNA.

For reduced stringency hybridization, the volume is adjusted with water following denaturation in formamide. High specific activity probes of $>10^9$ cpm/$\mu$g DNA can be routinely obtained with commercially available nick translation (i.e., Amersham, Arlington Heights, IL) or random primer kits. Under these conditions "prehybridization" of the filter is not required, as long as the filter is prewetted in nonradioactive hybridization solution prior to adding the probe.

5. After hybridization, filters are washed three times for 5 min each in 2× SSC, 0.01% SDS at room temperature. Two high-stringency washes of 20 min each are then conducted at 50° in 0.1× SSC/0.1% SDS and 0.1× SSC, respectively.

## Detection of Novel Tyrosine Kinase Gene Fragments by Genomic Hybridization Analysis at Reduced Stringencies

Cross-hybridization of related tyrosine kinase gene fragments is accomplished by reduced stringency hybridization facilitating hybrid formation and stability between partially mismatched sequences. While theoretical aspects of the formation of DNA hybrids and sequence homology have been extensively discussed in the literature,[7] most of these findings derive from liquid hybridization and may vary to some extent in filter hybridization where the template DNA is immobilized.

Thus, to determine experimentally the optimal stringency for cross-hybridization of a novel related tyrosine kinase gene, it proves useful to subject replica nitrocellulose filters containing genomic DNA to hybridization under stringencies incrementally decreased by 7°. Discrete intermediate levels of reduced stringency (Table I) may be useful in some cases. Since stringency is presumed to affect predominantly the formation of hybrids during hybridization and the stability of hybrids during washing, both hybridization and washing are conducted under equally reduced stringencies. Experimentally these are controlled by the formamide concentration during hybridization and salt concentration during washing, respectively, with otherwise identical conditions (Table I).

It is helpful to relate stringency in temperature degrees (°C) assuming that decrease of formamide concentration by 1% increases the $T_m$ of the hybrid by 0.7° and a logfold increase in ionic strength raises the stability of a hybrid by 18.5°.[7,8] Thus, considering incubation temperature ($T_i$),

[7] M. L. M. Anderson and B. D. Young, *in* "Nucleic Acid Hybridisation" (B. D. Hames and S. J. Higgins, eds.), p. 73. IRL Press, Oxford, 1985.

[8] G. A. Beltz, K. A. Jacobs, T. H. Eickbush, P. T. Cherbas, and F. C. Kafatos, *in* "Methods in Enzymology" (R. Wu, L. Grossman, and U. Moldave, eds.), p. 266. Academic Press, New York, 1983.

TABLE I
STRINGENCY REDUCTION[a]

| Stringency reduction ($\Delta T_s$) (°C) | Hydridization (42°/5× SSC) | Washing (50°) |
|---|---|---|
| **0** | **50% FA** | **0.1× SSC → 0.02 *M* Na⁺** |
| −3.5 | 45% FA | 0.15× SSC → 0.03 *M* $Na^+$ |
| − 7 | 40% FA | 0.25× SSC → 0.04 *M* $Na^+$ |
| −10.5 | 35% FA | 0.4× SSC → 0.08 *M* $Na^+$ |
| −14 | 30% FA | 0.6× SSC → 0.12 *M* $Na^+$ |
| −17.5 | 25% FA | 1× SSC → 0.2 *M* $Na^+$ |
| −21 | 20% FA | 1.5× SSC → 0.3 *M* $Na^+$ |

[a] $\Delta T_s$ indicates reduction of stringency relative to high-stringency conditions (bold type). FA, Formamide; 1× SSC = 0.15 *M* NaCl, 0.015 *M* sodium citrate, pH 7. For each stringency, hybridization is conducted at 42° in 5× SSC, while washing occurs at 50°.

formamide concentration (FA), and ionic strength ($\mu$), experimental stringency conditions relative to high-stringency hybridization can be approximated as

$$T_s = T_i + 0.7(\%\ \text{FA}) - 18.5 \log(\mu/\mu_0)$$

where $T_s$ is the stringency expressed as temperature degrees, $T_i$ the incubation temperature, and $\mu$ the ionic strength, while $\mu_0$ represents the ionic strength during hybridization conditions equalling 1 *M* $Na^+$ (5× SSC). At these high salt concentrations the hybrid stability is maximal and relatively unaffected by variations of ionic strength (log 1 = 0). Under high-stringency conditions the estimated $T_s$ of the hybridization ($T_s = 42° + 0.7° \times 50 - 18.5 \times \log 1$) is 77° and the $T_s$ during washing ($T_s = 50° + 0.7° \times 0 - 18.5° \times \log 0.02$) is 81°. Reducing the formamide concentration by 10% results in an estimated stringency of $T_s = 70°$, yielding a stringency reduction of $\Delta T_s = -7°$ in the hybridization. Equivalent reduction of the washing stringency requires an increase of the salt concentration to 0.25 × SSC (0.05 *M* $Na^+$). Table I lists experimental conditions with the respective estimated reduction in stringency achieved simultaneously during hybridization and washing. Since the rate of filter hybridization is not affected by formamide concentrations in the range from 50 to 30% and only slightly reduced at 20% formamide,[7] hybrid formation at different stringencies mainly depends on the degree of mismatches between probe and template DNA, eliminating hybridization kinetics as a major variable. Under these experimental conditions, we observe comparable signal/noise ratios at both high and reduced stringency.

The extent of stringency reduction required to detect a related se-

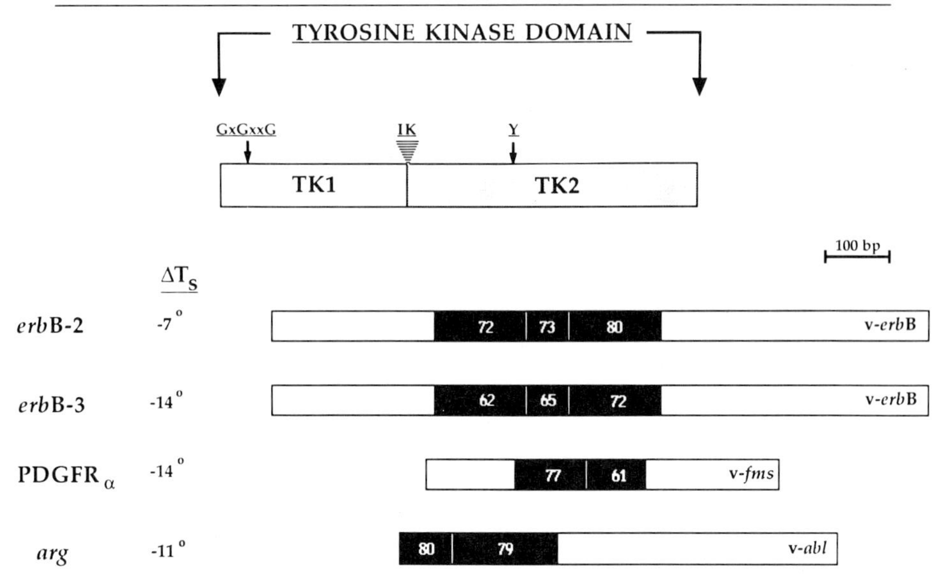


FIG. 1. Structural relationship of exons encoding novel receptor-like and cytoplasmic tyrosine kinases detected by reduced stringency hybridization. The junction between tyrosine kinase subdomains TK1 and TK2 is defined by a stretch of less well-conserved coding sequence, the interkinase region (IK), present in the *fms*/PDGFR family. The positions of ATP-binding site (GxGxxG), interkinase region (IK), and a tyrosine residue homologous to the autophosphorylation Tyr-416 (Y) of v-*src* are indicated. $\Delta T_s$ represents the reduction of stringency required for detection of related tyrosine kinases. Percentage nucleotide sequence identities of individual exons (solid boxes) with the homologous regions of the utilized probes (open bars) are indicated.

quence can be correlated to some degree with the nucleotide sequence identity score. Hybridization efficiency and hybrid stability of related sequences also depend on the distribution of nucleotide sequence homology and contiguous stretches of sequence identity with the probe (Fig. 1). Furthermore, in using genomic DNA, experimental conditions are influenced by the intron–exon structure. A closer homology between related tyrosine kinases tends to coincide with conserved exon structure. For instance, three exons that have been characterized in the tyrosine kinase domains of both *erbB-2* and *erbB-3* genes exhibit identical splice junctions, while these splice borders differ from those of tyrosine kinases of other subfamilies (Fig. 1). Thus, due to the conservation of nucleotide sequence and exon structure, the approach enhances the probability for detection of gene fragments most closely related to the gene from which the probe is derived. Figure 1 compares nucleotide sequence identity scores between exons of novel *PTK*s cross-hybridizing with the probe and the reduction of stringency required for the detection of single-copy gene

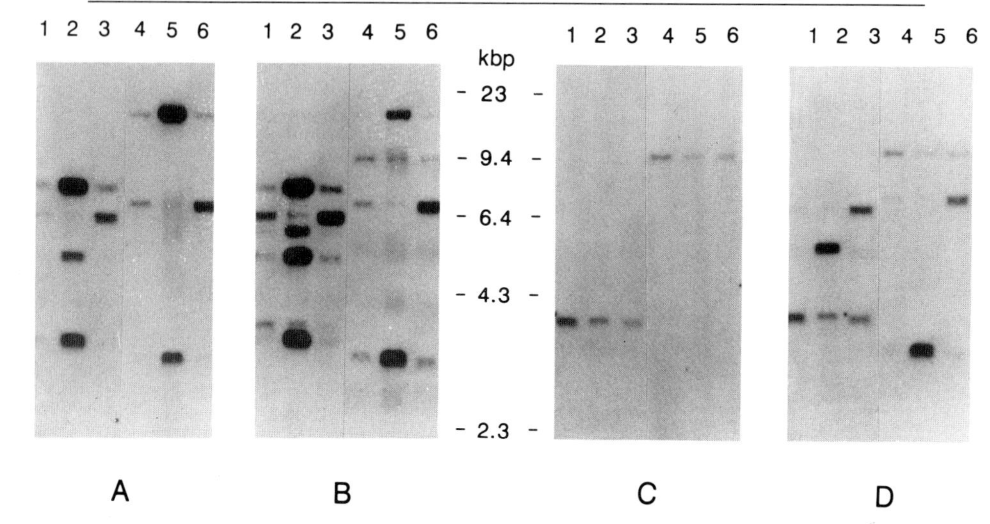


FIG. 2. *erbB*-Related sequences in the human genome. Replica Southern blots containing *Eco*RI (lanes 1–3)- or *Sac*I (lanes 4–6)-cleaved normal human genomic (1, 4), MDA-MB468 (2, 5), and SK-BR-3 (3, 6) DNAs were hybridized with v-*erbB* at intermediate ($\Delta T_s = -7°$) (A) and reduced ($\Delta T_s = -14°$) (B) or with an exon-specific *erbB*-3 probe at high (C) and reduced stringencies ($\Delta T_s = -14°$) (D). The migration positions of novel *erbB*-related *Eco*RI and *Sac*I restriction fragments identified under reduced stringency hybridization with v-*erbB* as a probe are indicated by arrowheads in (B).

fragments in normal DNA. This suggests that a reduction of stringency by 14° is adequate to detect related tyrosine kinase exons sharing in the range of 61 to 80% nucleotide sequence identity with the probe. In the EGF-R family, single gene copies of the more closely related *erbB-2* were detected at only 7° stringency reduction (Fig. 1).

A typical experimental approach is illustrated in Fig. 2. Replica nitrocellulose filters are generated containing several restriction digests of normal genomic DNA (lanes 1 and 4). The use of multiple enzyme digests increases the probability of detecting a novel *PTK* fragment distinct in size from those of the cognate gene. Furthermore, it is possible to exclude restriction fragment length polymorphisms. In this example, conducted for the identification of *erbB-3*,[3] we controlled for the gene fragments of known family members by their amplification in tumor cell DNAs. Alternatively, restriction fragments of known family members are identified by high-stringency hybridization with a perfectly matched probe. Using v-*erbB* as a probe at a stringency reduced by 7° (Fig. 2A), only restriction fragments specific to the EGF-R or *erbB-2* genes hybridized as determined by their amplification in MDA-MB468 (lanes 2 and 5) or SK-BR-3 (lanes 3 and 6), respectively. An additional 7° reduction of stringency to $\Delta T_s = -14°$ facilitated the detection of a novel v-*erbB*-related gene distinct from EGF-R and *erbB-2* genes. It was most noticeable in the *SacI* digest as a 9-kbp restriction fragment (Fig. 2B, lanes 4–6), but was also discerned as a 3.7-kbp *Eco*RI fragment (Fig. 2B, lanes 1–3). The specific detection of a novel *erbB*-related gene was inferred by absence of hybridization under the higher stringency conditions and the lack of gene amplification in MDA-MB468 and SK-BR-3 cells.

Evaluation of Southern blots and subsequent genomic cloning are facilitated by choosing restriction enzymes and probes such that single or few bands for individual *PTK* genes are generated. If exon structure and nucleotide sequence conservation are known for a subfamily of interest, a shorter probe is preferred, as the probability of multiple hybridizing fragments decreases for each member. Otherwise, longer probes spanning several exons are preferable, since the region of relatively highest conservation within the TK domain varies among distinct families (Fig. 1). We have successfully utilized probes extending from a single exon (135 bp) to most of the tyrosine kinase domain, respectively. Endonucleases cleaving the coding sequence in the probe region should be avoided. This is underscored by comparison of the *SacI* with the EcoRI digest in Fig. 2 using a 1.1-kbp v-*erbB* probe at reduced stringency. The presence of the novel 3.7-kbp *Eco*RI *erbB-3* fragment is obscured by four EGF-R and one *erbB-2* fragments in the range of 3–8 kbp, whereas the 9-kbp *erbB-3*-specific *Sac*I fragment is readily distinguished from two EGF-R- and one *erbB-2*-specific fragments.

## Characterization of Novel Tyrosine Kinase Exons

For further characterization, a novel related restriction fragment is cloned from a genomic library constructed by standard methodologies[9] in bacteriophage λ. At this step it is helpful to enrich the library source DNA for the novel restriction fragment (i.e., sucrose gradient centrifugation, DEAE membrane elution from agarose gel) with the attempt to exclude restriction fragment sizes of the cognate genes from the library. The clones of interest are identified by *in situ* hybridization at the reduced stringency initially determined by genomic Southern blot hybridization.

1. The bacteriophages are propagated according to standard procedures in 8 ml 0.7% top agarose in 150-mm plates on a 1.1% bottom agar layer at a density and plaque size that ensure that plaques are separated on the plates. After lifting the filters are air dried (15 min) and treated on 3MM Whatman sheets with the absorbed bacteriophages facing upward.

[9] T. Maniatis, E. F. Fritsch, and J. Sambrook, "Molecular Cloning: A Laboratory Manual." Cold Spring Harbor Lab., Cold Spring Harbor, New York, 1982.

Following denaturation (1 min) in 0.5 *M* NaOH/1.5 *M* NaCl the filters are briefly drained of excess alkali on dry 3MM paper, equilibrated with 1 *M* $NH_4OAc$/0.02 *M* NaOH (5 min), air dried, and baked for 2–3 hr at 80° in a vacuum oven.

2. For an efficient transfer of colonies of plasmid-based libraries from the plates onto nitrocellulose filters, 1.1% agar plates are used, where the melted agar has been cooled to 45° prior to pouring the plates. The marked filters are lifted gently, and with the colony side facing upward instantly placed onto 3MM paper saturated with denaturing solution. Sufficient lysis can be visually monitored, as the colonies become translucent (1–2 min). The filters are then treated essentially as described above. To minimize background, the filters following treatment with 1 *M* $NH_4OAc$/0.02 *M* NaOH are pressed while still wet between dry sheets of 3MM paper by firmly rolling a 25-ml pipette across the sandwich. By peeling the nitrocellulose filter off the 3MM paper, excess bacterial debris will adhere to the 3MM paper whereas DNA remains bound to the nitrocellulose membrane.

3. Hybridization and washing conditions are essentially as described for genomic Southern blot hybridization. Hybridization is conducted in 150-mm Petri dishes. Radioactive and nonradioactive hybridization solutions are prepared as described earlier and poured into separate dishes. Filters are individually wetted in cold hybridization solution and submerged in hybridization solution containing the probe ($1–3 \times 10^6$ cpm/ml). The covered dish enclosed in a bag or $\beta$ radiation-safe container is rotated slowly (<100 rpm) in a 42° orbital shaker for overnight hybridization. As a guideline, 30 large round filters can be hybridized in 1 dish containing ~100 ml of radioactive hybridization solution. An additional 100 ml of nonradioactive hybridization solution is required to wet the nitrocellulose filters.

4. Washing of the filters is carried out in 4-liter covered glass beakers throughout. As many as 50 filters can be processed in 1 container. For even washing, the filters must remain in motion and submerged in the solution. A volume of 1–1.5 liters of washing solution for each beaker is adequate. Four room-temperature washes of 15–20 min each are followed by two final stringency washes of 30 min each. It is advisable to monitor the actual temperature in the solution during the final stringency washes at 50° in order to ensure correct stringencies.

5. Clones for the cognate genes can be distinguished by hybridizing replica filters from the same plates with probes specific for those genes under high strigency. More conveniently, 1 $\mu$l of first cycle positive phage stocks are applied to gridded plates by piercing the surface of the top agarose that contains susceptible bacterial cells with a pointed pipette tip. Phage lysis will occur after a 10- to 12-hr incubation at 37°. These plates

are then used for the generation of replica nitrocellulose filters. The latter approach for distinguishing clones of the cognate genes is particularly useful, when many first round positives require further analysis.

For the characterization of *erbB-3,* a normal human genomic DNA library enriched for 8- to 12-kbp *Sac*I restriction fragments was constructed in bacteriophage. Among 4 × $10^5$ recombinants, 29 positives were identified using v-*erbB* as a probe under reduced hybridization stringency. One positive represented an EGF-R clone, whereas none was detected for *erbB-2*. Several positives plaques are purified and the phage DNAs are subjected to restriction and hybridization analysis at reduced stringency in order to map the region of homology within the phage DNA insert. Hybridizing insert fragments of suitable size (preferably 1 kpb or less) are then subcloned in plasmid vectors with high replication rate (pUC, Bluescript, etc.) and subjected to nucleotide sequence analysis by the dideoxy chain termination method using supercoiled DNA as template.

Gene-specific probes from the genomic clone can be used to investigate at high-stringency mRNA expression in various sources. In addition, genetic alterations including gene amplification or rearrangement as well as altered expression for a novel tyrosine kinase gene can be searched for in human disease. The predicted amino acid sequence of exons can be used to design peptides for the generation of polyclonal antisera in efforts to study the encoded gene product. For structural characterization and functional studies involving *in vitro* expresssion of the protein, the complete coding sequence can be isolated by cDNA cloning using gene-specific exon-containing probes at high stringency. The initial expression analysis is useful in detection of sources with relatively high transcript levels for cDNA cloning.

Utilizing a single genomic exon probe of *erbB-3* (Fig. 1, downstream exon) we rehybridized the initial genomic Southern blot under both high and reduced stringencies. This probe detected under high stringency the *erbB-3*-specific genomic fragments initially identified by v-*erbB* under reduced stringency (Fig. 2C). As expected at the same level of stringency reduction, the *erbB-3*-specific exon probe recognized only the homologous EGF-R- and *erbB-2*-specific gene fragments in addition to the endogenous restriction fragments (Fig. 2D), corroborating the specificity as well as reproducibility of this approach.

## Discussion

In this chapter, we have summarized a general approach for detecting novel *PTK*-related genes in genomic DNA. Based upon knowledge of structural conservation in a distinct *PTK* subfamily, the approach can be optimized for specific tasks by choice of restriction enzymes, probe design, and genomic DNA source. Several other approaches (see Chapters 44, 45, 47 in this volume), including reduced stringency hybridization of cDNA libraries with TK probes or degenerate oligonucleotides, have been employed in the successful isolation of novel *PTK*s. More recently, the polymerase chain reaction with degenerate oligonucleotides as primers has been utilized to isolate novel TK-coding sequences from DNA complementary to mRNA. Finally, screening of cDNA expression libraries with anti-phosphotyrosine antibodies has facilitated the successful isolation of *PTK*s based upon their expression at the protein level.

Approaches utilizing RNA as a source either combine criteria for structural relationship and levels of expression of a novel *PTK* or are strictly based on protein function. In the latter case this might allow the isolation of *PTK*s lacking sufficient sequence conservation for detection by the other methods described. Those techniques involving RNA as the source have the potential advantage of providing a more rapid determination of larger portions of coding sequence than the genomic method. On the other hand, the search for novel *PTK*s in mRNA-derived sources could be biased by the expression pattern. For example, a highly conserved, novel *PTK* could be missed, if not sufficiently expressed in a particular cell source, or inefficiently converted to cDNA due to its structure.

Since normal genomic DNA contains a complete set of genes encoding PTKs, the genomic approach is most comprehensive in the search for structurally conserved genes. Moreover, titration of stringency reduction on the genomic Southern blot can provide information about the number of different genes detected at a given stringency. A potential disadvantage of this method is the detection of nonfunctional genes, although in our search for novel *PTK*s we have not encountered such genes. While an intermediate genomic cloning step is required, knowledge of the expression pattern of a novel *PTK* prior to cDNA cloning may also be critical to the subsequent isolation of its complete coding sequence from the appropriate cDNA libraries. Hence, each of the described methods entails certain advantages, and the method of choice is determined by the specific goal. While those methods utilizing cDNAs may be particularly valuable in the isolation of tissue-specific expressed novel *PTK*s or molecules of more divergent genetic structure, genomic identification of novel *PTK*s appears most suitable for the systematic search for structurally related PTK coding sequences.

# [47] Isolation of cDNA Clones That Encode Active Protein-Tyrosine Kinases Using Antibodies against Phosphotyrosine

*By* RICHARD A. LINDBERG and ELENA B. PASQUALE

Protein-tyrosine kinases (PTKs) are involved in signal transduction pathways that control the proliferative status of the cell. They are also thought to play major roles in cell differentiation and embryonal development. In addition to these normal functions PTKs are often implicated in malignant transformation. Given the importance of this class of enzymes, it is of interest to identify and to characterize new members.

It is likely that only a fraction of the existing PTKs are known at this time. The number of identified PTKs has increased very rapidly but many more PTK genes are thought to exist. Molecular cloning of protein kinase genes, rather than enzyme purification, has proved to be the most successful strategy for identifying novel PTKs. cDNAs for several receptor PTKs have been cloned by screening libraries with oligonucleotides designed on the basis of protein sequence information. Other PTKs have been identified by transfection/transformation assays. Low-stringency screening with cDNAs encoding known PTKs has also been productive. Novel PTK genes have been cloned by utilizing the fact that they all share conserved motifs that can be used for the design of degenerate oligonucleotides capable of hybridizing to different PTK cDNAs. Such oligonucleotides have been used for screening libraries and for the polymerase chain reaction. Several of these methods are described in this volume.[1-3]

We have successfully used a method of cloning that does not rely on sequence homology, but rather on enzyme activity.[4,5] This approach consists of screening expression libraries with antibodies to phosphotyrosine (P-Tyr) and is based on the following rationale: (1) all protein kinases have a conserved catalytic domain capable of phosphotransferase activity even in the absence of adjacent sequences; (2) many PTKs display autophosphorylating activity and will also phosphorylate other proteins quite promiscuously; and (3) in bacteria endogenous PTK activity is undetectable by immunoblotting techniques. As we have found, PTKs are indeed active when expressed as $\beta$-galactosidase fusion proteins in bacteria and

[1] S. K. Hanks and R. A. Lindberg, this volume [44].

[2] A. F. Wilks, this volume [45].

[3] M. H. Kraus and S. A. Aaronson, this volume [46].

[4] R. A. Lindberg, D. P. Thompson, and T. Hunter, *Oncogene* **3,** 629 (1988).

[5] E. B. Pasquale and S. J. Singer, *Proc. Natl. Acad. Sci. U.S.A.* **86,** 5449 (1989).

PTK-containing plaques can be easily identified with antibodies to P-Tyr. In this chapter we describe the expression cloning methods that we have used to isolate and characterize cDNA clones encoding PTKs.

## Methods

### *Screening Procedure*

The two most important materials required for the screening are the anti-P-Tyr antibodies and the library. We have used antibodies raised against either bacterially produced v-*abl* kinase or randomly polymerized P-Tyr, alanine, and glycine.[4,5] The commercially available monoclonal from ICN (Lilse, IL), Py20, has also been used successfully (John Tan and James Spudich, personal communication, 1990; Martin McMahon and J. Michael Bishop, personal communication, 1990). Presumably any antibodies that have been affinity purified with P-Tyr or an analog should be adequate as long as they work well in immunoblotting experiments. The library should be complex and have an average insert size of at least 1 kilobase (kb), which is the minimal length of DNA needed to encode a protein kinase catalytic domain. Inserts should be longer in order to isolate clones with long 3′ untranslated regions. The specifications of the library will determine what subset of the PTKs amenable to this type of cloning will be isolated.

We chose to screen λgt11 cDNA expression libraries in *Escherichia coli* and performed the screening (originally described by Young and Davis[6]) as follows:

1. A stationary phase culture of Y1090 is incubated with the λ phage for 15 min at 37°, diluted 1 : 13 into top agarose, from which 8 ml is spread onto each 15-cm LB plate. The 8 ml should contain about 20,000 phage.

2. The plates are incubated at 42° for 3–4 hr (until plaques are easily visible).

3. Nitrocellulose filters are wetted in a 10 m*M* solution of the inducing agent, isopropyl-β-D-thiogalactoside (IPTG), and then air dried. They are laid on the plaques and the plates are incubated at 37° for 2–3 hr. A second filter can be incubated on each plate if duplicates are desired.

4. The filters should then be treated with a procedure that works for immunoblotting with the particular antibodies used. For poly(Ala, Gly, P-Tyr) antibodies (described by Kamps and Sefton[7]) the filters are blocked

[6] M. Snyder, S. Elledge, D. Sweetser, R. A. Young, and R. W. Davis, this series, Vol. 154, p. 107.

[7] M. P. Kamps and B. M. Sefton, *Oncogene* 2, 305 (1988).

in 5% bovine serum albumin (BSA) and 1% (w/v) ovalbumin in Tris–saline solution [10 m$M$ Tris, pH 7.4, 150 m$M$ NaCl, 0.01% (w/v) sodium azide] for several hours to overnight, incubated with antibody (in blocking solution) for 2 hr, rinsed two times (10 min each) in the saline, saline with 0.05% (v/v) Nonidet P-40, and saline two times again.

5. The bound immunoglobulins are detected with $^{125}$I-labeled protein A (0.1 $\mu$Ci/ml in blocking solution), followed by the same set of rinses performed after antibody incubation. The $^{125}$I-labeled protein A is 2–10 $\mu$Ci $\mu g^{-1}$ (NEN, Boston, MA); the filters are exposed 1–4 days at $-70°$ with intensifying screens.

### *Analysis of Positive Clones*

Once the recombinant clones have been plaque purified, the cloned cDNAs and the PTKs they encode can be analyzed in many ways; a few protocols we have used are outlined below.

*Characterization of Lysates.* Liquid lysates obtained from *E. coli* Y1090 infected with the PTK λgt11 recombinant phages can be used to purify DNA for restriction mapping, subcloning and sequencing. The *E. coli* lysates are also useful for the analysis of proteins. For instance, the proteins in the lysates can be fractionated by SDS-PAGE and probed by immunoblotting with anti-P-Tyr antibodies or with anti-$\beta$-galactosidase antibodies. These experiments will indicate the extent of protein tyrosine phosphorylation in *E. coli* after the expression of an active PTK, the level of autophosphorylation on tyrosine of the $\beta$-galactosidase–PTK fusion proteins, and their molecular weight (if the fusion proteins are not degraded by proteases). In some instances enough fusion protein is produced in a stable form to be purified from polyacrylamide gels and used as an immunogen to prepare antibodies (see below).

A protocol for the preparation of *E. coli* lysates for DNA and protein analysis is as follows[8]:

1. A culture of *E. coli* Y1090 is grown to saturation in LB medium + 0.4% (w/v) maltose + 50 $\mu$g/ml ampicillin at 37°.
2. Bacterial culture (1.2 ml) is centrifuged for 1 min at 10,000 $g$, resuspended in the same volume of 10 m$M$ $MgCl_2$, 20 m$M$ Tris-HCl, pH 8.0, and infected with $10^6$–$10^7$ phage/ml for 30 min at room temperature. Phage from a liquid lysate or a plate stock can be used.
3. The infected *E. coli* cells are transferred to 40 ml of LB medium +

[8] F. M. Ausubel, R. Brent, R. E. Kingston, D. D. Moore, J. G. Seidman, J. A. Smith, and K. Struhl, "Current Protocols in Molecular Biology." Wiley, New York, 1987.

5 m$M$ $CaCl_2$ + 50 $\mu$g/ml ampicillin, prewarmed at 37°, and shaken vigorously at 37° with good aeration for 1 hr.

4. The culture is then split in two portions (20 ml each) and IPTG is added to 1 m$M$ to one 20-ml portion, in order to induce the synthesis of fusion protein.

5. The cultures are grown for an additional 3–6 hr and harvested immediately upon clearing.

6. For the preparation of protein:

   a. The 20-ml aliquot of the bacterial lysate that was treated with IPTG is cooled on ice.
   b. Dry ammonium sulfate is added to 35–40% (w/v).
   c. The bacterial lysate is stirred on ice for 1 hr and centrifuged at 10,000 $g$ for 10 min.
   d. The pellet is resuspended in 0.5 ml of sodium dodecyl sulfate (SDS)-containing sample buffer, sonicated briefly, and boiled for 3 min.
   e. An aliquot (about 50 $\mu$l/lane) is used for SDS-PAGE. Polyacrylamide gels (7.5%, w/v) are adequate for the analysis of most $\beta$-galactosidase fusion proteins.

This procedure should yield 1 to 3 mg of total protein. The yield of fusion protein can vary, typically from a few to about 50 $\mu$g.

7. For the purification of $\lambda$ DNA:

   a. The 20-ml aliquot of the lysate that was not induced with IPTG is centrifuged at 10,000 $g$ for 10 min.
   b. The phage is purified using a commercially available phage immunoadsorbent [such as lambdaTRAP from Clontech Laboratories (Palo Alto, CA) or LambdaSorb from Promega (Madison, WI)] according to the manufacturer's specifications.

*Known or Novel?* The most direct way to determine the identity of a PTK is by sequence analysis. Any regime of subcloning and subsequent sequence analysis will suffice. However, the most efficient way is to start by sequencing the 5′ end of the cloned cDNAs in the original $\lambda$gt11 or in a plasmid vector, using appropriate primers. Translation of the DNA sequences obtained, followed by a comparison with published data or a search of computer databases, will indicate if the cloned kinases have been previously sequenced by others.

*Confirmation that Cloned cDNAs Code for PTKs*. The initial characterization of the clones isolated using anti-P-Tyr antibodies should establish whether they indeed encode PTKs. This can be done by analyzing the sequence of conserved regions in the catalytic domain. It is possible to target sequencing reactions to these regions by using degenerate oligonucleotides, as described in this volume.[1] The presence of P-Tyr in bacteria expressing the cloned protein should also be confirmed by phosphoamino acid analysis of $^{32}$P-labeled bacteria. Although it may be possible to use bacteria containing the original λgt11 vector for this purpose, we have routinely subcloned cDNA inserts into plasmid expression vectors. *Eco*RI fragments from λgt11, comprising the complete inserts, or catalytic domain-coding regions (if the sequence is known), can be subcloned and expressed from a plasmid promoter. We have expressed active catalytic domains in IPTG-inducible vectors (PKK233-2; Pharmacia, Piscataway, NJ), temperature-inducible vectors (pEX; Boehringer Mannheim, Indianapolis, IN), *trp*-repressible vectors (pATH[9]), and T7 polymerase-transcribed vectors (pGem; Promega, Madison, WI). The PTK-expressing bacteria can be metabolically labeled with [$^{32}$P]$P_i$ for phosphoamino acid analysis. The following protocol was used for the pATH vectors, but can be modified for any expression system.

1. A stationary phase culture grown in LB + 50 μg/ml ampicillin is diluted 1 : 100 into complete M9 medium [0.5% casamino acids (acid hydrolysate), 1 m*M* $MgSO_4$, 0.1 m*M* $CaCl_2$, 0.02% (w/v) glucose, 10 μg/ml thiamin, 6 g/liter $Na_2HPO_4 \cdot 7H_2O$, 3 g/liter $KH_2PO_4$, 0.5 g/liter NaCl, 1 g/liter $NH_4Cl$] + 50 μg/ml ampicillin. The cells are grown at 37° until expression begins (~1–2 hr).
2. The bacteria are resuspended in a small volume (i.e., 10 ml of culture in 0.1 ml) of phosphate-free M9 medium. [$^{32}$P]$P_i$ (500 μCi) is added and the bacteria are incubated at 37° for 30 min. After several washes in M9 medium the bacteria are resuspended in $H_2O$, then diluted 1 : 1 in SDS-containing sample buffer and boiled for 5 min. With IPTG-inducible systems we have added the inducer and [$^{32}$P]$P_i$ concurrently to log-phase bacteria in phosphate-free M9 medium and labeled for several hours.
3. The lysates are fractionated on SDS-polyacrylamide gels and transferred to Immobilon (Millipore, Bedford, MA). After autoradiography the labeled proteins can be subjected to phosphoamino acid analysis directly from the filter as described by Kamps and Sefton.[10] Briefly, the band is cut out, wetted in methanol, then washed in $H_2O$ and added directly to 5.7 *M* HCl for 1 hr at 110°. The phosphoamino acids are then separated by two-dimensional electrophoresis at pH 1.9 and 3.5 on cellulose thin-layer plates.[11]

[9] W. J. Boyle, J. S. Lipsick, and M. A. Baluda, *Proc. Natl. Acad. Sci. U.S.A.* **83,** 4685 (1986).

[10] M. P. Kamps and B. M. Sefton, *Anal. Biochem.* **176,** 22 (1989).

[11] J. A. Cooper and T. Hunter, this series, Vol. 99, p. 387.

It is critical to label the bacteria early during induction. P-Tyr appears to turn over very slowly in bacteria and the toxicity of the expressed kinases may rapidly reduce their synthesis. If [$^{32}$P]P$_i$ is added after most of the recombinant protein has been synthesized, the bacterial cells may contain high levels of unlabeled P-Tyr. The time course of induction can be monitored in immunoblotting experiments with anti-P-Tyr antibodies.

*Preparation of β-Galactosidase–PTK Fusion Proteins to Be Used as Antigens.* We have used several plasmid expression vectors to produce antigen, i.e., pATH and pEX. However, even before the corresponding cDNAs have been isolated and characterized, the β-galactosidase–PTK fusion proteins expressed in *E. coli* infected with the original λgt11 can be identified by immunoblotting the proteins from *E. coli* liquid lysates (see Characterization of Lysates) with anti-P-Tyr and anti-β-galactosidase antibodies. The PTK fusion proteins should be labeled with anti-β-galactosidase antibodies and are usually the proteins most heavily phosphorylated on tyrosine in the lysates due to autophosphorylation. If the fusion proteins present in the *E. coli* lysates appear stable and sufficiently abundant, they can be purified by electrophoresis on SDS-polyacrylamide gels, as described below, and used to immunize rabbits. They can also be coupled to a solid support and used for the affinity purification of specific antibodies from the immune sera.

The recombinant proteins, obtained from *E. coli* lysates as described above or using pEX and pATH expression vectors, can be purified using SDS-PAGE as follows:

1. After electrophoresis, the gel is stained with Coomassie Blue for 5 to 10 min or until the fusion protein band is visible.
2. The gel is then rinsed with $H_2O$.
3. A gel slice containing the fusion protein is removed.
4. The gel slice is rinsed with PBS and crushed into small fragments with a glass rod.
5. The proteins in the gel slice are eluted by shaking for 12–24 hr in about 15 ml/gel of elution buffer (50 m*M* $NH_4HCO_3$ + 0.5% SDS, pH 9.0). The pH of the elution buffer should be checked after addition to the crushed gel; we have found that proteins will not be extracted efficiently if the pH is lower than 9.0.
6. The crushed gel is centrifuged in a desk top centrifuge at 400 *g* for 5 min.
7. The supernatant is removed and the proteins eluted once more from the crushed gel.
8. The two supernatants are pooled and lyophilized.

9. The dried, eluted fusion protein is resuspended in about 0.3 ml $H_2O$/gel, dialyzed in PBS, mixed with adjuvant, and injected (0.3–0.6 ml/rabbit).

Antibodies to the $\beta$-galactosidase (or trpE) components of the antigen can be eliminated by absorption on $\beta$-galactosidase (or trpE) affinity columns. It is also possible to immunize rabbits with a $\beta$-galactosidase fusion protein and use the correspondent trpE fusion protein for affinity purification (or vice versa). In this case, only the antibodies specific for the PTK region of the fusion protein will be purified. If cDNAs encoding entire PTK catalytic domains are cloned in the expression vectors, the fusion proteins obtained may contain phosphorylated tyrosine residues. If, for this reason, anti-P-Tyr antibodies are produced, they can be absorbed on P-Tyr affinity columns or "neutralized" by the addition of free P-Tyr to the antibody preparations. These antisera can be used as a source of anti-P-Tyr antibodies after affinity purification.

Alternative procedures for most of the methods described in this chapter can be found in laboratory manuals.[8,12,13]

## Discussion

This method of screening appears to be of general application for the identification of PTKs and confirmation of their catalytic activity.[4,5,14,15] Unlike methods based on sequence homology, it allows the isolation of clones encoding at least the whole catalytic domain and allows for the possibility to isolate PTKs that are not closely homologous to known ones. For example, PTKs from microbial eukaryotes have been isolated with anti-P-Tyr antibodies (John Tan and Martin McMahon, personal communication, 1990).

Several unexpected results have been obtained with this type of screening. We originally assumed that cDNA inserts would have to be in the proper orientation and reading frame in order to be detected. However, many clones have been isolated with the insert in reverse orientation, or with stop codons upstream of the catalytic domain. Proteins encoded by such inserts are not fused to $\beta$-galactosidase and are expressed at very low levels.[6] It should be noted that the method is sensitive enough to detect these inefficiently expressed PTKs. There may even be a selection

[12] S. B. Carroll and A. Laughon, "DNA Cloning: A Practical Approach," Vol. 3, p. 89. IRL Press, Oxford, 1987.

[13] E. Harlow and D. Lane, "Antibodies: A Laboratory Manual." Cold Spring Harbor Lab., Cold Spring Harbor, New York, 1989.

[14] K. Letwin, S.-P. Yee, and T. Pawson, *Oncogene* **3,** 621 (1988).

[15] S. Kornbluth, K. E. Paulson, and H. Hanafusa, *Mol. Cell. Biol.* **8,** 5541 (1988).

against efficiently expressed fusion proteins due to the toxicity of tyrosine phosphorylation to bacteria. We also have isolated with anti-P-Tyr antibodies a protein kinase that has sequence homology with protein-serine/threonine kinases and phosphorylates proteins on threonine when expressed from plasmids. We have not yet mapped the epitope recognized by the anti-P-Tyr antibodies, but this finding reinforces the importance of determining the amino acid specificity of newly identified protein kinases. Using this method other laboratories have also isolated protein kinase cDNAs that have a primary structure that places them in the protein-serine/threonine kinase family (David Stern and Gordon Mills, personal communication, 1990).

### Acknowledgments

We thank Tony Hunter and Jon Singer for the encouragement and guidance that made this work possible. We are grateful to Bart Sefton and Mark Kamps for the anti-P-Tyr antibodies. The work was supported by USPHS Grant CA-39780 (R.A.L.) and by an Arthritis Investigator Award (E.B.P.).

## [48] cDNA Cloning and Properties of Glycogen Synthase Kinase-3

*By* JAMES ROBERT WOODGETT

Glycogen synthase kinase-3 (GSK-3) was first identified in 1980 as an enzyme, distinct from cyclic AMP-dependent protein kinase, that could phosphorylate glycogen synthase.[1,2] Extensive purification of GSK-3 from rabbit skeletal muscle demonstrated it to be identical to the proteinaceous activating factor ($F_A$) of the MgATP-dependent protein phosphatase, an inactive form of protein phosphatase-1.[3–5] cDNA cloning of GSK-3 has revealed the enzyme to comprise a multigene family in common with other signal transductory proteins, such as protein kinase C and cyclic AMP-

[1] N. Embi, D. B. Rylatt, and P. Cohen, *Eur. J. Biochem.* **107,** 519 (1980).

[2] D. B. Rylatt, A. Aitken, T. Bilham, G. D. Condon, N. Embi, and P. Cohen, *Eur. J. Biochem.* **107,** 529 (1980).

[3] B. A. Hemmings and P. Cohen, this series, Vol. 99, p. 337.

[4] B. A. Hemmings, D. Yellowlees, J. C. Kernohan, and P. Cohen, *Eur. J. Biochem.* **119,** 443 (1981).

[5] J. R. Vandenheede, S.-D. Yang, J. Goris, and W. Merlevede, *J. Biol. Chem.* **255,** 11768 (1980).

dependent protein kinase. In this chapter I summarize the structural and functional properties of this protein kinase family.

## Purification and Cloning

GSK-3 has been purified to homogeneity by four groups from rabbit skeletal muscle and bovine or porcine brain.[4,6–9] The procedure for isolation of the muscle enzyme using substrate peptide affinity columns is described in detail elsewhere in this volume.[10] GSK-3 has a very weak affinity for anion exchangers such as DEAE-cellulose but binds to a variety of cationic resins, hydroxylapatite, phosphocellulose, and heparin-Sepharose (Pharmacia, Piscataway, NJ); the enzyme is also retained by Cibacron Blue dye resins such as Blue-Sepharose (Pharmacia). Upon gel-permeation chromatography, the skeletal muscle enzyme elutes according to a mass of 75 kDa, whereas upon sedimentation equilibrium centrifugation in 500 m*M* NaCl a mass of 47 kDa is recorded.[6] In skeletal muscle, the concentration of GSK-3 is approximately 1 $\mu$g/ml, about 10-fold lower than cyclic AMP-dependent protein kinase. Consequently >50,000-fold purification is required, resulting in yields of 20–75 $\mu$g of homogeneous protein from 2–3 kg of muscle. Upon polyacrylamide gel electrophoresis in the presence of sodium dodecyl sulfate (SDS), the pure skeletal muscle enzyme migrates as a single band of 51 kDa.[7]

## cDNA Cloning

The method used for the isolation of GSK-3 cDNA is potentially applicable to the molecular cloning of other protein kinases of low abundance where the amount of peptide sequence information is limited, and is thus described in detail.

Tryptic peptides derived from 25 $\mu$g of muscle GSK-3, were resolved by reversed-phase high-performance liquid chromatography (HPLC) at pH 6.8 and selected peaks rechromatographed at pH 2.[11] The sequences of several peptides isolated by this procedure were determined at levels of 20–50 pmol using a gas–liquid sequenator (model 477A; Applied Biosystems, Foster City, CA). None of these sequences was suitable for the design of oligonucleotide probes for screening cDNA libraries by hybrid-

[6] J. R. Woodgett and P. Cohen, *Biochim. Biophys. Acta* **788,** 339 (1984).
[7] J. R. Woodgett, *Anal. Biochem.* **180,** 237 (1989).
[8] H. Y. L. Tung and L. J. Reed, *J. Biol. Chem.* **264,** 2985 (1989).
[9] S.-D. Yang, *J. Biol. Chem.* **261,** 11786 (1986).
[10] J. R. Woodgett, this volume [13].
[11] P. Tempste, D. D.-L. Woo, D. B. Teplow, R. Aebersold, L. E. Hood, and S. B. H. Kent, *J. Chromatogr.* **359,** 403 (1986).

ization, due to either their short length or their amino acid composition, which included residues encoded by four or more codons. The maximum usable degeneracy for oligonucleotide hybridization probes is around 100, governed by reductions in specific activity of a degenerate probe and the limits of signal detection.

All sequenced protein kinases share several structural features that are likely to be involved in the catalytic process.[12] Several of these "motifs" have been successfully used to isolate novel protein kinases by conventional hybridization and by use of the polymerase chain reaction.[13,14] This latter technique tolerates highly degenerate oligonucleotides, selectivity being governed by the requirement for two independent DNA sequences being proximal to each other and in the correct relative orientation.[15] To clone GSK-3, it was assumed that this enzyme contains protein kinase motifs. A generic oligonucleotide designed to represent one of these conserved sequences could then be paired with a GSK-3-specific oligonucleotide designed from one of the tryptic peptides. In general, the relative position of the peptide in the polypeptide chain with respect to the kinase motif will be unknown, requiring that both sense (relative to the RNA) and antisense forms of the oligonucleotides be synthesized.

In the case of GSK-3, the antisense form of a 17-mer representing all of the possible codons for the sequence Glu-Met-Asn-Pro-Asn-Tyr (32-fold degenerate) was initially synthesized. This was paired with the sense from of a 23-mer representing a number of sequence variations of the region VI protein kinase motif.[12] This generic region VI oligonucleotide was 60,000-fold degenerate.

## Amplification of cDNA

Sterile technique is used throughout to prevent contamination.

*Reagents*

10× Polymerase chain reaction (PCR) buffer: 100 m$M$ Tris-HCl, pH 8.4 (at 20°), 500 m$M$ KCl, 1% (v/v) gelatin, 0.5% (v/v) Nonidet P-40, 0.5% (w/v) Tween 20, 15 m$M$ $MgCl_2$. This solution should be autoclaved and aliquoted for storage at −20°

Deoxynucleotide mix: A mixture of highly purified deoxyATP,

[12] S. K. Hanks, A. M. Quinn, and T. Hunter, *Science* **241,** 42 (1987).
[13] S. K. Hanks, *Proc. Natl. Acad. Sci. U.S.A.* **84,** 388 (1987).
[14] A. Wilks, *Proc. Natl. Acad. Sci. U.S.A.* **86,** 1603 (1989).
[15] K. Mullis, F. Faloona, S. Scharf, R. Saiki, G. Horn, and H. Erlich, *Cold Spring Harbor Symp. Quant. Biol.* **51,** 263 (1986).

deoxyCTP, deoxyGTP, and deoxyTTP, each at 1.25 m$M$ and pH 7.0 in water, stored at $-20°$ in aliquots

*Thermophilus aquaticus* (*Taq*) DNA polymerase: Available from Perkin-Elmer Cetus (Emeryville, CA). Stored at $-20°$ at 5 U/$\mu$l

*Oligonucleotides.* The criteria for oligonucleotide design for PCR are different from those for hybridization to immobilized DNA. Degeneracy is not as important, but should be minimized. The G + C content should approximate 50% and inosine can be substituted for positions where the base could be A, T, or G, to further reduce degeneracy. Oligonucleotides of length 15–30 have been successfully used. A linker sequence encoding a restriction enzyme site can be appended to the 5′ end to aid in subsequent cloning, although this is not necessary and can result in loss of specificity. Deprotected and detritylated oligonucleotides should be extracted with phenol/chloroform/isoamyl alcohol (50 : 48 : 2), followed by chloroform extraction to remove residual organic compounds. Following precipitation with 2 vol of ethanol in the presence of 300 m$M$ sodium acetate, the oligonucleotides are dissolved in water to a final concentration of 0.5 mg/ml (1 $OD_{260}$ unit = 20 $\mu$g/ml) and stored frozen. Further purification is not usually required.

*Template DNA.* Due to the possibility of introns being present between the two oligonucleotide primers in genomic DNA, cDNA is the preferred template. cDNA purified from plasmid or bacteriophage $\lambda$ is suitable, although background amplification is lower using the former since the vector sequences are smaller. Plasmid DNA should be linearized before use. To reduce the probability of cleaving the DNA between the oligonucleotide-binding sites, two aliquots can be cut with different restriction enzymes and then combined. Linearized template DNA is stored frozen at a concentration of 1 mg/ml.

*Amplification*

Reactions are performed using 0.5-ml microcentrifuge tubes in a final volume of 50 $\mu$l. To each tube are added in order: 33.5 $\mu$l water, 5 $\mu$l 10× PCR buffer, 8 $\mu$l deoxynucleotide triphosphate mix, 1 $\mu$l sense oligonucleotide, 1 $\mu$l antisense oligonucleotide, 1 $\mu$l cDNA. The components are mixed and drawn to the bottom of the tube by brief centrifugation. Each tube is carefully overlaid with 50 $\mu$l of paraffin oil (IR spectroscopy grade; Fluka, Ronkonkoma, NY) and heated to 95° for 5 min. The tubes are then cooled to ~50° and 0.5 $\mu$l of *Taq* DNA polymerase is added. The tubes are tightly capped and subjected to 30 thermal cycles of 93° for 1 min, 50° for 2 min, and 71° for 5 min in a programmable heating block (Techne, Princeton, NJ). Following the cycles, 1 $\mu$l of DNA polymerase (Klenow

```
    1-GGGCACGCCGGAGCCCGAACGGCGCAGCCTGGAAGAGGCCGGAGCCCAAGGGAGGCGGCGGTAGAGGCAGCGGCGGGGGCAGCCCAGGCAGCCCGAGCCCCGCGGCCTGGGCCTGCGCTC

                                                                   T  S  S  F  A  E  P  G
             M  S  G  G  G  P  S  G  G  G  P  G  G  S  G  R  A  R  T  S  S  F  A  E  P  G  G  G  G  G  G  G  G  G  G  P  G  G - 38
  121-GGCGCCATGAGCGGCGGCGGGCCTTCGGGAGGTGGCCCTGGGGGCTCGGGCCGGGCGCGGACCAGCTCGTTCGCGGAGCCAGGCGGCGGAGGCGGAGGCGGTGGCGGCGGCCCCGGGGGC

                                           A  S  V  G  A  M  G  G  G  V  G  A  S  S  S  G  G  G  P
       S  A  S  G  P  G  G  T  G  G  G  K  A  S  V  G  A  M  G  G  G  V  G  A  S  S  S  G  G  G  P  S  G  S  G  G  G  G  S  G - 78
  241-TCGGCCTCCGGCCCAGGAGGCACTGGCGGCGGGAAGGCGTCAGTCGGGGCTATGGGTGGGGGCGTGGGAGCCTCGAGCTCTGGGGGTGGCCCCAGCGGCAGCGGCGGAGGAGGCAGCGGT

                                                                      V  T  T  V  V  A  T  L  G  Q  .  P
       G  P  G  A  G  T  S  F  P  P  P  G  V  K  L  G  R  D  S  G  K  V  T  T  V  V  A  T  L  G  Q  G  P  E  R  S  Q  E  V  A -118
  361-GGCCCCGGCGCGGGCACTAGCTTCCCGCCGCCCGGAGTGAAGCTGGGCCGTGACAGCGGGAAGGTGACCACAGTGGTAGCCACTCTAGGCCAAGGCCCAGAGCGTTCCCAAGAGGTGGCT

       Y  T  D  I  K  V  I  G  N  G  S  F  G  V  V  Y  Q  A  R  L  A  E  T  R  E  L  V  A  I  K  K  V  L  Q  D  K  R  F  K  N -158
  481-TACACCGACATCAAAGTGATTGGCAATGGCTCATTCGGAGTAGTGTACCAGGCACGGCTGGCAGAAACGAGGGAACTGGTGGCCATCAAGAAGGTTCTTCAGGACAAAAGGTTCAAGAAC

          E  L  Q  I  M  R                                 Y  F  F  Y  S
       R  E  L  Q  I  M  R  K  L  D  H  C  N  I  V  R  L  R  Y  F  F  Y  S  S  G  E  K  K  D  E  L  Y  L  N  L  V  L  E  Y  V -198
  601-CGAGAGCTGCAGATTATGCGTAAGCTGGACCACTGCAATATCGTGAGGCTGCGGTACTTTTTCTACTCCAGTGGGGAGAAGAAAGATGAGCTGTATTTAAATCTGGTGCTGGAATATGTG

       P  E  T  V  Y  R  V  A  R  H  F  T  K  A  K  L  I  I  P  I  I  Y  V  K  V  Y  M  Y  Q  L  F  R  S  L  A  Y  I  H  S  Q -238
  721-CCCGAGACCGTGTACCGAGTGGCCCGTCACTTTACCAAGGCCAAGTTGATCATCCCTATCATCTATGTCAAGGTGTACATGTACCAGCTCTTCCGGAGCTTGGCCTACATCCACTCCCAA

                      D  I  K  P  Q  N  L  L  V  D  P  D  T  A  V  L  K
       G  V  C  H  R  D  I  K  P  Q  N  L  L  V  D  P  D  T  A  V  L  K  L  C  D  F  G  S  A  K  Q  L  V  R  G  E  P  N  V  S -278
  841-GGTGTGTGTCACCGTGACATCAAGCCCCAGAATTTGCTTGTGGACCCTGACACTGCTGTCCTCAAGCTCTGCGACTTTGGCAGTGCAAAGCAGTTGGTTCGGGGGGAGCCCAACGTGTCC

       Y  I  C  S  R  Y  Y  R  A  P  E  L  I  F  G  A  T  D  Y  T  S  S  I  D  V  W  S  A  G  C  V  L  A  E  L  L  L  G  Q  P -318
  961-TACATCTGTTCTCGGTACTACCGTGCTCCGGAGCTCATCTTTGGAGCCACAGATTACACCTCGTCCATCGATGTGTGGTCAGCTGGCTGTGTACTGGCTGAGCTGCTTCTTGGCCAGCCC

                                                       V  L  G  T  P  T                 E  M  N  P  N  Y  T  E  F  K     P  Q
       I  F  P  G  D  S  G  V  D  Q  L  V  E  I  I  K  V  L  G  T  P  T  R  E  Q  I  R  E  M  N  P  N  Y  T  E  F  K  F  P  Q -358
 1081-ATCTTCCCTGGGGACAGTGGGGTAGACCAGCTTGTGGAGATCATCAAGGTACTAGGGACGCCAACCAGGGAGCAAATCCGAGAGATGAACCCTAACTACACAGAATTCAAGTTCCCCCAG
```

```
        I K A H P F T K
        I K A H P W T K V F K S R T P P E A I A L C S S L L E Y T P S S R L S P L E A C -398
1201-ATCAAAGCTCACCCCTGGACAAAGGTGTTCAAATCTCGGACACCACCTGAGGCCATCGCACTCTGCTCTAGCCTGCTGGAGTACACTCCATCCTCAAGGCTCTCCCCACTAGAGGCCTGT

        A H S F F D E L R S L G T Q L P N N R P L P P L F N F S P G E L S I Q P S L N A -438
1321-GCCCACAGTTTCTTTGATGAACTGCGGAGTCTCGGAACCCAGCTCCCCAACAACCGCCCGCTTCCCCCCCTCTTCAACTTCAGTCCTGGTGAACTTTCCATCCAACCGTCTCTCAATGCC

                        S P A G S T P L S Q S S Q A L T E A Q T G S D W Q S . E A T P Q
        I L I P P H L R S P S G P A T L T S S S Q A L T E T Q T G Q D W Q A P D A T P T -478
1441-ATTCTCATCCCTCCTCACTTGAGGTCCCCATCAGGCCCTGCTACCCTCACCTCGTCCTCACAAGCTTTAACTGAGACTCAGACTGGCCAAGACTGGCAGGCACCTGATGCCACACCTACC

        L
        L T N S S -483
1561-CTCACTAACTCTTCCTGAGGGCCCTACCGACTACCCCTCACGCTCACTCGGAAGGCTCAAGTGGGCTGGAAAGGGCCATAGCCCATCCAGCTCCTGCCTGCTGGCCCTAGACTGAGGGCA

1681-GAGGTAAATGAACTACCCAGCATCTAGGCCTCCCTCACCAGCCTCACCTTGTGGTGGCTTTTAAAGAGGATTTAACTGTTTGTGGGGGAGGGAAGGGAAGAGAAGGACGGACGGGGTTTG

1801-GGGTATGAGGACCTTCTACCCCCGTGGTCCCCTCCCCTCCCCCAGGCTACCACTCCTCCCCCCCCTCCCATGTCCCTTGTAAATAGAACCAGCCCAGCCGTATCCTCTTCCTGGCCCTTG

1921-GGTGTAAATAGATTGTTATAATTTTTTTCTTAAAGAAAACGTCGATTCACACCATCCAACCTGCCCTCCCCCTCAGCTGTACCCCCCTCTTGTCCTCTGCTCCCAAGGCTTCCTCCCTCT

2041-CCCCATCCCAAGGAGGGGAGTAGGGAGAGCCCCTGGTGTCTTAGTTTCCACAGTAAGGTTTGCCTGTGTACAGACCTCCGTTCAATAAATTATTGGCATGAAAAAAAAAAAAAAAA
```

FIG. 1. Amino acid sequence of rat GSK-3$\alpha$. The determined peptide sequences are aligned with the predicted open reading frame.

fragment; Pharmacia/LKB), 0.5 $\mu$l of polynucleotide kinase (Pharmacia/LKB), 1 $\mu$l of ATP (10 m$M$), and 1 $\mu$l of the deoxynucleotide mix are added and incubated at 37° for 30 min to blunt end and phosphorylate the termini of the DNA. The aqueous fraction is then extracted with phenol/chloroform/isoamyl alcohol (50 : 48 : 2), and the aqueous phase precipitated with 2 vol of ethanol in the presence of 300 m$M$ sodium acetate for 2 hr at −20°. After centrifugation at 20,000 $g$ for 10 min, the pellet is briefly dried under vacuum and resuspended in 20 $\mu$l 10 m$M$ Tris-HCl, 1 m$M$ EDTA, pH 8.0 and electrophoresed through 1% agarose in the presence of 5 $\mu$g/ml ethidium bromide. The amplified DNA is excised and purified by standard procedures for ligation into a suitable blunt-ended vector for sequence analysis.

## Isolation of Full-Length GSK-3 Clones

Amplification of HeLa cell cDNA with the two GSK-3 oligonucleotides generated a single fragment of 340 base pairs (bp), the nucleotide sequence of which contained an open reading frame that included a sequence determined to be from a second GSK-3 tryptic peptide, confirming that the DNA fragment was derived from GSK-3 cDNA.

The 340-bp amplified fragment was used to isolate full-length clones of GSK-3 from a rat brain cDNA library.[16] Two classes of cDNA were identified. One class contained an open reading frame of 483 amino acids (including the initiating methionine residue). Translation of synthetic RNA derived from the cDNA by a reticulocyte lysate *in vitro* generated a 51-kDa protein. The cDNA sequence contained all of the tryptic peptide sequences isolated from the muscle protein and was thus designated GSK-3$\alpha$ (see Fig. 1). The second type of cDNA contained an open reading frame of 420 amino acids related to the predicted sequence of the GSK-3$\alpha$ clone, and generated a 47-kDa protein upon translation of synthetic RNA *in vitro*.[16] This putative protein kinase was designated GSK-3$\beta$ (Fig. 2). Hybridization to genomic DNA revealed a distinct pattern of fragments for the two clones, which, combined with the nucleotide sequence differences, demonstrated the two cDNA clones to be derived from distinct genes.[16]

## Tissue Distribution

Although GSK-3 was initially purified from skeletal muscle, the RNA for both types of GSK-3 is widely expressed, with highest levels for $\alpha$ and $\beta$ in brain. GSK-3$\alpha$ RNA levels are approximately 3- to 10-fold higher than

[16] J. R. Woodgett, *EMBO J.* **9,** 2431 (1990).

```
                                                                                                                        M S -   2
   1-TTTTTCTTCGCGGGAGAACTTAATGCTGCATTTATTATTAACCTAGTACCCTAACATAAAACAAAAGGAAGAAAAGGACCAAGGAAGGAAAAGGTGAATCGAGAAGAGCCATCATGTCG

     G  R  P  R  T  T  S  F  A  E  S  C  K  P  V  Q  Q  P  S  A  F  G  S  M  K  V  S  R  D  K  D  G  S  K  V  T  T  V  V  A -  42
 121-GGGCGACCGAGAACCACCTCCTTTGCGGAGAGCTGCAAGCCAGTGCAGCAGCCTTCAGCTTTTGGTAGCATGAAAGTTAGCAGAGATAAAGATGGCAGCAAGGTAACCACAGTGGTGGCA

     T  P  G  Q  G  P  D  R  P  Q  E  V  S  Y  T  D  T  K  V  I  G  N  G  S  F  G  V  V  Y  Q  A  K  L  C  D  S  G  E  L  V -  82
 241-ACTCCAGGACAGGGTCCTGACAGGCCACAGGAAGTCAGTTACACAGACACTAAAGTCATTGGAAATGGGTCATTTGGTGTGGTATATCAAGCCAAACTTTGTGACTCAGGAGAACTGGTG

     A  I  K  K  V  L  Q  D  K  R  F  K  N  R  E  L  Q  I  M  R  K  L  D  H  C  N  I  V  R  L  R  Y  F  F  Y  S  S  G  E  K -122
 361-GCCATCAAGAAAGTTCTTCAGGACAAGCGATTTAAGAACCGAGAGCTCCAGATCATGAGAAAGCTAGATCACTGTAACATAGTCCGATTGCGGTATTTCTTCTACTCGAGTGGCGAGAAG

     K  D  E  V  Y  L  N  L  V  L  D  Y  V  P  E  T  V  Y  R  V  A  R  H  Y  S  R  A  K  Q  T  L  P  V  I  Y  V  K  L  Y  M -162
 481-AAAGATGAGGTCTACCTTAACCTGGTGCTGGACTATGTTCCGGAAACAGTGTACAGAGTCGCCAGACACTATAGTCGAGCCAAGCAGACACTCCCTGTGATCTATGTCAAGTTGTATATG

     Y  Q  L  F  R  S  L  A  Y  I  H  S  F  G  I  C  H  R  D  I  K  P  Q  N  L  L  L  D  P  D  T  A  V  L  K  L  C  D  F  G -202
 601-TACCAGCTGTTCAGAAGTCTAGCCTATATCCATTCCTTTGGGATCTGCCATCGAGACATTAAACCACAGAACCTCTTGCTGGATCCTGATACAGCTGTATTAAAACTCTGCGACTTTGGA

     S  A  K  Q  L  V  R  G  E  P  N  V  S  Y  I  C  S  R  Y  Y  R  A  P  E  L  I  F  G  A  T  D  Y  T  S  S  I  D  V  W  S -242
 721-AGTGCAAAGCAGCTGGTCCGAGGAGAGCCCAATGTTTCATATATCTGTTCTCGGTACTACAGGGCACCAGAGCTGATCTTTGGAGCCACCGATTACACGTCTAGTATAGATATGTGGTCT

     A  G  C  V  L  A  E  L  L  L  G  Q  P  I  F  P  G  D  S  G  V  D  Q  L  V  E  I  I  K  V  L  G  T  P  T  R  E  Q  I  R -282
 841-GCAGGCTGTGTGTTGGCTGAATTGTTGCTAGGACAACCAATATTTCCTGGGGACAGTGGTGTGGATCAGTTGGTGGAAATAATAAAGGTCCTAGGAACACCAACAAGGGAGCAAATTAGA

     E  M  N  P  N  Y  T  E  F  K  F  P  Q  I  K  A  H  P  W  T  K  V  F  R  P  R  T  P  P  E  A  I  A  L  C  S  R  L  L  E -322
 961-GAAATGAACCCAAATTATACAGAATTCAAATTCCCCCAAATCAAGGCACATCCTTGGACGAAGGTCTTTCGGCCCCGAACTCCACCAGAGGCAATCGCACTGTGTAGCCGTCTCCTGGAG

     Y  T  P  T  A  R  L  T  P  L  E  A  C  A  H  S  F  F  D  E  L  R  D  P  N  V  K  L  P  N  G  R  D  T  P  A  L  F  N  F -362
1081-TACACGCCGACCGCCCGGCTAACACCACTGGAAGCTTGTGCACATTCATTTTTTGATGAATTACGGGACCCAAATGTCAAACTACCAAATGGGCGAGACACACCTGCCCTCTTCAACTTT

     T  T  Q  E  L  S  S  N  P  P  L  A  T  I  L  I  P  P  H  A  R  I  Q  A  A  A  S  P  P  A  N  A  T  A  A  S  D  T  N  A -402
1201-ACCACTCAAGAACTGTCAAGTAACCCACCCCTGGCCACCATCCTTATCCCTCCTCACGCTCGGATTCAGGCAGCTGCTTCACCTCCTGCAAACGCCACAGCAGCCTCAGATACTAATGCT

     G  D  R  G  Q  T  N  N  A  A  S  A  S  A  S  N  S  T -420
1321-GGAGACCGTGGACAGACCAATAACGCCGCTTCTGCATCAGCCTCCAACTCTACCTGAACAGCCCCAAGTAGCCAGCTGCGCAGGGAAGACCAGCACTTACTTGAGTGCCACTCAGCAACA

1441-CTGGTCACGTTTGGAAAGAAAATTAAAAAAAAAA
```

FIG. 2. Amino acid sequence of rat GSK-3$\beta$.

GSK-3$\beta$. This difference in expression is not so marked at the level of protein. Immunoblotting analysis of rat tissues with antibodies to GSK-3$\alpha$ reveals expression of a 51-kDa protein in all tissues, with highest levels in brain, heart, and testis.[16] Antibodies directed to GSK-3$\beta$ detect expression of a 47-kDa band in all tissues at a lower level than GSK-3$\alpha$. The richest source for GSK-3$\beta$ is brain, where the ratio of GSK-3$\alpha$ to GSK-3$\beta$ is 1 : 1.

## Structure of Enzyme

Comparison of the predicted protein sequences of GSK-3$\alpha$ and GSK-3$\beta$ reveals a region of considerable identity in the central portion of the proteins which contains the protein kinase domain flanked by regions of much lower similarity (Fig. 3). The amino-terminal region of GSK-3$\alpha$ is very glycine/proline rich and accounts for the 4-kDa difference in the predicted molecular masses of the two proteins. This region is likely to be devoid of secondary structure and may extend from the protein, perhaps accounting for the anomalous behavior of GSK-3$\alpha$ upon gel-permeation chromatography (see above). The C-terminal variable region contains several differences in the amino acid sequences derived from the rabbit skeletal muscle protein and the rat GSK-3$\alpha$ cDNA, suggesting considerable variability in this region between species, as well as enzyme subtype.

Comparison of the sequence of the catalytic domain with other protein-serine kinases reveals highest homology with members of the cdc2/KIN28 subgroup {see phylogenetic relationship of GSK-3 to other protein-serine kinases ([2] in this volume)}.[16]

## Substrate Specificity

Preparations of GSK-3 from skeletal muscle contain only the $\alpha$ form, as determined by immunoblotting (Fig. 4). In contrast, purification of GSK-3 from bovine brain using a method similar to that described by Tung and Reed[8] results in the equimolar isolation of both GSK-3$\alpha$ and GSK-3$\beta$ as judged by antibody detection. Since this procedure requires several chromatographies on ion exchangers, phosphocellulose, heparin-Sepharose, and gel filtration, the physicochemical properties of these two GSK-3 proteins must be very similar. The two proteins in the preparation of Tung and Reed were suggested to represent a regulatory and a catalytic subunit of GSK-3 forming a heterodimer.[8] Expression of GSK-3$\beta$ in COS-1 cells clearly demonstrates this protein to have glycogen synthase kinase activity.[16] The two proteins are thus monomeric and each contains a protein kinase catalytic domain.

```
             MSGGGPSGGGPGGSGRARTSSFAEPGGGGGGGGGGPGGSASGPGGTG   α

GGKASVGAMGGGVGASSSGGGPSGSGGGGSGGPGAGTSFPPPGVKLGRDSGKVTTVVATL   α
                 ::     :             :    :    :  ::::::::
                MSGRPRTTSFAESCKPVQQPSAFGSMKVSRDKDGSKVTTVVATP   β

GQGPERSQEVAYTDIKVIGNGSFGVVYQARLAETRELVAIKKVLQDKRFKNRELQIMRKL   α
:::: : ::: ::: :::::::::::::: :    :::::::::::::::::::::::::
GQGPDRPQEVSYTDTKVIGNGSFGVVYQAKLCDSGELVAIKKVLQDKRFKNRELQIMRKL   β

DHCNIVRLRYFFYSSGEKKDELYLNLVLEYVPETVYRVARHFTKAKLIIPIIYVKVYMYQ   α
::::::::::::::::::::: :::::: ::::::::::::   ::   : :::: ::::
DHCNIVRLRYFFYSSGEKKDEVYLNLVLDYVPETVYRVARHYSRAKQTLPVIYVKLYMYQ   β

LFRSLAYIHSQGVCHRDIKPQNLLVDPDTAVLKLCDFGSAKQLVRGEPNVSYICSRYYRA   α
:::::::::: : ::::::::::: :::::::::::::::::::::::::::::::::::
LFRSLAYIHSFGICHRDIKPQNLLLDPDTAVLKLCDFGSAKQLVRGEPNVSYICSRYYRA   β

PELIFGATDYTSSIDVWSAGCVLAELLLGQPIFPGDSGVDQLVEIIKVLGTPTREQIREM   α
::::::::::::::::::::::::::::::::::::::::::::::::::::::::::::
PELIFGATDYTSSIDVWSAGCVLAELLLGQPIFPGDSGVDQLVEIIKVLGTPTREQIREM   β

NPNYTEFKFPQIKAHPWTKVFKSRTPPEAIALCSSLLEYTPSSRLSPLEACAHSFFDELR   α
:::::::::::::::::::::  ::::::::::: ::::::  :: ::::::::::::::
NPNYTEFKFPQIKAHPWTKVFRPRTPPEAIALCSRLLEYTPTARLTPLEACAHSFFDELR   β

SLGTQLPNNRPLPPLFNFSPGELSIQPSLNAILIPPHLRSPSGPATLTSSSQALTETQTG   α
     ::: :  : ::::   :::  : :  :::::: :             :
DPNVKLPNGRDTPALFNFTTQELSSNPPLATILIPPHARIQAAASPPANATAASDTNAGD   β

QDWQAPDATPTLTNSS         α
       :     ::
RGQTNNAASASASNST         β
```

FIG. 3. Comparison of the amino acid sequences of the two GSK-3 cDNAs. Identities are indicated by colons.

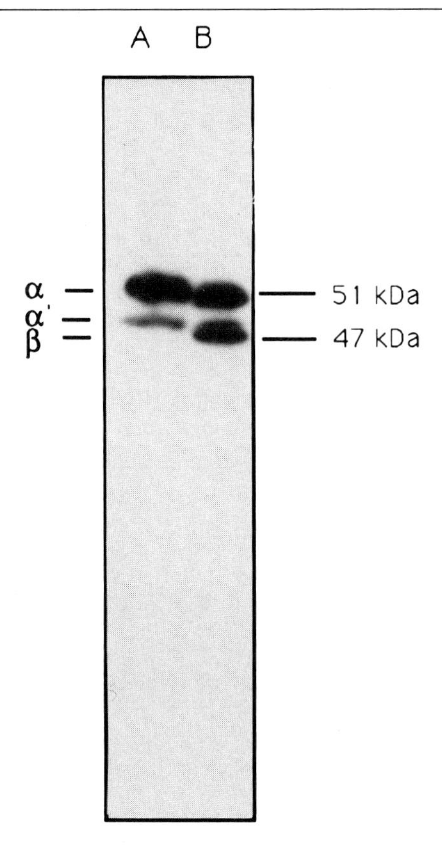


FIG. 4. Immunoblot of purified rabbit skeletal muscle (lane A) and bovine brain GSK-3 (lane B) preparations. Antibodies were raised against bacterial fusion proteins containing C-terminal portions of GSK-3$\alpha$ or GSK-3$\beta$ and blotted according to standard procedures.[16]

Since there are as yet no preparations of GSK-3$\beta$ devoid of GSK-3$\alpha$ (COS-1 cells have endogenous GSK-3$\alpha$ which copurifies with GSK-3$\beta$), the possibility that the $\beta$ form differs in its substrate preferences from GSK-3$\alpha$ must await its expression and purification to homogeneity.

GSK-3$\alpha$ phosphorylates a limited number of proteins (Table I). In the case of four of these, the enzyme exhibits a unique requisite for prior phosphorylation of the substrate at a neighboring site before it becomes a target for GSK-3. This "priming" phosphorylation is performed by casein kinase II in the case of glycogen synthase,[17] RII subunit of cyclic AMP-

[17] C. Picton, J. R. Woodgett, B. A. Hemmings, and P. Cohen, *FEBS Lett.* **150,** 191 (1982).

TABLE I
KNOWN SUBSTRATES OF GSK-3

| Protein | Site | Effect | Primer[a] | Ref. |
|---|---|---|---|---|
| Glycogen synthase | Ser-30, -34, -38, -42 | Inactivation | CK II | 29 |
| Inhibitor-2 | Thr-72 | Inactivation (of inhibitor-2) | CK II | 19 |
| RII subunit | Ser-44, -47 | ?[b] | CK II | 18 |
| G subunit | Serines | ? | A-kinase | 20 |
| c-jun | Ser-243, -250, Thr-239 | Inhibition of DNA binding | None | 22 |
| c-myb | Ser-357, -369, -371, Thr-355 | ? | None | 22 |

[a] Protein kinase that performs prephosphorylation of neighboring residue.
[b] ?, Unknown.

dependent protein kinase[18] and the inhibitor-2 subunit of protein phosphatase-1,[19] and cyclic AMP-dependent protein kinase in the case of the glycogen-binding subunit of phosphatase-1.[20] The molecular basis for this specificity has been examined in the case of glycogen synthase, resulting in a model in which phosphoserine acts as a specificity determinant for phosphorylation of a second serine four residues N terminal to it, and so on until four serines have been serially phosphorylated by GSK-3.[21] This requirement for prephosphorylation of glycogen synthase by casein kinase II has been exploited in the purification of GSK-3.[7,10]

However, this model does not account for the larger distance found between the priming and target residues in the other three substrates which exhibit prerequisite phosphorylation. Furthermore, in the cases of the c-jun and c-myb proteins,[22] there is no requirement for prephosphorylation, the function of the phosphoamino acid perhaps being substituted by a structural feature such as acidic residues. Comparison of the sequences around GSK-3 phosphorylation sites in these six proteins does not reveal

[18] B. A. Hemmings, A. Aitken, P. Cohen, M. Rymond, and F. Hofmann, *Eur. J. Biochem.* **127,** 473 (1982).
[19] A. A. DePaoli-Roach, *J. Biol. Chem.* **259,** 12144 (1984).
[20] P. Dent, D. G. Campbell, M. J. Hubbard, and P. Cohen, *FEBS Lett.* **248,** 67 (1989).
[21] C. J. Fiol, A. M. Mahrenholz, Y. Wang, R. W. Roeske, and P. J. Roach, *J. Biol. Chem.* **262,** 14042 (1987).
[22] W. B. Boyle, T. Smeal, L. H. K. Defize, P. Angel, J. R. Woodgett, M. Karin, and T. Hunter, *Cell,* **64,** 573 (1991).

TABLE II
SEQUENCES AROUND GSK-3 PHOSPHORYLATION SITES[a]

| Substrate | Sequence | Ref. |
|---|---|---|
| Glycogen synthase | VPRPASVPPSPSLSRHSSPHQSEDEE | 29 |
| Inhibitor-2 | IDEPSTPYHSMIGDDDDAYSDTE | 19 |
| RII subunit | LREARSRASTPPAAPPS | 18 |
| G subunit | FKPGFSPQPSRRGSESSEEV | 20 |
| c-myb | EHHHCTPSPPVDHGCLPEESASPAR | 22 |
| c-jun | EMPGETPPSLPIDMESQER | 22 |

[a] Glycogen synthase kinase-3 sites are in bold type. Casein kinase II and cyclic AMP-dependent protein kinase sites are underlined.

a simple common motif. Rather, the sites appear to be clustered and are in proline-rich acidic regions (Table II).

## Factor A Activity

On extraction of various tissues and assay of protein phosphatase activity, a portion of the type-1 phosphatase is detectable only on incubation of the crude extract with MgATP. This cryptic activity is due to a fraction of protein phosphatase being in an inactive complex with inhibitor-2. Activation occurs via phosphorylation of inhibitor-2 by GSK-3 on threonine-72,[23] which relieves the inhibition, although subsequent time-dependent "autodephosphorylation" of phospho inhibitor-2 by the active phosphatase recreates the inactive complex. This regulation of a protein phosphatase by a protein kinase presents a conundrum since GSK-3 both phosphorylates proteins and also activates one of the phosphatases that catalyzes removal of the phosphate.[24] Of course, the two types of reactions could be partitioned in some way, such as by compartmentalization, or be differentially regulated.

The physiological relevance of the MgATP-dependent protein phosphatase has been questioned, since no phosphothreonine has been detected in inhibitor-2 isolated under conditions designed to prevent dephosphorylation.[25,26] However, it is possible that dephosphorylation still occurs

[23] C. F. B. Holmes, D. G. Campbell, F. B. Caudwell, A. Aitken, and P. Cohen, *Eur. J. Biochem.* **155,** 173 (1986).

[24] P. Cohen, P. J. Parker, and J. R. Woodgett, *in* "Molecular Basis of Insulin Action" (M. P. Czech, ed.), p. 213. Plenum, New York, 1985.

[25] P. Cohen, *Annu. Rev. Biochem.* **58,** 453 (1989).

[26] C. F. B. Holmes, N. K. Tonks, H. Major, and P. Cohen, *Biochim. Biophys. Acta* **929,** 208 (1987).

during the initial stages of isolation due to the close apposition of the phosphatase with inhibitor-2. Immunoprecipitation of inhibitor-2 from $^{32}$P-labeled cells has detected phosphothreonine as well as phosphoserine, albeit at low levels.[27]

## Hormonal Signal Transduction and GSK-3

One of the first documented effects of insulin was the net dephosphorylation and catalytic activation of the rate-limiting enzyme of glycogen deposition, glycogen synthase. Although this enzyme is multiply phosphorylated on at least nine serine residues, the insulin-induced phosphate loss is concentrated in the region targeted by GSK-3.[24,28,29]

The nuclear oncoproteins jun and myb have been identified as substrates for GSK-3. This protein kinase phosphorylates sites on these proteins that are phosphorylated in cells. In the case of c-jun, at least one of these sites is specifically dephosphorylated upon exposure of the cells to phorbol esters. *In vitro,* phosphorylation of c-jun by GSK-3 prevents its interaction with DNA. Since phorbol esters activate transcription of certain genes through activation of c-jun, removal of inhibitory phosphate may be a necessary requisite for gene induction, with GSK-3 acting to suppress this process.

The effect of phorbol esters on c-jun phosphorylation parallels that of insulin on glycogen synthase phosphorylation, such that the target sites for GSK-3 are dephosphorylated.[22] Dephosphorylation of these residues in response to insulin or phorbol esters could be signaled via either activation of a site-specific protein phosphatase or inhibition of GSK-3: experiments have yet to discern the actual cellular mechanism.

GSK-3 is associated with intracellular membranes, including the plasma membrane, a localization consistent with a role in the transmission of signals (J.R.W., 1990, unpublished observations). The availability of cDNA clones allows manipulation of the levels of GSK-3 RNA and protein within cells, which should clarify the role of this particular protein kinase in cellular signal transduction.

## Acknowledgments

I am indebted to Nick Totty and Rob Philp for their protein-sequencing expertise, and to Bill Boyle, Tony Hunter, and numerous members of the Ludwig Institute for their help and encouragement.

[27] J. C. Lawrence, J. Hiken, B. Burnett, and A. A. DePaoli-Roach, *Biochem. Biophys. Res. Commun.* **150,** 197 (1988).

[28] P. J. Parker, F. B. Caudwell, and P. Cohen, *Eur. J. Biochem.* **130,** 227 (1983).

[29] L. Poulter, S. G-. Ang, B. W. Gibson, D. H. Williams, C. F. B. Holmes, F. B. Caudwell, J. Pitcher, and P. Cohen, *Eur. J. Biochem.* **175,** 497 (1988).

# Section VIII

# Expression and Analysis of Protein Kinases Using cDNA Clones

# [49] Prokaryotic Expression of Catalytic Subunit of Adenosine Cyclic Monophosphate-Dependent Protein Kinase

*By* WES M. YONEMOTO, MARIA L. MCGLONE, LEE W. SLICE, and SUSAN S. TAYLOR

cAMP-dependent protein kinase (cAPK) is a tetrameric holoenzyme composed of two catalytic and two regulatory subunits. Activation of cAPK occurs following the binding of cAMP to the regulatory subunit dimer and subsequent release of the monomeric catalytic subunits. Due to its small size, abundance, and relative ease of purification, the catalytic subunit of cAPK is one of the better characterized protein kinases in terms of the substrate requirements of the enzyme, the identification of residues involved in catalysis, and its mechanism of regulation.[1]

Protein kinases function as regulatory enzymes involved in signal transduction and cellular metabolism.[2] The activity of these proteins, in general, is tightly regulated and directed toward specific cellular targets *in vivo*. Since protein kinases are usually expressed at low levels in cells, overexpression of eukaryotic protein kinases has been used to aid in the purification and characterization of their qualitative properties *in vivo* and *in vitro*. In addition, the production of milligram quantities of protein is necessary for the identification of functional amino acid residues by chemical modification as well as for structural analysis by physical methods such as nuclear magnetic resonance (NMR) and crystallography.[3]

Three major eukaryotic expression systems have been utilized for the production of protein kinases: (1) mammalian cells, (2) baculovirus, and (3) yeast. Despite the importance and advantages of *in vivo* analysis of biological function and accurate posttranslational processing available with the use of mammalian cells, the level of protein expression is low relative to the other systems discussed here.[4] In addition, depending on the cell line used for expression, endogenous protein kinases may contaminate purified preparations of mutant recombinant enzyme. Baculovirus expression has been used for the overexpression of a variety of proteins with various degrees of success.[5-7] Higher levels of protein expression can be obtained compared to mammalian cells with reasonably similar posttrans-

[1] S. S. Taylor, J. A. Buechler, and W. Yonemoto, *Annu. Rev. Biochem.* **59,** 971 (1990).

[2] E. G. Krebs, *Enzymes* **17,** 3 (1986).

[3] S. S. Taylor, J. A. Buechler, and D. R. Knighton, *in* "Peptides and Protein Phosphorylation" (B.E. Kemp, ed.), p. 1. Uniscience CRC Press, Boca Ration, Florida, 1990.

[4] S. R. Olsen and M. D. Uhler, *J. Biol. Chem.* **264,** 20940 (1989).

[5] V. A. Luckow and M. D. Summers, *Bio/Technology* **6,** 47 (1988).

lational modification. There is some technical difficulty in identifying recombinant viral plaques necessary for expression and this system cannot be used to analyze biological function *in vivo*. Overall, however, baculovirus represents an attractive system for protein production with the drawbacks of the extensive labor and expense inherently involved in standard tissue culture techniques. The expression of protein kinases in yeast represents one alternative to the use of tissue culture cells. Genetic manipulation of yeast cells can allow for expression of certain mammalian enzymes in the absence of expression of the analogous yeast enzyme. Interestingly, a number of studies have been described in which the expression of mammalian gene is able to complement a loss of the analogous yeast gene.[8,9] Zoller and co-workers discuss in detail the methodology and optimization for the expression of protein kinases in yeast.[10]

In contrast to the eukaryotic systems described above, prokaryotic expression of foreign genes is a flexible and inexpensive system, using a variety of strong constitutive and inducible promoters. Milligram quantities of proteins can easily be produced for biochemical and physical studies with very little expense. In addition, expression in bacterial cells provides a method to produce recombinant eukaryotic proteins in the absence of endogenous wild-type enzymes. Finally, the complete lack of specific eukaryotic posttranslational modifications allows one to address the role of these modifications in the protein function. Expression in *Escherichia coli* is successfully used for high-level production of numerous eukaryotic proteins.[11,12] Unfortunately there has been limited success for the expression of protein kinases in bacteria. Although the expression of many protein kinases leads to the production of milligram quantities of protein, it is mainly in an insoluble, inactive form, such as seen for pp60$^{src}$.[13,14] The expression of truncated protein kinases (such as p62$^{v\text{-}abl}$) can produce a

---

[6] R. Herrera, D. Lebwohl, A. Garcia-de-Herreros, R. G. Kallen, and O. M. Rosen, *J. Biol. Chem.* **263,** 5560 (1988).

[7] H. Piwnica-Worms, N. G. Williams, S. H. Cheng, and T. M. Roberts, *J. Virol.* **64,** 61 (1990).

[8] T. Kataoka, S. Powers, S. Cameron, O. Fasano, M. Goldfarb, J. Broach, and M. Wigler, *Cell (Cambridge, Mass.)* **40,** 19 (1985).

[9] M. G. Lee and P. Nurse, *Nature (London)* **327,** 31 (1987).

[10] M. J. Zoller, K. E. Johnson, W. Yonemoto, and L. Levine, this volume [51].

[11] F. A. O. Marston, *Biochem. J.* **240,** 1 (1986).

[12] F. A. O. Marston, "DNA Cloning: A Practical Approach" (D. M. Glover, ed.), Vol. 3, p. 59. IRL Press, Oxford, 1987.

[13] T. M. Gilmer and R. L. Erikson, *Nature (London)* **294,** 771 (1981).

[14] J. D. McGrath and A. D. Levinson, *Nature (London)* **295,** 423 (1982).

protein that can be solubilized with intact enzymatic activity.[15] Other eukaryotic protein kinases have been expressed in prokaryotic cells. This chapter will deal with the expression and purification of soluble and active cAPK catalytic (C) subunit in *E. coli.*

## Materials

Reagents are purchased as follows: Bacto-tryptone and Bacto-yeast extract bacterial media (Difco, Detroit, MI); ampicillin (sodium salt), Boeringer Mannheim (Mannheim, Germany); kanamycin monosulfate (Sigma, St. Louis, MO); P-11 phosphocellulose (Whatman, Clifton, NJ); Q-Sepharose, CM-Sepharose, and Sephacryl-200 HR resins (Pharmacia/LKB, Piscataway, NJ); isopropyl-β-D-thiogalactopyranoside (IPTG), Sequenase, deoxy- and dideoxynucleotides (U.S. Biochemicals, Cleveland, OH). All chemicals not specifically cited are of reagent or molecular biology grade. All other DNA-modifying and restriction enzymes are obtained from Bethesda Research Laboratories (Gaithersburg, MD) or New England Bio Labs, Inc. (Boston, MA).

The following bacterial strains are used: *E. coli* RZ1032 (ATCC); *E. coli* DH5α (Bethesda Research Laboratories); *E. coli* BL21(DE-3) (W. Studier, Brookhaven National Laboratory, Upton, NY).

## Prokaryotic Expression of $C_\alpha$ Subunit of cAMP-Dependent Protein Kinase

Slice and Taylor[16] have described the expression of a cDNA clone for the murine $C_\alpha$ subunit of cAPK[17] in bacterial cells. A number of different expression vectors and bacterial strains were tested for their ability to express the $C_\alpha$ protein. These vectors included expression of the $C_\alpha$ cDNA from the *lac*UV5 promoter of pUC18 as well as coexpression with the cDNA of the wild-type bovine RI subunit.[18] Although expression was obtained from the different $C_\alpha$ subunit-producing *E. coli* cells examined, in general, the recombinant protein produced was inactive and found in insoluble inclusion bodies.[19]

The expression of the $C_\alpha$ subunit from the bacteriophage T7 gene 10

[15] J. Y. J. Wang, C. Queen, and D. Baltimore, *J. Biol. Chem.* **257,** 13181 (1982).

[16] L. W. Slice and S. S. Taylor, *J. Biol. Chem.* **264,** 20940 (1989).

[17] M. D. Uhler, D. F. Carmichael, D. C. Lee, J. C. Chrivia, E. G. Krebs, and G. S. McKnight, *Proc. Natl. Acad. Sci. U.S.A.* **83,** 1300 (1986).

[18] L. D. Saraswat, M. Filutowics, and S. S. Taylor, *J. Biol. Chem.* **261,** 11091 (1986).

[19] N. J. Darby and T. E. Creighton, *Nature* (*London*) **334,** 715 (1990).

($\phi$10) promotor[20] in *E. coli* BL21(DE-3) made significant amounts of soluble and active $C_\alpha$ subunit. *Escherichia coli* BL21(DE-3) contains the bacteriophage T7 gene *1* (T7 RNA polymerase) under inducible control of the *lac*UV5 promotor.[21] Following addition of IPTG, the expression of T7 RNA polymerase directs the expression of the $C_\alpha$ gene. When $C_\alpha$ expression was carried out at 37°, approximately 10 to 20% of the protein expressed was found in the soluble fraction. This finding was significant since it provided a method for the production of large amounts of recombinant $C_\alpha$ subunit with relative ease at little expense. In addition, this system provides a method with which to produce mutants of the C subunit for analysis of their biochemical and physical properties.

Interestingly, the use of *E. coli* BL21(DE-3) was important for the production of soluble enzyme, for when the same T7 gene *10* promotor/$C_\alpha$ cDNA expression plasmid vector (pLWS-3)[16] was expressed in *E. coli* K38, only insoluble C subunit was produced. *Escherichia coli* K38 contains the bacteriophage T7 RNA polymerase under control of a heat-inducible λ $P_L$ promotor and requires elevated temperature for gene expression.[20] It seems likely that this elevation in temperature during expression is at least partly responsible for the insolubility of the recombinant $C_\alpha$ protein (see below), although other variations between the two T7 expression *E. coli* strains may account for the difference in the ability to produce soluble recombinant protein.

In order to make oligonucleotide site-directed mutants, the $C_\alpha$ cDNA is subcloned into the phagemid vector, pUC119[22] and single-stranded DNA template is prepared as previously described in *E. coli* RZ1032.[23] Mutagenesis protocols are performed essentially as described by Zoller and Smith[24] and the mutants are verified by dideoxy-DNA sequencing with Sequenase (U.S. Biochemicals). Mutant $C_\alpha$ genes are resubcloned back into the multiple cloning site of the plasmid vector pT7-7, downstream from the bacteriophage T7 gene *10* promoter.[20] In order to obtain expression of the $C_\alpha$ cDNA, the pT7/$C_\alpha$ plasmid vectors are transformed into the *E. coli* BL21(DE-3) as previously described.[16] Colonies are screened for expression of $C_\alpha$ protein after growth in YT medium (8 g Bacto-tryptone, 5 g Bacto-yeast extract, 5 g NaCl/liter) with ampicillin (100 $\mu$g/ml) followed by induction with 0.4 m*M* IPTG for 3 hr at 37°. Total crude cell extracts are prepared by lysing the cells in boiling SDS sample buffer and analyzed

[20] S. Tabor and C. C. Richardson, *Proc. Natl. Acad. Sci. U.S.A.* **82,** 1074 (1985).

[21] F. W. Studier and B. A. Moffatt, *J. Mol. Biol.* **189,** 113 (1986).

[22] J. Viera and J. Messing, this series, Vol. 153, p. 3.

[23] T. A. Kunkel, J. D. Roberts, and R. A. Zakour, this series, Vol. 154, p. 367.

[24] M. J. Zoller and M. Smith, this series, Vol. 154, p. 329.

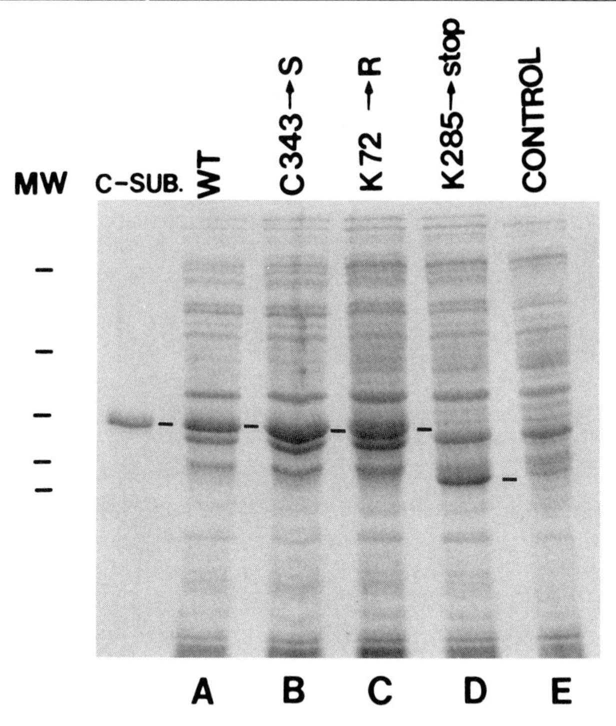


FIG. 1. Expression of wild-type and mutant $C_\alpha$ proteins in total bacterial cell extracts. Total extracts of *E. coli* BL21(DE-3) cells expressing $C_\alpha$ proteins were prepared as described in the text and analyzed by SDS–PAGE: (A) wild-type $C_\alpha$ subunit, (B) $C_\alpha$(Cys$^{343}$Ser) subunit, (C) $C_\alpha$(Lys$^{72}$Arg) subunit, (D) $C_\alpha$(Lys$^{285}$ stop) subunit, (E) BL21(DE-3) cells.

by SDS–PAGE.[25] Figure 1 shows the Coomassie Blue-stained SDS gel of a number of different recombinant $C_\alpha$ proteins. Lane A (Fig. 1) shows the relative expression of wild-type recombinant $C_\alpha$ subunit in the total bacterial lysate following induction with IPTG, while lanes B, C, and D (Fig. 1) show the relative level of expression of various mutant $C_\alpha$ proteins expressed in BL21(DE-3) cells. Lane E (Fig. 1) is a sample of total bacterial lysate from the parental BL21(DE-3) cells.

It is important to screen a number of different BL21(DE-3) transformants for expression of recombinant $C_\alpha$ subunit since we have noticed variations in the level of $C_\alpha$ expression in different colonies. Those bacterial colonies which express high levels of $C_\alpha$ subunit are grown to midlog phase (OD 0.5 to 0.6 at $A_{600}$). The cells are harvested by gentle centrifugation and resuspended in an equal volume of 2YT

[25] U. K. Laemmli, *Nature* (*London*) **227,** 680 (1970).

medium (16 g Bacto-tryptone, 10 g Bacto-yeast extract, 5 g NaCl/liter) containing 50% (v/v) glycerol (no antibiotics). The cell suspension is then divided into 200-$\mu$l aliquots in 1.5-ml Eppendorf tubes and frozen quickly over dry ice. The frozen bacterial stocks can be stored at $-70°$ indefinitely. We have found it is important not to grow the stock cell cultures to late log or stationary phase since the growth of cells producing the highest levels of $C_\alpha$ subunit tends to be unfavorable under those conditions.

*Effect of Temperature on the Production of Catalytic Subunit*

Previous studies have indicated that temperature can affect the solubility of proteins produced in *E. coli*.[26] This finding is of interest since many eukaryotic protein kinases produced in bacterial systems are made in an insoluble, inactive form. Generally only the fraction of soluble recombinant protein kinase, or enzyme resolubilized from the bacterial insoluble fraction, possesses functional kinase activity. Therefore, the effect of temperature on the solubility of the recombinant $C_\alpha$ subunit was examined. BL21(DE-3) cells containing the wild-type $C_\alpha$ gene (pLWS-3) were grown at 37° in YT medium with ampicillin (100 $\mu$g/ml). When the cell culture reached an OD of 0.8 at $A_{600}$, IPTG was added to a final concentration of 0.4 m*M*. The culture was divided into three samples which were individually shaken and incubated for 4 hr at the temperature indicated. After the indicated period, the cells were harvested and lysed using a French pressure cell. The cell lysates were separated into soluble and particulate fractions by differential centrifugation at 17,000 *g* for 30 min. The soluble and particulate fractions were collected and samples were analyzed by SDS–PAGE and then stained with Coomassie Blue. Figure 2 shows the analysis of the effect of temperature on the solubility of the recombinant C subunit. At 37°, a majority of the $C_\alpha$ subunit was found in the insoluble, particulate fraction of the cell lysate. At 31°, roughly 50% of the $C_\alpha$ subunit produced was observed in the soluble fraction. At room temperature (24°) essentially all of the $C_\alpha$ subunit produced was found in the soluble fraction of the cell lysate. The overall level of production of the $C_\alpha$ subunit was lower at 24 versus 37°, which reflects a slower rate of protein synthesis. Therefore, as the incubation temperature during induction was lowered, a greater proportion of the $C_\alpha$ subunit produced was found in the soluble fraction of the cell lysate.

[26] C. H. Schein and H. M. Noteborn, *Bio/Technology* **6,** 291 (1988).

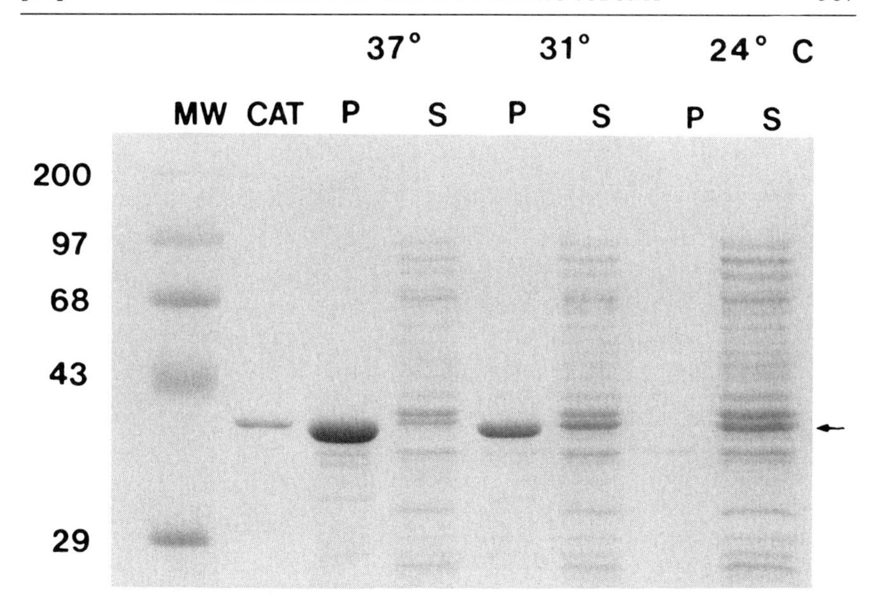


FIG. 2. Effect of temperature on solubility of the $C_\alpha$ subunit. Wild-type $C_\alpha$ subunit-expressing *E. coli* BL21(DE-3) cells were grown and induced under different temperature conditions as specified in the text. The bacterial cells were harvested, lysed, and fractionated into particulate (P) and soluble (S) by differential centrifugation. The samples induced at 37, 31, and 24° were analyzed by SDS–PAGE. Molecular weight ($\times 10^{-3}$).

### *Effect of Induction Time on Production of Catalytic Subunit*

Studier and Moffat have described the effect of induction time upon RNA and protein production under the control of bacteriophage T7 promoters in *E. coli* BL21(DE-3) cells.[21] At 37°, we have found maximal production of $C_\alpha$ subunit within 3 to 4 hr after the addition of IPTG and induction of bacteriophage T7 RNA polymerase. Since production of soluble $C_\alpha$ subunit was favorable at lower temperatures, the time course of protein synthesis and accumulation was reexamined at 24°. The pT7/$C_\alpha$–BL21(DE-3) cells were grown in YT with ampicillin until an OD at $A_{600}$ of 0.8 was reached, at which time the temperature was shifted to 24° and the cultures were induced by the addition of IPTG (0.4 m*M*). Samples were taken at the times indicated, harvested by centrifugation in a microfuge, and resuspended in SDS sample buffer. The cells were lysed by boiling and analyzed by SDS–PAGE and Coomassie Blue staining. Figure 3 shows the results of the induction time course. Detectable levels of the $C_\alpha$ subunit continued to accumulate until 6 to 8 hr after induction. After 8 hr, no change in the level of $C_\alpha$ subunit was observed up to 24 hr

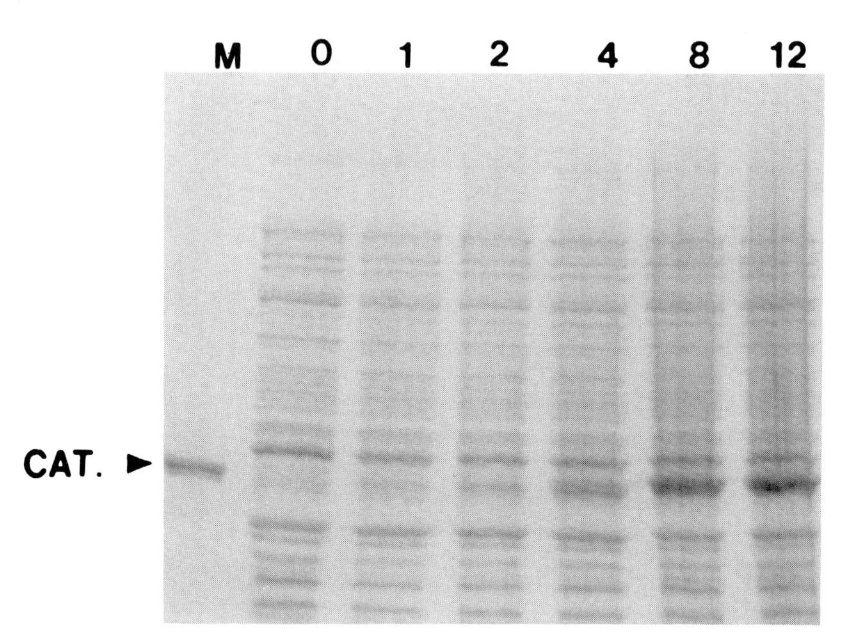


FIG. 3. Time course of expression of the $C_\alpha$ subunit. Wild-type $C_\alpha$ subunit-expressing *E. coli* BL21(DE-3) cells were grown and then induced at 25°. Samples were removed at the times indicated and samples were prepared and analyzed by SDS–PAGE as described in the text.

(data not shown). Therefore the use of lower temperature during induction requires a concomitant lengthening of the time required for protein synthesis and accumulation in order to achieve production of maximal levels of the soluble recombinant $C_\alpha$ subunit.

### Purification of the $C_\alpha$ Subunit

When cAPK is purified from mammalian cells or tissues, it is generally first isolated as holoenzyme complex on an anion-exchange resin, such as DEAE-Sepharose.[27] Following dissociation of the holoenzyme with the addition of cAMP, the C subunit is typically purified to homogeneity by CM-Sepharose chromatography.[28] Hydroxyapatite[29] and Blue dextran

[27] M. J. Zoller, A. R. Kerlavage, and S. S.Taylor, *J. Biol. Chem.* **254,** 2408 (1979).

[28] N. C. Nelson and S. S. Taylor, *J. Biol. Chem.* **256,** 3743 (1981).

[29] E. M. Reimann and R. A. Beham, this series, Vol. 99, p. 51.

chromatography[30] have also been used to purify the C subunit. Initial attempts at purification of the recombinant $C_\alpha$ subunit from the crude cell supernatant by standard anion, cation, hydrophobic, and hydroxyapatite chromatography all failed, possibly due to the aggregation of the C subunit with bacterial proteins and/or nucleic acids. Purification of the recombinant $C_\alpha$ subunit was finally achieved by the use of two novel protocols involving (1) the cation-exchange resin, phosphocellulose, and (2) the addition of a recombinant double mutant of the RI subunit which forms holoenzyme in the crude bacterial lysate.

*Phosphocellulose Chromatography/Gel Filtration*

In order to purify the $C_\alpha$ subunit, 12 liters of *E. coli* BL21(DE-3) cells transformed with a pT7/$C_\alpha$ plasmid are grown overnight at 37° in YT medium containing 100 $\mu$g/ml ampicillin. When the OD at $A_{600}$ reaches between 0.8 and 1.0, IPTG is added to a final concentration of 0.4 m*M* and the cells are grown for an additional 8 to 10 hr at 24°. The bacterial cells are harvested, resuspended in 260 ml buffer A [30 m*M* MES, pH 6.5, 50 m*M* KCl, 1 m*M* EDTA, 5 m*M* 2-mercaptoethanol(2-ME)] and lysed in a cold (4°) French pressure cell. Protease inhibitors may be added to the lysis buffer, but the recombinant $C_\alpha$ subunit does not appear to be very sensitive to proteolysis provided the lysate is kept cold (4°) at all subsequent purification steps. The cell lysate is clarified by centrifugation at 17,000 *g* for 30 min and the pellet is discarded.

The first purification method utilizes P-11 phosphocellulose and gel-filtration chromatography. This procedure can be used for wild-type recombinant $C_\alpha$ subunit, as well as for a number of mutant $C_\alpha$ subunits. P-11 phosphocellulose resin is prepared according to the manufacturer's instructions. Briefly, 15 g of resin is swollen in a 50× volume of 0.5 *M* NaOH for exactly 5 min. The resin is washed extensively with $H_2O$ in a sintered glass funnel until the eluate was at pH 11, and then is transferred into a 50× volume of 0.5 *M* HCl for 5 min. The resin is again washed with $H_2O$ until the eluate is at pH 3.0. The pretreated resin is washed with excess 300 m*M* MES, pH 6.5 until the washes equilibrate at pH 6.5. The resin then is poured into a column [3 cm (diameter) × 23 cm (length)] and preequilibrated with buffer B (30 m*M* MES, pH 6.5, 1 m*M* EDTA). This is a very important step since the recombinant $C_\alpha$ subunit binds over a very narrow pH range to phosphocellulose. The supernatant from the cell lysate is diluted with cold $H_2O$ to a conductivity of 1.1 m$\Omega$ and applied to the prepared 100 ml of P-11 column. After extensive washing with buffer

[30] J. J. Witt and R. Roskoski, *Biochemistry* **14,** 4503 (1975).

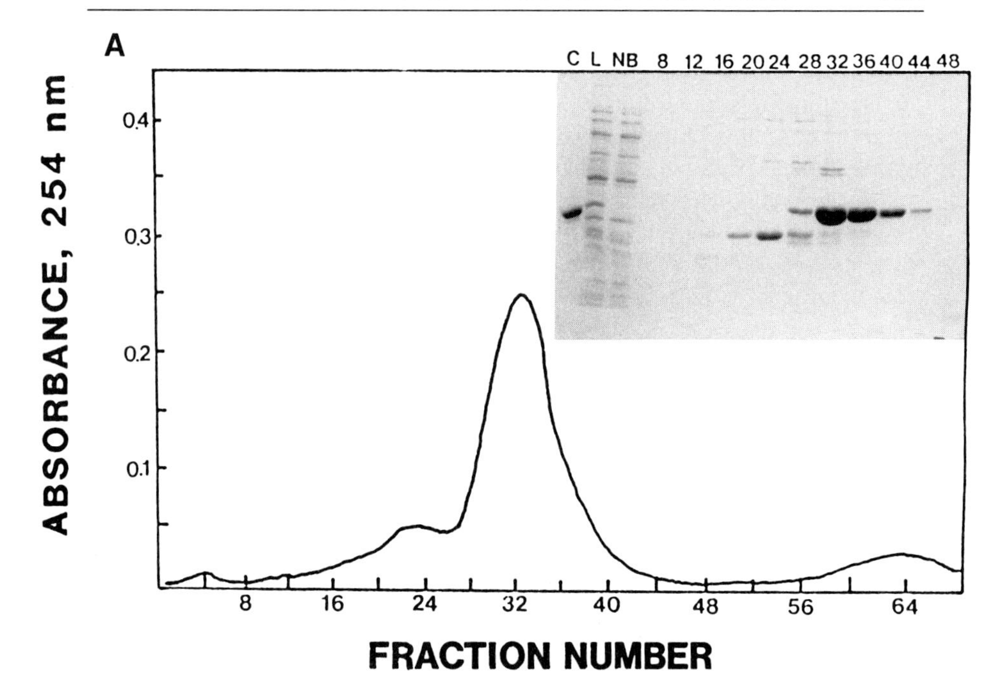


FIG. 4. (A) Phosphocellulose chromatography of $C_\alpha$ subunit. The column (3 × 23 cm) was packed with 15 g of prepared P-11 phosphocellulose equilibrated in buffer B. Twelve liters of *E. coli* BL21(DE-3) cells expressing $C_\alpha$(Cys$^{343}$Ser) subunit was processed as described in the text. The $C_\alpha$ subunit was eluted with a linear $KPO_4$ gradient (50–400 m*M*) and fractions of 3 ml were collected. Selected fractions were also subjected to SDS–PAGE (inset). (B) Sephacryl-200 HR chromatography of $C_\alpha$ subunit. The fractions containing $C_\alpha$(Cys$^{343}$Ser) subunit after phosphocellulose chromatography were pooled and concentrated by ammonium sulfate precipitation as described in text. The column (3 × 120 cm) was equilibrated with buffer C and eluted with a flow rate of 2 ml/min. Fractions of 3 ml were collected and the $C_\alpha$ protein was detected by SDS–PAGE (inset).

B, the proteins are eluted with a 300-ml linear gradient from 50 to 400 m*M* $KPO_4$, pH 6.5. Figure 4A shows the elution profile and SDS–PAGE analysis after P-11 chromatography for the mutant Cys$^{343}$Ser $C_\alpha$ protein. This mutant protein behaves identically to the wild-type $C_\alpha$ subunit during phosphocellulose chromatography. The Cys$^{343}$Ser $C_\alpha$ subunit is identified by its migration on the SDS–PAGE and by spectrophotometric protein kinase activity assays with the substrate peptide, kemptide.[31]

The P-11 fractions containing the $C_\alpha$ subunit are pooled and concentrated by precipitation with the addition of solid ammonium sulfate to a

[31] R. Roskoski, this series, Vol. 99, p. 1.

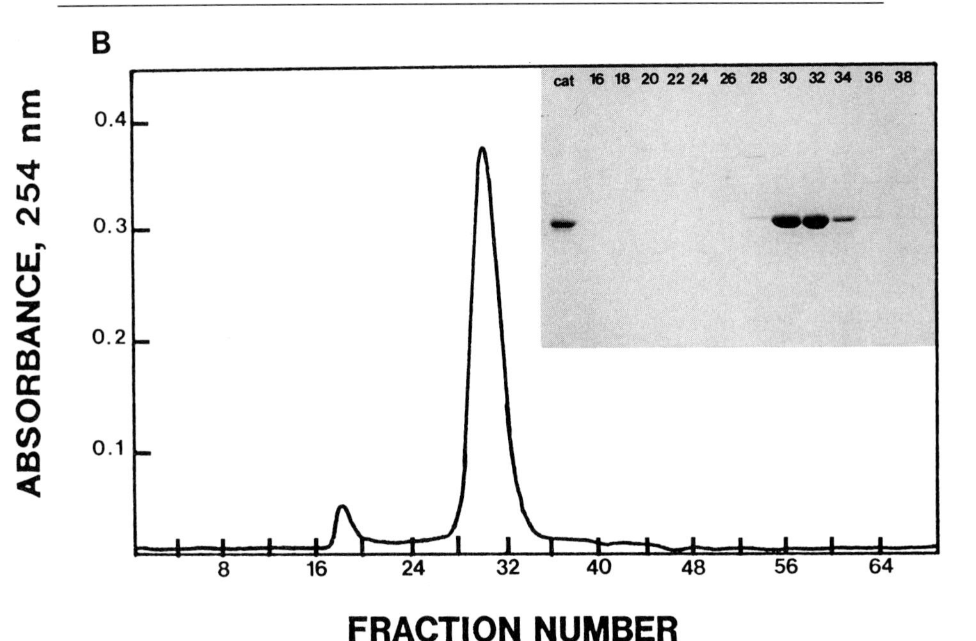


FIG. 4. (*continued*)

final concentration of 80%. The precipitate is collected and redissolved in approximately 2.0 ml of buffer C (25 m*M* $KPO_4$, pH 7.0, 200 m*M* KCl, 5 m*M* 2-mercaptoethanol) and applied to a Sephacryl-200 HR column equilibrated with buffer C. The $C_\alpha$ subunit is eluted with 300 ml of buffer C, pumped at a rate of 2 ml/min. Figure 4B shows the elution profile and SDS–PAGE analysis after gel-filtration chromatography. One major peak of protein elutes from the column. This protein is confirmed to be the $C_\alpha$ subunit by its migration on SDS–PAGE, Western blot analysis with antibodies directed against the $C_\alpha$ subunit, and *in vitro* protein kinase assays.

*Holoenzyme Purification with RI Subunit*

Although phosphocellulose chromatography provides an effective purification for the wild-type and some of the mutant recombinant $C_\alpha$ proteins, other mutant $C_\alpha$ subunits do not bind to P-11 directly from the crude bacterial lysate. Most of these mutant $C_\alpha$ subunits contain point mutations of highly conserved residues within the catalytic core region shared by all known eukaryotic protein kinases.[32] Therefore a second and novel proto-

[32] S. K. Hanks, A. M. Quinn, and T. Hunter, *Science* **241,** 42 (1988).

col was devised to aid in the purification of recombinant $C_\alpha$ proteins. This procedure employs the use of a mutant R subunit with an altered affinity for cAMP. This mutant R subunit contains a Lys residue at a single conserved Arg position within each tandem cAMP-binding domain.[33] This double RI subunit mutant, RI($Arg^{209}Lys : Arg^{333}Lys$), results in an increase of two orders of magnitude in the concentration of cAMP required for activation of the holoenzyme complex. This change in the affinity for cAMP allows the mutant RI subunit to bind to the catalytic subunit and form homoenzyme in the crude bacterial lysate. Wild-type RI subunit fails to form holoenzyme in crude cell lysates, due to the presence of endogenous cAMP.

In order to purify the recombinant $C_\alpha$ subunit by this alternative procedure, 12 liters of $C_\alpha$-expressing cells are harvested by centrifugation and combined with the bacterial pellet from 6 liters of *E. coli* 222 expressing the RI($Arg^{209}Lys : Arg^{333}Lys$) subunit. The cells are lysed together in 300 ml buffer A (without EDTA) using a French pressure cell. Holoenzyme formation occurs immediately upon lysis as monitored by cAMP-dependent protein kinase activity. The cell lysate is centrifuged at 17,000 *g* for 40 min at 4° and the supernatant is collected. The conductivity is adjusted to 1.1 mΩ with cold $H_2O$ and the diluted supernatant is applied to a column prepared with 100 ml of Q-Sepharose equilibrated in buffer D (10 m*M* MES, pH 6.5, 5 m*M* 2-mercaptoethanol). The column is washed extensively and eluted with a 500-ml linear gradient from 0 to 250 m*M* NaCl in buffer D. The fractions containing holoenzyme as monitored by SDS–PAGE and protein kinase activity are pooled and diluted to a conductivity of 1.1 mΩ with cold $H_2O$.

It has previously been shown that the C subunit binds well to CM-Sepharose, while the R subunits and the holoenzyme are not retained.[28] Therefore, the diluted, pooled Q-Sepharose-purified holoenzyme fractions are first passed over a 100-ml CM-Sepharose column, equilibrated in buffer E (17 m*M* $KPO_4$, pH 6.5, 5 m*M* 2-mercaptoethanol) in order to remove the bacterial proteins which bind to the resin. The flow-through fractions are collected and cAMP is added to a final concentration of 1 m*M* to dissociate the unbound holoenzyme complex. The dissociated holoenzyme is then reapplied to a fresh 100-ml column of CM-Sepharose equilibrated with buffer E, washed extensively with buffer E, and eluted with a 300-ml linear gradient from 0 to 400 m*M* NaCl in buffer E. Figure 5 shows the results of the CM-Sepharose purification of the $C_\alpha$ subunit. The peak fractions containing the $C_\alpha$ subunit are pooled and assayed for protein kinase activity.

[33] I. T. Weber, T. A. Steitz, J. Bubis, and S. S. Taylor, *Biochemistry* **26,** 343 (1987).

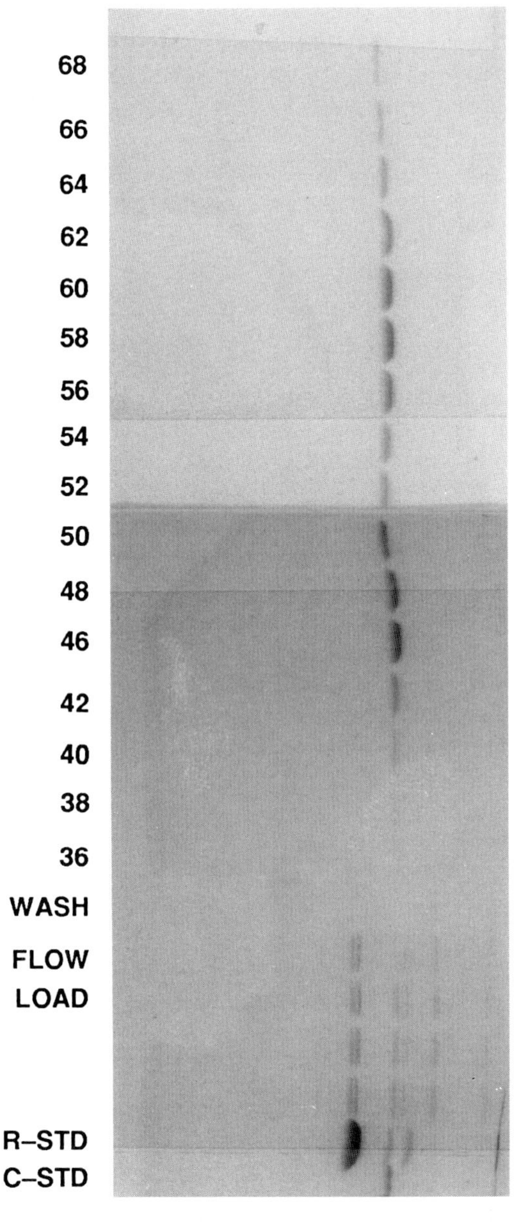


FIG. 5. CM-Sepharose chromatography of the $C_\alpha$ subunit. The column (3 × 23 cm) packed with 100 ml of CM-Sepharose was equilibrated with buffer E. The Q-Sepharose-purified holoenzyme was chromatographed before (not shown) and after dissociation as described in text. The $C_\alpha$ subunit was eluted with a linear NaCl gradient (0–400 m*M*). Fractions of 2.5 ml were collected and examined by SDS–PAGE.

Approximately 20 to 40 mg of $C_\alpha$ subunit is purified from 12 liters of bacterial cells by either the phosphocellulose or holoenzyme purification methods described above. The bacterial cells can be grown, induced, harvested, and frozen as a solid paste at $-70°$ for an indefinite period of time without any loss in yield of purified $C_\alpha$ subunit. A 12-liter preparation should take between 3 and 4 days from bacterial pellet to purified enzyme. A single operator can easily complete two purification runs in 1 week.

## Analysis of Properties of Recombinant $C_\alpha$ Subunit

### *Kinetic Parameters*

An initial characterization of the properties of the purified recombinant murine $C_\alpha$ subunit compared to the C subunit protein purified from porcine heart was completed.[16] Both the recombinant and mammalian C subunits were able to form holoenzyme with the $R^I$ subunit and the cAMP activation of the holoenzymes was indistinguishable. Both enzymes possess similar kinetic parameters with respect to the binding of the cosubstrates MgATP and the peptide, kemptide. The apparent $K_m$ values for MgATP and for kemptide were approximately 15 and 45 $\mu M$, respectively, for both the recombinant and mammalian enzyme. In addition, both proteins were efficiently inhibited by the 20-residue inhibitor peptide[34] with an apparent $K_i$ of approximately 12 n$M$. These results indicate that the recombinant $C_\alpha$ subunit produced in *E. coli* has roughly the same enzymatic properties as the protein isolated from mammalian tissues.

One clear distinction between the recombinant and mammalian C subunit was the thermostability of the enzymatic activity to heat denaturation. There is approximately a 5° difference in the temperature required to inactivate 50% of the recombinant murine $C_\alpha$ subunit after a 2.5-min incubation compared to the mammalian porcine enzyme. It is presently unclear why this difference in thermostability between these two forms of the C subunit exists; however, it has been proposed that myristoylation of the N terminus may influence the stability of the recombinant protein.[16]

### *Posttranslational Modifications*

The mammalian C subunit has been shown to contain two types of posttranslational modification: (1) phosphorylation[35] and (2) myristoylation.[36] Two sites of phosphorylation were detected by amino acid sequenc-

[34] D. A. Walsh, K. L. Angelos, S. M. Van Patten, D. B. Glass, and L. P. Garetto, "Peptides and Protein Phosphorylation" (B.E. Kemp, ed.). Uniscience CRC Press, Boca Raton, Florida, 1990.

[35] K. A. Peters, J. G. Demaille, and E. H. Fisher, *Biochemistry* **16,** 5691 (1977).

[36] S. A. Carr, K. Biemann, S. Shoji, D. C. Parmalee, and K. Titani, *Proc. Natl. Acad. Sci. U.S.A.* **79,** 6128 (1982).

ing of bovine C subunit, one at Thr-197, the other at Ser-338.[37] In addition, the porcine C subunit has been shown to contain one major site of *in vitro* phosphorylation at Ser-10.[38]

*In vivo* labeling of wild-type $C_\alpha$ subunit-expressing *E. coli* cells shows one major phosphoprotein which exactly comigrates with the C subunit. This protein was demonstrated to be the $C_\alpha$ subunit after purification by P-11 phosphocellulose chromatography, immunoblotting with polyclonal antibodies, and protein kinase activity assays. Phosphoamino acid analysis of the *in vivo* labeled recombinant $C_\alpha$ subunit identifies the presence of both phosphoserine and phosphothreonine. Preliminary sequence analysis of the high-performance liquid chromatography (HPLC)-purified phosphotryptic peptides indicates that both Thr-197 and Ser-338, residues previously identified in the mammalian C subunit, are also phosphorylated in the recombinant enzyme.

The amino terminus of the mammalian C subunit is blocked by the addition of a myristoyl group. Although the C subunit was the first protein discovered with this unique fatty acid modification, the biological function of this N-terminal myristoylation of the C subunit is not known.[36] Subsequently this posttranslational modification has been found on other cellular and viral proteins and in some cases has been shown to be necessary for association with membranes.[39] Since myristoylation is a eukaryotic protein modification, it was not expected that the recombinant $C_\alpha$ subunit would be acylated. Amino acid sequencing of the recombinant $C_\alpha$ subunit revealed a free N-terminal Gly residue.[16] Therefore, the major physical difference between the recombinant and mammalian enzymes, detected so far, is the lack of an N-terminal myristoylation.

Gordon and co-workers have cloned the gene for the yeast N-terminal myristoyltransferase and have shown that prokaryotic expression of the gene generates an active recombinant enzyme. Coexpression of the transferase together with the $C_\alpha$ subunit leads to acylation of the recombinant $C_\alpha$ subunit *in vivo,* thereby reconstituting this posttranslational modification system in *E. coli.*[40] Expression and purification of recombinant, myristoylated $C_\alpha$ subunit allows investigation of the physical properties of the modified $C_\alpha$ subunit. Specifically, the role of myristoylation in the thermostability of the recombinant $C_\alpha$ subunit can directly be readdressed.

[37] S. Shoji, K. Titani, J. G. Demaille, and E. H. Fisher, *J. Biol. Chem.* **254,** 6211 (1979).

[38] J. Toner-Webb and S. S. Taylor, *J. Biol. Chem.* (submitted for publication).

[39] D. A. Towler, S. R. Eubanks, D. S. Towery, S. P. Adams, and L. Glaser, *J. Biol. Chem.* **262,** 1030 (1987).

[40] R. J. Duronio, E. Jackson-Machelski, R. O. Heuckeroth, P. O. Olins, C. S. Devine, W. Yonemoto, L. W. Slice, S. S. Taylor, and J. I. Gordon, *Proc. Natl. Acad. Sci. U.S.A.* **87,** 1506 (1990).

## Summary

The prokaryotic expression of the $C_\alpha$ subunit of cAPK provides a system for the production of milligram quantities of wild-type and mutant protein with relative ease and little expense. The protocols described here have been optimized for the production of soluble and active protein kinase, with kinetic parameters similar to those found for the enzyme isolated from mammalian tissue. The $C_\alpha$ subunit expressed in *E. coli* is a phosphoprotein and appears to contain the sites of phosphorylation identified in the mammalian protein. The free N-terminal Gly indicates the lack of an N-terminal myristic acid; however, coexpression of the gene encoding the yeast *N*-myristoyltransferase allows for myristoylated $C_\alpha$ subunit to be produced in *E. coli*. Finally, the purified recombinant $C_\alpha$ subunit has been used in X-ray crystallographic studies which have yielded the crystals that will allow the first three-dimensional structure of a protein kinase to be solved. Therefore, although the employment of prokaryotic expression in the production of functional protein kinases has been limited, it is hoped that the methods presented here for the C subunit of cAPK will encourage the use of this simple and versatile expression system.

## Acknowledgments

This work was supported in part by the American Cancer Society Grant BC-678E to S.S.T. W.M.Y. is a postdoctoral fellow of the California Division of the American Cancer Society.

# [50] Expression and Purification of Active *abl* Protein-Tyrosine Kinase in *Escherichia coli*

*By* SYDONIA I. RAYTER

Protein-tyrosine kinase activity is the functional enzyme activity associated with many transforming retroviruses (such as those encoding *src* and *abl*) and growth factor receptors (such as epidermal growth factor, platelet-derived growth factor, insulin, insulin-related growth factor-1, and colony-stimulating factor-1). Approximately 30 proven or putative protein-tyrosine kinases have been identified, either as the transforming protein of retroviruses, the activity associated with growth factor receptors, by transfection of human tumor DNA into NIH3T3 cells, or as amplified sequences in primary human tumors.[1,2]

The protein-tyrosine kinase encoded by the Abelson murine leukemia virus (A-MuLV) induces the *in vitro* transformation of both fibroblasts and pre-B lymphocytes.[3,4] The gene encoding this protein is a fusion of coding sequences from the Moloney murine leukemia virus *gag* protein and a portion of the cellular c-*abl* gene.[5] A construct containing DNA sequences encoding the minimal kinase domain of A-MuLV has allowed expression of the pt*abl*50 kinase in *Escherichia coli,*[6] purification of the enzyme to homogeneity, and its enzymatic properties to be characterized.[7]

*Buffer Solutions*

Buffer A: 50 m*M* piperazine-*N,N*-bis-2-ethanesulfonic acid (PIPES), pH 7.5, 0.1 m*M* EDTA, 0.015% (w/v) Brij 35

Buffer B: 20 m*M* PIPES, pH 8.15, 0.1 m*M* EDTA, 0.015% (w/v) Brij 35, 20% (w/v) ethylene glycol, 0.2% (v/v) 2-mercaptoethanol

Buffer C: 20 m*M* Tris-HCl, pH 8.8 (0°), 0.015% (w/v) Brij 35, 20% (w/v) ethylene glycol, 0.2% (v/v) 2-mercaptoethanol, 100 m*M* NaCl

Buffer D: 50 m*M* PIPES, pH 7.5, 0.1 m*M* EDTA, 0.015% (w/v) Brij 35, 10% (w/v) ethylene glycol, 0.2% (v/v) 2-mercaptoethanol

Buffer E: 50 m*M* PIPES, pH 7.5, 0.1 m*M* EDTA, 0.015% (w/v) Brij 35, 20% (w/v) ethylene glycol, 2 m*M* dithiothreitol, 200 m*M* NaCl

Buffer F: 50 m*M* HEPES, pH 7.0, 0.1 m*M* EDTA, 0.015% (w/v) Brij 35, 20% (w/v) ethylene glycol, 0.2% (v/v) 2-mercaptoethanol

Buffer G: 50 m*M* PIPES, pH 7.5, 0.1 m*M* EDTA, 0.015% (w/v) Brij 35, 0.2% (v/v) 2-mercaptoethanol, 20 m*M* NaCl, 100 m*M* sucrose

Buffer H: 50 m*M* PIPES, pH 7.5, 0.1 m*M* EDTA, 0.015% (w/v) Brij 35, 40% (w/v) ethylene glycol, 2 m*M* dithiothreitol, 100 m*M* KCl

Kinase dilution buffer: 0.1 mg/ml bovine serum albumin, 0.2% (v/v) 2-mercaptoethanol in buffer A

Bacterial medium: 25 m*M* potassium phosphate, pH 7.5, containing (per liter) 4.37 g $NH_4Cl$, 20 g glucose, 10 g tryptone, 5 g yeast extract, 5

[1] S. I. Rayter, K. K. Iwata, R. W. Michitsch, J. M. Sorvillo, D. M. Valenzuela, and J. G. Foulkes, *in* "Oncogenes" (D. M. Glover and B. D. Hanes, eds.), p. 113. IRL Press, Oxford, 1989.

[2] T. Hunter, *Cell* (Cambridge, Mass.) **50,** 823 (1987).

[3] N. Rosenberg and D. Baltimore, *in* "Viral Oncology" (G. Klein, ed.), p. 187. Raven Press, New York, 1980.

[4] D. Baltimore, N. Rosenberg, and O. N. Whitte, *Immunol. Rev.* **48,** 1 (1979).

[5] S. Goff and D. Baltimore, *Adv. Viral Oncol.* **1,** 127 (1982).

[6] J. Y. J. Wang and D. Baltimore, *J. Biol. Chem.* **260,** 64 (1985).

[7] J. G. Foulkes, M. Chow, C. Gorka, A. R. Frackelton, Jr., and D. Baltimore, *J. Biol. Chem.* **260,** 8070 (1985).

g NaCl, 50 mg ampicillin, 232 mg $Mg^{2+}$, 11 mg $Ca^{2+}$, 10 mg $Fe^{2+}$, 0.4 mg $Mn^{2+}$, 2 mg $Zn^{2+}$, 0.04 mg $Co^{2+}$

Bacterial wash buffer: 100 m*M* NaCl, 20 m*M* Tris-HCl, pH 8.8 (0°), 2 m*M* EDTA

Bacterial sonication buffer: 40 m*M* NaCl, 1 m*M* EDTA, 100 m*M* sucrose, 0.015% (w/v) Brij 35, 20 m*M* Tris-HCl, pH 8.8 (0°), 0.2% (v/v) 2-mercaptoethanol

## Expression of Abelson (pt*abl*50) Kinase Activity in *E. coli*

Construction of the expression vector containing v-*abl* sequences has been described in detail elsewhere.[6] In brief, sequences coding for the kinase domain of A-MuLV were transferred onto a vector which allows expression in *E. coli*. This vector, pCS4, contains the $P_r$ promoter of λ bacteriophage, a ribosome-binding sequence, and an ATG codon, followed by 0.24 kilobase pairs (kbp) of sequences coding for small t of simian virus 40.[8] In addition, a temperature-sensitive *cI* gene is present such that transcription from the $P_r$ promoter is represssed at 30° and induced at 42°. To this vector, sequences coding for a specific N-terminal region of the A-MuLV kinase (from the *Hin*cII site at the *gag–abl* junction to the first *Pst* I site 1.2 kbp downstream) were placed in frame behind the small t sequences. Thus at the permissive temperature, this plasmid expresses a fusion protein, termed the pt*abl*50 kinase, which consists of 80 amino acids of small t, 403 amino acids from the N-terminal domain of the viral Abelson kinase, and 5 amino acids at its C terminus from pBR322.

Bacteria are grown in a New Brunswick Scientific (Edison, NJ) fermentor. Medium (9.5 liter) is inoculated with 100 ml of an overnight culture grown at 30°. Bacteria are grown for 7 hr at 30° (until $A_{660}$ = 4.0), followed by 4 hr at 42.5° (method 1) or 2.5 hr at 42.5° (method 2) (final $A_{660}$ is approximately 13). The pH is monitored during the fermentation and adjusted to pH 7.5 with 5 *M* NaOH, bacteria are harvested by centrifugation (12 min, 6500 *g* at 4°), quickfrozen, and stored at −70° in 6 × 40-g aliquots.

## Assays

### *Protein-Tyrosine Kinase Assay*

pt*abl*50 kinase activity is measured by the incorporation of $^{32}P$ from [$\gamma$-$^{32}P$]MgATP into the modified gastrin analog Raytide (Oncogene Science, Inc., Manhasset, NY). Raytide incorporates at least eightfold more

[8] C. Queen, *J. Mol. Appl. Genet.* **2,** 1 (1983).

$^{32}P$ than does angiotensin. The standard assay consists of 10 $\mu$l of pt*abl*50 kinase in kinase dilution buffer, 10 $\mu$l of Raytide (1 mg/ml in buffer A) and is initiated by the addition of 10 $\mu$l of 30 m*M* $Mg^{2+}$, 0.3 m*M* [$\gamma$-$^{32}P$]ATP (1500–10,000 cpm/pmol). After 30 min at 30°, the reaction is terminated by the addition of 120 $\mu$l of 10% (v/v) phosphoric acid. Each reaction tube is vortexed, and within 10 min, 120 $\mu$l of the reaction mixture is spotted onto phosphocellulose paper (2 $cm^2$). Papers are washed extensively in 4 liters 0.5% (v/v) phosphoric acid, washed once in acetone, dried, then counted. Controls incorporated into every assay are reaction mix minus peptide substrate and reaction mix minus enzyme extract. The pt*abl*50 kinase is activated upon dilution of bacterial extracts, and should therefore be diluted to at least 0.1 mg/ml in kinase dilution buffer before performing the assay. During the purification procedure kinase inhibitors are removed and activity becomes linear with dilution of the enzyme. Enzyme should be diluted in kinase dilution buffer immediately prior to performing the assay.

### *Definition of Unit*

One unit of pt*abl*50 kinase activity is defined as 1 pmol of phosphate transferred from ATP to Raytide/min at 30° in the standard assay.

### *Protein Determinations*

Protein concentrations are determined by the method of Bradford[9] using bovine serum albumin as standard.

## Purification of pt*abl*50 Kinase: Method 1[7]

All solutions and fractions are kept at 0–4° throughout the purification. The presence of ethylene glycol and the nonionic detergent Brij 35 in all the buffers ensures enzyme stability at 4° throughout the purification procedure. An 80-g aliquot of bacteria is thawed on ice, washed once in 100 m*M* NaCl, 20 m*M* Tris-HCl, pH 8.8 (0°), 2 m*M* EDTA, and then lysed by sonication (four times, 30 sec each) in 600 ml of 40 m*M* NaCl, 1 m*M* EDTA, 100 m*M* sucrose, 0.015% Brij 35, 20 m*M* Tris-HCl, pH 8.8 (0°), 0.2% 2-mercaptoethanol. Extracts are clarified by centrifugation (190,000 *g*, 45 min), and the supernatants are applied batchwise to DE-52 cellulose (440 g, wet wt) equilibrated in buffer B. The DE-52 cellulose (Whatman) is washed with buffer B, 20 m*M* NaCl, and pt*abl*50 kinase eluted by the application of buffer B, 120 m*M* NaCl, 2 m*M* $MgCl_2$.

This preparation is applied batchwise to hydroxylapatite (800 g, wet

[9] M. M. Bradford, *Anal. Biochem.* **72,** 248 (1976).

wt, preequilibrated in buffer C). The resin is washed with buffer C, 35 m*M* phosphate, and pt*abl*50 kinase eluted with buffer C, 120 m*M* phosphate. The preparation is concentrated using Amicon (Danvers, MA) stirred cells (YM10 membranes) and dialyzed overnight against buffer D. After each dialysis step particulate matter is removed by centrifugation at 12,000 *g*, 10 min, 4°C, and supernatant applied to the next step in the procedure. The next day, the material is applied to an Affi-Gel Blue column (50-ml bed volume) equilibrated in buffer D, the column is washed with buffer D, 220 m*M* NaCl, and the kinase eluted by the addition of buffer D, 860 m*M* NaCl. The 10% ethylene glycol concentration in buffer D allows the kinase activity to bind to the Affi-Gel Blue column. At 20% ethylene glycol, no binding occurs. Kinase activity is concentrated using an Amicon stirred cell (YM10 membranes) and, at the end of the second day, is applied to Sephadex G-100 superfine column (2.6 × 95 cm, flow rate 8 ml/hr) equilibrated in buffer E. Fractions eluting from the column which contain kinase activity are pooled and dialyzed against buffer F. The preparation is then applied to a Spherogel TSK DEAE-3SW high-pressure liquid chromatography column (7.5 × 75 mm). The column is washed with buffer F, 100 m*M* NaCl, and the kinase activity eluted with a 100 to 250 m*M* NaCl gradient (in buffer F), 0.7 ml/min flow rate, with a total gradient volume of 120 ml.

Fractions containing kinase activity are pooled and dialyzed against buffer G. The final purification step employs an anti-phosphotyrosine monoclonal antibody (1G2; Oncogene Science, Inc.) coupled covalently to Sepharose 4B. The preparation is applied to this column (2-ml bed volume), the column is washed with buffer G, and kinase activity eluted by the addition of buffer G, 2 m*M* phenyl phosphate (as a tyrosine phosphate analog).

The final preparation is dialyzed against buffer H and stored at −20°. The enzyme is stored in a dithiothreitol-containing buffer for long-term stability. A summary of the purification based on a starting preparation of 80 g of *E. coli* is given in Table I.

### *Comments on Purification Procedure: Method 1*

There are at least five forms of the pt*abl*50 kinase which separate at various stages of the purification. Although all forms migrate reproducibly when rechromatographed, the ratio of the forms varies with different fermentations. Therefore, Table I represents averaged data obtained from multiple fermentations, with the data normalized with respect to the two forms which are purified (by this procedure; Fig. 1). These two forms constitute 35% of the total activity in the extract so that the actual specific

TABLE I
PURIFICATION OF pt*abl*50 KINASE: METHOD 1

| Step | Volume (ml) | Units (pmol/min) | Protein (mg) | Specific activity (units/mg) | Yield (%) | Purification (-fold) |
|---|---|---|---|---|---|---|
| 1. Extract | 300 | 1,631,840 | 2,824 | 576 | 100 | 1 |
| 2. DE-52 cellulose | 3,000 | 1,533,920 | 786 | 1,952 | 94 | 3.4 |
| 3. Hydroxylapatite | 3,000 | 1,441,920 | 216 | 6,672 | 88 | 11.6 |
| 4. Affi-Gel Blue | 850 | 1,209,600 | 66 | 18,560 | 75 | 32 |
| 5. Sephadex G-100 superfine | 17 | 864,800 | 14 | 61,760 | 53 | 107 |
| 6. TSK DEAE-3SW | 10 | 605,440 | 2.74 | 220,960 | 37 | 384 |
| 7. Anti-phosphotyrosine | 6 | 505,920 | 0.234 | 2,160,000 | 31 | 3,750 |

activity of the extract is 206 ± 11 units/mg of protein. The various forms that are not purified can be found in the following fractions: 3–25% in the DE-52–20 m*M* NaCl fraction, 22–48% in the hydroxylapatite–35 m*M* phosphate fraction, 6–60% in the Affi-Gel Blue–210 m*M* NaCl fraction, and 10% in the breakthrough of the anti-phosphotyrosine column.

On gel filtration, the enzyme moves as a monomer [$M_r$ 60,300 ± 700 ($n$ = 3)]. In the absence of NaCl in the column buffer, 25% of the enzyme moves with an apparent $M_r$ 125,000, and this form of the enzyme can be partially dissociated (up to 50%) to the $M_r$ 60,300 form if rechromatographed in the presence of 200 m*M* NaCl.

The two forms of pt*abl*50 kinase, as isolated in step 5 and onward, prove to already be phosphorylated and, therefore, 80–90% of the kinase activity is recovered by antiphosphotyrosine MAb affinity chromatography. The 10% appearing in the breakthrough fraction of the column appears to represent the dephosphorylated forms of the enzyme, because on rechromatography, over 80% of this activity is again recovered in the breakthrough.

With respect to the two forms purified, the overall purification is 3750-fold with a 31% yield, such that 234 μg of pt*abl*50 kinase is obtained from 80 g of *E. coli* within 6–7 days. Assuming all forms have equal activity, the total pt*abl*50 kinase in *E. coli* extracts represents only 0.076% of the soluble protein. The most powerful purification step is the anti-phosphotyrosine monoclonal antibody column, where a 10-fold purification to homogeneity is obtained. This column also binds the platelet-derived growth factor receptor, the EGF receptor, and the p120$^{gag-abl}$ protein of A-MuLV. This step cannot be used earlier in our protocol because expression of the Abelson kinase in *E. coli* results in the phosphorylation of many bacterial proteins on tyrosine.

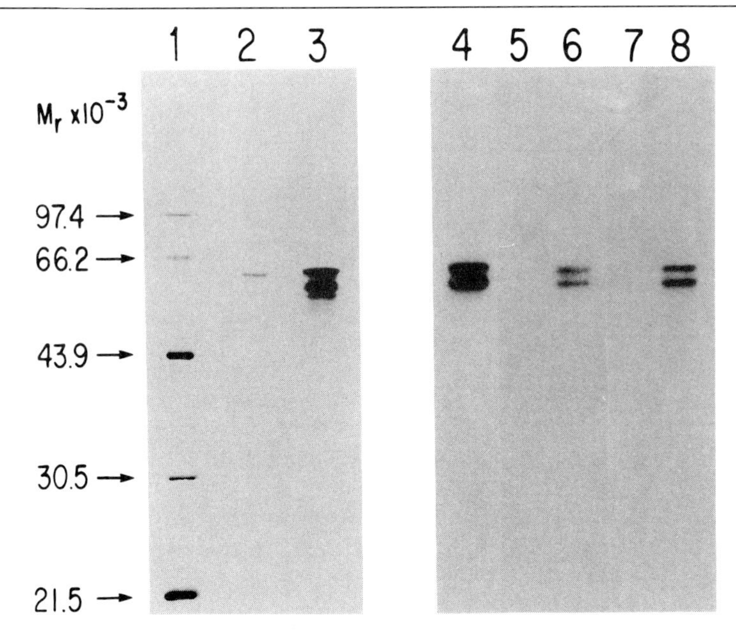


FIG. 1. SDS-polyacrylamide gel electrophoresis of purified pt*abl*50 kinase. Portions of the purified enzyme (from step 7) were subjected to SDS-polyacrylamide gel electrophoresis (10% gel), followed by silver staining. Lane 1 represents protein markers of known molecular weight, with 40 ng of each marker loaded; lane 2, 35 ng; and lane 3, 425 ng of the purified pt*abl*50 kinase. Lanes 4–8 represent an autoradiograph of the pt*abl*50 kinase, incubated with [$\gamma$-$^{32}$P]MgATP, and subjected to SDS-polyacrylamide gel electrophoresis. Lane 4, pt*abl*50 kinase incubated with MgATP; lanes 5–8, pt*abl*50 kinase $^{32}$P-labeled and immunoprecipitated with either preimmune serum (lane 5) followed by immune Abelson-specific serum (lane 6) or a nonspecific monoclonal antibody (lane 7) followed by specific small t monoclonal antibody (lane 8). Immunoprecipitates were collected with *Staphylococcus aureus* and washed in 1 m*M* EDTA, 1% deoxycholate, 0.1% SDS, 100 m*M* NaCl, 50 m*M* Tris-HCl, pH 8.0, prior to dissociation with sample buffer.

The purified preparation (step 7) appears as two major protein bands as judged by autophosphorylation followed by SDS-polyacrylamide gel electrophoresis (Fig. 1). Silver staining the gel reveals the presence of the two major protein bands as well as one minor and two very faint bands of lower molecular weight (Fig. 1). Whether these represent breakdown products or minor contaminants is unknown. The purification scheme, as summarized in Table I, allowed 234 $\mu$g of essentially homogeneous pt*abl*50 kinase to be obtained from 80 g of *E. coli* within 6–7 days.

The turnover number of 170 $\mu$mol/min/$\mu$mol of pt*abl*50 kinase, deter-

TABLE II
PURIFICATION OF pt*abl*50 KINASE: METHOD 2

| Step | Volume (ml) | Units (pmol/min) | Protein (mg) | Specific activity (units/mg) | Yield (%) | Purification (-fold) |
|---|---|---|---|---|---|---|
| 1. Extract | 306 | 23,216,832 | 4,835 | 4,800 | 100 | 1 |
| 2. 55% ammonium sulfate precipitation | 60 | 13,564,800 | 2,826 | 4,800 | 58 | 1 |
| 3. DE-52 80 m*M* NaCl Amicon concentration | 30 | 6,460,896 | 39 | 165,664 | 28 | 35 |
| 4. Red Sepharose 50 m*M* and 300 m*M* NaCl pool | 74.5 | 3,403,296 | 10.6 | 321,432 | 15 | 67 |
| 5. Anti-phosphotyrosine | 4 | 2,463,480 | 0.32 | 7,698,376 | 10.6 | 1,604 |

mined for ATP, is comparable to the specific activity of the most active protein-serine kinases.[10-12]

### *Alternative Steps in Purification Procedure: Method 2*

We have also attempted to simplify the previous purification procedure, and one scheme which has been applied will be outlined. After clarification (190,000 *g*, 45 min, 4°) of the sonicated bacteria (100 g in 300 ml sonication buffer), the cleared lysate is subjected to ammonium sulfate precipitation. A 30% ammonium sulfate cut is applied first to the bacterial extract by adding 34.26 g powder to 195 ml over 10 min, with stirring, at 4°. Stirring is continued for 2 hr, at 4°, then precipitated protein is pelleted by centrifugation at 10,000 rpm, 30 min, 4° (Sorvall, GSA rotor). The supernatant is then subjected to a 55% ammonium sulfate cut by adding 31.1 g ammonium sulfate powder to 195 ml supernatant as above. After stirring for 2 hr at 4°, the precipitated protein is recovered by centrifugation as above. Pellets from the 30 and 55% ammonium sulfate cuts are resuspended in buffer B and dialyzed overnight against 2 liters of buffer B. All fractions are assayed the next day for kinase activity, which is found in the 55% ammonium sulfate-precipitated cut. Although this step does not appear to be a purification procedure (see Table II), this may allow a greater fold purification to be achieved in the following DEAE chromatography step than in the scheme outlined in Table I (35-fold compared to

[10] P. H. Sugden, L. A. Holladay, E. M. Reimann, and J. D. Corbin, *Biochem. J.* **159,** 409 (1976).
[11] P. J. Bechtel, J. A. Beavo, and E. G. Krebs, *J. Biol. Chem.* **252,** 2691 (1977).
[12] G. W. Tessmer, J. R. Skuster, L. B. Tabatabai, and D. J. Graves, *J. Biol. Chem.* **252,** 5666 (1977).

3.4-fold, respectively). DE-52 cellulose (Whatman) (150 g) is allowed to swell at 4° overnight in 2 liters of buffer B.

The fraction containing kinase activity is applied batchwise to DE-52 as in method 1, but kinase activity is eluted in 80 m*M* NaCl in buffer B instead of 120 m*M* NaCl. The active fraction is centrifuged at 12,000 *g* (15 min, 4°) to remove any beads and the supernatant is Amicon concentrated (YM10 membrane) to give 10–30 ml, which is then dialyzed overnight at 4° in buffer D. The next step we have succeeded in using is a red-Sepharose (Pharmacia, Piscataway, NJ) affinity column which binds proteins with nucleotide-binding sites. One gram of matrix is swollen in 10 ml $H_2O$, washed with 300 ml $H_2O$, then with 30 ml buffer D. The pt*abl*50 kinase is first diluted with an equal volume of buffer D minus ethylene glycol, and then loaded onto the column (final concentration of ethylene glycol is 5% in the loading buffer). The sample is passed through a 2-ml column two or three times. The column is then washed batchwise with 10 ml buffer D, 10 ml buffer D + 50 m*M* NaCl, 10 ml buffer D + 300 m*M* NaCl, 10 ml buffer D + 500 m*M* NaCl, and finally 10 ml buffer D + 1 *M* NaCl. The kinase is found to be distributed as follows: 35% in the breakthrough and buffer wash, 13% in the 50 m*M* NaCl buffer, 46% in the 300 m*M* NaCl wash, and 6% in the final 500 m*M* and 1.3 *M* NaCl washes. The specific activities of pt*abl*50 were found to be identical in the 50 m*M* and 300 m*M* NaCl washes, and these two fractions are therefore pooled for the next step in the procedure.

The breakthrough and initial buffer wash are pooled and reapplied to the red-Sepharose column a second time, and eluted by 50 and 300 m*M* NaCl washes as described above. In this second round of purification only 28% of the kinase is recovered in the salt washes, while the remainder is accounted for in the breakthrough and initial buffer wash. All the 50 and 300 m*M* NaCl fractions are pooled and, after dialysis in buffer G overnight at 4°, applied to the 1G2 (Oncogene Science, Inc.) anti-phosphotyrosine MAb affinity column, as described in method 1. Based on a starting preparation of 100 g of *E. coli,* a summary of the purification is given in Table II.

*Comments on Purification Procedure: Method 2*

The advantages of this method compared to the first are mainly that fewer steps are involved (although the red-Sepharose step involves reapplying the breakthrough to the column for a second round of purification), and the fact that the final pt*abl*50 kinase consists of a single protein species as measured by autophosphorylation and silver staining (data not shown). An alternative step to DE-52 is Mono Q (Pharmacia) chromatography using the same buffer solutions described for DE52. This has given a 5- to 10-fold purification with insignificant loss of enzyme activity.

# [51] Functional Expression of Mammalian Adenosine Cyclic Monophosphate-Dependent Protein Kinase in *Saccharomyces cerevisiae*

*By* MARK J. ZOLLER, KAREN E. JOHNSON, WES M. YONEMOTO, and LONNY LEVIN

## Introduction

Phosphorylation and dephosphorylation of proteins is a primary cellular regulatory mechanism that functions to alter and reversibly control protein activity, association, localization, and stability.[1] The important biological questions regarding protein kinases are the identification of the signals that regulate their activities and the identification and localization of their target substrates. The important structural questions for this class of enzymes concern enzymatic mechanism, regulation of activity, and substrate specificity.

The expression of heterologous eukaryotic proteins in yeast provides an excellent system to probe the structure and function of proteins whose function is unknown.[2] In a number of cases, the yeast gene encoding the analogous protein can be cloned. Then, by manipulation of chromosomal sequences, the biological function of the heterologous gene can be studied in a genetic background that lacks the yeast homolog. In addition, mutant proteins can be expressed in yeast and studied in the absence of wild-type proteins. Genetic screens can be used to obtain protein variants with interesting or desired properties. Last, it is important to note that yeast is an eukaryote and that many of the biological and biochemical hallmarks of higher eukaryotes are also present in yeast. This chapter presents the functional expression in yeast of catalytic (C) subunit from mammalian cAMP-dependent protein kinase (cAdPK). In addition, we show that the mammalian C subunit can replace its yeast homolog and maintain the viability of a strain that lacks the yeast C subunits.

The best understood protein kinase in terms of structure and biochemistry is the cAMP-dependent protein kinase.[3] C subunit from cAdPK has been purified from a number of eukaryotic sources, including bovine

[1] E. G. Krebs, *Enzymes* **17,** 3 (1986).
[2] D. Botstein and G. Fink, *Science* **240,** 1439 (1988).
[3] S. S. Taylor, *BioEssays* **7,** 24 (1987).

heart,[1] porcine heart,[4] skeletal muscle,[5] and yeast.[6] The holoenzyme from most eukaryotic organisms is a tetramer composed of two regulatory (R) subunits and two catalytic (C) subunits. In the yeast *Saccharomyces cerevisiae,* three genes, *TPK1, TPK2,* and *TPK3*, encode distinct catalytic subunits of cAMP-dependent protein kinase, $C_1$, $C_2$, and $C_3$, respectively.[7] The three proteins exhibit approximately 75% amino acid identity with each other. Toda *et al.*[7] demonstrated genetically that the three C subunits were functionally similar and that at least one gene is necessary and sufficient for viability. The cDNAs and genes encoding mammalian C subunits from a number of sources have also been isolated and characterized. This has identified at least two distinct mammalian C subunits termed $C_\alpha$ and $C_\beta$.[8] Although the two proteins are nearly identical both structurally and biochemically, Uhler *et al.* demonstrated brain-specific expression of $C_\alpha$ mRNA.

The C subunits from *Drosophila*[9] and *Aplysia*[10] have been isolated and characterized. The amino acid sequences of C subunits from higher and lower eukaryotes provide some clues about important amino acid residues for this specific protein kinase as well as for all members of the protein kinase family.[11,12] However, it is not yet known whether these conserved residues are simply structural or play a functional role as well. The *Drosophila* and *Aplysia* C subunits are about 85% identical to the mammalian sequences. In contrast, the level of sequence conservation between the yeast and mammalian C subunits is only about 50%. With the isolation of cDNAs for the various subunits, the structure–function of cAMP-dependent protein kinase can now be approached by expression in yeast. In addition, it may be possible to dissect the numerous biological roles ascribed to this protein.

## Experimental Rationale

This chapter presents a procedure for introduction of expression vectors containing catalaytic subunits of cAdPK into the yeast *S. cerevisiae*.

[4] N. C. Nelson and S. S. Taylor, *J. Biol. Chem.* **256,** 3743 (1981).

[5] M. J. Zoller, A. R. Kerlavage, and S. S. Taylor, *J. Biol. Chem.* **254,** 2408 (1979).

[6] M. J. Zoller, J. Kuret, S. Cameron, L. Levin, and K. E. Johnson, *J. Biol. Chem.* **263,** 9242 (1988).

[7] T. Toda, S. Cameron, P. Sass, M. J. Zoller, and M. Wigler, *Cell (Cambridge, Mass.)* **50,** 277 (1987).

[8] J. C. Chrivia, M. D. Uhler, and G. S. McKnight, *J. Biol. Chem.* **263,** 5739 (1988).

[9] D. Kalderon and G. M. Rubin, *Genes Dev.* **2,** 1539 (1988).

[10] S. Beushausen, P. Bergold, S. Sturner, A. Elste, V. Roytenberg, J. H. Schwartz, and H. Bayley, *Neuron* **1,** 853 (1988).

[11] W. C. Barker, and M. O. Dayhoff, *Proc. Natl. Acad. Sci. U.S.A.* **79,** 2839 (1982).

[12] S. K. Hanks, A. M. Quinn, and T. Hunter, *Science* **241,** 42 (1988).

Upon introduction of the expression vector, strains can be constructed that contain only the heterologous catalytic subunit. In the current system, successful replacement of the yeast C subunit requires that the heterologous subunit be functional. Increased expression levels of C subunits can be obtained by using strong promoters but require coexpression of the R subunit. To demonstrate the system, the mouse $C_\alpha$ subunit is transformed into *S. cerevisiae* and expressed as the sole C subunit in the cell. The mouse $C_\alpha$ and yeast $C_1$, the product of the *TPK1* gene, exhibit approximately 50% sequence identity. The yeast enzyme, unlike all mammalian C subunits, does not contain a consensus sequence for N-terminal myristylation. The phosphorylation site at Thr-241 (equivalent to mammalian Thr-197) is conserved between yeast and mammals, but the residue in yeast $C_1$ analogous to Ser-338, the other known phosphorylation site of the mammalian C, is not conserved.

The $C_\alpha$ cDNA is cloned into two different yeast expression vectors (Fig. 1). YEp/*LEU2*/$C_\alpha$ expresses $C_\alpha$ using the promoter and terminator from the yeast C subunit gene, *TPK1*. YEp/*LEU2*/ADH-$C_\alpha$ expresses $C_\alpha$ from the *ADC1* promoter, a strong, constitutive promoter from the gene encoding yeast alcohol dehydrogenase I.[13] These vectors are multicopy episomal plasmids that are maintained by prototrophic selection for leucine. The plasmids were engineered using oligonucleotide-directed mutagenesis to express the natural $C_\alpha$ protein. The expressed protein would be expected to be myristylated on the N-terminal glycine, as is the protein in mammalian cells.[14] Production of a strain expressing only mammalian C is accomplished by substitution of the yeast *TPK1* gene with the $C_\alpha$ cDNA using a "plasmid swap" procedure depicted in Fig. 2. The starting strain, LL8, contains chromosomal disruptions of all three yeast C subunit (*TPK*) genes. The strain is kept alive by maintenance of a plasmid (YEp/*ADE8*/*TPK1*) that contains the wild-type *TPK1* gene. This vector is distinguished by the presence of the *ADE8* gene, which confers adenine prototrophy.

The experiment is done as follows: LL8 is transformed with the $C_\alpha$ plasmids, YEp/*LEU2*/$C_\alpha$ or YEp/*LEU2*/ADH-$C_\alpha$ or control plasmids YEp/*LEU2*/*TPK1* and YEp/*LEU2*. $Ade^+$,$Leu^+$ transformants are selected. These cells contain two plasmids, the newly transformed plasmid and the original plasmid expressing *TPK1*. Next, loss of one of the plasmids is encouraged by growth under nonselective conditions. Following growth overnight, the cultures are plated and individual colonies are tested for adenine and leucine prototrophy. The plasmid(s) present in the resulting

[13] G. Ammerer, *in* "Methods in Enzymology" (R. Wu, L. Grossman, and K. Moldare, eds.), Vol. 101, p. 192. Academic Press, New York, 1983.

[14] S. A. Carr, K. Biemann, S. Shoji, D. C. Parmelee, and K. Titani, *Proc. Natl. Acad. Sci. U.S.A.* **79,** 6128 (1982).

colonies are identified by their prototrophic markers. Viable cells must maintain at least one of the C subunit-expressing plasmids since at least one C subunit gene is required for viability. Thus, we predicted that following nonselective growth, $Ade^-$,$Leu^+$ cells would contain the swapped plasmid. These are the cells that require adenine in the medium for growth but not leucine. Having obtained strains that express only the mouse C subunit, we demonstrate that the protein produced in yeast contains myristic acid.

This chapter outlines a number of techniques: transformation of yeast, selection of plasmids, Western blotting proteins in yeast extracts, radiolabeling yeast cultures, and immunoprecipitation of proteins from yeast extracts. A number of manuals and articles are available that describe detailed protocols for yeast molecular genetics.[15,16]

*Materials*

cAMP, ATP, and kemptide (Leu-Arg-Arg-Ala-Ser-Leu-Gly) are purchased from Sigma (St. Louis, MO). [$\gamma$-$^{32}$P]ATP (>3000 Ci/mmol) and [$^3$H]myristic acid (40 Ci/mmol) are purchased from New England Nuclear (Boston, MA). Rabbit anti-mouse and goat anti-rabbit horseradish peroxidase (HRP)-coupled antibodies and 4-chloronaphthol are purchased from Bio-Rad (Richmond, CA). Rabbit antisera against mouse $C_\alpha$ is prepared using insoluble protein expressed in *Escherichia coli*. Yeast media are obtained from Difco.

*Media*

Yeast–*E. coli* shuttle vectors are propagated in *E. coli* strain MM294 using LB + ampicillin.[15] Preparation of yeast media is described in Sherman *et al.*,[16] except phosphate-free yeast media.[17] All percentages given below are weight/volume.

YPD: 2% yeast extract, 1% peptone, 2% glucose
SD: 0.67% yeast nitrogen base without amino acids, 2% glucose
SD + adenine: SD containing 20 mg/ml adenine
SD + leucine: SD containing 20 mg/ml leucine
Plates: 2% Bacto-agar is added to the above medium

[15] J. Sambrook, E. F. Fritsch, and T. Maniatis, "Molecular Cloning: A Laboratory Manual." Cold Spring Harbor Lab., Cold Spring Harbor, New York, 1982.

[16] F. Sherman, G. R. Fink, and J. B. Hicks, "Methods in Yeast Genetics." Cold Spring Harbor Press, Cold Spring Harbor, New York, 1983.

[17] G. Rubin, *J. Biol. Chem.* **248,** 3860 (1973).

*Yeast Strains*

Yeast strains are derivatives of SP1.[7] Strain LL8[18] contains chromosomal disruptions of all three *TPK* genes encoding catalytic subunits of yeast cAMP-dependent protein kinase and contains the *TPK1* gene on a multicopy vector marked by the *ADE8* gene.

SP1: *MATa TPK1 TPK2 TPK3 BCY1 his3 Leu2 ura3 trp1 ade8*[7]
LL8: *MATa TPK1 :: URA3 TPK2 :: HIS3 TPK3 :: TRP1 BCY1 leu2* + YEp/*ADE8*/*TPK1*[18]
LL8-1: Transformant of LL8 by YEp/*LEU2* (this work)
LL8-2: Transformant of LL8 by YEp/*LEU2*/*TPK1* (this work)
LL8-3: Transformant of LL8 by YEp/*LEU2*/$C_\alpha$ (this work)
MZY2: Derivative of LL8-2 that lost YEp/*ADE8*/*TPK1* by plasmid swap (this work)
MZY3: Derivative of LL8-3 that lost YEp/*ADE8*/*TPK1* by plasmid swap (this work)
MZY9: *MATa TPK1 :: URA3 TPK2 :: HIS3 TPK3 :: TRP1 BCY1 leu2* + YEp/*ADE8*/*BCY1*$^{\text{Ala-145}}$ + YEp/*LEU2*/ADH-$C_\alpha$ (this work)

*Buffers*

H buffer: 0.2 *M* potassium phosphate (pH 7.5), 4 m*M* ethylenediaminetetraacetic acid (EDTA), 10% (v/v) glycerol, 0.02% (w/v) Triton X-100, 1 m*M* phenylmethylsulfonyl fluoride (PMSF), 6 m*M* dithiothreitol, 8 m*M* leupeptin, 5 m*M* mercaptoethanol
RIPA-SDS: 50 m*M* Tris (pH 8.0), 1 m*M* dithiothreitol (DTT), 150 m*M* NaCl, 5 m*M* EDTA, 1% (w/v) Triton X-100, 0.5% (w/v) deoxycholate, 10 μ*M* cAMP, 50 m*M* NaF, 2 m*M* PMSF
PEG 3500: 50% (w/v) polyethylene glycol 3500 in water
Phosphate-buffered saline (PBS): 10 m*M* potassium phosphate (pH 7.2), 150 m*M* sodium chloride (w/v)
RIPA buffer: 50 m*M* Tris (pH 8.0), 1 m*M* DTT, 150 m*M* NaCl, 5 m*M* EDTA, 1% (w/v) Triton X-100, 0.5% (w/v) deoxycholate, 0.1% (w/v) sodium dodecyl sulfate (SDS)
S-300 buffer: 0.05 *M* potassium phosphate (pH 7.5), 0.1 *M* sodium chloride, 5 m*M* mercaptoethanol, 5% (v/v) glycerol, and 1 m*M* benzamidine
TE (10 : 1): 10 m*M* Tris-HCl, 1 m*M* EDTA (pH as indicated)
TE (10 : 0.1): 10 m*M* Tris-HCl, 0.1 m*M* EDTA (pH as indicated)
Tris-SDS buffer: 40 m*M* Tris (pH 7.4), 1% (w/v) SDS, 1 m*M* dithiothreitol, 5 m*M* EDTA, 50 m*M* NaF, and 2 m*M* PMSF

[18] L. R. Levin and M. J. Zoller, *Mol. Cell. Biol.* **10,** 1066 (1990).

Western transfer buffer: 50 m*M* glycine, 6 m*M* Tris-HCl, and 20% (v/v) methanol (pH 8.3)

YPD-$PO_4$ medium: Prepared according to Rubin[17]

2× Laemmli sample buffer: 0.1 *M* Tris-HCl (pH 6.8), 20% (v/v) glycerol, 4% (w/v) SDS, 5% (v/v), mercaptoethanol, and 0.02% (w/v) Bromphenol Blue

*Expression Vectors*

1. YEp/*LEU2* is the parent vector for the expression plasmids used. It contains the *LEU2* gene, a fragment of the 2-μm plasmid containing an origin of replication and the pUC119 backbone[18]
2. YEp/*ADE8/TPK1* expresses the yeast $C_1$ from its natural promoter and contains the *ADE8* gene for adenine selection, a yeast 2-μm origin, and the pUC18 backbone
3. YEp/*LEU2/TPK1* contains the *TPK1* gene, expressing $C_1$, cloned into YEp/*LEU2*
4. YEp/*LEU2*/$C_\alpha$ contains the mouse $C_\alpha$ coding sequence[8] cloned between the *TPK1* promoter and terminator sequences[7] in YEp/*LEU2*
5. YEp/*LEU2*/ADH-$C_\alpha$ contains the mouse $C_\alpha$ coding sequence cloned between the yeast *ADCI* promoter and terminator sequences[13] in YEp/*LEU2*
6. YEp/*ADE8/BCY1*$^{\text{Ala-145}}$ contains a mutant derivative of the yeast *BCY1* gene cloned in YEp/*LEU2*. This gene expresses a mutant yeast R subunit that contains a serine-to-alanine substitution at amino acid-145. This mutation increases the affinity for yeast C subunit 10× compared with wild-type R.[19] Expression of this R subunit variant increases the cellular level of C subunit and facilitates its purification. In this plasmid, *BCY1* is expressed from its own promoter

## *Yeast Transformation*

Yeast transformation of strain LL8 is accomplished by the lithium acetate procedure of Ito *et al.*[16,20]

### *Preparation of Competent Yeast Cells*

1. A single colony of LL8 is inoculated into 5 ml of YPD in a 50-ml polypropylene Falcon tube and grown at 30° for about 20 hr.

2. YPD (50 ml) is inoculated with 1 ml from step 1 overnight and the

[19] J. Kuret, K. E. Johnson, C. Nicolette, and M. J. Zoller, *J. Biol. Chem.* **263,** 9149 (1988).

[20] H. Ito, Y. Fukuda, K. Murata, and A. Kimura, *J. Bacteriol.* **153,** 163 (1983).

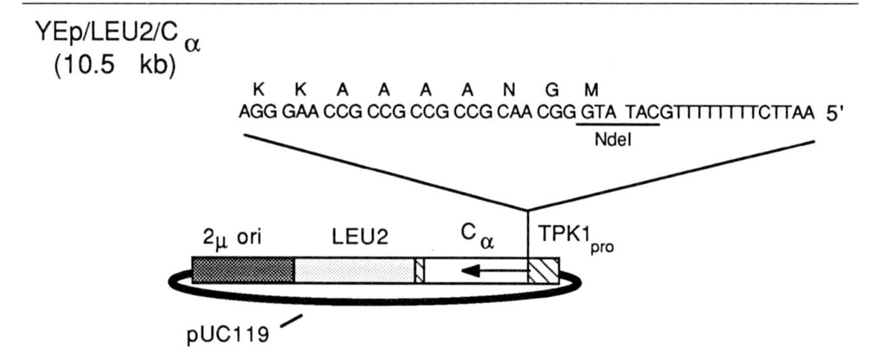


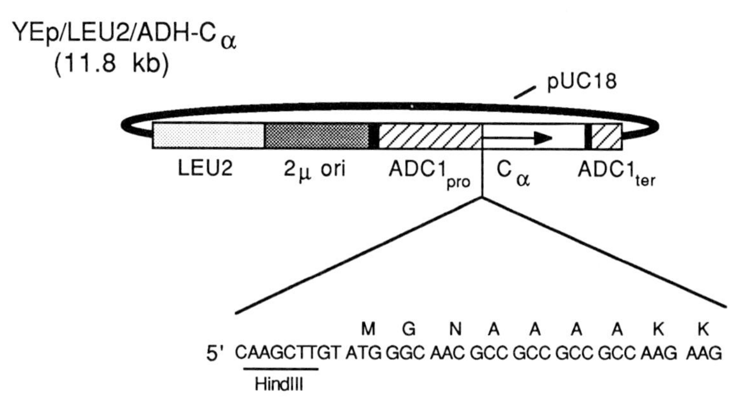


FIG. 1. Vectors for expression of mammalian C subunit in yeast. *Top:* YEp/*LEU2*/$C_\alpha$ expresses mouse $C_\alpha$ using the promoter and terminator sequences from *TPK1*. *Bottom:* YEp/*LEU2*/ADH-$C_\alpha$ expresses mouse $C_\alpha$ using the strong constitutive promoter and terminator from yeast alcohol dehydrogenase I (*ADC1*). Both plasmids contain a 2-$\mu$m origin of replication sequence and the *LEU2* gene for selection. The direction of transcription is indicated by the arrow. The constructions were made to produce the exact amino acid sequence of the natural protein. The expected N-terminal nine amino acids are shown. The sequence Gly-Asn-Ala-Ala-Ala-Ala-Lys-Lys encodes a signal for myristylation on glycine by the enzyme myristoyl-CoA:glycylpeptide *N*-myristoyltransferase (NMT).

cells are grown at 30° until $A_{600}$ is 0.7–1.0 (approximately $2 \times 10^7$ cells/ml).

3. The cells are harvested by centrifugation (5 min at 2000 rpm at 23°) and washed twice with 10 ml of TE (10 : 0.1; pH 7.5).

4. The washed cells are resuspended in 5 ml 0.1 *M* lithium acetate in TE (10 : 0.1; pH 7.5) and incubated at 30° with agitation for 1 hr.

5. The cells are harvested by centrifugation, then resuspended in 5 ml 0.1 *M* lithium acetate/15% glycerol. Aliquots (300 μl) are dispensed into individual 1.5-ml Eppendorf tubes, then placed at −70° until use.

*Transformation of Competent Yeast Cells*

1. Plasmid DNA (2 μg) and 30 μg of salmon sperm DNA (as carrier) are added to 300 μl of competent cells.
2. PEG 3500 (50%, w/v), 0.7 ml, is added and the contents are mixed by inverting three times.
3. The tube is incubated at 30° for 1 hr without agitation.
4. Following incubation, the cells are placed at 42° for 3 min, after which the cells are gently pelleted by centrifugation for 2 sec.
5. The supernatant is removed by aspiration and the cells are resuspended in 0.5 ml of TE (10 : 0.1; pH 7.5).
6. Two 200-μl aliquots of the transformation mixture are plated onto SD plates. After 3–4 days transformants appear and several transformants are restreaked for single colonies on SD.

*Plasmid Swap*

This is based on a procedure for switching plasmids by manipulation of the media.[21] Since we use the *LEU2* and *ADE8* genes as selectable markers, we do not use 5-fluoroorotic acid (5FOA) selection, which selects against the presence of the *URA3* gene.

1. A colony (from the transformation above) is inoculated into 5 ml YPD medium in a 50-ml Falcon tube and grown overnight at 30°.
2. The culture is diluted $10^3$-fold, $10^4$-fold, and $10^5$-fold in water, then 100-μl aliquots are spread onto YPD plates.
3. After 2 days, a plate containing about 50–100 colonies is replicated onto YPD, SD + adenine, and SD + leucine plates. (Alternatively, 50 colonies can be toothpicked in series onto the plates.)
4. Colonies that grow on SD + adenine but not on SD + leucine are predicted to contain only the plasmid containing $C_\alpha$. These strains are streaked out for single colonies onto SD + adenine and SD + leucine plates to retest the markers. The biochemical presence of yeast or mammalian C subunits can be determined by either protein kinase assay or Western blotting (see below). A plasmid expressing the yeast R subunit can be subsequently introduced by transformation as described above.
5. To prepare a frozen stock of these cells, a colony is inoculated into

[21] J. D. Boeke, J. Trueheart, G. Natsoulis, and G. Fink, *in* "Methods in Enzymology" (R. Wu and L. Grossman, eds.), Vol. 154, p. 164. Academic Press, San Diego, California, 1987.

5 ml SD + adenine and the culture is grown to midlog phase ($OD_{660}$ = 0.5–1). Culture (750 $\mu$l) is then added to a sterile cryovial containing 150 $\mu$l sterile glycerol (so final glycerol concentration is 15%), the vial is mixed well, then stored at $-70°$.

6. To obtain a fresh culture, a sterile wooden stick or Eppendorf tip is scraped on the top of the frozen culture, and the cells are streaked on a plate containing the appropriate medium. Frozen cells generally take 2–4 days to yield large, single colonies. The strains carrying episomal plasmids should not be maintained on plates for prolonged periods of time.

LL8 was transformed with one of four plasmids, YEp/*LEU2*, YEp/*LEU2*/*TPK1*, YEp/*LEU2*/$C_\alpha$, or YEp/*LEU2*/ADH-$C_\alpha$, and $Ade^+$,$Leu^+$ transformants were selected. No transformants were obtained with YEp/*LEU2*/ADH-$C_\alpha$. By analogy with the yeast C subunit, overexpression of an unregulated C subunit appears to be lethal. A colony from each of the transformation experiments was isolated: LL8-1, LL8-2, and LL8-3, which contain YEp/*ADE8*/*TPK1* and either YEp/*LEU2*, YEp/*LEU2*/*TPK1*, or YEp/*LEU2*/$C_\alpha$, respectively. These three strains were subjected to the plasmid swap protocol described above and the results of the swap are summarized in Table I. Since yeast requires at least one functional cAMP-dependent protein kinase catalytic subunit for viable growth, all of the strains resulting from nonselective growth maintained at least one kinase-expressing plasmid. For example, the resulting colonies from LL8-1, which contains the wild-type *TPK1* gene and a control plasmid YEp/*LEU2*, all contain the *TPK1* plasmid (all 106 colonies were $Ade^+$). Since the YEp/*LEU2* plasmid is not needed, it was often lost during the nonselective growth (27/106 colonies had lost the YEp/*LEU2* plasmid). The resulting colonies from LL8-2, which contains two wild-type *TPK1*-expressing plasmids, lost one or the other plasmid but not both (there were no $Ade^-$,$Leu^-$ colonies). Last, some of the resulting colonies from LL8-3 contained both the *TPK1*-expressing and $C_\alpha$-expressing plasmids, some contained only the *TPK1*-expressing plasmid, and a third class contained only the $C_\alpha$-expressing plasmid. The isolation of this last class of cells demonstrated that the mammalian C subunit was capable of replacing its yeast homolog.

### *Expression of $C_\alpha$ Using ADH Promoter*

Biochemical analysis of the strain expressing $C_\alpha$ from the *TPK1* promoter demonstrated that the expression was low. The level of C subunit in the cells was increased about fivefold by subsequently introducing a plasmid expressing the yeast R subunit. However, in order to achieve levels of expression so that physical and kinetic properties of the C subunit

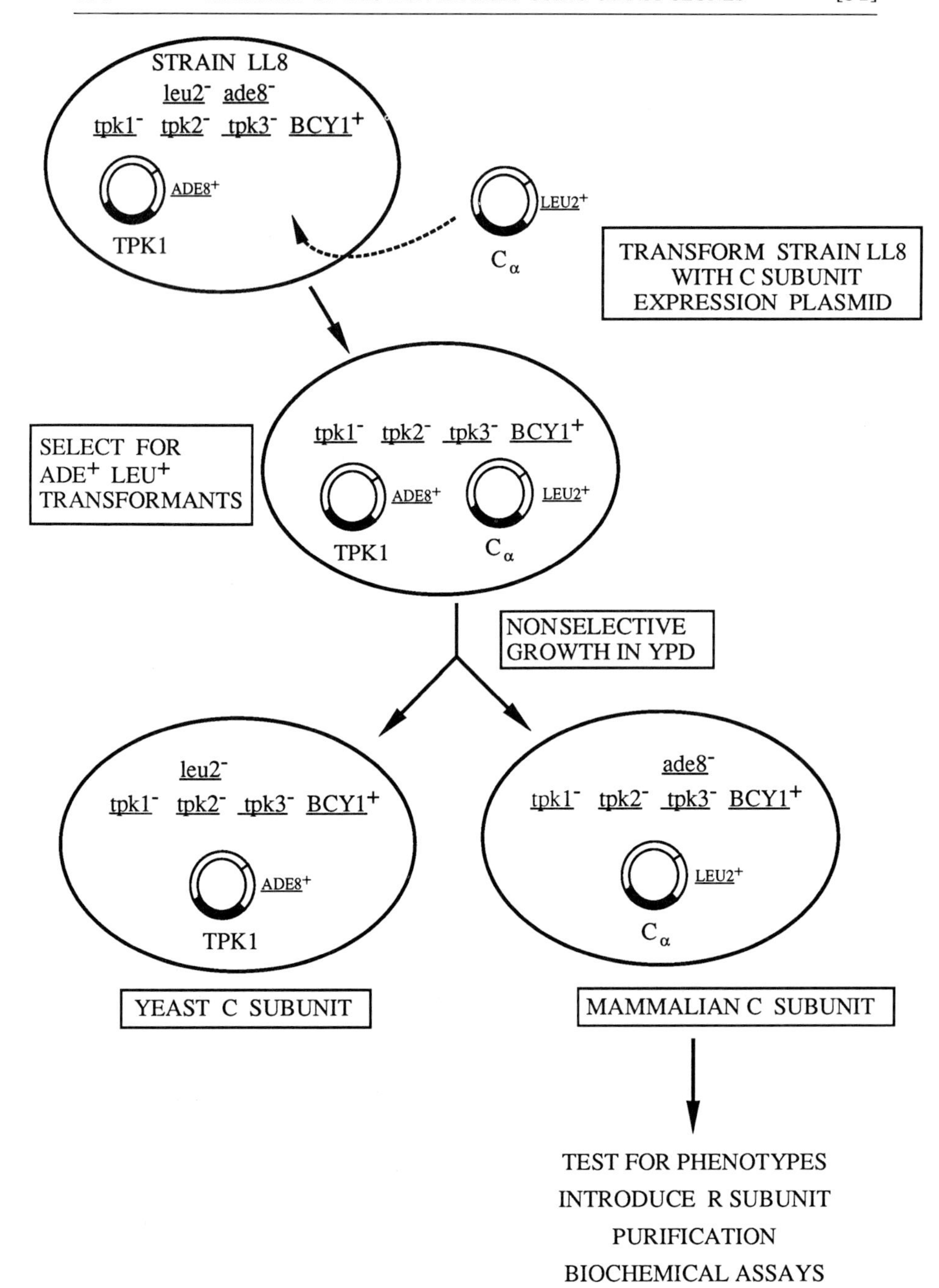
STRAIN LL8
leu2⁻ ade8⁻
tpk1⁻ tpk2⁻ tpk3⁻ BCY1⁺
ADE8⁺
TPK1
LEU2⁺
Cα
TRANSFORM STRAIN LL8 WITH C SUBUNIT EXPRESSION PLASMID
SELECT FOR ADE⁺ LEU⁺ TRANSFORMANTS
tpk1⁻ tpk2⁻ tpk3⁻ BCY1⁺
ADE8⁺
LEU2⁺
TPK1
Cα
NONSELECTIVE GROWTH IN YPD
leu2⁻
tpk1⁻ tpk2⁻ tpk3⁻ BCY1⁺
ADE8⁺
TPK1
ade8⁻
tpk1⁻ tpk2⁻ tpk3⁻ BCY1⁺
LEU2⁺
Cα
YEAST C SUBUNIT
MAMMALIAN C SUBUNIT
TEST FOR PHENOTYPES
INTRODUCE R SUBUNIT
PURIFICATION
BIOCHEMICAL ASSAYS

could be analyzed, we needed to utilize a stronger promoter than the *TPK1* promoter. Although our initial attempts to transform LL8 with only YEp/*LEU2*/ADH-$C_\alpha$ were unsuccessful, we found that we were able to overexpress $C_\alpha$ if we also overexpressed the yeast R subunit. Below is a procedure for cotransformation of LL8 with YEp/*LEU2*/ADH-$C_\alpha$ and YEp/*ADE8*/*BCY1*$^{\text{Ala-145}}$.

1. Strain LL8 is transformed with 2 $\mu$g each of YEp/*LEU2*/ADH-$C_\alpha$ and YEp/*ADE8*/*BCY1*$^{\text{Ala-145}}$ using the lithium acetate procedure described above.
2. Transformants are selected by plating on SD plates. Since LL8 is auxotrophic for leucine, cells that grow on SD are predicted to contain YEp/*LEU2*/ADH-$C_\alpha$. However, these cells still contain the plasmid that expressed the yeast $C_1$.
3. In order to lose the *TPK1* plasmid, a colony is inoculated into 5 ml SD + adenine liquid medium and grown overnight.
4. Aliquots of the culture are diluted, plated onto SD plates, and grown for 2–4 days at 30°.
5. Ten randomly picked colonies are inoculated into 5 ml of SD medium and the 10 individual cultures are grown for 24 hr at 30°.
6. One milliliter from each culture is used to prepare a frozen stock (described above). The remainder of each culture is harvested by centrifugation for Western blot analysis.

A strain that had lost the plasmid expressing yeast C is identified by Western blot analysis using anti-yeast C and anti-mouse C antibodies (see the next section). Generally, 10–20% of the strains had lost the yeast C and now overexpressed only the mouse C as well as the yeast R. One of these was chosen for further studies (MZY9). It should be noted that we used such an indirect method because we did not have a strain with enough selectable markers so that the plasmid expressing the yeast R could be added first. This situation may not be typical since there are three genes encoding yeast C subunits and each must be genetically deleted in

FIG. 2. The protein kinase plasmid swap procedure. The starting strain LL8 contains genetic disruptions of all three *TPK* genes. Viability is maintained by the presence of YEp/*ADE8*/*TPK1* that expresses wild-type $C_1$. The *ADE8* gene confers adenine prototrophy. A plasmid that expresses mouse C and confers leucine prototrophy is introduced into LL8 by transformation of competent cells. $\text{Ade}^+$,$\text{Leu}^+$ are selected by growth on SD plates. Plasmid loss is forced by growth without selection in YPD medium, then individual colonies are tested for leucine and adenine prototrophy. Colonies that are $\text{Ade}^-$,$\text{Leu}^+$ are predicted to express only the mammalian C subunit. To obtain higher expression of C subunit, plasmids expressing R subunit can be introduced subsequently by transformation.

TABLE I
CONSTRUCTION BY PLASMID SWAP OF STRAIN MZY3 THAT EXPRESSES ONLY $C_\alpha$[a]

| Strain[a] | Number of colonies with each phenotype after growth in YPD[b] | | |
|---|---|---|---|
| | $Ade^+,Leu^+$ | $Ade^+,Leu^-$ | $Ade^-,Leu^+$ |
| LL8-1 | 79 | 27 | 0 |
| LL8-2 | 0 | 7 | 73 |
| LL8-3 | 150 | 34 | 111 |

[a] Yeast strain LL8 was transformed with C subunit-expressing plasmids YEp/*LEU2*/$C_\alpha$ or YEp/*LEU2*/ADH-$C_\alpha$, or with control plasmids YEp/*LEU2* or YEp/*LEU2*/TPK1, by the lithium acetate procedure. Transformants were selected on SD plates and grown for 2 days at 30°. Three strains were generated: LL8-1, LL8-2, and LL8-3, the resulting transformants of LL8 by YEp/*LEU2*, YEp/*LEU2*/*TPK1*, and YEp/*LEU2*/$C_\alpha$, respectively. These stains contain two plasmids, YEp/*ADE8*/*TPK1*, the vector resident in LL8, and one of the vectors listed above. No transformants were obtained from YEp/*LEU2*/ADH-$C_\alpha$.

[b] A plasmid swap was performed in order to produce a strain containing only the mammalian $C_\alpha$ expression vector. LL8-3 was grown under nonselective conditions in YPD, and individual colonies were tested for growth in minimal media lacking either leucine or adenine, the selectable markers on the two plasmids in the cell. Following growth overnight at 30°, an aliquot of the culture was spread onto a YPD plate. After 2 days, a plate containing 100–200 colonies was replicated onto SD, SD plus adenine, and SD plus leucine plates. Growth on each plate was scored after 2 days at 30°. Control experiments using strains LL8-1 and LL8-2 were accomplished in the same manner. For each strain, colonies were scored for the requirement of adenine and/or leucine. In this way, the plasmid(s) contained in that colony could be inferred from its phenotype. Cells were scored having an $Ade^+,Leu^+$ phenotype if they grew on SD, $Ade^+,Leu^-$ if they required leucine for growth, and $Ade^-,Leu^+$ if they required adenine for growth. Thus, $Ade^+,Leu^+$ cells contain both *ADE8*- and *LEU2*-marked plasmids. $Ade^+,Leu^-$ cells contain only the original *ADE8*-marked plasmid. $Ade^-,Leu^+$ cells contain only the *LEU2*-marked plasmid. A strain, MZY3, containing only the mammalian $C_\alpha$ expression vector, was picked from one of the colonies with an $Ade^-,Leu^+$ phenotype derived from LL8-3.

order to produce a strain that can be used to express mutants in the absence of any wild-type protein. For experiments where there is only a single yeast homolog, the common laboratory strains should contain a sufficient number of selectable markers so that this scheme would not be necessary.

### *Purification of $C_\alpha$ from Yeast*

Purification of the C subunit from mammalian cAdPK is generally accomplished by a procedure in which holoenzyme is bound to a DEAE column, then C subunit is specifically eluted by application of cAMP.[4] We have developed an analogous immunoaffinity procedure for purification of yeast C in which the homoenzyme is bound to an antibody column that recognizes the yeast R subunit.[6] Yeast-expressed $C_\alpha$ was purified using this protocol.

1. SD (5 ml) is inoculated with a colony of strain MZY9 and grown overnight at 30°.
2. The 5-ml culture is then used to inoculate 1 liter SD. After growth for 18 hr, 10 ml of 40% glucose is added and growth is continued for an additional 8 hr, at which time the $A_{660}$ is generally between 2 and 6.
3. The cells are harvested by centrifugation, washed with 100 ml H buffer, harvested in two 50-ml polypropylene tubes (Falcon, Oxnard, CA), then stored overnight at −70°.
4. The cells are resuspended in 6 ml of cold H buffer (total volume about 13 ml), then transferred to two 25-ml Corex tubes (Corning Glass Works, Corning, NY). All subsequent steps are carried out at 4°, with the extract kept on ice.
5. Glass beads (Sigma; 400–600 $\mu$m) are added until only a small meniscus of liquid is observed. The cells are broken by vortexing on maximum for 10 bursts of 30 sec, with 1 min of rest on ice between bursts. Completeness of lysis is monitored by phase-contrast microscopy in 1% SDS. Lysed cells look dark and unlysed cells appear bright and refractile.
6. The broken cell extract is centrifuged for 10 min in an SS-34 rotor (8000 rpm, 4°), then the supernatant is transferred to a fresh 30-ml Corex tube on ice. The beads are washed with 3 ml of H buffer, centrifuged, and the supernatants are combined (total volume, 10 ml).
7. The extract is brought to 5% ammonium sulfate and 0.4% polymin-P, incubated on ice for 10 min, then centrifuged at 12,000 rpm for 10 min in an SS-34 rotor at 4°.
8. The supernatant is brought to 1 m*M* benzamidine and 35% ammonium sulfate, incubated on ice for 30 min, then centrifuged at 12,000 rpm for 30 min in an SS-34 rotor.

9. The supernatant is brought to 65% ammonium sulfate, incubated on ice for 30 min, and centrifuged at 12,000 rpm for 30 min in an SS-34 rotor.

10. The pellet is resuspended with 3 ml of H buffer to a final volume of 4.5 ml, and applied to a Sephacryl-300 column ($2.3 \times 47$ cm) equilibrated in S-300 buffer and run at 80 ml/hr. Fractions (4 ml) are collected and monitored for absorbance at 280 nm. Aliquots (5 $\mu$l) are assayed for phosphotransferase activity in the absence and presence of cAMP by the method of Roskoski.[22]

11. The peak fractions containing cAMP-dependent kinase activity are pooled, then passed through a 1-ml anti-yeast R/MAb427 immunoaffinity column at a rate of 1 ml/min. The column is prepared as described previously.[6] The sample containing cAdPK is passed through the column three times, then the antibody column is washed with 100 ml of S-300 buffer to remove nonspecifically bound proteins.

12. The catalytic subunit is eluted batchwise from the immunoaffinity column by addition of six 1-ml aliquots 0.8 m*M* cAMP in S-300 buffer.

13. Aliquots from each fraction are assayed for protein kinase activity. The purity is assessed by SDS–PAGE. Generally fractions 1–3 contain the majority of the kinase activity.

14. The peak fractions are combined and dialyzed at 4° versus S-300 buffer (three times, 2 liters each) to remove cAMP. Then the sample is dialyzed at 4° for 18 hr versus 250 ml of 50 m*M* potassium phosphate, 100 m*M* sodium chloride, 1 m*M* EDTA, 0.02% Triton X-100, 5 m*M* mercaptoethanol, and 50% glycerol, then stored at −20°.

The physical and biochemical properties of the yeast-expressed $C_\alpha$ compared to the yeast $C_1$, and $C_\alpha$ expressed in *E. coli*[23] and mammalian[24] cells are listed in Table II.[24a] The molecular weight of $C_\alpha$ by SDS–PAGE differs slightly from the other sources and may reflect posttranslational modifications. The kinetic properties of the yeast- and *E. coli*-expressed proteins are similar but differ from the properties of the mammalian-expressed protein. This may be due to differences in assay conditions or purification schemes. The yeast- and mammalian-expressed $C_\alpha$ proteins exhibit similar kinetics of inhibition by a peptide derivative of the heat-

[22] R. Roskoski, *in* "Methods in Enzymology" (J. D. Corbin and J. D. Hardman, eds.), Vol. 99, p. 3. Academic Press, New York, 1983.

[23] L. W. Slice and S. S. Taylor, *J. Biol. Chem.* **264,** 20940 (1989).

[24] S. R. Olsen and M. D. Uhler, *J. Biol. Chem.* **31,** 18662 (1989).

[24a] M. D. Uhler, D. F. Carmichael, D. C. Lee, J. C. Chrivia, E. G. Krebs, and G. S. McKnight, *Proc. Natl. Acad. Sci. U.S.A.* **83,** 1300 (1986).

TABLE II
STRUCTURAL AND BIOCHEMICAL PROPERTIES OF MAMMALIAN VERSUS YEAST C SUBUNIT

| | Protein kinase subunits | | | |
|---|---|---|---|---|
| Property[a] | Yeast $C_1$[b] | $C_\alpha$ (yeast)[c] | $C_\alpha$ (*E. coli*)[d] | $C_\alpha$ (mammalian)[e] |
| $M_r$ calculated (kDa) | 45,945 | 40,590 | 40,590 | 40,590 |
| $M_r$ SDS (kDa) | 52,000 | 42,600 | 42,000 | 42,000 |
| $K_m$ ATP ($\mu M$) | 33 | 37 | 18.5 | 4 |
| $K_m$ kemptide ($\mu M$) | 101 | 53 | 43 | 5.6 |
| $K_i$ $PKI_{5-24}$ (n$M$) | 280 | 12 | 14 | nd |

[a] $M_r$ calculated, the calculated molecular weight based on the amino acid sequence; $M_r$ SDS, the molecular weight determined by SDS-PAGE; $K_m$ values were determined from the slope of Eadie–Hofstee plots; $K_i$ $PKI_{5-24}$ is the inhibition constant of phosphotransferase activity determined from the value of the $IC_{50}$.[25]

[b] Values for purified yeast $C_1$ are taken from Zoller *et al.*[6] The calculated molecular weight for yeast $C_1$ was obtained from Toda *et al.*[7]

[c] Values for $C_\alpha$ produced in yeast are taken from this work. The calculated molecular weight for $C_\alpha$ was obtained from Uhler *et al.*[24a]

[d] Values for $C_\alpha$ produced in *E. coli* are taken from Slice and Taylor.[23]

[e] Values for $C_\alpha$ produced in a mammalian cell line are taken from Olsen and Uhler.[24] nd, Not determined.

stable inhibitor (PKI).[25,26] The interaction between mouse $C_\alpha$ and yeast R subunit is approximately five times weaker than that between yeast C and yeast R, as judged by $IC_{50}$ of catalytic activity (data not shown).

### *SDS–PAGE and Immunoblot (Western) Analysis*

1. To analyze total cellular proteins, the cells in 0.5–1.0 ml of yeast culture (in minimal medium) are pelleted by centrifugation. The supernatant is removed and the cells are resuspended in 30 $\mu$l of 1× Laemmli sample buffer.

2. The samples are heated at 100° for 3 min, then subjected to SDS–PAGE.[27] Generally 20 $\mu$l is loaded in one lane of a 0.75-mm thick 12% polyacrylamide gel.

3. Following electrophoresis, the unstained gel is soaked in 100 ml of

[25] J. D. Scott, M. B. Glaccum, E. H. Fischer, and E. G. Krebs, *Proc. Natl. Acad. Sci. U.S.A.* **83,** 1613 (1986).

[26] D. B. Glass, H.-C. Cheng, L. Mende-Mueller, J. Reed, and D. A. Walsh, *J. Biol. Chem.* **264,** 12166 (1989).

[27] E. Harlow and D. Lane, "Antibodies: A Laboratory Manual." Cold Spring Harbor Lab., Cold Spring Harbor, New York, 1988.

Western transfer buffer for 15 min at 4°, then placed into a Hoeffer electroblot transfer apparatus.

4. Electrophoretic transfer is conducted by a modification of Towbin *et al.*,[28] using Western transfer buffer at 30 V for 16 hr at 4°.

5. The gel is removed from the nitrocellulose filter, then the filter is incubated with 100 ml in PBS + 3% nonfat dried milk (w/v) for 1 hr at room temperature.[29]

6. Antibody is diluted in 15 ml PBS/3% nonfat milk and incubated with the filter for 1–8 hr at room temperature.[27]

7. Unbound primary antibody is removed by washing the filter with PBS three times (100 ml each) over 30 min.

8. HRP-coupled antibodies are diluted in 15 ml PBS/3% nonfat milk and incubated with each filter for 1 hr at room temperature.

9. Excess reagent is washed off with PBS and the color development is accomplished using 4CN (Bio-Rad), using the manufacturer's directions.

Figure 3 shows a Western blot of $C_\alpha$ from strain MZY9 compared with purified yeast $C_1$. For detection of yeast C subunit, hybridoma tissue culture containing anti-yeast C monoclonal antibody (MAb Z9) is diluted 1 : 100. For detection of mouse $C_\alpha$, a rabbit polyclonal antiserum (#306) is diluted 1 : 750. Since the yeast and mammalian C subunits would be expected to copurify, this analysis demonstrates that the $C_\alpha$ is the sole C subunit in this strain.

### *Radiolabeling and Immunoprecipitation*

Since the natural amino acid sequence of $C_\alpha$ was expressed, we predicted that the protein would be N-myristylated on the amino terminus.[14] It is known that *S. cerevisiae* contains an enzyme that catalyzes this modification and that the substrate specificity is similar to that of the

[28] H. Towbin, T. Staehelin, and J. Gordon, *Proc. Natl. Acad. Sci. U.S.A.* **76,** 4350 (1979).

[29] D. A. Johnson, J. W. Gautsch, J. R. Sportsman, and J. H. Elder, *Gene Anal. Technol.* **1,** 3 (1984).

FIG. 3. Electrophoretic mobility and immunological reactivity of mammalian $C_\alpha$ subunit and yeast $C_1$ purified from yeast. Yeast $C_1$ and mouse $C_\alpha$ purified from yeast (Y) and from *E. coli* (E) were subjected to SDS-polyacrylamide gel electrophoresis in duplicate. Approximately 50 ng of each protein is applied to the gel. The proteins were electrophoretically transferred to nitrocellulose, then one-half is incubated with anti-$C_\alpha$ polyclonal antibodies (#306) and the other half is incubated with anti-yeast $C_1$ MAb (Z9). The positions of immune complexes were observed by horseradish peroxidase-coupled second antibody. The positions and sizes of prestained molecular weight ($\times 10^{-3}$) standards are shown on the right.

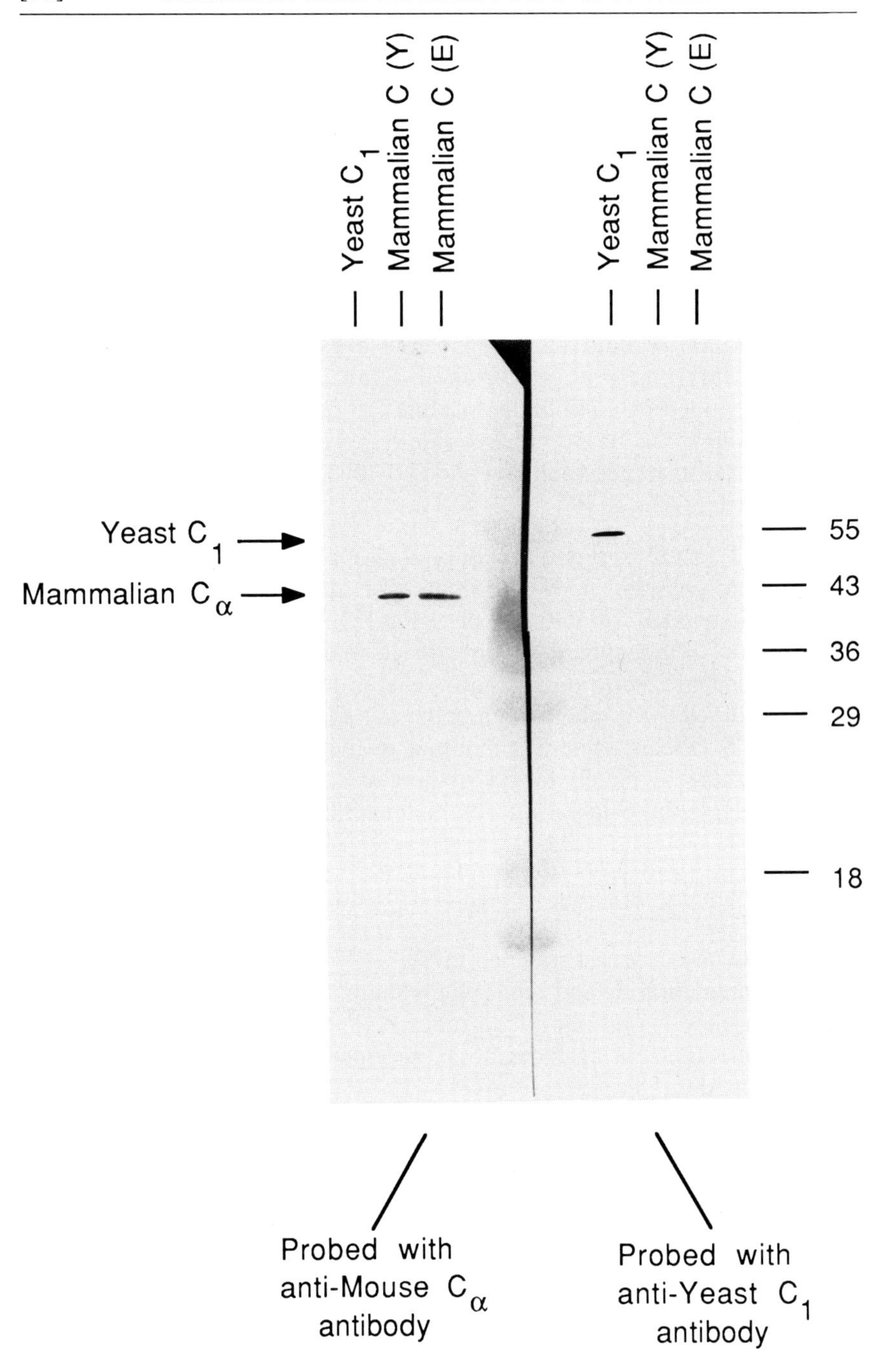
Yeast $C_1$
Mammalian C (Y)
Mammalian C (E)
Yeast $C_1$
Mammalian C (Y)
Mammalian C (E)
Yeast $C_1$
Mammalian $C_\alpha$
55
43
36
29
18
Probed with anti-Mouse $C_\alpha$ antibody
Probed with anti-Yeast $C_1$ antibody

mammalian enzyme.[30] To investigate the modification of $C_\alpha$ expressed in yeast, strain MZY9 is labeled with [$^3$H]myristic acid, the cells are broken, then immunoprecipitated C subunit is subjected to SDS–PAGE and analyzed by autoradiography. This procedure has been used to analyze $^{32}$P- and $^{35}$S-labeled proteins as well.[18]

1. A colony is inoculated into 5 ml SD medium and grown overnight.
2. One milliliter is diluted into 50 ml fresh SD medium and grown to midlog phase (approximately $2 \times 10^7$ cells/ml).
3. The cells are harvested by centrifugation at 5000 rpm for 10 min, the supernatant is removed, and the cells are resuspended in 0.1 ml SD.
4. [$^3$H]Myristic acid (40 Ci/mmol) (2 mCi) is evaporated to dryness and resuspended in a minimum volume (10–20 $\mu$l) of dimethylsulfoxide (DMSO).
5. The resuspended cells are added to the myristic acid and the culture is incubated 3–4 hr at 30°.

For $^{32}$P labeling, a culture containing approximately $10^8$ cells is pelleted, washed with phosphate-free medium (YPD-$PO_4^{2-}$), and resuspended in 100 $\mu$l of YPD-$PO_4^{2-}$. We use 1–3 mCi of ortho[$^{32}$P]phosphate (ICN; carrier free). Generally, the level of phosphate incorporation into yeast proteins is much less than that obtained from mammalian cells.

For $^{35}$S labeling, use SD minimal medium lacking methionine and cysteine. Typically for $^{35}$S labeling we use 1 mCi of [$^{35}$S]methionine (ICN; Tran$^{35}$S-label, approximately 1000 Ci/mmol).

6. Cells are pelleted by centrifugation in a microfuge for 30 sec, and the supernatant containing free label is carefully removed and discarded appropriately.
7. The cells are resuspended in 500 $\mu$l TE (10 : 1, pH 7.5) containing 50 m*M* sodium fluoride and 2 m*M* phenylmethylsulfonyl fluoride and transferred to a new 1.5-ml microfuge tube.
8. Pellet the cells again, remove the supernatant, and then resuspend the cells in 100 $\mu$l Tris-SDS buffer.
9. Add an equal volume of glass beads, then break the cells open by vortexing four times (30-sec bursts) with 1-min rests on ice in between.
10. Heat the lysate at 100° for 10 min to denature proteins and complete the cell lysis. (*Hint:* Poke a small hole into the top of the tube with a 22-gauge needle so pressure does not build up and pop the top during heating.)
11. Add 400 $\mu$l of RIPA-SDS to the tube, mix carefully, let the glass

[30] D. A. Towler, S. R. Eubanks, D. S. Towery, S. P. Adams, and L. Glaser, *J. Biol. Chem.* **262,** 1030 (1987).

beads settle, then remove the supernatant to a fresh microcentrifuge tube.

12. Wash the glass beads with 500 $\mu$l of RIPA-SDS + 10 $\mu M$ cAMP, 50 m$M$ NaF, 2 m$M$ PMSF. Combine the two supernatants.

13. Clarify the supernatant by centrifugation for 10 min at 4° in an Eppendorf centrifuge. Remove the supernatant and place it into a fresh microcentrifuge tube. Total volume is about 0.8 ml.

14. For immunoprecipitation of yeast C, 0.5 $\mu$l of monoclonal antibody MAb/Z9 (anti-*TPK1*)[6] ascitic fluid (or 100 $\mu$l hybridoma tissue culture fluid) is added to the cell extract and incubated with the cell extract for 1 hr at 4° with rotation. For immunoprecipitation of mouse $C_\alpha$, 10 $\mu$l rabbit anti-C antiserum is incubated with the cell extract for 1 hr.

15. Remove nonspecific aggregates by centrifugation for 10 min at 4°. Carefully remove the supernatant and add it to a fresh tube. This step reduces precipitation of unrelated proteins.

16. Add 100 $\mu$l of 50% protein A-Sepharose (Pharmacia/LKB Biotechnology, Inc., Piscataway, NJ) to the supernatant and incubate by rotation at 4° for 20–60 min.

17. Centrifuge the tube for 30 sec, remove the supernatant, then wash the beads four times with 0.5 ml RIPA buffer.[27] (*Optional:* For $^{32}$P labeling, treat beads with 25 $\mu$g RNase and 50 $\mu$g DNase in 0.5 ml RIPA buffer for 5 min on ice. This reduces background of $^{32}$P-labeled nucleic acids on the SDS gel. Caution should be taken with this, since we once had a protease contamination of the DNase.)

18. Add 50 $\mu$l of 1 × Laemmli sample buffer to each tube and denature the proteins by boiling for 3 min. Subject 25 $\mu$l to SDS–polyacrylamide gel electrophoresis.

Figure 4 shows that an immunoprecipitated protein with a relative molecular weight of $C_\alpha$ appears to contain myristic acid. This protein comigrates with an $^3$H-labeled protein in the total extract (lane 1, Fig. 4). Further studies are in progress to demonstrate that the modification is myristate and that it is attached to the N terminus.

## Summary

The heterologous expression of protein kinases in *E. coli* has proved difficult and unpredictable. Although the v-*abl* protein kinase is successfully expressed in *E. coli,*[31] our experiments on expression of yeast C subunits in *E. coli* produced large amounts of predominantly insoluble and inactive protein.[6] Attempts to refold the protein proved unsuccessful. In contrast, a major fraction of mouse $C_\alpha$ expressed in *E. coli* is soluble and

[31] J. Y. Wang and D. Baltimore, *J. Biol. Chem.* **260,** 64 (1985).

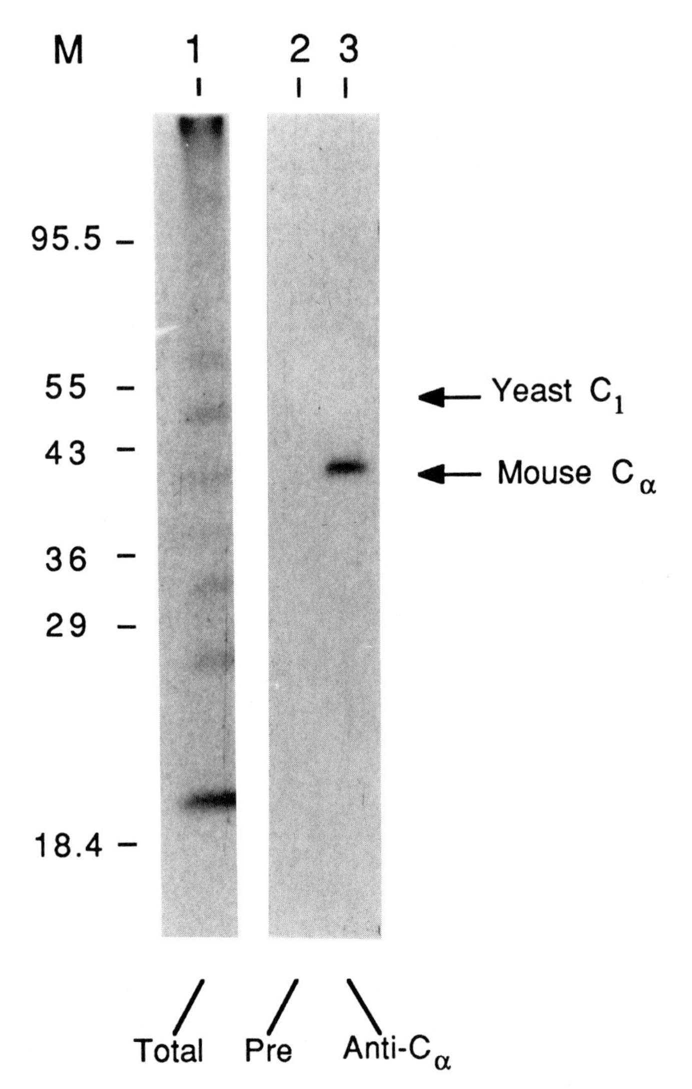


FIG. 4. Mammalian $C_\alpha$ subunit is myristylated in yeast. Strain MZY9 was labeled with [$^3$H]myristic acid, then the cells were lysed using glass beads (see Radiolabeling and Immunoprecipitation). A sample of the total cell extract (20 $\mu$l out of the total 800 $\mu$l) was removed. About 390 $\mu$l of the sample was incubated with 10 $\mu$l of preimmune rabbit serum and the other 390 $\mu$l was incubated with 10 $\mu$l rabbit anti-$C_\alpha$ serum (#306). The immune complex was precipitated using protein A-Sepharose and washed (see Radiolabeling and Immunoprecipitation). The samples were mixed with Laemmli sample buffer and subjected to SDS-PAGE. Lane 1 shows an aliquot of the total cell extract applied to the gel. Lanes 2 and 3 show the immunoprecipitate from the preimmune and anti-$C_\alpha$ immune sera, respectively. Following electrophoresis the gel was fixed for 30 min, treated with En$^3$Hance (New England Nuclear), then dried. Lane 1 was autoradiographed for 5 days. Lanes 2 and 3 were autoradiographed for 4 weeks. The localization of mouse $C_\alpha$ was identified by Western blotting purified mouse C that was coelectrophoresed (data not shown).

the enzyme in the soluble fraction is active[23]; however, certain mutant forms have proved to be unstable, difficult to purify, or insoluble. In addition, the *E. coli* system cannot be used to study the biological role of posttranslational modifications specific to eukaryotic systems. Several protein kinases have been expressed in soluble form in insect cells using baculovirus,[32] suggesting that this system is generally more reliable than *E. coli*. However, the presence and nature of posttranslational modifications in insect cells may be different from that found in the natural source and may affect the biochemical function. In addition, baculovirus expression is not particularly useful for studying biological questions. Mouse $C_\alpha$ and $C_\beta$ have been overexpressed in NIH3T3 cells.[24] This approach is useful in characterizing the biochemical properties of $C_\alpha$ versus $C_\beta$, but it may not be an ideal system for studying mutant proteins since wild-type C subunits are still expressed from the chromosomal copies in this genetic background. This small level of wild type may make it difficult to analyze weakly functional mutants, which have activities less than 10% that of wild type. Several cell lines with altered subunits of cAMP-dependent protein kinase have been identified[33] but a strain completely devoid of C subunit has not been adequately characterized for protein structure/function studies. Disruption of the genes encoding cAMP-dependent protein kinase in mammalian cells has not yet been accomplished.

This chapter describes a method to express a C subunit of mammalian cAMP-dependent kinase in yeast. We have demonstrated that the mouse $C_\alpha$ subunit can substitute for its yeast counterpart. Since at least one functional C subunit is required for viability, these results suggest that the yeast substrates important for viability are recognized by the mammalian C subunit. Although the sequence conservation between yeast and mouse C subunit is only about 50%, these results demonstrate that heterologous proteins with relatively low sequence conservation with their yeast counterparts can be functional in yeast. This is an important result because it validates the use of yeast to identify the biological role of newly cloned genes from heterologous systems, a key tenet of the Human Genome Project.[2,34] There are a number of examples of the functional expression in yeast of genes from higher eukaryotes. Among these are mammalian *ras*,[35] *Drosophila* topoisomerase II,[36] and the mammalian homolog of the

[32] R. Herrera, D. Lebwohl, A. G. de Herreos, R. G. Kallen, and O. M. Rosen, *J. Biol. Chem.* **263,** 5560 (1988).

[33] M. Gottesman, *in* "Methods in Enzymology" (J. D. Corbin and J. D. Hardman, eds.), Vol. 99, p. 197. Academic Press, New York, 1983.

[34] J. D. Watson, *Science* **248,** 44 (1990).

[35] T. Kataoka, S. Powers, S. Cameron, O. Fasano, M. Goldfarb, J. Broach, and M. Wigler, *Cell (Cambridge, Mass.)* **40,** 19 (1985).

[36] E. Wyckoff and T.-S. Hsieh, *Proc. Natl. Acad. Sci. U.S.A.* **85,** 6272 (1988).

cell cycle protein kinase cdc2.[37] These genes complemented conditional mutants or lethal gene knockouts. In addition to these examples, the mammalian transcription factors AP1/jun[38] and TFIID,[39] cytoskeletal proteins,[40] and proteins involved in secretion[41] share sequence and/or functional similarities with yeast proteins.

The approach described in this chapter could be used to study site-directed mutants of mammalian C subunits as well as to investigate the structure and function of other protein kinases. For example, yeast protein kinase C has been cloned[42] and has been shown to be required for viable growth of yeast cells. Thus, a similar "plasmid swap" system can be engineered for the analysis of wild-type and mutant forms of PK-C. The expression of site-directed mutants in the absence of the wild-type protein is a powerful tool that complements experiments using affinity and chemical labeling[43] and X-ray crystallography.[44] In addition, hypotheses concerning the structure or enzymatic mechanism of the protein that stem from the three-dimensional structure can be tested by expression of mutants in this system. Protein kinase mutants can also be identified in yeast by genetic screens and selections.[18,45] Last, the expression of the mammalian cAMP-dependent protein kinase in yeast provides a genetic system to study the biological role of cAMP-dependent protein kinase in gene expression, signal transduction, and cellular proliferation.[46]

## Acknowledgments

This work is performed at Cold Spring Harbor Laboratory and is conducted in collaboration with the laboratory of Dr. Susan Taylor (University of California, San Diego). We thank Dr. Taylor for encouragement and interest in this work. In addition, we acknowledge Dr. Stan McKnight (Department of Biochemistry, University of Washington) for providing a

[37] M. G. Lee and P. Nurse, *Nature (London)* **327,** 31 (1987).

[38] K. D. Harshman, W. S. Moye-Rowley, and C. S. Parker, *Cell (Cambridge, Mass.)* **53,** 321 (1988).

[39] S. Buratowski, S. Hahn, P. Sharp, and L. Guarente, *Nature (London)* **334,** 37 (1988).

[40] J. Huffaker, A. Hoyt, and D. Botstein, *Annu. Rev. Genet.* **21,** 259 (1987).

[41] G. S. Payne, T. B. Hasson, M. S. Hasson, and R. Sheckman, *Mol. Cell. Biol.* **7,** 3888 (1987).

[42] D. E. Levin, F. O. Fields, R. Kunisawa, J. M. Bishop, and J. Thorner, *Cell (Cambridge, Mass.)* **62,** 213 (1990).

[43] J. Buechler, J. Toner-Webb, and S. S. Taylor, this volume [41].

[44] J. M. Sowadski, N. Xuong, D. Anderson, and S. S. Taylor. *J. Mol. Biol.* **182,** 617 (1985).

[45] L. R. Levin, J. Kuret, K. E. Johnson, S. Powers, S. Cameron, T. Michaeli, M. Wigler, and M. J. Zoller, *Science* **240,** 69 (1988).

[46] R. H. Jones and N. C. Jones, *Proc. Natl. Acad. Sci. U.S.A.* **86,** 2176 (1989).

cloned cDNA of mouse $C_\alpha$ to Dr. Taylor. We also thank Dr. Lee Slice for providing pLWS-3, from which we made our constructs. This work was funded by a grant from the National Science Foundation to M.J.Z. (DMB-8918788). L.R.L. was a graduate student in the Genetics Program at SUNY, Stony Brook. W.Y. is a postdoctoral fellow in the laboratory of Dr. Susan Taylor (Department of Chemistry, University of California, San Diego).

# [52] Expression and Properties of Epidermal Growth Factor Receptor Expressed from Baculovirus Vectors

*By* MICHAEL D. WATERFIELD and CHARLES GREENFIELD

## Introduction

The biological effects of epidermal growth factor (EGF) and transforming growth factor $\alpha$ (TGF$\alpha$) are mediated by interaction with a specific cell surface glycoprotein of 175,000 kDa. The EGF receptor has an external ligand-binding domain that is separated by a hydrophobic transmembrane region from a cytoplasmic domain possessing a ligand-stimulatable tyrosine kinase activity. This receptor is closely related to the erb-B2 (Neu or HER-2) and erb-B3 putative receptors and more distantly to the insulin and IGF I receptors.[1] The EGF receptor is encoded by a protooncogene located on human chromosome 7. When the receptor is overexpressed, in an environment where its ligand is produced, transformation of cultured cells can occur.[2,3] Overexpression may also contribute to the etiology of human tumors. Gross rearrangments of the gene, involving deletions of the region encoding the external domain, have been detected in human brain tumors and in the oncogene v-*erbB* of avian erythroblastosis virus (AEV).[4,5] In both cases the truncated receptor may function independently of ligand, thereby transforming cells and producing tumors.

To explore the molecular mechanisms of signal transduction by the EGF receptor, expression of both normal and mutant receptors in large amounts is desirable. To achieve this, we have employed baculovirus

[1] A. Ullrich and J. Schlessinger, *Cell (Cambridge, Mass.)* **61,** 203 (1990).

[2] P. P. Di Fiore, J. H. Pierce, T. P. Fleming, R. Mazan, A. Ullrich, C. R. King, J. Schlessinger, and S. A. Aaronsen, *Cell (Cambridge, Mass.)* **51,** 1063 (1987).

[3] H. Riedel, S. Massoglia, J. Schlessinger, and A. Ullrich, *Proc. Natl. Acad. Sci. U.S.A.* **85,** 1477 (1988).

[4] P. Collins, personal communication.

[5] J. Downward, Y. Yarden, E. Mayes, G. Scrace, N. Totty, P. Stockwell, A. Ullrich, J. Schlessinger, and M. D. Waterfield, *Nature (London)* **307,** 521 (1984).

vectors to produce intact functional receptor, and its domains, in insect cells.[6,7] This protein is being used for crystallization and determination of the three-dimensional structure of the receptor domains.

The baculoviruses occlude and hence protect their genomes in large protein crystals of a 29-kDa polyhedron protein. This provides a means for horizontal transmission of the virus since when infected larvae die, large numbers of occluded virus particles are left in the decomposing tissue. When larvae feed on contaminated plants, they ingest the polyhedra and the whole cycle begins again. It is clear that polyhedrin production is essential for transmission of the virus in the natural environment. However, for the infection of insect cells in tissue culture the production of polyhedrin is dispensible. Because high levels of polyhedron protein are made by these viruses, a system has been developed to allow insertion of foreign genes into the polyhedron gene. In most cases the recombinant virus has been derived from the *Autographa californica* nuclear polyhedrosis virus (AcNPV), a prototype virus of the family Baculoviridae, which has then been used to infect *Spodoptera frugiperda* Sf9 cells in tissue culture. The genetic engineering of this virus is usually accomplished by allelic replacement. This involves cell-mediated homologous recombination of cotransfected wild-type viral DNA and a transfer plasmid containing the gene of interest attached to the polyhedrin promoter flanked by viral sequences. Recombination results in the replacement of wild-type sequences with plasmid sequences located between the crossover points. Many different plasmids have been developed which facilitate the insertion of foreign genes at appropriate sites within the polyhedrin region. These transfer vectors are constructed to enable the cloning of the foreign gene into a site either before or after the polyhedrin initiator codon ATG. Following homologous recombination, cells infected solely with the recombinant virus can be plaque purified, as these cells show cytopathic effects but no polyhedra occlusion bodies. Budded extracellular virus (ECV) recovered from these occlusion negative cells can be propogated to produce additional recombinant virus and to direct expression of the heterologous protein.

There are many major advantages in the use of this invertebrate virus expression vector over bacterial, yeast, and mammalian systems. One is that functional proteins are produced at high levels. A second is that the baculovirus is not pathogenic to vertebrates or plants and their propagation does not utilize transformed cells or transforming elements. Third, the

[6] C. Greenfield, G. Patel, N. Jones, and M. D. Waterfield, *EMBO J.* **7,** 139 (1988).

[7] C. Greenfield, I. Hiles, M. D. Waterfield, M. Federwisch, A. Wollmer, T. L. Blundell, and N. McDonald, *EMBO J.* **8,** 4115 (1989).

baculovirus expression system is suited to the production of toxic products. Furthermore, recombinant virus preparation and purification are relatively rapid, requiring about 8 to 12 weeks for production of the recombinant protein, versus 6 months for dihydrofolate reductase (DHFR) amplification in Chinese hamster ovary (CHO) cells. Finally, recombinant proteins produced in insect cells with baculovirus vectors are usually biologically active and for the most part appear to undergo similar post-translational processing to those proteins produced in higher eukaryotic cells. Recombinant proteins synthesized by insect cells can be secreted, transported to the nucleus, transported to the cell surface, proteolytically processed, phosphorylated, N-glycosylated, O-glycosylated, myristylated, or palmitylated. There is also evdience that insect cells can assemble heterologous proteins with correct secondary and tertiary conformations. Proteins have been produced at levels ranging from 1 to 500 mg/liter. The factors which determine how well a foreign gene is expressed in this system are not yet well characterized, and so it is difficult to predict how efficiently different genes will be expressed.

The application of the baculovirus expression system has been limited by difficulties in the scale-up of insect cell culture. Insect cells are reported to be unusually shear sensitive, to be damaged by gas sparging, and to have an unusually high oxygen demand. Taken together, these factors have limited insect cultures to typically 3 liters or less.

The system we have used employs the *Autographica californica* nuclear polyhedrosis virus (AcNPV) and the cells of *Spodoptera frugiperda* in tissue culture. The methodologies for producing recombinant viruses and for growing and infecting insect cells have been extensively documented by the Summers laboratory, which was responsible for establishing the system as a laboratory tool.[8] Indeed, laboratory manuals which describe in great detail many aspects of the system can be obtained from Summers.[9]

The essential first step in the preparation of a recombinant virus expressing the intact EGF receptor involves the construction of an expression vector where the receptor cDNA is fused to the polyhedron gene translational control sequences so that the full-length receptor polypeptide, including its signal sequence, can be expressed. The plasmid containing the receptor cDNA is then transfected together with viral DNA

[8] M. D. Summers and G. E. Smith, *in* "Genetically-Altered Viruses and Their Environment: The Banley Report" (B. Fields, M. A. Martin, and D. Kamley, eds.), Vol. 22. p. 319. Cold Spring Harbor Lab., Cold Spring Harbor, New York, 1985.

[9] M. D.Summers and G. E. Smith, eds., "A Manual of Methods for Baculovirus Vectors and Insect Culture Procedures," Tex. Exp. Stn. Bull. No. 155. Texas A & M Agric. Coll., College Station, 1987.

into *Spodoptera* cells. Recombinant viruses, which no longer express the polyhedron protein and hence cannot form occlusion bodies, can be selected through several rounds of screening—either visually or by filter hydridization. Purified recombinant virus, free of wild-type virus, can then be used to study the expression of the receptor in *Spodoptera* cells employing reagents that have been previously defined in studies of mammalian cells, in particular the A431 human vulval carcinoma cell line.

*Materials*

Grace's insect cell culture medium (product 041-01590), gentamycin (product 043-05750 D), and amphotericin B (Fungizone, product 043-05290) are obtained from GIBCO (Grand Island, NY). Tryptose phosphate broth is from Oxoid (product CM 283), Sea Plaque agarose from FMC Bioproducts (Rockland, MD), and TC yeastolate (product 5577-15-5) and TC lactalbumin hydrolysate (product 5996-01) are from Difco Laboratories. Fluorescein-conjugation goat anti-mouse IgG is from Cappel (Turnhout, Belgium) and 1.4-diazobicyclo [2.2.2]octane from Sigma (St. Louis, MO). Pluronic F68 is from DSF Wyandotte (Parsittany, NJ). *Spodoptera frugiperda* cells (Sf9) are from the American Type Culture Collection (Rockville, MD) (20852-1776, Accession Number CRL1711). Transfer vector pAc 373 is the gift of Max Summers of the Department of Entomology, Texas A&M University College Station, TX).

## Preparation of Tissue Culture Medium

*TNM-FH Medium.* Grace's insect cell culture medium is converted to TNM-FH by the addition of TC yeastolate (3.3 g/liter) and TC lactalbumin hydrolysate, (3.3 g/liter). The solution is then filter sterilized using a 0.2-$\mu$m filter and incubated at 37° to check sterility before it is stored at 4°. For complete growth medium, 10% fetal calf serum together with antibiotics (50 mg/ml gentamycin, 2.5 mg/ml amphotericin B) is added before use as required.

## Insect Cell Culture Technique

*Spodoptera frugiperda* cells have a doubling time of 18–24 hr in TNM-FH medium containing 10% fetal calf serum. Optimal growth at 27° in the absence of $CO_2$ is obtained by subculturing the cells two to three times a week. For transfections or plaque assays Sf9 cells are propagated in flasks at 27° in complete TNM-FH medium supplemented with 10% fetal calf serum without antibiotics. For routine culture the following seeding densities are tabulated below:

| Type vessel | Cell density | Minimum virus volume | Final volume |
|---|---|---|---|
| 96-well plate | $2.0 \times 10^4$/well | 10 μl | 100 μl |
| 24-well plate | $3.0 \times 10^5$/well | 200 μl | 500 μl |
| 60-mm² dish | $2.5 \times 10^6$/dish | 1 ml | 3 ml |
| 25-cm² flask | $3.0 \times 10^6$/dish | 1 ml | 5 ml |
| 75-cm² flask | $9.0 \times 10^6$/flask | 2 ml | 10 ml |
| 150-cm² flask | $1.8 \times 10^7$/flask | 4 ml | 20 ml |
| Spinner flasks | $1.0 \times 10^5$/ml | Based on multiplicity of infection | Based on size |

*Culture in Flasks.* Cells are seeded at the appropriate density (see above) and allowed to attach to the plastic; the medium is then replaced with fresh medium. When nearing confluence, the cells are detached from the plastic by knocking the flasks fairly vigorously with the palm of the hand. The contents are then subdivided among four fresh flasks of equal dimensions.

*Culture in Suspension.* Cells are generally propagated in 1-liter Techne stirrers in water vath vessels (Techne MCS104L and MWB-10L, Princeton, NJ). For a 200-ml culture, the cells from about five 75-cm² confluent flasks are harvested and then centrifuged using a TECHNOSPIN R (Sorvall instruments, Du Pont, Wilmington, DE) at 800 rpm for 10 min at room temperature. The cell pellet is then resuspended in fresh medium and the cell density measured using a hemacytometer. The size of the inoculum is adjusted so that the final cell density of Sf9 cells seeded into spinner flasks is about $2–4 \times 10^5$ cells/ml. The culture is propagated at 27° with constant stirring at 70 rpm until the cells reach a density of $2–4 \times 10^6$ when they are infected with virus.

*Infection of Sf9 Cells.* Cells are seeded into flasks at the appropriate density (see above) and allowed to attach for 1 hr. The medium is then removed and the appropriate amount of virus added. After a 1-hr incubation the virus inoculum is removed and replaced with fresh medium. The cells or medium are then harvested 2–4 days postinfection. For cells grown in suspension culture, the required amount of virus stock is added to the total suspension culture to achieve a multiplicity of infection of 5–10.

*Frozen cell stocks:* Cells are taken for freezing in the logarithmic phase of growth. They are detached from the plastic of the tissue culture flask and centrifuged at 800 rpm for 10 min at room temperature. The cell pellet is suspended in fresh medium to achieve a final cell density of $2 \times 10^7$ cells/ml. The cell suspension is kept on ice and to it is added, drop by drop, an equal volume of previously chilled new growth medium containing 15% (w/v) dimethysulfoxide and 20% (v/v) fetal calf serum with 100 mg/ml gentamycin and 5 mg/ml amphotericin B (Fungizone). Aliquots (1 ml) of the resulting cell suspension are dispensed into cryostat tubes and placed in a $-80°$ freezer overnight. The next day the tubes are immersed in liquid nitrogen for longer term storage.

## Preparation of Wild-Type Viral DNA

*Spodoptera frugiperda* cells in flasks or suspension culture (100–1000 ml of culture medium) are infected with wild-type baculovirus. At 36–48 hr postinfection, the infected cells are removed by pelleting at 2500 rpm for 10 min and the supernatant transferred to ultracentrifuge tubes. The extracellular virus (ECV) is then pelleted by spinning for 30 min at 100,000 *g* (24,000 rpm in a Beckman SW-27 rotor). The medium is then decanted and the tube inverted on paper towels for a few minutes. Any residual medium is wiped from the sides of the tube. The ECV is then resuspended in 4.5 ml of extraction buffer (12.1 g/liter Tris, 33.6 g/liter $Na_2EDTA \cdot 2H_2O$, 14.9 g/liter KCl, pH 7.5) and digested for 2 hr at 50° with proteinase K (40 mg/liter), after which 0.5 ml of 10% Sarkosyl is added and the incubation continued overnight. The next day, a phenol/chloroform/isoamyl alcohol (25 : 24 : 1) extraction is performed three times on the solution containing the ECV, taking care to adequately mix the two phases while avoiding vigorous shaking, as the viral DNA (AcMNPV) is relatively sensitive to shearing. The DNA-containing aqueous phase is then transferred to another tube and the DNA precipitated by the addition of 10 ml of absolute alcohol. After storing overnight at $-20°$, the DNA is pelleted by low-speed centrifugation (2500 rpm for 20 min) and then briefly washed in 90% ethanol before being resuspended in 500 ml of $0.1\times$ TE. To aid solubilization of the DNA, the sample is heated for about 15 min at 65°. The optical density is then measured at 260 nm, and the sample stored at 4°.

## Preparation of Transfer Vectors

One of the main considerations in constructing an expression vector is whether to fuse a foreign gene sequence in the vector before or after the polyhedrin translational start codon, ATG, where the A is nucleotide +1. Since we wanted to express a mature protein rather than a fusion product, a transfer vector with a cloning site(s) upstream (in the 5′ direction) from the polyhedrin translation initiator codon was chosen. This was the transfer vector pAc373, in which artificial *Bam*HI linkers have been inserted at a point 8 bases before the first ATG of the polyhedrin transcription start

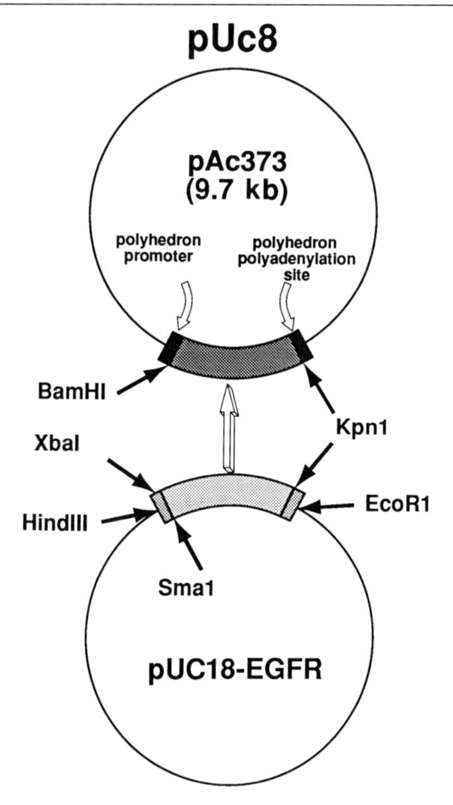


FIG. 1. Construction of pAc373–EGFR. The coding sequence of the polyhedron protein was isolated from pAc373 using a *Bam*HI–*Kpn* digest, leaving the polyhedron promoter (−8 bp) and polyadenylation sequences intact. The left and right arms of this vector are homologous to the flanking sequences of the polyhedron gene in wild-type baculovirus, thus allowing recombination to take place between the transfer vector and wild-type virus at a later stage. The cDNA encoding the full-length human EGFR was excised from pUC18–EGFR using an *Xba*I–*Kpn*I digest and transferred to the modified pAc373. The resulting construct contained all but 8 bp of the polyhedron 5′ untranslated leader sequence joined to 20 nucleotides from the polylinker of pUC18 which was linked to 17 nucleotides of the EGFR 5′ noncoding sequence.

site.[10] This vector can be used to insert a foreign gene sequence such that the first ATG is the initiator ATG of the foreign protein. Plasmid pUC18-EGFR contains the full-length cDNA of the human EGF receptor cloned into the polylinker of pUC18 (see Fig. 1). To construct the recombinant baculovirus vAc-EFGR, the segment of DNA corresponding to the full-

[10] G. E. Smith, M. D. Summers, and M. J. Fraser, *Mol. Cell. Biol.* **3,** 2156 (1983).

length EGF receptor (constructed in our laboratory from three overlapping partial clones) is removed from pUC18–EGFR by first digesting with *Xba*I, filling in with the Klenow fragment of DNA polymerase I, and then digesting with *Kpn*I.[11] This segment is then transferred into the baculovirus polyhedrin plasmid pAc 373, which had been digested with *Bam*HI, filled in with Klenow polymerase, and then digested with *Kpn*I. The resulting construct pAc373–EGFR therefore contained all but eight nucleotides of the polyhedrin 5′ untranslated leader sequence fused to 20 bp from the polylinker of pUC18 followed by 17 bp of receptor 5′ noncoding sequences. At least 10 mg of highly purified pAc373–EGFR plasmid DNA was prepared. Impure preparations of plasmid DNA are toxic to *Spodoptera frugiperda* cells, with many cells undergoing lysis shortly after transfection. This results in a lower recombination frequency and increased difficulty in detecting recombinant viruses. To test the quality of a plasmid DNA preparation, a control with AcMNPV DNA alone is included in all transfection experiments and the viability of these cells is compared to those transfected with AcMNPV DNA plus plasmid DNA.

Preparation of Recombinant

The plasmid pAc373–EGFR was cotransfected with wild-type AcNPV DNA into Sf9 cells.[10,12] *In vivo* recombination events yielded a small proportion of viruses with the EGF receptor cDNA incorporated into the original viral genome. Recombinant baculoviruses were identified in infected cell monolayers by DNA hybridization using the cDNA of the EGF receptor as a probe. The recombinant virus was then plaque purified by screening for the occlusion-negative phenotype.[9]

To produce a recombinant virus, *S. frugiperda* cells are seeded into 25-$cm^2$ flasks at a density of $2.5 \times 10^6$ cells/flask and are allowed to attach for 1 hr. The medium is then removed and replaced with 0.75 ml of Grace's medium with 10% (Fetal calf serum (FCS) and antibiotics. Meanwhile, 1 mg AcMNPV viral DNA and 2 mg plasmid DNA are added to 0.75 ml of filter-sterilized transfection buffer (25 m*M* HEPES, pH 7.1, 140 m*M* NaCl, 125 m*M* $CaCl_2$) and the mixture briefly vortexed. The DNA solution is then added dropwise to the Grace's solution already in the cell culture flasks. The cells are incubated for 4 hr at 27°, after which the medium is removed from each flask and replaced with TNM-FH containing 10% FCS

[11] A. Ullrich, L., Coussens, J. S. Mayflick, T. J. Dull, A. Gray, A. W. Tam, J. Lee, J. Yarden, T. A. Libermann, J. Schlessinger, J. Downward, E. L. Mayes, N. Whittle, M. D. Waterfield, and P. H. Seeberg, *Nature* (*London*) **309,** 418 (1984).

[12] C. Miyamato, G. E. Smith, J. Farrell-Towt, R. Chizzmite, M. D. Summers, and J. Ju, *Mol. Cell. Biol.* **5,** 2860 (1985).

and antibiotics. This procedure is repeated four times to remove any traces of calcium phosphate precipitate. The cells are then incubated for 4–6 days and observed for signs of infection, which include the appearance of polyhedra, an increase in the cell size, and cell lysis. When most of the cells appear to be infected, the medium is removed and cleared of cells and debri by low-speed centrifugation. The virus (ECV in the medium) is then serially diluted into fresh medium to obtain 2-ml aliquots of $10^{-3}$, $10^{-4}$, and $10^{-5}$ dilutions. These dilutions of virus are then used for the isolation and purification of the recombinant virus.

## Detection of Recombinant Virus Using Plaque Hybridization Assay

There are two ways to detect recombinant viruses: visual screening for plaques formed by occlusion-negative viruses or the use of filter hybridization. Although visual screening can be the most rapid approach, the ease with which this can be carried out depends on many factors. For example, it is very difficult, without previous experience, to discriminate between plaques formed from occlusion-negative and from occlusion-positive (wild-type) viruses. In our experience it is preferable to initially detect recombinants using a plaque hybridization procedure and then later, after enrichment of the recombinant virus, to select for the occlusion-negative plaques by eye. The plaque hybridization procedure is similar to colony hybridization techniques and is as follows: Sf9 cells (97–98% viable in logarithmic growth phase) are seeded into 60-mm culture plates at a density of $2.5 \times 10^6$ viable cells/ml and allowed to attach for 1 hr. (The cells must be subconfluent as they need to grow and divide during the course of the plaque assay.) After the cells attach, the medium is removed and 1 ml of virus inoculum is added to each plate. The inoculum is either derived from transfection mixes diluted to the concentration of approximately $10^{-3}$ to $10^{-5}$, or viruses picked from a plaque and diluted to a concentration of $10^{-1}$ and $10^{-2}$. Samples of each diluted stock (1 ml) are stored at 4° until the nature of the recombinant virus has been determined. While the plates are incubating at room temperature, 50 ml of 1.5% low melting point agarose is prepared (1.5 g agarose in 50 ml $H_2O$), autoclaved, and allowed to cool to 37° before being added to 50 ml 2× TNM-FH containing 20% fetal calf serum plus 100 mg/ml gentamicin and 5 mg/ml amphotericin at 37°. These media are maintained at 37–40° in a water bath until ready for use. After a 1-hr incubation period the virus inoculum is completely removed from each plate. Slowly, 4 ml of overlay is added to the edge of each plate and the plate rocked to spread the agarose evenly. The plates are then left for at least 1 hr to allow the overlay to solidify before being placed in a humidified incubator at 27°. After 3–4 days plaques are usually beginning to form and at this stage the plates are allowed to dry overnight by leaving the plates in a nonhumid environment ready for plaque hybridization.

For the hybridization procedure, four blotting solutions are prepared:

Solution A: 0.05 *M* Tris-HCl (pH 7.4), 0.15 *M* NaCl
Solution B: 0.5 *N* NaOH, 1.5 *M* NaCl
Solution C: 1.0 *M* Tris-HCl (pH 7.4), 1.5 *M* NaCl
Solution D: 0.3 *M* NaCl, 0.03*M* sodium citrate

The agarose and plates are marked for orientation and the edges of the agarose loosened away from the plate and removed to sterile Petri dishes for subsequent storage. A dry nitrocellulose 47-mm filter (Whatman 7184004) is then placed on top of the cells remaining in each plate. A drop of solution A is added to secure the filter to the plate and the orientation markings from the plate are traced onto the nitrocellulose filter. Whatman 3MM paper (47-mm circles) is saturated with solution A and placed on top of the nitrocellulose. The filters are firmly applied to the tissue culture plates, moving from the center outward, with a large rubber stopper so as to bring the nitrocellulose filter in direct contact with the cells and to force out any air pockets. After a minute, the filters are carefully removed and placed cell side up for 3 min on a Whatman filter saturated with solution B. The filters are then dried on paper towels before they are transferred to a Whatman filter saturated with solution C. After 3 min the filters are again dried on paper towels before washing them in solution D. When all of the filters have been prepared in this manner, they are baked for 2 hr at 80° under vacuum. The filters are then hybridized with a receptor cDNA probe, washed under stringent conditions, and autoradiographed. Suitable controls are included (wild-type infected and uninfected cells) to indicate the specificity of the hybridization. After developing the autoradiogram of the hybridized filters, the positive areas are aligned to the original agarose layer (to determine which location to pick) and a fairly large area of agarose (several millimeters in diameter) picked. This plug is then placed in 1 ml of medium in preparation for the next round of plaque purification.

### Detection of Recombinant Virus Using Visual Plaque-Screening Assay

Recombinant viruses obtained through the filter hybridization assay are further plaque purified by screening for the occlusion-negative phenotype as described in detail by Summers and Smith in their manual of methods for baculovirus culture.[9]

### Determination of Virus Titer Using a Plaque Assay

Ten-fold dilutions of the virus stock from $10^{-1}$ to $10^{-9}$ are prepared in complete cell culture medium and these dilutions are used to infect, in duplicate, Sf9 cells. Following infection, a plaque assay is performed. After 4–6 days postinfection the number of plaques on each plate is counted, optimally around 100 plaques/dish, and the viral titer calculated.

### Detection of Expression of Recombinant Protein

Purified recombinant virus is used to infect monolayers of cells adherent to poly(L-lysine) (Sigma)-treated glass coverslips. At 24, 36, 48, 60, and 72 hr postinfection the coverslips are washed in PBSA (phosphate buffered saline consisting of 140 m*M* NaCl, 4 m*M* KCl, 1.7 m*M* $KH_2PO_4$, 8 m*M* $Na_2HPO_4$, pH 7.4) containing 10% FCS and then treated for 15 min with 5% formaldehyde with or without 0.2% Triton X-100. The cells are then washed twice in PBSA/10% FCS before a 1 : 100 dilution of monoclonal antibody in PBSA/10% FCS is added and cells incubated for 30 min at 37°. The presence of the EGF receptor is detected using immunofluorescence detected by monoclonal antibodies R1 and F4 (20 mg/ml),[13,14] which recognize, respectively, the external and internal domains of the receptor. Binding is detected by a fluorescein-conjugated goat anti-mouse IgG. The coverslips are mounted in a 10% solution of 1,4-diazobicyclo[2.2.2]octane in glycerol (50%) and examined under UV light. An example of such analysis is shown in Fig. 2.

### Purification of Epidermal Growth Factor and Transforming Growth Factor α

Mouse epidermal growth factor (EGF) is prepared from submaxillary glands obtained from adult male mice (Parkes strain, NIMR), as described by Savage and Cohen.[15] Human TGFα and EGF are recombinant products and are the gift of H. Gregory ICI Pharmaceuticals. Mouse EGF and recombinant human TGFα and EGF undergo final purification by high-performance liquid chromatography (HPLC).

Cells are removed from the dish by washing in Eagle's buffered salt solution with 25 m*M* HEPES, 0.5% BSA, pH 7.5 (binding buffer). Aliquots ($4 \times 10^5$ cells) are incubated in a final volume of 500 μl with $^{125}$I-labeled

[13] W. J. Gullick, J. J. Marsden, N. Whittle, B. Ward, L. Bowbrow, and M. D. Watefield, *Cancer Res.* **46,** 285 (1986).

[14] W. J. Gullick, J. Downward, P. J. Parker, N. Whittle, R. Kris, J. Schlessinger, A. Ullrich, and M. D. Waterfield, *Proc. R. Soc. London, Ser. B* **226,** 127 (1985).

[15] C. R. Savage, Jr. and S. Cohen, *J. Biol. Chem.* **247,** 7609 (1972).

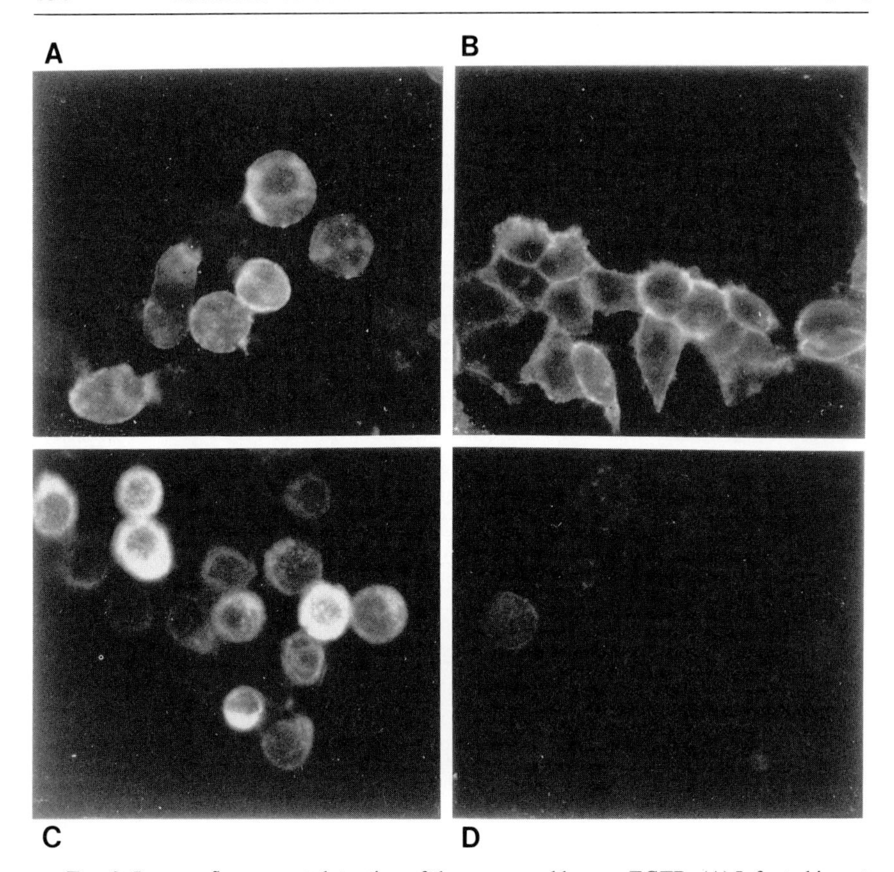


FIG. 2. Immunofluorescent detection of the expressed human EGFR. (A) Infected insect cells and (B) A431 cells were treated with MAb R1 (which recognizes the external domain[19]) and then with goat anti-mouse fluorescein-labeled IgG. (C) and (D) Infected insect cells were permeabilized with Triton and then treated with MAb F4 (which recognizes the kinase domain[13,14]) followed by goat anti-mouse fluorescein-labeled IgG. Competing peptide antigen was included in the treated cells (D).

EGF (120 pg) and increasing concentrations of unlabeled EGF (0–100 ng) for 4 hr at 4°. The cells are then centrifuged at 4° and washed twice in binding buffer before being counted in an $\gamma$ counter. A431 cells are included as a control.[16] The data are corrected for nonspecific binding (always <10% total counts) and analyzed by the method of Scatchard as shown in Fig. 3.[17]

[16] D. Barnes, *in* "Methods in Enzymology" (D. Barnes and D. A. Sirbasku, eds.), Vol. 146, p. 88. Academic Press, San Diego, California, 1987.

[17] G. Scatchard, *Ann. N. Y. Acad. Sci.* **51,** 660 (1949).

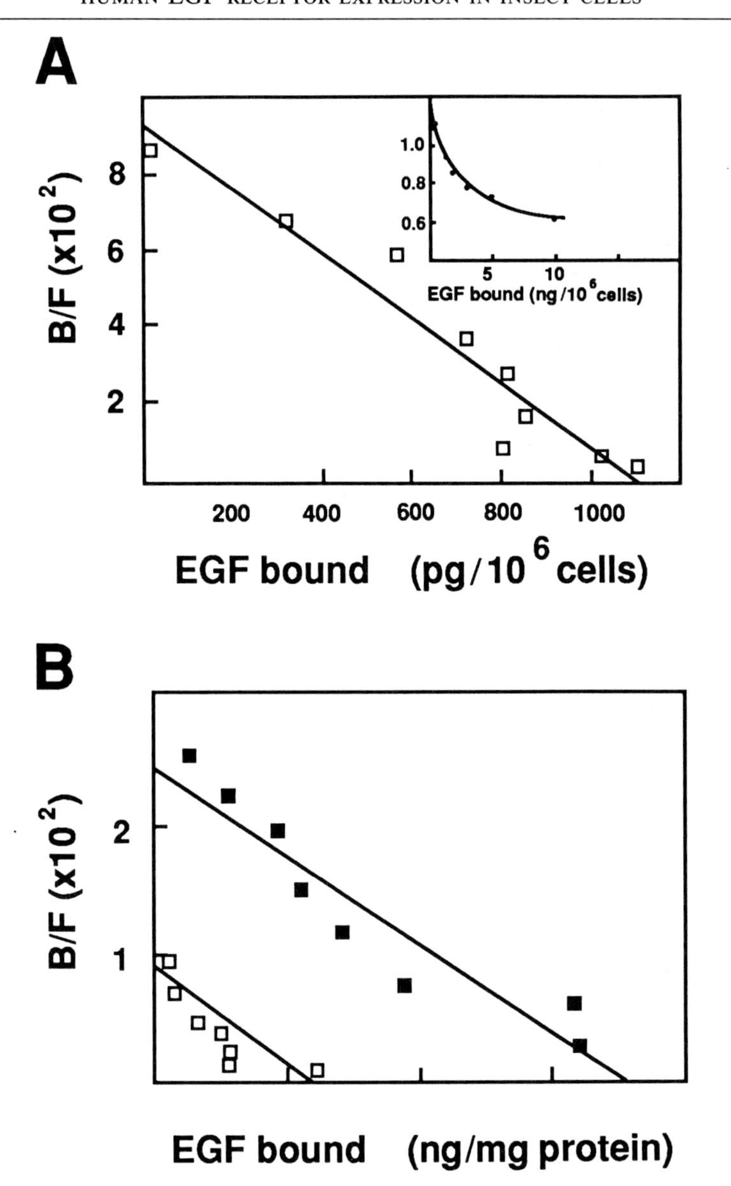


FIG. 3. EGF binding to human receptor. [125]I-labeled EGF and increasing concentrations of unlabeled EGF were used to derive Scatchard plots of EGF binding to insect and A431 cells. (A) Intact infected insect and (inset) intact A431 cells. (B) Solubilized infected inset (□) and A431 cells (■).

### Epidermal Growth Factor Binding to Solubilized Insect Cells

The cells are solubilized in lysis buffer [100 m*M* HEPES, pH 7.4, 5 m*M* EGTA, 150 m*M* NaCl, 0.1% (w/v) BSA, 1% (v/v) Nonidet P-40, to which 0.1% phenylmethylsulfonyl fluoride (PMSF, 50 mg/ml), 25 m*M* benzamidine, leupeptin (25 $\mu$g/ml), and aprotonin (10 $\mu$g/ml) are added] and cleared at 10,000 *g* for 10 min at 4°. A final reaction volume of 40 ml contains 40–100 mg of cell lysate, 120 pg $^{125}$I-labeled EGF, and increasing concentrations of unlabeled EGF (2.5–100 ng) and is incubated for 1 hr at 24°. At this time 10 ml EGF-$R_1$ antibody (0.2 mg/ml) is added for 10 min at 4° followed by 10 ml 0.24% $\gamma$-globulin and 50 ml polyethylene glycol 8000 (PEG) for a further 10 min at 4°. The insoluble material is pelleted and washed in 25% PEG. The data are collected and analyzed[18] (see Fig. 4).

### Immunoblotting

Proteins are electroblotted on nitrocellulose ("Hybond C"; Amersham, Bucks, England). Nonspecific binding sites are blocked using 5% Marvel. The blot is then incubated for 4 hr at 20° with anti-15E antiserum, then extensively washed (seven times) in PBSA/Tween 20 (0.05%) and the bound antibody is then detected using iodinated protein A (Amersham) ($5 \times 10^4$ cpm/ml).[13,14,19]

### Immunoprecipitation

Infected Sf9 cells ($10^7$) are washed with PBSA, the cell pellet is solubilized in lysis buffer, and the receptor protein immunoprecipitated with antibodies Mab $R_1$, anti-15E, or anti-14E.[13,14,19] Antisera 15E and 14E are used with or without competing peptide antigen as a control. Analysis of isotopically labeled cells is shown in Fig. 5.

### Isotopic Labeling

Sf9 cells ($2.5 \times 10^6$) are plated onto 6-cm dishes and infected with the recombinant virus. Thirty hours later the medium is replaced with medium deficient in methionine supplemented with 10% dialyzed FCS. After 2 hr [$^{35}$S]methionine (>800 Ci/mmol; Amersham) is added to a final concentra-

[18] W. Gullick, J. H. Downward, J. J. Marsden, and M. D. Waterfield, *Anal. Biochem.* **141,** 253 (1984).

[19] M. D. Waterfield, E. L. V. Mayes, P. Stroobant, P. L. P. Bennett, S. Young, P. N. Goodfellow, G. S. Bunting, and B. Ozanne, *J. Cell. Biochem.* **20,** 149 (1982).

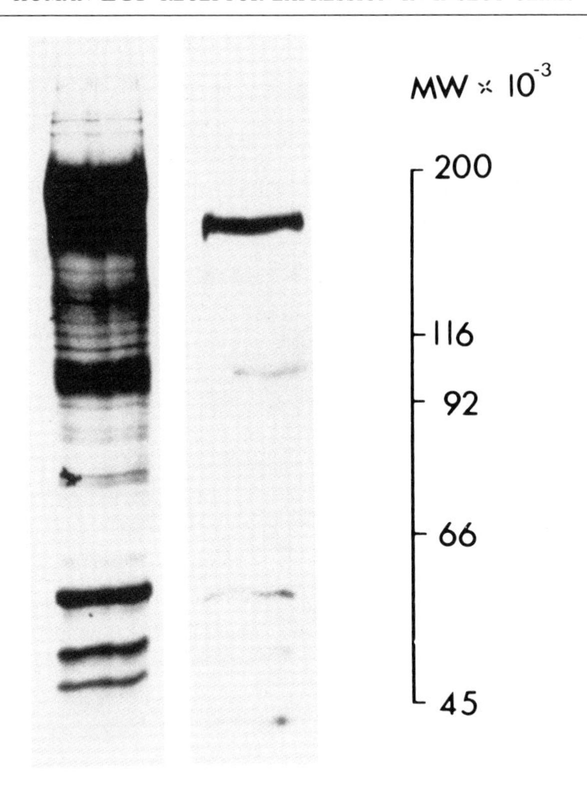


FIG. 4. Immunoprecipitation analysis of human EGFR expression. A431 and infected insect cells were labeled with [$^{35}$S]methionine and immunoprecipitated with MAb R1.

tion of 500 $\mu$Ci/ml. Thirty minutes later the cells are washed and recultured in complete medium. The cells are lysed 48 hr postinfection. As a control study, A431 cells are labeled with [$^{35}$S]methionine as described by Mayes and Waterfield (see Fig. 5).[20]

## Autophosphorylation

Lysates of A431 or Sf9 cells ($10^7$ cells/500 $\mu$l of lysate) are incubated at 4° with or without EGF ($10^{-7}$ *M*) for 10 min. The EGF receptor is then immunoprecipiated as above, but after washing the pellet is suspended in 50 $\mu$l of phosphorylation buffer (10 $\mu M$ ATP, 2 m*M* $Mn^{2+}$, 12 m*M* $Mg^{2+}$, 100 m*M* sodium orthovanadate, 40 m*M* HEPES, pH 7.4, 5% glycerol, 0.2% Triton X-100, 150 m*M* NaCl) containing [$\gamma$-$^{32}$P]ATP (5000 cpm/

[20] E. L. V. Mayes and M. D. Waterfield, *EMBO J.* **3,** 531 (1984).

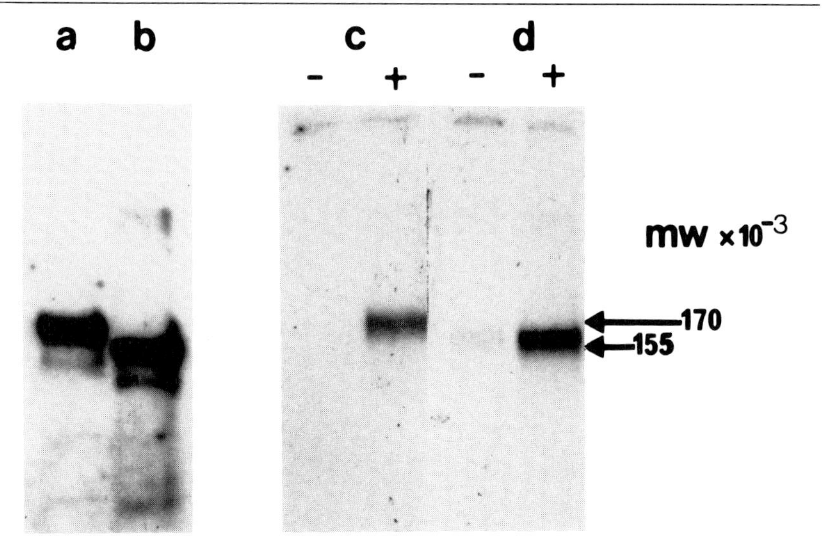


FIG. 5. Analysis of autophosphorylation activity of human EGFR. (a) Aliquots of A431 and infected insect cell lysates were immunoblotted using anti-15E antiserum and equal aliquots were immunoprecipitated with (+) and without EGF (−), autophosphorylation was carried out, and samples run on SDS-PAGE were autoradiographed. Bands were quantified by scanning densitometry.

pmol). After 10 min at 4° the reaction is stopped by the addition of sample buffer. The samples are then analyzed on 7% SDS-PAGE gels, which are then dried and exposed to Kodak (Rochester, NY) X-AR5 film at −70° using intensifying screens (see Fig. 6).

## Comments on the Characterization of Human Epidermal Growth Factor Receptor in Insect Cells

The expression of the human EGF receptor in insect cells can be best detected initially by immunofluorescence and immunoprecipitation studies of infected cells labeled with [$^{35}$S]methionine using monoclonal anti-receptor antibodies such as MAb EGF-R1, which is available commercially (Amersham) (see Figs. 2 and 4.)[6] The expressed receptor can be detected in intact cells by immunofluorescence using the external domain MAb EGFR as a probe in conjunction with a fluorescent tagged second antibody.[6] It is useful to analyze A431 cells which express high levels of receptor as a control.[6,7,18,19] Results of immunoprecipitation analysis are shown in Fig. 4, where the receptor is detected as a 170-kDa protein in A431 cells and a 155-kDa protein in insect cells. Treatment of such

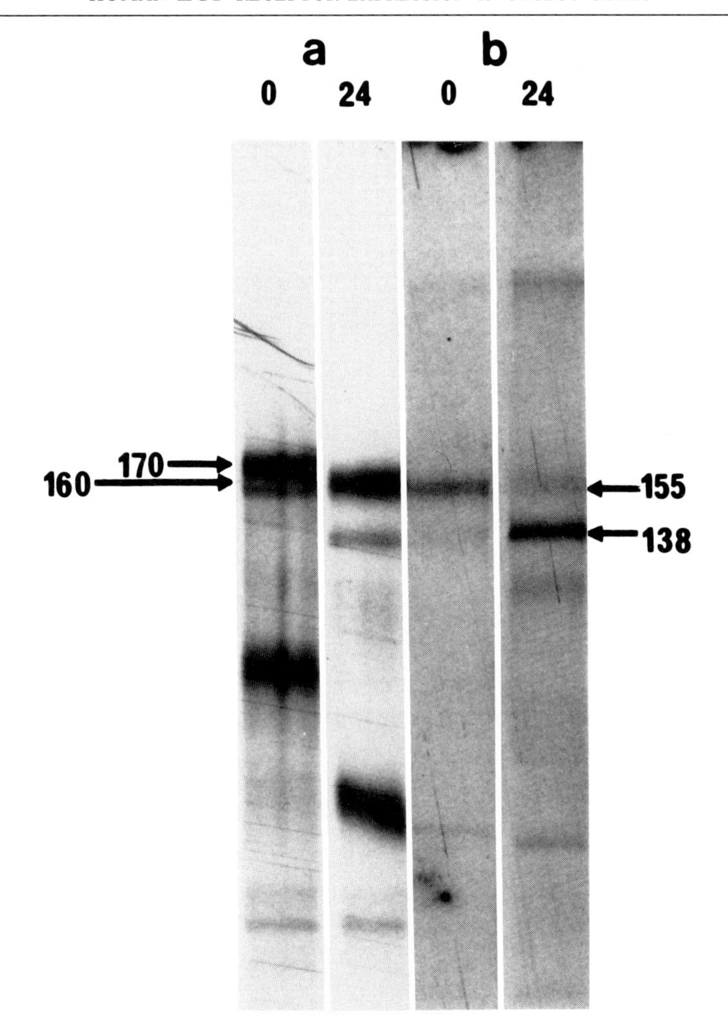


FIG. 6. Analysis of endoglycosidase sensitivity of the insect-expressed human EGFR. Immunoprecipitates with MAb R1 of infected insect and A431 cells were treated for 24 hr with 5 mU of endoglycosidase H. (a) A431 receptor and (b) infected insect cells after a 0- and 24-hr digestion. Molecular weight ($\times 10^{-3}$).

immunoprecipitates with endoglycosidase H shows the difference in molecular weight to be due to differential glycosylation since the mature glycoproteins from both cell types were convertable to partially glycosylated receptor precursors of 138 kDa following enzyme treatement (see Fig. 6). Analysis of receptors expressed in monensin-treated cells supports the concept that only high mannose-type glycosaccharides are present on the insect cell-expressed receptor. Processing to generate complex carbohydrate side chains takes place in A431 cells.[6] After permeabilization of the plasma membrane with Triton X-100 detergent, antibodies which recognize the cytoplasmic domain reveal large amounts of internal receptor in addition to that seen at the cell surface (see Fig. 3). The levels of receptor expressed can be quantified by analysis of ligand binding using EGF or TGF$\alpha$. With A431 cells as controls, Scatchard analysis of ligand-binding data[17] (see Fig. 3) usually reveals two classes of binding affinities of approximately $3.7 \times 10^{-10}$ and $3.1 \times 10^{-9}$ $M$[6] on intact cells. In detergent lysates of A431 cells only a single affinity class of receptors with a $K_d$ of $2.9 \times 10^{-7}$ $M$ are detectable. Similar analysis of intact insect cells shows only a single class of binding affinity, $2 \times 10^{-9}$ $M$, which was reduced to $2 \times 10^{-7}$ $M$ upon solubilization in detergent. It is not clear why the receptor shows a single class of binding sites in insect cells.

The insect cell-expressed human receptor possessed a ligand-stimulatable tyrosine kinase activity which can be detected by autophosphorylation using the techniques described above (see Fig. 5). Detailed phosphopeptide analysis has shown that identical sites in the receptor are phosphorylated to those which are used in A431 cells.[6] Analysis of the degree to which the kinase is ligand stimulatable in the insect cells shows that at late times (3 days) postinfection the receptor posseses an increasing level of ligand independent kinase activity. The precise reason for this is unclear but similar effects have been reported in studies of baculovirus-expressed PDGF receptor.[21]

### *Purification of the Epidermal Growth Factor Receptor*

The purification of human EGF receptor from insect cells with an active tyrosine kinase cannot be accomplished using monoclonal antibody columns because the conditions required to elute receptor from the columns inactivate the kinase. Consequently EGF affinity columns offer the best method to achieve purification. Suitable methods have been described by Gill and Weber.[22]

### *Large-Scale Production of Receptor in Insect Cells*

The availability of serum-free medium at a reasonable cost (Excel from J.R. Scientific, Woodland, CA) has made it possible to process 400-ml batches of insect cells in 1-liter spinner flasks using magnetic stirrers and

[21] L. Williams, personal communication, 1990.

[22] G. N. Gill and W. Weber, *in* "Methods in Enzymology" (D. Barnes and D. A. Sirbasku, eds.), Vol. 146, p. 82. Academic Press, San Diego, California, 1987.

simple water baths. It is also possible to use 2-liter plastic roller bottles containing 500 ml of medium in conventional roller machines employing a rotation speed of 3 rpm. Further scale up is problematic and requires very careful optimization of the stirring speed, oxygenation requirements, infection conditions (such as MOI and cell density), and of the harvesting time for each type of vessel or bioreactor. Although some laboratories have managed to grow cells in 30-liter vessels with overhead stirrers gassed with oxygen, this cannot easily be achieved without a great number of frustrating failures. At present, large-scale production can best be accomplished by setting up large numbers of 1-liter stirred vessels. In our own laboratory we can easily process 12 1-liter Techne vessels/week. The insect system has also been used to produce milligrams of secreted external domain of the human EGF receptor.[7]

# [53] Production of p60$^{c\text{-}src}$ by Baculovirus Expression and Immunoaffinity Purification

*By* DAVID O. MORGAN, JOSHUA M. KAPLAN, J. MICHAEL BISHOP, and HAROLD E. VARMUS

Many features of the structure and function of an enzyme can be explored only if large quantities of the native protein are available in highly purified form. The availability of a pure enzyme makes it possible to pursue X-ray crystallographic studies of protein structure, develop specific antibodies to native epitopes, and perform detailed analyses of interactions between the enzyme and its substrates, regulators, and other proteins.

This chapter describes an effective method for the large-scale production of p60$^{c\text{-}src}$, a 60-kDa protein-tyrosine kinase associated with the inner face of cellular membranes.[1] A brief account of this method has been published previously.[2] In this chapter we emphasize general considerations that should be useful for applying this technique to the study of any protein kinase or other enzyme. The description is divided into three parts: (1) expression, (2) purification, and (3) protein characterization.

[1] J. A. Cooper, *in* "Peptides and Protein Phosphorylation" (B. Kemp ed.), p. 85. CRC Press, Boca Raton, Florida, 1990.

[2] D. O. Morgan, J. M. Kaplan, M. J. Bishop, and H. E. Varmus, *Cell* (*Cambridge, Mass.*) **57,** 775 (1989).

### Expression of $p60^{c\text{-}src}$ with the Baculovirus System

The purification of a rare regulatory enzyme can be greatly simplified if a rich source of the protein is developed by overexpression in a heterologous system. In our studies of $p60^{c\text{-}src}$, the baculovirus system[3] has been particularly useful for this purpose. This approach employs the *Autographa californica* nuclear polyhedrosis virus (AcMNPV), a baculovirus that can be used to infect *Spodoptera frugiperda* (Sf9) insect cells in culture. Late in the lytic virus life cycle, the viral polyhedrin protein is expressed at extremely high levels (over 50% of cell protein), resulting in the formation of crystalline viral occlusions in the infected cell. Polyhedrin synthesis is not essential for infection or replication of the virus; thus, it is possible to replace the polyhedrin coding sequence in the viral genome with a foreign cDNA. Construction of this recombinant baculovirus simply involves cotransfection of insect cells with baculovirus genomic DNA and a transfer vector, a bacterial plasmid containing the foreign cDNA flanked by the polyhedrin promoter and other polyhedrin gene-flanking sequences. The foreign gene is transferred to the viral genome by homologous recombination. Wild-type and recombinant viruses are produced by the transfected cells and are distinguished by their plaque morphology on cell monolayers: wild-type plaques contain readily visible polyhedrin occlusions, while recombinant plaques are occlusion negative.

The baculovirus system frequently provides very high levels of recombinant protein expression. In addition (unlike expression in *Escherichia coli*), expression in the higher eukaryotic insect cell generally results in normal folding and posttranslational processing of the foreign protein. It should be noted that certain proteins are not highly expressed in this system, perhaps due to their intrinsic activities or due to problems in their posttranslational processing; thus, high-level expression with this system is not guaranteed. Nevertheless, many proteins (including several protein kinases[2,4–8]) have been successfully overexpressed from recombinant baculovirus vectors.

[3] V. A. Luckow and M. D. Summers, *Bio/Technology* **6,** 47 (1988).

[4] L. Ellis, A. Levitan, M. H. Cobb, and P. Ramos, *J. Virol.* **62,** 1634 (1988).

[5] R. Herrera, D. Lebwohl, A. Garcia de Herreros, R. G. Kallen, and O. M. Rosen, *J. Biol. Chem.* **263,** 5560 (1988).

[6] C. Greenfield, G. Patel, S. Clark, N. Jones, and M. D. Waterfield, *EMBO J.* **7,** 139 (1988).

[7] A. M. Pendergast, R. Clark, E. S. Kawasaki, F. P. McCormick, and O. N. Witte, *Oncogene* **4,** 759 (1989).

[8] H. Piwnica-Worms, N. G. Williams, S. H. Cheng, and T. M. Roberts, *J. Virol.* **64,** 61 (1990).

*Transfer Vector Construction*

Baculovirus transfer vectors are bacterial plasmids (often pUC8) containing the AcMNPV polyhedrin gene and flanking sequences.[3] In most cases, large regions of the polyhedrin coding sequence have been deleted. The desired foreign gene is inserted at cloning sites either a short distance downstream of the polyhedrin start codon (to produce a fused protein) or upstream (to produce a nonfused protein). Although fused proteins are generally expressed at higher levels than nonfused proteins, the analysis of protein structure and function is best performed with the normal nonfused protein.

To express p60$^{c\text{-}src}$ as a nonfused protein, the chicken c-*src* cDNA was inserted in the pVL941 transfer vector. This vector was developed during studies[9] suggesting that sequences surrounding the polyhedrin start codon may be partly responsible for high levels of polyhedrin expression. In pVL941, the polyhedrin start codon has been mutated to ATT, and the foreign gene is inserted at a *Bam*HI site about 35 nucleotides downstream. Thus, sequences proximal to the polyhedrin start codon are retained, but translation is initiated at the first start codon in the inserted foreign cDNA. The usefulness of the pVL941 transfer vector has been enhanced by the development of new vectors (pVL1392, pVL1393) that contain multiple cloning sites instead of the single *Bam*HI site.

To further optimize expression levels with nonfused transfer vectors, it is useful to reduce the size of the noncoding 5′ leader sequence of the inserted foreign cDNA. For this reason, the c-*src* cDNA inserted in pVL941 contained minimal noncoding sequences. A *Bgl*II fragment encompassing the chicken c-*src* coding region (originally derived from plasmid p5H; a gift of Hanafusa[10]) was obtained using a *Bgl*II site about 20 bp downstream of the termination codon, plus a *Bgl*II site generated by linker insertion 35 bp upsteam of the start codon. This 1.65-kb fragment was inserted in the *Bam*HI site of pVL941 to generate the pVL–c-*src* baculovirus transfer vector.

Baculovirus vectors expressing several mutant forms of p60$^{c\text{-}src}$ were also developed by replacing fragments of the c-*src* cDNA with previously constructed fragments bearing various mutations. These mutants include the following:

1. p60$^{c\text{-}srcM}$, a kinase-deficient mutant with a lysine-295 to methionine

[9] V. A. Luckow and M. D. Summers, *Virology* **170,** 31 (1989).

[10] J. B. Levy, H. Iba, and H. Hanafusa, *Proc. Natl. Acad. Sci. U.S.A.* **83,** 4228 (1986).

mutation in the ATP-binding site of the kinase domain (constructed with a mutant v-*src* fragment[11])

2. p60$^{c\text{-}src\text{A}}$, an activated mutant in which tyrosine-527 is changed to phenylalanine (constructed with a fragment provided by Shalloway[12]). In fibroblasts, the kinase activity of wild-type p60$^{c\text{-}src}$ is inhibited severalfold by constitutive phosphorylation at this site (probably by a separate kinase); thus, mutation of this site activates p60$^{c\text{-}src}$ in fibroblasts. A second mutant protein, p60$^{c\text{-}src\text{AM}}$, is a double mutant with methionine at position 295

3. p60$^{c\text{-}src\text{SD}}$, a mutant lacking N-terminal myristylation due to replacement of glycine-2 with alanine (constructed from a mutant v-*src* fragment provided by Kamps and Sefton[13]). A double mutant containing methionine-295, p60$^{c\text{-}src\text{SDM}}$, was also constructed

4. p60$^{c\text{-}src\text{N}}$, the neural-specific form of p60$^{c\text{-}src}$ that contains a six-amino acid insertion (constructed with a chicken neural c-*src* cDNA provided by Brugge[14])

### *Preparation of Recombinant Baculovirus*

Generation of the recombinant baculovirus involves a variety of techniques that are thoroughly described in a laboratory manual written by Summers and Smith.[15] This section emphasizes those aspects of our techniques that differ from the standard approach.

*Insect Cell Culture.* Sf9 insect cells are grown in Grace's insect cell medium (GIBCO, Grand Island, NY) supplemented with 10% fetal calf serum plus yeastolate and lactalbumin hydrolysate (each 3.3 g/liter). Cell stocks are grown in humidified air at 28° in stirred suspension cultures. To grow these cells to high density, aeration must be maximized by using minimal volumes: we generally use 150 ml in a standard 1-liter suspension culture flask, or about 2–3 cm depth. Every 2 days the culture is split by discarding 75% and adding fresh medium; every few weeks the medium is completely replaced. Under these conditions the cells are maintained at densities between $1 \times 10^6$ and $4 \times 10^6$ cells/ml. After 2 or 3 months, cells may exhibit signs of ill health (slower growth, poor recombinant protein production, etc.); in this case a new culture is started from stocks in liquid nitrogen.

[11] M. A. Snyder, M. J. Bishop, J. P. McGrath, and A. D. Levinson, *Mol. Cell. Biol.* **5,** 1772 (1987).

[12] T. E. Kmiecik and D. Shalloway, *Cell (Cambridge, Mass.)* **49,** 64 (1987).

[13] M. P. Kamps, J. E. Buss, and B. M. Sefton, *Proc. Natl. Acad. Sci. U.S.A.* **82,** 4625 (1985).

[14] J. B. Levy, T. Dorai, L.-H. Wang, and J. S. Brugge, *Mol. Cell. Biol.* **7,** 4142 (1987).

[15] M. D. Summers and G. E. Smith, "A Manual of Methods for Baculovirus Vectors and Insect Cell Culture Procedures," Tex. Agric. Exp. Stn. Bull. No. 1555. Texas A & M Agric. Coll., College Station, 1987.

For transfections and small-scale infections, small amounts of cells are removed from the stock suspension culture and plated for 1 hr on tissue culture dishes. It is essential that these and all other infection experiments be performed with cells that are completely viable and growing rapidly (i.e., split the day before).

*Transfection.* Freshly plated cell monolayers in 25-cm$^2$ flasks are transfected by calcium phosphate precipitation of the transfer vector (2 $\mu$g, twice banded on cesium chloride gradients) and baculovirus genomic DNA (1 $\mu$g, method 1 of Summers and Smith[15]). We have found that total nucleic acids prepared from infected cells are an adequate and convenient source of baculovirus genomic DNA for transfections. Roughly 25% of the total nucleic acids prepared from infected cells is baculovirus DNA, and so 4 $\mu$g of this preparation can be used to provide 1 $\mu$g baculovirus DNA for transfections. Four to 5 days after transfection, most cells contain polyhedra, and the culture medium contains wild-type viruses and a small proportion of recombinant viruses that can be isolated in plaque assays.

*Isolation of Recombinant Plaques.* Cell monolayers on 60-mm dishes (LUX, Nunc, Naperville, IL) are infected with 1 ml diluted medium from transfected cells as described,[15] and then overlaid with 1% agarose in complete medium. Plaques become apparent 5–6 days later, and are mainly wild-type plaques containing polyhedral occlusions barely visible to the naked eye. Less than 1% of the plaques are recombinants lacking polyhedral occlusions. Visualization of occlusion-negative plaques is greatly simplified if the dish is overlaid with 1.5 ml 1% agarose containing 0.04% Trypan Blue in complete medium. After an overnight incubation with the stain, plaques are readily apparent (especially under low-power brightfield microscopy) as clusters of blue cell ghosts surrounded by clear viable cells. Wild-type plaques are identified by their dark polyhedral occlusions; recombinants are completely free of polyhedra, even when analyzed under high-power phase-contrast microscopy. In some cases, recombinant plaques are considerably smaller than wild-type plaques, perhaps due to toxic effects of the expressed foreign protein (see below). Generally, four to six occlusion-negative plaques are chosen for each transfer vector construct. Agarose plugs containing these plaques are picked with a 200-$\mu$l pipettor and placed in 1 ml complete medium. Two more rounds of plaque assays are required to obtain pure recombinant plaques.

### *Characterization of Recombinant Protein Production*

A significant fraction (10–20%) of occlusion-negative viruses do not express the foreign protein. Thus, several occlusion-negative viruses are plaque purified for each transfer vector construct. Once these viruses are prepared, it is then necessary to identify recombinant viruses that provide high levels of foreign protein expression.

Several straightforward techniques can be employed to measure the amount of foreign protein being expressed in the infected insect cell. The simplest approach, which has worked well in our studies of $p60^{c\text{-}src}$, is to run crude lysates of infected cells on SDS–polyacrylamide gels and then stain with Coomassie Blue. If the protein is expressed at high levels (and is not the same size as a major insect cell protein), then a distinct band in the expected molecular weight range should be apparent. This band should be absent from lysates of cells infected with a negative control virus that does not express the protein. This approach is not only very simple but also allows a quick estimate of the absolute amount of protein expressed in the infected cell. If the protein is not expressed at such high levels, but if specific antibodies to the protein are available, then it is possible to analyze levels of protein expression by immunoblotting of cell lysates or immunoprecipitation of $[^{35}S]$methionine-labeled cell lysates.

Typically, the initial screen of protein expression is performed as follows. Viable, log-phase cells are plated in wells of six-well culture dishes at $1 \times 10^6$ cells/well. After 1 hr, the medium is removed and 200 $\mu$l of first passage viral stock (plus 300 $\mu$l medium) is added. As in all infections where maximal protein production is desired, the multiplicity of infection in these experiments is roughly 10 plaque-forming units (pfu)/insect cell. After a 1-hr infection, an additional 2 ml complete medium is added and the cells are incubated at 28° for 48 hr. Cells are then washed once with phosphate-buffered saline and lysed on ice with 300 $\mu$l lysis buffer [50 m*M* HEPES pH 7.5, 1% Triton X-100, 5 m*M* EDTA, 1 m*M* phenylmethylsulfonyl fluoride (PMSF), and 1 $\mu$g/ml leupeptin]. After 10 min, lysates are transferred to 1.5-ml microfuge tubes for centrifugation (10 min, 12,000 rpm, 4°). Thirty to 40 $\mu$l of these lysates is then analyzed by SDS–polyacrylamide gel electrophoresis and Coomassie Blue staining. In addition, smaller aliquots of these lysates are analyzed by immunoblotting with monoclonal antibody (MAb) 327[16] (Promega Protoblot Western blot AP system, Promega, Madison, WI). Under these mild lysis conditions, about 80% of the $p60^{c\text{-}src}$ in the infected cell is solubilized, and the kinase activity of the protein is maintained (see below).

Coomassie Blue-stained lysates of cells infected with various VL–c-*src* vectors are shown in Fig. 1. Wild-type and mutant forms of the protein are all expressed at high levels (~0.5 to 2% of the solubilized protein in these lysates), although different forms of the protein are consistently expressed at different levels. Interestingly, the level of expression may be

[16] L. A. Lipsich, A. J. Lewis, and J. S. Brugge, *J. Virol.* **48,** 352 (1983).

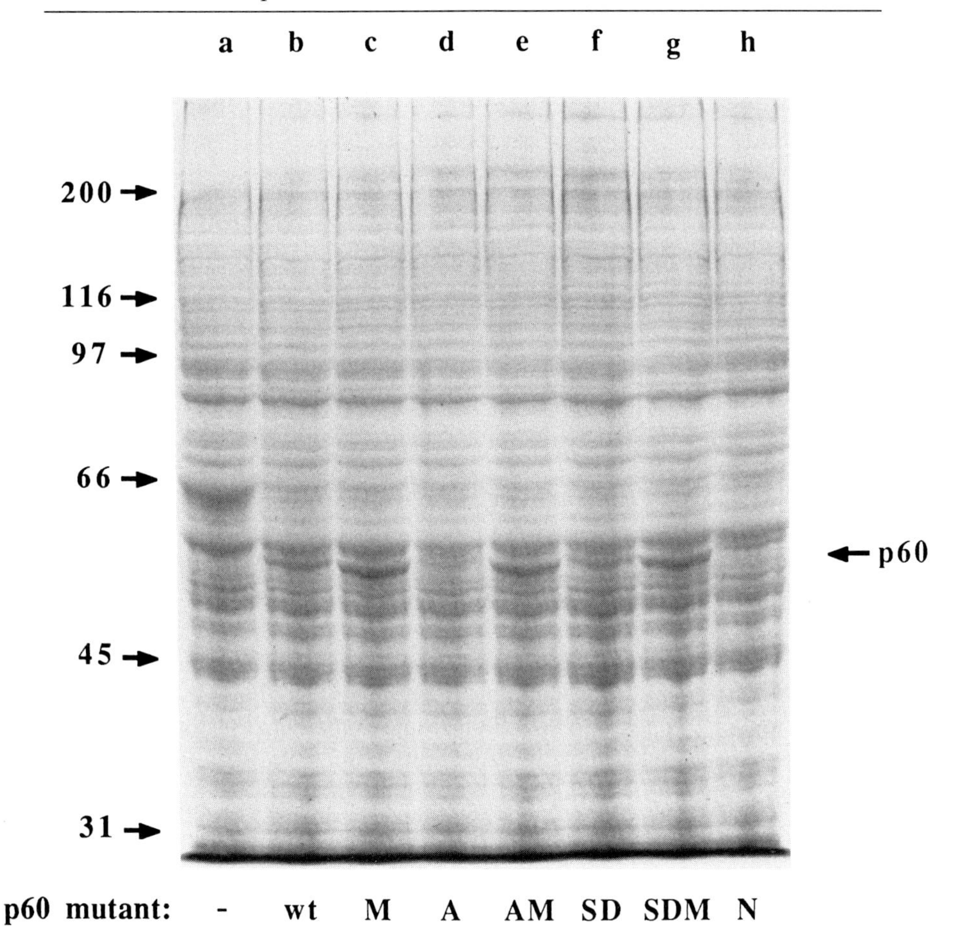


FIG. 1. Detergent lysates of insect cells infected with various recombinant baculoviruses. Lysates prepared from insect cells infected with various baculovirus vectors were analyzed by electrophoresis on a 7.5% polyacrylamide gel, followed by Coomassie Blue staining (as described in the text). The baculoviruses encode the following proteins (p60$^{c\text{-}src}$ mutants are explained in the text): (a) no p60$^{c\text{-}src}$ protein is encoded by this virus (a negative control); (b) wild-type p60$^{c\text{-}src}$; (c)p60$^{c\text{-}srcM}$; (d) p60$^{c\text{-}srcA}$; (e) p60$^{c\text{-}srcAM}$; (f) p60$^{c\text{-}srcSD}$; (g) p60$^{c\text{-}srcSDM}$; (h) p60$^{c\text{-}srcN}$. Positions of molecular weight markers ($\times 10^{-3}$) are indicated.

affected by the kinase activity of the p60$^{c\text{-}src}$ protein. Kinase-deficient mutants bearing the lysine-295 mutation in the ATP-binding site are expressed at two- to threefold higher levels than is the wild-type protein. It is thus possible that the tyrosine kinase activity of p60$^{c\text{-}src}$ somehow leads to restricted expression of the protein. This potential toxicity of the active p60$^{c\text{-}src}$ kinase may have implications for those interested in overexpressing any active enzyme in this system.

## Purification of p60$^{c\text{-}src}$ by Immunoaffinity Chromatography

The production of insect cell lysates containing large amounts of a foreign protein makes it possible to develop purification procedures that provide pure protein in a few simple steps. If 1–2% of insect cell protein is the desired foreign protein (as is the case for p60$^{c\text{-}src}$), then extensive purification of the protein should be possible in a few standard chromatographic steps. Standard chromatography is particularly effective if the protein is sensitive to the extreme conditions that may be required for elution from immunoaffinity columns. On the other hand, if an immunoaffinity or other affinity method can be used without harm to the protein, then purification becomes very simple. Fortunately, in our studies of p60$^{c\text{-}src}$, an immunoaffinity method has been found that has little effect on the kinase activity of the protein and provides essentially homogeneous p60$^{c\text{-}src}$ protein in a simple, one-step procedure.

Immunoaffinity chromatography is a well-established technique that has been useful in the purification of many proteins, including several protein tyrosine kinases (e.g., the receptors for insulin and insulin-like growth factor I[17,18]). A thorough discussion of the concepts and methods involved in the development of an immunoaffinity approach is provided elsewhere,[19] and so the present discussion will focus on methods that we have used for the purification of p60$^{c\text{-}src}$. Many aspects of this procedure will be useful in the purification of other protein kinases, while certain aspects (e.g., column washing and elution conditions) may not apply in approaches using other antibodies and antigens.

### *Column Construction*

Our approach employs columns containing MAb 327, a monoclonal antibody that binds with moderately high affinity and high specificity to an amino-terminal epitope on p60$^{c\text{-}src}$ from several species.[16] For large-scale purification (up to 10 mg of p60$^{c\text{-}src}$), a column was developed by binding 80 mg of protein A-purified MAb 327 to 10 ml protein A-Sepharose. The antibody was then covalently coupled to the protein A with dimethyl

[17] R. A. Roth, D. O. Morgan, J. Beaudoin, and V. Sara, *J. Biol. Chem.* **261,** 3753 (1986).

[18] D. O. Morgan, K. Jarnagin, and R. A. Roth, *Biochemistry* **25,** 5560 (1986).

[19] E. Harlow and D. Lane, "Antibodies: A Laboratory Manual." Cold Spring Harbor Lab., Cold Spring Harbor, New York, 1988.

pimelimidate by the procedure of Schneider *et al.*[20] This method, unlike methods where the antibody is simply coupled to an activated support (e.g., Bio-Rad Affi-Gel), provides maximum binding capacity, since the antibody on the column is theoretically oriented with all of its antigen binding sites accessible to the antigen. Prior to use, the new column should be treated with the same washing and elution conditions that will be used in the purification, since uncoupled antibody may be present. Columns are stored between use at 4° in phosphate-buffered saline containing 0.02% sodium azide: under these conditions, the same column can be used 10 to 20 times over a period of several months without significant loss of capacity.

*Large-Scale Insect Cell Infection*

For the preparation of 5–10 mg pure p60$^{c\text{-}src}$, Sf9 cell cultures are scaled up as follows. Standard 150-ml cultures (described earlier) are grown to high density in a well-aerated 1-liter suspension flask (3–4 × $10^6$ cells/ml). These cells are transferred to a standard 2.8-liter Fernbach flask [Corning (Corning, NY) No. 4420], and 700 ml medium is added. The culture is then incubated at 28° and stirred at 80 rpm with a Teflon floating stir bar (Nalgene, Rochester, NY, 92 × 30 mm). About 2 days later, when the culture reaches a density of 2 × $10^6$ cells/ml, the cells are collected by centrifugation in 250-ml bottles (10 min, 1200 rpm, Sorvall GSA rotor) and gently resuspended in 100 ml viral stock (multiplicity of infection should be 5–10 pfu/cell). After a 1-hr infection at 28°, 800 ml medium is added to the cells (including 400 ml used medium saved from the original culture). The cells are then stirred for 48 hr at 28°, and then harvested for p60$^{c\text{-}src}$ purification.

*Purification of p60$^{c\text{-}src}$*

The general approach to p60$^{c\text{-}src}$ purification begins with overnight incubation of the column with the cell lysate, followed by extensive washes in high salt and a chaotropic agent (potassium thiocyanate, KSCN). The purified protein is then eluted in an alkaline buffer. KSCN wash and alkaline elution conditions are based on conditions originally designed by J. Brugge and co-workers (personal communication, 1987).

The following procedure is the result of various preliminary studies in which conditions were refined to provide the maximum yield of active, soluble, and highly purified p60$^{c\text{-}src}$ protein. We have found that extensive washing is crucial for removing nonspecifically bound proteins from the

[20] C. Schneider, R. A. Newman, D. R. Sutherland, U. Asser, and M. F. Greaves, *J. Biol. Chem.* **257,** 10766 (1982).

column; omission of any wash step results in contamination of the final preparation. Other components that are crucial include the 10% glycerol in all solutions; this is essential for maintenance of kinase activity. Maintenance of kinase activity and protein solubility also requires the use of sodium cholate and dithiothreitol in the elution buffer.

*Solutions*

Lysis: 50 m*M* HEPES, pH 7.4, 1% (w/v) Triton X-100, 5 m*M* EDTA, 1 m*M* PMSF, 1 μg/ml leupeptin

Salt wash: 50 m*M* HEPES, pH 7.4, 1% (w/v) sodium cholate, 500 m*M* NaCl, 10% (v/v) glycerol

KSCN wash: 1 *M* potassium thiocyanate (KSCN), pH 8.0, 1% (w/v) sodium cholate, 10% (v/v) glycerol

Elution: 10 m*M* diethylamine, pH 11.0, 1% (w/v) sodium cholate, 150 m*M* NaCl, 10% (v/v) glycerol, 1 m*M* dithiothreitol (DTT)

*Method.* The 900-ml culture of infected cells is centrifuged and washed by gentle resuspension in 100 ml phosphate-buffered saline (infected cells are fragile). After a second centrifugation, the cells are thoroughly resuspended in 200 ml lysis buffer on ice. All subsequent procedures are performed at 4°. After 20 min with frequent swirling, the lysate is centrifuged (20,000 *g*, 30 min, Sorvall GSA). Glycerol (20 ml) is added to the supernatant, which is then loaded over the MAb 327 column overnight (about 16 hr at 14 ml/hr). The column is then washed with 50 ml lysis buffer, 50 ml salt wash, and 50 ml KSCN wash (all at 30 ml/hr). The column is eluted with 40 ml elution buffer at 10 ml/hr, and 3-ml fractions are collected in polypropylene tubes. Fractions are neutralized immediately after collection by addition of 300 μl of 0.5 *M* Tricine, pH 8.5. Fractions are then analyzed by SDS–polyacrylamide gel electrophoresis (Fig. 2) or by protein assay (Bradford). Peak fractions are pooled, aliquotted, frozen in liquid nitrogen, and stored at −80° until use.

*Protein Solubility.* These studies were initially hampered by problems of $p60^{c\text{-}src}$ solubility. In early experiments, much of the purified $p60^{c\text{-}src}$ eluting from the column precipitated and could not be redissolved. Much of this aggregation could be avoided after consideration of the following factors:

1. Incubation of purified $p60^{c\text{-}src}$ at different pH values (pH 6 to 11) revealed that the protein precipitated mainly at pH values below 7.5. Isoelectric focusing of the purified protein (S. Basu, unpublished results) indicated an isoelectric point of about pH 5.5–6.0. Thus, to avoid the aggregation that apparently results from incubation near the isoelectric point, fractions are neutralized to pH 8.5.

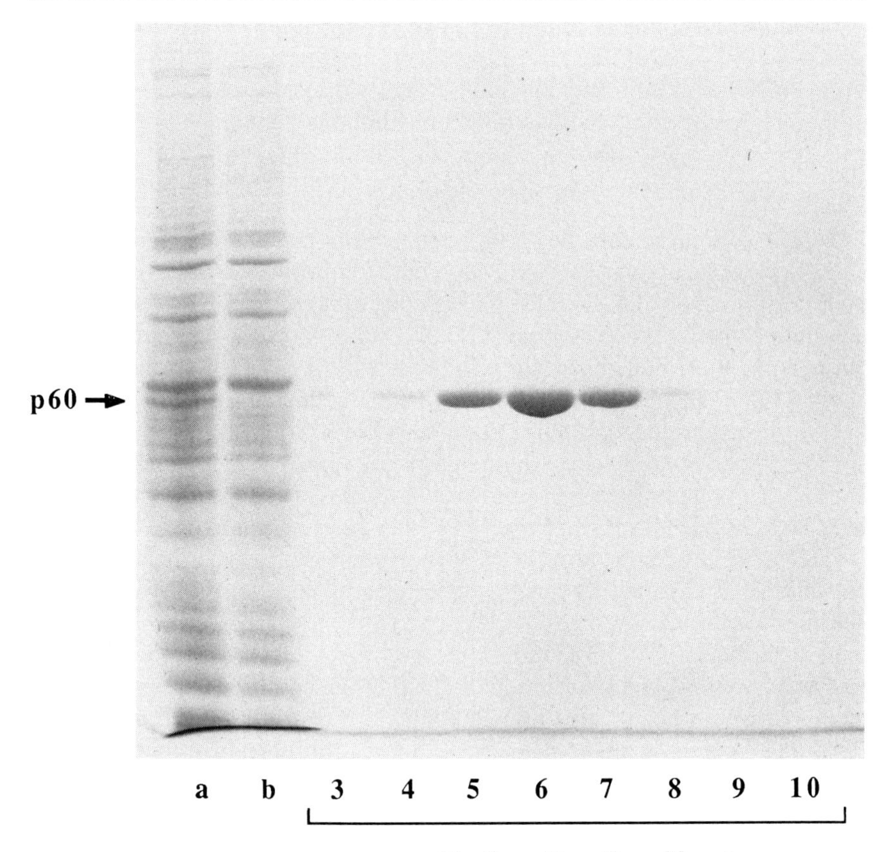


FIG. 2. Immunoaffinity purification of p60$^{c\text{-}src}$ from insect cell lysates. A large-scale preparation of p60$^{c\text{-}src}$ was carried out as described in the text. Samples of the original infected cell lysate (20 $\mu$l, lane a), the lysate after passage over the immunoaffinity column (20 $\mu$l, lane b), and fractions from the final alkaline elution (10 $\mu$l of each) were analyzed on a 7.5% polyacrylamide gel (stained with Coomassie Blue).

2. Some aggregation of p60$^{c\text{-}src}$ is probably due to disulfide cross-linking, since the inclusion of a reducing agent (DTT) in the elution buffer also decreased precipitation.

3. Initially, eluted fractions were collected in tubes already containing the neutralizing buffer. However, much less p60$^{c\text{-}src}$ precipitation occurred if the neutralizing buffer was added after the fraction was completely collected. It is possible that protein aggregates form when the first drops of purified p60$^{c\text{-}src}$ at pH 11 mix with the concentrated neutralizing buffer in the tube.

4. Precipitation was found to occur mainly at p60$^{c\text{-}src}$ concentrations over 1 mg/ml, a level that is sometimes reached in peak elution fractions. To avoid precipitation, the protein is slightly diluted (to about 0.5 mg/ml) by pooling peak fractions with adjacent fractions containing less protein.

### *Purification of p60$^{c\text{-}src}$ from Other Sources*

The above procedure has also been used to purify p60$^{c\text{-}src}$ from other sources: (1) human platelets[2] and (2) NIH 3T3 fibroblasts that overexpress p60$^{c\text{-}src}$ from a retroviral vector (these cells express about 20-fold greater amounts of p60$^{c\text{-}src}$ than normal NIH 3T3 fibroblasts; H. Hirai and H. E. Varmus, unpublished observations). These cell types express considerably less p60$^{c\text{-}src}$ than baculovirus-infected insect cells (roughly 0.05–0.1% of detergent-soluble cell protein). As a result, p60$^{c\text{-}src}$ prepared from these sources by the single-step immunoaffinity procedure contains minor amounts of contaminating protein. Nevertheless, roughly 90% of the protein in these preparations is p60$^{c\text{-}src}$. This is certainly adequate for many purposes. If homogeneous p60$^{c\text{-}src}$ from these sources is desired, then an additional chromatographic step (e.g., gel filtration) can be added. Additional purification steps also reduce p60$^{c\text{-}src}$ degradation by the abundant proteases that are found in platelet lysates (D. Feder and J. M. Bishop, unpublished observations). Despite the additional steps required to prepare pure p60$^{c\text{-}src}$ from these sources, they do have certain advantages as a source of the protein. For example, fibroblasts are particularly useful if a p60$^{c\text{-}src}$ protein is desired that is phosphorylated normally at Tyr-527. Similarly, if cells are synchronized in mitosis prior to lysis, it is possible to purify p60$^{c\text{-}src}$ that is phosphorylated at mitotic phosphorylation sites.

## Characterization of Protein Structure and Function

In most cases it is important to characterize certain aspects of the structure and function of a protein prepared by baculovirus expression and affinity purification. Clearly, the extent of this analysis depends on the purpose for which the protein is being produced. For example, extensive analysis may not be necessary if the protein is simply being used as a general purpose immunogen. For most purposes, however, it is worthwhile to carry out a few basic studies of the antigenic structure, posttranslational modifications, and functional activities of the protein.

The techniques used in these studies are standard biochemical methods, and so the following is mainly a general discussion of our approach in the analysis of the p60$^{c\text{-}src}$ protein in the infected insect cell and after purification. A more detailed analysis of p60$^{c\text{-}src}$ structure in insect cells has been presented elsewhere.[8]

*Protein Structure*

Various approaches can be taken to confirm that the structure of a foreign protein in insect cells is equivalent to that in higher cells. Specific antibodies should recognize the foreign protein from insect cells. The tertiary structure of the protein can be crudely assessed by testing the binding of antibodies that recognize conformation-specific epitopes on the protein. Subunit structure, if any, can be revealed by standard protein sizing methods. Finally, if the protein is known to undergo posttranslational modifications in its normal enviroment, then the extent of these modifications can be analyzed in insect cells by standard cell labeling and immunoprecipitation techniques.

Although the insect cell provides the environment necessary for the normal translation and folding of proteins from higher eukaryotic cell types, it does not always contain adequate machinery for normal posttranslational processing. Posttranslational modification (e.g., phosphorylation) of a foreign protein in the insect cell will depend on several factors, including (1) the presence of the required modifying enzyme (e.g., kinase) in the insect cell, (2) the ability of the insect cell modifying enzyme to recognize a foreign protein from a different species, (3) the activity or concentration of the insect cell enzyme during the late stages of baculovirus infection, when cell division is arrested, synthesis of host proteins is inhibited, and cell metabolism is directed almost entirely to viral production, and (4) the capacity of the insect cell enzyme to modify the extremely high levels of foreign substrate that are present in the insect cell.

The major form of posttranslational modification in p60$^{c\text{-}src}$ is phosphorylation (reviewed in Ref. 1). In interphase mammalian fibroblasts, p60$^{c\text{-}src}$ is mainly phosphorylated at Ser-17 (probably by cAMP-dependent kinase) and Tyr-527 (probably by a separate, unknown kinase). Autophosphorylation *in vivo* (at Tyr-416) is usually minimal (due to the inhibitory effects of phosphorylation at Tyr-527).[21] Under some conditions, other residues are also phosphorylated: for example, Ser-12 is phosphorylated by C kinase, and a group of N-terminal sites is phosphorylated during mitosis (probably by p34$^{cdc2}$-associated kinase activity[2,22]).

We were primarily interested in measuring the state of Tyr-527 phosphorylation in the insect cell, since we planned to use the protein as a

[21] T. Hunter, *Cell (Cambridge, Mass.)* **49,** 1 (1987).

[22] S. Shenoy, J.-K. Choi, S. Bagrodia, T. D. Copeland, J. L. Maller, and D. Shalloway, *Cell (Cambridge, Mass.)* **57,** 761 (1989).

substrate to pursue the kinase that phosphorylates this residue. Protein phosphorylation was analyzed by labeling infected insect cell monolayers (40 hr after infection) with ortho[$^{32}$P]phosphate in phosphate-free Grace's medium plus 5% dialyzed fetal calf serum. After 4 hr, cells were washed and lysed (as above) and p60$^{c\text{-}src}$ was immunoprecipitated with MAb327 and cleaved with cyanogen bromide.[23] This treatment yields various distinct C-terminal fragments, including one that contains Tyr-416 and another that contains Tyr-527. This analysis revealed that p60$^{c\text{-}src}$ from insect cells, unlike that from chicken and mammalian fibroblasts, is mainly phosphorylated at Tyr-416 and relatively poorly (less than 5%) at Tyr-527 (data not shown). Similar results were obtained by Piwnica-Worms *et al.*,[8] who also used mapping of tryptic phosphopeptides from p60$^{c\text{-}src}$ to show that phosphorylation at Ser-17 may also be reduced in the insect cell.

In chicken and mammalian cells, p60$^{c\text{-}src}$ is also modified by N-terminal myristylation. To determine if this modification also occurs in chicken p60$^{c\text{-}src}$ overexpressed in insect cells from a baculovirus vector, infected insect cells were labeled with [$^{3}$H]myristate. Immunoprecipitation of p60$^{c\text{-}src}$ from detergent lysates of these cells indicated that p60$^{c\text{-}src}$ is myristylated, although the exact stoichiometry of the modification has not been calculated (not shown).

### *Protein Function*

It is also important to demonstrate that a foreign protein expressed in insect cells possesses normal functional activities. In the case of p60$^{c\text{-}src}$ and other protein kinases, the protein should contain the appropriate kinase activity. p60$^{c\text{-}src}$ kinase activity is most conveniently measured by incubating the protein in an immune complex with [$\gamma$-$^{32}$P]ATP and an exogenous substrate such as acid-denatured enolase.[24] The rate of enolase phosphorylation, which increases in a linear fashion under the conditions used, yields an estimate of p60$^{c\text{-}src}$ kinase activity.

The kinase domain of p60$^{c\text{-}src}$expressed in the insect cell appears to behave normally. The protein possesses tyrosine-specific kinase activity toward enolase and other substrates such as poly(Glu : Tyr), and autophosphorylates *in vitro* at Tyr-416 (as judged by tryptic phosphopeptide mapping, not shown). The enolase kinase activity of wild-type p60$^{c\text{-}src}$ immunoprecipitated from insect cells is significantly higher than that of wild-type p60$^{c\text{-}src}$ from mammalian fibroblasts, possibly because the protein from insect cells is poorly phosphorylated at the negative regulatory site Tyr-527. Nevertheless, the activity of p60$^{c\text{-}src}$ in the insect cell is not

[23] R. Jove, S. Kornbluth, and H. Hanafusa, *Cell (Cambridge, Mass.)* **50,** 937 (1987).

[24] J. A. Cooper, F. S. Esch, S. S. Taylor, and T. Hunter, *J. Biol. Chem.* **259,** 7835 (1984).

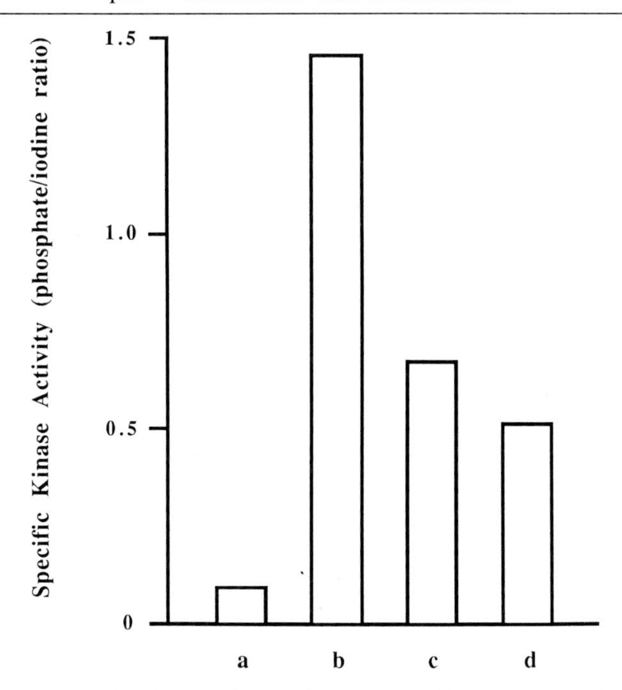


FIG. 3. Kinase activity of p60$^{c\text{-}src}$ from various sources. Similar amounts of p60$^{c\text{-}src}$ (10–50 ng/reaction) were immunoprecipitated from the following sources with MAb 327: (a) a detergent lysate (prepared with the lysis buffer described in the text) of NIH 3T3 cells overexpressing p60$^{c\text{-}src}$, (b) a similar lysate of NIH 3T3 cells overexpressing activated p60$^{c\text{-}src}$ in which Tyr-527 is replaced by Phe, (c) a similar lysate of Sf9 insect cells after a 48-hr infection with the p60$^{c\text{-}src}$ baculovirus, (d) p60$^{c\text{-}src}$ purified from baculovirus-infected insect cells by the method described in the text. Immune complexes were washed three times with lysis buffer and once in kinase buffer (20 m*M* Tris, pH 7.4, 0.1% Triton X-100, 10 m*M* $MgCl_2$). Half of each sample was then analyzed in kinase assays to determine p60$^{c\text{-}src}$ kinase activity, and the other half was analyzed by immunoblotting to determine the amount of p60$^{c\text{-}src}$ protein. For the kinase assay, the immune complex was combined with 30 $\mu$l kinase buffer containing 5 $\mu M$ ATP, 2 $\mu$Ci [$\gamma$-$^{32}$P]ATP (3000 Ci/mmol), and 2 $\mu$g acid-denatured rabbit muscle enolase. After a 1-min reaction at room temperature, reactions were stopped and run on an SDS–polyacrylamide gel. Radiolabeled enolase bands were excised and counted. To determine p60$^{c\text{-}src}$ levels in these reactions, immune complexes prepared in parallel were run on polyacrylamide gels, transferred to nitrocellulose and immunoblotted with MAb 327. These blots were then incubated with $^{125}$I-labeled protein A, and the resulting $^{125}$I-labeled p60 bands were excised and counted. Specific kinase activity was then calculated by determining the ratio of kinase activity ($^{32}$P counts) to p60$^{c\text{-}src}$ quantity ($^{125}$I counts). Values shown represent the average of duplicates.

as high as that of activated (Tyr-527 to Phe) $p60^{c\text{-}src}$ from mammalian fibroblasts (Fig. 3).

A major concern in the purification of $p60^{c\text{-}src}$ is the effect of brief alkaline elution on the structural and functional integrity of the protein. It is clear from several experiments with many different preparations of purified $p60^{c\text{-}src}$ that the enolase kinase activity of $p60^{c\text{-}src}$ is not greatly affected by the purification procedure. In the experiment shown in Fig. 3, the specific kinase activity of a purified $p60^{c\text{-}src}$ preparation was about 25% lower than the activity of $p60^{c\text{-}src}$ freshly immunoprecipitated from an infected insect cell lysate.

In addition to its carboxy-terminal kinase domain, $p60^{c\text{-}src}$ contains amino-terminal regions thought to be involved in the regulation and substrate specificity of the kinase.[1] The function of these regions is unclear and thus not easily measured. In particular, it has not been possible to demonstrate that the alkaline elution procedure has not affected this region of the protein. The only evidence to suggest that this region is intact comes from studies showing that purified cAMP-dependent kinase and $p34^{cdc2}$-associated kinase will phosphorylate purified $p60^{c\text{-}src}$ at the appropriate amino-terminal sites *in vitro*.[2] These experiments may suggest that this region of the purified protein retains certain aspects of its structure.

## Acknowledgments

We thank Jan Kitajewski and Mario Chamorro for advice on insect cell culture techniques, Joan Brugge for generously providing MAb 327, the neural c-*src* cDNA, and advice on $p60^{c\text{-}src}$ purification, Helen Piwnica-Worms and Thomas Roberts for assistance with the baculovirus system, David Shalloway, Mark Kamps, and Bart Sefton for *src* mutants, and Verne Luckow, Max Summers, and Hidesaburo Hanafusa for plasmids.

# [54] Use of Recombinant Baculoviruses and $^1$H Nuclear Magnetic Resonance to Study Tyrosine Phosphorylation by a Soluble Insulin Receptor Protein-Tyrosine Kinase

*By* LELAND ELLIS and BARRY A. LEVINE

## Introduction

The insulin receptor (IR), like a number of transmembrane receptors and transforming proteins, has intrinsic protein-tyrosine kinase (PTK) activity. Autophosphorylation occurs at multiple tyrosine sites, with the rapid phosphorylation of three tyrosines (1158, 1162, and 1163; receptor residues are numbered herein according to Ebina and colleagues) which reside in the kinase homology region of the receptor primary sequence, and two tyrosines (1328 and 1334) in the carboxy-terminal tail. One or more additional tyrosines (965, 972, and 984) in the juxtamembrane region of the cytoplasmic domain may serve as additional sites.[1–3] While autophosphorylation of tyrosine-1158, -1162, and -1163 seems to correlate with an increase in the activity of the enzyme toward exogenous substrates,[1–5] the functional consequence(s) of utilizing three closely spaced tyrosines (including twin tyrosines at 1162/3) for autophosphorylation and activation is not understood at present (i.e., why not use one, or does some fine degree of control of autophosphorylation kinetics derive from having three sites?). Given that no physiologically relevant endogenous substrates for the IR have been characterized, the requirements for a good substrate for this receptor PTK are also not known.

The wild-type IR is a transmembrane glycoprotein, composed of two $\alpha$ subunits (~135 kDa) which bind insulin, and two $\beta$ subunits (~95 kDa) which each span the membrane a single time, and whose cytoplasmic domains include the PTK.[6,7] Despite the size (1355 residues per half-receptor) and overall complexity of this receptor (the extracellular domain contains 41 cysteines and 16 potential N-linked glycosylation sites per half-receptor), its deduced (and subsequently experimentally demonstrated) transmembrane topology is rather simple. Truncation prior to the deduced transmembrane domain results in the efficient secretion of a heterotetrameric soluble ectodomain [each half-receptor is composed of 923 amino acids, including all of the $\alpha$ subunit and the extracellular one-third (approximately) of the $\beta$ subunit] that binds insulin with wild-type affinity,[8–11] while the deduced cytoplasmic domain can be expressed as a soluble,

[1] H. E. Tornqvist, M. W. Pierce, A. R. Frackelton, R. A. Nemenoff, and J. Avruch, *J. Biol. Chem.* **262,** 10212 (1987).

[2] J. M. Tavaré and R. M. Denton, *Biochem. J.* **252,** 607 (1988).

[3] M. F. White, S. E. Shoelson, H. Keutmann, and C. R. Kahn, *J. Biol. Chem.* **263,** 2969 (1988).

[4] O. M. Rosen, R. Herrera, Y. Olowe, L. M. Petruzzelli, and M. H. Cobb, *Proc. Natl. Acad. Sci. U.S.A.* **80,** 3237 (1983).

[5] K.-T. Yu and M. P. Czech, *J. Biol. Chem.* **259,** 5277 (1984).

[6] Y. Ebina, L. Ellis, K. Jarnigan, M. Edery, L. Graf, E. Clauser, J. Ou, F. Masiarz, Y. W. Kan, I. D. Goldfine, R. A. Roth, and W. J. Rutter, *Cell (Cambridge, Mass.)* **40,** 747 (1985).

[7] A. Ullrich, J. R. Bell, R. Y. Chen, R. Herrera, L. M. Petruzzelli, T. J. Dull, A. Gray, L. Coussens, Y.-C. Liao, M. Tsubokawa, A. Mason, P. H. Seeburg, C. Grunfeld, O. M. Rosen, and J. Ramachandran, *Nature (London)* **313,** 756 (1985).

[8] L. Ellis, J. Sissom, and A. Levitan, *J. Mol. Recognition* **1,** 25 (1988).

[9] J. D. Johnson, M. L. Wong, and W. J. Rutter, *Proc. Natl. Acad. Sci. U.S.A.* **85,** 7516 (1988).

[10] J. Whittaker and A. Okamoto, *J. Biol. Chem.* **263,** 3063 (1988).

[11] J. Sissom and L. Ellis, *Biochem. J.* **261,** 119 (1989).

monomeric, and enzymatically active PTK[12–14] [401 amino acids, the cytoplasmic two-thirds (approximately) of the β subunit, referred to herein as the hIR PTK]. The fact that the two functional domains of the IR can be expressed and therefore studied independently as soluble proteins has prompted further studies directed at the overexpression of each domain, with the long-term goal of providing sufficient quantites (tens of milligrams) of each protein for detailed biochemical and biophysical studies.

## Baculovirus Expression

The expression of recombinant proteins in insect Sf9 cells via baculovirus vectors has found widespread application for most classes of proteins (including secreted, transmembrane, cytoplasmic, and nuclear[15]), and the experimental details for each step of the methodology (from vector design to production of recombinant baculovirus to characterization of expressed proteins) has been thoroughly described by Summers and Smith.[16] Following unsuccessful initial attempts to express the hIR PTK in bacteria (which resulted in the inefficient production of an inactive enzyme; unpublished observations), an expression plasmid (pAchIRPTK) was engineered such that an ~1.4-bp cDNA encoding all but the first six amino acids (RKRQPD) of the cytoplasmic domain of the hIR [thus including residue G-959 (bp 3087) to S-1355 (bp 4284), plus 3′ untranslated sequence to bp 4443] was placed under the transcriptional control of the late viral polyhedrin promoter (Fig. 1). Following cotransfection of Sf9 cells, a recombinant baculovirus (designated AchIRPTK) was derived by homologous recombination between the introduced pAchIRPTK plasmid and wild-type viral DNA, as described.[12,16] As so engineered, the AchIRPTK virus includes the complete 5′ untranslated sequence of the polyhedrin gene and utilizes the polyhedrin AUG for initiation of translation, and is followed by three heterologous amino acids (RPM) which derive from the polylinker sequence (thus the amino terminus of the protein is MRPMG. . . ; cf. RKRQPDG. . . for the wild-type receptor).[12] The protein is efficiently expressed in this heterologous cell system as a soluble, monomeric, and

[12] L. Ellis, A. Levitan, M. H. Cobb, and P. Ramos, *J. Virol.* **62,** 1634 (1988).

[13] R. Herrera, D. Lebwohl, A. G. de Herreros, R. G. Kallen, and O. M. Rosen, *J. Biol. Chem.* **263,** 5560 (1988).

[14] M. H. Cobb, B.-C. Sang, R. Gonzalez, E. Goldsmith, and L. Ellis, *J. Biol. Chem.* **264,** 18701 (1989).

[15] V. A. Luckow and M. D. Summers, *Bio/Technology* **6,** 47 (1988).

[16] M. D. Summers and G. E. Smith, "A Manual of Methods for Baculovirus Vectors and Insect Cell Culture Procedures," Tex. Agric. Exp. Stn. Bull. No. 1555. Texas A & M Agric. Coll., College Station, 1987.

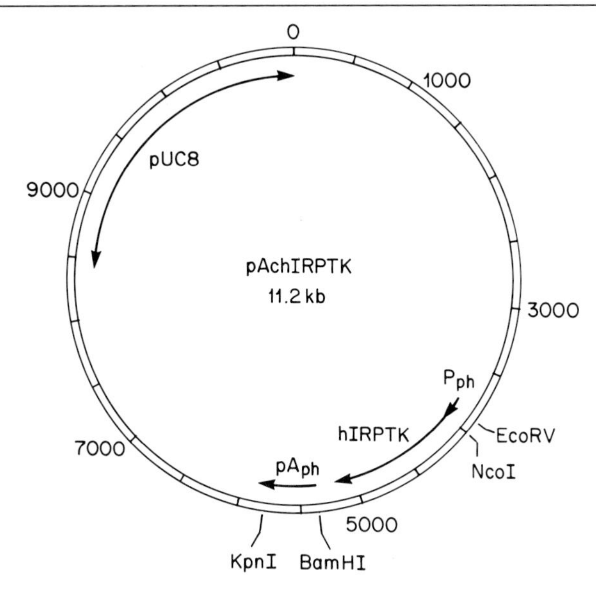


FIG. 1. The 11.2-kb expression plasmid pAchIRPTK. The cDNA encoding the cytoplasmic protein-tyrosine kinase domain of the human insulin receptor (hIRPTK) is placed under the transcriptional control of the polyhedrin promoter ($P_{ph}$). Poly(A) addition signals are provided by 3′ untranslated sequences of the polyhedrin gene ($pA_{ph}$).

stable enzyme (401 amino acids, $M_r$ 45,297), which exhibits a number of the properties of the kinase of the wild-type receptor[12,14,17]: (1) the enzyme reacts with four classes of conformation-sensitive monoclonal antibodies; (2) as isolated from infected cells, the enzyme exhibits a low basal level of kinase activity, which increases during autophosphorylation; (3) two-dimensional tryptic phosphopeptide maps of the autophosphorylated soluble enzyme are entirely typical of those observed for the intact wild-type receptor following insulin stimulation (i.e., at sites corresponding to tyrosine-1158, -1162, -1163, -1328, and -1334 of the native receptor); (4) the enzyme phosphorylates a number of exogenous substrates on tyrosine residues. Furthermore, the scaleup of expression has proved efficient, as 10 to 20 mg of protein is now routinely purified from 800 ml of cells grown in suspension culture by the use of a single chromatographic step [high-resolution Mono Q fast protein liquid chromatography (FPLC)].[14–19] This level of expression and relative ease of purification derives not only from the use of the strong polyhedrin promoter, but also follows from the

[17] J. M. Tavaré, B. Clack, and L. Ellis, *J. Biol. Chem.* **266,** 1390 (1991).
[18] B. A. Levine, B. Clack, and L. Ellis, *J. Biol. Chem.* **266,** 3565 (1991).
[19] L. Ellis, J. M. Tavaré, and B. A. Levine, *Biochem. Soc. Trans.* **19,** 426 (1991).

fact that the host insect cell protein synthesis is largely shut down as a consequence of the late lytic baculoviral infection.[15,16] Thus, this expression system has proved to be very efficient with respect to this domain of the hIR. In contrast, the secretion of the extracellular domain is much less efficient: a high-mannose nonprocessed $\alpha\beta$ precursor accumulates intracellularly, and both precursor and processed receptor are secreted into the medium.[11]

### $^1$H Nuclear Magnetic Resonance of Tyrosine-Containing Peptides

Previous studies have demonstrated that a dodecapeptide based on the sequence of a major site of autophosphorylation of the IR (RRDIYETDY-YRK, designated YYY; the tyrosines at positions 5, 9, and 10 of the peptide correspond to receptor tyrosine-1158, -1162, and -1163, respectively) is phosphorylated by both the wild-type receptor and the hIR PTK, with an apparent $K_m$ of ~100 to 240 $\mu M$.[14,20,21] While an initial study concluded that only the tyrosine at position 9 (Y-9) is phosphorylated,[20] the analysis of phosphorylation reactions both by thin-layer chromatography and autoradiography,[22] and by $^1$H nuclear magnetic resonance (NMR)[18] reveals that phosphorylation of the YYY peptide occurs at multiple tyrosines. Given this multiple stoichiometry, does the phosphorylation reaction proceed in an ordered and progressive manner at successive sites, or at random? Ideally, one would like to be able to monitor such a phosphorylation reaction in real time, with each of the three tyrosines resolved, so that occupation of each site can be directly followed as a function of time. Intuitively, one might imagine attempting to do so by $^{31}$P NMR, but the sensitivity of this method is only 6.63% that of $^1$H NMR,[23] and these studies will always be limited by enzyme availability. The latter method, with its greater sensitivity, has proved to also have the required resolution as well.

Tyrosine residues have two pairs of equivalent aromatic symmetry-related protons (at ring positions 2 and 6, and 3 and 5: designated 2,6-H and 3,5-H, respectively[24]; Fig. 2), which have distinct chemical shift positions and are well resolved in $^1$H NMR spectra (trace Y of Fig. 3). Furthermore, as illustrated in Fig. 3 (trace YYY), each of the three tyrosines of the YYY peptide are well resolved, and have been assigned directly by

[20] L. Stadtmauer and O. M. Rosen, *J. Biol. Chem.* **261,** 10000 (1986).

[21] S. E. Shoelson, M. F. White, and C. R. Kahn, *J. Biol. Chem.* **264,** 7831 (1989).

[22] M. Dickens, J. M. Tavaré, B. Clack, L. Ellis, and R. M. Denton, *Biochem. Biophys. Res. Commun.* **174,** 772 (1991).

[23] R. A. Dwek, "Nuclear Magnetic Resonance in Biochemistry." Oxford Univ. Press (Clarendon), London and New York, 1973.

[24] K. Wüthrich, "NMR of Proteins and Nucleic Acids." Wiley, New York, 1985.

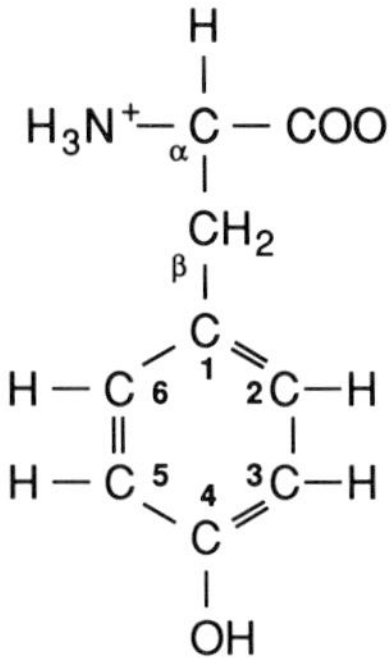

```
              H
              |
     H3N+ — αC — COO−
              |
             CH2
             β|
              C
           /  1  \\
     H — C6        2C — H
         ||         |
     H — C5   4    3C — H
           \\   //
              C
              |
              OH
```

FIG. 2. Diagram of the amino acid tyrosine. Equivalent symmetry-related protons of the aromatic ring correspond to positions 2 and 6, and 3 and 5.

the use of two-dimensional NMR methods, and the examination of a series of mono- and disubstituted peptides, with phenylalanine(s) in the place of tyrosine(s).[18] This resolution can be achieved with peptide concentrations as low as 50 to 100 $\mu M$ at 500 to 600 MHz.

Phosphorylation at position 4 of the tyrosine ring results in a shift downfield (to the left in these spectra) in the positions of the resonances of the 2,6-H and 3,5-H protons (Fig. 3, trace Y · P: note that for free tyrosine, the 3,5-H protons shift further downfield than the 2,6-H protons). Thus one can monitor the phosphorylation of a particular tyrosine by monitoring the downfield shift of the resonances of its aromatic ring protons, as illustrated in the spectrum of peptide YYY following phosphorylation at position Y-9 [Fig. 3, trace Y(Y · P)Y].

## Phosphorylation of Peptide YYY Is Highly Ordered and Progressive

Phosphorylation reactions are carried out in 0.5-ml volumes of $D_2O$ directly in a 5-mm i.d. quartz NMR tube, so that the time course of the reaction (i.e., the rate of substrate utilization and the rate of product formation) can be monitored directly as a function of time. The solution is buffered with sodium bicarbonate (20 m$M$, pH 7.5), as the more commonly employed biological buffers (e.g., Tris or HEPES) themselves give rise to signals in the proton spectra (this limitation has been overcome with the availability of deuterated Tris, which we now utilize for experiments such as those summarized herein). The reaction is initiated with the addition of activated (i.e., autophosphorylated) enzyme to a final concentration of 5 $\mu M$ [the molecular weight of the enzyme is 45,297, thus 5 $\mu M$ in 0.5 ml is 113 $\mu$g or 2.4 nmol of enzyme; we have examined enzyme to peptide ratios ranging from 1 : 80 (as in Figs. 4 and 5) to 1 : 5] to the buffered solution of peptide YYY (340 $\mu M$ = 170 nmol) and excess $MgCl_2$ and

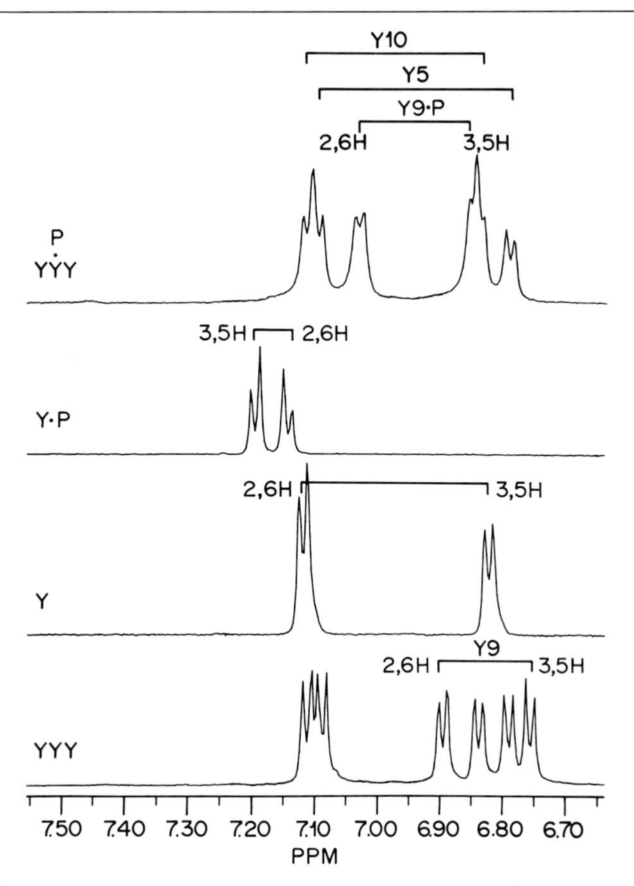


FIG. 3. $^{1}H$ NMR spectra of the following: YYY, the dodecapeptide RRDIYETDYYRK; Y, the amino acid tyrosine; Y · P, tyrosine phosphate; Y(Y · P)Y, the YYY peptide following monophosphorylation at Y-9 (peptide was phosphorylated by the hIR PTK *in vitro* and purified by thin-layer chromatography and HPLC; courtesy of Dr. Jeremy Tavaré, University of Bristol, England). The positions of resonances derived from the 3,5H and 2,6H ring protons (cf. Fig. 2) are indicated. Pulse-and-collect spectra were recorded at 500 or 600 MHz at 300 K (27°) on Brucker instruments with a 60° pulse width (~5 μsec) over a 10-ppm sweep width in 8K data blocks (acquisition time ~0.8 sec). An interpulse delay time of 0.4 sec was typically used. During peptide phosphorylation (e.g., Figs. 4 and 5), the number of transients collected per spectrum ranged from 16 to 256, and was dictated by the rate of the reaction, as monitored during accumulation of spectra by transformation in a second memory block. Typical accumulation times were ~2 min.

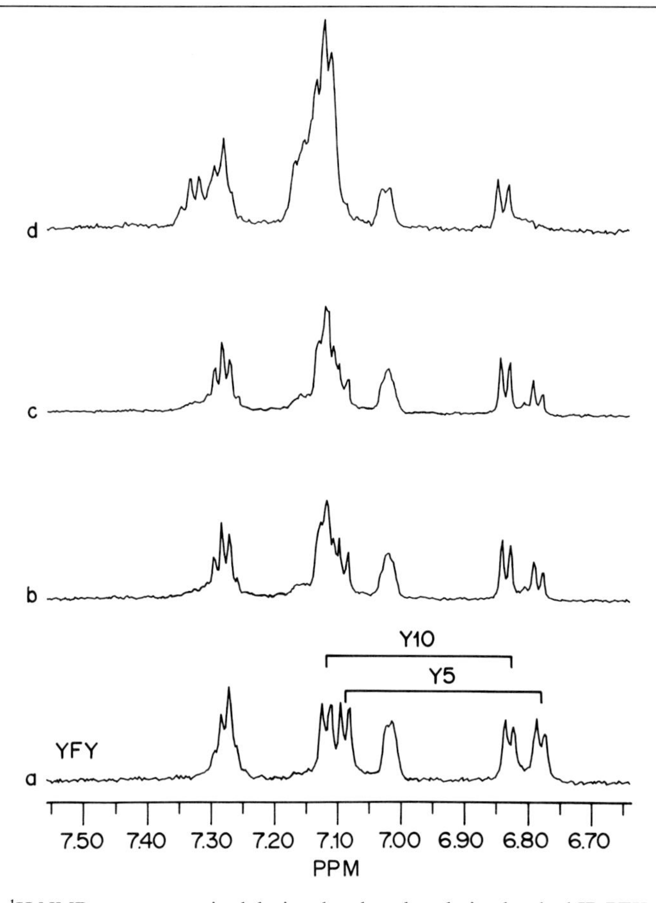


FIG. 4. $^1$H NMR spectra acquired during the phosphorylation by the hIR PTK of peptide YFY, in which the tyrosine at position 9 is substituted with phenylalanine. Phosphorylation at Y-5 proceeds from trace a to d. Trace a: $t$ = 0 min, prior to the addition of enzyme. Traces b–d: $t$ = 15, 20, and 500 min following the addition of enzyme, respectively. In trace d, Y-10 has just begun to be phosphorylated.

ATP (6 and 3 m$M$, respectively, in the experiments of Figs. 4 and 5). Phosphorylation of the YYY peptide proceeds in a highly ordered and progressive manner, with phosphorylation first at Y-9 (at a rate of ~275 nmol/min/mg enzyme). The reaction proceeds to completion, prior to phosphorylation at Y-10, which proceeds at a somewhat slower rate (~75 nmol/min/mg). Last, Y-5 is phosphorylated, although slowly, and has never been observed to proceed to completion.[18] Thus the enzyme can clearly discriminate among the three tyrosines of this dodecapeptide sub-

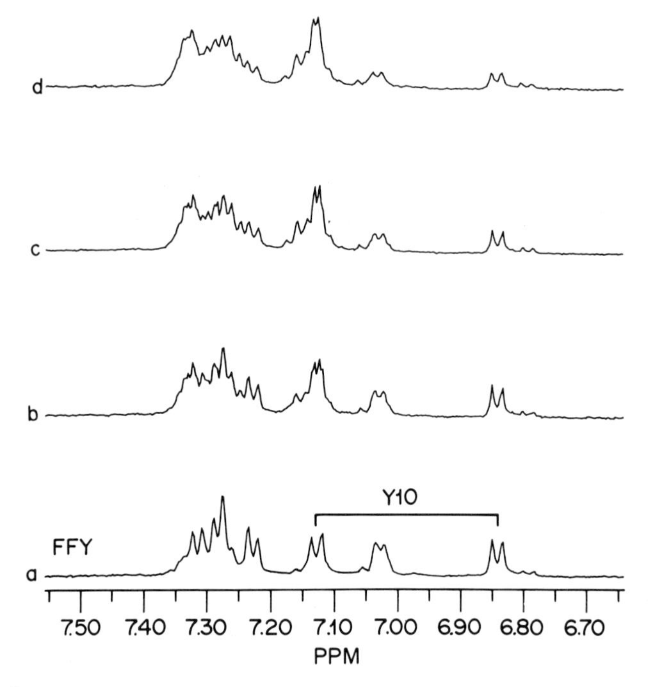


FIG. 5. $^{1}H$ NMR spectra acquired during the phosphorylation by the hIR PTK of peptide FFY, in which the tyrosines at positions 5 and 9 are substituted with phenylalanine. Phosphorylation at Y-10 proceeds from trace a to d. Trace a: $t = 0$ min, prior to the addition of enzyme. Traces b–d: 40, 85, and 190 min following the addition of enzyme, respectively.

strate. The examination of the phosphorylation of a series of Y to F-substituted peptides reveals that Y-5 and/or Y-10 are not required for phosphorylation at Y-9.[18] However, a Y to F substitution at position 9 results in the phosphorylation of Y-5 (at ~44 nmol/min/mg), with the very slow further phosphorylation at Y-10 (Fig. 4). Phosphorylation first at Y-10 occurs only in the absence of Y-5 and Y-9 (at ~15 nmol/min/mg; Fig. 5). Thus, for the peptide YYY, phosphorylation initiates at Y-9, which is flanked on its amino-terminal side by a negatively charged residue (D-8), a frequent but not steadfast requirement at known sites of tyrosine phosphorylation.[25] Phosphorylation at Y-9 seems then to activate Y-10 for phosphorylation, we surmise based on present evidence by now creating a negative environment amino proximal to Y-10. In the absence of Y-9, phosphorylation first occurs at Y-5, which is flanked on its carboxy-termi-

[25] T. Hunter and J. A. Cooper, *Annu. Rev. Biochem.* **54,** 897 (1985).

nal side by a negatively charged residue (E-6)—thus Y-5 seems to be the default site of phosphorylation, in the absence of Y-9.

These studies demonstrate that $^1$H NMR provides an experimental method with sufficient resolution to explore the peptide substrate preferences of this member of the PTK family in solution and in real time, and thus provides a means to directly assess the influence of neighboring residues on the order and kinetics of phosphorylation at multiple sites.

## Prospects for Directly Observing Autophosphorylation of Protein-Tyrosine Kinase

Given that the sequence of the YYY peptide (apart from R-1) is found in the primary sequence of the hIR itself,[6,7] where the three tyrosines are major sites of hIR PTK[17] (as well as wild-type receptor[1-3]) autophosphorylation, and that autophosphorylation of the soluble enzyme in the presence of $MgCl_2$ occurs *in trans* between enzyme monomers,[14] we wish to extend this analysis to the direct examination of autophosphorylation of soluble hIR PTKs. Changes in the aromatic spectrum of ~800 $\mu M$ hIR PTK are in fact observed during autophosphorylation of the enzyme (unpublished observations), but the size of the protein (401 amino acids, $M_r$ 45,297) and the number of tyrosines therein (13) render the sequence-specific assignment of the enzyme beyond the capability of current NMR methods.[24] However, the progress being reported with the introduction of heteronuclear isotopes (e.g., $^{15}$N-, $^{13}$C-, or $^2$H-labeled amino acids) into recombinant proteins by biosynthetic labeling,[26-28] together with conventional protein engineering (truncations and site-directed point mutations), provide a multifaceted approach to simplify the overall complexity of the problem. These experiments represent the next hurdle to be cleared toward the long-term goal of visualizing the temporal and spatial properties of tyrosine autophosphorylation in the context of an intact enzyme.

## Acknowledgments

We would like to thank our colleagues in Dallas and Oxford for their experimental contributions and discussions to this ongoing study, especially Drs. Andrew Atkinson, Beatrice Clark, Wayne Fairbrother, Dan Raleigh, Nadira Sayed, Jeremy Tavaré, and Professor R.J.P. Williams. The work in the authors' laboratories is supported by the NIH (L.E.), the Howard Hughes Medical Institute (L.E.), and the Medical Research Council (B.A.L.). B.A.L. is an associate member of the Oxford Center for Molecular Sciences.

[26] R. H. Griffey and A. G. Redfield, *Q. Rev. Biophys.* **19,** 51 (1987).

[27] B. H. Oh, W. M. Westler, P. Darba, and J. L. Markley, *Science* **240,** 908 (1988).

[28] H. Oschkinat, C. Griesinger, P. J. Kraulis, O. W. Srensen, R. R. Ernst, A. M. Gronenborn, and G. M. Clore, *Nature (London)* **332,** 374 (1988).

# [55] Expression of Protein Kinase C Isotypes Using Baculovirus Vectors

*By* S. STABEL, D. SCHAAP,[1] and P. J. PARKER

## Introduction

The isolation of protein kinase C (PKC) cDNAs has led to the identification of seven related gene products of which three—$\alpha$, $\beta_{(1+2)}$, and $\gamma$—have been separated by standard chromatographic procedures. However, PKC-$\beta_1$ and -$\beta_2$ and other members of this family have not been separated and purified. Furthermore, an assessment of purity for any isotype preparation is limited since unknown isotypes may be present. To this end we have investigated the expression of recombinant PKC proteins using the baculovirus expression system.[2] The advantages of this system are the potential for high yields and the production of defined, isotype-pure PKCs. Using the pAcC4 vector (F. McCormick, Cetus Corp., Emeryville, CA) high levels of expression have been obtained. We describe below the expression and purification procedure for PKC-$\alpha$, -$\beta_1$, -$\beta_2$, and -$\gamma$; a separate procedure is described for PKC-$\varepsilon$.

## Procedure for $\alpha$, $\beta_1$, $\beta_2$, $\gamma$ Expression and Purification

Full-length PKC cDNAs are constructed and inserted into the pAcC4 vector by standard procedures.[3] This vector has an in-frame *Nco*I site at the polyhedrin initiator ATG, allowing for direct cloning of the required cDNA (also carrying an *Nco*I site at the initiator ATG) with the viral 5′ untranslated context. Recombinant virus is obtained by cotransfection of Sf9 cells (grown in Grace's medium at 27° with 10% (v/v) fetal calf serum (FCS), 3.3 g/liter lactalbumin, 3.3 g/liter yeastolate) with wild-type AcNPV DNA and the pAcC4-PKC construct.[2] Initially recombinant virus is detected by hybridization to the relevant PKC cDNA probe and subsequently recombinant plaques are identified visually.[2]

Purified recombinant virus is passaged through infection of Sf9 cells in supplemented Grace's medium to $\sim 10^8$ pfu/ml. For infection, 400 ml of

[1] D.S. is a fellow of the Netherlands Cancer Foundation.

[2] M. D. Summers and G. E. Smith, "A Manual of Methods for Baculovirus and Insect Cell Culture Procedures," Tex. Exp. Stn. Bull. No. 1555. Texas A & M Agric. Coll., College Station, 1987.

[3] G. Patel and S. Stabel, *Cell. Signal.* **1**, 227–240 (1989).

Sf9 cells (1–1.5 × $10^6$ cells/ml) is pelleted at 1000 rpm in a sterile 1-liter centrifuge bottle using a Beckman (Palo Alto, CA) J6 centrifuge. The medium is removed and virus (50–100 ml) is added to the cell pellet and the suspension gently agitated at room temperature for 1 hr. The cells are repelleted and the virus removed. The cells are resuspended in Ex-Cell 400 medium (J. R. Scientific, Woodland, CA) containing amphotericin (2.5 $\mu$g/ml) and gentamicin (50 $\mu$g/ml).

Cells are monitored for infection by visual inspection of wet mounts and by sampling for PKC activity determination. For the latter, 1-ml samples are taken (~$10^6$ cells) and the cells pelleted in a microfuge. The pellet is frozen in liquid $N_2$ (these can be stored at −80°). Frozen pellets are extracted in 100 $\mu$l of 20 m*M* Tris-HCl, pH 7.5, 5 m*M* EDTA, 10 m*M* benzamidine, 50 $\mu$g/ml phenylmethylsulfonyl fluoride (PMSF), 0.3% (v/v) 2-mercaptoethanol, 0.5% (v/v) Triton X-100 (buffer A + PMSF). The extract is diluted (25- to 50-fold) and assayed with histone III-S substrate as described by Parker and Marais ([19], this volume). The peak of PKC expression is usually 3–4 days postinfection. At this time cells are harvsted by centrifugation and pellets frozen at −80° until required.

Frozen cell pellets are extracted with buffer A + PMSF (25 ml/400 ml cell culture) and a 40,000 *g* supernatant is loaded onto a Sephacryl S-200 column (5 × 60 cm) at a flow rate of 2 ml/min; the column is developed in buffer B [buffer A with Triton X-100 reduced to 0.02% (v/v)] and 15-ml fractions are collected. The PKC activity peak (which elutes with an apparent molecular weight of 70,000) is pooled and concentrated on a 25-ml DEAE-cellulose column equilibrated in buffer B and step eluted with buffer B + 0.15 *M* NaCl. The eluted activity is then made 0.44 *M* in NaCl and loaded onto a 50-ml threonine-Sepharose column (1.6 × 25 cm); this is washed with two column volumes of buffer B + 0.44 *M* NaCl and developed with a 50-ml linear gradient from 0.44 to 1.2 *M* NaCl in buffer B followed by 100 ml of buffer B + 1.2 *M* NaCl. Fractions are assayed for PKC, which normally elutes at 700–800 m*M* NaCl and samples are routinely run on a Phast gel system (Pharmacia, Piscataway, NJ) to monitor purity. Pooled fractions are dialyzed into buffer B + 50% (v/v) glycerol and stored at −20° with 1 m*M* dithiothreitol.

### Expression and Purification of PKC-ε

The generation of recombinant virus and infection of Sf9 cells are as described above for other PKC isotypes with the exception that the pAcC3 vector is employed. Also, in monitoring infections, a synthetic peptide substrate (ERMRPRKRQGSVRRRV) is employed since histone is a very poor PKC-ε substrate.[4]

[4] D. Schaap and P. J. Parker, *J. Biol. Chem.* **265,** 7301 (1990).

For purification a frozen pellet (from a 500-ml infection) is thawed into 20 m*M* Tris-HCl, pH 7.5, 5 m*M* EDTA, 0.3% (v/v) 2-mercaptoethanol, 250 $\mu$g/ml leupeptin, 50 $\mu$g/ml PMSF, 10 m*M* benzamidine, and 1% (v/v) Triton X-100 (buffer C + 1% Triton X-100). The extract is centrifuged at 15,000 *g* for 20 min at 4° and the supernatant loaded onto a 100-ml DEAE-cellulose column. The column is washed with buffer C containing 0.02% (v/v) Triton X-100 (buffer D) and developed with a linear gradient from 0 to 1 *M* NaCl in the same buffer. Active fractions eluting at ~0.1 *M* NaCl are pooled and loaded onto a serine-Sepharose column (prepared as for threonine-Sepharose) preequilibrated in buffer D + 0.15 *M* NaCl. Active fractions eluted on a linear 0.15 to 1.0 *M* NaCl gradient in buffer D (activity elutes at ~0.4 *M* NaCl) are pooled, dialyzed for 2 hr against buffer D, and loaded onto an 8-ml Mono Q column attached to a fast protein liquid chromatography (FPLC) unit. PKC-$\varepsilon$ elutes at 0.15 *M* NaCl on a linear gradient (0 to 0.5 *M* NaCl in buffer D). At this stage the enzyme is ≥90% pure, but minor contaminants can be removed by gel filtration (employing a Superose 12 column). The enzyme is dialyzed and stored at −20° as above.

## Comments

The use of the defined Ex-Cell 400 medium for cell culture postinfection prevents contamination of extracts and subsequent purification steps with serum albumin. However, routine passage of Sf9 cells in Ex-Cell 400 medium has not been found to be efficient.

The first procedure is routinely used for PKC-$\alpha$, -$\beta_1$, -$\beta_2$, and -$\gamma$. The initial level of expression is 10–20 mg/liter culture and the purification yields 2–6 mg/liter culture of purified PKC (90–95% purity). The enzymes are stable on storage at −20°; however, we find that PKC-$\alpha$ undergoes a slow, apparently irreversible, conversion to a lipid/$Ca^{2+}$/phorbol ester-independent form (the reason for this is not understood). Similar purification results are obtained if extraction is carried out in the absence of Triton X-100; under these conditions effective purification can be obtained by sequential chromatography on DEAE-cellulose and phenyl-Sepharose (S. Stabel and D. Frith, unpublished).

For PKC-$\varepsilon$ the purification is necessarily somewhat different (for example, the enzyme binds less well to threonine-Sepharose compared to other isotypes) and yields of purified protein are slightly lower (~2 mg/liter culture). This is partly due to the four-step procedure and in part due to an increased susceptibility to degradation.

The procedures described are for a single frozen pellet of 4–5 × $10^8$

cells; however, the purifications have been linearly scaled up so that multiple frozen pellets can be extracted, pooled, and worked up. The individual recombinant proteins appear to have the same biochemical properties as the isotypes purified from bovine brain and as such these recombinant proteins can be employed for various PKC studies.

# [56] Construction and Expression of Linker Insertion and Site-Directed Mutants of v-*fps* Protein-Tyrosine Kinase

*By* JAMES C. STONE, MICHAEL F. MORAN, and TONY PAWSON

## Introduction

Several strategies are available for creating mutations in molecularly cloned oncogenes and protein kinase cDNAs,[1] each with its attendant strengths and weaknesses. The approach must depend on the specific questions being asked: What protein determinants are required for biological activity and which residues are dispensable? What are the limits of individual functional domains? What is the significance of a particular residue or sequence motif that has been conserved in evolution? How do the domains interact? Perhaps most important, with what other cellular proteins (such as substrates and regulators) does the protein kinase interact to induce a cellular response?

To identify the functional domains of the $P130^{gag\text{-}fps}$ protein-tyrosine kinase, encoded by the v-*fps* oncogene of Fujinami avian sarcoma virus, we exploited a simple and generally applicable method to generate numerous defined lesions throughout the v-*fps* coding sequence.[2] Mutants were then expressed in mammalian or bacterial cells. The elements of this approach are defined below.

## In-Phase Linker-Insertion Mutagenesis: General Strategy for Genetic Dissection of Complex Polypeptides

### *General Strategy*

The basic approach is to generate linear plasmids by partial digestion with restriction enzymes that recognize numerous positions throughout

[1] M. Smith, *Annu. Rev. Genet.* **19,** 423 (1985).

[2] J. C. Stone, T. Atkinson, M. Smith, and T. Pawson, *Cell* (*Cambridge, Mass.*) **37,** 549 (1984).

the region of interest, to attach small duplex DNA molecules designed to leave the reading frame of the target coding sequence unaltered, to circularize these variant plasmids *in vitro,* and to recover and map these insertion mutations in *Escherichia coli* prior to functional analysis in vertebrate cells. The net effect of these manipulations is to create many different mutant proteins, each with a small peptide insertion at a distinct site in the polypeptide chain. Such insertions can have profound effects on protein function, and can give clues as to the limits, activities, and interactions of polypeptide domains. This technique serves as a basis for subsequent, more detailed analysis by oligonucleotide-directed deletion, point and cassette mutagenesis. Although originally developed to investigate an oncogenic tyrosine kinase, this approach has been successfully applied by us to the c-*raf* serine/threonine protein kinase[3] and to the c-*myc* nuclear protein,[4] and by others to a variety of proteins, including protein kinases[5–9] and proteins involved in cellular or viral transcription and replication.[10–13]

### *Construction of the Vector for Mutagenesis*

Substantial effort should be invested in design of the vector. The relevant cDNA should be inserted into a high-copy plasmid such as pUC.[14] Construction of mutations in a plasmid that serves as a eukaryotic expression vector facilitates subsequent gene transfer and analysis without recourse to subcloning. Incorporation of a simian virus 40 (SV40) replication origin into the plasmid allows mutant proteins to be transiently expressed to high levels in COS monkey cells (see below). This type of analysis is extremely valuable, since one can rapidly verify that a full-length protein is made, and assess its stability and kinase activity. Useful vectors for this

[3] U. R. Rapp, G. Heidecker, M. Huleihel, J. L. Cleveland, W. C. Choi, T. Pawson, J. N. Ihle, and W. B. Anderson, *Cold Spring Harbor Symp. Quant. Biol.* **53,** 173 (1988).

[4] J. C. Stone, T. deLange, G. Ramsey, E. Jakobovits, J. M. Bishop, H. E. Varmus, and W. Lee, *Mol. Cell. Biol.* **7,** 1697 (1987).

[5] M. Ng and M. L. Privalsky, *J. Virol.* **58,** 542 (1986).

[6] R. Prywes, J. G. Foulkes, and D. Baltimore, *J. Virol.* **54,** 114 (1985).

[7] E. T. Kipreos, G. J. Lee, and J. Y. J. Wang, *Proc. Natl. Acad. Sci. U.S.A.* **84,** 1345 (1987).

[8] J. E. DeClue and G. S. Martin, *J. Virol.* **63,** 542 (1989).

[9] S. D. Lyman and L. R. Rohrschneider, *Mol. Cell. Biol.* **7,** 3287 (1987).

[10] V. Giguerre, S. M. Hollenberg, M. G. Rosenfeld, and R. M. Evans, *Cell (Cambridge, Mass.)* **46,** 645 (1986).

[11] M. L. Privalsky, P. Boucher, A. Koning, and C. Judelson, *Mol. Cell. Biol.* **8,** 4510 (1988).

[12] L. Zumstein and J. C. Wang, *J. Mol. Biol.* **191,** 333 (1986).

[13] V. R. Prasad and S. P. Goff, *Proc. Natl. Acad. Sci. U.S.A.* **86,** 3104 (1989).

[14] C. Yanisch-Perron, J. Vieira, and J. Messing, *Gene* **33,** 103 (1985).

purpose include pECE[15] and pCMV.[16] The vector may also include a selectable marker, such as *neo*$^r$,[17] to allow for selection of stably transfected mammalian cells (i.e., pIV2.3 in Fig. 2). However, the efficiency with which mutations are recovered in the target coding sequence depends on the relative proportion of this sequence in the plasmid, which will obviously be decreased by inclusion of a selectable marker. If possible, we recommend the use of cotransfection with a selectable marker plasmid such as pSV2*neo*[17] for the isolation of stably transfected cells. If one anticipates using techniques such as electroporation or microinjection, a plasmid vector including an internal selectable marker is more suitable, especially in the analysis of biologically defective mutants. For retroviral transfer, it may be simplest to do the mutagenesis using a cassette containing the cDNA of interest, and to insert the individual mutants into a retroviral vector, as described for linker insertion mutagenesis of v-*src* by DeClue and Martin.[8] Since the actual construction and mapping of mutations in the plasmid is straightforward, emphasis should be placed on constructing the vector that best suits the functional tests anticipated. Since tyrosine kinases and many other eukaryotic enzymes are active following expression in *E. coli,* it can be advantageous to do the initial mutagenesis and screening in a bacterial system, and then to transfer interesting mutants into a mammalian expression vector (see Refs. 7 and 13, and Fig. 2). Using pUC-based plasmids we rarely obtain mutations within the bacterial plasmid backbone, presumably because these tend to abolish antibiotic resistance or plasmid replication.

### *Design of the Mutagenic Oligonucleotide*

Mutagenic oligonucleotides are synthesized based on the following criteria. The linker should insert a multiple of 3 bases so as not to disturb the reading frame, and must contain a restriction site for an enzyme that does not cut the parental plasmid. Some enzymes have unfavorable properties, such as inhibition by contaminants commonly found in crude plasmid preparations, and should be avoided. In principle a complementary hexanucleotide can be employed,[18] although we have been unsuccessful with several of these. A dimer of a 6-bp recognition sequence, on the other hand, can be attached to the blunt ends of linear plasmids with high efficiency. For example, the 12-mer 5′-CTCGAGCTCGAG-3′ has two

[15] L. Ellis, E. Clauser, D. O. Morgan, M. Edery, R. A. Roth, and W. J. Rutter, *Cell (Cambridge, Mass.)* **45,** 721 (1986).

[16] D. Russell, personal communication.

[17] P. J. Southern and P. Berg, *J. Mol. Appl. Genet.* **1,** 327 (1982).

[18] F. Barany, *Proc. Natl. Acad. Sci. U.S.A.* **82,** 4202 (1985).

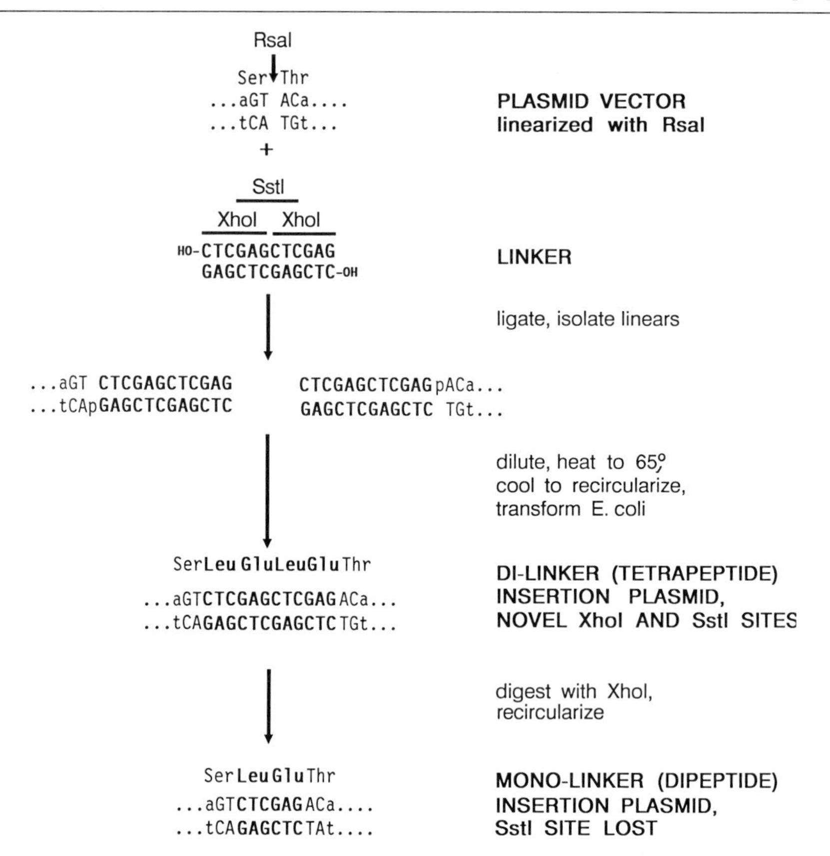


FIG. 1. Linker-insertion mutagenesis. A plasmid is linearized with a blunt-end cutter (*Rsa*I), and a nonphosphorylated dodecameric linker with novel *Xho*I and *Sst*I sites is introduced, creating an insertion of four codons. The insertion can be reduced to a hexamer (monolinker) encoding only two extra codons, with the loss of the *Sst*I site, by digestion with *Xho*I. Variations on this theme are discussed in the text.

*Xho*I recognition sites (5′-CTCGAG-3′), and an internal *Sst*I site (5′-GAGCTC-3′) where the two *Xho*I sites meet (Fig. 1). Depending on the exact protocol followed, one can selectively recover mutations with an insertion of one, two, or multiple copies of the hexameric (5′-CTCGAG-3′) *Xho*I sequence, creating what we refer to as mono-, di-, and polylinker mutations, respectively. The absence of the *Sst*I site can generally be used to distinguish monolinker mutations from other alleles.

A second type of mutagenic linker we have successfully employed is typified by the partially self-complementary sequence 5′-CGACCTCGAGGT-3′. This dodecamer is designed to attach to protruding 5′-CG ends, and to specify an *Xho*I site (underlined). In this case a single 12-bp insertion would constitute a monolinker mutation, and a novel *Sal*I site is diagnostic of multiple insertions. While hexamers can be efficiently attached to recessed ends,[18] the use of a 12-base mutagenic oligonucleotide allow one to insert the same recognition sequence (i.e., *Xho*I) into all sites.

The linker should be designed to avoid stop codons in any reading frame. While large linker insertions might generally be expected to give a more severe phenotype, there may be cases where this is not so; a tetrapeptide insertion in an $\alpha$-helical region might give less angular distortion than a dipeptide, for example. Clearly one is not limited to the linkers described above. The sequence of the linker can be tailored to suit any enzyme used to linearize the DNA, or to introduce any novel restriction site (with the caveat that some such linkers may encode translational termination signals in one or more reading frames). Linkers encoding specific structural motifs, rather than random oligopeptides, can also be designed.[19] Obviously the quality of the chemistry employed in the synthesis of the mutagenic oligonucleotide is crucial if one is to avoid mutations of unexpected structure.

### *Partial Digestion of the Plasmid*

Starting with a plasmid preparation which contains predominantly covalently closed circles and virtually no linear molecules, conditions are established to optimize the yield of linear plasmid by partial digestion. *Alu*I, *Dpn*I, *Fnu*DII, *Hae*III, and *Rsa*I are enzymes with tetranucleotide recognition sequences which generate blunt ends suitable for attaching the *Xho*I dimer. *Hin*PI, *Taq*I, and *Msp*I all generate 5′-CG overhangs, which can be attached to the partially self-complementary dodecamer described above. When initial functional tests indicate that a particular region of the protein is interesting, it is often possible to increase the local density of mutations by locating sites for single-cutting enzymes. In theory, and usually in practice, it is possible to create an in-phase insertion at any restriction site by modification of the ends of the linear plasmid or by synthesizing appropriate oligonucleotides.

Partial digestion giving adequate yields of linear plasmid DNA can be achieved either by performing the digestion in the presence of ethidium bromide, or by limiting both the duration of the incubation and the concentration of the restriction enzyme. There seems to be no advantage to the ethidium method, and the latter technique is simpler. In either case the

[19] D. Periman and H. O. Halvorson, *Nucleic Acids Res.* **14,** 2139 (1986).

conditions are empirically determined as described in the protocol below. To monitor plasmid digestion it is imperative to use electrophoresis conditions that adequately resolve open circles, covalently closed circles, full-length linear species, and linear species of less than full length that arise from multiple cleavages. It is not necessary to isolate linear species from the partially digested plasmid prior to linker attachment. Ligations are performed on the mixed products of a partial enzyme digestions, and the linear plasmid with attached linkers is subsequently isolated and introduced into *E. coli*.

*Protocol for Linker-Insertion Mutagenesis*

For detailed descriptions of recombinant DNA procedures the reader is referred to Sambrook *et al.*[20]

1. Restriction enzyme (2–4 units) is added to a tube containing 50–100 μg of plasmid in 100 μl of the appropriate buffer. If the enzyme is very concentrated it may be advantageous to dilute 1–2 μl of the enzyme into ice-cold reaction buffer immediately before use. In this way the use of very small volumes can be avoided and the method can be readily reproduced.
2. Incubate the reaction at the recommended temperature for enzyme digestion.
3. At 2, 4, 8, 16, 32, and 64 min, remove 15 μl of the reaction to tubes containing 85 μl of TNE (10 m*M* Tris-HCl, pH 8.0, 100 m*M* NaCl, 1 m*M* EDTA) and 100 μl of Tris-buffered phenol–chloroform–isoamyl alcohol (25 : 24 : 1) at room temperature. Vortex each sample briefly after the DNA sample is added.
4. After the last sample is removed from the digestion tube vortex all the samples vigorously, centrifuge at room temperature, and remove the aqueous phase to fresh tubes.
5. Precipitate each sample with ethanol. Add 0.1 vol of 3 *M* sodium acetate and 2 vol of 95% ethanol, vortex, and leave at −20° for 1 hr. Centrifuge each sample at maximum speed in a microfuge and decant the supernatants. Rinse the tubes with 70% ethanol and then with 95% ethanol. Spin the tubes briefly after each rinse if the pellets are not visibly secure on the bottoms of the tubes. Dry the pellets and dissolve each in 20 μl of TE (10 m*M* Tris-HCl, pH 8.0, 1 m*M* EDTA).
6. Mix 1 μl of each sample with 9 μl of TE and 2 μl of sample buffer, and electrophorese in a 20-cm 0.5% agarose gel in Tris–acetate–EDTA buffer for 12–18 hr at 20 V. Run uncut plasmid and plasmid cut with an

[20] J. Sambrook, E. F. Fritsch, and T. Maniatis, "Molecular Cloning: A Laboratory Manual." Cold Spring Harbor Lab., Cold Spring Harbor, New York, 1989.

enzyme that cleaves only once as controls. It may be necessary to allow the gel to solidify for 1 hr at 4° prior to loading. Ethidium bromide can be included in the gel at 1.0 μg/ml or the gels can be stained after electrophoresis.

7. Fractions of partially digested DNA that are rich in linear species and which show no signs of multiply cut plasmid are identified and can be pooled.

8. Linker is prepared as described by Lathe *et al.*[21] Briefly, the nonphosphorylated oligonucleotide is dissolved in TE at a concentration of 0.5–1.0 μg/ml, heated to 65°, and allowed to cool slowly to room temperature to form duplex DNA. These preparations may be stored at −20° until use.

9. An aliquot of partially digested DNA corresponding to 1.0 μg of linear species is mixed with 1.0 μg of linker and adjusted to 9.0 μl with TE. DNA species are then coprecipitated by addition of 0.1 vol of 3 *M* sodium acetate and 3 vol of 95% ethanol.

10. The coprecipitated DNA is dissolved in 5.0 μl of TE by pipetting up and down and then allowing the tube to sit on ice for several minutes. Solution (4.5 μl) is transferred to a fresh tube and 0.5 μl of 10× T4 DNA ligase buffer (200 m*M* Tris-HCl, pH 7.6, 50 m*M* $MgCl_2$, 50 m*M* dithiothreitol, 500 μg/ml bovine serum albumin, 5 m*M* ATP) is added. T4 ligase is added (0.5 μl; 2–3 Weiss units), and the reaction is incubated at room temperature for 4 hr. Blunt-end ligation proceeds much more efficiently at room temperature than at 4°. This reaction requires a high concentration of ligase, but the glycerol from the ligase storage buffer should not exceed 5% of the final volume. Since the linkers have 5′ hydroxyl groups they cannot be ligated to one another, and can only be covalently attached to the phosphorylated 5′ end of the linearized plasmid DNA.

11. After addition of sample buffer, the ligation reaction is heated to 65° for a few minutes, and then loaded on a 0.5% agarose gel as described above to separate linear molecules which should now have linkers attached.

12. The linear species is excised from the gel, and recovered by electroelution followed by ethanol precipitation.

13. After estimating the recovery of linear species by running a small sample on an agarose gel and staining with ethidium bromide, an aliquot corresponding to about 0.1 to 0.2 μg of linear DNA is diluted into 20 μl of TE. The sample is heated to 65° and allowed to cool slowly to promote recircularization. In this step, the covalently attached strands of the linker

[21] R. Lathe, M. P. Kieny, S. Skory, and J. P. Lecocq, *DNA* **3,** 173 (1984).

will hybridize, although the plasmid is not yet covalently closed. The plasmid DNA molecules are then transformed into *E. coli* using standard procedures, where the circles are closed and the plasmids replicate. The choice of *E. coli* strain does not seem critical. This procedure will result in the insertion of a single linker. In the case of the *Xho*I dimer dodecameric linker attached to blunt-ended DNA, the size of the linker can be reduced to a hexameric ("monolinker") 5′-CTCGAG-3′ insertion by subsequent digestion of mutant plasmid DNA with *Xho*I, followed by religation and transformation of *E. coli* (Fig. 1).

*Variations*

1. If polylinker mutations are desired, the linker must be phosphorylated before the coprecipitation step. Immediately after ligation, digestion with the linker-cleaving enzyme (i.e., *Xho*I) is required to convert plasmid-linker concatamers to species of plasmid unit length, to reduce the number of attached linkers, and to generate cohesive ends. If the digestion is thorough, monolinker mutations are generated by this method. In practice the digestion of linker is incomplete and polylinker insertions are recovered. After isolating the linear species from the gel, an aliquot of this DNA is converted to covalently closed circular form using T4 DNA ligase, followed by transformation into *E. coli.*

2. Ethidium bromide can be included in the initial enzymatic digestion to inhibit multiple cutting, and thereby promote the formation of linear molecules. Plasmid DNA (2 $\mu$g), 1 unit restriction enzyme, and 0.1 to 20 $\mu$g ethidium bromide are incubated in 20 $\mu$l containing 10 m$M$ Tris-HCl, pH 8.0, 50 m$M$ NaCl, 10 m$M$ $MgCl_2$ at 37° for 60 min. The concentration of ethidium bromide and enzyme that gives predominantly linear molecules is empirically determined as above. This procedure has been extensively described.[2,5]

*Identification and Mapping of Linker Mutations*

Linker-containing plasmids are identified and roughly mapped within the target sequence by a pair of double restriction enzyme digests, each using single cutters that cleave outside the region of interest, in addition to the enzyme that cuts within the linker. Mutants will contain a novel restriction site specified by the linker, whose approximate position is given by the sizes of the digestion products. Although rapid and simple, the following problems must be considered in mapping the mutations.

1. A cautionary point is that the original preparation of linear species used to transform *E. coli* is invariably contaminated with plasmid cleavage

products of slightly less than full length, arising from double digestion at two neighboring sites in the plasmid. The consequence of this double cleavage, if followed by linker attachment, is a small deletion that may not be apparent from the initial mapping experiments described above. Digestion of all putative linker insertion mutations with an enzyme that does not cut the linker, but rather cuts the parental plasmid into several small fragments in the 3.0- to 0.5-kb range, allows identification of such unwanted deletion plasmids.

2. Small amounts of covalently closed circles contaminating the original linear preparation will seriously decrease the percentage yield of mutant plasmids, since uncut plasmid is recovered in *E. coli* with high efficiency relative to mutated plasmid.

Both these difficulties can be reduced or eliminated by carefully controlling the partial digestion and preparative electrophoresis conditions. High-resolution mapping using molecules end labeled on the *Xho*I-generated termini allows one to deduce the exact positions of the mutations. Alternatively, the mutant structures can be determined directly by DNA sequence analysis, using a set of oligonucleotide primers for dideoxynucleotide sequencing spaced at 200-bp intervals throughout the region of interest. We find that the sequence of the mutants invariably conforms to that expected from the initial mapping. Ultimately the quality of the synthetic oligonucleotide and the enzymes employed determine the predictability of the insertion mutations.

A plasmid that contains di- or polylinker insertions can be converted to the monolinker form by exhaustive digestion with the linker-cleaving enzyme. The linear plasmid is then circularized using ligase, and introduced back into *E. coli*. It is simple to screen for the reduction in linker size by DNA sequencing, although the loss of the internal restriction site (*Sst*I or *Sal*I in the examples above) is generally diagnostic of a monolinker. A rare complication in this latter test arises when the restriction site diagnostic of di- or polylinker mutations is created at the boundary of the insertion. For example, when the *Xho*I dimer described above is attached to an *Alu*I site, an *Sst*I site should be created 6/16 times on a random basis, and hence the monolinker insertion mutant will still retain an *Sst*I recognition sequence at the mutation site.

### *Generating Deletion Mutants*

An advantage of linker insertion mutants is that they can be used to generate additional mutants. Deletions can readily be constructed by digesting two different linker insertion mutants with the enzyme specified by the linker, and a single cutter elsewhere in the plasmid. If the two linker insertions are in frame, appropriate DNA fragments from the mutants can simply be recombined. If they are in different reading frames, one can generally modify the ends of the plasmid cleaved at the linker to generate an in-phase deletion. For example, the 3′ recessed ends of *Xho*I-generated linear DNA are filled out using Klenow polymerase; ligation of two such ends results in a +1 shift in reading frame. Digestion with single-strand-specific mung bean nuclease generates a −1 shift. Thus any pair of *Xho*I sites can be recombined in frame. Alternatively one can cut at the novel restriction site provided by the linker and digest with the slow isozyme of *Bal*31 exonuclease to generate a series of deletion variants.[22] These deletion methods were used with particular advantage in the definition of functional protein domains crucial to c-*myc* transforming activity.[4]

### Alternative Strategies for Insertion Mutagenesis

Several other attractive insertion mutagenesis methods have been developed. One approach employs a DNA fragment containing a marker which can be scored in *E. coli,* such as *supF*,[23] rather than a linker. Subsequently the marker is removed, leaving a small insert of foreign DNA. This method has some drawbacks, including (1) the preparation of the mutagen is much more laborious, (2) the procedure requires two rounds of plasmid manipulation, and (3) the minimal insertion is generally larger than 6 bp. The advantage of this technique is that virtually 100% of the plasmids analyzed contain inserts. However, since the protocol detailed above often yields greater than 80% mutant plasmids, the need for a selectable marker is obviated.

For genes that are small relative to the size of the plasmid it can be difficult to obtain a collection of linker insertions by the methods described here. The more cumbersome linker-scanning method, whereby a set of 5′ and a set of 3′ deletions are recombined using a common linker, achieves nearly the same net result.[22] In this case the reading frame is not dictated by the enzymology, however, but rather is controlled by recombining appropriate pairs of plasmids after DNA sequence analyis.

## Transient Expression of Wild-Type and Mutant v-*fps* Proteins in COS Monkey Cells

### *Transient Expression in COS Monkey Cells*

If the linker insertion mutagenesis has been done in a plasmid vector containing an SV40 origin of replication, the activities of mutant proteins

[22] L. I. Lobel and S. P. Goff, *Proc. Natl. Acad. Sci. U.S.A.* **81,** 4149 (1984).

[23] B. M. Willumsen, A. Christensen, N. L. Hubbert, A. G. Papageorge, and D. R. Lowy, *Nature (London)* **310,** 583 (1984).

can be rapidly assessed in COS monkey cells. COS cells constitutively synthesize the SV40 large T antigen, which provokes the extrachromosomal replication of circular DNA molecules with an SV40 replication origin, and hence the high level synthesis of any protein encoded by the plasmid. Transfection can be carried out using a variety of procedures, including calcium phosphate coprecipitation,[24,25] DEAE-dextran-mediated transfection,[26] or lipid-mediated transfection[27] (lipofection). Two of these techniques are described below. See Sambrook *et al.*[20] and Freshey[28] for further details and discussion on transfection and tissue culture, respectively.

*Protocol for Calcium Phosphate Transfection of COS Cells*

COS-1 cells[29] are maintained in Dulbecco's modified Eagle's medium (DMEM) containing 10% fetal bovine serum (FBS). They are seeded at approximately 5–10 $\times$ $10^5$ cells/100-mm dish (approximately one-tenth of a confluent plate) and then transfected the next day essentially according to Graham and Van der Eb.[24] Use 10 $\mu$g of plasmid DNA that has been purified by equilibrium centrifugation in a CsCl–ethidium bromide density gradient. It is important to optimize the pH and DNA concentration which give the best yield.

*Reagents*

2$\times$ HEPES-buffered saline (2$\times$ HBS): 1.64 g NaCl (280 m*M*), 1.19 g HEPES (58 m*M*), 21.3 mg $Na_2HPO_4$ (1.5 m*M*); dissolve in distilled $H_2O$ to a volume of 100 ml and adjust the pH to 7.10 with NaOH. Sterilize through a 0.22-$\mu$m filter

$CaCl_2$ (2.5 *M*): Dissolve 18.4 g $CaCl_2 \cdot 2H_2O$ in distilled $H_2O$ to a volume of 50 ml and sterilize through a 0.22-$\mu$m filter

1. In a 5-ml sterile polystyrene tube, combine

   Tris-HCl (1 m*M*) (0.5 ml), 0.1 m*M* EDTA, pH 8.0 containing 10 $\mu$g plasmid DNA
   $CaCl_2$ (2.5 *M*) (50 $\mu$l)

In a 15-ml sterile conical polystyrene tube place 0.5 ml 2$\times$ HBS.

2. Bubble sterile air into the 2$\times$ HBS through a plastic 1-ml pipette

[24] F. L. Graham and A. J. Van der Eb, *Virology* **52,** 456 (1973).
[25] C. Chen and H. Okayama, *Mol. Cell. Biol.* **7,** 2745 (1987).
[26] J. H. McCutchan and J. S. Pagano, *J. Natl. Cancer Inst.* (*U.S.*) **41,** 351 (1968).
[27] P. L. Felgner, T. R. Gadek, M. Holm, R. Roman, H. W. Chan, M. Wenz, J. P. Northrop, G. M. Ringold, and M. Danielsen, *Proc. Natl. Acad. Sci. U.S.A.* **84,** 7413 (1987).
[28] R. I. Freshney, "Culture of Animal Cells." Alan R. Liss, New York, 1983.
[29] Y. Gluzman, *Cell* (*Cambridge, Mass.*) **23,** 175 (1981).

attached to a Pipet-Aid (Drummond Scientific). Slowly, and dropwise, add the $Ca^{2+}$–DNA solution into the phosphate solution using a 1-ml blue-tipped pipettor. Let the precipitate accumulate at room temperature for 20 min (the precipitate may not be very obvious).

3. Pipette the mixture onto the COS-1 cells using a plastic pipette and gently swirl the plate to evenly distribute the precipitate.

4. The next day (16 hr later) rinse the cells once with warm phosphate-buffered saline (PBS) and then feed with fresh 10% FBS-DMEM.

5. Two days later the cells should be expressing the plasmid-encoded protein in substantial amounts. The transfected gene product can be detected by Western blotting or by immunoprecipitation, and analyzed for kinase activity following immunoprecipitation or partial purification.

We have not routinely employed dimethylsulfoxide (DMSO) or glycerol "shocking" or chloroquine treatment to enhance DNA uptake by COS cells or Rat-2 cells,[20] or the modified calcium phosphate technique of Chen and Okayama.[25] Such additional steps may ultimately increase the expression of the plasmid-encoded protein when using the calcium-phosphate method.

*Lipofection of COS Cells*

As an alternative to introducing plasmid DNA into COS cells as a calcium–phosphate precipitate, we have used Lipofectin reagent (BRL, Inc., Bethesda, MD). This is a 1 : 1 solution of *N*-[1-(2,3-dioleyloxy)propyl]-*N*,*N*,*N*-trimethylammonium chloride and dioleoylphosphatidylethanolamine as described by Felgner *et al.*[27] In this method, the DNA is first mixed with the lipid, giving a lipid–DNA complex that can apparently fuse with the cells. We have found this method to be very simple and reproducible, although it is expensive. However, we have experienced batch-to-batch variations in the Lipofectin reagent which can affect is cytotoxicity. New lots of the reagent should be optimized for the amount of Lipofectin and the duration of exposure which give maximal transfection efficiency and minimal cell death.

*Method*

1. COS-1 cells are seeded at approximately $2 \times 10^5$/60-mm dish the day before the transfection.

2. Plasmid DNA (1.0 $\mu$g) is combined with 1.5 ml of serum-free, growth factor-supplemented medium (Opti MEM I; BRL, Inc.) containing 55 $\mu M$ 2-mercaptoethanol in a sterile polystyrene tube (polypropylene tubes should be avoided). In a second polystyrene tube combine 25 $\mu$g of the

lipid reagent in 1.5 ml of the above medium. Combine the diluted DNA and lipid solutions and mix gently (do not vortex) and let stand at room temperature for 15 min.

3. Wash cells three times with warm PBS lacking $Ca^{2+}$ and $Mg^{2+}$. It is essential that all traces of serum are removed from the cells. Add the 3 ml of DNA–lipid mixture to the cells and incubate for 5 hr.

4. After the 5-hr incubation add to the cells 3 ml of DMEM containing 20% FBS such that the final serum concentration is brought to 10%.

5. The next day (approximately 16 hr later) change to 5 ml of fresh DMEM containing 10% FBS.

6. Two days later the cells may be harvested for analysis of protein kinase expression and activity.

### *Stable Expression of Mutant Protein Kinases in Rat-2 Fibroblasts*

To obtain stable cell lines expressing wild-type or mutant forms of $P130^{gag-fps}$, we have generally employed the Rat-2 fibroblast cell line.[30] This line has several useful characteristics. It lacks endogenous thymidine kinase activity, meaning that the *tk* gene can be used as a coselectable marker if required. Rat-2 cells can be transfected with good efficiencies (approximately 1 : $10^3$ stable transfectants), arrest their growth when confluent, and virtually never give rise to spontaneously transformed colonies. This is an important consideration when transforming activity is being measured. With the *gag–fps* oncogene, it is obviously possible (and desirable) to measure transforming efficiency in monolayers of transfected Rat-2 cells. Transformed foci can be readily picked, expanded, and cloned by growth in soft agar or by limiting dilution. Such cells invariably express the transfected oncogene at high levels. However, to study mutants with little or no transforming activity, or to express protein kinases or other signaling proteins with no dramatic effects on cell morphology or growth properties, it is necessary to select transfectants on the basis of a cotransfected marker and to screen for synthesis of the desired gene product. Resistance to the antibiotic G418 conferred by the *neo*$^r$ gene is generally the selection procedure of choice.

### *Protocol for Isolation of Stably Transfected Cell Lines Using Coselection for G418 Resistance*

This method allows for the screening of several G418-resistant colonies for expression of the transfected gene product with a minimum delay. Colonies are transferred into 12-well plates and analyzed by Western

[30] W. C. Topp, *Virology* **113,** 408 (1981).

blotting of whole-cell lysates. By screening for expression soon after the initial picking, positive and negative clones can be identified and negative isolates can be discarded before they have been needlessly expanded.

1. Approximately $2 \times 10^6$ Rat-2 cells are plated in a 100-mm dish in DMEM + 10% FBS and left overnight, or until about 30% confluent.

2. Cells are transfected by either of the two procedures described above. If the plasmid does not contain an internal selectable marker, a 10 : 1 ratio of the plasmid encoding the kinase to pSV2*neo* (which encodes the aminoglycoside phosphotransferase that confers resistance to the antibiotic G418) is employed. With the calcium–phosphate method, 1 $\mu$g of plasmid (combined with 0.1 $\mu$g of pSV2*neo* if necessary) and 9 $\mu$g of sheared Rat-2 genomic DNA (as carrier) are used. The Rat-2 carrier DNA is not included when using lipid-mediated transfection. It is recommended that these DNAs be sterilized by ethanol precipitation before transfection when attempting to establish cell lines.

3. Two days after the transfection the cells are trypsinized and reseeded onto five plates in DMEM + 10% FBS and 400 $\mu$g/ml G418 (1× concentration) (Geneticin; GIBCO, Grand Island, NY).

4. After 2–3 days cells will begin to slough off of the monolayer. The G418-containing medium is changed every 2 days. Ten to 14 days after the transfection, colonies containing approximately 500–1000 G418-resistant cells should be visible.

5. At this point individual colonies can be isolated using cloning cylinders. The positions of up to 12 well-isolated colonies are marked on the underside of a plate with a marker pen. In a sterile tissue culture hood remove the medium and then rinse the cells gently with 5 ml of PBS, which is then aspirated from the plate. Using sterile forceps, place cloning cylinders onto the individual colonies according to the markings on the underside of the plate. Using a pipettor fitted with a sterile yellow tip, aliquot 50 $\mu$l of 0.25% trypsin into each of the cylinders. Two to 4 min later, pipette the trypsin up and down within the cylinder to completely dislodge the cells from the surface of the plate. Care must be taken not to disturb the cylinder during this step. Finally, the cells are taken from the cylinder and placed into individual wells of a 12-well plate containing 1.5 ml of DMEM + 10% FBS + 1× G418.

6. Once the cells are confluent (approximately 4 days), they are rinsed with 1 ml of warm PBS, and then approximately 150 $\mu$l (three drops) of 0.25% trypsin is added. Two to 4 min later 1 ml of DMEM + 10% FBS is added to the cells to inhibit the trypsin. A sterile blue-tipped pipettor is used to gently disperse the cells into the medium. One drop of the cell suspension is transferred into a fresh 12-well plate containing DMEM + 10% FBS + 1× G418 to propagate the clone. The rest of the cells are

placed into a snap-cap tube and pelleted by centrifugation for 2 min at 3000 rpm in a microfuge. (Note that this protocol is not suitable for transmembrane proteins that will be exposed directly to trypsin).

7. The supernatant is completely removed, and 30 $\mu$l of SDS gel sample buffer is added to the cells. At this point the lysates may be stored at $-20°$ for the extra 1–2 days that may be required for slower growing clones to accumulate.

8. Lysates are heated to 100° for 2 min, followed by vigorous vortexing to reduce their viscosity prior to sodium dodecyl sulfate–polyacrylamide gel electrophoresis (SDS–PAGE) and Western blotting. By using a mini-SDS–PAGE apparatus, and an antibody-alkaline phosphatase conjugate and color reaction to develop the Western blot, clones expressing the desired protein can be identified in 1 day. Positive clones can be identified in less than 3 weeks following the initial DNA transfection.

9. Positive clones are expanded from the second 12-well plate for further analysis and freezing. Cell lines should be maintained in medium containing at least 40 $\mu$g/ml G418 to maintain the selective pressure on protein expression.

*Analysis of Transforming Activity*

1. To measure transforming activity cells are plated as above, and transfected using the calcium–phosphate procedure with 150 ng plasmid DNA and 10 $\mu$g Rat-2 DNA/10-cm plate as described above.

2. Cells are allowed to grow to confluence, and the medium is switched to DMEM + 5% calf serum. The lower serum concentration favors the overgrowth of transformed cells with reduced growth factor requirements.

3. Cells are maintained for 2 weeks to 2 months. Cultures are monitored for the appearance of transformed foci, which can be quantitated as to foci formed per microgram of DNA transfected. For reasons that are unclear, poorly transforming alleles may induce foci on Rat-2 cells with a longer latent period than the wild-type gene, and hence it is worth keeping cells for a protracted period.

4. Foci of transformed cells can be picked either using a cloning cylinder (as above), or more readily with a sterile drawn-out Pasteur pipette. Picked cells are grown up, and can be cloned either by limiting dilution or by colony formation in soft agar (a property of many but not all morphologically transformed cells).

## Specific Oligonucleotide-Directed Mutagenesis and Rapid Expression of v-*fps* Kinase Domain

Once a general view of the functional domains within a protein kinase has been obtained, it is possible to ask more specific questions about the functions of individual residues or defined domains. For this purpose it is useful to create predetermined mutations using synthetic oligonucleotides. We have produced a variety of point, deletion, and insertion mutations in the viral and cellular *fps* and *fms* tyrosine kinases. These procedures have been described in detail elsewhere.[31-35] Two advances have made this procedure more rapid and efficient, especially if the mutant sequence must subsequently be expressed in mammalian cells. First, the coding sequence to be mutated can be placed into a plasmid containing an f1 or M13 replication origin (phagemid), which can be propagated as a plasmid but rescued as single-stranded DNA at will following superinfection with a helper bacteriophage such as M13K07,[31] providing a template for mutagenesis. The stability of inserts within such phagemids is generally greater than in M13 itself. Obviously it is helpful if the phagemid is designed to effect subsequent expression of the insert in mammalian cells (i.e., pCMV), since no subcloning of mutants is then necessary. Second, the template DNA can be prepared by growth of the bacteriophage or plasmid in $dut^-ung^-$ cells (i.e., CJ236), which yield uracil-substituted DNA.[32] On *in vitro* mutagenesis and introduction into wild-type *E. coli* the parental strand is selected against, and a very high frequency of mutants is produced. Usually mutants can be identified directly by DNA sequencing without recourse to prior screening procedures. For rapid analysis, the mutated sequences are subcloned into both bacterial and mammalian expression vectors (Fig. 2).

## Discussion

### *Linker Insertion Mutagenesis Suggests a Domain Structure for the v-fps Tyrosine Kinase*

In analyzing a collection of linker insertion mutations in the *gag–fps* coding sequence, some of which are shown in Fig. 3, we have recovered mutant proteins with the following phenotypes:

1. Indistinguishable from wild type

[31] J. Viera and J. Messing, *in* "Methods in Enzymology" (R. Wu and L. Grossman, eds.), Vol. 153, p. 3. Academic Press, San Diego, California, 1987.

[32] T. A. Kunkel, J. D. Roberts, and R. A. Zakour, *in* "Methods in Enzymology" (R. Wu and L. Grossman, eds.), Vol. 154, p. 367. Academic Press, San Diego, California, 1987.

[33] M. J. Zoller and M. Smith, *in* "Methods in Enzymology" (R. Wu and L. Grossman, eds.), Vol. 154, p. 329. Academic Press, San Diego, California, 1987.

[34] M. Moran, C. A. Koch, I. Sadowski, and T. Pawson, *Oncogene* **3,** 665 (1988).

[35] G. R. Taylor, M. Reedijk, V. Rothwell, L. R. Rohrschneider, and T. Pawson, *EMBO J.* **8,** 2029 (1989).

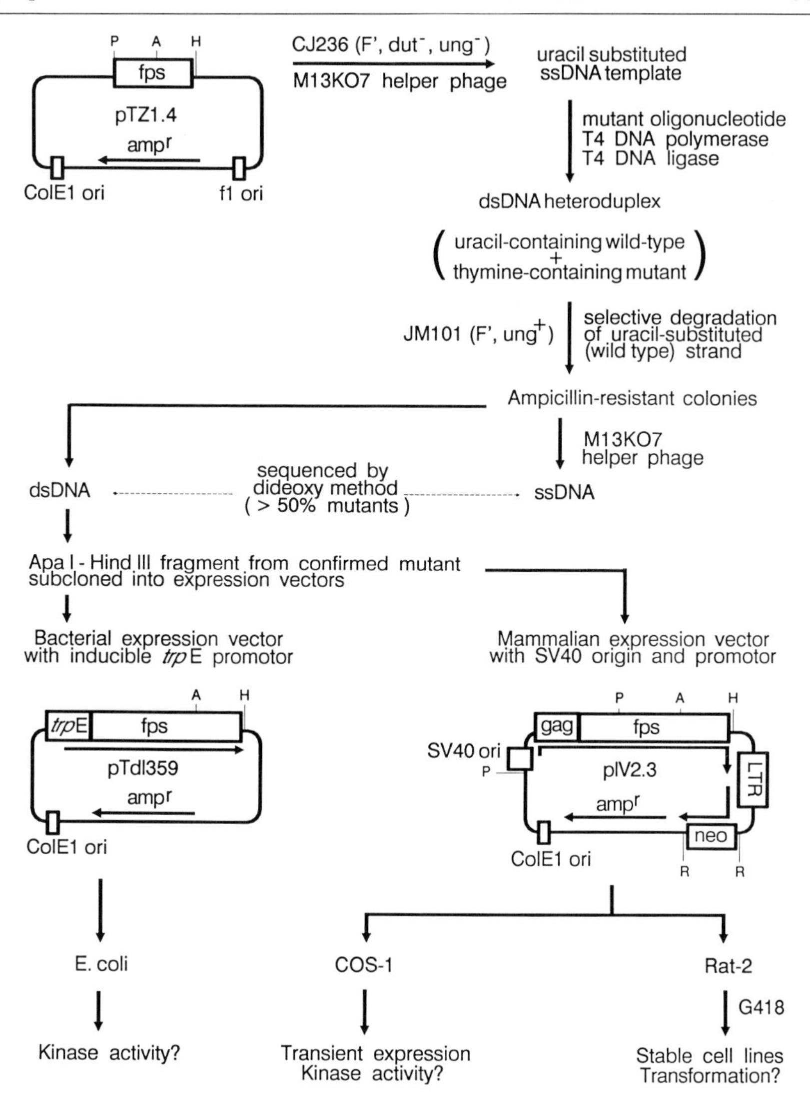


FIG. 2. Oligonucleotide-directed mutagenesis and expression of the v-*fps* tyrosine kinase. The v-*fps*-containing phagemid pTZ1.4 are passaged through the dut$^-$, ung$^-$ *E. coli* strain CJ236, and ssDNA is prepared by superinfection with M13K07. Following *in vitro* mutagenesis using the uracil-substituted single-stranded template and a mutant oligonucleotide, the double-stranded heteroduplex is used to transform the ung$^+$ strain JM101 to ampicillin resistance. Mutants are identified by dideoxy sequencing. The mutations are then subcloned into bacterial or mammalian expression vectors. Mutant proteins are rapidly screened for kinase activity in bacteria and COS monkey cells, followed by analysis of transforming activity and recovery of stably expressing lines using Rat-2 fibroblasts.

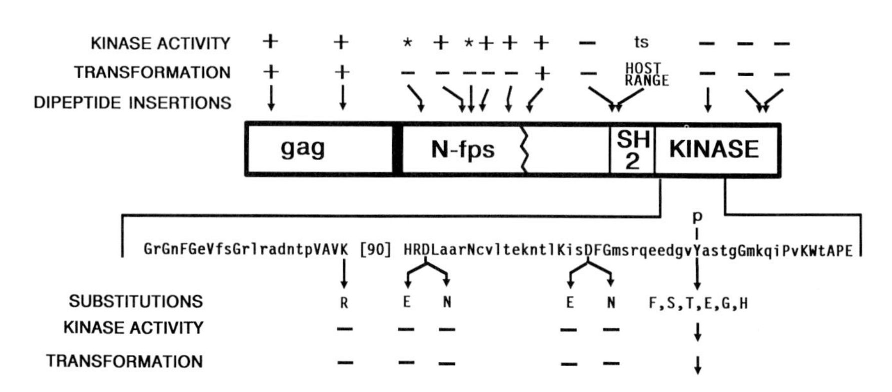


FIG. 3. *gag–fps* linker insertion and point mutants generated using the techniques shown in Figs. 1 and 2. The kinase and transforming activities of the mutants, and the positions of the domains, are indicated (*, kinase activity not tested). Residues within the kinase domain which are highly conserved among protein kinases are in upper case, as is the major autophosphorylation site (Tyr-1073), which has been changed to a variety of different amino acids,[34,38,39] none of which can functionally replace tyrosine. Two aspartic acid residues are conserved in all protein kinases and some aminoglycoside phosphotransferases,[40] and cannot be substituted with glutamic acid or asparagine without loss of activity.[34]

2. Transformation defective, but positive for *in vitro* kinase activity
3. Temperature sensitive for both kinase and transforming activities
4. Host dependent for transforming activity
5. Transformation defective and kinase inactive

This broad range of mutant phenotypes, taken in conjunction with the locations of the insertion mutations and sequence comparisons of related gene family members, has allowed us to propose a domain structure for P130$^{gag-fps}$ and has revealed previously unappreciated aspects of tyrosine kinase structure and function.

The most informative mutants are those that retain some aspect of their biological activity, or which lose activity unexpectedly. A series of adjacent mutations which induce similar phenotypes suggest a discrete functional domain made up of a contiguous stretch of the polypeptide backbone. For example, we described a series of insertions within the N-*fps* region of P130$^{gag-fps}$ (Fig. 3) that dramatically decrease transforming activity in Rat-2 cells without any overt effect of kinase activity.[2] Subsequent analysis has suggested a role in subcellular localization.[36] We found an excellent correlation between mutations that perturbed the hypothetical N-*fps* functional domain and the failure of the encoded mutant proteins to react with two different anti-Fps monoclonal antibodies.[37] Apparently the antibodies recognize conformational epitopes within a domain whose structure is perturbed by any one of several insertions, resulting in a loss of biological, although not catalytic, activity.

The SH2 domain (Fig. 3)[38–40] was also first identified and defined on the basis of linker insertions that abolished transforming activity.[41] In contrast to the amino-terminal domain, however, these SH2 mutations inhibited kinase activity. This was a curious observation, since biochemical and structural evidence indicated that the catalytic domain is composed of residues entirely carboxy-terminal to SH2. We postulated that the SH2 region influences kinase activity through an intramolecular interaction with the catalytic domain. Consistent with this notion, SH2 insertions which affect kinase activity increase the sensitivity of the kinase domain to limited proteolysis, suggesting that they cause a substantial conformational change within the catalytic domain. Some SH2 mutations in avian v-*fps* and v-*src* induce a host range phenotype, transforming chicken but not rat cells.[42] These and other observations have led to the speculation that the SH2 domain also influences substrate recognition, and may associate directly with other cellular proteins. The subsequent discoveries that phospholipase C-$\gamma$ and p21$^{ras}$ GTPase activating protein (GAP) contain SH2 domains, and are substrates for tyrosine phosphorylation, support the view that this protein substructure may play a key role in normal as well as oncogenic signal transduction.[43,44]

Insertions in the v-*fps* or v-*src* catalytic domain most frequently result in a loss of kinase activity.[2,8] Null mutations presumably identify critical regions or residues, although they tend to be otherwise rather uninformative. The identification of further mutations which restore activity to null alleles (pseudorevertants) may be useful in defining sites of intramolecular interactions. A host range v-*src* mutant with an insertion at the principal

[36] A. R. Brooks-Wilson, E. Ball, and T. Pawson, *Mol. Cell. Biol.* **9,** 2214 (1989).

[37] J. C. Stone and T. Pawson, *J. Virol.* **55,** 721 (1985).

[38] G. Weinmaster, M. J. Zoller, M. Smith, E. Hinze, and T. Pawson, *Cell (Cambridge, Mass.)* **37,** 559 (1984).

[39] G. Weinmaster and T. Pawson, *J. Biol. Chem.* **261,** 328 (1986).

[40] S. Brenner, *Nature (London)* **329,** 21 (1987).

[41] I. Sadowski, J. C. Stone, and T. Pawson, *Mol. Cell. Biol.* **6,** 4396 (1986).

[42] J. E. Declue, I. Sadowski, G. S. Martin, and T. Pawson, *Proc. Natl. Acad. Sci. U.S.A.* **84,** 9064 (1987).

[43] T. Pawson, *Oncogene* **3,** 491 (1988).

[44] C. Ellis, M. Moran, F. McCormick, and T. Pawson, *Nature (London)* **343,** 377 (1990).

autophosphorylation site within the kinase domain has been identified.[8] Clearly, insertions or other mutations in the kinase domain which induce a partial or complete loss of biological activities, but do not affect enzymatic activity per se (measured for example by autophosphorylation), may be useful in identifying important substrates. Ultimately it may be possible through the isolation of appropriate mutants to identify the kinase residues that actually contact substrates, allowing the definition of an "effector domain" as has been proposed from the analysis of oncogenic *ras* genes. Insertion mutations in oncogenic tyrosine kinases seem to be a particularly rich source of temperature-sensitive (ts) mutants.[7,42] Some of these may be specifically defective in their interactions with key substrates at the nonpermissive temperature. The majority of ts mutants, however, lose both kinase and transforming activities at the nonpermissive temperatures. These latter mutants are valuable tools in following the time course of events in kinase-generated signals.

Insertions with no apparent effect on activity may indicate the position of hinge structures, or of regions that are irrelevant to the properties under examination. They can be useful starting points for the construction of chimeric proteins.[36]

## Acknowledgments

Work in the authors' laboratories is supported by grants from the National Cancer Institute of Canada and the Medical Research Council of Canada to T.P., and from the National Institutes of Health (CA44900 and 44012) to J.C.S. T.P. is a Terry Fox Cancer Research Scientist and M. F. M. is a postdoctoral fellow of the National Cancer Institute of Canada.

# AUTHOR INDEX

Numbers in parentheses are footnote reference numbers and indicate that an author's work is referred to although the name is not cited in the text.

## A

C

## D

E

F

## H

# I



# J

## K

L

## M

## N

## O

P

S

## T

X



Y

## Z

# Subject Index

## A

## D

## E

## F

## G

## J



## K



## L

## N

## R

## U



## V



## W



## X



## Y

## Z